Microbiome Innovations

This book explores the diversity of microbiomes, ecological significance, and applications in agriculture, human health, and environmental sustainability. The introductory chapter provides state-of-the-art methods for studying microbiomes, including microscopy, culturing techniques, and molecular biology tools, while examining how artificial intelligence (AI), machine learning (ML), and computational biology are revolutionizing microbiome research. It also covers advancements in next-generation sequencing (NGS) technologies, various metagenomic approaches, and their significance in studying microbiomes and identifying genes of agricultural importance. Practical applications are explored, such as enhancing plant nutrient utilization, improving crop yields using biofertilizers, and leveraging rhizosphere microbiomes for better nutrient uptake and drought tolerance. The potential of microbiomes as biocontrol agents to manage pests and diseases is also discussed. Advanced techniques such as CRISPR–Cas systems for genome editing within microbiomes are examined, along with the role of microbiomes in improving crop performance in high-altitude, high-saline, and high-temperature conditions. The book concludes by addressing current challenges in microbiome research and application, highlighting future opportunities for innovation in agriculture. This book is intended for researchers and students in microbiology, agriculture, and environmental science.

Key Features:

- Explores the diversity, ecological significance, and applications of microbiomes in agriculture, human health, and environmental sustainability
- Covers advanced methods for studying microbiomes, including microscopy, culturing techniques, molecular biology tools, AI, ML, and computational biology
- Discusses practical applications of microbiomes in enhancing plant nutrient utilization, improving crop yields with biofertilizers, and using microbiomes as biocontrol agents
- Includes insights into cutting-edge technologies such as CRISPR–Cas systems for genome editing and their applications in microbiome research
- Addresses current challenges in microbiome research and highlights potential future opportunities for innovation and application

Microbiome Innovations

Advancing Sustainable Agriculture

Edited by
Indra Mani and Vijai Singh

CRC Press
Taylor & Francis Group
Boca Raton London New York

CRC Press is an imprint of the
Taylor & Francis Group, an **informa** business

Designed cover image: Shutterstock

First edition published 2027
by CRC Press
2385 NW Executive Center Drive, Suite 320, Boca Raton FL 33431

and by CRC Press
4 Park Square, Milton Park, Abingdon, Oxon, OX14 4RN

CRC Press is an imprint of Taylor & Francis Group, LLC

ISBN: 978-1-032-89749-3 (hbk)
ISBN: 978-1-032-89750-9 (pbk)
ISBN: 978-1-003-54442-5 (ebk)

DOI: 10.1201/9781003544425

Typeset in Times
by Apex CoVantage, LLC

Contents

Foreword

I am pleased to write the foreword for the book *Microbiome Innovations: Advancing Sustainable Agriculture*, a timely and significant contribution to the rapidly evolving landscape of life sciences. This book provides basic and applied knowledge of the microbiome and microbiota for better use in the agricultural sector.

A group of microorganisms residing in a particular niche is considered a microbiota. Microbiota includes many microorganisms such as viruses, archaea, bacteria, fungi, and parasites. Soil consists of diverse communities of microorganisms that are crucial for enabling the soil to sustain plant growth and development by means of nitrogen fixation, the solubilization of phosphorus and zinc, the production of siderophores, the creation of substances that promote plant growth, disease resistance, enhanced drought tolerance, increased root expansion, and the breakdown of organic matter. Therefore, from an agricultural standpoint, knowledge about the presence of particular microorganisms and their functional activities is crucial. Changes in soil pH, temperature, water content, crop cultivation, fertilizer use, and so on all affect the microbial properties of soils. It is interesting to note that biofertilizers, also known as microbial inoculants, are employed in agriculture to lessen the need of inorganic fertilizers and show encouraging results. Because they can fix nitrogen from the air, a variety of microorganisms can act as biofertilizers, assisting plants in obtaining nutrients from organic fertilizers. Additionally, some microorganisms can boost salt tolerance, enhance plant health, or improve drought tolerance. Microorganisms are ubiquitous. These microbes play a significant role in nature.

Culture-independent molecular techniques are being used to assess the microbiome and its function in particular crop soils. This allows us to get an idea of microbes—like who is there? And/or what are they doing? The answers to these questions will be interesting for understanding where microbes reside and their functions in particular agricultural lands. It may vary due to differences in the soil's physical and chemical characteristics. This information can improve soil quality and enhance crop yield. Further, agriculturally important genes/pathways can be prospected and help improve microbial strains using engineering biology tools, such as genome editing techniques (e.g., CRISPR–Cas). This book highlights a wide range of applications of microbiomes, which will be helpful for students, researchers, investigators, stakeholders, and industries.

I express my sincere appreciation to Dr. Indra Mani and Prof. Vijai Singh for their dedication and effort in bringing this fantastic book to our scientific audience, in collaboration with CRC Press, Taylor & Francis Group. This resource will serve as a valuable guide for students, researchers, scientists, policymakers, and other stakeholders.

Rodrigo Ledesma-Amaro
Department of Bioengineering
Imperial College London, London, United Kingdom

Preface

This book elucidates the composition and the role of beneficial or harmful microorganisms, offering a promising strategy to optimize plant yield, enhance plant health, and improve productivity. Until recently, the understanding of soil microbiome and plants remained largely inaccessible. However, technological advancements such as metagenomics, high-throughput sequencing (next-generation sequencing, NGS), and bioinformatics (*in silico* tools and biological databases) have made it possible to analyze the microbiome directly from samples. It can be utilized to screen and select a potential microorganism to develop a microbial consortium that harnesses the full potential of microbiota for agricultural sustainability. This book presents numerous methodologies for analyzing soil microbiomes and their uses in diverse areas, including crop improvement, the development of biofertilizers, and the bioremediation of contaminated agricultural soils.

The book is structured to cover both fundamental and applied aspects of microbiome research.

Chapter 1 introduces the concept of microbiomes and their applications. Chapter 2 discusses advances in techniques for studying the microbiome. Chapter 3 highlights the role of artificial intelligence and machine learning for analyzing the microbiome. Chapter 4 focuses on computational biology for microbiome data generation and analysis. Chapters 5 and 6 discuss advancements in NGS and metagenomics approaches. Chapter 7 emphasizes metagenomics in the development of microbiome consortia. Chapter 8 explores metagenomics approaches for prospecting genes of agricultural importance. Chapters 9 and 10 address the use of the microbiome to enhance nutrient utilization and crop yield, utilizing biofertilizers. Chapter 11 explores the role of rhizosphere microbiomes in the improvement of nutrient uptake. Chapters 12 and 13 discuss the mechanisms and perspectives of the microbiome in improving salinity tolerance and drought tolerance. Chapter 14 highlights the use of microbiomes as biocontrol agents for pest and disease management. Chapter 15 discusses CRISPR–Cas systems in microbiome genome editing. Chapters 16 and 17 highlight microbiome-based crop improvement strategies under high-altitude and high-temperature conditions. Chapter 18 concludes with current challenges and opportunities in microbiome research.

Overall, *Microbiome Innovations: Advancing Sustainable Agriculture* serves as a comprehensive resource on the recent development and applications in microbiome research and innovations. This book will be valuable not only for beginners in the field but also for students, researchers, stakeholders, and policymakers engaged in advancing agricultural sustainability through microbiome science.

Acknowledgment

I would like to express my sincere gratitude and deep appreciation to Prof. Sangeeta Bhatia, Principal, Gargi College, the University of Delhi, India, for extending her outstanding support and motivation to complete this book. I am delighted to thank all the authors for their excellent contributions to this book.

I would like to thank Dr. Bhavik Sawhney (Senior Editor—Life Science) and the entire Taylor and Francis team for their excellent management of this project. I would like to thank Prof. Kailash N. Pandey, Prof. Vijai Singh, Prof. P. K. Srivastava, Dr. Ravindra Kumar, Prof. Janardan Yadav, Dr. Satya Prakash, Dr. Vipin Chauhan, Dr. Anurag Tiwari, Dr. Jay Prakash Verma, Dr. Ajay Kumar Singh, Dr. Santosh Kumar Mishra, Dr. Ashok Saini, Dr. Vimal C. Pandey, and those who have directly or indirectly helped in shaping this project.

I wish to express my heartfelt gratitude to my parents, and all members of my family who always filled my mind with affirmations and crowded out the negative thoughts. Most of all, I thank my wife, Reena Gond, sons, Ankur Kumar Gond and Ankit Kumar Gond, and daughter-in-law, Shivangi Gond, who have helped me during this project. I would like to warmly thank the Faculty and Staff of the Gargi College, University of Delhi for their encouragement and support.

And finally, I express my sincere gratitude to all my friends and well-wishers for providing me moral support during the process of writing this book. Finally, I would like to express my heartfelt gratitude to GOD for His immense POWER, allowing me to experience joy and triumph embodied in this book.

Indra Mani

I am pleased to extend my heartfelt thanks to all the contributing authors for their excellent contributions to this book. I would like to give many thanks to Dr. Bhavik Sawhney (Senior Editor—Life Science), and the entire Taylor and Francis team for their excellent support and seamless management throughout this project.

My sincere appreciation goes to Prof. Rakesh Rawal, Prof. Sudhir P. Singh, and Prof. Bharat Maitreya. Dr. Poonam Bhargava, Dr. Ravindra P. Singh, Prof. Bhabatosh Das, Dr. Indra Mani, Dr. Pablo Carbonell, Dr. Satya Prakash, Prof. Suresh Ramakrishna, Dr. Dinh-Toi Chu, Prof. Antonia Sagona, Dr. Rodrigo Amaro, and all others—named or unnamed—whose direct or indirect contributions have played a key role in shaping this book.

I am deeply thankful to my PhD students—Mr. Khushal Khambhati, Mr. Karan Murjani, Ms. Khushbu Panchal, and Ms. Dharmisha Solanki—whose thoughtful discussions and valuable feedback have helped to shape this book.

Despite our best efforts, there may be some errors that may have inadvertently made their way into this first edition. I welcome feedback from readers that can be useful to improve in future editions. Finally, I offer my heartfelt thanks to God for his supreme grace, which enables me to complete this journey with joy and fulfilment in the form of this book.

Vijai Singh

About the Editors

Indra Mani is Assistant Professor in the Department of Microbiology, Gargi College, University of Delhi (DU), New Delhi, India. He received his MSc in microbiology from Bundelkhand University (BU), Jhansi, Uttar Pradesh, India, and PhD in biochemistry (2010) from Banaras Hindu University (BHU), Varanasi, Uttar Pradesh, India. He completed postdoctoral training in the Department of Physiology, School of Medicine, Tulane University, New Orleans, LA, USA. He has worked as Senior Research Officer (SRO) in the Department of Medicine, All India Institute of Medical Sciences (AIIMS), New Delhi, India. He has more than 13 years of research and teaching experience in microbiology, molecular genetics, cell signaling, and bioinformatics. He has published more than 50 research articles in peer-reviewed journals, 58 book chapters, and eight books. He received the MBSI (Microbiologist Society, India) Senior Teacher award—2024 and the Agathiyar Chemical Biology award—2024 from the Society of Chemical and Synthetic Biology, India. He has extensive experience in microbial characterization, molecular genetics, cell culture, fluorescence/confocal microscopy, and in silico analysis. He is a life member of the Association of Microbiologists of India (AMI), the Microbiologists Society, India (MBSI), the Society for Chemical and Synthetic Biology (SCSB), the Indian Science Congress Association (ISCA), the Indian Network for Soil Contamination Research (INSCR), and the Aquatic Biodiversity and Conservation Society (ABCS), India. He has presented several papers at different national and international conferences. He serves as an editorial board member, a reviewer of peer-reviewed journals, and a member of national and international societies.

Vijai Singh serves as Professor and Head of the Department of Biosciences, School of Science, and Dean (Research and Innovation) at Indrashil University, Rajpur, Mehsana, India. Prior to this, he held the position of Associate Professor in the same department at Indrashil University. He has also served as Assistant Professor in the Department of Biological Sciences and Biotechnology at the Institute of Advanced Research, Gandhinagar, and in the Department of Biotechnology at Invertis University, Bareilly, India. Earlier in his career, Dr. Singh worked as a postdoctoral fellow in the Synthetic Biology Group at the Institute of Systems and Synthetic Biology, Paris, France, and at the School of Energy and Chemical Engineering at the Ulsan National Institute of Science and Technology, Ulsan, South Korea. He earned his PhD in biotechnology in 2009 from the ICAR-National Bureau of Fish Genetic Resources, Uttar Pradesh Technical University, Lucknow, India. His doctoral research focused on the development of molecular and immunoassay for the diagnosis of *Aeromonas hydrophila*. Dr. Singh's research interests lies in the design and construction of novel biosynthetic pathways for the production of medically and industrially important biomolecules. His laboratory is also actively engaged in developing CRISPR–Cas systems for gene knock-out, knock-in, gene regulation, and diagnostics. With over 15 years of research and teaching experience, his areas of expertise include synthetic biology, metabolic engineering, bioinformatics, microbiology, and industrial microbiology. He has an extensive academic record, with over 113 articles, 97 book chapters, 37 books, and three patents to his credit. Dr. Singh has received a number of prestigious awards, including the Bioclues Innovation, Research and Development Award (2023) from the BIOCLUES Society, the Vice Chancellor's Best Research Award (2023) from Indrashil University, and the Agathiyar Chemical Biology Award (2023) from the Society of Chemical and Synthetic Biology. He serves as an associate editor, editorial board member, and reviewer of several reputed peer-reviewed journals.

1 Insight into the Soil Microbiome and Its Applications

Jagriti Saraswat[1], *Ingita Dey Munshi*[2], *Mehak*[3], *Reet Arora*[4], *Anshu Mathuria*[5], *Debolina Das*[1], *Eshani Choudhury*[1], *Nikita Bisht*[1], *Kritika Jain*[6], *Vijai Singh*[7], *Ashok Saini*[1]*, *and Indra Mani*[5]*

[1]Department of Microbiology, Institute of Home Economics, University of Delhi, New Delhi. India

[2]School of Computational and Integrative Sciences, Jawaharlal Nehru University, New Delhi. India

[3]University College Dublin, Belfield, Dublin 4. Ireland

[4]Medybiz Pharma Pvt Ltd., Bangalore, 560071, India

[5]Department of Microbiology, Gargi College, University of Delhi, New Delhi. India

[6]Laboratory Medicine, Manipal Hospitals, Dwarka, New Delhi, 110075, India

[7]Department of Biosciences, School of Science, Indrashil University, Rajpur, Gujarat. India

1.1 INTRODUCTION

The microbiome is the collective genetic material of microorganisms and their surrounding environment, whose study has enabled us to understand the microbial ecosystems and their impact on human health, agriculture, and environmental systems. Diverse types of microbiomes exist across different environments, including the gastrointestinal tract, skin, oral, and vaginal microbiomes in humans and animals, as well as soil, marine, freshwater, and plant-associated microbiomes in nature (de Jonge et al., 2022). Each microbiome type harbors distinct microbial compositions adapted to its specific ecological niche and environmental conditions.

Another term, "Microbiota" specifically refers to the pool of all the living microorganisms constituting a microbial community that inhabits a specific environment (Berg et al., 2020). These microbial communities are remarkably diverse, including eubacteria, archaea, various fungi, viruses, and protists that coexist in complex ecological relationships (Gupta et al., 2022a). From an evolutionary perspective, microbiota has co-evolved along with their hosts over the period of millions of years, forming intricate symbiotic relationships that have shaped the development of both microbes and hosts. Microbiomes are found in plants, animals, soil, water, and humans, and interact with them in various ways. In plants, microbiomes influence various physiological processes such as tolerance to stress, nutrient uptake, and resistance toward various diseases (Liu et al., 2020), while animal microbiomes across species exhibit remarkable adaptations to their hosts' diets and environments; for example, ruminants harbor specialized microbial communities that enable the digestion of cellulose (Weimer et al., 2022).

Soils play a significant role in various ecosystem functions such as the production of biomass, safeguarding biodiversity, filtering water for drinking, and managing insects (Brevik et al., 2019). Soil is observed to have different kinds of microorganisms, which constitute the soil microbiome and play a crucial role in various functions of soil, including agriculture, decay of organic matter in soil, nutrient recycling, and pest/insect population regulation (Fierer, 2017; Brevik et al., 2019). Research done in recent years has brought about the importance of the soil microbiome in ecosystems and soil quality, as well as in human and animal health. Studies suggest that 1 gram of soil may contain over 50,000 microbial species belonging to diverse genetic backgrounds (Bardgett and van der Putten, 2014; Fierer, 2017; Kendzior et al., 2022).

1.2 COMPOSITION OF SOIL MICROBIOME

Microbial communities found in soil are complex and dynamic in nature (Neher et al., 2022). Most microbes present in the soil are usually found in greater numbers in the topsoil, especially in the rhizosphere around the plant roots, where nutrient concentration is very high. However, some microorganisms can be present in deep layers of soil, but with the increase in depth, microbial activity decreases (Van Leeuwen et al., 2017).

Furthermore, soil microorganisms primarily make up less than 0.4–0.5% (w/w) of the overall soil mass and play a significant role in influencing soil characteristics (Chakraborty et al., 2022). Soil microbiome consists of microorganisms from different types, including bacteria, fungi, archaea, and protists (Bardgett and van der Putten, 2014). Although it

***Corresponding authors:** ashok.saini@ihe.du.ac.in, drashoksaini22@gmail.com (Dr. Ashok Saini) and indra.mani@gargi.du.ac.in, indramanibhu@gmail.com (Dr. Indra Mani)

DOI: 10.1201/9781003544425-1

has been observed that soil contains a greater number of bacterial and fungal species than protists, archaea, and viral particles (Fierer, 2017).

Table 1.1 summarizes the different types of microbes along with their function in the soil ecosystem.

1.3 FACTORS AFFECTING THE SOIL MICROBIOME

Soil microbiome consists of various microbial communities, whose composition is influenced by various abiotic and biotic factors, which are controlled by geographical and seasonal parameters. Biotic factors include various plants, microbial competitors, and insects, while abiotic factors include the type of soil and its structure, water availability in the soil, pH, availability and composition of nutrients, and other geographical factors (Blum et al., 2019; Islam et al., 2020). The constitution of the soil microbiome is determined by the interaction of these factors and is not dependent on any single factor (Fierer, 2017). Some factors affect particular microbial communities present in the microbiota as compared with other microbes, such as soil moisture, which has been observed to stimulate protist composition, which is interpreted in terms of microbial biomass (Mishra et al., 2023). Water availability, pH (acidity), and texture of soil have been considered as main factors that affect the composition of microbiome (Brockett et al., 2012; Rousk et al., 2010).

Additionally, soil microbiome shows sensitivity toward various soil contaminants such as heavy metals (e.g., added from fertilizers or emitted from industries), microplastics, or pesticides. Heavy metals are usually recalcitrant and persist rigidly in soil for longer periods as they are resistant toward microbial decomposition and thus disrupt the stability of the soil microbiome by affecting various biochemical processes. Moreover, accumulation of microplastics into soils also adversely affects the composition and function of the soil microbiome (Guo et al., 2020).

1.3.1 Effect of Soil Types on the Composition of the Soil Microbiome

Soil microbiomes are an integral part of soil and play various crucial roles in agroecosystems. The microbiome is closely associated with soil structure, including aspects like aggregation and pore connectivity, as this structure governs the movement of water, oxygen, and nutrients within the ecosystem (Hartmann and Six, 2023).

The diversity of the microbial species found in the soil is greatly determined by the characteristics of the soil (Table 1.2).

1.4 TYPES OF SOIL MICROBIOME

Soil is a rich source of diverse kinds of microorganisms, which became an integral component of plant microbiomes and play important roles in the ecosystem, plant health, and one health for all other living organisms (Banerjee & van der Heijden, 2023); Bardgett and van der Putten, 2014; Fierer, 2017). Soil microbiomes can be classified on the basis of their location or interaction with plants (Figure 1.1). The main categories of soil microbiome are plant growth-promoting soil bacteria (PGPB), rhizospheric microbes, genetically engineered microbes in soil, epiphytic and phyllospheric microbes, and endophytic microbes.

Thus, the soil microbiome can be classified into various groups as discussed below.

1.4.1 Plant Growth-Promoting Soil Bacteria

PGPB are the beneficial microorganisms found in abundance mainly in the underground environment, where they make remarkable associations with plants to increase their resilience and performance (Qadir et al., 2024).

PGPB plays an important role in the development of plants and also stimulates various processes occurring in the soil, such as bioremediation in various ways, such as secretion of metabolites and various types of hormones, fixation of nitrogen, and increasing bioavailability of nutrients via mineral solubilization (Poria et al., 2022).

For instance, specialized bacteria like *Rhizobium* develop intricate root nodules in legumes, establishing nutrient factories that capture nitrogen from the air (Hayat et al., 2010). Meanwhile, free-living species such as *Azospirillum* alter root architecture and enhance nutrient scavenging capabilities (Poria et al., 2022).

The biochemical repertoire of PGPB extends to the synthesis of vital plant hormones such as gibberellins, auxins, and CKs, which fine-tune plant growth patterns. These remarkable organisms also produce 1-aminocyclopropane-1-carboxylic acid (ACC) deaminase, which is an enzyme that dismantles stress-induced ethylene (ET) precursors, enabling plants to maintain robust root development even under challenging conditions (Glick, 2010). Beyond nurturing plant growth, PGPB serve as natural defenders, manufacturing an arsenal of protective compounds—including siderophores, antibiotics, and hydrogen cyanide—that keep pathogens away. Their contribution to soil health manifests through the release of sticky exopolysaccharides, which hold soil particles together, creating a moisture-retentive environment resistant to erosion (Hayat et al., 2010).

PGPB exhibit strategic colonization patterns, with distinct populations establishing themselves either in the dynamic root-soil interface (rhizosphere) or within plant tissues as endophytes. This dual distribution creates a comprehensive support network that strengthens plant performance across diverse environmental conditions. The agricultural community increasingly recognizes these bacteria as game-changers in sustainable farming practices, offering alternatives to synthetic fertilizers, maximizing crop output, and empowering plants to flourish in nutritionally challenged or contaminated soils (Poria et al., 2022).

TABLE 1.1
Different Categories of Microbes Constitute the Soil Microbiome and Their Functions

Microorganism	Function in Soil	References
Bacteria		
Rhizobium	Symbiotic N_2 fixation in legume roots converts atmospheric nitrogen into bioavailable ammonia.	Fierer (2017)
Pseudomonas	Phosphate solubilization, promoting plant growth and Induced Systemic Resistance (ISR) Production of- 1. Siderophores 2. NH_3 3. Antibiotics 4. Enzymes aiding cell wall degradation, like chitinases and cellulases.	Kumar et al. (2017)
Streptomyces	Phosphate solubilization, disease suppression via antibiosis, mycoparasitism, and nutrient competition, decomposition of biopolymers like chitin, lignocellulose, and starch. Production of- 1. Plant growth-promoting hormones like indole-3-acetic acid (IAA), cytokinin (CK), 2. Essential minerals like iron and copper. 3. Hydrolytic enzymes like proteases, glucanases, and chitinases. 4. Osmolytes 5. Antibiotics 6. Antioxidants	Al-Quwaie (2023)
Bacillus	N_2 fixation, solubilization, phosphorus mineralization, and mineralization of other nutrients aid in ISR. Production of- 1. Phytohormones like IAA 2. Antimicrobial metabolites, including lytic enzymes, VOCs, i.e., volatile organic compounds, and toxins like cry proteins.	Saxena et al. (2019)
Azospirillum	Free-living N_2 fixer, solubilization of inorganic phosphorus, degradation of xenobiotics like benzoate and phenols. Production of- 1. Phytohormones like IAA and CKs 2. PGRs like development controllers' putrescine, spermine, spermidine, and cadaverine 3. Siderophores	Fukami et al. (2018), Cruz-Hernández et al. (2022), and Uma Devi et al. (2020)
Archaea		
Nitrososphaera	It oxidises NH_3 to NO_2^-, hence it is an essential agent for nitrification.	Tourna et al. (2011)
Methanobacterium	It converts CO_2 and H_2 to CH_3 (methane) gas in anaerobic soils, hence performing methanogenesis.	Thauer et al. (2008)
Fungi		
Glomeromycota (AMF)	AMF enhances water and nutrient uptake, particularly P and N, improves soil aggregation, and confers resistance to the plant.	Khanum et al. (2023)
Trichoderma	Biocontrol agent that produces antibiotic agents and suppresses soil-borne pathogens, enhancing plant and root growth, and organic matter turnover.	Harman et al. (2004)
Penicillium	Phosphate solubilization, production of 1. Bioactive metabolites 2. Phytohormones IAA and gibberelins 3. Siderophores	Altaf et al. (2017)
Aspergillus	Phosphate solubilization, chelation of metal ions. Production of 1. Low mol. wt. organic acids 2. Complex compounds like siderophores, OH-ions, CO_2, and protons	Cruz-Hernández et al. (2022)
Viruses		
Bacteriophages	Nutrient metabolism (P and C), degradation of organic matter, helps control the bacterial population.	Zhang et al. (2022)
RNA viruses (Partitiviridae)	Infect fungi and protozoa in soil, which affect the soil microbiome and ecosystem function.	Starr et al. (2019)
Protozoa		
Amoeba	Act as microbial predators using chemotaxis and enhance nutrient cycling.	Hu et al. (2025)
Ciliates	Predates on bacterial populations, performs nutrient cycling, and enhances soil structure.	Lara and Acosta-Mercado (2011)
Algae		
Chlorella spp.	Enhance nutrient levels in soil (C, K, Cu, Mn, Zn, etc.), and improve physicochemical characteristics of soil.	Gougoulias et al. (2017)
Nostoc spp. (Cyanobacteria)	Helps in fixing atmospheric nitrogen, restores biocrust functioning in case of sandy soils.	Román et al. (2018)

TABLE 1.2
Different Types of Soil and Their Dominant Microbial Taxa

Soil Type	Characteristics	Dominant Microbial Taxa	References
Sandy soil	Reduced water retention: moisture availability impacts microbial diversity.	*Pseudomonadota* (34%), *Actinomycetes* (26%)	Al-Awthan et al. (2024)
Loamy soil	Even texture fosters variable microbial taxa.	*Proteobacteria*, *Acidobacteria*, *Actinobacteria*	Frąc et al. (2020)
Clayey soil	Elevated nutrient levels facilitate anaerobic microbial processes.	*Firmicutes*, *Bacteroidetes*, *Methanogens*	Xia et al. (2020)
Arid soil	Less moisture and organic content; microflora adapted to a harsh environment.	*Actinobacteria*, *Proteobacteria*, *Firmicutes*	Vásquez-Dean et al. (2020)
Alkaline soil	Elevated pH levels promote certain bacterial groups.	*Bacillus*, *Pseudomonas*, *Rhizobium*	Yu et al. (2024)
Acidic soil	Low pH, favoring acidophilic microbes.	*Acidobacteria*, *Fungi* (e.g., *Ascomycota*)	Yuan et al. (2024)
Saline soil	Elevated salt concentration; suitable for halophiles.	*Halobacteria*, *Salinibacter*, *Halomonas*	Cao et al. (2025)

In addition, PGPB can be spread vertically through seeds to other parts of plants, such as leaves and roots (Abdelfattah et al., 2021).

1.4.2 Rhizosphere Microbiome

The term "Rhizosphere" is used to define the interface between plant roots and soil (Hiltner, 1904; Hartmann et al., 2008). It represents a highly specialized soil zone adjoining the plant root systems, distinguished by its exceptionally rich microbial diversity and metabolic activity. This microenvironment serves as a biological interface where diverse prokaryotic and eukaryotic microorganisms establish intricate molecular interactions with host plants (McNear, 2013). Within this complex ecological framework, various types of bacteria and fungi interact with plant roots, including PGPB, which serve as critical regulators of various physiological processes including nutrient homeostasis, developmental programming, and defensive mechanisms (Gupta et al., 2022b).

In the rhizosphere, bacterial-mediated plant nutrition enhancement relies heavily on their diazotrophic activity. Advanced nitrogenase complexes enable the transformation of molecular nitrogen into physiologically accessible ammonium compounds in specialized root endosymbiotic structures that are mainly composed of specialized prokaryotes like *Rhizobium* and *Bradyrhizobium* species. Thus, the need for industrial nitrogen replenishment is greatly reduced by this enzymatic reaction (Hayat et al., 2010). Further, rhizoplane-colonizing diazotrophs like *Azospirillum* simultaneously enhance nitrogen bioavailability and complement these symbiotic associations, while modifying root system architecture through localized auxin gradients (Glick, 2010).

Another rhizospheric bacterial consortium consisting of *Pseudomonas* and *Bacillus* species exhibits remarkable phosphorus mobilization capabilities through the coordinated secretion of organic acid metabolites and phosphohydrolase enzymes. These molecular mechanisms facilitate the dissolution of mineral-sequestered phosphates, enhancing their bioavailability for plant absorption (Poria et al., 2022). Such biochemical interactions prove particularly significant in phosphorus-deficient substrates, where bacterial solubilization activities substantially augment plant phosphorus acquisition efficiency.

The metabolic repertoire of rhizosphere bacteria, especially plant growth-promoting rhizobia (PGPR), encompasses the biosynthesis of multiple phytohormone classes, including gibberellins and CKs, which modulate plant developmental pathways. These microorganisms simultaneously function as biological control agents, synthesizing high-affinity ferric chelators that establish competitive exclusion of pathogenic organisms (Chepsergon and Moleleki, 2023). Their antimicrobial arsenal encompasses diverse secondary metabolites and compounds that are volatile in nature, creating antagonistic gradients against harmful pathogenic microorganisms (Hayat et al., 2010).

Under environmental stress conditions, rhizospheric bacteria modulate plant adaptive responses through sophisticated biochemical pathways. The enzymatic degradation of ACC via bacterial deaminases regulates stress-induced ET accumulation, preserving root elongation potential under suboptimal conditions (Glick, 2010). In contaminated substrates, these microbes enhance their bioremediation efficiency by modifying pollutant bioavailability and facilitating plant-mediated detoxification mechanisms (Poria et al., 2022).

Through the biosynthesis of complex polysaccharide compounds, these microorganisms enhance soil structural integrity and moisture retention properties. This dynamic rhizosphere microbiome constitutes an essential component of sustainable agricultural systems, providing biological alternatives to synthetic amendments while maximizing crop productivity and environmental resilience. Future agricultural advancement necessitates the optimization of

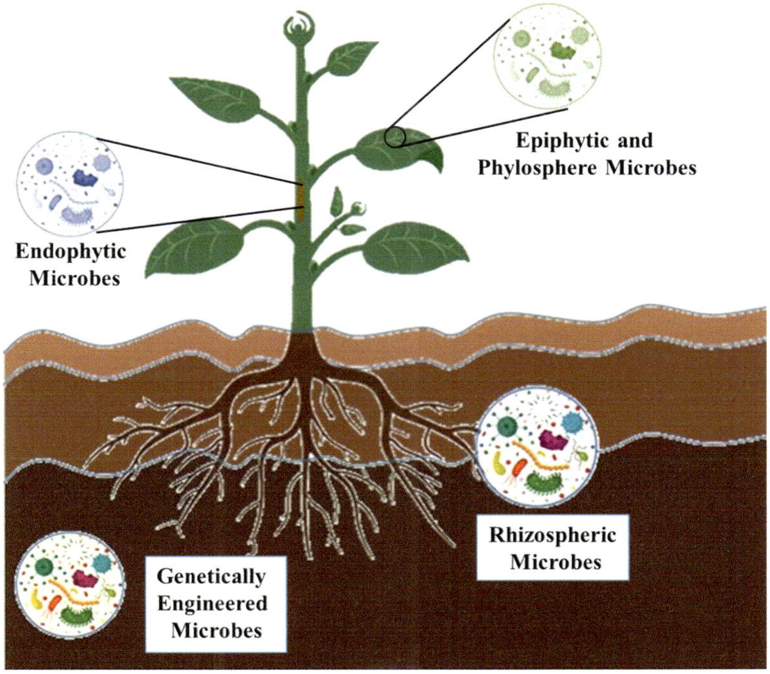

FIGURE 1.1 Different categories of soil microbiome.

these sophisticated plant–microbe interactions to ensure both food security and ecological sustainability.

1.4.3 Phyllosphere Microbiome

Phyllosphere refers to the components of the plants that are above the ground. They encompass flowers, fruits, leaves, stems, and pollen, each exhibiting distinct morphological and physical characteristics (Sohrabi et al., 2023b).

Significant variations in water content, temperature, and ultraviolet (UV) radiation make the phyllosphere a difficult and unpredictable habitat that hinders the growth of many microbial species. In addition to seed-mediated transmission, the phyllosphere microbiome can be established through a variety of dispersal methods, including precipitation, aerosol deposition via wind, activities of the insects, and soil (Koskella, 2020).

During the germination of seeds and the subsequent growth of plants, there is a significant transformation in their phyllosphere microbiome, primarily driven by two key factors:

1. The phyllosphere's inherently open environment facilitates the continuous influx of new microbial taxa.
2. The alterations in plant physiology and leaf defensive chemistry that occur throughout development thereby modify the conditions that microbial organisms must navigate to survive and compete (Koskella, 2020).

Microorganisms residing in the phyllosphere perform numerous roles, including the degradation or breakdown of environmental pollutants, nitrogen fixation, enhancement of plant growth, suppression of pathogenic microbes, and bolstering plant defenses against pathogens. For instance, biocontrol bacteria sourced from the phyllosphere can be employed to mitigate and manage diseases affecting plants in this region. Additionally, the functional characteristics of the phyllosphere can be modulated by its microbial inhabitants to enhance the adaptability of plants (Huang et al., 2023).

Despite the extensive taxonomical diversity observed in nature, the bacterial communities inhabiting the seed plant phyllosphere are predominantly composed of a limited number of phyla, namely

1. Pseudomonadota,
2. Bacteroidota,
3. Actinomycota, and
4. Bacillota (Laforest-Lapointe et al., 2016).

Microbial communities that inhabit the surface of the phyllosphere are known as epiphytes, while those that exist within the tissues of the plant, either in the intercellular spaces (apoplast) or inside the plant cells, are termed endophytes. A single microbial species may function as both an endophyte and an epiphyte, thereby occupying both of the ecological niches (Sohrabi et al., 2023b).

The surface of leaves presents a challenging environment for the epiphytes residing there, exposing them to various physicochemical stresses such as intense light and UV radiation, fluctuations in temperature and moisture, as well as nutrient scarcity. Consequently, phyllosphere microbial inhabitants employ multiple strategies to withstand and mitigate the adverse impacts of these abiotic stressors, thereby facilitating their successful colonization of the phyllosphere (Sohrabi et al., 2023b).

1.4.4 Mycorrhiza

Mycorrhiza essentially represents a symbiotic relationship between fungi and the plant roots, in which both of them benefit in a manner that the fungi help plants in nutrient and water uptake, and the plant, in return, supplies the fungi with sugars and other essential compounds. Mycorrhizal fungi make available various nutrients to plants, such as up to 75% of nitrogen (N), 90% of phosphorus (P), and in return they get about 20% of plant carbohydrate (Shi et al., 2023).

1.4.4.1 Types of Mycorrhiza

Mycorrhizal associations can be broadly classified into two types.

Type	Characteristics	Host Plants	Fungal Species	References
Endomycorrhiza	Non-differentiated, coil-like or spherical intracellular hyphae infiltrating the plant cells, presence of vesicles and arbuscles	Angiosperms-Green, Leafy Plants, fruit trees, commercial crops, orchids	Glomeromycota (VAM or arbuscular mycorrhiza)	Roth-Bejerano et al. (2013) and Pandey et al. (2019)
Ectomycorrhiza	The existence of an intercellular network composed of hyphae, known as the Hartig Net, encircles the cells without infiltrating the cortical cells of plant roots.	Gymnospermic and some angiospermic trees	Basidiomycota, Zygomycota, and Ascomycota phyla.	Pandey et al. (2019)

Endomycorrhiza is the most prevalent type, which is associated with most of the plants that belong to angiosperms, as well as crops. Out of its various types, the most prevalent is arbuscular mycorrhiza (AM) (Wang and Chen, 2024). AM fungi belong to the Glomeromycotina subphylum. It forms a symbiotic association with over 80% of plant species (Smith and Read, 2008). In this association, AM fungi form arbuscules, which are extensively branched intracellular structures. These arbuscules are formed within the cells of the root cortex of host plant cells, the place where nutrient exchange takes place (Rich et al., 2021).

Moreover, these fungi also take part in various other processes such as nutrient recycling, various biogeochemical cycles, maintaining soil fertility, soil aggregation, and decomposition (Powell and Rillig, 2018; Tedersoo et al., 2020; Zhang et al., 2024).

Other roles of mycorrhiza include protection of plants from disease and strengthening plants' resistance against them. During any pathogen attack, plants stimulate their tolerance and resistance mechanisms to fight against the pathogen either by inhibiting their growth and multiplication or reducing their pathogenicity (Mikaberidze and McDonald, 2020).

The AM fungi inhibit the colonization and invasion of soil-borne pathogens into the root cortex by using various strategies, like competing with the pathogen for nutrients, photosynthates, limited space, and alteration of the secretion of plant root exudates. For example, *Phytophthora nicotianae* (a plant pathogen) hyphae were observed to decline in the AM colonized as compared with the uncolonized root cortex (Cordier et al., 1996, 1998). AM fungi also alter the production of root exudates and secondary metabolites like organic acids, amino acids (Harrier and Watson, 2004; Kaur et al., 2022), phenolic compounds and flavonoids (Eke et al., 2016; Bencherif et al., 2019), lignans (Sanmartín et al., 2020), phytoalexins (Yao et al., 2003; Song et al., 2011), stilbenoids (Bruisson et al., 2016), etc. Mycorrhizal fungi influence the carbon sequestration capacity of a plant by modifying both the quality and quantity of root exudates (Sarathambal et al., 2023).

The composition of microbiome present around the soil has the potential to antagonize the root pathogens present, such as *Penicilium* spp. and *Trichoderma* spp. (Jamiołkowska et al., 2019), and *Pseudomonas* spp. (Bharadwaj et al., 2008).

1.4.5 Endophytic Microbiome

The endophytic microbiome encompasses a specialized consortium of microorganisms, predominantly bacteria, which establish their habitat within plant tissue matrices without inducing pathogenic responses. These internal colonizers establish sophisticated molecular partnerships with their host organisms, fundamentally modulating plant developmental pathways and adaptive responses. In the context of endophytes, the rhizosphere, which is the soil region influenced by roots, plays a paramount role in determining the diversity of root endophytes, which can subsequently transition to become shoot endophytes (Pandey et al., 2022)

Unlike their rhizosphere-dwelling counterparts, endophytic bacteria penetrate and colonize diverse plant organs, from root systems to reproductive structures, establishing precise biochemical interfaces within the host cellular environment. This localization confers distinct ecological advantages, including homeostatic microenvironments, protection from external perturbations, and direct access to host-derived metabolites (Hayat et al., 2010).

In reference to plant mineral nutrition, endophytic bacteria execute critical metabolic functions. These microorganisms enhance nutrient bioavailability through multiple biochemical pathways, including phosphate solubilization, atmospheric nitrogen fixation, and micronutrient mobilization. Specifically, endophytic populations of *Azospirillum* and *Pseudomonas* demonstrate significant diazotrophic activity, facilitating direct nitrogen transfer to host tissues and reducing external nitrogen requirements (Poria et al., 2022). These prokaryotes secrete an array of organic acids and hydrolytic enzymes that transform recalcitrant phosphate complexes into bioavailable forms, a mechanism particularly crucial in nutrient-limited substrates.

The metabolic versatility of endophytes extends to phytohormone biosynthesis within host tissues. Their molecular machinery produces a variety of plant growth regulators, including auxins, CKs, and gibberellins, which create specific hormonal gradients that influence cellular differentiation and the development of organs. The intracellular production of these compounds provides spatially targeted growth regulation superior to exogenous applications. For instance, auxin biosynthesis by endophytic *Bacillus* populations demonstrates significant correlation with enhanced root system architecture, facilitating improved resource acquisition from deeper soil horizons (Glick, 2010).

Endophytic bacteria exhibit remarkable capabilities in modulating plant stress responses through sophisticated biochemical mechanisms. These microorganisms enhance host tolerance to diverse abiotic stresses through the synthesis of specialized metabolites (Kamran et al., 2022). They generate osmolytes such as proline and trehalose that maintain cellular integrity under stress conditions, while simultaneously producing antioxidant compounds that neutralize stress-induced reactive oxygen molecules. For example, *Bacillus cereus*, *Enterobacter tabaci*, *Curtobacterium oceanosedimentum*, etc., were observed to have improved saline tolerance of rice by deploying defense enzymes like IAA, ABA, and gibberellin. (Khan et al., 2020). Notably, endophytic strains expressing ACC deaminase activity modify ET-mediated stress responses, preserving root elongation potential under adverse conditions (Poria et al., 2022). Furthermore, some endophytes, such as *Trichoderma harzianum* Th-56, alter superoxide dismutase (SOD), proline, and lipid peroxidation products in rice plants, which results in an increased ability of rice to withstand drought conditions (Singh et al., 2023; Ameen et al., 2024).

In environmental remediation contexts, endophytic bacteria facilitate sophisticated detoxification processes within plant tissues. These specialized microorganisms mediate the transformation and sequestration of metallic pollutants, reducing their phytotoxicity through various biochemical pathways. Additionally, they possess extensive metabolic capabilities for degrading complex organic contaminants, converting these compounds into simplified, non-toxic derivatives suitable for plant assimilation or compartmentalization (Hayat et al., 2010).

The endophytic microbiome significantly augments plant immune responses through multiple mechanisms. These bacteria synthesize diverse antimicrobial compounds, including bioactive lipopeptides and volatile metabolites, establishing chemical barriers against pathogenic invasion. Furthermore, they modulate host immune system programming, enhancing defensive response efficiency against potential pathogens (Glick, 2010). For example, *Pseudomonas aeruginosa* and *Aneurinibacillus aneurinilyticus*, isolated from *Ocimum tenuiflorum* plants, could induce ISR in *Pisum sativum* against pathogenic *Fusarium* wilt (*Fusarium oxysporum* f.sp. pisi) (Gupta et al., 2022b).

This specialized bacterial community represents a cornerstone of agricultural sustainability, providing sophisticated biological mechanisms for enhancing crop performance and environmental resilience. Their intimate association with plant tissues enables precise regulation of growth parameters and stress responses, reducing dependence on synthetic agricultural inputs. Contemporary biotechnological approaches are facilitating the optimization of endophytic strains for specific agricultural applications, positioning these microorganisms as crucial elements in addressing global food security challenges and environmental sustainability imperatives.

1.4.6 Genetically Engineered Soil Microbiomes

The manipulation of soil microbiomes through genetic engineering represents a cutting-edge frontier in agricultural sustainability and environmental remediation efforts. These complex microbial ecosystems, encompassing diverse bacterial and fungal communities, serve as important regulators of essential ecological processes, from nutrient transformation to xenobiotic compound degradation. While native microbial populations demonstrate remarkable capabilities, their performance often falls short under extreme conditions or when faced with novel environmental challenges, highlighting the need for genetic enhancement strategies (Rebello et al., 2021). The inherent limitations of wild-type microorganisms, particularly in hostile environments, underscore the importance of genetic modification approaches (Kaul et al., 2021).

Environmental decontamination stands as a primary application for genetically modified soil microbiomes. Traditional microbial populations frequently exhibit suboptimal degradation rates when confronting recalcitrant pollutants, including agrochemicals, petroleum derivatives, and metallic contaminants. Through targeted genetic modifications, researchers have successfully augmented metabolic pathways, enabling enhanced degradation of aromatic compounds such as toluene and naphthalene (Rebello et al., 2021). Similarly, the incorporation of metal-sequestering protein sequences has dramatically improved microbial capacities for heavy metal extraction, particularly for elements like cadmium and mercury (Jansson et al., 2023).

Agricultural enhancement through engineered microbiomes extends beyond remediation applications. Modified strains exhibiting enhanced capabilities for atmospheric nitrogen fixation and mineral phosphate solubilization present sustainable alternatives to conventional fertilization practices, which often result in environmental deterioration (Kaul et al., 2021). The engineering of microorganisms to amplify phytohormone production, particularly IAA, has demonstrated significant improvements in plant root architecture and nutrient acquisition efficiency, leading to enhanced crop performance in suboptimal soil conditions (Rebello et al., 2021; Jansson et al., 2023). These innovations prove particularly valuable in addressing climate-induced stressors, including soil salinization and moisture deficiency (Kaul et al., 2021).

The development of engineered soil microbiomes leverages sophisticated biotechnological approaches, incorporating synthetic biology principles and precise genome editing methodologies. CRISPR–Cas-based modifications enable targeted genetic alterations to enhance environmental adaptation and metabolic efficiency. Research has demonstrated that engineered microbial communities, designed to execute complementary ecological functions, exhibit superior resilience and effectiveness compared to monoculture approaches (Jansson et al., 2023). These synthetic communities demonstrate remarkable stability in fluctuating environmental conditions (Kaul et al., 2021).

The implementation of genetically engineered microbiomes presents significant ecological considerations. The introduction of modified organisms may potentially disrupt indigenous microbial communities and ecosystem dynamics. Research has documented cases where modified rhizobial strains altered existing bacterial and fungal community structures, with a potential tendency of long-term soil ecosystem stability (Rebello et al., 2021). Environmental persistence and biosafety considerations necessitate comprehensive regulatory frameworks to minimize potential ecological disruptions (Jansson et al., 2023).

Despite these challenges, engineered soil microbiomes offer transformative potential for addressing contemporary environmental and agricultural challenges. These innovations show particular promise in climate change mitigation through enhanced carbon fixation and reduced emissions of greenhouse gases (Kaul et al., 2021; Jansson et al., 2023). Additionally, engineered microorganisms demonstrate significant potential for improving agricultural productivity in degraded soils, offering solutions for food security challenges in regions affected by environmental deterioration and climate instability (Rebello et al., 2021).

The advancement of this field necessitates integrated research approaches. The convergence of expertise from microbiology, synthetic biology, ecological sciences, and agricultural research enables comprehensive development of engineered soil microbiome technologies. Recent technological advancements, particularly in the field of metagenomics and metatranscriptomics, have facilitated detailed characterization of microbial functions and interactions, enabling more precise engineering strategies (Jansson et al., 2023). This technological progression ensures the development of engineered microbiomes that balance efficacy with environmental sustainability.

The field of genetically engineered soil microbiomes represents a revolutionary approach to enhancing terrestrial ecosystem health and agricultural productivity. These biotechnological innovations offer solutions to pressing environmental challenges, including soil contamination, ecosystem degradation, and agricultural sustainability. However, successful implementation requires continued scientific investigation, careful ecological assessment, and robust regulatory oversight to ensure responsible deployment and long-term sustainability (Rebello et al., 2021; Kaul et al., 2021; Jansson et al., 2023).

1.5 ROLE OF SOIL MICROBIOME AS BIOCONTROL AGENTS AGAINST PESTS AND DISEASES

Diseases and pests affecting plants present significant challenges to global food security, and the widespread utilization of chemical pesticides remains a common approach to mitigating damage caused by bacterial and fungal pathogens. These plant pathogens and pests significantly impact agricultural productivity, with plant diseases reducing the agricultural produce by 21–30% in key crops globally (Savary et al., 2019). Furthermore, numerous pathogens have acquired resistance to chemical control strategies (Lucas, 2011), leading to a rise in the occurrence of specific plant diseases. The reliance on chemical pesticides, while effective, poses substantial environmental and public health risks, as many insecticides persist in soil, leading to toxic residues (Gilden et al., 2010). There is an increasing recognition of these risks, and it is advised to transition toward more sustainable agricultural practices that utilize fewer chemicals (Donley, 2019). In response to the limitations of synthetic agrochemicals, there has been an increasing interest in environmentally friendly alternatives. With the increase in the global population, there is an urgent necessity to find sustainable alternatives to chemical pesticide use in agriculture. In recent years, microbial biological control agents (MBCAs) have emerged as a very powerful and eco-friendly sustainable solution to plant disease management, enhancing plant growth and crop production (El-Saadony et al., 2022). In addition to promoting plant development, PGPR and plant growth-promoting fungi (PGPF) can also combat plant diseases by producing bioactive compounds

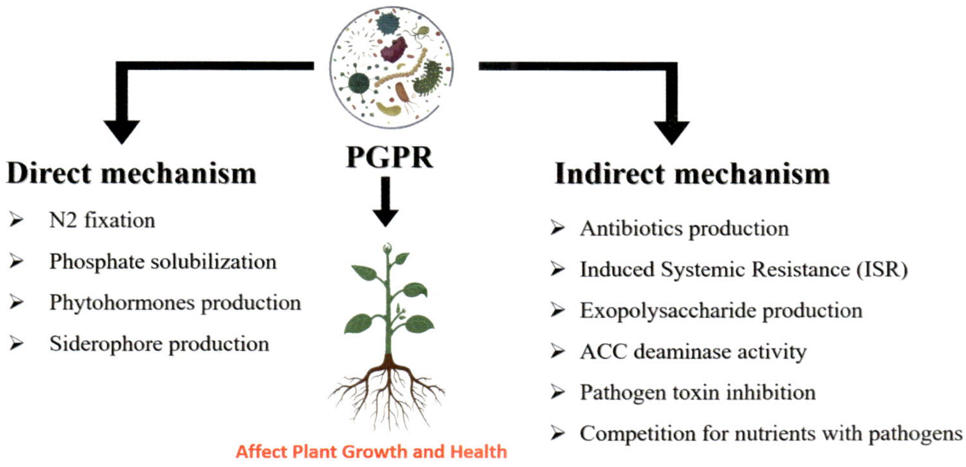

FIGURE 1.2 Mechanisms used by PGPR to control pests and stimulate growth of plants—direct and indirect mechanisms.

that inhibit pathogens and by triggering the plant's immune responses (Fones et al., 2020). The microbiome serves as a biocontrol agent, where the microbiota associated with plants can facilitate the synthesis of siderophores, volatile compounds, enzymes, and antibiotics, while also modulating levels of phytohormones (Vandana et al., 2021).

Additionally, the plant microbiome can suppress the growth and activity of pathogens by competing for nutrients and microenvironments, employing strategies such as parasitism, antibiosis, and enhancing the plants' immune system resistance. Microbes that protect against pathogens are primarily from the genera *Pseudomonas*, *Streptomyces*, *Bacillus*, *Paenibacillus*, *Enterobacter*, *Pantoea*, *Burkholderia*, and *Paraburkholderia* (Boro et al., 2022).

The soil microbiome facilitates plant growth through biocontrol mechanisms by enhancing the availability and absorption of vital nutrients required for plant development. Consequently, modifying or manipulating the plant microbiome could decrease the occurrence of plant diseases and improve agricultural yield. Therefore, changes in the microbiome significantly diminish the reliance on and use of chemical pesticides, which in turn reduces greenhouse gas emissions, promoting sustainable agricultural practices. In developed nations, biological control has emerged as a crucial strategy for achieving more sustainable, cost-efficient, and safer pest management, offering benefits over conventional synthetic chemical pest control (Fan et al., 2020).

However, for widespread adoption and registration for commercial use, MBCAs must address several challenges (Lucas, 2011). Different classes of microorganisms evolve with different mechanisms that control pests and diseases. A few of them are discussed below.

1.5.1 Biocontrol Mechanism using PGPR

PGPR offers a range of advantageous effects on plants, like enhancing the availability of nutrients, production of phytohormones, and improving stress tolerance. They aid phosphate solubilization and N_2 fixation, as well as synthesize plant hormones like jasmonic acid (JA), IAA, CKs, gibberellic acid (GA), and ET, which contribute to enhancing plants' growth and fitness (Gouda et al., 2018; Sham et al., 2019). PGPR promotes plant growth through two mechanisms: direct and indirect (Kundan et al., 2015), as illustrated in Figure 1.2.

Direct Effective Mechanisms: In these mechanisms, PGPR directly influence plant growth through the production of phytohormones, nitrogen fixation, the solubilization of nutrients such as phosphate and potassium, and the enhancement of iron uptake (Kundan et al., 2015).

Indirect Effective Mechanisms: These mechanisms encompass the production of antibiotics, induced systemic resistance (ISR), competition for nutrients against pathogens, and the inhibition of pathogen toxins (Kundan et al., 2015; Elnahal et al., 2022). Additionally, PGPR generate signaling metabolites that assist plants in coping with various abiotic stresses caused by drought or salinity (Jochum et al., 2019). Moreover, PGPR can modify root morphology, thereby increasing root surface area and enhancing nutrient absorption (Abbas et al., 2019). They can effectively outcompete detrimental bacteria by rapidly colonizing the root system and securing available nutrients, which prevents the proliferation of other organisms (Lyu et al., 2020; Abd El-Mageed et al., 2020).

PGPR employ diverse strategies to colonize specific plant hosts, including the synthesis of antimicrobial compounds such as bacteriocins, antibiotics, and siderophores that inhibit microbial pathogens by targeting their cellular structures and metabolic functions (Abriouel et al., 2011).

Bacteriocins are secondary metabolites secreted by microorganisms to eliminate closely related strains (Riley and Wertz, 2002). The majority of bacteriocins produced by gram-negative bacteria are referred to as "colicins," which are proteins generated by certain strains of *Escherichia coli* that are lethal to related organisms.

Other bacteriocins, such as pyocins from *Pseudomonas aeruginosa* strains, cloacins from *Enterobacter cloacae*, marcescins from *Serratia marcescens*, and megacins from *Bacillus megaterium*, have also been identified and named similarly to colicin, a term derived from *E. coli* (Cascales et al., 2007). Notably, bacteriocins from *Bacillus* spp. are gaining increasing importance due to their often broader spectrum of inhibition (Table 1.3), which includes gram-negative bacteria, yeasts. Siderophores are another type of molecules produced by PGPR, which solubilize the iron present in the surrounding environment to form a complex (ferric-siderophore complex), which is then transported inside or outside of the cell through diffusion (Andrews et al., 2003).

Bacterial siderophores are classified into four major classes—carboxylate, phenol catecholates, hydroxamates, and pyoverdines, according to their structural characteristics, iron-coordinating functional groups, and ligand types (Crowley, 2006). For cultivable bacteria, hundreds of siderophores have been found and documented; some are species-specific, while others are broad range and inhibit many microbes (Crowley, 2006; Sandy and Butler, 2009). The potential of certain microbes to enhance plant development in soil is mostly determined by their siderophore-producing activity.

PGPR can colonize roots and exclude other microorganisms from this ecological niche due to the competitive advantages conferred by siderophore synthesis (Haas and Défago, 2005).

Among the various bacterial siderophores studied, those produced by pseudomonads are particularly noted for their strong affinity for ferric ions. In iron-deficient soils, a pseudobactin siderophore produced by the *P. putida* B10 strain effectively suppressed *Fusarium oxysporum*; however, this suppression was negated when iron was reintroduced into the soil, which inhibits the microbial synthesis of iron chelators (Kloepper et al., 1980b).

In addition to siderophore production, the generation of low-molecular-weight compounds known as antibiotics are secreted by many PGPB which act as antagonistic agents against phytopathogens and serve as biocontrol agents (Duffy, 2003; Haas and Keel, 2003; Glick et al., 2007). These antibiotics can be classified into six distinct categories: hydrogen cyanide (HCN; which is volatile), cyclic lipopeptides, pyrrolnitrin, pyoluteorin, phenazines, and phloroglucinols (Haas and Défago, 2005). Numerous bacterial and fungal strains produce a variety of antibiotics that target their respective hosts (Table 1.4) in multiple ways, such as affecting transcription and translation, altering cell membrane structures, and inhibiting the synthesis of cell walls in pathogens (Maksimov et al., 2011). For instance, antibiotics secreted by various *Bacillus* species (including polymyxin, circulin, and colistin) inhibit a range of bacteria (both gram-positive and gram-negative) as well as numerous pathogenic fungi (Maksimov et al., 2011).

Moreover, PGPR are involved in reducing the harmful effects of various plant pathogens by producing a plethora of lytic enzymes like chitinases, cellulases, 1,3-glucanases, proteases, and lipases, which are involved in the cell wall

TABLE 1.3
List of Bacteriocins Produced by Various Microbes

Bacteriocin/Compound	Producing Microbe	Target/Inhibited Species	Additional Features	References
Malacidin A	Soil-derived microbial community (metagenomic discovery)	Gram-positive pathogens, including *Staphylococcus aureus* (incl. MRSA)	Calcium-dependent activity, a novel class discovered through culture-independent soil screening	Hover et al. (2018)
Desotamides B, C, D	*Streptomyces scopuliridis*	Gram-positive bacteria	Cyclic hexapeptides, bacteriocin-like; novel antimicrobial peptides from soil *Streptomycete*	Fazal et al. (2020)
Bacillocin Bb	*Bacillus brevis Bb (soil)*	Gram-positive *Bacillus* strain BC31	Approx. <10 kDa, thermostable, pH stability pH 1–9, detergent/solvent resistant, titer 5280 AU/mL	Saleem et al. (2009)
Pyocin Pa (bacteriocin)	*Pseudomonas aeruginosa* Pa (soil)	Gram-positive *Bacillus* BC31	olubil <10 kDa, highly stable pH 1–11, thermostable, titer 640 AU/mL, activity lost with proteinase K (Proteinaceous)	Saleem et al. (2009)
Cerein 7	*Bacillus cereus* (rhizosphere soil)	Other Gram-positive bacteria	Approx 3.94 kDa peptide	Subramanian and Smith (2015)
Subtilin	*Bacillus subtilis* (rhizosphere/soil)	Gram-positive species	Widely studied antibiotic; ribosomally synthesized peptide; used as a model bacteriocin.	Subramanian and Smith (2015)
Brevibacillus laterosporus bacteriocin-like compound	*B. laterosporus* (soil isolate)	~83% of tested bacteria, including *E. coli*, MRSA, VRE	Broad-spectrum antibacterial; non-hemolytic; high coverage of clinical pathogens	Pino-Hurtado et al. (2024)

TABLE 1.4
List of Antibiotics Produced by Various Microbes

Antibiotic Produced	Producing Organism	Against	Reference
Pyrrolnitrin	*P. fluorescens* BL915	*Rhizoctonia solani*	Hill et al. (1994)
2,4-Diacetylphloroglucinol (DAPG)	Pseudomonads	*Pythium* spp.	de Souza et al. (2003)
Phenazine	Pseudomonads	*Gaeumannomyces graminis*, *F. oxysporum*	Chin-A-Woeng et al. (2003)
Phenazine-1-carboxamide	*P. chlororaphis* PCL1391	Tomato	Hernandez et al. (2004) and Haas and Défago (2005)
Surfactin	*Bacillus subtilis* strain 6051	*P. syringae* pv *DC3000*	Bais et al. (2004)

lysis of many pathogenic fungi. Along with these, PGPR also produce different antibiotics for inhibition of the growth of plant pathogens; however, sometimes bacteria may develop resistance against these antibiotics (Schippers et al., 1987; El-Saadony et al., 2022).

Furthermore, a plant's specific microbiome triggers induced systemic resistance (ISR) when it interacts with the plant and improves its resistance against diseases by inducing defense mechanisms (Raza et al., 2016). In ISR, the resistance has been induced after the attack of a pathogen at a specific site of the plant. Although it is a non-specific mechanism, it provides increased resistance against plant pathogens to control diseases using jasmonate and ET as signaling molecules (Suslow et al., 1979).

Thus, biocontrol agents are one of the most promising approaches for sustainable agriculture and offer an eco-friendly method for pest management. This approach depends on the utilization of live microorganisms to effectively diminish pest populations in a sustainable, dependable, and eco-friendly manner.

1.6 ROLE OF SOIL MICROBIOME IN NUTRIENT CYCLING AND SOIL FERTILITY

Soil health plays a fundamental role in ensuring productive agriculture, climate change, population growth, and soil degradation. Maintenance of healthy soils has become even more critical for food security. Historically, healthy soil has been the foundation of successful farming, with its composition of sand, silt, and clay influencing nutrient cycling and water retention (Xing et al., 2025). Over centuries, agricultural techniques like crop rotation and the usage of nitrogen-fixing legumes have been key to maintaining soil fertility. Furthermore, the creation of synthetic fertilizers, particularly following the advancement of the Haber–Bosch process in the early 20th century, significantly enhanced agricultural productivity, particularly during the Green Revolution (Hayat et al., 2025).

The soil microbiome primarily consists of bacteria, various fungal species, protists, and other microorganisms that play a crucial role in the development of soil structure, fertility, and plant health (Suman et al., 2022). Healthy soils, with diverse microbial communities, form stable, water-retaining macroaggregates. These aggregates help the soil retain moisture, which reduces runoff and flooding, while improving nutrient availability and fertilizer efficiency.

The soil microbiome comprises taxonomically diverse prokaryotic and eukaryotic microorganisms that collectively influence soil aggregation through various mechanisms:

1. *Secretion of Exopolysaccharides (EPS)*: Various bacteria secrete exopolysaccharides, which function as biological adhesives that bind primary soil particles into microaggregates (20–250 μm) (Costa et al., 2018).
2. *Presence of Fungal Hyphae*: A vast range of fungal species present in the soil belong to the ascomycetes and basidiomycetes classes and interact with plant roots as mycorrhiza (Bowles et al., 2018). The mycelium of the fungi physically entangles the soil particles due to their adhesive nature and forms microaggregates, which further assemble to form macroaggregates (>250 μm), creating a hierarchical soil structure with enhanced porosity (Totsche et al., 2018).
3. *Production of Glomalin-Related Soil Proteins (GRSPs)*: GRSPs are glycoproteins found in soil, usually secreted by AM fungi, which form a complex mixture with other substances such as humic and soil organic matter. These GRSPs are hydrophobic in nature and capable of forming stable soil aggregates (Ling et al., 2025; Wang et al., 2021)

These biologically mediated structural improvements create a favorable soil architecture with 40–60% pore space that optimizes water infiltration rates, gas exchange, and root penetration while simultaneously reducing bulk density.

The soil microbiome orchestrates essential biogeochemical transformations that underpin ecosystem functioning and agricultural productivity. Within the nitrogen cycle, diazotrophic bacteria such as *Rhizobium*, *Bradyrhizobium*, and *Azospirillum* convert atmospheric N_2 to plant-available NH_3 through specialized nitrogenase enzyme complexes, contributing 40–60 kg N/ha/year in legume-based agricultural systems (McNeill and Unkovich, 2007). This biological nitrogen fixation is complemented by nitrification, a two-step process where ammonia-oxidizing bacteria and archaea transform NH_4^+ to NO_2^-, followed by nitrite-oxidizing bacteria converting NO_2^- to NO_3^-. Denitrification serves as a possible route for nitrogen loss, occurring via the anaerobic conversion of NO_3^- to N_2 by facultative anaerobes. In contrast, the dissimilatory nitrate reduction to ammonium (DNRA) pathway plays a crucial role in retaining nitrogen within the soil ecosystem (Liao et al., 2025).

Phosphorus cycling is similarly dependent on microbial activities, with phosphate solubilization performed by *Pseudomonas*, *Bacillus*, and *Penicillium* species that produce organic acids and phosphatases to convert insoluble phosphorus forms into plant-available orthophosphates. AM fungi establish extensive hyphal networks that extend up to 100 times beyond plant root zones, effectively mining soil for phosphorus and other immobile nutrients otherwise inaccessible to plant roots (Bhantana et al., 2021). These fungi transfer essential minerals to their plant hosts through their roots. In turn, they got carbon compounds formed by the process of photosynthesis by plants, creating a bidirectional nutrient exchange system that optimizes resource acquisition (Bowles et al., 2018).

1.7 MICROBIOTA-MEDIATED NITROGEN FIXATION

Microbes associated with plants are crucial for enhancing nutrient absorption, improving stress resilience, and promoting overall plant vitality. Notably, nitrogen-fixing bacteria, especially those that form symbiotic and associative relationships with different plants/crops (Table 1.5), have attracted considerable interest because of their capacity to increase the nitrogen availability for plants by transforming atmospheric nitrogen into forms that are usable by plants (Mahmud et al., 2020). Research has primarily focused on soil-associated diazotrophs, but there has been increasing interest in other plant-associated niches like xylem, rhizomes, and aerial roots.

TABLE 1.5
Example of a Few N_2-Fixing Bacteria Associated with Different Plants

Bacterial N_2 Fixer	Crop	References
Bradyrhizobium spp. and *Azorhizobium* sp. *Beijerinckia* sp.	Sugarcane, sweet potato	Dobereiner (1961) and Hill et al. (1983)
Azospirillum brasilense	Foxtail millet	Okon et al. (1983)
Klebsiella pneumonia	Wheat	Iniguez et al. (2004)
Azotobacter and *Clostridia*	Paddy	Yoshida and Ancajas (1973)
H. seropedicae, *A. brasilense*, and *B. unamae*.	Maize	Van Deynze et al. (2018)
Burkholderia spp., and *Ralstonia taiwanensis*	Switch grass	Compant et al. (2008)

The nitrogen-fixer microbiome can be divided into three types:

1. *Symbiotic Nitrogen Fixers*: These microorganisms engage directly with the roots or stems of plants, creating specialized structures referred to as nodules, which function as the site for nitrogen fixation. The bacterial partner provides nitrogen to the plant in exchange for sugars and carbohydrates from plants. For example, *Rhizobium* forms symbiotic relationships with a variety of legume plants (peas, beans, etc.) and forms root nodules. The composition of the microbiome differs as per crop (Qiao et al., 2024).
2. *Associative Nitrogen Fixers*: In this type, N_2 is fixed in a non-symbiotic mechanism in which various types of rhizosphere-associative N_2-fixing bacteria fix nitrogen for the plant by utilizing carbon along with energy sources available in the surrounding environment, rather than from the plant. These microbes do not form any kind of nodules and usually release the fixed nitrogen for the plants after their lysis, for example, *Azospirillum*, *Herbaspirillum*, and *Burkholderia* (White et al., 2012; Delwiche and Wijler, 1956; Grover et al., 2023).
3. *Free-Living Nitrogen Fixers*: The soil microbiome contains various heterotrophic bacteria that can fix a substantial amount of nitrogen without any direct interaction with plants.

These microbes live chemolithotrophically in soil, where they use inorganic compounds as their energy source and use exudates of plants as carbon and energy sources. Examples—*Azotobacter*, *Bacillus*, *Clostridium*, and *Klebsiella* (Wagner, 2011).

Exudates released from the roots have a complex composition consisting of various simple as well as complex sugars along with different fatty acids, organic acids, vitamins, and amino acids, which play a crucial role in the attraction and assembling of various types of microbes to the roots (Neumann and Romheld, 2007; Turner et al., 2013).

Recent studies revealed the purpose of mucilage in nitrogen fixation in aerial roots by various diazotrophs. The microbial community within the mucilage was found to be enriched in N_2-fixing bacteria such as *Burkholderia*, *Herbaspirillum*, and *Azospirillum*, highlighting the crucial role of the mucilage–microbiota system in meeting the nitrogen needs of the plant.

For example, maize aerial roots exude a carbohydrate-rich mucilage that attracts N_2-fixing bacteria and plays an instrumental role in supplying nitrogen to the plant from the atmosphere (Van Deynze et al., 2018; Pang et al., 2023).

1.8 MICROBIOTA-MEDIATED PHOSPHORUS SOLUBILIZATION

Phosphorus is a critical nutrient essential for plant growth, but since it is not readily available in soils, it limits plant growth, making it essential for agricultural productivity (Kavanagh et al., 2018). However, the widespread usage of high-concentration phosphate fertilizers has raised concerns about the sustainability of phosphorus resources, as natural phosphorus reserves are non-renewable. Moreover, much of the phosphorus applied through fertilizers becomes unavailable to plants or leads to environmental imbalances in soil and water. In response to these challenges, microorganisms like bacteria, fungi, and mycorrhizal fungi offer promising solutions (Bilal et al., 2021). These microorganisms engage directly with the roots or stems of plants which improves phosphorus cycling in the soil and supports the growth of crops while minimizing the environmental impact of traditional fertilizer use. Therefore, microorganisms play a crucial role in advancing sustainable agricultural methods by enhancing phosphorus bioavailability and reducing the need for synthetic fertilizers (Doilom et al., 2020).

Phosphate-solubilizing microorganisms (PSM) encompass a diverse assembly of microbes, including actinobacteria, bacteria, mycorrhizal fungi, and blue-green algae, which can transform both organic and inorganic insoluble forms of phosphorus into soluble forms that plants can absorb (Rawat et al., 2021). These microbes are found abundantly in soil, especially in the rhizosphere, where they play an essential role in enhancing phosphorus availability. A study conducted by Djuuna et al. in 2022 in Indonesia showed that significant populations of solubilizing bacteria and fungi are available in agricultural soils. The bacterial populations ranged between 25×10^3 and 550×10^3 CFU per gram of soil, while fungal populations ranged between 2.0×10^3 and 5.0×10^3 CFU per gram of soil. PSM exhibits a wide diversity across different genera (Kalayu, 2019). Bacteria from genera including *Azospirillum*, *Bacillus*, *Pseudomonas*, *Rhizobium*, and *Serratia* are common solubilizers. Fungal genera like *Penicillium*, *Fusarium*, and *Trichoderma* also play significant roles, while mycorrhizal fungi, like *Rhizophagus irregularis* and *Glomus mossea*, form symbiotic relationships with plant roots and increase the uptake of phosphorus (Shrivastava et al., 2018). Additionally, actinobacteria from genera like *Streptomyces* and *Thermobifida*, as well as cyanobacteria such as *Anabaena variabilis* and *Scytonema* sp., contribute to phosphorus solubilization. This diversity of PSM highlights the complex interactions in the soil environment that contribute to phosphorus availability and support plant growth (Behera et al., 2014).

PSMs mobilize phosphorus through various mechanisms (Sharma et al., 2013), including solubilizing inorganic phosphorus, mineralizing organic phosphorus, and releasing phosphorus bound in soil complexes (Silva et al., 2023).

Key mechanisms include:

1. *Organic Acid Production*: PSMs secrete a variety of organic acids like gluconic acid and citric acid, which lower soil pH and chelate cations (e.g., calcium, iron, and aluminum) that bind phosphorus, making it soluble and hence enhancing its availability to plants (McGill et al., 1981; Tate, 1984).
2. *Enzymatic Activity*: Enzymes like phosphatases and phytases produced by PSMs hydrolyze organic phosphorus compounds into inorganic forms that plants can absorb (Zhu et al., 2011).
3. *Proton Release*: Microorganisms release protons during their metabolic processes, acidifying the rhizosphere and enhancing phosphorus solubilization.
4. *Siderophore Production*: Siderophores are iron-chelating compounds, which indirectly mobilize phosphorus bound to iron complexes (McGill et al., 1981; Cui et al., 2022).
5. *Direct Oxidation of Glucose*: In this process, gluconic acid is generated through the activity of the glucose dehydrogenase enzyme, which additionally aids in the solubilization of phosphorus (Song et al., 2008).

To summarize, microorganisms, especially those that solubilize phosphate, offer a viable and sustainable approach to enhancing soil quality and plant vitality (Chen et al., 2021). By increasing the availability of nutrients, these microbes can diminish reliance on chemical fertilizers, thereby fostering more eco-friendly and resource-efficient agricultural methods.

1.9 ROLE OF SOIL MICROBIOME IN DECOMPOSITION OF ORGANIC MATTER

Soil organic matter (SOM) is predominantly found in the upper approximately 20–30 cm of the soil profile and is composed of intricate organic macromolecules that include elements such as carbon, oxygen, nitrogen, phosphorus, and sulfur (Senesi and Loffredo, 1999).

These organic substances primarily originate from plants via photosynthesis and play a critical role in plant growth and functioning, as well as in sustaining soil fertility (Manlay et al., 2007).

The decomposition of organic matter in the soil is largely facilitated by microorganisms, including bacteria and fungi that inhabit the soil (Table 1.6). This decomposition process is essential as it transforms organic matter into simpler substances like CO_2, water, and various nutrients such as nitrogen, phosphorus, and sulfur, which are then made available to plants.

Nevertheless, the rate and extent of decomposition are influenced by a range of environmental factors such as:

- *Soil temperature*: Temperature severely affects the growth of microbes in any environment. It has been observed that warmer conditions often accelerate microbial growth and their activity, speeding up decomposition.
- *Soil moisture*: Water availability is essential for microbial life and decomposition processes. Microbial activity decreases as the water activity decreases.

TABLE 1.6
Examples of Microorganisms Involved in SOM Degradation

Microorganism Involved	Role in SOM Degradation	Reference
Bacillus subtilis *Bacillus licheniformis*	Break down organic matter into essential nutrients.	Wang et al. (2024), Bais et al. (2004), and Lee et al. (2022)
Cellulomonas uda	Decompose intricate macromolecular organic substances (cellulose) into more basic compounds.	Pérez-Avalos et al. (2008) and Ontañon et al. (2021)
Pseudomonas sp	Degrade polycyclic aromatic hydrocarbons (PAH)	Saeed et al. (2022)
Flavobacterium	Breaks down organic contaminants such as phenolic compounds	Larsbrink et al. (2016)
Bacillus mucilaginous	Decompose feldspar and convert the insoluble potassium components into soluble components.	Ahmad et al. (2016) and Tallapragada and Matthew (2021)
Pseudomonas aeruginosa *Chromobacterium violaceum* *Bacillus subtilis*	Collaborate to discharge organic compounds, including polysaccharides and adhesives, into the soil to enhance its health.	Martens and Frankenberger (1992)
Archaea (Methanogenic archaea and ammonia-oxidising archaea)	The breakdown of organic matter, the nitrogen cycle, and the utilization of methane contribute to the promotion of plant growth and overall health.	Jung et al. (2020) and Ni et al. (2023)

- *Oxygen availability*: Microbial decomposition needs oxygen, especially for aerobic microbes, while degradation done by anaerobic microbes does not require oxygen.
- *Nitrogen content*: The nitrogen-to-carbon ratio (N:C ratio) in organic material influences decomposition. High N:C ratio limits the microbial activity due to a deficiency of carbon required for growth and thus slows down the decomposition, while in low N:C ratio microorganisms grow rapidly, leading to much quicker decomposition, but sometimes this may lead to ammonia toxicity due to nitrogen accumulation (Yang et al., 2020; Ekinci et al., 2019).
- *Quality and quantity of available carbon substrates*: Different types of organic matter (e.g., cellulose, lignin, lipids) decompose at different rates. The availability and quality of carbon substrates significantly influence the rate and extent of decomposition and degradation processes, impacting both soil health and the overall carbon cycle. Higher quality, readily decomposable carbon sources generally promote faster degradation rates compared to complex, recalcitrant substrates. Additionally, the quantity of available carbon can either stimulate or inhibit decomposition, depending on the specific context and the nature of the soil ecosystem (Jayasekara et al., 2025).

The breakdown of soil organic carbon (SOC) is crucial in the global carbon cycle, affecting the manner in which carbon is either stored or emitted into the atmosphere as carbon dioxide. This has significant consequences for carbon sequestration, defined as the process of capturing and storing atmospheric carbon dioxide to alleviate climate change. Precise modeling of SOC decomposition rates is vital for comprehending and forecasting the effects of land use, forest management, and global climate change on carbon storage.

1.10 ROLE OF SOIL MICROBIOME IN STRESS RESISTANCE AND RESILIENCE

Plants encounter a variety of biotic and abiotic stresses throughout their lifespan. Abiotic stress encompasses conditions such as drought, salinity, metal toxicity, and other adverse environmental factors, while biotic stress involves pests and pathogens that frequently lead to disease manifestation in plants (Chandra et al., 2024). These stresses trigger alterations in plants, resulting in the activation of their defense mechanisms, as well as changes in transcriptomics and metabolomics, which culminate in the release of leaf and root exudates. The production of these exudates not only stimulates the plant's defense system but also influences the microbial communities associated with the plants. Nevertheless, the plant immune system alone is insufficient to safeguard against pathogens. Various microbes present in the soil and the rhizosphere of plants assist in combating these pathogens (Liu et al., 2020). Members of the plant microbiome can provide the host plant with enhanced immunological capabilities at both individual and community levels. The traits conferred by these microbial communities are contingent upon the interplay between the nutritional properties of the soil and the plant's immune response (Vannier et al., 2019; Berg and Koskella, 2018; Castrillo et al., 2017). There exist multiple defense strategies, including the secretion of antibiotics, siderophores, and other compounds by plants to address biotic stress and diseases, as previously mentioned. In response to abiotic stresses such as salinity and drought, plants activate additional mechanisms in conjunction with their associated microbiome.

1.10.1 "Cry for Help" Mechanism

Plants use a defense mechanism called the "cry for help" in response to abiotic stressors, including drought, salt, or pest and disease attacks (Bakker et al., 2018). Through this mechanism, plants recruit beneficial soil microbes by secreting a varied range of chemical signals, thereby enhancing their capacity to reduce the damaging effects of these kinds of stresses (Rizaludin et al., 2021; Rolfe et al., 2019; Bakker et al., 2018; Carrión et al., 2019). This strategy is particularly crucial because abiotic stresses not only affect plant health but also have lasting impacts on the surrounding microenvironment and future plant generations. The interactions between plants and recruited beneficial microbes can be heritable, leading to a phenomenon termed soil-borne legacy, where offspring benefit from enhanced resilience due to inherited microbial associations (Yuan et al., 2018; Bakker et al., 2018).

Plants adopt several specific strategies as part of the "cry for help" response:

- **Co-adaptive strategies**: Under drought conditions, within the rhizosphere and endosphere, plants experience an increase in beneficial microbial populations, such as *Acinetobacter* species and other monoderms (Santos-Medellín et al., 2017; Fitzpatrick et al., 2018).
- **Symbiotic nutrient acquisition**: During phosphate or nitrogen starvation, plants enhance symbiotic relationships with beneficial microbes to optimize nutrient uptake (Van Deynze et al., 2018; Ham et al., 2018).
- **Selective recruitment via root exudates**: Plants modulate root exudates to selectively promote chemotaxis, swarming, and biofilm formation by beneficial microbes. These microbes, in turn, evolve mechanisms to activate plant defense signaling pathways and inhibit pathogen proliferation (Dudenhöffer et al., 2016; Xie et al., 2022).

Several Examples Illustrate This Phenomenon

- Leaves of *Arabidopsis* infected with *Pseudomonas syringae* pv. Tomato (Pst) was found to attract the beneficial bacterium *Bacillus subtilis* FB17 to the roots, which in turn initiated systemic resistance against subsequent Pst infections (Rudrappa et al., 2008).
- The rhizosphere of *Arabidopsis* was also found to be enriched with various bacterial taxa, including *Microbacterium* sp., *Stenotrophomonas* sp., and *Xanthomonas* sp., due to infection by the downy mildew pathogen *Hyaloperonospora arabidopsidis*.

A recent field investigation found that durum wheat (*Triticum turgidum* L. var. durum) naturally infected with the crown-rot pathogen *Fusarium graminearum* attracted *Stenotrophomonas rhizophila* (SR80) to its rhizosphere and root endosphere, providing additional evidence in favor of the "cry for help" theory (Liu et al., 2021). Remarkably, reintroduction of strain SR80 into wheat plants induced resistance against the crown-rot disease and promoted plant growth (Liu et al., 2021).

1.11 ROLE OF SOIL MICROBIOME FOR IMPROVING CROP QUALITY

Enhancing crop quality is crucial for advancing agricultural practices and boosting economic development. As sustainable development becomes a global priority, there is a rising need for environmentally friendly and non-toxic methods to improve agriculture. Soil bacteria, especially those in the rhizosphere and topsoil, play an essential role in soil development and fertility, creating conditions conducive to agriculture.

A classic example of beneficial interaction is the lichen symbiosis, where fungi collaborate with algal partners. Similarly, leguminous plants develop root nodules that accommodate nitrogen-fixing bacteria, which transform atmospheric nitrogen into a form that can be utilized by plants.

Furthermore, genetic engineering has allowed the transfer of beneficial traits, such as nitrogen fixation, through vectors like the Ti plasmid from *Agrobacterium tumefaciens*. This approach has led to the development of GMOs such as Bt cotton and Bt brinjal for insect resistance, as well as biofortified crops such as Golden Rice enriched with provitamin A (East, 2013; Zhang et al., 2015). Additionally, genes from blue-green algae, purple sulfur bacteria, and microalgae like *Spirulina* have been explored as sources of lipopolysaccharides (LPS) and protein-rich biomolecules.

Thus, these microorganisms increase the crop quality by acting as biofertilizers and biostimulants, which enhance the fertility of the soil and plant health.

1.12 MICROBES AS BIOFERTILIZERS

The Green Revolution aimed to enhance agricultural productivity through the use of high-yielding varieties (HYVs), chemical fertilizers, pesticides, and advanced irrigation techniques to address hunger issues (Fitzpatrick et al., 2018). Nevertheless, the over-reliance on agrochemicals has resulted in soil degradation, water pollution, increased greenhouse gas emissions, and negative impacts on human health (Funabashi, 2018; Fanelli, 2020). Overuse of nitrogen-based fertilizers can cause groundwater pollution, eutrophication, and atmospheric contamination by ammonia (NH_3) and nitrogen oxides (NOx), contributing to ozone layer depletion (Kanter and Searchinger, 2018). Therefore, alternative sustainable soil inputs are necessary.

Research indicates that the soil microbiome plays a crucial role in enhancing soil productivity (East, 2013; Zhang et al., 2015; Imran et al., 2021). As biofertilizers, PGPB offer a sustainable alternative to GMOs and synthetic fertilizers (Rozier et al., 2017; Itelima et al., 2018). By maintaining appropriate levels of macro- and micronutrients and facilitating the production of plant hormones, biofertilizers not only boost crop yields but also help plants combat biotic and abiotic stresses (Xu et al., 2014). A variety of beneficial microorganisms, such as phosphate-solubilizing and nitrogen-fixing bacteria, present in the rhizosphere can

TABLE 1.7
Variety of Microorganisms Used as Biofertilizers and Biostimulants to Support Plant Health and Productivity

Biofertilizer Type	Microbial Source
N_2 fixing—Symbiotic	*Rhizobium, Frankia, Anabaena azollae*
N_2 fixing—Free-living	*Clostridium, Azotobacter, Nostoc, Beijerinkia*
Phosphate solubilizers	*Penicillium, Bacillus subtilis, Aspergillus, Pseudomonas*
Plant growth-promoting rhizobacteria (PGPR)	*Pseudomonas, Bacillus*
Bacterial-based biostimulants	*Bacillus licheniformis, Pseudomonas fluorescens, Streptomyces* spp.
Fungal-based biostimulants	*Trichoderma harzianum, Rhizophagus intraradices*
Cyanobacteria-based biostimulants	*Anabaena* spp., *Nostoc* spp.
Yeast-based biostimulants	*Saccharomyces cerevisiae*

serve as biofertilizers (Battini et al., 2017; Yasmeen et al., 2021). Some of the microbial-based biofertilizers are mentioned in Table 1.7. Furthermore, PGPB transforms unavailable phosphorus into soluble forms that plants can readily absorb (Umesha et al., 2018). These microorganisms also synthesize IAA, which enhances plant growth by stimulating cell division, elongation, fruiting, and flowering (Kumar et al., 2018).

Biofertilizers can mitigate abiotic stresses such as drought, salinity, and heavy metal contamination by modulating ET levels through the activity of ACC deaminase, which converts ET into α-ketobutyrate and ammonium (Tiwari et al., 2018; Van Oosten et al., 2018). This process also diminishes reactive oxygen species (ROS) levels, enhances potassium uptake, and improves photosynthesis and water absorption (Safdarian et al., 2019; Yoolong et al., 2019).

Additionally, biofertilizers can safeguard plants by functioning as biocontrol agents, producing antimicrobial compounds, or strengthening the plant's immune system (Tilman et al., 2002).

1.13 MICROBES AS BIOSTIMULANTS

Biostimulants encompass a diverse array of substances, both natural and synthetic, that can be utilized on soils or plants to improve their growth, development, and resilience against various stresses. This category includes a wide variety of products such as protein hydrolysates, humic substances, seaweed extracts, and microbial inoculants (Bulgari et al., 2015).

Microbial inoculants, which consist of live microorganisms like bacteria, fungi, and yeasts, are introduced to soils or plants to foster resilience and promote growth (Table 1.7). They function by generating chemicals that enhance plant development, fixing nitrogen, and solubilizing phosphate (Vejan et al., 2016; Bargaz et al., 2018; Canellas et al., 2015).

As noted by Battacharyya et al. (2015), organic matter decomposes into humic substances that are crucial for improving soil structure and nutrient availability, thereby facilitating plant growth and nutrient absorption.

Seaweed extracts, derived from various marine algae, are rich in bioactive substances such as polysaccharides, amino acids, and plant growth regulators. These extracts have been shown to support plant development and enhance stress tolerance (Colla et al., 2015).

Protein hydrolysates, produced through the chemical or enzymatic hydrolysis of proteins from microorganisms, plants, or animals, can stimulate defense mechanisms and promote plant growth due to the bioactive peptides and amino acids they contain (Le Mire et al., 2016).

The increasing demand for environmentally sustainable agricultural practices has led to a heightened interest in the application of biostimulants in agriculture (Rouphael and Colla, 2018; Nardi et al., 2016; Mącik et al., 2020).

One of the primary advantages of biostimulants is their ability to enhance plant growth and stress tolerance without the harmful effects on the environment and human health that are often associated with chemical fertilizers and pesticides (Rouphael and Colla, 2018).

Biostimulants improve the quality and yield of crops by facilitating nutrient absorption, promoting optimal plant growth, and enhancing resistance to abiotic stresses such as salinity, drought, and extreme temperatures (Calvo et al., 2014).

Moreover, by activating natural defense mechanisms, encouraging the proliferation of beneficial microbial communities, and increasing the production of secondary metabolites with antimicrobial properties, biostimulants can enhance a plant's resilience against pests and diseases (Van Oosten et al., 2017). Their application fosters the establishment of more robust and sustainable agricultural systems (Povero et al., 2016; Pichyangkura and Chadchawan, 2015).

Crucially, biostimulants also contribute to the preservation of biodiversity, soil health, and water quality by minimizing dependence on chemical inputs (Pylak et al., 2019; Calvo et al., 2014).

Furthermore, by enhancing soil structure and fertility, biostimulants play a significant role in the long-term sustainability and productivity of agricultural lands (Sharma et al., 2014).

1.14 ROLE OF THE SOIL MICROBIOME IN SUSTAINABLE AGRICULTURE

Sustainable agriculture relies not only on plant genetics and farming techniques but also on the unseen yet essential realm beneath our feet—the soil microbiome. This intricate ecosystem, which includes bacteria, fungi, archaea, protozoa, and various other microorganisms, plays a crucial role in plant health, crop yield, and the overall sustainability of agriculture. These microbial communities improve soil fertility, aid in nutrient absorption, and promote vigorous plant growth, ultimately leading to increased yields and more resilient agroecosystems (Chen et al., 2024). A flourishing and diverse microbiome decreases the dependence on synthetic agrochemicals, such as chemical fertilizers and pesticides, which frequently result in long-term detrimental ecological effects. By encouraging natural nutrient cycling and enhancing plant defense mechanisms, beneficial microbes contribute to the reduction of environmental degradation. Consequently, leveraging microbial potential is increasingly recognized as a fundamental aspect of sustainable agriculture, essential for ensuring food security and ecological equilibrium for future generations (Suman et al., 2022).

To comprehend the composition and dynamics of the soil microbiome, it is necessary to examine the factors that influence it. Metagenomic research indicates that the makeup of microbial communities is significantly affected by both biotic and abiotic stresses. Key factors include the genotype of the host plant and its surrounding geographical context (Dastogeer et al., 2020).

The plant genotype influences microbiome assembly through differences in immune signaling, root exudates, age, morphology, and physiological characteristics. These elements collectively shape the microbial communities present in both the rhizosphere and phyllosphere, resulting in species-specific microbial profiles. For instance, research comparing Indica and Japonica rice varieties has demonstrated that Indica roots support a greater abundance and diversity of nitrogen-fixing bacteria (Noman et al., 2021).

Environmental conditions introduce an additional layer of complexity. Factors such as soil pH, moisture, texture, and temperature are all essential in determining which microbes can establish and flourish in a specific environment. Abiotic stresses, including drought, salinity, flooding, and heat, can significantly alter the structures of microbial communities. Nevertheless, certain genera—like *Pseudomonas, Rhizobium, Bacillus*, and *Enterobacter*—exhibit remarkable adaptability in the face of these challenging conditions.

Microorganisms utilize a variety of survival strategies to manage environmental changes. These strategies encompass biofilm formation, exopolysaccharide (EPS) production, osmolyte accumulation, and dormancy. Concurrently, plants modify their hormonal signaling to better endure these stresses. For example, *Azospirillum* species are recognized for increasing ABA levels in plants during drought, thereby enhancing their stress tolerance.

In environments heavily impacted by pollution, such as oil-contaminated soils, the interdependence between plants and their microbiomes becomes even more evident. In these contexts, microbes not only facilitate detoxification but also promote plant growth, nutrient absorption, and biomass production. This dynamic adaptability underscores the vital role of microbiomes in maintaining plant health under extreme conditions. At the heart of these interactions is a vibrant, interconnected soil ecosystem comprising bacteria, actinomycetes, fungi, and algae. These organisms establish complex relationships with plants that can vary from beneficial to neutral or even antagonistic. A well-known beneficial interaction is the symbiotic relationship between *Rhizobium* and leguminous plants, which results in effective biological nitrogen fixation.

The release of 5–20% of a plant's photosynthates into the soil as root exudates underscores the importance of the rhizosphere. These exudates, which are abundant in organic acids, sugars, amino acids, fatty acids, vitamins, growth factors, hormones, and antimicrobial compounds, shape the microbial ecology surrounding the root zone (Köhl et al., 2014). The advent of advanced technologies such as high-throughput sequencing has allowed scientists to conceptualize a core microbiome—a stable microbial community that is consistently found across various plant species and generations (Hamonts et al., 2018). Although the microbiome may be limited in diversity, it is characterized by its abundance and stability, typically comprising genera such as *Sphingomonas*, *Burkholderia*, *Pseudomonas*, *Bradyrhizobium*, *Rhizobium*, and *Mesorhizobium* mainly related to citrus plants (Xu et al., 2018).

Further enhancing our comprehension is the identification of "hub microorganisms"—essential species that uphold the structure and stability of microbial networks. The elimination of these hubs, such as *E. cloacae*, can precipitate the disintegration of entire microbial communities (Niu et al., 2017). Thus, the preservation of these hub species is crucial for maintaining microbiome functionality, particularly under stress conditions.

The role of the microbiome extends beyond merely stabilizing communities; it also significantly contributes to nutrient acquisition. Nitrogen-fixing bacteria—including Rhizobia, cyanobacteria such as *Anabaena*, *Nostoc*, *Calothrix*, and free-living diazotrophs like *Beijerinckia*, *Klebsiella*, and *Bacillus*—are instrumental in enhancing the nitrogen content of the soil (Ladha et al., 2016). Likewise, phosphate-solubilizing microbes such as *Azotobacter*, *Bacillus*, and *Pseudomonas* facilitate the conversion of inaccessible phosphorus into forms that are more readily available to plants (Malik et al., 2001).

Furthermore, microorganisms generate a variety of plant hormones, including IAA, gibberellins, CKs, and salicylic acid, which promote plant growth and enhance stress responses. Additionally, some of these microbes synthesize ACC deaminase, an enzyme that lowers ET levels in plants and encourages vigor. For instance, IAA-producing microbes found in the tomato rhizosphere have been demonstrated to significantly improve plant growth (Egamberdieva et al., 2017).

Even in challenging environments such as high salinity or exposure to heavy metals, beneficial microbes assist in regulating plant hormone levels and stress responses, thereby increasing resilience and ensuring sustained productivity (Vannier et al., 2019).

The existence and variety of microorganisms in the soil rhizosphere guarantee sustainable agriculture. Therefore, plant composition, productivity, and sustainability are significantly impacted by the diversity and abundance of soil microorganisms. Because they guarantee low-input, sustainable agriculture, microorganisms provide a unique and noteworthy strategy for reestablishing soil health and agricultural output. Understanding how the microbiome can be utilized in agriculture enables the development of inoculants or the manipulation of microbial communities to select for more effective groups that enhance plant growth.

Grasping the significance of the plant microbiome, which plays a crucial role in advancing sustainable agricultural practices, will result in a significant reduction in the reliance on chemical fertilizers and pesticides.

1.15 CONCLUSION AND FUTURE PERSPECTIVES

The increasing number of studies on microbiomes emphasizes how important they are to the advancement of biocontrol, environmental remediation, and sustainable farming methods. Our knowledge of microbial interactions has greatly increased as a result of the integration of metagenomics and multi-omics techniques, which have made it possible to create customized microbiome-based interventions for agricultural productivity, disease prevention, and soil health. While naturally existing microbiomes offer resilience against biotic and abiotic challenges, genetically modified microbial communities present viable solutions for environmental detoxification, pollutant degradation, and nutrient mobilization. Moreover, microbial applications in organic farming can reduce dependence on chemical inputs, hence boosting long-term soil health and ecological balance. Current agricultural methods could be completely transformed by more study and use of microbiomes, promoting food security and lessening environmental impacts. Optimizing microbiome engineering, utilizing synthetic biology, and examining ecological concerns related to microbial changes should be the top priorities of future studies. Furthermore, creating synergistic microbial consortia can increase plant–microbe interactions, resulting in more resilient crops and sustainable harvests. To guarantee the ethical use of microbial technologies, policies and regulations must also change to strike a balance between innovation, safety, and ecological preservation.

Microbiome-based approaches present a viable way to achieve long-term sustainability in agriculture as long as global issues like soil erosion, climate change, and food scarcity continue to exist. A new age of environmentally friendly agricultural innovations can be sparked by combining cutting-edge microbial technologies with conventional farming methods. We can create resilient ecosystems, slow down environmental deterioration, and guarantee food security for future generations by utilizing the potential of microbiomes. Microbiome-based approaches present a viable way to achieve long-term sustainability in agriculture as long as global issues like soil erosion, climate change, and food scarcity continue to exist. A new age of environmentally friendly agricultural innovations can be sparked by combining cutting-edge microbial technologies with conventional farming methods. The creation of microbial bioinformatics tools to forecast and control microbial activity, improvements in precision agriculture, and customized microbiome applications suited to particular ecosystems should be the highlights of future microbiome research. More research into artificial microbiomes and synthetic microbial communities will improve their use in enhancing soil fertility and fending against environmental pressures. We can create resilient ecosystems, slow down environmental deterioration, and guarantee food security for upcoming generations by utilizing the ability of microbiomes.

REFERENCES

Abbas, R., Rasul, S., Aslam, K., Baber, M., Shahid, M., Mubeen, F., & Naqqash, T. (2019). Halotolerant PGPR: A hope for cultivation of saline soils. *Journal of King Saud University-Science*, 31(4), 1195–1201.

Abd El-Mageed, T. A., Rady, M. M., Taha, R. S., Abd El Azeam, S., Simpson, C. R., & Semida, W. M. (2020). Effects of integrated use of residual sulfur-enhanced biochar with effective microorganisms on soil properties, plant growth and short-term productivity of *Capsicum annuum* under salt stress. *Scientia Horticulturae*, 261, 108930.

Abdelfattah, A., Wisniewski, M., Schena, L., & Tack, A. J. M. (2021). Experimental evidence of microbial inheritance in plants and transmission routes from seed to phyllosphere and root. *Environmental Microbiology*, 23(4), 2199–2214.

Abriouel, H., Franz, C. M., Omar, N. B., & Gálvez, A. (2011). Diversity and applications of *Bacillus* bacteriocins. *FEMS Microbiology Reviews*, 35(1), 201–232.

Ahmad, M., Nadeem, S. M., Naveed, M., & Zahir, Z. A. (2016). Potassium-solubilizing bacteria and their application in agriculture. In Meena, V., Maurya, B., Verma, J., Meena, R. (Eds.), *Potassium Solubilizing Microorganisms for Sustainable Agriculture*. New Delhi: Springer. https://doi.org/10.1007/978-81-322-2776-2_21

Al-Awthan, Y. S., Mir, R., Alatawi, F. A., Alatawi, A. S., Almutairi, F. M., Khafaga, T., Shohdi, W. M., Fakhry, A. M., & Alharbi, B. M. (2024). Metagenome analysis identified novel microbial diversity of sandy soils surrounded by natural lakes and artificial water points in King Salman Bin Abdulaziz Royal Natural Reserve, Saudi Arabia. *Life (Basel)*, 14(12), 1692.

Al-Quwaie, D. A. (2023). The role of *Streptomyces* species in controlling plant diseases: A comprehensive review. *Australasian Plant Pathology*, 53(1), 1–14. https://doi.org/10.1007/s13313-023-00959-z

Altaf, M. M., Imran, M., Abulreesh, H. H., Khan, M. S., & Ahmad, I. (2017). Diversity and applications of *Penicillium* spp. In plant-growth promotion. *Elsevier eBooks*, 261–276. https://doi.org/10.1016/b978-0-444-63501-3.00015-6

Ameen, M., Mahmood, A., Sahkoor, A., Zia, M. A., & Ullah, M. S. (2024). The role of endophytes to combat abiotic stress in plants. *Plant Stress*, 12, 100435. https://doi.org/10.1016/j.stress.2024.100435

Andrews, S. C., Robinson, A. K., & Rodríguez-Quiñones, F. (2003). Bacterial iron homeostasis. *FEMS Microbiology Reviews*, 27, 215–237. https://doi.org/10.1016/S0168-6445(03)00055-X

Bais, H. P., Fall, R., & Vivanco, J. M. (2004). Biocontrol of *Bacillus subtilis* against infection of *Arabidopsis* roots by *Pseudomonas syringae* is facilitated by biofilm formation and surfactin production. *Plant Physiology*, 134(1), 307–319. https://doi.org/10.1104/pp.103.028712

Bakker, P. A., Pieterse, C. M., de Jonge, R., & Berendsen, R. L. (2018). The soil-borne legacy. *Cell*, 172(6), 1178–1180.

Banerjee, S., & van der Heijden, M. G. A. (2023). Soil microbiomes and one health. *Nature Reviews Microbiology*, 21(1), 6–20. https://doi.org/10.1038/s41579-022-00779-w

Bardgett, R. D., & van der Putten, W. H. (2014). Below ground biodiversity and ecosystem functioning. *Nature*, 515, 505–511.

Bargaz, A., Lyamlouli, K., Chtouki, M., Zeroual, Y., & Dhiba, D. (2018). Soil microbial resources for improving fertilizers efficiency in an integrated plant nutrient management system. *Frontiers in Microbiology*, 9, 1606.

Battacharyya, D., Zamani Babgohari, M., Rathor, P., & Prithiviraj, B. (2015). Seaweed extracts as biostimulants in horticulture. *Scientia Horticulturae*, 196, 39–48. https://doi.org/10.1016/j.scienta.2015.09.012

Battini, F., Grønlund, M., Agnolucci, M., Giovannetti, M., & Jakobsen, I. (2017). Facilitation of phosphorus uptake in maize plants by mycorrhizosphere bacteria. *Scientific Reports*, 7(1), 4686.

Behera, B. C., Singdevsachan, S. K., Mishra, R. R., Dutta, S. K., & Thatoi, H. N. (2014). Diversity, mechanism and biotechnology of phosphate solubilising microorganism in mangrove: A review. *Biocatalysis and Agricultural Biotechnology*, 3(2), 97–110.

Bencherif, K., Djaballah, Z., Brahimi, F., Boutekrabt, A., Dalpè, Y., & Lounès-Hadj Sahraoui, A. (2019). Arbuscular mycorrhizal fungi affect total phenolic content and antimicrobial activity of *Tamarix gallica* in natural semi-arid Algerian areas. *South African Journal of Botany*, 125, 39–45.

Berg, G., Rybakova, D., Fischer, D., Cernava, T., Vergès, M. C. C., Charles, T., Chen, X., Cocolin, L., Eversole, K., Corral, G. H., & Kazou, M. (2020). Microbiome definition re-visited: Old concepts and new challenges. *Microbiome*, 8, 103.

Berg, M., & Koskella, B. (2018). Nutrient- and dose-dependent microbiome-mediated protection against a plant pathogen. *Current Biology*, 28(15), 2487–2492.

Bhantana, P., Rana, M. S., Sun, X. C., Moussa, M. G., Saleem, M. H., Syaifudin, M., Shah, A., Poudel, A., Pun, A. B., Bhat, M. A., & Mandal, D. L. (2021). Arbuscular mycorrhizal fungi and its major role in plant growth, zinc nutrition, phosphorous regulation and phytoremediation. *Symbiosis*, 84, 19–37.

Bharadwaj, D. P., Lundquist, P.-O., & Alström, S. (2008). Arbuscular mycorrhizal fungal spore-associated bacteria affect mycorrhizal colonization, plant growth and potato pathogens. *Soil Biology and Biochemistry*, 40(10), 2494–2501.

Bilal, S., Hazafa, A., Ashraf, I., Alamri, S., Siddiqui, M. H., Ramzan, A., Qamar, N., Sher, F., & Naeem, M. (2021). Comparative effect of inoculation of phosphorus-solubilizing bacteria and phosphorus as sustainable fertilizer on yield and quality of mung bean (*Vigna radiata* L.). *Plants*, 10(10), 2079.

Blum, W. E. H., Zechmeister-Boltenstern, S., & Keiblinger, K. M. (2019). Does soil contribute to the human gut microbiome? *Microorganisms*, 7(9), Article 287. https://doi.org/10.3390/microorganisms7090287

Boro, R. C., Nath, D. J., Bordoloi, N., & Das, R. (2022). Plant growth-promoting rhizobacteria as biocontrol agents against plant pathogens: Mechanisms and applications. *Frontiers in Microbiology*, 13, 1023456. https://doi.org/10.3389/fmicb.2022.1023456

Boros-Lajszner, E., Wyszkowska, J., Borowik, A., & Kucharski, J. (2021). The response of the soil microbiome to contamination with cadmium, cobalt and nickel in soil sown with *Brassica napus*. *Minerals*, 11(5), Article 498. https://doi.org/10.3390/min11050498

Bowles, T. M., Jackson, L. E., & Cavagnaro, T. R. (2018). Mycorrhizal fungi enhance plant nutrient acquisition and modulate nitrogen loss with variable water regimes. *Global Change Biology*, 24(1), e171–e182. https://doi.org/10.1111/gcb.13884

Brevik, E. C., Steffan, J. J., Rodrigo-Comino, J., Neubert, D., Burgess, L. C., & Cerdà, A. (2019). Connecting the public with soil to improve human health. *European Journal of Soil Science*, 70(4), 898–910.

Brockett, B. F. T., Prescott, C. E., & Grayston, S. J. (2012). Soil moisture is the major factor influencing microbial community structure and enzyme activities across seven biogeoclimatic zones in western Canada. *Soil Biology and Biochemistry*, 44(1), 9–20.

Bruisson, S., Maillot, P., Schellenbaum, P., Walter, B., Gindro, K., & Deglène-Benbrahim, L. (2016). Arbuscular mycorrhizal symbiosis stimulates key genes of the phenylpropanoid biosynthesis and stilbenoid production in grapevine leaves in response to downy mildew and grey mould infection. *Phytochemistry*, 131, 92–99.

Bulgari, R., Cocetta, G., Trivellini, A., Vernieri, P., & Ferrante, A. (2015). Biostimulants and crop responses: A review. *Biological Agriculture & Horticulture*, 31(1), 1–17.

Calvo, P., Nelson, L., & Kloepper, J. W. (2014). Agricultural uses of plant biostimulants. *Plant and Soil*, 383, 3–41.

Canellas, L. P., Olivares, F. L., Aguiar, N. O., Jones, D. L., Nebbioso, A., Mazzei, P., & Piccolo, A. (2015). Humic and fulvic acids as biostimulants in horticulture. *Scientia Horticulturae*, 196, 15–27.

Cao, X., Cui, Q., Li, D., Liu, Y., Liu, K., & Li, Z. (2025). Characteristics of soil microbial community structure in different land use types of the Huanghe Alluvial Plain. *Microorganisms*, 13(2), 273. https://doi.org/10.3390/microorganisms13020273

Carrión, V. J., Perez-Jaramillo, J., Cordovez, V., Tracanna, V., De Hollander, M., Ruiz-Buck, D., Mendes, L. W., van Ijcken, W. F., Gomez-Exposito, R., Elsayed, S. S., & Mohanraju, P. (2019). Pathogen-induced activation of disease-suppressive functions in the endophytic root microbiome. *Science*, 366(6465), 606–612.

Cascales, E., Buchanan, S. K., Duché, D., Kleanthous, C., Lloubès, R., Postle, K., Riley, M., Slatin, S., & Cavard, D. (2007). Colicin biology. *Microbiology and Molecular Biology Reviews*, 71, 158–229. https://doi.org/10.1128/MMBR.00036-06

Castrillo, G., Teixeira, P. J. P. L., Paredes, S. H., Law, T. F., De Lorenzo, L., Feltcher, M. E., Finkel, O. M., Breakfield, N. W., Mieczkowski, P., Jones, C. D., & Paz-Ares, J. (2017). Root microbiota drive direct integration of phosphate stress and immunity. *Nature*, 543(7646), 513–518.

Chakraborty, S., Kumari, M., Lal, A., & Rawat, N. (2022). Types of microorganisms in soil. In Singh, Y. V. (Ed.), *Emerging Trends in Agriculture Sciences* (Vol. 7, pp. 1–146). Integrated Publications. https://doi.org/10.22271/int.book.173

Chandra, P., Egamberdieva, D., Basak, N., Rai, A. K., & Sonth, B. (2024). Editorial: Plants surviving in extreme environment: Harnessing soil-plant-microbial relationship to enhance crop health and productivity. *Frontiers in Microbiology*, 15, 1388802. https://doi.org/10.3389/fmicb.2024.1388802

Chen, J., Tian, F., Ge, F., Zhang, D., Deng, S., & Liu, X. (2021). Roles of phosphate-solubilizing microorganisms from managing soil phosphorus deficiency to mediating biogeochemical P cycle. *Biology*, 10(2), 158. https://doi.org/10.3390/biology10020158

Chen, Q., Song, Y., An, Y., Lu, Y., & Zhong, G. (2024). Soil microorganisms: Their role in enhancing crop nutrition and health. *Diversity*, 16(12), 734. https://doi.org/10.3390/d16120734

Chepsergon, J., & Moleleki, L. N. (2023). Rhizosphere bacterial interactions and impact on plant health. *Current Opinion in Microbiology*, 73, 102297. https://doi.org/10.1016/j.mib.2023.102297

Chin-A-Woeng, T. F., Bloemberg, G. V., & Lugtenberg, B. J. (2003). Phenazines and their role in biocontrol by *Pseudomonas* bacteria. *New Phytologist*, 157, 503–523. https://doi.org/10.1046/j.1469-8137.2003.00686.x

Colla, G., Nardi, S., Cardarelli, M., Ertani, A., Lucini, L., Canaguier, R., & Rouphael, Y. (2015). Protein hydrolysates as biostimulants in horticulture. *Scientia Horticulturae*, 196, 8–38. https://doi.org/10.1016/j.scienta.2015.08.037

Compant, S., Nowak, J., Coenye, T., Clément, C., & Ait Barka, E. (2008). Diversity and occurrence of Burkholderia spp. In the natural environment. *FEMS Microbiology Reviews*, 32, 607–626. https://doi.org/10.1111/j.1574-6976.2008.00113.x

Cordier, C., Gianinazzi, S., & Gianinazzi-Pearson, V. (1996). Colonisation patterns of root tissues by *Phytophthora nicotianae* var. *parasitica* related to reduced disease in mycorrhizal tomato. *Plant and Soil*, 185(2), 223–232.

Cordier, C., Pozo, M. J., Barea, J. M., Gianinazzi, S., & Gianinazzi-Pearson, V. (1998). Cell defense responses associated with localized and systemic resistance to *Phytophthora parasitica* induced in tomato by an arbuscular mycorrhizal fungus. *Molecular Plant-Microbe Interactions*, 11(10), pp. 1017–1028.

Costa, O. Y. A., Raaijmakers, J. M., & Kuramae, E. E. (2018). Microbial extracellular polymeric substances: Ecological function and impact on soil aggregation. *Frontiers In Microbiology*, 9, 1636. https://doi.org/10.3389/fmicb.2018.01636

Crowley, D. E. (2006). Microbial Siderophores in the plant rhizosphere. In Barton, L. L. & Abadia, J. (Eds.), *Iron Nutrition in Plants and Rhizospheric Microorganisms*. Dordrecht: Springer. https://doi.org/10.1007/1-4020-4743-6_8

Cruz-Hernández, M. A., Mendoza-Herrera, A., Bocanegra-García, V., & Rivera, G. (2022). *Azospirillum* spp. From plant growth-promoting bacteria to their use in bioremediation. *Microorganisms*, 10(5), 1057. https://doi.org/10.3390/microorganisms10051057

Cui, K., Xu, T., Chen, J., Yang, H., Liu, X., Zhuo, R., Peng, Y., Tang, W., Wang, R., Chen, L., & Zhang, X. (2022). Siderophores, a potential phosphate solubilizer from the endophyte *Streptomyces* sp. CoT10, improved phosphorus mobilization for host plant growth and rhizosphere modulation. *Journal of Cleaner Production*, 367, 133110.

Dastogeer, K. M., Tumpa, F. H., Sultana, A., Akter, M. A., & Chakraborty, A. (2020). Plant microbiome–an account of the factors that shape community composition and diversity. *Current Plant Biology*, 23, Article 100161. https://doi.org/10.1016/j.cpb.2020.100161

De Jonge, N., Carlsen, B., Christensen, M. H., Pertoldi, C., & Nielsen, J. L. (2022). The gut microbiome of 54 mammalian species. *Frontiers in Microbiology*, 13, 886252. https://doi.org/10.3389/fmicb.2022.886252

de Souza, J. T., Arnould, C., Deulvot, C., Lemanceau, P., Gianinazzi-Pearson, V., & Raaijmakers, J. M. (2003). Effect of 2,4-diacetylphloroglucinol on Pythium: Cellular responses and variation in sensitivity among propagules and species. *Phytopathology*, 93(8), 966–975. https://doi.org/10.1094/PHYTO.2003.93.8.966

Delwiche, C., & Wijler, J. J. (1956). Non-symbiotic nitrogen fixation in soil. *Plant Soil*, 7, 113–129. https://doi.org/10.1007/BF01343722

Djuuna, I. A. F., Prabawardani, S., & Massora, M. (2022). Population distribution of phosphate-solubilizing microorganisms in agricultural soil. *Microbes and Environments*, 37(1), ME21041.

Dobereiner, J. (1961). Nitrogen-fixing bacteria of the genus Beijerinckia Derx in the rhizosphere of sugar cane. *Plant Soil*, 15, 211–216. https://doi.org/10.1007/BF01400455

Doilom, M., Guo, J. W., Phookamsak, R., Mortimer, P. E., Karunarathna, S. C., Dong, W., Liao, C. F., Yan, K., Pem, D., Suwannarach, N., & Promputtha, I. (2020). Screening of phosphate-solubilizing fungi from air and soil in Yunnan, China: Four novel species in Aspergillus, Gongronella, Penicillium, and Talaromyces. *Frontiers In Microbiology*, 11, 585215.

Donley, N. (2019). The USA lags behind other agricultural nations in banning harmful pesticides. *Environmental Health*, Full Text. https://ehjournal.biomedcentral.com/articles/10.1186/s12940-019-0488-0

Dudenhöffer, J. H., Scheu, S., & Jousset, A. (2016). Systemic enrichment of antifungal traits in the rhizosphere microbiome after pathogen attack. *Journal of Ecology*, 104(6), 1566–1575.

Duffy, B. (2003). Pathogen self-defense: Mechanisms to counteract microbial antagonism. *Annual Review of Phytopathology*, 41, 501–538. https://doi.org/10.1146/annurev.phyto.41.052002.095606

East, R. (2013). Microbiome: Soil science comes to life. *Nature*, 501(7468), S18–S19.

Egamberdieva, D., Wirth, S. J., Alqarawi, A. A., Abd_Allah, E. F., & Hashem, A. (2017). Phytohormones and beneficial microbes: Essential components for plants to balance stress and fitness. *Frontiers in Microbiology*, 8, 2104.

Eke, P., Chatue Chatue, G., Wakam, L. N., Kouipou, R. M. T., Fokou, P. V. T., & Boyom, F. F. (2016). Mycorrhiza consortia suppress the fusarium root rot (*Fusarium solani* f. sp. *Phaseoli*) in common bean (*Phaseolus vulgaris* L.). *Biological Control*, 103, 240–250.

Ekinci, K., Tosun, İ., & Bıtrak, B., et al. (2019). Effects of initial C/N ratio on organic matter degradation of composting of rose oil processing solid wastes. *International Journal of Environmental Science and Technology*, 16, 5131–5140. https://doi.org/10.1007/s13762-018-1947-9

Elnahal, A. S., El-Saadony, M. T., Saad, A. M., Desoky, E. S. M., El-Tahan, A. M., Rady, M. M., AbuQamar, S. F., & El-Tarabily, K. A. (2022). The use of microbial inoculants for biological control, plant growth promotion, and sustainable agriculture: A review. *European Journal of Plant Pathology*, 162(4), 759–792.

El-Saadony, M. T., Saad, A. M., Soliman, S. M., Salem, H. M., Ahmed, A. I., Mahmood, M., El-Tahan, A. M., Ebrahim, A. A.

M., Abd El-Mageed, T. A., Negm, S. H., Selim, S., Babalghith, A. O., Elrys, A. S., El-Tarabily, K. A., & AbuQamar, S. F. (2022). Plant growth-promoting microorganisms as biocontrol agents of plant diseases: Mechanisms, challenges and future perspectives. *Frontiers in Plant Science*, 13, 923880. https://doi.org/10.3389/fpls.2022.923880

Fan, D., Subramanian, S., & Smith, D. L. (2020). Plant endophytes promote growth and alleviate salt stress in *Arabidopsis thaliana*. *Scientific Reports*, 10(1), 12740.

Fanelli, R. M. (2020). The spatial and temporal variability of the effects of agricultural practices on the environment. *Environments*, 7(4), 33.

Fazal, A., Webb, M. E., & Seipke, R. F. (2020). The desotamide family of antibiotics. *Antibiotics (Basel, Switzerland)*, 9(8), 452. https://doi.org/10.3390/antibiotics9080452

Fierer, N. (2017). Embracing the unknown: Disentangling the complexities of the soil microbiome. *Nature Reviews Microbiology*, 15, 579–590.

Fitzpatrick, C. R., Copeland, J., Wang, P. W., Guttman, D. S., Kotanen, P. M., & Johnson, M. T. (2018). Assembly and ecological function of the root microbiome across angiosperm plant species. *Proceedings of the National Academy of Sciences*, 115(6), E1157–E1165.

Fones, H. N., Bebber, D. P., Chaloner, T. M., Kay, W. T., Steinberg, G., & Gurr, S. J. (2020). Threats to global food security from emerging fungal and oomycete crop pathogens. *Nature Food*, 1(6), 332–342.

Frąc, M., Lipiec, J., Usowicz, B., Oszust, K., & Brzezińska, M. (2020). Structural and functional microbial diversity of sandy soil under cropland and grassland. *PeerJ*, 8, e9501. https://doi.org/10.7717/peerj.9501

Fukami, J., Cerezini, P., & Hungria, M. (2018). *Azospirillum*: Benefits that go far beyond biological nitrogen fixation. *AMB Express*, 8(1), 73. https://doi.org/10.1186/s13568-018-0608-1

Funabashi, M. (2018). Human augmentation of ecosystems: Objectives for food production and science by 2045. *NPJ Science of Food*, 2(1), 16.

Gilden, R. C., Huffling, K., & Sattler, B. (2010). Pesticides and health risks. *Journal of Obstetric, Gynecologic & Neonatal Nursing*, 39(1), 103–110.

Glick, B. R. (2010). Using soil bacteria to facilitate phytoremediation. *Biotechnology Advances*, 28(3), 367–374.

Glick, B. R., Cheng, Z., Czarny, J., & Duan, J. (2007). Promotion of plant growth by ACC deaminase-producing soil bacteria. *European Journal of Plant Pathology*, 119, 329–339.

Gouda, S., Kerry, R. G., Das, G., Paramithiotis, S., Shin, H. S., & Patra, J. K. (2018). Revitalization of plant growth promoting rhizobacteria for sustainable development in agriculture. *Microbiological Research*, 206, 131–140.

Gougoulias, N., Papapolymerou, G., Karayannis, V., Spiliotis, X., & Chouliaras, N. (2017). Effects of manure enriched with algae *Chlorella vulgaris* on soil chemical properties. *Soil and Water Research*, 13(1), 51–59. https://doi.org/10.17221/260/2016-swr

Grover, M., Yaadesh, S., & Jayasurya, A. (2023). Associative nitrogen fixers-options for mitigating climate change. In Singh, S., Prasanna, R., & Pranaw, K. (Eds.), *Bioinoculants: Biological Option for Mitigating Global Climate Change*. Singapore: Springer. https://doi.org/10.1007/978-981-99-2973-3_10

Guo, J.-J., Huang, X.-P., Xiang, L., Wang, Y.-Z., Li, Y.-W., Li, H., Cai, Q.-Y., Mo, C.-H., & Wong, M.-H. (2020). Source, migration and toxicology of microplastics in soil. *Environment International*, 137, 105263.

Gupta, A., Mishra, R., Rai, S., Bano, A., Pathak, N., Fujita, M., Kumar, M., & Hasanuzzaman, M. (2022b). Mechanistic insights of plant growth-promoting bacteria-mediated drought and salt stress tolerance in plants for sustainable agriculture. *International Journal of Molecular Sciences*, 23(7), 3741.

Gupta, A., Singh, V., & Mani, I. (2022a). Dysbiosis of human microbiome and infectious diseases. *Progress in Molecular Biology and Translational Science*, 192(1), 33–51. https://doi.org/10.1016/bs.pmbts.2022.06.016

Haas, D., & Défago, G. (2005). Biological control of soil-borne pathogens by fluorescent pseudomonads. *Nature Reviews Microbiology*, 3, 307–319. https://doi.org/10.1038/nrmicro1129

Haas, D., & Keel, C. (2003). Regulation of antibiotic production in root-colonizing *Pseudomonas* spp. And relevance for biological control of plant disease. *Annual Review of Phytopathology*, 41, 117–153. https://doi.org/10.1146/annurev.phyto.41.052002.095656

Ham, B. K., Chen, J., Yan, Y., & Lucas, W. J. (2018). Insights into plant phosphate sensing and signaling. *Current Opinion in Biotechnology*, 49, 1–9.

Hamonts, K., Trivedi, P., Garg, A., Janitz, C., Grinyer, J., Holford, P., Botha, F. C., Anderson, I. C., & Singh, B. K. (2018). Field study reveals core plant microbiota and relative importance of their drivers. *Environmental Microbiology*, 20(1), 124–140. https://doi.org/10.1111/1462-2920.14031

Harman, G. E., Howell, C. R., Viterbo, A., Chet, I., & Lorito, M. (2004). *Trichoderma* species—opportunistic, avirulent plant symbionts. *Nature Reviews Microbiology*, 2(1), 43–56. https://doi.org/10.1038/nrmicro797

Harrier, L. A., & Watson, C. A. (2004). The potential role of arbuscular mycorrhizal (AM) fungi in the bioprotection of plants against soil-borne pathogens in organic and/or other sustainable farming systems, *Pest Management Science*, 60(2), 149–157.

Hartmann, A., Rothballer, M., & Schmid, M. (2008). Lorenz Hiltner, a pioneer in rhizosphere microbial ecology and soil bacteriology research. *Plant Soil*, 312, 7–14. Hartmann et al., 2008.

Hartmann, M., & Six, J. (2023). Soil structure and microbiome functions in agroecosystems. *Nature Reviews Earth & Environment*, 4(1), 4–18.

Hayat, M. K., Khan, T. Y., Azam, M. A., Imran, F., Ifra, I., Shan, M., . . . & Oumar, B. A. (2025). Introduction to organic agriculture: From Haber-Bosch to howard: A century of agricultural transformation. *World News of Natural Sciences*, 60, 40–64.

Hayat, R., Ali, S., Amara, U., Khalid, R., & Ahmed, I. (2010). Soil beneficial bacteria and their role in plant growth promotion: A review. *Annals of Microbiology*, 60, 579–598.

Hernandez, M. E., Kappler, A., & Newman, D. K. (2004). Phenazines and other redox-active antibiotics promote microbial mineral reduction. *Applied and Environmental Microbiology*, 70, 921–928. https://doi.org/10.1128/AEM.70.2.921-928.2004

Hill, D. S., Stein, J. I., Torkewitz, N. R., Morse, A. M., Howell, C. R., Pachlatko, J. P., Becker, J. O., & Ligon, J. M. (1994).

Cloning of genes involved in the synthesis of pyrrolnitrin from *Pseudomonas fluorescens* and role of pyrrolnitrin synthesis in biological control of plant disease. *Applied and Environmental Microbiology*, 60, 78–85. https://doi.org/10.1128/aem.60.1.78-85.1994

Hill, W. A., Bacon-Hill, P., Crossman, S. M., & Stevens, C. (1983). Characterization of N_2-fixing bacteria associated with sweet potato roots. *Canadian Journal Of Microbiology*, 29, 860–862. https://doi.org/10.1139/m83-139

Hiltner, L. (1904). Ueber neuere Erfahrungen und Probleme auf dem Gebiete der Bodenbakteriologie und unter besonderer BerUcksichtigung der Grundungung und Brache. *Arbetian Der Deutschen Landwirtschafts–Gesellschafts*, 98, 59–78.

Hover, B. M., Kim, S. H., Katz, M., et al. (2018). Culture-independent discovery of the malacidins as calcium-dependent antibiotics with activity against multidrug-resistant Gram-positive pathogens. *Nature Microbiology*, 3, 415–422. https://doi.org/10.1038/s41564-018-0110-1

Hu, H., Luan, F., Liu, X., Jiang, Y., Hu, D., Long, H., Zhai, Z., Yan, J., Cui, C., Song, H., Zhou, J., He, G., Chen, S., Tennakoon, D. S., & Gao, Y. (2025). Soil amoeba richness drives local-scale bacterial and fungal community assembly. *Soil Biology and Biochemistry*, 211, 109972. ISSN 0038-0717. https://doi.org/10.1016/j.soilbio.2025.109972

Huang, S., Zha, X., & Fu, G. (2023). Affecting factors of plant phyllosphere microbial community and their responses to climatic warming: A review. *Plants*, 12(16), 2891. https://doi.org/10.3390/plants12162891

Imran, A., Hakim, S., Tariq, M., Nawaz, M. S., Laraib, I., Gulzar, U., Hanif, M. K., Siddique, M. J., Hayat, M., Fraz, A., & Ahmad, M. (2021). Diazotrophs for lowering nitrogen pollution crises: Looking deep into the roots. *Frontiers in Microbiology*, 12, 637815.

Iniguez, A. L., Dong, Y., & Triplett, E. W. (2004). Nitrogen fixation in wheat provided by Klebsiella pneumoniae 342. *Molecular Plant-Microbe Interactions*, 17, 1078–1085. https://doi.org/10.1094/MPMI.2004.17.10.1078

Islam, W., Noman, A., Naveed, H., Huang, Z., & Chen, H. Y. H. (2020). Role of environmental factors in shaping the soil microbiome. *Environmental Science and Pollution Research*, 27(33), 41225–41247.

Itelima, J. U., Bang, W. J., Onyimba, I. A., Sila, M. D., & Egbere, O. J. (2018). Bio-fertilizers as key player in enhancing soil fertility and crop productivity: A review. *Direct Research Journal of Agriculture and Food Science*, 6(3), 73–83.

Jamiołkowska, A., Thanoon, A. H., Patkowska, E., & Grządziel, J. (2019). Impact of AMF *Claroideoglomus etunicatum* on the structure of fungal communities in the tomato rhizosphere. *Acta Mycologica*, 54(1), 11–20.

Jansson, J. K., McClure, R., & Egbert, R. G. (2023). Soil microbiome engineering for sustainability in a changing environment. *Nature Biotechnology*, 41(12), 1716–1728.

Jayasekara, C., Leigh, C., Shimeta, J., et al. (2025). Organic matter decomposition in mountain peatlands: Effects of substrate quality and peatland degradation. *Plant and Soil*, 506, 639–654. https://doi.org/10.1007/s11104-024-06725-4

Jochum, M. D., McWilliams, K. L., Borrego, E. J., Kolomiets, M. V., Niu, G., Pierson, E. A., & Jo, Y.-K. (2019). Bioprospecting plant growth-promoting rhizobacteria that mitigate drought stress in grasses. *Frontiers In Microbiology*, 10, 2106. https://doi.org/10.3389/fmicb.2019.02106

Jung, J., Kim, J. S., Taffner, J., Berg, G., & Ryu, C. M. (2020). Archaea, tiny helpers of land plants. *Computational and Structural Biotechnology Journal*, 18, 2494–2500. https://doi.org/10.1016/J.CSBJ.2020.09.005

Kalayu, G. (2019). Phosphate solubilizing microorganisms: Promising approach as biofertilizers. *International Journal of Agronomy*, 2019, 4917256. https://doi.org/10.1155/2019/4917256

Kamran, M., Imran, Q. M., Ahmed, M. B., Falak, N., Khatoon, A., & Yun, B. W. (2022). Endophyte-mediated stress tolerance in plants: A sustainable strategy to enhance resilience and assist crop improvement. *Cells*, 11(20), 3292. https://doi.org/10.3390/cells11203292

Kanter, D. R., & Searchinger, T. D. (2018). A technology-forcing approach to reduce nitrogen pollution. *Nature Sustainability*, 1(10), 544–552.

Kaul, S., Choudhary, M., Gupta, S., & Dhar, M. K. (2021). Engineering host microbiome for crop improvement and sustainable agriculture. *Frontiers in Microbiology*, 12, 635917.

Kaur, S., Campbell, B. J., & Suseela, V. (2022). Root metabolome of plant–arbuscular mycorrhizal symbiosis mirrors the mutualistic or parasitic mycorrhizal phenotype. *New Phytologist*, 234(2), 672–687.

Kavanagh, P. H., Vilela, B., Haynie, H. J., Tuff, T., Lima-Ribeiro, M., Gray, R. D., Botero, C. A., & Gavin, M. C. (2018). Hindcasting global population densities reveals forces enabling the origin of agriculture. *Nature Human Behaviour*, 2(7), 478–484.

Kendzior, J., Raffa, D. W., & Bogdanski, A. (2022). *A Review of the Impacts of Crop Production on the Soil Microbiome*. Food and Agriculture Organization of the United Nations.

Khan, M., Asaf, S., Khan, A., Adhikari, A., Jan, R., Ali, S., Imran, M., Kim, K. M., & Lee, I. J. (2020). Plant growth–promoting endophytic bacteria augment growth and salinity tolerance in rice plants. *Plant Biology*, 22, 850–862. https://doi.org/10.1111/plb.13124

Khanum, S., Al-Tawaha, A. R. M., Al-Tawaha, A. R., Alatrash, H., Rauf, A., Karnwal, A., Dey, A., Alimad, N., Thangadurai, D., Sangeetha, J., Islam, S., Imran, N., Amanullah, N., Khalid, S., Saranraj, P., & Gunal, E. (2023). Role of AMF in sustainable agriculture. *Apple Academic Press eBooks*, 219–236. https://doi.org/10.1201/9781003429708-15

Kloepper, J. W., Leong, J., Teintze, M., & Schroth, M. N. (1980). Pseudomonas siderophores: A mechanism explaining disease-suppressive soils. *Current Microbiology*, 4, 317–320.

Köhl, J., Kolnaar, R., & Ravensberg, W. J. (2014). Mode of action of microbial biological control agents against plant diseases: Relevance beyond efficacy. *Frontiers in Plant Science*, 5, 504. https://doi.org/10.3389/fpls.2014.00504

Köhl, L., Oehl, F., & van der Heijden, M. G. A. (2014). Agricultural practices indirectly influence plant productivity and ecosystem services through effects on soil biota. Ecological Applications, 24(7), 1842–1853. https://doi.org/10.1890/13-1821.1

Koskella, B. (2020). The phyllosphere. *Current Biology*, 30(19), R1143–R1146. https://doi.org/10.1016/j.cub.2020.07.037

Kumar, R., Kumawat, N., & Sahu, Y. K. (2017). Role of biofertilizers in agriculture. *Popular Kheti*, 4, 4.

Kumar, V., Wani, S. H., Suprasanna, P., & Tran, L. S. P. (2018). *Salinity Responses and Tolerance in Plants, Volume 1*. Cham: Springer. https://doi.org/10.1007/978-3-319-75671-4_6

Kundan, R., Pant, G., Jadon, N., & Agrawal, P. K. (2015). Plant growth promoting rhizobacteria: Mechanism and current prospective. *Journal of Biofertilizers & Biopesticides*, 6, 155.

Ladha, J. K., Tirol-Padre, A., Reddy, C. K., Cassman, K. G., Verma, S., Powlson, D. S., Van Kessel, C., Richter, D. B., Chakraborty, D., & Pathak, H. (2016). Global nitrogen budgets in cereals: A 50-year assessment for maize, rice and wheat production systems. *Scientific Reports*, 6(1), 19355.

Laforest-Lapointe, I., Messier, C., & Kembel, S. W. (2016). Host species identity, site and time drive temperate tree phyllosphere bacterial community structure. *Microbiome*, 4(1). https://doi.org/10.1186/s40168-016-0174-1

Lara, E., & Acosta-Mercado, D. (2011). A molecular perspective on ciliates as soil bioindicators. *European Journal of Soil Biology*, 49, 107–111. https://doi.org/10.1016/j.ejsobi.2011.11.001

Larsbrink, J., Zhu, Y., Kharade, S. S., Kwiatkowski, K. J., Eijsink, V. G. H., Koropatkin, N. M., McBride, M. J., & Pope, P. B. (2016). A polysaccharide utilization locus from *Flavobacterium johnsoniae* enables conversion of recalcitrant chitin. *Biotechnology for Biofuels*, 9(1). https://doi.org/10.1186/S13068-016-0674-Z

Le Mire, G., Nguyen, M., Fassotte, B., du Jardin, P., Verheggen, F., Delaplace, P., & Jijakli, H. (2016). Implementing biostimulants and biocontrol strategies in the agroecological management of cultivated ecosystems. *Biotechnologie, Agronomie, Société et Environnement*, 20(3), 299–31.

Lee, D., Oh, T., Kang, B., Ahn, J. S., & Cho, Y. (2022). Throughput screening of *Bacillus subtilis* strains that abundantly secrete surfactin in vitro identifies effective probiotic candidates. *PloS One*, 17(11 November). https://doi.org/10.1371/journal.pone.0277412

Liao, Y., He, T., Wang, C., Zheng, C., & Zhang, M. (2025). Dissimilatory nitrate reduction to ammonium has a competitive advantage over denitrification under nitrate-limited conditions. *Reviews in Environmental Science and Bio/Technology*, 1–18.

Ling, Q., Wu, H., Xie, L., Zhao, Y., Huang, Q., Zhang, Q., Liu, J., Hu, P., Tang, T., Xiao, J., Du, H., Zhao, J., Zhang, W., Chen, H., & Wang, K. (2025). Advances, challenges, and perspectives in glomalin-related soil protein research. *Microorganisms*, 13(4), 740. https://doi.org/10.3390/microorganisms13040740

Liu, H., Brettell, L. E., Qiu, Z., & Singh, B. K. (2020). Microbiome-mediated stress resistance in plants. *Trends in Plant Science*, 25(8), 733–743.

Liu, H., Carvalhais, L. C., Crawford, M., Singh, E., Dennis, P. G., Pieterse, C. M. J., & Singh, B. K. (2021). Inner plant values: Diversity, colonization and benefits from endophytic bacteria. *Frontiers in Microbiology*, 12, 650698.

Lucas, J. A. (2011). Advances in plant disease and pest management. *The Journal of Agricultural Science*, 149(S1), 91–114.

Lyu, D., Backer, R., Subramanian, S., & Smith, D. L. (2020). Phytomicrobiome coordination signals hold potential for climate change-resilient agriculture. *Frontiers in Plant Science*, 11, 634.

Mącik, M., Gryta, A., & Frąc, M. (2020). Biofertilizers in agriculture: An overview on concepts, strategies and effects on soil microorganisms. *Advances in Agronomy*, 162, 31–87.

Mahmud, K., Makaju, S., Ibrahim, R., & Missaoui, A. (2020). Current progress in nitrogen fixing plants and microbiome research. *Plants (Basel, Switzerland)*, 9(1), 97. https://doi.org/10.3390/plants9010097

Maksimov, I. V., Abizgil'dina, R. R., & Pusenkova, L. I. (2011). Plant growth promoting rhizobacteria as alternative to chemical crop protectors from pathogens (Review). *Applied Biochemistry and Microbiology*, 47, 333–345.

Malik, A. I., Colmer, T. D., Lambers, H., Setter, T. L., & Schortemeyer, M. (2002). Short-term waterlogging has long-term effects on the growth and physiology of wheat. *New Phytologist*, 153(2), 225–236. https://doi.org/10.1046/j.0028-646X.2001.00318.x

Malik, K. A., Mehnaz, S., Rasul, G., Mirza, M. S., & Ali, S. (2001). Association of nitrogen-fixing, plant-growth-promoting rhizobacteria with kallar grass and rice. *Plant and Soil*, 194, 37–44.

Manlay, R., Feller, C., & Swift, M. J. (2007). Historical evolution of soil organic matter concepts and their relationships with the fertility and sustainability of cropping systems. *Agriculture, Ecosystems & Environment*, 119, 217–233. https://doi.org/10.1016/j.agee.2006.07.011

Martens, D. A., & Frankenberger, W. T. (1992). Decomposition of bacterial polymers in soil and their influence on soil structure. *Biology and Fertility of Soils*, 13(2), 65–73. https://doi.org/10.1007/BF00337337

McGill, W. B., & Cole, C. V. (1981). Comparative aspects of cycling of organic C, N, S and P through soil organic matter. *Geoderma*, 26(4), 267–286. https://doi.org/10.1016/0016-7061(81)90024-0.

McGill, W. B., Cole, C. V., & Steward, J. W. B. (1981). Forms of phosphorus and transformations in soils, plants, and microorganisms. In *The Role of Phosphorus in Agriculture* (pp. 201–238). American Society of Agronomy.

McNear, D. H., Jr. (2013). The rhizosphere—roots, soil and everything in between. *Nature Education Knowledge*, 4(3), 1.

McNeill, A., & Unkovich, M. (2007). The nitrogen cycle in terrestrial ecosystems. In Marschner, P. & Rengel, Z. (Eds.), *Nutrient Cycling in Terrestrial Ecosystems* (pp. 37–64). Springer. https://doi.org/10.1007/978-3-540-68027-7_3

Mikaberidze, A., & McDonald, B. A. (2020). A tradeoff between tolerance and resistance to a major fungal pathogen in elite wheat cultivars. *New Phytologist*, 226(3), 879–890.

Mishra, A., Singh, L., & Singh, D. (2023). Unboxing the black box—one step forward to understand the soil microbiome: A systematic review. *Microbial Ecology*, 85(2), 669–683.

Nardi, S., Pizzeghello, D., Schiavon, M., & Ertani, A. (2016). Plant biostimulants: Physiological responses induced by protein hydrolyzed-based products and humic substances in plant metabolism. *Scientia Agricola*, 73(1), 18–23.

Neher, D. A., Harris, J. M., Horner, C. E., Scarborough, M. J., Badireddy, A. R., Faulkner, J. W., White, A. C., Darby, H. M., Farley, J. C., & Bishop-von Wettberg, E. J. (2022). Resilient soils for resilient farms: An integrative approach to assess, promote, and value soil health for small- and medium-size farms. *Phytobiomes Journal*, 6(3), 201–206.

Neumann, G., & Romheld V. (2007). The release of root exudates as affected by the plant physiological status. In Pinton, R., Varanini, Z., & Nannipieri, P. (Eds.), *The Rhizosphere: Biochemistry and Organic Substances at the Soil-Plant Interface*. CRC Press.

Ni, R., Wang, S., Lin, X., & Song, L. (2023). Antibiotics inhibit methanogenesis during municipal solid waste

decomposition. *Science of the Total Environment*, 905. https://doi.org/10.1016/j.scitotenv.2023.167397
Niu, Y., Bainard, L. D., Bandara, M., Hamel, C., & Gan, Y. (2017). Soil residual water and nutrients explain about 30% of the rotational effect in 4-yr pulse-intensified rotation systems. *Canadian Journal of Plant Science*, 97(5), 852–864.
Noman, M., Ahmed, T., Ijaz, U., Shahid, M., Azizullah, Li, D., Manzoor, I., & Song, F. (2021). Plant–microbiome crosstalk: Dawning from composition and assembly of microbial community to improvement of disease resilience in plants. *International Journal of Molecular Sciences*, 22(13), 6852. https://doi.org/10.3390/ijms22136852
Okon, Y., Heytler, P., & Hardy, R. J. (1983). N_2 fixation by *Azospirillum brasilense* and its incorporation into host Setaria olubi. *Applied And Environmental Microbiology*, 46, 694–697. https://doi.org/10.1128/AEM.46.3.694-697.1983
Ontañon, O. M., Bedő, S., Ghio, S., Garrido, M. M., Topalian, J., Jahola, D., Fehér, A., Valacco, M. P., Campos, E., & Fehér, C. (2021). Optimisation of xylanases production by two *Cellulomonas* strains and their use for biomass deconstruction. *Applied Microbiology and Biotechnology*, 105(11), 4577–4588. https://doi.org/10.1007/S00253-021-11305-Y
Pandey, D., Kehri, H. K., Zoomi, I., Akhtar, O., & Singh, A. K. (2019). Mycorrhizal fungi: Biodiversity, ecological significance, and industrial applications. In Yadav, A., Mishra, S., Singh, S., & Gupta, A. (Eds.), *Recent Advancement in White Biotechnology Through Fungi. Fungal Biology*. Cham: Springer. https://doi.org/10.1007/978-3-030-10480-1_5
Pandey, S. S., Jain, R., Bhardwaj, P., Thakur, A., Kumari, M., Bhushan, S., & Kumar, S. (2022). Plant probiotics: Endophytes pivotal to plant health. *Microbiological Research*, 263, 127148. https://doi.org/10.1016/j.micres.2022.127148
Pang, Z., Mao, X., Zhou, S., Yu, S., Liu, G., Lu, C., Wan, J., Hu, L., & Xu, P. (2023). Microbiota-mediated nitrogen fixation and microhabitat homeostasis in aerial root-mucilage. *Microbiome*, 11(1), 85.
Pérez-Avalos, O., Sánchez-Herrera, L. M., Salgado, L. M., & Ponce-Noyola, T. (2008). A bifunctional endoglucanase/endoxylanase from *Cellulomonas flavigena* with potential use in industrial processes at different pH. *Current Microbiology*, 57(1), 39–44. https://doi.org/10.1007/s00284-008-9149-1
Pichyangkura, R., & Chadchawan, S. (2015). Biostimulant activity of chitosan in horticulture. *Scientia Horticulturae*, 196, 49–65.
Pino-Hurtado, A., Martínez-García, P. M., Rodríguez-Rubio, L., García, P., & Martínez, B. (2024). A bacteriocin-like compound produced by Brevibacillus laterosporus shows broad-spectrum antibacterial activity against clinically relevant pathogens. *Microbiology Spectrum*, 12(2), e03124-23. https://doi.org/10.1128/spectrum.03124-23
Poria, V., Dębiec-Andrzejewska, K., Fiodor, A., Lyzohub, M., Ajijah, N., Singh, S., & Pranaw, K. (2022). Plant growth-promoting bacteria (PGPB) integrated phytotechnology: A sustainable approach for remediation of marginal lands. *Frontiers in Plant Science*, 13, 999866.
Povero, G., Mejia, J. F., Di Tommaso, D., Piaggesi, A., & Warrior, P. (2016). A systematic approach to discover and characterize natural plant biostimulants. *Frontiers in Plant Science*, 7, 435.
Powell, J. R., & Rillig, M. C. (2018). Biodiversity of arbuscular mycorrhizal fungi and ecosystem function. *New Phytologist*, 220(4), 1059–1075.
Pylak, M., Oszust, K., & Frąc, M. (2019). Review report on the role of bioproducts, biopreparations, biostimulants and microbial inoculants in organic production of fruit. *Reviews in Environmental Science and Bio/Technology*, 18(3), 597–616.
Qadir, M., Iqbal, A., Hussain, A., Hussain, A., Shah, F., Yun, B. W., & Mun, B. G. (2024). Exploring plant–bacterial symbiosis for eco-friendly agriculture and enhanced resilience. *International Journal of Molecular Sciences*, 25(22), 12198.
Qiao, M., Sun, R., Wang, Z., et al. (2024). Legume rhizodeposition promotes nitrogen fixation by soil microbiota under crop diversification. *Nature Communications*, 15, 2924. https://doi.org/10.1038/s41467-024-47159-x
Rawat, P., Das, S., Shankhdhar, D., & Shankhdhar, S. C. (2021). Phosphate-solubilizing microorganisms: Mechanism and their role in phosphate solubilization and uptake. *Journal Of Soil Science And Plant Nutrition*, 21(1), 49–68.
Raza, W., Ling, N., Yang, L., Huang, Q., & Shen, Q. (2016). Response of tomato wilt pathogen *Ralstonia solanacearum* to the volatile organic compounds produced by a biocontrol strain *Bacillus amyloliquefaciens* SQR–9. *Scientific Reports*, 6(1), 24856.
Rebello, S., Nathan, V. K., Sindhu, R., Binod, P., Awasthi, M. K., & Pandey, A. (2021). Bioengineered microbes for soil health restoration: Present status and future. *Bioengineered*, 12(2), 12839–12853.
Rich, M. K., Vigneron, N., Libourel, C., Keller, J., Xue, L., et al. (2021). Lipid exchanges drove the evolution of mutualism during plant terrestrialization. *Science*, 372(6544), 864–868.
Riley, M. A., & Wertz, J. E. (2002). Bacteriocins: Evolution, ecology, and application. *Annual Review of Microbiology*, 56, 117–137. https://doi.org/10.1146/annurev.micro.56.012302.161024
Rizaludin, M. S., Stopnisek, N., Raaijmakers, J. M., & Garbeva, P. (2021). The chemistry of stress: Understanding the 'cry for help' of plant roots. *Metabolites*, 11(6), 357.
Rolfe, S. A., Griffiths, J., & Ton, J. (2019). Crying out for help with root exudates: Adaptive mechanisms by which stressed plants assemble health-promoting soil microbiomes. *Current Opinion in Microbiology*, 49, 73–82.
Román, J. R., Roncero-Ramos, B., Chamizo, S., Rodríguez-Caballero, E., & Cantón, Y. (2018). Restoring soil functions by means of cyanobacteria inoculation: Importance of soil conditions and species selection. *Land Degradation and Development*, 29(9), 3184–3193. https://doi.org/10.1002/ldr.3064
Roth-Bejerano, N., Navarro-Ródenas, A., Gutiérrez, A. (2014). Types of Mycorrhizal association. In Kagan-Zur, V., Roth-Bejerano, N., Sitrit, Y., Morte, A. (Eds.), *Desert Truffles. Soil Biology* (Vol 38). Berlin, Heidelberg: Springer. https://doi.org/10.1007/978-3-642-40096-4_5
Rouphael, Y., & Colla, G. (2018). Synergistic biostimulatory action: Designing the next generation of plant biostimulants for sustainable agriculture. *Frontiers in Plant Science*, 9, 1655. https://doi.org/10.3389/fpls.2018.01655
Rousk, J., Bååth, E., Brookes, P. C., Lauber, C. L., Lozupone, C., Caporaso, J. G., Knight, R., & Fierer, N. (2010). Soil bacterial and fungal communities across a pH gradient in an arable soil. *The ISME Journal*, 4(10), 1340–1351.
Rozier, C., Hamzaoui, J., Lemoine, D., Czarnes, S., & Legendre, L. (2017). Field-based assessment of the mechanism of maize yield enhancement by *Azospirillum lipoferum*

CRT1. *Scientific Reports*, 7, 1–12. https://doi.org/10.1038/s41598-017-07929-8

Rudrappa, T., Czymmek, K. J., Paré, P. W., & Bais, H. P. (2008). Root-secreted malic acid recruits beneficial soil bacteria. *Plant Physiology*, 148(3), 1547–1556.

Saeed, M., Ilyas, N., Bibi, F., Jayachandran, K., Dattamudi, S., & Elgorban, A. M. (2022). Biodegradation of PAHs by *Bacillus marsiflavi*, genome analysis and its plant growth promoting potential. *Environmental Pollution*, 292(Pt B). https://doi.org/10.1016/j.envpol.2021.118343

Safdarian, M., Askari, H., Shariati, J. V., & Nematzadeh, G. (2019). Transcriptional responses of wheat roots inoculated with *Arthrobacter nitroguajacolicus* to salt stress. *Scientific Reports*, 9(1), 1792.

Saleem, F., Ahmad, S., Yaqoob, Z., & Rasool, S. A. (2009). Comparative study of two bacteriocins produced by representative indigenous soil bacteria. *Pakistan Journal of Pharmaceutical Sciences*, 22(3), 252–258. PMID: 19553169.

Sandy, M., & Butler, A. (2009). Microbial iron acquisition: Marine and terrestrial siderophores. *Chemical Reviews*, 109, 4580–4595. https://doi.org/10.1021/cr9002787

Sanmartín, N., Sánchez-Bel, P., Pastor, V., Pastor-Fernández, J., Mateu, D., Pozo, M. J., Cerezo, M., & Flors, V. (2020). Root-to-shoot signalling in mycorrhizal tomato plants upon *Botrytis cinerea* infection. *Plant Science*, 298, Article 110595.

Santos-Medellín, C., Edwards, J., Liechty, Z., Nguyen, B., & Sundaresan, V. (2017). Drought stress results in a compartment-specific restructuring of the rice root-associated microbiomes. *Mbio*, 8(4), e00764-17.

Sarathambal, C., Jeevalatha, A., Sivaranjani, R., Biju, C. N., Charles, S., Srinivasan, V., George, P., Peter, B., & Radhika, R. (2023). Arbuscular mycorrhizal colonization alters biochemical, molecular defense responses and root exudate composition against *Phytophthora capsici* infection in black pepper. *Rhizosphere*, 25, Article 100651.

Savary, S., Willocquet, L., Pethybridge, S. J., Esker, P., McRoberts, N., & Nelson, A. (2019). The global burden of pathogens and pests on major food crops. *Nature Ecology & Evolution*, 3(3), 430–439.

Saxena, A., Kumar, M., Chakdar, H., Anuroopa, N., & Bagyaraj, D. (2019). *Bacillus* species in soil as a natural resource for plant health and nutrition. *Journal of Applied Microbiology*, 128(6), 1583–1594. https://doi.org/10.1111/jam.14506

Schippers, B., Bakker, A. W., & Bakker, A. H. M. (1987). Interactions of deleterious and beneficial rhizosphere microorganisms and the effect of cropping practice. *Annual Review of Phytopathology*, 25, 339–358.

Senesi, N., & Loffredo, E. (1999). The chemistry of soil organic matter. In Sparks, D. L. (Ed.), *Soil Physical Chemistry*. CRC Press.

Sham, A., Al-Ashram, H., Whitley, K., Iratni, R., El-Tarabily, K. A., & AbuQamar, S. F. (2019). Metatranscriptomic analysis of multiple environmental stresses identifies RAP2.4 gene associated with *Arabidopsis* immunity to *Botrytis cinerea*. *Scientific Reports*, 9(1), 17010.

Sharma, H. S., Fleming, C., Selby, C., Rao, J. R., & Martin, T. (2014). Plant biostimulants: A review on the processing of macroalgae and use of extracts for crop management to reduce abiotic and biotic stresses. *Journal of Applied Phycology*, 26, 465–490.

Sharma, S. B., Sayyed, R. Z., Trivedi, M. H., & Gobi, T. A. (2013). Phosphate solubilizing microbes: Sustainable approach for managing phosphorus deficiency in agricultural soils. *SpringerPlus*, 2, 1–14.

Shi, J., Wang, X., & Wang, E. (2023). Mycorrhizal symbiosis in plant growth and stress adaptation: From genes to ecosystems. *Annual Review of Plant Biology*, 74, 569–607.

Shrivastava, M., Srivastava, P. C., & D'Souza, S. F. (2018). Phosphate-solubilizing microbes: Diversity and phosphates solubilization mechanism. In Meena, V. (Ed.), *Role of Rhizospheric Microbes in Soil*. Singapore: Springer. https://doi.org/10.1007/978-981-13-0044-8_5

Silva, U. C., Medeiros, J. D., Leite, L. R., Morais, D. K., Cuadros-Orellana, S., Oliveira, C. A., & Silva, I. R. (2023). Phosphate-solubilizing microorganisms as a sustainable strategy for improving phosphorus availability in agricultural soils. *Frontiers in Microbiology*, 14, 1182457. https://doi.org/10.3389/fmicb.2023.1182457

Singh, P., Singh, R., Madhu, G. S., & Singh, V. P. (2023). Seed biopriming with *Trichoderma harzianum* for growth promotion and drought tolerance in rice (*Oryza sativa*). *Agricultural Research*, 12(2), 154–162. https://doi.org/10.1007/s40003-022-00641-8

Smith, S. E., & Read, D. (2008). *Mycorrhizal Symbiosis* (Third Edition). Academic Press. https://doi.org/10.1016/B978-012370526-6.50003-9.

Sohrabi, R., Paasch, B. C., Liber, J. A., & He, S. Y. (2023). Phyllosphere microbiome. *Annual Review of Plant Biology*, 74(1), 539–568. https://doi.org/10.1146/annurev-arplant-102820-032704

Song, O. R., Lee, S. J., Lee, Y. S., Lee, S. C., Kim, K. K., & Choi, Y. L. (2008). Solubilization of insoluble inorganic phosphate by *Burkholderia cepacia* DA23 isolated from cultivated soil. *Brazilian Journal of Microbiology*, 39, 151–156.

Song, Y. Y., Cao, M., Xie, L. J. et al. (2011). Induction of DIMBOA accumulation and systemic defense responses as a mechanism of enhanced resistance of mycorrhizal corn (*Zea mays* L.) to sheath blight. *Mycorrhiza*, 21, 721–731. https://doi.org/10.1007/s00572-011-0380-4

Starr, E. P., Nuccio, E. E., Pett-Ridge, J., Banfield, J. F., & Firestone, M. K. (2019). Metatranscriptomic reconstruction reveals RNA viruses with the potential to shape carbon cycling in soil. *Proceedings of the National Academy of Sciences*, 116(51), 25900–25908. https://doi.org/10.1073/pnas.1908291116

Subramanian, S., & Smith, D. L. (2015). Bacteriocins from the rhizosphere microbiome—from an agriculture perspective. *Frontiers In Plant Science*, 6, 909. https://doi.org/10.3389/fpls.2015.00909

Suman, J., Rakshit, A., Ogireddy, S. D., Singh, S., Gupta, C., & Chandrakala, J. (2022). Microbiome as a key player in sustainable agriculture and human health. *Frontiers in Soil Science*, 2, 821589. https://doi.org/10.3389/fsoil.2022.821589

Suslow, T. V., Kloepper, J. W., Schroth, M. N., & Burr, T. J. (1979). Beneficial bacteria enhance plant growth. *California Agriculture*, 33, 15–17.

Tallapragada, P., & Matthew, T. (2021). Potassium solubilizing microorganisms (KSM) a very promising biofertilizers. In Mishra, B. B., Nayak, S. K., & Pahari, A. (Eds.), *Agriculturally Important Microorganisms: Mechanisms and Applications for Sustainable Agriculture*. CRC Press.

Tate, K. R. (1984). The biological transformation of P in soil. *Plant and Soil*, 76(1–3), 245–256. https://doi.org/10.1007/BF02205584

Tedersoo, L., Bahram, M., & Zobel, M. (2020). How mycorrhizal associations drive plant population and community biology. *Science*, 367(6480), eaba1223.

Thauer, R. K., Kaster, A.-K., Seedorf, H., & Buckel, W. (2008). Methanogenic archaea: Ecologically relevant differences in energy conservation. *Nature Reviews Microbiology*, 6, 579–591. https://doi.org/10.1038/nrmicro1931

Tilman, D., Cassman, K. G., Matson, P. A., Naylor, R., & Polasky, S. (2002). Agricultural sustainability and intensive production practices. *Nature*, 418(6898), 671–677.

Tiwari, G., Duraivadivel, P., Sharma, S., & P. H. (2018). 1-Aminocyclopropane-1-carboxylic acid deaminase producing beneficial rhizobacteria ameliorate the biomass characters of Panicum maximum Jacq. By mitigating drought and salt stress. *Scientific Reports*, 8(1), 17513.

Totsche, K. U., Amelung, W., Gerzabek, M. H., Guggenberger, G., Klumpp, E., Knief, C., Lehndorff, E., Mikutta, R., Peth, S., Prechtel, A., & Ray, N. (2018). Microaggregates in soils. *Journal of Plant Nutrition and Soil Science*, 181(1), 104–136.

Tourna, M., Stieglmeier, M., Spang, A., Könneke, M., Schintlmeister, A., Urich, T., Engel, M., Schloter, M., Wagner, M., Richter, A., & Schleper, C. (2011). *Nitrososphaera viennensis*, an ammonia oxidizing archaeon from soil. *Proceedings of the National Academy of Sciences*, 108(20), 8420–8425. https://doi.org/10.1073/pnas.1013488108

Turner, T. R., Ramakrishnan, K., Walshaw, J., Heavens, D., Alston, M., Swarbreck, D., Osbourn, A., Grant, A., & Poole, P. S. (2013). Comparative metatranscriptomics reveals kingdom level changes in the rhizosphere microbiome of plants. *ISME Journal*, 7, 2248–2258. https://doi.org/10.1038/ismej.2013.119

Uma Devi, S., Shyamala Gowri, R., & Kumar, S. D. (2020). Utilization of *Azospirillum* as a biofertilizer: An overview. *International Journal of Pharmaceutical Sciences Review and Research*, 2, 141–145.

Umesha, S., Singh, P. K., & Singh, R. P. (2018). Microbial biotechnology and sustainable agriculture. In Singh R. l. & Mondal, S. (Eds.), *Biotechnology for Sustainable Agriculture*. Woodhead Publishing. https://doi.org/10.1016/B978-0-12-812160-3.00006-4.

Van der Heijden, M. G. A., Bardgett, R. D., & Van Straalen, N. M. (2008). The unseen majority: Soil microbes as drivers of plant diversity and productivity in terrestrial ecosystems. *Ecology Letters*, 11, 296–310.

Van Deynze, A., Zamora, P., Delaux, P. M., Heitmann, C., Jayaraman, D., Rajasekar, S., Graham, D., Maeda, J., Gibson, D., Schwartz, K. D., & Berry, A. M. (2018). Nitrogen fixation in a landrace of maize is supported by a mucilage-associated diazotrophic microbiota. *PloS Biology*, 16(8), e2006352.

Van Leeuwen, J. P., Djukic, I., Bloem, J., Lehtinen, T., Hemerik, L., de Ruiter, P. C., & Lair, G. J. (2017). Effects of land use on soil microbial biomass, activity and community structure at different soil depths in the Danube floodplain. *European Journal of Soil Biology*, 79, 14–20. https://doi.org/10.1016/j.ejsobi.2017.02.001

Van Oosten, M. J., Di Stasio, E., Cirillo, V., Silletti, S., Ventorino, V., Pepe, O., Raimondi, G., & Maggio, A. (2018). Root inoculation with Azotobacter chroococcum 76A enhances tomato plants adaptation to salt stress under low N conditions. *BMC Plant Biology*, 18, 1–12.

Van Oosten, M. J., Pepe, O., De Pascale, S., et al. (2017). The role of biostimulants and bioeffectors as alleviators of abiotic stress in crop plants. *Chemical And Biological Technologies In Agriculture*, 4(5). https://doi.org/10.1186/s40538-017-0089-5

Vandana, U. K., Rajkumari, J., Singha, L. P., Satish, L., Alavilli, H., Sudheer, P. D., Chauhan, S., Ratnala, R., Satturu, V., Mazumder, P. B., & Pandey, P. (2021). The endophytic microbiome as a hotspot of synergistic interactions, with prospects of plant growth promotion. *Biology*, 10(2), 101.

Vannier, N., Agler, M., & Hacquard, S. (2019). Microbiota-mediated disease resistance in plants. *PloS Pathogens*, 15(6), e1007740.

Vásquez-Dean, J., Maza, F., Morel, I., Pulgar, R., & González, M. (2020). Microbial communities from arid environments on a global scale: A systematic review. *Biological Research*, 53(1). https://doi.org/10.1186/s40659-020-00296-1

Vejan, P., Abdullah, R., Khadiran, T., Ismail, S., & Nasrulhaq Boyce, A. (2016). Role of plant growth promoting rhizobacteria in agricultural sustainability: A review. *Molecules*, 21(5), 573.

Wagner, S. C. (2011). Biological nitrogen fixation. *Nature Education Knowledge*, 3(10), 15.

Wang, H., & Chen, Y. (2024). Protecting plants from pathogens through arbuscular mycorrhiza: Role of fungal diversity. *Microbiological Research*, 289, 127919. https://doi.org/10.1016/j.micres.2024.127919

Wang, Q., Hong, H., Liao, R., Yuan, B., Li, H., Lu, H., Liu, J., & Yan, C. (2021). Glomalin-related soil protein: The particle aggregation mechanism and its insight into coastal environment improvement. *Ecotoxicology and Environmental Safety*, 227, 112940. https://doi.org/10.1016/j.ecoenv.2021.112940

Wang, X., Chi, Y., & Song, S. (2024). Important soil microbiota's effects on plants and soils: A comprehensive 30-year systematic literature review. *Frontiers in Microbiology*, 15, 1347745. https://doi.org/10.3389/FMICB.2024.1347745

Weimer, P. J. (2022). Degradation of cellulose and hemicellulose by ruminal microorganisms. *Microorganisms*, 10(12), Article 2345. https://doi.org/10.3390/microorganisms10122345

Weimer, P. J., Russell, J. B., & Muck, R. E. (2022). Lessons from the cow: What the ruminant animal can teach us about consolidated bioprocessing of cellulosic biomass. *Bioresource Technology*, 353, 127116. https://doi.org/10.1016/j.biortech.2022.127116

White, J. F., Crawford, H., Torres, M. S., Mattera, R., Irizarry, I., & Bergen, M. (2012). A proposed mechanism for nitrogen acquisition by grass seedlings through oxidation of symbiotic bacteria. *Symbiosis*, 57, 161–171. https://doi.org/10.1007/s13199-012-0189-8

Xia, Q., Rufty, T., & Shi, W. (2020). Soil microbial diversity and composition: Links to soil texture and associated properties. *Soil Biology and Biochemistry*, 149, 107953. https://doi.org/10.1016/j.soilbio.2020.107953

Xie, S., Jiang, L., Wu, Q., Wan, W., Gan, Y., Zhao, L., & Wen, J. (2022). Maize root exudates recruit *Bacillus amyloliquefaciens* OR2–30 to inhibit Fusarium graminearum infection. *Phytopathology®*, 112(9), 1886–1893.

Xing, Y., Wang, X., & Mustafa, A. (2025). Exploring the link between soil health and crop productivity. *Ecotoxicology and Environmental Safety*, 289, 117703. ISSN 0147-6513. https://doi.org/10.1016/j.ecoenv.2025.117703

Xu, J., Zhang, Y., Zhang, P., Trivedi, P., Riera, N., Wang, Y., Liu, X., Fan, G., Tang, J., Coletta-Filho, H. D., & Cubero, J. (2018). The structure and function of the global citrus rhizosphere microbiome. *Nature Communications*, 9(1), 4894.

Xu, S., Bai, Z., Jin, B., Xiao, R., & Zhuang, G. (2014). Bioconversion of wastewater from sweet potato starch production to *Paenibacillus polymyxa* biofertilizer for tea plants. *Scientific Reports*, 4(1), 4131.

Yang, Y., Du, W., Ren, X., Cui, Z., Zhou, W., & Lv, J. (2020). Effect of bean dregs amendment on the organic matter degradation, humification, maturity and stability of pig manure composting. *Science of The Total Environment*, 708, 134623.

Yao, M. K., Désilets, H., Charles, M. T., Boulanger, R., & Tweddell, R. J. (2003). Effect of mycorrhization on the accumulation of rishitin and solavetivone in potato plantlets challenged with *Rhizoctonia solani*. *Mycorrhiza*, 13(6), 333–336.

Yasmeen, T. et al. (2021). *Pseudomonas* as plant growth-promoting bacteria and its role in alleviation of abiotic stress. In Mohamed, H. I., El-Beltagi, H. E. D. S., & Abd-Elsalam, K. A. (Eds.), *Plant Growth-Promoting Microbes for Sustainable Biotic and Abiotic Stress Management*. Cham: Springer. https://doi.org/10.1007/978-3-030-66587-6_7

Yoolong, S., Kruasuwan, W., Thanh Phạm, H. T., Jaemsaeng, R., Jantasuriyarat, C., & Thamchaipenet, A. (2019). Modulation of salt tolerance in Thai jasmine rice (*Oryza sativa* L. cv. KDML105) by *Streptomyces venezuelae* ATCC 10712 expressing ACC deaminase. *Scientific Reports*, 9(1), 1275.

Yoshida, T., & Ancajas, R. R. (1973). Nitrogen-fixing activity in upland and flooded rice fields. *Soil Science Society Of America Journal*, 37, 42–46. https://doi.org/10.2136/sssaj1973.03615995003700010017x

Yu, Z., Chen, C., Li, Z., Song, Y., Yan, C., Jiang, X., Jia, H., Shang, Y., & Tian, M. (2024). Potential biofertilizers for alkaline soil: Bacteria isolated from the rhizosphere of potatoes. *Agronomy*, 14(6), 1241. https://doi.org/10.3390/agronomy14061241

Yuan, J., Zhao, J., Wen, T., Zhao, M., Li, R., Goossens, P., Huang, Q., Bai, Y., Vivanco, J. M., Kowalchuk, G. A., & Berendsen, R. L. (2018). Root exudates drive the soil-borne legacy of aboveground pathogen infection. *Microbiome*, 6, 1–12.

Yuan, Y., Li, X., Liu, F., Tian, X., Shao, Y., Yuan, Z., & Chen, Y. (2024). Differences in soil microbial communities across soil types in China's temperate forests. *Forests*, 15(7), 1110. https://doi.org/10.3390/f15071110

Zhang, J., Zhao, R., Li, X., & Zhang, J. (2024). Potential of arbuscular mycorrhizal fungi for soil health: A review. *Pedosphere*, 34(2), 279–288.

Zhang, X., Davidson, E. A., Mauzerall, D. L., Searchinger, T. D., Dumas, P., & Shen, Y. (2015). Managing nitrogen for sustainable development. *Nature*, 528(7580), 51–59.

Zhang, Y., Guo, Y., Qiu, T., Gao, M., & Wang, X. (2022). Bacteriophages: Underestimated vehicles of antibiotic resistance genes in the soil. *Frontiers in Microbiology*, 13, 936267. https://doi.org/10.3389/fmicb.2022.936267

Zhu, F., Qu, L., Hong, X., & Sun, X. (2011). Isolation and characterization of a phosphate-solubilizing halophilic bacterium *Kushneria* sp. YCWA18 from Daqiao Saltern on the coast of Yellow Sea of China. *Evidence-based Complementary and Alternative Medicine*, 2011(1), 615032.

2 Advances in Techniques for Studying Microbiome

*Pál Salamon, Emese Bálint, and Gyöngyvér Mara**
Faculty of Economics, Socio-Human Sciences and Engineering, Sapientia Hungarian University of Transylvania, Libertății Sq. 1, 530104 Miercurea Ciuc, Romania.

2.1 INTRODUCTION

Soil ecosystems are intricate and diverse, containing over 100 million microbial cells per gram, including bacteria, fungi, archaea, and protists. To gain a deeper comprehension of the microbial activities implicated in key ecological soil processes, researchers have analyzed the microbial community structure, biomass, and function (Widmer et al., 2001; Nannipieri 2020). The functions of the soil microbiome influence microbe–microbe and plant–microbe relationships, leading to ecological impacts that range from localized effects on soil nutrient levels to global consequences, such as greenhouse gas emissions. Microbial diversity strengthens an ecosystem's resilience to disturbances and helps control harmful organisms, including soil-borne pathogens. However, the correlation between the microbiome diversity and the resilience of soil environment is based on the presence of microorganisms with specific traits (Berihu et al., 2023; Garg et al., 2024).

The initial approach for investigating the soil microbiome involves culture-based techniques. Despite the vast number of microbes in the soil, only a small percentage (0.1–51%) can be grown in laboratory settings (Garg et al., 2024). Microorganisms that cannot be cultivated represent an untapped source of species with distinctive biological and chemical traits. These uncultivable bacteria play a pivotal role in their natural environment, and our lack of understanding of them creates a knowledge gap (Garg et al., 2024). Consequently, more advanced and labor-intensive methods have been developed to comprehensively examine the microbiome. In the mid- to late 20th century, biochemical and chemotaxonomic techniques based on fatty acid methyl ester (FAME) and phospholipid fatty acid (PLFA) analysis, along with Biolog systems, were employed for taxonomic classification and microbial biomass measurement (Figure 2.1). The Biolog system has been extensively used to estimate the metabolic diversity of soil microbial populations (Lawlora et al., 2000). Subsequently, molecular-based methods, such as amplicon sequencing (16S rDNA) and other PCR-based techniques, have emerged, enabling community-level studies of microbial diversity. The development of high-throughput sequencing in the 21st century marked a significant breakthrough in exploring unknown, hidden, and uncultivable portions of the soil microbiome. Each of these approaches provides insight into aspects of soil microbiological features and offers a standalone analysis of changes in the soil microbial community structure and function. The integration of omics-based methods for documenting community composition and fluctuations, combined with metabolic modeling techniques, is advancing the structured explanation of ecosystem architecture. Projecting microbial taxa-related functions onto the meta-network, which represents the complete set of soil enzymes, provides insights into the core and specialized functions. This community-level perspective on metabolism will enhance our understanding of in-situ microbial activities, which are key determinants of ecosystem structure and microbial community functioning (Berihu et al., 2023).

In the following, we will discuss the methods used to discover soil microbiome composition and function in order to understand the occurrence of the method, giving a broader overview of how today's knowledge and understanding have evolved.

2.2 BIOCHEMICAL AND CHEMOTAXONOMICAL METHODS

2.2.1 Fatty Acid Analysis

Fatty acid (FA) analyses were the first biochemical and chemotaxonomic methods used to understand soil microbial composition. FAs are widely employed as biomarkers to assess microbial community composition in soil ecosystems subjected to various management practices and vegetation types. The PubMed database (National Center for Biotechnology Information, U.S. National Library of Medicine) search on 13 February 2025 using the keyword combination fatty acids AND soil resulted in 10871 hits dated back to 1970–1980, when the first articles were published regarding soil FA analysis.

This method is based on the hypothesis that alterations in overall FA profiles primarily reflect changes in microbial populations. The abundance and chemical diversity of FAs in cells make them suitable for this purpose. Certain FAs

***Corresponding author:** maragyongyver@uni.sapientia.ro (Gyöngyvér Mara)

 DOI: 10.1201/9781003544425-2

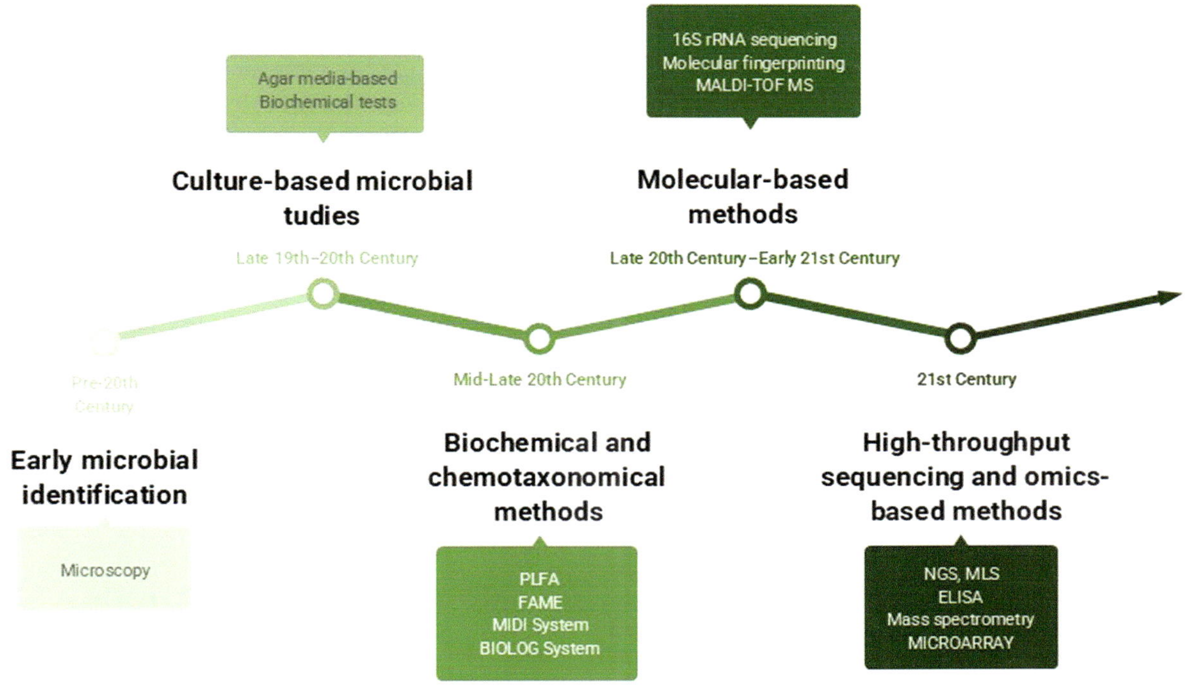

FIGURE 2.1 Timeline of techniques for studying the microbiome.

have been identified as characteristic indicators of specific microbial functional groups, enabling researchers to interpret observed changes in community structure. This method assumes that shifts in FA profiles can be attributed to the corresponding shifts in microbial community composition (Fernandes et al., 2013).

One of the most widely employed methods for evaluating soil microbial communities is examination of PLFAs. This technique allows researchers to measure the quantity of microbes present and to investigate the composition of their communities. The initial application of PLFA biomarker analysis can be traced back to 1979 when researchers employed it to assess microbial biomass in sediments from marine and estuarine environments (White et al., 1979).

The analytical process typically involves four sequential steps. Initially, a two-phase extraction method is employed to isolate lipids from soil samples. The extracted lipids are then separated using solid-phase extraction. This separation is accomplished by sequentially eluting lipids with increasing polarity solvents (e.g., chloroform, acetone, and methanol), resulting in the isolation of lipid fractions (neutral lipids, glycolipids, phospholipids). Subsequently, the separated lipid fractions undergo derivatization to form their corresponding FAMEs: neutral lipid fatty acids (NLFAs), glycolipid fatty acids (GLFAs), and PLFAs. All these fractions may contain notable supplementary data about the microbial populations in the soil (Gorka et al., 2023).

Many laboratories employ gas chromatographic whole-cell fatty acid methyl ester (FAME) analysis for bacterial species identification because of its cost-effectiveness, simplicity, and automation. This chemotaxonomic method typically relies on commercially available systems, such as the Sherlock Microbial Identification System (MIS, Microbial ID Inc. software [MIDI], Newark, DE, USA) (Slabbinck et al., 2009).

PLFA analysis can yield similar indicators for microbial community composition as other methods, but may show varying degrees of sensitivity to soil characteristics. Although PLFA profiling is an effective method for examining soil microbiomes, it is crucial to acknowledge its constraints and consider incorporating additional lipid analyses, such as NLFA and GLFA. This comprehensive approach offers a more thorough insight into microbial community structure and dynamics. By combining these techniques, researchers can obtain more extensive datasets, enabling them to better evaluate the consequences of environmental shifts and management strategies on soil health. FAME–MIDI systems also have a significant drawback: their identification libraries are updated infrequently and only cover a portion of the known species. This update latency severely compromises the reliability of the bacterial species identification. To address this issue, the development of a modern FAME-based system for bacterial species identification utilizing artificial intelligence-based techniques has been proposed (Slabbinck et al., 2009).

Nevertheless, despite its limitations, this technique continues to be a useful approach in soil microbiome studies, offering reproducible results, cost-efficiency, and quick assessments of microbial community structures.

2.2.2 Biolog System

Among the biochemical and chemotaxonomical methods, Biolog systems should be mentioned. The Biolog system was introduced in 1991 and is widely used in medical and general microbiology to identify bacteria. The PubMed database (National Center for Biotechnology Information, U.S. National Library of Medicine) search on 13 February 2025 using the keyword combination Biolog system AND soil resulted in 184 hits dated back to 1992, when the first article, regarding a *Pseudomonas* sp. STAIN characterization was published (Hickey et al., 1992).

The identification method relies on carbon source utilization by bacteria, and a given profile is compared to a database. Low-molecular-weight carbonaceous compounds are used as substrates, and tetrazolium salts are used as indicators of substrate utilization and microbial growth. GenIII microbial identification plates consist of 94 simultaneous phenotypic tests (71 carbon source and 23 chemical sensitivity analysis) and provide a unique profile of the selected strain suitable for identification. It has been broadly used for the identification of soil and rhizobacteria, and the outcomes were evaluated against those obtained from alternative methods, such as 16S rDNA sequencing. The success of identification varies and is generally higher at the genus level and lower at the species level. In Woźniak et al.'s (2019) study, 78% of the plant growth promoting rhizobacterial isolates were accurately identified at the genus level, whereas 48% at the species level (Woźniak et al., 2019).

It has been used not only for identification, but also for phenotype assays and community analysis in various environments related to environmental factors and agricultural practices.

Biolog Phenotype MicroArray plates were developed to assess the utilization of carbon, nitrogen, and phosphorus sources, biosynthetic pathways, osmosis, pH response, and chemical sensitivity of bacteria and yeast. In the case of *Pseudomonas* species from the sugarcane rhizosphere, H.-B. Li et al. (2017) used phenotypic assays to study metabolic diversity and define chemical guilds. Based on these results, they highlighted the main factors accounting for metabolic variation (nitrogen utilization, osmolytes, and pH) in *Pseudomonas* sp. (H.-B. Li et al., 2017).

EcoPlates with three repeated sets of 31 carbon sources have been developed and used extensively in applied ecological research. These plates can be used to study the stability of a standard microbial population and determine and evaluate changes related to environmental variables. EcoPlates are suitable for soil microbiome analysis because at least 9 of the 31 environmentally applicable substrates are relevant components of plant root exudates (Preston-Mafham et al., 2002).

However, GN and GP plates were developed mainly for bacterial identification, and Widmer et al. (2001) also attempted to obtain community-level substrate utilization (CLSU) fingerprints. They compared the results obtained with DNA- and PLFA-based datasets and observed a different dendrogram topology for Biolog data.

Although Biolog systems have been used successfully in many cases, there are several limitations that must be considered. In the case of community analysis, due to the complexity and species diversity, microorganisms will be subjected to competition, and the short incubation period will favor the R strategist species, reflecting the growth of these species more than the whole community reaction to different parameters (Preston-Mafham et al., 2002; Widmer et al., 2001; Kruglov 2016). Other synergistic or antagonistic interactions among species may have influenced the data obtained (Kruglov 2016). The usefulness of the Biolog system is uncontested and must be used and interpreted with care. According to earlier studies, it is more beneficial for comparing communities than for describing them (Preston-Mafham et al., 2002), and it is particularly effective in investigating the influence of natural and human-induced factors on soil microbial communities (Kruglov 2016).

2.3 MOLECULAR-BASED METHODS

2.3.1 Ribosomal RNA Genes

This method dates back to 1977, when Woese and Fox (1977) developed 16S Ribosomal RNA (rRNA) sequencing, which revolutionized bacterial phylogenetics. rRNA genes are commonly employed as molecular markers in the investigation of soil microbial phylogeny. A PubMed database (National Center for Biotechnology Information, U.S. National Library of Medicine) search on 13 February 2025 using the keyword combination 16S rRNA gene AND soil resulted in 11707 hits, with articles published since 1991, and the largest number of articles, around 600, were published between 2009 and 2018, annually.

These genes are preferred due to their widespread presence, conservation of function, and the existence of variable regions that enable taxonomic differentiation. An extensive and growing digital repository of small subunit rRNA gene sequences (16S rRNA for prokaryotes, 18S rRNA for eukaryotes) enables the identification of numerous soil archaea, bacteria, and fungi. The 16S rRNA gene carries nine hypervariable regions, labeled V1 through V9, which exhibit sufficient diversity to enable taxonomic classification of bacteria (H. Li et al., 2009). This identification often reaches the genus or species level and occasionally extends to the subspecies level. However, many organisms remain unclassified within the studied phyla and are categorized as "uncultured" (Hirsch et al., 2010). When estimating the relative abundance of specific bacterial groups based on the quantity of different 16S rRNA genes in a trial, it is crucial to take into consideration that the number of rRNA gene copies fluctuates from 1 to >15 among bacterial species and brings possible biases to microbial diversity analyses (Gao and Wu 2023). This information can be found in the ribosomal rRNA operon copy number database, known as rrnDB (Hirsch et al., 2010), and a novel method and software called RasperGade16S has emerged to maximize the reliability of 16S rRNA-based community prediction (Gao and Wu, 2023).

rRNA analysis methods offer a wealth of information about the different taxa present in a given environment, yet they offer limited understanding of the functional roles played by each phylogenetic group (Torsvik and Øvreås, 2002). Besides, the choice of primers may selectively amplify specific taxa, which can lead to biases, as certain hypervariable regions and the number of rRNA copies can alter community structure prediction (Gao and Wu, 2023; H. Li et al., 2009).

2.3.2 PCR-Based Techniques

A molecular method commonly applied for the detection of microbial species in soil and leaf samples is the polymerase chain reaction (PCR) (Barghouthi 2011). Amplification of targeted sequences from nucleic acid PCR can identify bacterial species and establish phylogenetic relationships between microorganisms (Garibyan and Avashia, 2013; Mirmajlessi et al., 2015). A PubMed database (National Center for Biotechnology Information, U.S. National Library of Medicine) (National Center for Biotechnology Information, U.S. National Library of Medicine) search on 13 February 2025 using the keyword combination PCR AND soil resulted in 10076 hits, with articles published since 1988.

Different PCR techniques can simultaneously enable the isolation and measurement of viable bacterial counts. Multiplex PCR is a technique that can be employed to isolate mixed bacterial pathogens from a sample and distinguish between various species within the same genus (Franco-Duarte et al., 2019). RT-PCR enables rapid and high-volume detection and measurement of specific DNA sequences, and it minimizes the risk of cross-contamination as no additional handling of samples is necessary after amplification (Kralik and Ricchi 2017).

The advantages of PCR methods include rapid turnaround time, simplicity of use, sensitivity, and repeatability. They are relatively low-cost and can be considered as routine procedures for the detection and quantification of microorganisms. Among these disadvantages, one can enumerate the expensive equipment cost, the necessity of high-level skills, and non-specific amplification. Primer dimer formation is a huge disadvantage; in the case of a multiplex PCR, it can be formed between the various primers, which could result in reduced sensitivity and the selective amplification of specific targets (Markoulatos et al., 2002).

A major drawback is that only 0.1–1% of microbes can be detected (Barghouthi, 2011). Classic PCR cannot differentiate between viable and non-viable microorganisms, and is affected by PCR inhibitors that might be in the sample, leading to false-negative outcomes (Venbrux et al., 2023).

A PCR-based technique that analyzes the amplified region between 16S and 23S rRNA operons is called Ribosomal Intergenic Spacer Analysis (RISA). Depending on the species, this region has sequence and length variability, which makes it a useful tool for identifying microorganisms at a taxonomic level. By analyzing the results of RISA, we could easily differentiate various microbial strains and closely related species from an environmental sample. Although it is a quick and extremely easy rRNA fingerprinting method, its utilization is limited by the lack of a database of ribosomal intergenic spacer sequences (Malik et al., 2008).

Microarray techniques are used for concomitant assay of many genes, consisting of 500–5000 nucleotides in length of complementary DNA probes on a solid support, permitting the synchronous hybridization of a multiple set of genes to their corresponding targets in an environmental sample. This method is often used for pure cultures; however, during a functional gene array (FGA), the identification of genes encoding key enzymes can be measured directly from a metabolic process. The concept of a community genome array (CGA) is similar in that it can assess the genomic similarity of isolated bacteria based on the temperature at which hybridization occurs. As in the case of the previously mentioned techniques, culturable organisms are needed. These methodological variations only detect the dominant population in every environment; probes that are intended to target specific known sequences can sometimes bind to similar or unidentified sequences, potentially leading to inaccurate signals. Alternatively, soil samples can contain humic acid and other organic compounds, which inhibit the hybridization of microarrays (Gentry et al., 2006).

However, these methods do not provide sufficient information regarding the microbiome–host plant connections. Activity and functional capacity can be examined using metagenomics (Lundberg et al., 2012).

Host–microbiome relationships can be analyzed more thoroughly by introducing the concept of holo-omics, getting answers and connecting plant microbiome data with plant physiology, genetics, metabolism, and other host functions. This approach includes the simultaneous examination of host plant and microbiome interactions through amplicon sequencing, shotgun meta-genomics, meta-transcriptomics, and exometabolomics (Xu et al., 2021).

2.3.3 MALDI-TOF MS

In microbiome research, Matrix-Assisted Laser Desorption Ionization Time-of-Flight Mass Spectrometry (MALDI-TOF MS) is essential for rapidly and accurately identifying microbial communities, even in complex environmental samples. The unique ability of MALDI-TOF MS to identify organisms lies in the ability to distinguish and analyze specific protein fingerprints that can be mapped in mass spectra to different members of the same organism, from specific genera to species to specific strains. It allows for the analysis of the microbiome of soil, water, and air, where the microorganisms' diversity is critical for ecosystem functioning and biogeochemical cycling. The PubMed database (National Center for Biotechnology Information,

U.S. National Library of Medicine) search on 13 February 2025 using the keyword combination MALDI-TOF MS AND soil resulted in 277 hits, with the first articles published in 2003, and growing article numbers since 2010.

MALDI-TOF MS allows for the identification of complex microbial matrices. MALDI-TOF MS in agricultural and mining areas resulted in 30.2% species and 69.8% genus identification (Surányi et al., 2023). This shows that, despite being a practical technology, it is essential to study ecological diversity and its polymicrobial components. In addition to reliable results, high throughput and speed allow the simultaneous assessment of a great number of environmental samples, which can be used to track the biological origin and contamination of natural resources (water and soil), especially in ecologically sensitive regions (Avanzi et al., 2017).

MALDI-TOF MS can be used in soil microbiome research despite the complexity of soil microbial communities. MALDI-TOF MS was applied to distinguish between individual groups of soil microbes. Recent studies have shown that this method can be employed to determine the impacts of farming techniques (fertilization, crop rotation, and biochar), allowing the detection of the impact of different treatments on soil health and productivity. MALDI-TOF MS is suitable for monitoring microbial responses to soil contamination (heavy metal contamination or pesticide use), facilitating bioremediation, and developing sustainable land management strategies. The utilization of the MALDI-TOF MS method can reduce the time for identification to less than 24 h, allowing for the mapping of the microbiome across bacterial, fungal, viral, and parasitic communities (Tsuchida et al., 2020; Torres-Sangiao et al., 2021), which is invaluable for environmental monitoring, where timely data on microbial composition can inform ecosystem management and interventions.

In addition to its application in taxonomic profiling, MALDI-TOF MS enables biomolecular profiling in proteomics and lipidomics studies. Elucidation of microbial interactions and the ecological roles of various biomolecules (proteins and lipids) is essential for microbial function and community stability. These insights enhance our comprehension of microbial dynamics and provide essential viewpoints on the impact of microbial communities on environmental health and human well-being (Tarfeen et al., 2022; Alizadeh et al., 2021). Biomolecular profiling in soil using MALDI-TOF MS allows for a detailed analysis of microbial metabolic pathways involved in carbon cycling, phosphorus solubilization, and the production of secondary metabolites (e.g., antibiotics, plant growth-promoting compounds). These results help to clarify the role of soil microbiomes in ecosystem resilience and agricultural sustainability.

Overall, MALDI-TOF MS represents a transformative tool for microbiome research because of its speed and accuracy in studying microbial diversity and functional roles, enabling a deeper understanding of microbial interactions and their ecological impact.

2.4 HIGH-THROUGHPUT SEQUENCING AND OMICS-BASED METHODS

2.4.1 Genomics

Investigating plant–microbe interactions has been the breeding ground for the development of several molecular tools over the past three decades. The PubMed database (National Center for Biotechnology Information, U.S. National Library of Medicine) search on 13 February 2025 using the keyword combination genomics AND soil keywords resulted in 26131 hits, with the first two articles published in 1973, and growing article numbers since 2004. When the search was repeated with a more precise genomic tool, such as shotgun metagenome sequencing AND soil, there were 465 articles found in the database, with a growing number of articles since 2020.

Extracting microbial information from these plant–microbe systems and mapping their interactions have been demonstrated to be crucial in controlling plant development and enhancing resilience to both biotic and abiotic stresses. DNA sequencing is mostly used in the case of unculturable bacteria; however, for functional studies, culture-based sequencing is often used to understand the functions of genes, metabolic activity, or behavior.

The traditional Sanger sequencing method has its own limitations, such as low genomic coverage, with an error rate of 0.3%, which is laborious, requires the use of radioisotopes, and is expensive. They are predominantly used for sequencing the genome, exome, and transcriptome (Kumar et al., 2024). Fluorescent dye-based Sanger sequencing has predominantly supplanted radioisotope-based methods due to its enhanced safety and efficiency. The use of fluorescent dyes permits the incorporation of all four dideoxynucleotides within a single reaction. This advancement considerably accelerated the sequencing process and simplified the complexity of data analysis. At present, it continues to be used for a limited number of samples, especially in the analysis of microbial community structure (Malik et al., 2008) through single-gene or multilocus sequencing (MLS).

Through these methods, we gathered information about the stimulation of plant growth, biocontrol activity, and discoveries, including the production of useful metabolites by related microorganisms.

Next-generation sequencing (NGS) systems enable high-throughput parallel DNA or RNA sequencing, making them ideal for different genomic and transcriptomic utilizations.

The advent of NGS platforms (Mardis 2008; Novais and Thorstenson 2011) has facilitated high-throughput comparative metagenomics, sparking a surge in research aimed at deciphering the composition and roles of bacterial communities in a wide array of environments (Bolhuis et al., 2014; Ley et al., 2006; Garrett et al., 2010; Yoon et al., 2015; Caporaso et al., 2010; Norman et al., 2014; Huttenhower et al., 2014).

Whole-genome sequencing offers significant benefits, including the ability to identify novel functional genes, metabolisms, and draft genomes of uncultivated organisms, encompassing members of the microbial dark matter (Ma et al., 2023; Mlesnita, 2024). To the same extent, this level of discovery is not achievable through taxonomic marker sequencing. Furthermore, whole-genome sequencing eliminates potential PCR biases associated with marker gene amplification (Nam et al., 2023). These techniques use bioinformatics tools to examine the taxonomy and function of soil microbes.

Shotgun metagenomics offers a more direct evaluation and a wider understanding of the activities and functional characteristics of rhizosphere organisms compared to the optimization of crop-specific techniques. Additionally, a study explored microbial communities, including bacteria, fungi, and archaea at various levels, their functional capabilities, and how environmental factors affect maize rhizosphere soils, particularly in semi-arid areas of South Africa. Therefore, through metagenomics, the research aimed to analyze the structure and function of microbial communities in both the rhizosphere and bulk soils of maize, as well as the impact of environmental factors on these microbial communities (Omotayo et al., 2021).

Metagenomic data were further refined into metagenome-assembled genomes (MAGs) using an advanced metagenomic technique. The development of MAGs has contributed significantly to uncovering bacterial diversity, particularly in revealing microbial dark matter (Quince et al., 2017), and is a powerful tool for cataloguing microbial diversity and function. A MAG is a microbial genome at the species level that is created using metagenomic data from a community within an environmental sample. Shotgun sequencing is followed by raw data filtering and processing, metagenome assembly, contig binning, and quality assessment. High-quality MAGs can be used for taxonomic identification and functional analysis. MAGs facilitate the discovery of previously unknown species and metabolic pathways, thereby enhancing a more thorough grasp of microbial ecology and evolution.

Extensive metagenomic excavation has led to the reconstruction of genome bins, uncovering a multitude of previously unknown species-level genomes that have considerably increased our knowledge of soil microbial diversity and functionality. In a study by Ma et al. (2023), researchers delved into the soil microbial dark matter using 3304 metagenome datasets. Their efforts resulted in the reconstruction of over 40,000 MAG bins, from which 21,077 species-level genome bins were identified. Notably, nearly 80% of these were previously unidentified species-level genome bins. This research revealed numerous uncharacterized genes requiring further investigation, as well as a substantial number of potential biosynthetic gene clusters that may encode valuable secondary metabolites. The study also explored virus–host relationships, describing a wide array of viruses that infect various bacterial hosts, with prophages emerging as the most reliable indicators of these associations. Additionally, researchers have examined the defense mechanisms of the microbial community, identifying more than 8,500 CRISPR–Cas genes, highlighting the soil microbiome as a rich source of Cas proteins (Ma et al., 2023).

Although there are some obstacles, metagenomics holds significant potential for enhancing agricultural techniques and environmental stewardship. By uncovering the complete spectrum of microbial diversity, functions, and relationships, metagenomics is set to transform soil microbiology research. Metagenomic techniques offer the potential to create a vast repository of countless living species and enable the study of interactions between microorganisms, soil, and crops. As a result, soil microbial communities can be harnessed to develop more resilient crops and identify new biomolecules (Garg et al., 2024).

2.4.2 Proteomics

Proteomics involves the comprehensive study of proteins. It is a fundamental tool for understanding microbiome function and host–microbiome interactions. Proteomic analyses of the proteins and enzymes produced by microorganisms have revealed their roles in various biological processes. The PubMed database (National Center for Biotechnology Information, U.S. National Library of Medicine) search on 13 February 2025 using the keyword combination proteomics AND soil resulted in 1855 articles, with the first article published in 1999, and growing article numbers since 2016.

Multiple techniques, such as liquid chromatography-tandem mass spectrometry (LC-MS/MS), enable the detection and measurement of proteins, even in complex environments, such as soil. These methods can be used to map microbial contributions to the nutrient cycle, including the decomposition of soil carbon, nitrogen fixation, and phosphorus solubilization. In microbial ecosystems, nitrogenase enzymes, which are essential for nitrogen fixation, and cellulases, which degrade plant polysaccharides, are often identified and quantified (Chourey et al., 2010).

Proteomics is one of the key methods to decipher plant–microbe interactions. Recent proteomics studies have focused on soil microbial communities, identifying enzymes that promote phosphate solubilization and enhance plant nutrient uptake (Parray et al., 2023). Proteomics allows for the monitoring of microbial responses to soil amendments (e.g., biofertilizers).

Targeted proteomics approaches, such as enzyme-linked immunosorbent assays (ELISAs), offer highly sensitive methods for quantifying specific microbial enzymes and stress-related proteins. These techniques are particularly valuable in complex microbial environments, where traditional sequencing methods may fall short. Targeted proteomics provide greater sensitivity, throughput, and selectivity by focusing on individual proteins, making it ideal for studying microbial enzymatic functions and stress responses in diverse environments. For example, enzymes such as nitrate reductases, which play key roles in biogeochemical processes such as nitrate reduction, can be accurately quantified using these methods. By targeting the peptides conserved across diverse microorganisms, researchers can estimate the abundance of these enzymes in environmental samples. Similarly, enzymes involved in other biogeochemical transformations can be quantified, helping to link enzyme activity to environmental shifts (M. Li et al., 2017).

Proteomics is also a valuable tool for studying stress-related proteins produced by bacteria. These proteins serve as biomarkers of environmental stress and are sensitive indicators of contamination levels in soil bacteria (Kohan and Morgan 2007). ELISA is particularly effective for environmental monitoring, including the detection of proteins in genetically modified organisms (GMOs) (Settlage et al., 2017). In contrast, LC-MS/MS complements ELISA by allowing high-throughput quantification of a broader range of proteins (potential biomarkers, both in clinical and environmental samples) (Kim et al., 2013).

A disadvantage of targeted proteomics is that it requires prior knowledge of proteins and their peptide sequences, which can be challenging in complex microbial communities. However, advances in peptide selection tools and high-resolution mass spectrometry are helping to overcome these challenges, increasing the utility of targeted proteomics in environmental and clinical research (Medvecky and Mandalakis 2021; Kim et al., 2013).

Beyond microbiome research, proteomics is essential for understanding the dynamic interactions between host and microbiome.

Despite these advantages, proteomics faces several challenges, including the extraction of proteins from complex soil matrices and distinguishing microbial proteins from plant or host proteins. In addition, quantifying low-abundance proteins is a significant obstacle, but the continuous development of mass spectrometry and refined protein extraction protocols is helping overcome these challenges.

2.4.3 Lipidomics

Lipids predominantly support the essential functions of the plasma membrane and its capacity to adapt to abiotic stress, facilitate intercellular and intracellular communication, and enable nutrient exchange during interactions. These molecules not only constitute the plasma membrane structure but also play a critical role in maintaining their integrity. Living organisms can adjust the fluidity of their cell membranes in response to a changing environment. This is especially valid for soil microorganisms; therefore, changes in the lipid structure of the soil environment indicate perturbation. Monitoring the soil lipidome is a valuable tool for observing changes in the soil environment. The PubMed database (National Center for Biotechnology Information, U.S. National Library of Medicine) search on 13 February 2025 using the keyword combination lipidomics AND soil resulted in 105 articles, with the first article published in 2009 regarding the shotgun lipidomics approach in *Sinorhizobium meliloti* (Basconcillo et al., 2009), and growing article numbers since 2021.

Among the plasma membrane lipid components, glycerophospholipids, glycerolipids, and sphingolipids can be distinguished. Glycerophospholipids with different head groups, such as phosphatidylcholine and phosphatidylethanolamine, play a role in membrane fluidity and flexibility. Phosphatidylglycerol acts as a structural constituent, as well as a precursor (Couvillion et al., 2023; Macabuhay et al., 2022). The main types of lipids studied in the case of soil microorganisms are PLFAs and FAMEs, which are mainly used for microbial biomass measurements and microbial guilds. These techniques provide no data on the intact lipid molecular structure (Couvillion et al., 2023). The simultaneous extraction of neutral and GLFAs along with phospholipids contributes to a more holistic understanding of the soil microbial communities. This approach reveals distinct community characteristics and attributes that respond to environmental changes (Gorka et al., 2023).

It has been reported in different studies that there is a change in membrane phospholipids under stress and nutrient-limited conditions. Betaine lipids replace phospholipids in phosphate-limited conditions in *Rhodobacter sphaeroides* (Benning et al., 1995) and under drought conditions in the whole soil lipidome (Couvillion et al., 2023). Other sulfur-containing glycolipids were more abundant in dry soil, indicating their role in microbial adaptation to environmental stress (Couvillion et al., 2023). The increased presence of polyunsaturated fatty acids (PUFA) and sphingolipids in the soil under drought stress suggests their role in stress tolerance (Couvillion et al., 2023). Chwastek et al. (2020) examined the lipidome of plant-associated *Methylobacterium extorquens* and observed that only 11 lipids accounted for the main variability caused by varying environmental conditions. They observed that acyl chain remodeling was unevenly distributed, while the remodeling of acyl chains specific to headgroups served as an adaptive strategy (Chwastek et al., 2020).

Although lipid profiles provide valuable information regarding microbial community structure and function, analytical techniques struggle with challenges in obtaining accurate data because of the complex matrices of the soil. Moreover, the ever-changing nature of microbial populations and their lipid profiles necessitate ongoing improvements in methodology to effectively monitor real-time shifts (Gorka et al., 2023).

2.4.4 Metabolomics

Metabolomics involves the thorough examination of small-molecule metabolites within biological systems, which is crucial for understanding microbiome functions and host–microbiome interactions. Metabolomics provides an opportunity to map biochemical activities and metabolic pathways by identifying and quantifying the metabolites produced by microorganisms and their hosts. The metabolomics toolbox offers an opportunity to evaluate the functional capabilities of microbial communities, including their role in nutrient metabolism, signal transduction, and modulation of host physiology. The PubMed database (National Center for Biotechnology Information, U.S. National Library of Medicine) search on 13 February 2025 using the keyword combination metabolomics AND soil resulted in 2363 hits, with the first article published in 2003, and growing article numbers since 2020.

Mass spectrometry (MS) and nuclear magnetic resonance (NMR) spectroscopy are highly sensitive and specific techniques for identifying metabolites in complex mixtures (feces, plasma, and tissue) (Saputra et al., 2024). When integrated with other omics approaches (genomics, transcriptomics, proteomics), metabolomics enables system-wide investigations.

In agriculture, metabolomics clarifies the functionality of the soil microbiome, identifies metabolites associated with nutrient (nitrogen and phosphorus) cycling, and detects microbial stress responses to environmental changes. Applications include optimizing biofertilizers, monitoring soil health, and improving plant pathogen and drought resistance (Parray et al., 2023). In addition, metabolomics contributes to the detection of biomarkers for disease diagnosis and provides tools for assessing dysbiosis and its metabolic consequences.

Metabolomics is popularly used to identify key metabolites that mediate plant–microbe interactions, thus facilitating the development of microbial vaccines to promote plant growth. Advances in multi-omics integration and computational tools have enabled comprehensive metabolite pathway mapping from the microbiome, revealing novel mechanisms underlying host–microbiome interactions (Shtossel et al., 2024).

Along with the advantages of metabolomics, it is necessary to list its disadvantages, such as the need to standardize sample preparation protocols and the reproducibility of results. Although the structure of such models is biologically well understood, the inverse coupling of metabolomic data with environmental microbial activities is challenging. The detection limits and potential matrix interference in different agricultural samples can affect the accurate quantification of metabolites. However, increased mass spectrometric resolution and machine-learning algorithms are among the current technological developments that offer solutions to these limitations and enable the application of this method in modern microbiome studies.

2.5 CONCLUSIONS AND FUTURE PERSPECTIVES

This chapter, through chronology, illustrates the progression of microbiome research methodologies, demonstrating how advances in technology have influenced our comprehension of microbial communities and their utilization in agriculture and soil management practices. The techniques for studying microbiomes are continuously evolving, due to multi-omics integration (genomics, transcriptomics, proteomics, metabolomics), which provide holistic insights into microbial functions and interactions. Advanced analytical methods, including those based on machine learning, in combination with publicly available sequence databases, are essential for the upcoming stage of research in microbiome studies. The analysis of individual cell genomes will allow researchers to examine uncultivable microorganisms with unprecedented detail, while advancements in synthetic biology and microbiome manipulation are creating opportunities for precise microbial interventions in agricultural settings. Emerging knowledge will enable us to create predictive models for various purposes, including microbial community engineering, optimization of ecosystem services, strategies for disease suppression, and schemes for resource management.

REFERENCES

Alizadeh, Maryam, Leila Yousefi, Farzaneh Pakdel, Reza Ghotaslou, Mohammad Ahangarzadeh Rezaee, Ehsaneh Khodadadi, Mahin Ahangar Oskouei, Mohammad Hossein Soroush Barhaghi, and Hossein Samadi Kafil. 2021. "MALDI-TOF Mass Spectroscopy Applications in Clinical Microbiology." *Advances in Pharmacological and Pharmaceutical Sciences* 2021 (May): 9928238. https://doi.org/10.1155/2021/9928238.

Avanzi, Ingrid Regina, Louise Hase Gracioso, Marcela Dos Passos Galluzzi Baltazar, Bruno Karolski, Elen Aquino Perpetuo, and Claudio Augusto Oller do Nascimento. 2017. "Rapid Bacteria Identification from Environmental Mining Samples Using MALDI-TOF MS Analysis." *Environmental Science and Pollution Research International* 24 (4): 3717–26. https://doi.org/10.1007/s11356-016-8125-8.

Barghouthi, Sameer A. 2011. "A Universal Method for the Identification of Bacteria Based on General PCR Primers." *Indian Journal of Microbiology* 51 (4): 430–44. https://doi.org/10.1007/s12088-011-0122-5.

Basconcillo, Libia Saborido, Rahat Zaheer, Turlough M. Finan, and Brian E. McCarry. 2009. "A Shotgun Lipidomics Approach in Sinorhizobium Meliloti as a Tool in Functional Genomics." *Journal of Lipid Research* 50 (6): 1120–32. https://doi.org/10.1194/jlr.M800443-JLR200.

Benning, C., Z. H. Huang, and D. A. Gage. 1995. "Accumulation of a Novel Glycolipid and a Betaine Lipid in Cells of Rhodobacter Sphaeroides Grown under Phosphate Limitation." *Archives of Biochemistry and Biophysics* 317 (1): 103–11. https://doi.org/10.1006/abbi.1995.1141.

Berihu, Maria, Tracey S. Somera, Assaf Malik, Shlomit Medina, Edoardo Piombo, Ofir Tal, Matan Cohen, et al. 2023. "A Framework for the Targeted Recruitment of Crop-Beneficial Soil Taxa Based on Network Analysis of Metagenomics Data." *Microbiome* 11 (1): 8. https://doi.org/10.1186/s40168-022-01438-1.

Bolhuis, Henk, Mariana Silvia Cretoiu, and Lucas J. Stal. 2014. "Molecular Ecology of Microbial Mats." *FEMS Microbiology Ecology* 90 (2): 335–50. https://doi.org/10.1111/1574-6941.12408.

Caporaso, J. Gregory, Justin Kuczynski, Jesse Stombaugh, Kyle Bittinger, Frederic D. Bushman, Elizabeth K. Costello, Noah Fierer, et al. 2010. "QIIME Allows Analysis of High-Throughput Community Sequencing Data." *Nature Methods* 7 (5): 335–36. https://doi.org/10.1038/nmeth.f.303.

Chourey, Karuna, Janet Jansson, Nathan VerBerkmoes, Manesh Shah, Krystle L. Chavarria, Lauren M. Tom, Eoin L. Brodie, and Robert L. Hettich. 2010. "Direct Cellular Lysis/Protein Extraction Protocol for Soil Metaproteomics." *Journal of Proteome Research* 9 (12): 6615–22. https://doi.org/10.1021/pr100787q.

Chwastek, Grzegorz, Michal A. Surma, Sandra Rizk, Daniel Grosser, Oksana Lavrynenko, Magdalena Rucińska, Helena Jambor, and James Sáenz. 2020. "Principles of Membrane Adaptation Revealed through Environmentally Induced Bacterial Lipidome Remodeling." *Cell Reports* 32 (12): 108165. https://doi.org/10.1016/j.celrep.2020.108165.

Couvillion, Sneha P., Robert E. Danczak, Dan Naylor, Montana L. Smith, Kelly G. Stratton, Vanessa L. Paurus, Kent J. Bloodsworth, et al. 2023. "Rapid Remodeling of the Soil Lipidome in Response to a Drying-Rewetting Event." *Microbiome* 11 (1): 34. https://doi.org/10.1186/s40168-022-01427-4.

Fernandes, Marcelo F., Jyotisna Saxena, and Richard P. Dick. 2013. "Comparison of Whole-Cell Fatty Acid (MIDI) or Phospholipid Fatty Acid (PLFA) Extractants as Biomarkers to Profile Soil Microbial Communities." *Microbial Ecology* 66 (1): 145–57. https://doi.org/10.1007/s00248-013-0195-2.

Franco-Duarte, Ricardo, Lucia Černáková, Snehal Kadam, Karishma S. Kaushik, Bahare Salehi, Antonio Bevilacqua, Maria Rosaria Corbo, et al. 2019. "Advances in Chemical and Biological Methods to Identify Microorganisms-from Past to Present." *Microorganisms* 7 (5). https://doi.org/10.3390/microorganisms7050130.

Gao, Yingnan, and Martin Wu. 2023. "Accounting for 16S rRNA Copy Number Prediction Uncertainty and Its Implications in Bacterial Diversity Analyses." *ISME Communications* 3 (1): 59. https://doi.org/10.1038/s43705-023-00266-0.

Garg, Diksha, Niketan Patel, Anamika Rawat, and Alexandre Soares Rosado. 2024. "Cutting Edge Tools in the Field of Soil Microbiology." *Current Research in Microbial Sciences* 6 (February): 100226. https://doi.org/10.1016/j.crmicr.2024.100226.

Garibyan, Lilit, and Nidhi Avashia. 2013. "Polymerase Chain Reaction." *The Journal of Investigative Dermatology* 133 (3): 1–4. https://doi.org/10.1038/jid.2013.1.

Garrett, Wendy S., Jeffrey I. Gordon, and Laurie H. Glimcher. 2010. "Homeostasis and Inflammation in the Intestine." *Cell* 140 (6): 859–70. https://doi.org/10.1016/j.cell.2010.01.023.

Gentry, T. J., G. S. Wickham, C. W. Schadt, Z. He, and J. Zhou. 2006. "Microarray Applications in Microbial Ecology Research." *Microbial Ecology* 52 (2): 159–75. https://doi.org/10.1007/s00248-006-9072-6.

Gorka, Stefan, Sean Darcy, Julia Horak, Bruna Imai, Moritz Mohrlok, Erika Salas, Andreas Richter, et al. 2023. "Beyond PLFA: Concurrent Extraction of Neutral and Glycolipid Fatty Acids Provides New Insights into Soil Microbial Communities." *Soil Biology & Biochemistry* 187 (109205): 109205. https://doi.org/10.1016/j.soilbio.2023.109205.

Hickey, W. J., V. Brenner, and D. D. Focht. 1992. "Mineralization of 2-Chloro- and 2,5-Dichlorobiphenyl by Pseudomonas Sp. Strain UCR2." *FEMS Microbiology Letters* 77 (1–3): 175–80. https://doi.org/10.1016/0378-1097(92)90151-d.

Hirsch, Penny R., Tim H. Mauchline, and Ian M. Clark. 2010. "Culture-Independent Molecular Techniques for Soil Microbial Ecology." *Soil Biology & Biochemistry* 42 (6): 878–87. https://doi.org/10.1016/j.soilbio.2010.02.019.

Huttenhower, Curtis, Rob Knight, C. Titus Brown, J. Gregory Caporaso, Jose C. Clemente, Dirk Gevers, Eric A. Franzosa, et al. 2014. "Advancing the Microbiome Research Community." *Cell* 159 (2): 227–30. https://doi.org/10.1016/j.cell.2014.09.022.

Kim, Yeoun Jin, Sebastien Gallien, Jan van Oostrum, and Bruno Domon. 2013. "Targeted Proteomics Strategy Applied to Biomarker Evaluation." *Proteomics Clinical Applications* 7 (11–12): 739–47. https://doi.org/10.1002/prca.201300070.

Kohan, K., and W. Morgan. 2007. "Stress Proteins as a Suitable Biomarker of Environmental Pollution." *International Journal of Environmental Research* 1 (4): 290–5. https://doi.org/10.22059/ijer.2010.139.

Kralik, Petr, and Matteo Ricchi. 2017. "A Basic Guide to Real Time PCR in Microbial Diagnostics: Definitions, Parameters, and Everything." *Frontiers in Microbiology* 8 (February): 108. https://doi.org/10.3389/fmicb.2017.00108.

Kruglov, Yu V. 2016. "Microbial Community of Soil: Physiological Diversity Patterns and Assessment (Review)." *Agricultural Biology* 51 (1): 46–59. https://doi.org/10.15389/agrobiology.2016.1.46eng.

Kumar, Ajay, Manoj Kumar Solanki, Zhen Wang, Anjali Chandrol Solanki, Vipin Kumar Singh, and Praveen Kumar Divvela. 2024. "Revealing the Seed Microbiome: Navigating Sequencing Tools, Microbial Assembly, and Functions to Amplify Plant Fitness." *Microbiological Research* 279 (February): 127549. https://doi.org/10.1016/j.micres.2023.127549.

Lawlora, K., B. P. Knighta, V. L. Barbosa-Jeffersona, P. W. Laneb, A. K. Lilleyc, G. I. Patond, S. P. McGratha, S. M. O'Flahertya, and P. R. Hirscha. 2000. "Comparison of Methods to Investigate Microbial Populations in Soils under Different Agricultural Management." *FEMS Microbiology Ecology* 33 (2): 129–37. https://doi.org/10.1111/j.1574-6941.2000.tb00735.x.

Ley, Ruth E., Daniel A. Peterson, and Jeffrey I. Gordon. 2006. "Ecological and Evolutionary Forces Shaping Microbial Diversity in the Human Intestine." *Cell* 124 (4): 837–48. https://doi.org/10.1016/j.cell.2006.02.017.

Li, Hai-Bi, Rajesh K. Singh, Pratiksha Singh, Qi-Qi Song, Yong-Xiu Xing, Li-Tao Yang, and Yang-Rui Li. 2017. "Genetic Diversity of Nitrogen-Fixing and Plant Growth Promoting Species Isolated from Sugarcane Rhizosphere." *Frontiers in Microbiology* 8 (July): 1268. https://doi.org/10.3389/fmicb.2017.01268.

Li, Hui, Ying Zhang, Dong-Sheng Li, Hui Xu, Guan-Xiong Chen, and Cheng-Gang Zhang. 2009. "Comparisons of Different Hypervariable Regions of Rrs Genes for Fingerprinting of Microbial Communities in Paddy Soils." *Soil Biology & Biochemistry* 41 (5): 954–68. https://doi.org/10.1016/j.soilbio.2008.10.030.

Li, Minjing, Yuqian Gao, Wei-Jun Qian, Liang Shi, Yuanyuan Liu, William C. Nelson, Carrie D. Nicora, et al. 2017. "Targeted Quantification of Functional Enzyme Dynamics in Environmental Samples for Microbially Mediated Biogeochemical Processes." *Environmental Microbiology Reports* 9 (5): 512–21. https://doi.org/10.1111/1758-2229.12558.

Lundberg, Derek S., Sarah L. Lebeis, Sur Herrera Paredes, Scott Yourstone, Jase Gehring, Stephanie Malfatti, Julien Tremblay, et al. 2012. "Defining the Core *Arabidopsis* Thaliana Root Microbiome." *Nature* 488 (7409): 86–90. https://doi.org/10.1038/nature11237.

Ma, Bin, Caiyu Lu, Yiling Wang, Jingwen Yu, Kankan Zhao, Ran Xue, Hao Ren, et al. 2023. "A Genomic Catalogue of Soil Microbiomes Boosts Mining of Biodiversity and Genetic

Resources." *Nature Communications* 14 (1): 7318. https://doi.org/10.1038/s41467-023-43000-z.

Macabuhay, Allene, Borjana Arsova, Robert Walker, Alexander Johnson, Michelle Watt, and Ute Roessner. 2022. "Modulators or Facilitators? Roles of Lipids in Plant Root-Microbe Interactions." *Trends in Plant Science* 27 (2): 180–90. https://doi.org/10.1016/j.tplants.2021.08.004.

Malik, Seidu, Michael Beer, Mallavarapu Megharaj, and Ravi Naidu. 2008. "The Use of Molecular Techniques to Characterize the Microbial Communities in Contaminated Soil and Water." *Environment International* 34 (2): 265–76. https://doi.org/10.1016/j.envint.2007.09.001.

Mardis, Elaine R. 2008. "Next-Generation DNA Sequencing Methods." *Annual Review of Genomics and Human Genetics* 9 (1): 387–402. https://doi.org/10.1146/annurev.genom.9.081307.164359.

Markoulatos, P., N. Siafakas, and M. Moncany. 2002. "Multiplex Polymerase Chain Reaction: A Practical Approach." *Journal of Clinical Laboratory Analysis* 16 (1): 47–51. https://doi.org/10.1002/jcla.2058.

Medvecky, Matej, and Manolis Mandalakis. 2021. "PepMANDIS: A Peptide Selection Tool for Designing Function-Based Targeted Proteomic Assays in Complex Microbial Systems." *Frontiers in Chemistry* 9 (August): 722087. https://doi.org/10.3389/fchem.2021.722087.

Mirmajlessi, Seyed Mahyar, Marialaura Destefanis, Richard Alexander Gottsberger, Marika Mänd, and Evelin Loit. 2015. "PCR-Based Specific Techniques Used for Detecting the Most Important Pathogens on Strawberry: A Systematic Review." *Systematic Reviews* 4 (1): 9. https://doi.org/10.1186/2046-4053-4-9.

Mlesnita, Andreea-Mihaela. 2024. "Tackling the Soil Microbiome—Challenges and Oportunities." *Journal of Experimental and Molecular Biology* (December). https://doi.org/10.47743/jemb-2025-209.

Nam, Nguyen Nhat, Hoang Dang Khoa Do, Kieu The Loan Trinh, and Nae Yoon Lee. 2023. "Metagenomics: An Effective Approach for Exploring Microbial Diversity and Functions." *Foods (Basel, Switzerland)* 12 (11). https://doi.org/10.3390/foods12112140.

Nannipieri, Paolo. 2020. "Soil Is Still an Unknown Biological System." *Applied Sciences (Basel, Switzerland)* 10 (11): 3717. https://doi.org/10.3390/app10113717.

National Library of Medicine. 2025. PubMed. National Center for Biotechnology Information. https://pubmed.ncbi.nlm.nih.gov/ (Accessed February 13, 2025).

Norman, Jason M., Scott A. Handley, and Herbert W. Virgin. 2014. "Kingdom-Agnostic Metagenomics and the Importance of Complete Characterization of Enteric Microbial Communities." *Gastroenterology* 146 (6): 1459–69. https://doi.org/10.1053/j.gastro.2014.02.001.

Novais, R. C., and Y. R. Thorstenson. 2011. "The Evolution of Pyrosequencing® for Microbiology: From Genes to Genomes." *Journal of Microbiological Methods* 86 (1): 1–7. https://doi.org/10.1016/j.mimet.2011.04.006.

Omotayo, Oluwadara, Ozede Igiehon, and Olubukola Babalola. 2021. "Metagenomic Study of the Community Structure and Functional Potentials in Maize Rhizosphere Microbiome: Elucidation of Mechanisms behind the Improvement in Plants under Normal and Stress Conditions." *Sustainability* 13 (14): 8079. https://doi.org/10.3390/su13148079.

Parray, Javid A., Dilfuza Egamberdieva, Elsayed Fathi Abd_Allah, and R. Z. Sayyed. 2023. "Editorial: Soil Microbiome Metabolomics: A Way Forward to Sustainable Intensification." *Frontiers in Sustainable Food Systems* 7 (July). https://doi.org/10.3389/fsufs.2023.1251054.

Preston-Mafham, Juliet, Lynne Boddy, and Peter F. Randerson. 2002. "Analysis of Microbial Community Functional Diversity Using Sole-Carbon-Source Utilisation Profiles: A Critique." *FEMS Microbiology Ecology* 42 (1): 1–14. https://doi.org/10.1111/j.1574-6941.2002.tb00990.x.

Quince, Christopher, Alan W. Walker, Jared T. Simpson, Nicholas J. Loman, and Nicola Segata. 2017. "Shotgun Metagenomics, from Sampling to Analysis." *Nature Biotechnology* 35 (9): 833–44. https://doi.org/10.1038/nbt.3935.

Saputra, Rizki Rachmad, Mokhamat Ariefin, Meiyanti Ratna Kumalasari, Junita Dongoran, Mulani Jeni Lestari Tampubolon, Putri Sulistiawati, Sri Yulandari Simangunsong, et al. 2024. "Advancements in NMR and IR Spectroscopy: Enhancing Metabolomics and Disease Diagnostics in the Health Sector: A Comprehensive Review." *Indonesian Journal of Chemical Analysis (IJCA)* 7 (2): 77–88. https://doi.org/10.20885/ijca.vol7.iss2.art6.

Settlage, Sharon B., Julie E. Eble, Jayant K. Bhanushali, Matthew L. Cheever, Ai-Guo Gao, David A. Goldstrohm, Ryan Hill, et al. 2017. "Validation Parameters for Quantitating Specific Proteins Using ELISA or LC-MS/MS: Survey Results." *Food Analytical Methods* 10 (5): 1339–48. https://doi.org/10.1007/s12161-016-0689-x.

Shtossel, Oshrit, Omry Koren, Iris Shai, Ehud Rinott, and Yoram Louzoun. 2024. "Gut Microbiome-Metabolome Interactions Predict Host Condition." *Microbiome* 12 (1): 24. https://doi.org/10.1186/s40168-023-01737-1.

Slabbinck, Bram, Bernard De Baets, Peter Dawyndt, and Paul De Vos. 2009. "Towards Large-Scale FAME-Based Bacterial Species Identification Using Machine Learning Techniques." *Systematic and Applied Microbiology* 32 (3): 163–76. https://doi.org/10.1016/j.syapm.2009.01.003.

Surányi, B. B., A. Taczman-Brückner, Cs Mohácsi-Farkas, and T. Engelhardt. 2023. "Rapid Identification of Bacteria from Agricultural Environment Using MALDI-TOF MS." *Acta Alimentaria* 52 (1): 113–20. https://doi.org/10.1556/066.2022.00202.

Tarfeen, Najeebul, Khair Ul Nisa, and Qadrul Nisa. 2022. "MALDI-TOF MS: Application in Diagnosis, Dereplication, Biomolecule Profiling and Microbial Ecology." *Proceedings of the Indian National Science Academy* 88 (3): 277–91. https://doi.org/10.1007/s43538-022-00085-2.

Torres-Sangiao, Eva, Cristina Leal Rodriguez, and Carlos García-Riestra. 2021. "Application and Perspectives of MALDI-TOF Mass Spectrometry in Clinical Microbiology Laboratories." *Microorganisms* 9 (7). https://doi.org/10.3390/microorganisms9071539.

Torsvik, Vigdis, and Lise Øvreås. 2002. "Microbial Diversity and Function in Soil: From Genes to Ecosystems." *Current Opinion in Microbiology* 5 (3): 240–5. https://doi.org/10.1016/s1369-5274(02)00324-7.

Tsuchida, Sachio, Hiroshi Umemura, and Tomohiro Nakayama. 2020. "Current Status of Matrix-Assisted Laser Desorption/Ionization-Time-of-Flight Mass Spectrometry (MALDI-TOF MS) in Clinical Diagnostic Microbiology." *Molecules (Basel, Switzerland)* 25 (20). https://doi.org/10.3390/molecules25204775.

Venbrux, Marc, Sam Crauwels, and Hans Rediers. 2023. "Current and Emerging Trends in Techniques for Plant Pathogen Detection." *Frontiers in Plant Science* 14 (May): 1120968. https://doi.org/10.3389/fpls.2023.1120968.

White, D. C., W. M. Davis, J. S. Nickels, J. D. King, and R. J. Bobbie. 1979. "Determination of the Sedimentary Microbial Biomass by Extractible Lipid Phosphate." *Oecologia* 40 (1): 51–62. https://doi.org/10.1007/BF00388810.

Widmer, F., A. Fließbach, E. Laczkó, J. Schulze-Aurich, and J. Zeyer. 2001. "Assessing Soil Biological Characteristics: A Comparison of Bulk Soil Community DNA-, PLFA-, and Biolog™-Analyses." *Soil Biology & Biochemistry* 33 (7–8): 1029–36. https://doi.org/10.1016/s0038-0717(01)00006-2.

Woese, C. R., and G. E. Fox. 1977. "Phylogenetic Structure of the Prokaryotic Domain: The Primary Kingdoms." *Proceedings of the National Academy of Sciences of the United States of America* 74 (11): 5088–90. https://doi.org/10.1073/pnas.74.11.5088.

Woźniak, Małgorzata, Anna Gałązka, Renata Tyśkiewicz, and Jolanta Jaroszuk-Ściseł. 2019. "Endophytic Bacteria Potentially Promote Plant Growth by Synthesizing Different Metabolites and Their Phenotypic/Physiological Profiles in the Biolog GEN III MicroPlate Test." *International Journal of Molecular Sciences* 20 (21). https://doi.org/10.3390/ijms20215283.

Xu, Ling, Grady Pierroz, Heidi M.-L. Wipf, Cheng Gao, John W. Taylor, Peggy G. Lemaux, and Devin Coleman-Derr. 2021. "Holo-Omics for Deciphering Plant-Microbiome Interactions." *Microbiome* 9 (1): 69. https://doi.org/10.1186/s40168-021-01014-z.

Yoon, Sang Sun, Eun-Kyoung Kim, and Won-Jae Lee. 2015. "Functional Genomic and Metagenomic Approaches to Understanding Gut Microbiota-Animal Mutualism." *Current Opinion in Microbiology* 24 (April): 38–46. https://doi.org/10.1016/j.mib.2015.01.007.

3 Artificial Intelligence and Machine Learning for Analysis of Microbiomes

Ankita Shukla[*,a], *and Ankur Choudhary*[b]
[a]University Institute of Biotechnology (UIBT), Chandigarh University, Mohali, Punjab, India
[b]Chitkara University School of Engineering & Technology, Chitkara University, Punjab

3.1 INTRODUCTION

In recent years, artificial intelligence (AI) and machine learning (ML) fields have revolutionized the study of microbiomes, offering powerful tools to analyze complex biological data with remarkable accuracy and efficiency. The microbiome—consisting of the diverse communities of microorganisms is categorized into two major classes, i.e., human microbiome (gut, oral, skin, vaginal, respiratory, urogenital, etc.) and the environmental microbiome (soil, marine, fresh water, and plant) that exist within plants, soil, and the environment—plays a critical role in agricultural ecosystems (Panthee et al., 2022) (Figure 3.1) The terms microbiota and microbiome are often used interchangeably but they represent distinct concepts. The microbiota refers to the community of microorganisms—including bacteria, archaea, fungi, and viruses—that inhabit a specific environment, such as the human gut, skin, or oral cavity. In contrast, the microbiome encompasses not only these microbial communities but also their collective genomes and interactions with the host and environment (Hou et al., 2022; Berg et al., 2020).

Understanding the distinction between these terms is crucial, as it shapes how we study host–microbe interactions and their implications in health and disease. The microbiome plays a significant role in regulating immune responses, metabolic pathways, and even neurological functions, whereas the microbiota is central to ecological composition and diversity analysis within specific niches.

The microbiome presents in the human body such as bacteria, fungi, and viruses and their genes play a crucial role in maintaining human health while the environmental microbiome influences the human microbiome makeup (Panthee et al., 2022). Understanding and optimizing these microbial communities, researchers are uncovering new ways to improve crop resilience, enhance soil health, reduce dependence on chemicals, increase crop production, and also disease outbreaks. The trending AI and ML techniques aid in analyzing high-dimensional microbiome datasets, uncovering previously inaccessible insights into microbial interactions, nutrient cycling, and disease dynamics (Hernández Medina et al., 2022).

In the context of sustainable agriculture, AI-driven microbiome analysis has the potential to transform practices that can uplift agriculture into a new era. Advanced algorithms make use of historical data to identify patterns in microbial communities, predict disease outbreaks, and anticipate environmental stressors this helps minimize crop losses and reduce resource use (Mana et al., 2024). Nowadays, ensemble and hybrid AI models are highly used to boost accuracy and interpretability, thus empowering collaborators to make data-driven decisions that align with ecological and economic sustainability (Jabed and Azmi Murad, 2024). As these technologies continue to advance, they hold the promise of fostering agricultural practices that are both productive and environmentally responsible, paving the way for a more resilient global food system and thus controlling human diseases through environmental factors (Patil et al., 2024).

3.2 ROLE OF AI AND ML IN MICROBIOME ANALYSIS

Microbiomes consist of diverse microbial species and their activity region with distinct physio-chemical properties derive specific ecological niches (Berg et al., 2020). Traditional methods of microbiome analysis, such as whole-genome shotgun sequencing (WGS) and 16S rRNA sequencing, produce vast amounts of data that are difficult to handle manually (Kim et al., 2024). AI and ML techniques analyze this sequencing data more efficiently thus uncovering various patterns, interrelations, and trends that would otherwise be omitted (Rahmani et al., 2021). These technologies thus help speed up research by providing more accurate insights by permitting the automation of data processing, classification, prediction, and interpretation. It, therefore, aids in providing more inclusive insights regarding microbial diversity and their functionality in respective niches. If we talk about microbial behavior their interactions and their impact on a host such as in the case of plants or soil these methods provide better predictions. Deep learning methods have proved to be more able to predict the functional roles of microbes in an ecosystem depending on their genomic sequences, even in those conditions where functions have not been previously documented ("Identifying Keystone Species in Microbial Communities Using Deep Learning | Nature Ecology and Evolution" n.d.). Thus, it is useful in agriculture

***Corresponding author:** shukla.ankita39@gmail.com (Ankita Shukla)

DOI: 10.1201/9781003544425-3

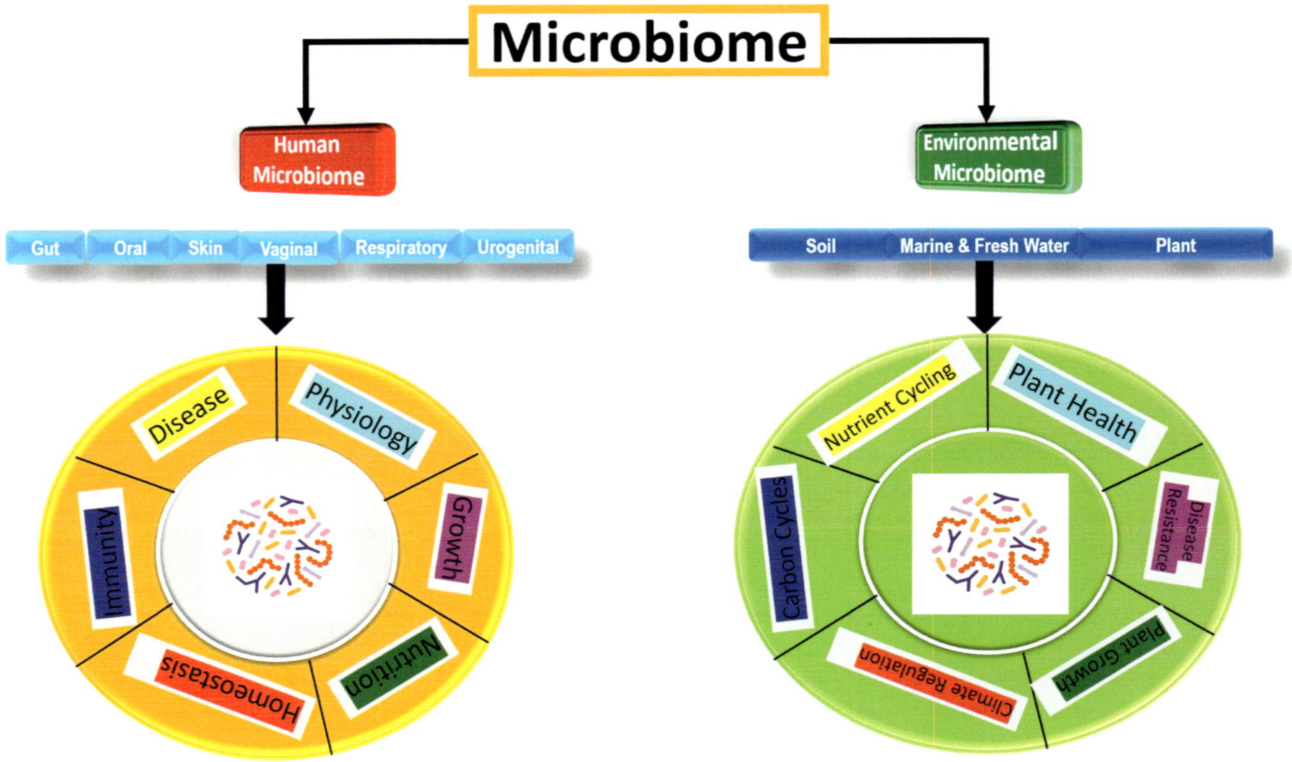

FIGURE 3.1 Types of microbiomes and their roles. This figure categorizes microbiomes into human and environmental microbiomes and illustrates their distinct roles. (1) *Human microbiome*: Includes seven types—gut, oral, skin, vaginal, respiratory, urogenital, and environmental microbiomes. These play critical roles in disease prevention, physiological regulation, growth, nutrition, homeostasis, and immunity within the human body. (2) *Environmental microbiome*: Encompasses soil, marine and freshwater, and plant microbiomes. These contribute to nutrient cycling, plant health, disease resistance, plant growth, climate regulation, and carbon cycling within natural ecosystems

for identifying beneficial or harmful microbes in the soil or plants.

AI methods, such as network analysis, help in mapping microbial interactions within a community. Considering these interactions provides insight into associations in agriculture such as nitrogen-fixing bacteria or pathogens that may perhaps prove to be dangerous threats to crop health (Matchado et al., 2021). In another category of ML methods, supervised learning, algorithms learn from a labeled set of data to make predictions this could be useful for predicting soil health, disease outbreaks, or crop yield depending on microbial composition (Thompson et al., 2019). This technique could be further used to analyze unlabeled data, for clustering microbial species or identifying patterns in community composition. In unsupervised learning, unknown microbial communities can be identified and their behavior can be traced depending on changes in environment or agricultural practices (Saeed et al., 2012). In addition, ensemble learning methods combine several ML models that help improve prediction accuracy and reduce bias. This approach is especially useful where the complexity of data can lead to overfitting or underfitting. Their application involves how changes in agricultural practices, such as crop rotation or irrigation methods, might affect the microbiome and crop yield (Sharma et al., 2020).

Imaging techniques (e.g., sequencing technologies or fluorescence microscopy), produce visual data that can be analyzed better with deep learning methods ("Applications, Promises, and Pitfalls of Deep Learning for Fluorescence Image Reconstruction | Nature Methods" n.d.). This imaging technique takes images of plant roots or leaves and with the help of deep learning algorithms, it helps understand the set of microbes most beneficial for plant health and growth (Deng et al., 2021). Also, methods are available for integrating AI with sensor networks in agriculture to monitor environmental factors like temperature, soil moisture, and microbial activity (Al-Qudah et al., 2025). This real-time data can thus be processed to optimize irrigation, fertilization, and pest control, leading to more sustainable farming practices (Md. A. Ali et al., 2023). The three-dimensional (3D) imaging techniques can uncover patterns not directly noticed from raw data. ML methods in microbiome research often include "black-box" algorithms, thus creating a concern regarding data interpretability (Topçuoğlu et al., 2020). However, recent advancements in explainable AI (XAI) are proving to be very helpful for researchers in the decision-making process behind AI models (S. Ali et al., 2023). This ensures that the insights are transparent and actionable for agricultural decision-making.

3.3 MICROBIOME DATASET ANALYSIS

This involves the following steps (Figure 3.2):

1. *Input*: It involves raw microbiome data (e.g., 16S rRNA sequences, whole-genome sequencing, metagenomic data).
2. *Data Preprocessing:* The input is raw sequence data that undergoes quality control, feature selection, and normalization processes to ensure high-quality normalized data.
3. *Feature Extraction:* It uses processed microbiome data that has undergone dimensionality reduction and feature selection to extract meaningful features for classification.
4. *Model Selection:* It uses selected features that undergo supervised or unsupervised learning and a cross-validation process, helping to choose and fine-tune ML models.
5. *Training and Testing*: It splits the data into training and testing sets and performs model training on the training dataset. This helps evaluate model performance on unseen data. In Microbiome data cross-validation techniques, such as *k*-fold cross-validation, are used to confirm that the model generalizes well and is not overfitted.
6. *Model Evaluation*: This generates the model output, performs metric calculations, and applies hyperparameter tuning. Model uses parameter optimization techniques to fine-tune model performance and improve predictive accuracy. Common metrics include accuracy, precision, recall, ROC-AUC, and F1-score for classification tasks, and mean squared error (MSE) or *R*-squared for regression tasks.
7. *Prediction and Interpretation*: This generates the final model, enabling the prediction of new microbiome samples and facilitating the biological interpretation of features. In microbiome research, interpreting the model output is crucial for biological relevance. Explainable AI techniques, are used to understand which microbial features are most influential in predictions, making the models more actionable in agricultural contexts.
8. *Visualization*: This visualizes predictions and microbial interactions, visually representing the obtained output.

3.4 CORE AI AND ML TECHNIQUES IN MICROBIOME ANALYSIS

3.4.1 Supervised Learning

The algorithm works on labeled input and predicts beneficial microbes that can enhance crop resilience and reduce dependence on chemical inputs (Huang et al., 2023). It uses a diversity of algorithms such as random forests (RFs), support vector machines (SVMs), neural networks (NNs), and gradient boosting methods (GBMs) which are as follows:

1. *RF*: This algorithm is useful in handling complex microbiome data as it is based on multiple decision trees to create a particularly effective ensemble model (Acharjee et al., 2020). This uses a ranking-based method to determine important microbial features depending on their importance thus helping to identify key microbial species or communities.
2. *SVMs*: SVMs work effectively with clear separable classes and small datasets. This is useful in microbiome research to distinguish between healthy and diseased states (Ghannam and Techtmann 2021).
3. *NNs*: Deep learning models can be trained on large microbiome datasets to capture complex relationships among microbial communities and predict multiple outcomes, such as soil properties or disease susceptibility (Ghannam and Techtmann 2021).
4. *GBMs*: These algorithms consecutively build ensemble models and perform weight adjustment based on previous errors to improve precision. They are extensively used in microbiome analysis for high-accuracy classification and regression tasks (e.g., XGBoost, LightGBM) (Ghannam and Techtmann 2021).

3.4.2 Unsupervised Learning

The algorithm can classify microbial communities based on their functional roles, thus categorizing them into growth promoters, nitrogen-fixers, pathogens, etc. ("Performance Determinants of Unsupervised Clustering Methods for Microbiome Data | Microbiome | Full Text" n.d.). By examining co-occurrence data, the algorithm can build networks that reveal relationships between microbial species, showing which microbes frequently appear together or have an influence on each other. A variety of algorithms are used in unsupervised learning, such as k-means clustering, hierarchical clustering, and (Density-Based Spatial Clustering of Applications with Noise (DBSCAN). These are as follows:

1. *K-Means Clustering*: This algorithm clusters microbes into k clusters based on their likeness. Thus, it helps to categorize microbial taxa into comparable groups that reveal primary patterns within the microbial communities ("Performance Determinants of Unsupervised Clustering Methods for Microbiome Data | Microbiome | Full Text" n.d.).
2. *Hierarchical Clustering*: It organizes microbial data into a dendrogram that helps visualize relationships between microbial communities ("Performance Determinants of Unsupervised

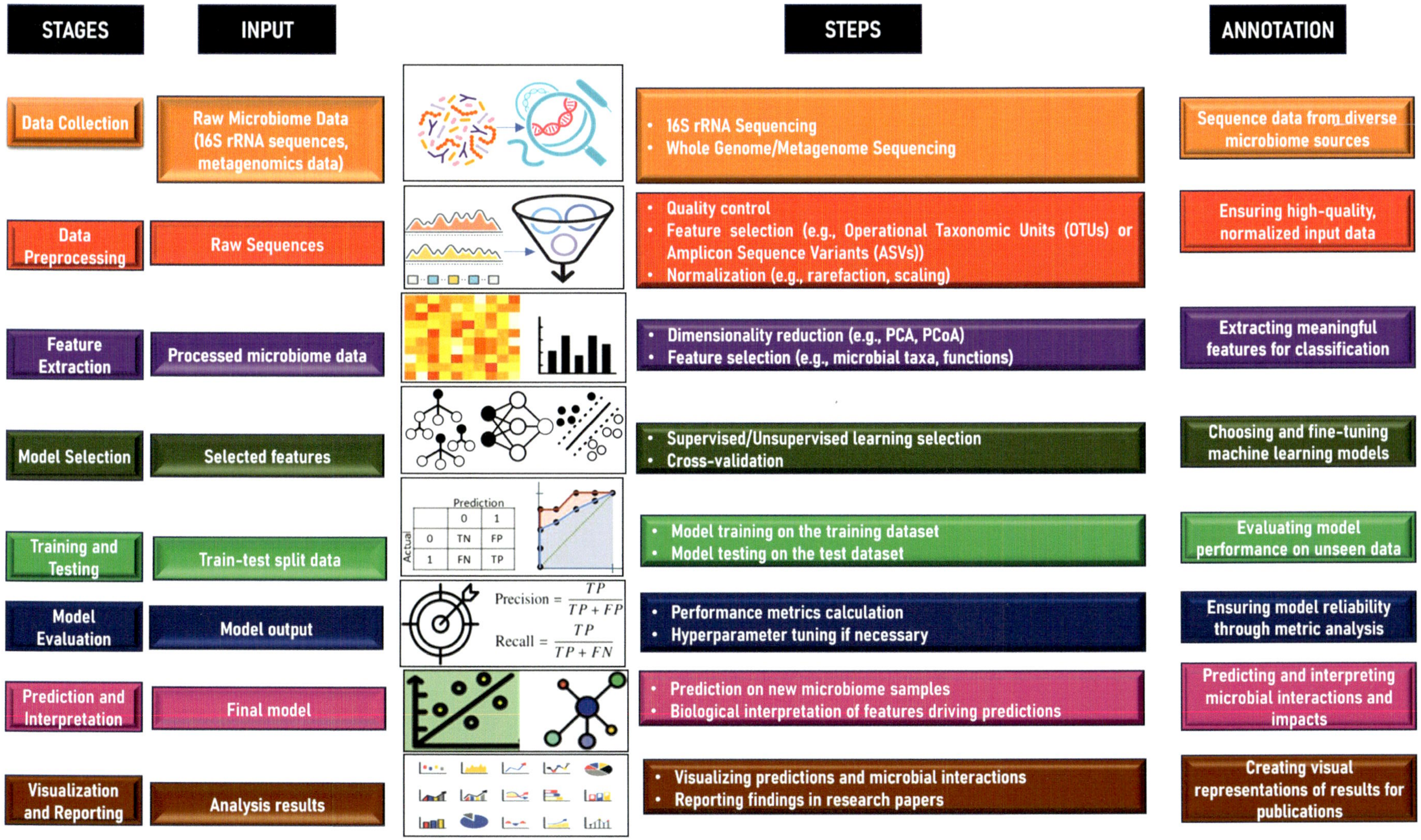

FIGURE 3.2 ML workflow for microbiome analysis. This workflow illustrates the eight key stages involved in microbiome analysis: Data Collection, Data Preprocessing, Feature Extraction, Model Selection, Training and Testing, Model Evaluation, Prediction and Interpretation, and Visualization and Reporting. First, sequencing data is collected from diverse microbiome sources and undergoes preprocessing to produce high-quality, normalized data. This refined data is then used to extract meaningful features, which are input into various ML models. The data is divided into training and testing sets, allowing the models to be trained and evaluated on unseen data. Model evaluation ensures reliability and accuracy. The final model is subsequently used to predict microbial interactions and their impacts, with results visualized and reported for further analysis.

Clustering Methods for Microbiome Data | Microbiome | Full Text" n.d.).

3. *Density-Based Spatial Clustering of Applications with Noise (DBSCAN):* It is effective for identifying clusters of microbes along with outliers, which may characterize unique microbial taxa that might be important for specific environmental functions. Also, it uses dimensionality reduction techniques such as principal component analysis (PCA), t-distributed Stochastic Neighbor Embedding (t-SNE), and Uniform Manifold Approximation and Projection (UMAP) that work effectively in high-dimensional microbiome data and is often used to visualize clusters of microbes (Liu et al., 2012).

3.4.3 Deep Learning

It is an ML type that includes multiple layers of NN, and that is why it is called a deep neural network (Hernández Medina et al., 2022). Deep learning can predict patterns from raw microbial data, which enables more precise predictions and deeper insights into microbial community dynamics (Hernández Medina et al., 2022). This method helps predict disease, microbial interaction networks, functional annotation, and decipher new biomarkers. Deep learning models that are generally used are as follows:

1. *Convolutional Neural Networks (CNNs)*: They are widely applied in pattern recognition, and these are proven to be more effective in data visualization, such as generating heatmaps and co-occurrence networks (Hernández Medina et al., 2022).
2. *Recurrent Neural Networks (RNNs) and Long Short-Term Memory networks (LSTMs)*: These are predominantly suitable for time-series data such as in microbiome analysis, they can trace changes in the microbiome during disease progression or crop growth cycles (Metwally et al., 2019).
3. *Autoencoders*: These are mainly used in case of data compression while preserving critical information. It benefits finding hidden patterns, clustering, and anomaly detection within microbial communities (Metwally et al., 2019).
4. *Deep Belief Networks (DBNs)*: This method is used to model the hierarchical data structures they can be applied to model microbial communities where fine-grained classification or pattern recognition is needed (Lee et al., 2020).
5. *Generative Adversarial Networks (GANs)*: It is mainly used when labeled data is scarce, it generates synthetic microbiome data, which can be used to generate new microbial compositions (Choi et al., 2023).

3.4.4 Reinforcement Learning

This is another important ML type where an agent (e.g., model or algorithm) learns to make decisions by interacting with an environment (Ghadermazi and Chan 2024). The following algorithm takes actions, observes the outcomes, and receives rewards or penalties based on the result. The key components are: (1) algorithm or model—that makes decisions; (2) environment—e.g., the microbial community or agriculture environment; (3) action—a decision made by the algorithm to select certain microbes, apply specific treatments, or manage environmental conditions; (4) reward—feedback from the environment to know how successful the action was to attain required outcome; (5) policy—strategy the agent made to perform the next action depending upon the current state of the environment; (6) value function—a function that estimates the reward an algorithm or model can obtain from the given state (Ghadermazi and Chan 2024). Reinforcement learning (RL) could guide personalized microbial interventions that improve plant growth and productivity. RL algorithms are as follows:

1. *Q-Learning*: In this algorithm, the model learns a value function that evaluates the best action to take from each state. In microbiome applications, it could be used to make decisions about microbial treatments or interventions that maximize crop yield or long-term health (Treloar et al., 2020).
2. *Deep Q-Networks (DQN)*: In case of high-dimensional data (e.g., large microbial datasets), this algorithm could be used to augment microbial community composition over the period and by considering a wide range of microbial features and environmental factors (Fahid et al., 2021).
3. *Policy Gradient Methods*: It uses the "reinforce" method in situations where the action space is continuous, it provides a smooth, optimized strategy for controlling microbial populations (Mendiola-Rodriguez and Ricardez-Sandoval 2023).
4. *Proximal Policy Optimization (PPO)*: It has applications in real-time data as it provides efficiency and stability. In microbiome analysis, it could be used to enhance complex microbiome systems, such as when environmental conditions need to be changed or adjustments made regarding microbial treatments (Ghadermazi and Chan 2024).

3.5 ENSEMBLE AND HYBRID MODELS IN MICROBIOME ANALYSIS

Ensemble methods comprise predictions from multiple base models to enhance overall accuracy (Miller et al., 2023). The idea of ensemble/hybrid methods existed as individual algorithms were insufficient for large data sets. Thus, to overcome the problem of overfitting, biases, and reducing the variance, methods have been improved by aggregating the predictions from different models (Miller et al., 2023). Hybrid models are designed to handle complex data by combining ML, deep learning, or supervised and unsupervised methods to enhance the prediction (Miller et al., 2023). Thus, these models helped handle different microbiome data features, like disease conditions, key interactions in a network, and classifying diverse microbial communities (Hernández Medina et al., 2022). There are a variety of methods which include:

1. *Bagging (also known as Bootstrap Aggregating)*: These methods inculcate a random set of data, (performing replacement each time) by combining multiple copies of the same model (Lindström and Oleandersson 2024). The final prediction is done by combining the predictions from each model trained on random datasets (Lindström and Oleandersson 2024). RF is a commonly used bagging method widely applied in microbiome profiles (Mo et al., 2024). It has applications in classifying microbiome samples into different communities and predicting disease outcomes.
2. *Boosting*: Another method called boosting is highly used in training a series of weak learners, and every new model corrects the errors of the previous one (Beck et al., 2024). Popular boosting methods include AdaBoost, XGBoost, and Gradient Boosting (Beck et al., 2024). In microbiome studies, it has applications in predicting plant health, identifying disease biomarkers, and classifying microbial communities. For example, XGBoost can identify microbial features associated with specific disease conditions.
3. *Stacking:* Last but not least, there is a method called stacking that combines multiple models and performs training on a meta-model and final predictions are based on the outputs of the base models (Ding et al., 2023). This method helps in improving the accuracy as the meta-model learns how to best combine the predictions from the base models. Stacked models combine the outputs of multiple ML algorithms such as SVMs, Decision Trees, and NNs and can be used in microbiome analysis to improve classification performance (Ding et al., 2023).

Merging multiple models in ensemble methods aids in decreasing overfitting and also refines generalization, which leads to more efficient and reliable predictions, especially in the case of high-dimensional microbiome data. The deep learning models can often be seen as "black boxes," while hybrid and ensemble models offer a better way to combine interpretable models with more complex models, thus helping the interpretability of microbiome analysis (Khamparia et al., 2024). One of the advantages of using ensemble and hybrid models is that they offer flexibility in picking different algorithms that best suit various parts of the analysis pipeline (Khamparia et al., 2024). For example, one portion of the analysis may benefit from supervised clustering, while another may require supervised classification or prediction. Depending upon the type of algorithm used for the analysis purpose there are a variety of tools available to perform analysis on plant microbiome data as shown in Table 3.1.

3.6 CHALLENGES AND FUTURE PROSPECTS

Although there are extreme application of AI and ML methods in analyzing the microbiome data but along with this it has different challenges too. If we talk about the supervised learning methods where it takes care of labeled data, this could be difficult to gain in the case of microbiome data, exclusively in agricultural settings. Also, there is one more difficulty as a supervised learning model trained on microbiome data from one location or crop type may not fit well with other contexts. Therefore, future work may emphasize making more robust models that can better cope with different environments and agricultural conditions. These supervised techniques permit learning from limited labeled data thus contributing new opportunities for microbiome research with controlled datasets. Unsupervised learning methods can produce complex networks that are often difficult to interpret biologically. Future work can focus on providing the domain knowledge to expand the biological importance of clustering microbial networks. In addition, merging microbiome data with functional data can improve the understanding obtained from unsupervised learning. Future directions also include creating multimodal unsupervised learning methods that not only consider microbiome composition but also functional, phenotypic, and environmental evidence. Though deep learning methods seem to be opportunistic; in this case, microbiome data obtained through sequencing can be variable, and thus affects the performance of deep learning models. One of the main challenges with deep learning is interpreting the models by considering which microbial features contribute to what type of a model's predictions which seems to be crucial, especially in agricultural contexts. Techniques such as SHapley Additive exPlanations (SHAP) values, Local Interpretable Model-agnostic Explanations (LIME), and attention mechanisms are highly used to infer deep learning models in microbiome research. Deep learning models often require large datasets, however, in agriculture large and diverse datasets may not always be available so this could lead to overfitting. Deep learning models are computationally affluent, demanding substantial hardware resources and time for training. Introducing multi-omics data such as those obtained from transcriptomics, metabolomics, or proteomics, can provide a more holistic view of microbiome dynamics and its impact on host and environmental health. Designing RL algorithms that can efficiently handle sparse feedback and long-term reward delays is a key challenge. As RL models continue to evolve, they can be utilized for personalized microbiome-based interventions in healthcare, agriculture, and environmental management, leading to

TABLE 3.1
ML and AI Tools for Analyzing Agriculture Microbiome Data

Tools	USE	AI/ML Integration	Website
		General Bioinformatics Tools with AI/ML Integration	
QIIME2	Microbial community analysis (e.g., diversity analysis, taxonomic classification)	q2-sample-classifier	https://qiime2.org
DADA2	Sequence quality filtering, error correction, and feature table generation for 16S rRNA data	Can output data for ML workflows using other tools like Python or R	https://www.bioconductor.org/packages/release/bioc/html/dada2.html
MEGAN (Metagenome Analyzer)	Functional and taxonomic analysis of microbiome datasets	Can preprocess data for downstream ML models	https://software-ab.cs.uni-tuebingen.de/download/megan6/welcome.html
Phyloseq (R Package)	Data exploration and visualization of microbiome datasets	Can integrate with ML libraries in R for predictive modeling	https://www.bioconductor.org/packages/release/bioc/html/phyloseq.html
		AI/ML-Specific Platforms for Microbiome Analysis	
MicrobiomeAnalyst	Statistical analysis and functional profiling of microbiome data	Features supervised and unsupervised learning options for classification and clustering	https://www.microbiomeanalyst.ca/
H2O.ai	General-purpose AI platform for developing ML models	Can analyze microbiome data by building classification or regression models using tools like RFs, gradient boosting, or deep learning	https://github.com/h2oai
Weka	ML software with a focus on biological datasets	Includes a wide range of ML algorithms for microbial dataset classification, clustering, and regression	https://sourceforge.net/projects/weka/
TensorFlow and PyTorch	Building and training custom deep learning models	Ideal for advanced ML tasks such as feature extraction, dimensionality reduction, and predictive modeling in microbiome data	https://www.tensorflow.org/ https://pytorch.org/
		Data Preprocessing and Feature Selection Tools	
scikit-learn	Feature selection, dimensionality reduction (e.g., PCA, t-SNE), and ML-based classification or regression	Widely used for building and evaluating ML models for microbiome datasets	https://scikit-learn.org/stable/index.html
Bioconductor (R)	Comprehensive tools for statistical analysis and feature selection	Provides ML-friendly outputs for supervised learning tasks	https://www.bioconductor.org/
		Tools for Multi-Omics and Microbiome Integration	
MetaboAnalyst	Integration of metabolomic and microbiome datasets	Features classification, clustering, and regression tools powered by ML	https://www.metaboanalyst.ca/
ML-Omics	A framework for multi-omics analysis using ML techniques	Customizable ML workflows for microbiome and multi-omics data	ML-Omics
Critical Assessment of Metagenome Interpretation (CAMI)	Benchmarking tools and workflows for microbiome metagenomics	Outputs data suitable for ML applications	https://cami-challenge.org/
		Visualization and Interaction Analysis Tools	
Gephi	Network analysis and visualization of microbial interactions	Can integrate ML outputs to highlight important microbial relationships	https://gephi.org/
Cytoscape	Visualization of microbial networks and interaction data	Can visualize ML-predicted microbial interaction networks or functional relationships	https://cytoscape.org/
		Cloud-Based Platforms	
Google Cloud AutoML	Automated ML for building ML models	Useful for predictive modeling with microbiome datasets	https://cloud.google.com/automl/
AWS ML Services	Scalable cloud-based ML tools	Ideal for microbiome studies requiring high computational power	https://aws.amazon.com/Free/MachineLearning
Microsoft Azure ML	End-to-end ML model development and deployment	Can handle microbiome dataset preprocessing, model building, and visualization	https://azure.microsoft.com/Data_Analytics/MachineLearning

more effective and efficient approaches for improving animal health, plant growth, or disease prevention. Ensemble and hybrid models that include multiple sets of models to make final predictions are generally expensive and may entail more resources and time to train than individual models. In addition, microbiome data comes in various forms (e.g., 16S rRNA sequences, metabolomics, metagenomics), and inculcating these diverse data types into a hybrid model can be challenging. Ensemble models generally cope with overfitting but still, there is a chance of overfitting if datasets are not properly tuned. Expanding computerized ensemble learning techniques could support microbiome researchers to quickly select and combine apt base models for specific tasks.

3.7 CONCLUSION

Combining AI and ML with microbiome research has significantly enhanced our ability to analyze, comprehend, and predict the complex dynamics of microbial communities. These sophisticated computational methods are no longer limited by conventional statistical approaches, which can detect non-linear patterns, reveal hidden attributes, and build predictive models from large and intricate microbiome datasets. AI and ML software have enabled us to know much more regarding the interactions between hosts and microbiomes, ranging from taxonomic profiling and functional annotation to disease prediction and biomarker discovery. Application of methods such as deep learning, support vector machines, RFs, and NNs to metagenomic, metatranscriptomic, and multi-omics data has unlocked new avenues in environmental microbiology, gut–brain axis studies, and precision medicine. However, there are still issues with the data quality, standardization, and interpretation. It is also important to watch out for ethical issues in clinical use, especially because many deep learning models work like "black boxes" and are hard to understand.

The future of microbiome research will likely be characterized by the convergence of AI and systems biology, network science, and integrative omics. As algorithms get better, more individuals work across fields and more are able to access computing, AI and ML can become indispensable tools to comprehend the role of the microbiome in health, disease, and ecosystems.

REFERENCES

Acharjee, Animesh, Joseph Larkman, Yuanwei Xu, Victor Roth Cardoso, and Georgios V. Gkoutos. 2020. "A Random Forest Based Biomarker Discovery and Power Analysis Framework for Diagnostics Research." *BMC Medical Genomics* 13 (1): 178. https://doi.org/10.1186/s12920-020-00826-6.

Ali, Md. Akkas, Rajesh Kumar Dhanaraj, and Anand Nayyar. 2023. "A High Performance-Oriented AI-Enabled IoT-Based Pest Detection System Using Sound Analytics in Large Agricultural Field." *Microprocessors and Microsystems* 103 (November): 104946. https://doi.org/10.1016/j.micpro.2023.104946.

Ali, Sajid, Tamer Abuhmed, Shaker El-Sappagh, et al. 2023. "Explainable Artificial Intelligence (XAI): What We Know and What Is Left to Attain Trustworthy Artificial Intelligence." *Information Fusion* 99 (April): 101805. https://doi.org/10.1016/j.inffus.2023.101805.

Al-Qudah, Rabia, Mrouj Almuhajri, and Ching Y. Suen. 2025. "Unveiling the Potential of Sustainable Agriculture: A Comprehensive Survey on the Advancement of AI and Sensory Data for Smart Greenhouses." *Computers and Electronics in Agriculture* 229 (February): 109721. https://doi.org/10.1016/j.compag.2024.109721.

"Applications, Promises, and Pitfalls of Deep Learning for Fluorescence Image Reconstruction | Nature Methods." n.d. Accessed December 18, 2024. https://www.nature.com/articles/s41592-019-0458-z.

Beck, Kristen L., Niina Haiminen, Akshay Agarwal, et al. 2024. "Development and Evaluation of Statistical and Artificial Intelligence Approaches with Microbial Shotgun Metagenomics Data as an Untargeted Screening Tool for Use in Food Production." *mSystems* 9 (11): e0084024. https://doi.org/10.1128/msystems.00840-24.

Berg, Gabriele, Daria Rybakova, Doreen Fischer, et al. 2020. "Microbiome Definition Re-Visited: Old Concepts and New Challenges." *Microbiome* 8 (1): 103. https://doi.org/10.1186/s40168-020-00875-0.

Choi, Joung Min, Ming Ji, Layne T. Watson, and Liqing Zhang. 2023. "DeepMicroGen: A Generative Adversarial Network-Based Method for Longitudinal Microbiome Data Imputation." *Bioinformatics (Oxford, England)* 39 (5): btad286. https://doi.org/10.1093/bioinformatics/btad286.

Deng, Zhiyu, Jinming Zhang, Junya Li, and Xiujun Zhang. 2021. "Application of Deep Learning in Plant–Microbiota Association Analysis." *Frontiers in Genetics* 12 (October). https://doi.org/10.3389/fgene.2021.697090.

Ding, Shang, Liyan Jiang, Jiyuan Hu, Wuji Huang, and Liping Lou. 2023. "Microbiome Data Analysis via Machine Learning Models: Exploring Vital Players to Optimize Kitchen Waste Composting System." *Bioresource Technology* 388 (November): 129731. https://doi.org/10.1016/j.biortech.2023.129731.

Fahid, Fahmid Morshed, Jonathan P. Rowe, Randall D. Spain, Benjamin S. Goldberg, Robert Pokorny, and James Lester. 2021. "Adaptively Scaffolding Cognitive Engagement with Batch Constrained Deep Q-Networks." In *Artificial Intelligence in Education*, edited by Ido Roll, Danielle McNamara, Sergey Sosnovsky, Rose Luckin, and Vania Dimitrova. Springer International Publishing. https://doi.org/10.1007/978-3-030-78292-4_10.

Ghadermazi, Parsa, and Siu Hung Joshua Chan. 2024. "Microbial Interactions from a New Perspective: Reinforcement Learning Reveals New Insights into Microbiome Evolution." *Bioinformatics (Oxford, England)* 40 (1): btae003. https://doi.org/10.1093/bioinformatics/btae003.

Ghannam, Ryan B., and Stephen M. Techtmann. 2021. "Machine Learning Applications in Microbial Ecology, Human Microbiome Studies, and Environmental Monitoring." *Computational and Structural Biotechnology Journal* 19: 1092–107. https://doi.org/10.1016/j.csbj.2021.01.028.

Hernández Medina, Ricardo, Svetlana Kutuzova, Knud Nor Nielsen, et al. 2022. "Machine Learning and Deep Learning Applications in Microbiome Research." *ISME Communications* 2 (1): 98. https://doi.org/10.1038/s43705-022-00182-9.

Hou, Kaijian, Zhuo-Xun Wu, Xuan-Yu Chen, et al. 2022. "Microbiota in Health and Diseases." *Signal Transduction and Targeted Therapy* 7 (1): 135. https://doi.org/10.1038/s41392-022-00974-4.

Huang, Shimeng, Elisabeth Ailer, Niki Kilbertus, and Niklas Pfister. 2023. "Supervised Learning and Model Analysis with Compositional Data." *PLOS Computational Biology* 19 (6): e1011240. https://doi.org/10.1371/journal.pcbi.1011240.

"Identifying Keystone Species in Microbial Communities Using Deep Learning | Nature Ecology & Evolution." n.d. Accessed December 18, 2024. https://www.nature.com/articles/s41559-023-02250-2.

Jabed, Md. Abu, and Masrah Azrifah Azmi Murad. 2024. "Crop Yield Prediction in Agriculture: A Comprehensive Review of Machine Learning and Deep Learning Approaches, with Insights for Future Research and Sustainability." *Heliyon* 10 (24): e40836. https://doi.org/10.1016/j.heliyon.2024.e40836.

Khamparia, Aditya, Babita Pandey, Devendra Kumar Pandey, and Deepak Gupta. 2024. *Microbial Data Intelligence and Computational Techniques for Sustainable Computing*. Singapore: Springer Nature.

Kim, Nayeon, Junyeong Ma, Wonjong Kim, Jungyeon Kim, Peter Belenky, and Insuk Lee. 2024. "Genome-Resolved Metagenomics: A Game Changer for Microbiome Medicine." *Experimental & Molecular Medicine* 56 (7): 1501–12. https://doi.org/10.1038/s12276-024-01262-7.

Lee, Joon-Yong, Natalie C. Sadler, Robert G. Egbert, et al. 2020. "Deep Learning Predicts Microbial Interactions from Self-Organized Spatiotemporal Patterns." *Computational and Structural Biotechnology Journal* 18: 1259–69. https://doi.org/10.1016/j.csbj.2020.05.023.

Lindström, Felix, and Robin Oleandersson. 2024. *Analysis of High-Dimensional Compositional Microbiome Data Using PERMANOVA and Machine Learning Classifiers*. https://urn.kb.se/resolve?urn=urn:nbn:se:uu:diva-530378.

Liu, Qiliang, Min Deng, Yan Shi, and Jiaqiu Wang. 2012. "A Density-Based Spatial Clustering Algorithm Considering Both Spatial Proximity and Attribute Similarity." *Computers & Geosciences* 46 (September): 296–309. https://doi.org/10.1016/j.cageo.2011.12.017.

Mana, A. A., A. Allouhi, A. Hamrani, S. Rehman, I. el Jamaoui, and K. Jayachandran. 2024. "Sustainable AI-Based Production Agriculture: Exploring AI Applications and Implications in Agricultural Practices." *Smart Agricultural Technology* 7 (March): 100416. https://doi.org/10.1016/j.atech.2024.100416.

Matchado, Monica Steffi, Michael Lauber, Sandra Reitmeier, et al. 2021. "Network Analysis Methods for Studying Microbial Communities: A Mini Review." *Computational and Structural Biotechnology Journal* 19: 2687–98. https://doi.org/10.1016/j.csbj.2021.05.001.

Mendiola-Rodriguez, Tannia A., and Luis A. Ricardez-Sandoval. 2023. "Integration of Design and Control for Renewable Energy Systems with an Application to Anaerobic Digestion: A Deep Deterministic Policy Gradient Framework." *Energy* 274 (July): 127212. https://doi.org/10.1016/j.energy.2023.127212.

Metwally, Ahmed A., Philip S. Yu, Derek Reiman, Yang Dai, Patricia W. Finn, and David L. Perkins. 2019. "Utilizing Longitudinal Microbiome Taxonomic Profiles to Predict Food Allergy via Long Short-Term Memory Networks." *PLoS Computational Biology* 15 (2): e1006693. https://doi.org/10.1371/journal.pcbi.1006693.

Miller, Tymoteusz, Grzegorz Mikiciuk, Anna Kisiel, et al. 2023. "Machine Learning Approaches for Forecasting the Best Microbial Strains to Alleviate Drought Impact in Agriculture." *Agriculture* 13 (8): 8. https://doi.org/10.3390/agriculture13081622.

Mo, Yujie, Raven Bier, Xiaolin Li, et al. 2024. "Agricultural Practices Influence Soil Microbiome Assembly and Interactions at Different Depths Identified by Machine Learning." *Communications Biology* 7 (1): 1–16. https://doi.org/10.1038/s42003-024-07059-8.

Panthee, Bimala, Saroj Gyawali, Pratiksha Panthee, and Kuaanan Techato. 2022. "Environmental and Human Microbiome for Health." *Life (Basel, Switzerland)* 12 (3): 456. https://doi.org/10.3390/life12030456.

Patil, Abhinandan, Neha Singh, Mohsina Patwekar, et al. 2024. "AI-Driven Insights into the Microbiota: Figuring out the Mysterious World of the Gut." *Intelligent Pharmacy*, ahead of print, August 19. https://doi.org/10.1016/j.ipha.2024.08.003.

"Performance Determinants of Unsupervised Clustering Methods for Microbiome Data | Microbiome | Full Text." n.d. Accessed December 18, 2024. https://microbiomejournal.biomedcentral.com/articles/10.1186/s40168-021-01199-3.

Rahmani, Amir Masoud, Elham Azhir, Saqib Ali, et al. 2021. "Artificial Intelligence Approaches and Mechanisms for Big Data Analytics: A Systematic Study." *PeerJ. Computer Science* 7: e488. https://doi.org/10.7717/peerj-cs.488.

Saeed, Isaam, Sen-Lin Tang, and Saman K. Halgamuge. 2012. "Unsupervised Discovery of Microbial Population Structure within Metagenomes Using Nucleotide Base Composition." *Nucleic Acids Research* 40 (5): e34. https://doi.org/10.1093/nar/gkr1204.

Sharma, Divya, Andrew D. Paterson, and Wei Xu. 2020. "TaxoNN: Ensemble of Neural Networks on Stratified Microbiome Data for Disease Prediction." *Bioinformatics (Oxford, England)* 36 (17): 4544–50. https://doi.org/10.1093/bioinformatics/btaa542.

Thompson, Jaron, Renee Johansen, John Dunbar, and Brian Munsky. 2019. "Machine Learning to Predict Microbial Community Functions: An Analysis of Dissolved Organic Carbon from Litter Decomposition." *PLoS One* 14 (7): e0215502. https://doi.org/10.1371/journal.pone.0215502.

Topçuoğlu, Begüm D., Nicholas A. Lesniak, Mack T. Ruffin, Jenna Wiens, and Patrick D. Schloss. 2020. "A Framework for Effective Application of Machine Learning to Microbiome-Based Classification Problems." *mBio* 11 (3): e00434-20. https://doi.org/10.1128/mBio.00434-20.

Treloar, Neythen J., Alex J. H. Fedorec, Brian Ingalls, and Chris P. Barnes. 2020. "Deep Reinforcement Learning for the Control of Microbial Co-Cultures in Bioreactors." *PLoS Computational Biology* 16 (4): e1007783. https://doi.org/10.1371/journal.pcbi.1007783.

4 Next-Generation Sequencing Technology and Bioinformatics Pipelines for Generation and Analysis of Microbiome Data

Megha Chaudhari[1†], *Sunita Verma*[1†], *and Sushanta Deb*[1,2#]

[1]Department of Microbiology, All India Institute of Medical Sciences, Ansari Nagar, New Delhi, 110029

[2]Department of Veterinary Microbiology and Pathology, College of Veterinary Medicine, Washington State University, Pullman 99164, WA, United States

4.1 INTRODUCTION

Amplicon sequencing and shotgun sequencing are effective methods to study microbial diversity and their function. Amplicon sequencing is the process of targeting and amplifying universal genetic markers (e.g., 16S rRNA) to identify bacteria and their quantities in a sample using tools like QIIME2, DADA2, and databases like SILVA and Greengenes (Straub et al., 2020). Database constraints present challenges such as sequencing mistakes, contamination, amplification bias, and taxonomic misclassification. On the basis of taxonomic profiles, functional analysis techniques such as FAPROTAX and PICRUSt2 predict the metabolic potential of microorganisms (Wang et al., 2022). Large datasets, host DNA contamination, and computationally demanding tasks like assembly and annotation using Bowtie2, Kraken2, and MetaPhlAn are all necessary for shotgun sequencing, which offers more comprehensive taxonomic and functional insights. Tools such as DeepMicro and Microbiome Analyst, machine learning (ML) has recently revolutionized microbiome research by facilitating high-dimensional data visualization, biomarker discovery, and predictive modeling. The need for strong methodology and integrated approaches is highlighted by the persistence of issues like standardization, computational demands, and interpretability notwithstanding progress.

4.2 UNDERSTANDING MICROBIOME

A microbiome is a group of microorganisms that coexist in a certain habitat, like the soil, the ocean, or the human body. The intricate ecosystems known as microbiomes are essential to the well-being of living systems. More than one trillion microorganisms, including a vast range of commensal bacteria, inhabit inside humans and are essential to an individual's health. These microorganisms inhabit a variety of human body organs, including the gut, skin, vagina, and mouth. Different organs host different types and numbers of microorganisms, and these can also differ between individuals. The environment, food, and host genetics are some of the potential causes of the high diversity of microbes. Numerous investigations into the human microbiome have shown that the microbiota of healthy and sick people differ (Rajpoot et al., 2018). Recent advancements in next-generation sequencing (NGS) technology have substantially enhanced the analysis of microbial communities across diverse ecosystems, including aquatic animal environments (Deb et al., 2020), freshwater habitats (Das et al., 2020), and hot spring ecosystems (Deb & Das, 2022; Debnath et al., 2022). Some of the important microbiome projects are summarized below:

4.2.1 Earth Microbiome Project

With the audacious objective of compiling a worldwide inventory of the many uncultured microorganisms that exist on Earth, the Earth Microbiome Project was launched in August 2010. Analyzing the microbial diversity and functional capacities of more than 200,000 environmental samples is the primary goal of the Earth Microbiome Project. In 2012, Jonathan Eisen was described in Nature as saying, "Knight and Gilbert literally talk about sampling the entire planet." This ambitious project was initially viewed as impractical. In Snowbird, Utah, the first concept was developed at a Department of the Environment (DOE) workshop on the possibilities of terabase-scale sequencing. This event aimed to alter microbial ecology and deepen our understanding of the microbial domain by generating research

†Both authors contributed equally

#**Corresponding author:** sushantadeb26@gmail.com (Sushanta Deb)

DOI: 10.1201/9781003544425-4

ideas with state-of-the-art technology. That meeting gave rise to a number of other exciting projects, like the Microbial Earth Project, which aims to improve the sequencing of type strains of cultivated bacterial species (Gilbert et al., 2014). Researchers from a wide range of microbial ecology disciplines, including humans, plants, animals, freshwater, aquatic, terrestrial, air, sediment, built environments, and all of these ecosystems' intersections, collaborated to create the Earth Microbiome Project (Earth Microbiome Project data, 2024).

The John Templeton Foundation and the WM Keck Foundation presented the EMP awards in June 2013 to promote efforts aimed at raising awareness of the catalogue. They had processed more than 30,000 samples as of July 2014, compared to 50,000 in the Human Microbiome Project (HMP)'s phase 1 amplicon screening of 5,771 samples (Huttenhower et al., 2012). The EMP suggested sample co-analysis using metagenomics and metabolic modeling of ecosystems during its development phase. These goals remain feasible, but up to this point, these efforts have been primarily focused on particular settings and research.

4.2.2 The Tara Oceans Project

This project was started in September 2009 and is overseen by the EMBL. This study has gathered over 30,000 samples from over 200 ocean sites over the last 3 years, which contain millions of tiny species. The 110-foot research schooner Tara collected most of the samples. From 2009 to 2013, the ship made trips to all of the major marine basins, including the North Atlantic Ocean, to study microscopic planktons found at depths reaching 2000 meters (Bork et al., 2015). Since 2009, this topic has attracted interest from a large number of academic groups worldwide and has been covered by the media on multiple occasions (Ainsworth, 2013).

Three notable studies on microbial big-data mining have focused on the data and samples gathered during the Tara Oceans project. The first study examined the composition and functions of aquatic microorganisms (Bork et al., 2015). A small portion of the experiment, namely 7.2 TB of metagenomic data from 243 Tara Oceans samples collected across 68 locations, was examined by Peer Bork's study team. They found millions of new sequences from viruses, prokaryotes, and picoeukaryotes, and they showed that temperature has a greater impact on the makeup of microbial communities than other environmental parameters. The second study focused on the ocean virome (Brum et al., 2015). Brum and colleagues used 43 samples from the upper-ocean viral environment to identify the pan and core gene sets. The seed-bank hypothesis states that"The viral community is a reservoir of dormant individuals that may be resuscitated in the future under different environmental conditions," which is supported by their assessments (Lennon and Jones, 2011). This approach could explain why oceanic viral communities maintain high levels of local diversity. The interpretation of the ocean plankton interactome is described in the third paper.

By combining state-of-the-art picture technology with modern sequencing, Tara Oceans is leading the way in marine science (Karsenti et al., 2011). The amount of data generated by these technologies is unmatched in the field of plankton ecology, requiring specialized storage infrastructures and collaborative platforms to annotate high-throughput photos and sequences both manually and automatically. These free scientific resources are being developed by Tara Oceans. Additionally, this project offered a wealth of information on biological elements in ocean models and ocean food webs (Lima-Mendez et al., 2015). Additional published research includes evaluations of eukaryotic plankton diversity (Kaneko et al., 2023; De Vargas et al., 2015) a summary of the global ocean microbiome derived from the Tara Oceans project (Sunagawa et al., 2015), and ways in which Agulhas rings' environmental features impact the movement of plankton across oceans (Villar et al., 2015).

4.2.3 Human Microbiome Project

Enhancing our understanding of the human microbiome is the goal of the HMP, a broad community endeavor funded by the NIH Roadmap for Biomedical Research (http://nihroadmap.nih.gov). The HMP intends to: (1) use cutting-edge high-throughput technologies to analyze samples from at least 250 "normal" people across different body sites; (2) look into a range of medical conditions to determine whether changes in the microbiome are linked to health or disease; and (3) offer a standardized data resource and creative technological approaches to support such studies within the scientific community. Additionally, rigorous research on the ethical, legal, and societal implications of this kind of research is underway (Peterson et al., 2009).

Since an interdisciplinary endeavor is being conducted concurrently in multiple nations, including the US (as part of the next phase of the National Institutes of Health's Roadmap for Medical Research), Europe, and Asia, the HMP makes conceptual and experimental sense. A revolutionary era in microbiology is being ushered in by the introduction of highly parallel DNA sequencers and high-throughput mass spectrometers that provide remarkable mass precision and sensitivity. The focus of the science is changing due to these techniques from studying individual species to comprehending the functions of large microbial ecosystems. The goal of the new field of metagenomics is to describe these communities' genomes, as well as the mRNA and proteins that are linked to them. Researchers have recently used 16S rRNA sequencing technologies to publish studies on the human microbiome under a variety of biological settings. Research on the gut microbiome at the 16S rRNA gene level, for instance, has shown notable differences in individual flora, tracked the development of the microbiome in infants (Palmer et al., 2007), and revealed significant differences in the microbial communities of lean versus obese individuals. Additionally, the 16S rRNA gene has been used in research as a marker for microbiomes in the skin (Gao et al., 2007), vagina (Hyman et al., 2005), and oral cavity (Faveri et al., 2008).

The program's data is publicly accessible through the NIH NCBI BioProject, where users can examine the different kinds of data produced by the research. Based on the four categories of data generated, the HMP program classifies four different projects: (1) targeted 16S ribosomal RNA gene sequences, obtained from the healthy adult cohort study and serving as a taxonomic marker; (2) whole-genome shotgun metagenomic sequences (metagenome: encompassing all sequences of genes from a single microbial group); (3) datasets produced from individual Demonstration Project activities; and (4) sequences of reference strain microbial genomes. The HMP's conceptual framework demonstrates how the program's six elements interact together through consortium activities (Proctor, 2011).

The HMP's focused initiatives to develop standards for employing next-generation technologies in metagenomic research and to provide precise, quality-assured data will build the groundwork for future investigations into the human microbiome.

4.2.4 The Global Coral Microbiome Project

Coral reefs, often called "rainforests of the sea," are the most productive ecosystems on the planet due to their incredible diversity (Knowlton, 2001). Tropical reefs often have low nutrient levels, despite their extraordinarily high primary output; effective nutrient retention mechanisms are required. Because they filter large amounts of seawater (up to thousands of liters per day) (Hentschel et al., 2006) retain the organic matter, and convert it into biomass that detritivores can consume, marine sponges are, therefore, a crucial part of reef ecosystems. The "sponge-loop," as this mechanism is called, recycles the water back into the reef system (De Goeij, 2013).

Coral reefs, known for being among the most biologically diverse marine ecosystems on Earth, provide coastal communities with considerable ecological and economic benefits. Corals are made up of intricate communities of unusual and little-studied microbiological species in addition to the cnidarian animal host. These natural wonders are currently in danger of going extinct due to the combined effects of several stressors, such as pollution, overfishing, and climate change. Corals may become more vulnerable to disease or algal competition as a result of these stresses' ability to change the microbial populations within corals. As a result, understanding the connections between reefs and their microbiota could aid in preventing coral disease and protecting reef ecosystems.

Most research on the microbial communities associated with corals has concentrated on a limited range of coral diversity, focusing on both well-explored corals and various reefs globally. Despite ongoing efforts to enhance our understanding of coral reef ecosystems and to safeguard them for future generations, this project intends to identify the different types of microbes present in all major coral groups that contribute to reef-building across various ecological systems worldwide.

Recent advancements in DNA sequencing technology and advanced computer techniques will enable a more comprehensive analysis of microbial diversity by optimizing the value of read bacterial genomes. By employing enhanced prediction models that consider both coral phylogenetic and microbial function, we can better inform conservation strategies and generate predictive indicators for coral susceptibility to disease or bleaching. Connecting the diversity of corals to that of their bacteria will also provide crucial insights into how intimate symbiotic connections with microorganisms evolved and are maintained in a range of species, including humans.

4.3 OVERVIEW ON AMPLICON SEQUENCING TECHNIQUES

4.3.1 Bioinformatics Tools and Databases for Amplicon Sequence Data Analysis

Amplicon sequencing includes sampling a community, DNA extraction from all cells in the sampled community, target and amplify a taxonomically informative universal genomic marker through PCR, and sequencing. The bioinformatics tools analyze the amplicons and identify types of microbes found in the sample and in what proportions (Beaudry et al., 2021). Some of the bioinformatics tools used are listed in Table 4.1.

4.3.2 Challenges Associated with Amplicon Sequence Data Analysis

The analysis of amplicon sequence data, which is commonly generated in microbial diversity investigations or targeted sequencing applications, faces a lot of obstacles. These difficulties can affect the quality of data, reliability, and interpretability (Hugerth & Andersson, 2017). NGS technologies may include errors such as substitutions, insertions, and deletions, which can have an impact on downstream analysis. Improper trimming of sequencing adapters and primer sequences might produce misleading results. Reads with poor Phred quality scores may result in false findings if not filtered properly (Deakin et al., 2014). Laboratory or reagent contamination can introduce non-target DNA. Inefficient sample preparation or indexing errors can lead to misassigned reads. Many microbial taxa are poorly represented in databases, leading to misclassifications or unclassified sequences. Closely related species or strains can have nearly identical sequences in the targeted region, making it challenging to resolve at the species level (Clarridge, 2004).

PCR amplification can produce chimera readings that are unrelated to any genuine organism. Inefficient or non-universal primers may not amplify all intended species equally, skewing community composition. Amplification bias during PCR can favor some sequences over others, altering abundance estimations. It requires precise clipping, filtering, and merging of paired-end readings, which can be computationally intensive. The choice between operational

TABLE 4.1
Bioinformatics Tools and Databases for Amplicon Sequence Data Analysis

Tools	Description	Web Link	References
QIIME 2 (Quantitative Insights Into Microbial Ecology):	• Analyze and interpret amplicon sequencing data • Particularly 16S rRNA gene sequences • Use for quality control • Diversity analysis.	https://qiime2.org/	Chen et al. (2025)
DADA2	• Focuses on high-resolution amplicon sequence variant (ASV) inference from raw sequencing data. • Corrects errors in sequenced data • Produces a table of ASVs	https://github.com/benjjneb/dada2	Callahan et al. (2016)
USEARCH	• Process and analyze amplicon sequences • Perform clustering, chimera detection, and taxonomic assignment	http://drive5.com/usearch/	Kang et al. (2021)
VSEARCH	• Offer similar functionalities for clustering, chimera detection • Useful for users looking for a free tool	https://github.com/torognes/vsearch	Rognes et al. (2016)
Mothur	• Analyze 16S rRNA gene sequences • Provide a wide range of tools for sequence processing • Clustering and diversity analysis	https://www.mothur.org/	Chappidi et al. (2019)
Databases for Amplicon Sequence Data			
Greengenes	• Database of 16S rRNA gene sequences. • Provides a reference for taxonomic classification.	http://greengenes.lbl.gov/	DeSantis et al. (2006)
SILVA	• Database of ribosomal RNA sequences, including 16S and 18S rRNA genes. • It offers high-quality alignments. • Perform taxonomic classifications • A valuable resource for microbial ecology studies.	https://www.arb-silva.de/	Quast et al. (2013)
RDP (Ribosomal Database Project)	• Use for the classification and analysis of ribosomal RNA sequences. • It includes a taxonomic classifier and a sequence alignment tool.	http://rdp.cme.msu.edu/	Cole et al. (2014)
NCBI (National Center for Biotechnology Information)	• A vast repository of biological data, including nucleotide sequences. • Provide wide range of amplicon sequences and related metadata	https://www.ncbi.nlm.nih.gov/	Jenuth (1999)
UNITE	• A database focused on fungal ITS sequences, providing a resource to study fungal diversity and ecology.	https://unite.ut.ee/	Abarenkov et al. (2010)

taxonomic units (OTUs) and amplicon sequence variations (ASVs) for analysis affects resolution and interpretability. Amplicon data frequently contain a large number of features (for example, taxa) that are sparsely represented, complicating statistical studies (Thorsen et al., 2016). Processing batches introduce variability, which can mask genuine biological trends. Large dataset analysis demands a lot of processing resources, especially for phylogeny, alignment, and clustering. Big data handling can be prohibitively expensive for academics with little funding. The results can be inconsistent due to differences in pipelines and parameter selections. Meta-analyses and reproducibility are hampered by a lack of standardization in the reporting and analysis of results (Polanin et al., 2020).

4.3.3 Tools for Functional Analysis of Amplicon Sequence Data

By connecting taxonomic designations to established functional capabilities, amplicon sequencing can be used to predict functional potential even though its primary purpose is to offer taxonomic classification. The following lists a few popular methods and instruments for functional analysis of amplicon sequencing data (Table 4.2).

4.4 OVERVIEW OF SHOTGUN SEQUENCING TECHNIQUES

Sequences from the host organism will typically contaminate shotgun metagenome sequencing data collected from a host environment. To prevent biases, lessen the computing load downstream, or protect privacy, host sequences should be eliminated prior to additional analysis. A useful technique for the taxonomic and functional characterization of microbial populations is shotgun metagenomic sequencing. This method is difficult to use where large quantities of host DNA and low microbe abundance lead to ineffective and expensive sequencing (Sedlar et al., 2017).

TABLE 4.2
Major Tools for the Functional Analysis of Amplicon Sequence Data

S. No.	Tools	Description	References
1	PICRUSt2 (Phylogenetic Investigation of Communities by Reconstruction of Unobserved States)	• Uses 16S rRNA gene marker gene sequences for the prediction of metagenomic functional profiles. • Maps taxonomic characteristics to functional pathways using a reference genome database (like KEGG). • Also sheds light on functional variety and metabolic capacity	Chen et al. (2021)
2	FAPROTAX (Functional Analysis for Prokaryotic Taxa)	• Give taxonomic classifications functional profiles directly. • It makes use of a carefully curated library of bacterial taxa and the roles they play. • Also offers a simple and rapid way to determine functional potential	Yang et al. (2022)
3	Tax4Fun2	• It is comparable to PICRUSt2, but it makes use of sophisticated statistical techniques and a larger reference genome database. • Applicable on both 16S and 18S rRNA gene data and increased the accuracy of functional predictions	Toole et al. (2021)
4	Phylogenetic Placement Tools (e.g., PPlacer, EPA-NG)	• In order to deduce evolutionary links and functional potential, amplicon sequences are placed onto a reference tree. • it can offer more accurate functional predictions	Hasan et al. (2022)
5	Metagenomics Analysis Tools (e.g., MetaPhlAn, Kraken2)	• Identify amplicon data in order to estimate functional potential, even though they were first created for metagenomic data. • Uses the sequence data to directly identify particular genes and pathways	Joris (2022)

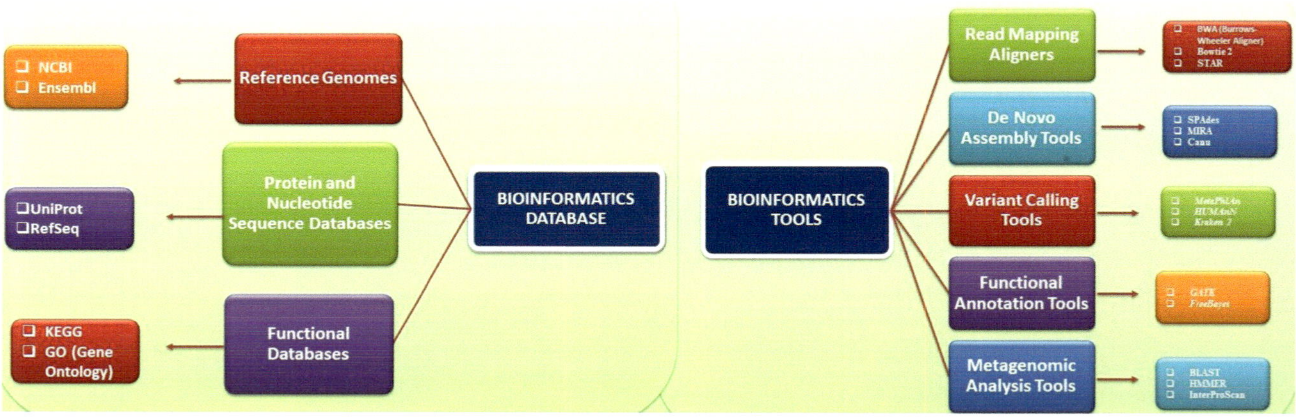

FIGURE 4.1 Bioinformatics tools and database used for shotgun sequence data analysis.

4.4.1 Bioinformatics Tools and Database Used for Shotgun Sequence Data Analysis

NGS has revolutionized the genome sequencing of microbial isolates, including those derived from humans, aquatic animals, plants, and various ecological niches (Das et al., 2021; Kumari et al., 2023; Das et al., 2022; Dash et al., 2023; Deb et al., 2020; Singh et al., 2024; Deb et al., 2020a, 2020b). Researchers have developed a wide range of computational tools and techniques for analyzing bacterial whole-genome sequences (Deb et al., 2024). Similarly, there has been a significant increase in computational biology approaches tailored for metagenome analysis. To comprehend the underlying biological information, shotgun sequencing produces a large quantity of short sequence reads that must be assembled and examined. To perform this, a range of bioinformatics databases and technologies are mentioned in Figure 4.1.

4.4.2 Functional Analysis of Shotgun Sequence Data

Interpreting raw DNA sequence reads to deduce the biological functions and activities of the organisms and genes contained in the sample is known as functional analysis of shotgun sequencing data. In environmental genomics, metagenomics, and microbiome research, shotgun sequencing is frequently employed to analyze microbial communities without the need for cultivation (Silva et al., 2016;

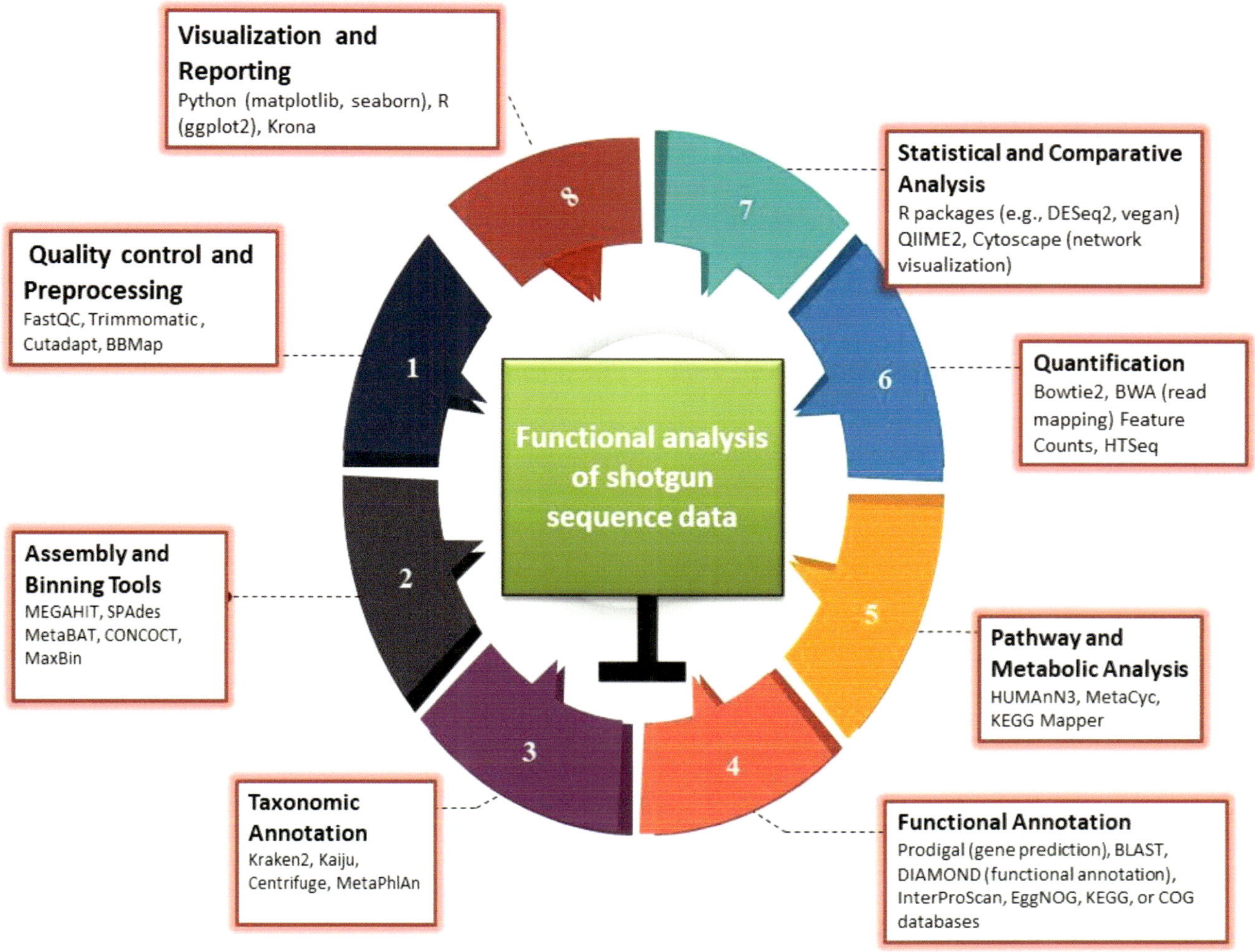

FIGURE 4.2 Functional analysis of shotgun sequences.

Sharpton, 2014). The functional analysis of the shotgun sequence is illustrated in Figure 4.2.

In environmental genomics, metagenomics, and microbiome research, shotgun sequencing is frequently employed to identify intricate microbial communities without cultivation. The best speed and accuracy combination for decontaminating a human gut microbiome using short readings with the help of HoCoRT was discovered utilizing BioBloom, Bowtie2 in end-to-end mode, and HISAT2. Although accuracy was sacrificed, Kraken2 continuously showed the fastest speed. Similarly, Bowtie2 was significantly slower than the other techniques in the case of an oral microbiota. It seems tough to detect human host reads for long reads. The best accuracy in this instance was obtained by combining Kraken2 and Minimap2, which identified 59% of human reads (Rumbavicius et al., 2023; Li, 2018). In end-to-end mode, HoCoRT employing Bowtie2 demonstrated to be significantly quicker and marginally more precise than the specialized DeconSeq tool. This method is difficult to use, though, in samples where large quantities of host DNA and low microbe abundance lead to ineffective and expensive sequencing (Yap et al., 2020).

4.4.3 Challenges Associated with Shotgun Sequence Data Analysis

The examination of shotgun sequence data poses a number of difficulties. Massive volumes of short sequence reads are produced by shotgun sequencing, which increases the computational demands of data processing and storage (Quince et al., 2017). Accurate assembly and analysis may be hampered by short read lengths and possible sequencing mistakes. It can be challenging to correctly assemble repetitive portions of the genome, which can result in gaps and mistakes in the finished assembly. It can be challenging to put short readings together into continuous sequences (contigs), particularly for large and intricate genomes (Logsdon et al., 2020). A high-quality reference genome, which may not be available for every organism, is frequently necessary for accurate analysis. Accurately identifying genetic changes can be difficult, particularly in complicated locations. It takes specific knowledge and bioinformatics tools to interpret big datasets and derive biologically significant findings (Merelli et al., 2014).

Large dataset analysis requires a substantial amount of processing power and storage space. To correctly assemble and interpret the vast number of short sequencing reads produced by shotgun sequencing, complex computing procedures are needed. Shotgun sequencing of complex genomes, including a large number of repetitive DNA sequences, can result in misassembly (Pop, 2004). Significant structural differences may be difficult for short-read sequencing to identify. The breadth and scale of research efforts may be constrained by the high costs of shotgun sequencing. Results from different studies can be inconsistent because different bioinformatics pipelines handle raw sequencing data slightly differently (SoRelle et al., 2020).

4.5 RECENTLY DEVELOPED ML TOOLS AND TECHNIQUES USED FOR MICROBIOME ANALYSIS

By allowing researchers to handle the complexity and high dimensionality of microbiome datasets, recent ML tools and approaches have greatly increased microbiome analysis (Abbasi et al., 2024). Microbiological biomarkers are associated with diseases identified using deep learning models like convolutional neural networks (CNNs) and recurrent neural networks (RNNs). SVMs and Random Forests help with classification and predictive modeling. For a comprehensive understanding of host–microbiome interactions, ML techniques combine data from metagenomics, metabolomics, and transcriptomics.

For the purpose of visualizing complicated datasets, dimension reduction methods like principal component analysis (PCA) and uniform manifold approximation and projection (UMAP) are essential (Allaoui et al., 2020). Predictive algorithms based on ML associate the composition of microbiomes with conditions like ASD, diabetes, and inflammatory bowel disease (IBD). Autoencoders assist in revealing obscure patterns unique to medical disorders. ML makes it possible to construct probiotics or synthetic microbial consortia for focused therapeutic interventions, especially in the treatment of dysbiosis. To stratify microbial populations and find new patterns, clustering methods such as k-means and hierarchical clustering are used. For sophisticated taxonomic and functional analysis, ML is integrated into tools such as DeepMicro, Microbiome Analyst, and upgraded versions of QIIME 2 (Ghannam and Techtmann, 2021). With continuous efforts to enhance the interpretability of ML models and their applicability in clinical and ecological microbiome research, these developments place a strong emphasis on data integration, visualization, and predictive modeling.

4.6 HOST GENOME CONTAMINATION IN SHOTGUN SEQUENCE

In shotgun metagenomic sequencing, host genome contamination is a common issue, especially when analyzing microbial communities linked to a host organism (such as the soil microbiome or the human gut microbiome) (Quince et al., 2017). Because this contamination causes bias and obscures the microbial signal, it can significantly affect downstream analytical accuracy. Poor DNA extraction techniques, residual host cells or DNA in the sample, or cross-contamination during sample processing could all be the cause of the contamination.

Additionally, it results in false-positive detection of microbial species, skewed functional profiling, incorrect taxonomic classification, and lower sequencing depth for microbial genomes. Bowtie2, BWA-MEM, Kraken2, and bbmap are a few of the bioinformatics tools that can be effectively utilized to eliminate host reads. By carefully assessing the aforementioned criteria and applying pertinent methodologies, it is possible to decrease host genome contamination and obtain precise insights into microbial populations.

4.6.1 Tools and Techniques to Remove Contamination from Shotgun Sequence

The comprehensive sequencing of the genomes of the microorganisms present in a host's surroundings has facilitated innovative approaches to studying the interactions between hosts and microbes. In order to avoid redundant calculations, efficient processing is vital because of the typically large volume of data involved. Moreover, non-microbial sequences could skew further analysis. As a result, eliminating host sequences ought to be given top priority as soon as possible (Bharti & Grimm, 2021).

Alignment tools are frequently used to scan readings against a host genome database in an ad hoc manner for decontamination. Ad hoc methods could result in less-than-ideal performance and be needlessly complex. It is difficult to compare research because there are no established best practices for this process. Below, we have outlined several specific tools designed for eliminating contaminating sequences (Figure 4.3).

DeconSeq: A command-line tool called DeconSeq is intended to identify and remove sequence contamination from genomic and metagenomic datasets. Because DeconSeq's source code contains a modified version of the BWA-SW classifier, it is challenging to modify. Since 2013, the program's code has not been updated, and it only supports single-end Illumina reads (Schmieder & Edwards, 2011).

GenCoF: A graphical user interface called GenCoF allows contaminants to be quickly eliminated from the human genome when metagenomic datasets are restricted to short reads. Because of Bowtie2's close interaction with its source code, extending GenCoF is challenging, and its GUI design precludes scripting. Because of their clear drawbacks, both tools are less suitable for most large-scale datasets (Czajkowski et al., 2019).

HoCoRT: An open-source Python 3 command-line tool called HoCoRT was created. It is designed

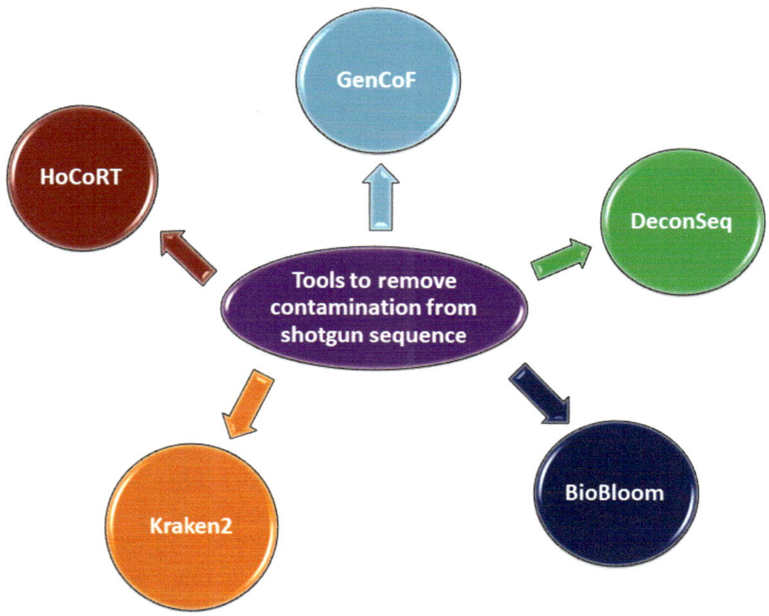

FIGURE 4.3 Major tools for removing contamination from shotgun sequence.

to be user-friendly and can be installed as a Docker container or as a package using Bioconda (Grüning et al., 2018). To distinguish between host and non-host (microbial) sequences, HoCoRT uses a modular pipeline method in addition to well-established classification, mapping, and alignment approaches. Creating new modules that incorporate extra tools can improve HoCoRT. It facilitates both writing and reading, and its integrated alarm systems and assistance features make documentation easy to use (Li et al., 2009).

The classification accuracy and performance of HoCoRT were assessed using a variety of underlying algorithms and configurations on both synthetic and real datasets. The required scripts for generating synthetic datasets and performing performance assessments can be found in the GitHub repository. The synthetic datasets for the human oral microbiome consisted of 50% human host and 50% microbial sequences, while the artificial datasets for the human gut microbiome contained 1% human host and 99% microbial sequences (Benson et al., 2012).

Kraken2: To tackle the problem of high memory usage, Kraken 2 implements two modifications to the data structures and techniques utilized in Kraken 1. While Kraken 1 employed a sorted array of k-mer/LCA pairs categorized by minimizers, Kraken 2 utilizes a compressed hash table that incorporates uncertainty to link minimizers to LCAs (Roberts et al., 2004). This new table requires only a third of the storage space typical of traditional hash tables, though it sacrifices some precision and detail.

Users have the option to select a desired hash table size when using Kraken 2's subsampling method based on hashes, which also reduces the collection of minimizer/LCA pairs in the table. Choosing smaller hash tables leads to reduced classification accuracy, but it also results in decreased memory consumption and increased classification speed.

Other runtime and accuracy enhancements are also included in Kraken 2. Kraken, a new translated search mode compared to nucleotide-based search, 2X, employs a smaller alphabet of amino acids and is more sensitive to viral datasets. Both batch and block-based parsing are used in the like in more current Bowtie 2 versions, the critical portion is employed to enhance thread scaling (Langmead et al., 2019).

BioBloom: BioBloom Tool uses Bloom filtration systems, which are probabilistic data structures with constant time access, to ascertain if components belong to a set (Bloom, 1970). Similar to hash tables, bloom filters deposit a predetermined number of bits for each component into a shared bit array rather than the actual contents. As a result, they consume less memory; however, due to hash collisions in the shared bit array, queries to the filter may result in false membership (hits). By making the filter larger, the false-positive rate (FPR) brought on by these false positives can be controlled. The fast and accurate classification of sequences (FACS) program first employed bloom filters for sequence classification (Stranneheim et al., 2010).

4.7 NANOPORE ADAPTIVE SAMPLING TECHNOLOGY AND MICROBIAL ENRICHMENT

Nanopore sequencing is helping to create a portable and affordable genetic surveillance tool. In contrast to other popular sequencing techniques that depend on DNA synthesis, this approach detects changes in current when DNA strands go through nanopores on the flow cell to determine genetic sequences (Jain et al., 2016). Quicker library preparation using less resources is made possible by the technology. Additionally, it permits special features like adaptive sampling (Loose et al., 2016; Payne et al., 2021).

Before sequencing, traditional technologies like Illumina employ PCR and other techniques for enrichment. The adaptive sampling concept from Oxford Nanopore Technologies (ONT), also referred to as selective sequencing, represents a software-driven approach to enrichment that offers an alternative to traditional lab-based depletion or enrichment techniques. The "ReadUntil" programming interface provides software with the capability to request the expulsion of the molecule currently being sequenced in a specific pore and allows for control over individual pores. This makes it possible to assess whether a molecule is "on target" by examining its initial few hundred bases. By reversing the current across the pore, "off target" molecules can be removed, enabling the pore to take in a new molecule (Martin et al., 2022).

A recent study used two different adaptive sampling approaches to evaluate the efficacy of adapted selection for amplifying low-abundance plasmids in known bacterial isolates. The results showed that even on flow cells that had expired, considerable enrichment could be obtained. Adaptive sampling also speeds up sequencing and improves the quality of freshly constructed plasmids. But when targeted and non-target sequences are located in similar geographic areas, this work also highlights the difficulties with adaptive sampling (Jens-Uwe et al., 2024).

4.8 APPLICATIONS OF MICROBIOME STUDIES IN LIVESTOCK AND WILDLIFE ANIMALS

As the importance of animal medicine in the thorough study of animal diseases is becoming more widely recognized, researchers are now concentrating more on the prevention and management of infectious diseases from an animal health viewpoint rather than just considering animals as tools in clinical medicine. The idea of "One Medicine" was first introduced in the 19th century as a result of Rudolf Virchow's support for veterinary pathology, which strengthened the connection between animal and human medicine (Bardgett et al., 2008). In addition to acknowledging the interdependence of human, animal, and ecosystem health, Karl Friedrich Meyer helped to establish the "One Health and One Medicine" concepts, emphasizing the value of cooperation among medical professionals to advance the health and well-being of all species (Pospischil, 2015).

A notable application of microbes in animal health is the creation of vaccines. Edward Jenner, often referred to as the "father of immunology," developed a strategy that considered animal health, successfully preventing and ultimately eradicating smallpox in the 18th century using vaccinia vaccination (Saied et al., 2021). In 1880, Louis Pasteur began his research on rabies, which led to the development of a canine rabies vaccine as a preventative measure after an effective human rabies vaccine was created (Atlas, 2012). Pasteurella was added to the cholera vaccine to successfully stop the spread of cholera in chickens (Atlas, 2012). In order to stop malaria from spreading, Koch used bed nets to keep mosquitoes away and quinine to stop *Plasmodium* from synthesizing proteins (Vonaesch et al., 2018).

Only a limited number of animal species cows, goats, chickens, pigs, and sheep, account for the majority of the animal products consumed by humans, such as meat, milk, and eggs (FAO, 2009; Teferra, 2022; Pulina et al., 2018). When comparing the chicken, a bird, to the other four species, which are mammals, it is evident that each has its own distinct evolutionary history that may be quite extensive. Even among the mammals, there are noticeable differences in their digestive systems, as well as in physiological traits such as lifespan, growth rates, behavior, and reproduction (Reece & Rowe, 2017).

Ruminant animals possess a multi-chambered gastrointestinal tract (reticulum, omasum, and abomasum) that functions to ferment plant materials, while monogastric mammals have a straightforward stomach composed of a single chamber. There are significant differences between these two categories. Since much of an individual's performance is influenced by the health and functioning of their digestive system (Celi et al., 2017), it is indeed a critical foundational element in animal production.

Consequently, most microbiota research has focused primarily on this aspect, with research on the reproductive tract following in importance (Heil et al., 2019). But in recent years, a growing number of these studies have also addressed public health and environmental concerns, specifically those pertaining to the spread of food-borne illnesses (Black et al., 2021), greenhouse gas emissions (Min et al., 2020), and the emergence of antibiotic resistance (Zalewska et al., 2021).

4.9 PROSPECTS AND POTENTIALS OF MICROBIOME STUDIES

Human microbiota is varied, differs from person to person and from site to site of the body, and is vital to human health. The human body's bacteria are vital to disease processes, general health, and immune system performance. The study of the human microbiome has been made easier by NGS and its metagenomic uses. This development makes it possible for scientists to examine the makeup of microorganisms in the human body and identify the genes these germs express for different purposes.

Since the human gut is the most diverse and densely inhabited place, it has been the subject of more research than other areas. According to preliminary studies, IBD, colon cancer, diabetes, and atopy are among the illnesses that may be linked to disturbances in the gut microbiota (Kim et al., 2013). In therapeutic contexts, bacteria are vital to human health. The value of the bacteria that inhabit the body has been brought to light by international metagenomics investigations of the human microbiome, which could also result in new medicinal applications. The development of bioinformatic algorithms and metagenome sequencing technology will soon ensure easy access to vast amounts of data. However, research would be needed to elucidate, for instance, the diseases brought on by microbiota dysbiosis, the advantages of microbiome control, and the degree of inter- and between-individual microbial heterogeneity before this data could be used clinically.

4.10 CONCLUSION

Computational biology is emerging as a significant tool for studying microbiomes due to its ability to generate, analyze, and understand complex biological data. It plays a crucial role in understanding the microbiome, the vast diversity of microbes, and the influence of such on health, sickness, and general physiological function in diverse habitats, including the human body. Computational biology has revolutionized microbiome research by allowing the massive-scale investigation of microbial communities at previously unprecedented scales in light of the advances experienced in sequencing technologies and developing bioinformatics tools. The complex functional potential and dynamics within microbiomes have been better revealed by combining advanced computational techniques with metagenomics, transcriptomics, and other methods of omics. More and more methods, such as statistical modeling, network analysis, and ML, are being applied to understand the complex interplay between microbial species, host variables, and environmental influences. Biomarker identification, disease prediction associated with the microbiome, and the creation of personalized treatment plans, including microbiome-based therapies, are made possible by computational biology. Standardizing analytical techniques, improving data integration across platforms, and accounting for the tremendous diversity and heterogeneity of microbiomes between individuals and populations are all challenging tasks, however. Notwithstanding these challenges, however, there is much scope to enhance our understanding of microbiome biology and its applications in environmental sciences, agriculture, and medicine through the continuing development of computational tools and algorithms. In summary, computational biology leads the field of microbiome research and is bringing about breakthroughs that will be able to revolutionize everything from sustainable agriculture to customized medicine. As technology advances and our knowledge grows, this combination of computational techniques and microbiome science will surely open up new avenues for improving environmental and human health.

REFERENCES

Abarenkov, K., Nilsson, R. H., Larsson, K. H., Alexander, I. J., Eberhardt, U., Erland, S., . . . & Kõljalg, U. (2010). The UNITE database for molecular identification of fun-gi–recent updates and future perspectives. The New Phytologist, 186(2), 281–285.

Abbasi, N., Nizamullah, F. N. U., Zeb, S., Fahad, M., & Qayyum, M. U. (2024). Ma-chine learning models for predicting susceptibility to infectious diseases based on microbiome profiles. Journal of Knowledge Learning and Science Technology ISSN: 2959–6386 (online), 3(4), 35–47.

Ainsworth, C. (2013). Biology on the high seas: How Eric Karsenti's quest to understand the cell launched a trip around the world. Nature, 501(7465), 20–24.

Allaoui, M., Kherfi, M. L., & Cheriet, A. (2020, June). Considerably improving clustering algorithms using UMAP dimensionality reduction technique: A comparative study. In International conference on image and signal processing (pp. 317–325). Cham: Springer International Publishing.

Atlas, R. M. (2012). One health: Its origins and future. One Health: The Human-Animal-Environment Interfaces in Emerging Infectious Diseases: The Concept and Examples of a One Health Approach, 1–13.

Bardgett, R. D., Freeman, C., & Ostle, N. J. (2008). Microbial contributions to climate change through carbon cycle feedbacks. The ISME Journal, 2(8), 805–814.

Beaudry, M. S., Wang, J., Kieran, T. J., Thomas, J., Bayona-Vásquez, N. J., Gao, B., Devault, A., Brunelle, B., Lu, K., Wang, J. S., Rhodes, O. E., Jr, & Glenn, T. C. (2021). Improved microbial community characterization of 16S rRNA via meta-genome hybridization capture enrichment. Frontiers in Microbiology, 12, 644662. https://doi.org/10.3389/fmicb.2021.644662

Benson, D. A., Karsch-Mizrachi, I., Clark, K., Lipman, D. J., Ostell, J., & Sayers, E. W. (2012). GenBank. Nucleic Acids Res. 40. D48–53. https://doi.org/10.1093/nar/gkr1202

Bharti, R., & Grimm, D. G. (2021). Current challenges and best-practice protocols for microbiome analysis. Briefings in Bioinformatics, 22(1), 178–193.

Black, Z., Balta, I., Black, L., Naughton, P. J., Dooley, J. S. G., & Corcionivoschi, N. (2021). The fate of foodborne pathogens in manure treated soil. Frontiers in Microbiology, 12, 781357.

Bloom, B. H. (1970). Space/time trade-offs in hash coding with allowable errors. Communications of the ACM, 13(7), 422–426.

Bork, P., Bowler, C., de Vargas, C., Gorsky, G., Karsenti, E., & Wincker, P. (2015). Tara Oceans studies plankton at planetary scale. Science, 348(6237), 873–873.

Brum, J. R., Ignacio-Espinoza, J. C., Roux, S., Doulcier, G., Acinas, S. G., Alberti, A., . . . & Sullivan, M. B. (2015). Patterns and ecological drivers of ocean viral communities. Science, 348(6237), 1261498.

Callahan, B., McMurdie, P., Rosen, M. et al. (2016). DADA2: High-resolution sample infer-ence from Illumina amplicon data. Nature Methods, 13, 581–583. https://doi.org/10.1038/nmeth.3869

Celi, P., Cowieson, A. J., Fru-Nji, F., Steinert, R. E., Kluenter, A. M., & Verlhac, V. (2017). Gastrointestinal functionality in animal nutrition and health: New opportunities for sustainable animal production. Animal Feed Science and Technology, 234, 88–100.

Chappidi, S., Villa, E. C., & Cantarel, B. L. (2019). Using Mothur to determine bacterial community composition and structure in 16S ribosomal RNA datasets. Current Protocols in Bio-Informatics, September, 67(1), e83.

Chen, J. W., Wu, J. H., Chiang, W. F., Chen, Y. L., Wu, W. S., & Wu, L. W. (2021). Taxonomic and functional dysregulation in salivary microbiomes during oral car-cinogenesis. Frontiers in Cellular and Infection Microbiology, 11, 663068.

Chen, Y., Sun, C., Yan, Y., Jiang, D., Huangfu, S., Tian, L. (2025). Impact of arbuscular mycorrhizal fungi on maize rhizosphere microbiome stability under moderate drought conditions. Microbiological Research, 290, 127957. https://doi.org/10.1016/j.micres.2024.127957

Clarridge, J. E., III. (2004). Impact of 16S rRNA gene sequence analysis for identification of bacteria on clinical microbiology and infectious diseases. Clinical Micro-Biology Reviews, 17(4), 840–862.

Cole, J. R., Wang, Q., Fish, J. A., Chai, B., McGarrell, D. M., Sun, Y., Brown, C. T., Porras-Alfaro, A., Kuske, C. R., Tiedje, J. M. (2014). Ribosomal database project: Data and tools for high throughput rRNA analysis. Nucleic Acids Research, January, 42(Database issue), D633–D642. https://doi.org/10.1093/nar/gkt1244. Epub 2013 Nov 27. PMID: 24288368; PMCID: PMC3965039.

Czajkowski, M. D., Vance, D. P., Frese, S. A., & Casaburi, G. (2019). GenCoF: A graphical user interface to rapidly remove human genome contaminants from metagenomic datasets. Bioinformatics (Oxford, England), 35(13), 2318–2319. https://doi.org/10.1093/bioinformatics/bty963

Das, L., Deb, S., Arakawa, E., Yamasaki, S., & Das, S. K. (2022). Pufferfish (Tetraodon cutcutia) sampled from a freshwater river serves as an intermediate reservoir of a sucrose nonfermenting variant of violubillerae PS-4. Microbiology Spectrum, February 23, 10(1), e0122121. https://doi.org/10.1128/spectrum.01221-21. Epub 2022 Feb 16. PMID: 35171023; PMCID: PMC8849079.

Das, L., Deb, S., & Das, S. K. (2020). Glutamicibacter mishrai sp. nov., isolated from the coral Favia veroni from Andaman Sea. Archives of Microbiology, 202, 733–745.

Das, L., Deb, S., & Das, S. K. (2021). Description of Acinetobacter kanungonis sp. nov., based on phylogenomic analysis. International Journal of Systematic and Evolutionary Microbiology, 71(6), 004833.

Dash, J., Sethi, M., Deb, S., Parida, D., Kar, S., Mahapatra, S., Minz, A. P., Pradhan, B., Prasad, P., & Senapati, S. (2023). Biochemical, functional and genomic characterization of a new probiotic *Ligilactobacillus salivarius* F14 from the gut of tribes of Odisha. World Journal of Microbiology Biotechnology, April 27, 39(7), 171. https://doi.org/10.1007/s11274-023-03626-z. PMID: 37101059.

De Goeij, J. M., Van Oevelen, D., Vermeij, M. J., Osinga, R., Middelburg, J. J., De Goeij, A. F., & Admiraal, W. (2013). Surviving in a marine desI The sponge loop retains resources within coral reefs. Science, 342(6154), 108–110.

De Vargas, C., Audic, S., Henry, N., Decelle, J., Mahé, F., Logares, R., . . . & Velayoudon, D. (2015). Eukaryotic plankton diversity in the sunlit ocean. Science, 348(6237), 1261605.

Deakin, C. T., Deakin, J. J., Ginn, S. L., Young, P., Humphreys, D., Suter, C. M., Al-exander, I. E., & Hallwirth, C. V. (2014). Impact of next-generation sequencing error on analysis of barcoded plasmid libraries of known complexity and sequence. Nucle-ic Acids Research, 42(16), e129. https://doi.org/10.1093/nar/gku607

Deb, S., Badhai, J., & Das, S. K. (2020a). Draft genome sequences of two vibrio fortis strains isolated fromcoral (Fungia sp.) from the Andaman Sea. Microbiology Resource Announcements, January 2, 9(1), e01225-19. https://doi.org/10.1128/MRA.01225-19. PMID: 31896635; PMCID: PMC6940287.

Deb, S., Badhai, J., & Das, S. K. (2020b). Draft genome sequences of vibrio alginolyticus strain S6-61 and vibrio diabolicus strain S7-71, isolated from corals in the Andaman Sea. Microbiology Resource Announcements, February 20, 9(8), e01465-19. https://doi.org/10.1128/MRA.01465-19. PMID: 32079632; PMCID: PMC7033269.

Deb, S., Basu, J., & Choudhary, M. (2024). An overview of next generation sequencing strategies and genomics tools used for tuberculosis research. Journal of Applied Microbiology, July 2, 135(7), lxae174. https://doi.org/10.1093/jambio/lxae174. PMID: 39003248.

Deb, S., Das, L., & Das, S. K. (2020). Composition and functional characterization of the gut microbiome of freshwater pufferfish (Tetraodon cutcutia). Archives of Microbiology, 202, 2761–2770.

Deb, S., & Das, S. K. (2020a). Draft genome sequence of idiomarina sp: Strain W-5T, isolated from the Andaman Sea. Microbiology Resource Announcements, January 30, 9(5), e01510-19. https://doi.org/10.1128/MRA.01510-19. PMID: 32001569; PMCID: PMC6992873.

Deb, S., & Das, S. K. (2020b). Draft genome sequence of microbacterium oryzae strain MB-10, isolated from a rice field in India. Microbiology Resource Announcements, February 6, 9(6), e01532-19. https://doi.org/10.1128/MRA.01532-19. PMID: 32029561; PMCID: PMC7005126.

Deb, S., & Das, S. K. (2022). Phylogenomic analysis of metagenome-assembled genomes deciphered novel acetogenic nitrogen-fixing Bathyarchaeota from hot spring sediments. Microbiology Spectrum, 10(3), e00352-22. https://doi.org/10.1128/spectrum.00352-22

Debnath, T., Deb, S., & Das, S. K. (2022). Influence of geochemistry in the tropical hot springs on microbial community structure and function. Current Microbiology, 80(1), 4. https://doi.org/10.1007/s00284-022-03118-7

DeSantis, T. Z., Hugenholtz, P., Larsen, N., Rojas, M., Brodie, E. L., Keller, K., Huber, T., Dalevi, D., Hu, P., & Andersen, G. L. (2006). Greengenes, a chimera-checked 16S rRNA gene database and workbench compatible with ARB. Applied and Environemental Microbiology, 72. https://doi.org/10.1128/AEM.03006-05

Faveri, M., Mayer, M. P. A., Feres, M., De Figueiredo, L. C., Dewhirst, F. E., & Paster, B. J. (2008). Microbiological diversity of generalized aggressive periodontitis by 16S rRNA clonal analysis. Oral Microbiology and Immunology, 23(2), 112–118.

FAO. (2009). How to feed the world in 2050. In Proceedings of the expert meeting on how to feed the world in 2050. Rome, Italy: Food and Agriculture Organization of the United Nations.

Gao, Z., Tseng, C. H., Pei, Z., & Blaser, M. J. (2007). Molecular analysis of human forearm superficial skin bacterial biota. Proceedings of the National Academy of Sciences, 104(8), 2927–2932.

Ghannam, R. B., & Techtmann, S. M. (2021). Machine learning applications in mi-crobial ecology, human microbiome studies, and environmental monitoring. Computational and Structural Biotechnology Journal, 19, 1092–1107.

Gilbert, J. A., Jansson, J. K., & Knight, R. (2014). The Earth microbiome project: Successes and aspirations. BMC Biology, 12, 1–4.

Grüning, B., Dale, R., Sjödin, A., Chapman, B. A., Rowe, J., Tomkins-Tinch, C. H., . . . & Bioconda Team. (2018). Bioconda: Sustainable and comprehensive software distribution for the life sciences. Nature Methods, 15(7), 475–476.

Hasan, N. B., Biswas, A., Balaban, M., Mirarab, S., & Bayzid, M. S. (2022). Fast and accurate branch support calculation for distance-based phylogenetic place-ments. In RECOMB international workshop on comparative genomics, May (pp. 33–51). Cham: Springer International Publishing.

Heil, B. A., Paccamonti, D. L., & Sones, J. L. (2019). Role for the mammalian female reproductive tract microbiome in pregnancy outcomes. Physiological Genomics, 51(8), 390–399.

Hentschel, U., Usher, K. M., & Taylor, M. W. (2006). Marine sponges as microbial fermenters. FEMS Microbiology Ecology, 55(2), 167–177.

Hugerth, L. W., & Andersson, A. F. (2017). Analysing microbial community composition through amplicon sequencing: From sampling to hypothesis testing. Frontiers in Microbiology, 8, 1561.

Huttenhower, C., Gevers, D., Knight, R., Abubucker, S., Badger, J. H., Chinwalla, A. T., . . . & McCorrison, J. M. (2012). Structure, function and diversity of the healthy human microbiome. Nature, 486(7402), 207–214.

Hyman, R. W., Fukushima, M., Diamond, L., Kumm, J., Giudice, L. C., & Davis, R. W. (2005). Microbes on the human vaginal epithelium. Proceedings of the National Academy of Sciences, 102(22), 7952–7957.

Jain, M., Olsen, H. E., Paten, B., & Akeson, M. (2016). The Oxford Nanopore MinION: Delivery of nanopore sequencing to the genomics community. Genome Biology, 17, 1–11.

Jens-Uwe, U., Lennard, E., Tanja, P., Birgit, W., Kerstin, S., Torsten, S., & Renard, B. Y. (2024). Nanopore adaptive sampling effectively enriches bacterial plasmids. MSystems, 9(3).

Jenuth, J. P. (1999). The NCBI: Publicly available tools and resources on the web. Bioinformatics Methods and Protocols, 301–312.

Joris, B. R. (2022). Towards more complete metagenomic analyses through circular-ized genomes and conjugative elements (Doctoral dissertation, The University of Western Ontario (Canada)).

Kang, X., Deng, D. M., Crielaard, W., & Brandt, B. W. (2021). Reprocessing 16S rRNA gene amplicon sequencing studies: (Meta) data issues, robustness, and repro-ducibility. Frontiers in Cellular and Infection Microbiology, 11, 720637.

Kaneko, H., Endo, H., Henry, N. *et al.* (2023). Predicting global distributions of eukaryotic plankton communities from satellite data. ISME *Communications,* 3, 101. https://doi.org/10.1038/s43705-023-00308-7

Karsenti, E., Acinas, S. G., Bork, P., Bowler, C., De Vargas, C., Raes, J., . . . & Tara Oceans Consortium. (2011). A holistic approach to marine eco-systems biology. PLoS Biology, 9(10), e1001177.

Kim, B. S., Jeon, Y. S., & Chun, J. (2013). Current status and future promise of the human microbiome. Pediatric Gastroenterology, Hepatology & Nutrition, 16(2), 71–79.

Knowlton, N. (2001). Coral reef biodiversity—habitat size matters. Science, 292(5521), 1493–1495.

Kumari, K., Rawat, V., Shadan, A., Sharma, P. K., Deb, S., & Singh, R. P. (2023). In-depth genome and pan-genome analysis of a metal-resistant bacterium Pseudomonas parafulva OS-1. Frontiers in Microbiology, June 20, 14, 1140249. https://doi.org/10.3389/fmicb.2023.1140249. PMID: 37408640; PMCID: PMC10318148.

Langmead, B., Wilks, C., Antonescu, V., & Charles, R. (2019). Scaling read aligners to hundreds of threads on general-purpose processors. Bioinformatics, 35(3), 421–432.

Lennon, J. T., & Jones, S. E. (2011). Microbialld banks: The ecological and evolutionary implications of dormancy. Nature Reviews Microbiology, 9(2), 119–130.

Li, H. (2018). Minimap2: Pairwise alignment for nucleotide sequences. Bioinformatics, 34(18), 3094–3100.

Li, H., Handsaker, B., Wysoker, A., Fennell, T., Ruan, J., Homer, N., . . . & 1000 Genome Project Data Processing Subgroup. (2009). The sequence alignment/map format and SAMtools. Bioinformatics, 25(16), 2078–2079.

Lima-Mendez, G., Faust, K., Henry, N., Decelle, J., Colin, S., Carcillo, F., . . . & Raes, J. (2015). Determinants of community structure in the global plankton interactome. Science, 348(6237), 1262073.

Logsdon, G. A., Vollger, M. R., & Eichler, E. E. (2020). Long-read human genome sequencing and its applications. Nature Reviews Genetics, 21(10), 597–614.

Loose, M., Malla, S., & Stout, M. (2016). Real-time selective sequencing using nanopore technology. Nature Methods, 13(9), 751–754. https://doi.org/10.1038/nmeth.3930

Martin, S., Heavens, D., Lan, Y., Horsfield, S., Clark, M. D., & Leggett, R. M. (2022). Nanopore adaptive sampling: A tool for enrichment of low abundance species in metagenomic samples. Genome Biology, 23(1), 11.

Merelli, I., Pérez-Sánchez, H., Gesing, S., & D'Agostino, D. (2014). Managing, analysing, and integrating big data in medical bioinformatics: Open problems and future perspectives. BioMed Research International, 2014(1), 134023.

Min, B. R., Solaiman, S., Waldrip, H. M., Parker, D., Todd, R. W., & Brauer, D. (2020). Dietary mitigation of enteric methane emissions fromb ruminants: A review of plant tannin mitigation options. Animal Nutrition, 6, 231–246.

Palmer, C., Bik, E. M., DiGiulio, D. B., Relman, D. A., & Brown, P. O. (2007). Development of the human infant intestinal microbiota. PLoS Biology, 5(7), e177.

Payne, A., Holmes, N., Clarke, T., Munro, R., Debebe, B. J., & Loose, M. (2021). Readfish enables targeted nanopore sequencing of gigabase-sized genomes. Nature Biotechnology, 39(4), 442–450.

Peterson, J., Garges, S., Giovanni, M., McInnes, P., Wang, L., Schloss, J. A., . . . & NIH HMP Working Group. (2009). The NIH human microbiome project. Genome Research, 19(12), 2317–2323.

Polanin, J. R., Hennessy, E. A., & Tsuji, S. (2020). Transparency and reproducibility of meta-analyses in psychology: A meta-review. Perspectives on Psychological Science, 15(4), 1026–1041.

Pop, M. (2004). Shotgun sequence assembly. Advanced Computing, 60, 193–248.

Pospischil, A. (2015). Human and animal health on three continents: A biography of the early life of Karl Friedrich Meyer (1884–1974). Pathogens and Disease, 73(6), ftv039.

Proctor, L. M. (2011). The human microbiome project in 2011 and beyond. Cell Host & Microbe, 10(4), 287–291.

Pulina, G., Milán, M. J., Lavín, M. P., Theodoridis, A., Morin, E., Capote, J., . . . & Caja, G. (2018). Invited review: Current production trends, farm structures, and economics of the dairy sheep and goat sectors. Journal of Dairy Science, 101(8), 6715–6729.

Quast, C., Pruesse, E., Yilmaz, P., Gerken, J., Schweer, T., Yarza, P., Peplies, J., & Glöckner, F. O. (2013). The SILVA ribosomal RNA gene database project: Improved data processing and web-based tools. Nucleic Acids Research, 41(D1), D590–D596.

Quince, C., Walker, A. W., Simpson, J. T., Loman, N. J., & Segata, N. (2017). Shotgun metagenomics, from sampling to analysis. Nature Biotechnology, 35(9), 833–844.

Rajpoot, M., Sharma, A. K., Sharma, A., & Gupta, G. K. (2018). Understanding the microbiome: Emerging biomarkers for exploiting the microbiota for personalized medicine against cancer. In Seminars in cancer biology, October (Vol. 52, pp. 1–8). Academic Press.

Reece, W. O., & Rowe, E. W. (2017). Functional anatomy and physiology of domestic animals. John Wiley & Sons.

Roberts, M., Hayes, W., Hunt, B. R., Mount, S. M., & Yorke, J. A. (2004). Reducing storage requirements for biological sequence comparison. Bioinformatics, 20(18), 3363–3369.

Rognes, T., Flouri, T., Nichols, B., Quince, C., Mahé, F. (2016). VSEARCH: A versatile open source tool for metagenomics. PeerJ, 4, e2584. https://doi.org/10.7717/peerj.2584

Rumbavicius, I., Rounge, T. B., & Rognes, T. (2023). HoCoRT: Host contamination removal tool. BMC Bioinformatics, 24(1), 371.

Saied, A. A., Metwally, A. A., Mohamed, H. M., & Haridy, M. A. (2021). The contribution of bovines to human health against viral infections. Environmental Science and Pollution Research, 28(34), 46999–47023.

Schmieder, R., & Edwards, R. (2011). Fast identification and removal of sequence contamination from genomic and metagenomic datasets. PloS One, 6(3), e17288.

Sedlar, K., Kupkova, K., & Provaznik, I. (2017). Bioinformatics strategies for taxonomy independent binning and visualization of sequences in shotgun metagenomics. Computational and Structural Biotechnology Journal, 15, 48–55.

Sharpton, T. J. (2014). An introduction to the analysis of shotgun metagenomic data. Frontiers in Plant Science, 5, 209.

Silva, G. G. Z., Green, K. T., Dutilh, B. E., & Edwards, R. A. (2016). Super-Focus: A tool for agile functional analysis of shotgun metagenomic data. Bioinformatics, 32(3), 354–361.

Singh, R. P., Sinha, A., Deb, S., & Kumari, K. (2024). First report on in-depth genome and comparative genome analysis of a metal-resistant bacterium Acinetobacter pittii S-30, isolated from environmental sample. Frontiers in Microbiology, April 29, 15, 1351161. https://doi.org/10.3389/fmicb.2024.1351161. PMID: 38741743; PMCID: PMC11089254.

SoRelle, J. A., Wachsmann, M., & Cantarel, B. L. (2020). Assembling and validating bioinformatic pipelines for next-generation sequencing clinical assays. Archives of Pathology & Laboratory Medicine, 144(9), 1118–1130.

Stranneheim, H., Käller, M., Allander, T., Andersson, B., Arvestad, L., & Lundeberg, J. (2010). Classification of DNA sequences using Bloom filters. Bioinformatics, 26(13), 1595–1600.

Straub, D., Blackwell, N., Langarica-Fuentes, A., Peltzer, A., Nahnsen, S., & Kleindienst, S. (2020). Interpretations of environmental microbial community studies are biased by the selected 16S rRNA (gene) amplicon sequencing pipeline. Frontiers in Microbiology, 11, 550420.

Sunagawa, S., Coelho, L. P., Chaffron, S., Kultima, J. R., Labadie, K., Salazar, G., . . . & Velayoudon, D. (2015). Structure and function of the global ocean microbiome. Science, 348(6237), 1261359.

Teferra, T. F. (2022). The cost of postharvest losses in Ethiopia: Economic and food security implications. Heliyon, 8(3).

Thorsen, J., Brejnrod, A., Mortensen, M., Rasmussen, M. A., Stokholm, J., Al-Soud, W. A., . . . & Waage, J. (2016). Large-scale benchmarking reveals false discoveries and count transformation sensitivity in 16S rRNA gene amplicon data analysis methods used in microbiome studies. Microbiome, 4, 1–14.

Toole, D. R., Zhao, J., Martens-Habbena, W., & Strauss, S. L. (2021). Bacterial func-tional prediction tools detect but underestimate metabolic diversity compared to shotgun metagenomics in southwest Florida soils. Applied Soil Ecology, 168, 104129.

Villar, E., Farrant, G. K., Follows, M., Garczarek, L., Speich, S., Audic, S., . . . & Iudicone, D. (2015). Environmental characteristics of Agulhas rings affect interocean plankton transport. Science, 348(6237), 1261447.

Vonaesch, P., Anderson, M., & Sansonetti, P. J. (2018). Pathogens, microbiome and the host: Emergence of the ecological Koch's postulates. FEMS Microbiology Reviews, 42(3), 273–292.

Wang, S., Wang, Y., Liu, H., Li, X., Zhao, J., Dong, Z., . . . & Shao, T. (2022). Using PICRUSt2 to explore the functional potential of bacterial community in alfalfa silage harvested at different growth stages. Chemical and Biological Technologies in Agriculture, 9(1), 98.

Yang, Z., Peng, C., Cao, H., Song, J., Gong, B., Li, L., . . . & Lu, L. (2022). Microbial functional assemblages predicted by the FAPROTAX analysis are impacted by physicochemical properties, but C, N and S cycling genes are not in mangrove soil in the Beibu Gulf, China. Ecological Indicators, 139, 108887.

Yap, M., Feehily, C., Walsh, C. J., Fenelon, M., Murphy, E. F., McAuliffe, F. M., . . . & Cotter, P. D. (2020). Evaluation of methods for the reduction of contaminating host reads when performing shotgun metagenomic sequenolubil the milk microbiome. Scientific Reports, 10(1), 21665.

Zalewska, M., Blazejewska, A., Czapko, A., & Popowska, M. (2021). Antibiotics and antibiotic resistance genes in animal manure consequences of its application in agriculture. Frontiers in Microbiology, 12, 610656.

5 Advancements in Next-Generation Sequencing

A Deep Dive into the Microbiome Research

Jhinuk Basu[1], *Madhusmita Nayak*[2], *Sulata Kumari Sahu*[3], *Amrapali Nandi*[4], *Badrinath Khadanga*[5], *and Sushanta Deb*[6#]

[1]Department of Cancer Chemoprevention, Chittaranjan National Cancer Institute, Kolkata, India

[2]Department of Botany, Government Women's College, Dhenkanal, India

[3]Ravenshaw University, Cuttack, India

[4]Department of Biotechnology and Microbiology, Sister Nivedita University

[5]Department of Botany, Berhampur University, Odisha-760007 INDIA

[6]Department of Veterinary Microbiology and Pathology, College of Veterinary Medicine, Washington State University, Pullman 99164, WA, United States

5.1 INTRODUCTION

The balance of ecosystems and human health is dependent on microbial communities, which are made up of various populations of bacteria, viruses, fungi, archaea, and other microorganisms (Fierer, 2017). With the development of next-generation sequencing (NGS) tools, the study of these communities has undergone substantial change. Conventional methods, such as culture-based techniques, are unable to provide the same level of insight into the structure, function, and dynamics of microbial communities as NGS (Bardgett and Van Der Putten, 2014). Historically, techniques based on culture have been used to examine microbial communities. By using these methods, scientists could isolate and examine microorganisms in carefully regulated lab environments. Since most microorganisms are difficult to culture in the lab, the application of culture-based methods is limited to a few members of microbes present in various ecosystems. Up to 99% of environmental microorganisms are thought to be unculturable using conventional methods, leaving a large information gap about microbial diversity and community structure (Amann et al., 1995).

NGS eliminated the necessity for culture, revolutionizing the field of microbial research. It enables the direct sequencing of genetic material from environmental samples by scientists, offering a thorough picture of the diversity of microbes, including those that cannot be cultured. Our knowledge of the human microbiome has expanded substantially due to the upgradation of NGS technology, which has made it possible to identify and characterize unculturable microorganisms and predict their functions (Scholz et al., 2012). Shotgun metagenomic sequencing, RNA sequencing, and 16S rRNA gene sequencing are important NGS techniques. The selection of NGS methodology for a particular objective might be confusing for researchers and physicians. Strategies like 16S rRNA amplicon sequencing, shotgun metagenomics, and meta-transcriptomics are deemed as vital resources for the analysis of microbial communities in a variety of settings, including the human gut, soil, and seas (Jansson and Hofmockel, 2018; Wensel et al., 2022).

The benefits of NGS techniques include high throughput and resolution, cost effectiveness, independence from culture, and the ability to perform quantitative and comparative analyses. On the contrary, a few challenges are associated with the NGS approach; the intricacy of data analysis is one of the main problems (Van Dijk et al., 2018). Additionally, large-scale NGS data generation necessitates the processing and interpretation of sophisticated bioinformatics tools and knowledge. The accuracy of NGS data can also be impacted by biases in DNA extraction, PCR amplification, and sequencing errors; however, advances in bioinformatics and sequencing technology are continuously addressing these problems (Scholz et al., 2012).

5.2 BRIEF INTRODUCTION TO THE ROLE OF MICROBIOMES IN VARIOUS ECOSYSTEMS (HUMAN HEALTH, AGRICULTURE, ENVIRONMENT)

Ubiquitous in nature, microbes can be found in nearly every habitat and are essential to human existence. The term "microbiome" describes the microbial community and its genetic makeup in a certain setting (Malla et al., 2019). In

#**Corresponding author:** Sushantadeb26@gmail.com (Sushanta Deb)

DOI: 10.1201/9781003544425-5

many ecosystems, microbial communities are essential for maintaining environmental stability, cycling nutrients, and even human health. A broad range of scientific disciplines, including ecology, agriculture, medicine, and biotechnology, depend on an understanding of the makeup and operation of these communities (Bardgett and Van Der Putten, 2014).

By promoting healthy digestion, controlling the immune system, sustaining metabolic processes, and even having an impact on mental health, the human microbiome is essential to preserving health. The microbiome is an essential target for therapeutic interventions and preventive health measures since imbalances in this microbial ecosystem can result in a range of health problems, including infections, metabolic diseases, and mental health concerns (Lynch and Pedersen, 2016). Digestion depends especially on the gut microbiome. Short-chain fatty acids (butyrate, acetate, and propionate) supply energy and control metabolic processes through catabolization of complex proteins, lipids, and carbohydrates that are indigestible. Maintaining general nutritional health also depends on a healthy microbiome, which facilitates the absorption of minerals and vitamins (such as B12 and K) (Flint et al., 2012). The state of the metabolism is significantly impacted by the gut flora. Specific microbial compositions are associated with the development of obesity, insulin resistance, and type 2 diabetes. Research has demonstrated that the microbiomes of fat people differ from those of lean people, and they are better able to absorb energy from food. Altering the gut microbiome with food, probiotics, and prebiotics can have an impact on metabolic health and weight management (Turnbaugh et al., 2006). The immune system is significantly shaped and regulated by the human microbiome. The presence of microbes in the gut lowers the risk of autoimmune illnesses by teaching the immune system to distinguish between foreign and benign molecules. Allergies, asthma, and inflammatory bowel disease (IBD) have all been linked to dysbiosis, or an imbalance in the microbiome. According to Round et al. (2009), the gut microbiota collaborates with immune cells to support anti-inflammatory responses, so as to avert chronic inflammation that may otherwise result in illnesses such as cancer (Round and Mazmanian, 2009).

The gut–brain axis, a term used to describe the enduring relationship between brain function and the gut microbiome, has been revealed by recent research. Neurotransmitters like dopamine and serotonin, which affect mood, stress response, and cognitive function, are produced by microbes in the gut. Disruptions in the microbiota have been related to mental health illnesses such as depression, anxiety, and even neurodegenerative diseases like Alzheimer's (Cryan and Dinan, 2012). As a healthy microbiome produces antimicrobial compounds and outcompetes harmful bacteria for resources, it acts as a barrier against them. Also, it adjusts the immune system to ward off infections. People with dysbiosis may be more prone to infections like *Helicobacter pylori* and *Clostridium difficile*, which have been connected to stomach cancer and ulcerative disorders. Treating such infections by altering the microbiome through dietary modifications or fecal microbiota transplantation (FMT) has gained importance as a therapeutic approach (Smits et al., 2013).

In order to improve soil fertility, boost crop yield, and support plant health, the agricultural microbiome is essential. Plant growth, nutrient uptake, and resistance to environmental stressors are all influenced by microbes, including bacteria, fungi, and archaea, that are found in the soil, in plant roots (rhizosphere), and inside plants (endosphere). It is becoming more widely accepted that using these microbes' power can provide a sustainable substitute for chemical pesticides and fertilizers (Graham et al., 2016).

Soil microbes play a crucial role in the cycling of nutrients by converting organic matter into forms that are absorbed by plants. As they break down organic matter, vital nutrients, including potassium (K), phosphorus (P), and nitrogen (N), are released. For example, ammonia, which plants can use for growth, is created when nitrogen-fixing bacteria like Rhizobium transform atmospheric nitrogen. Similar to this, phosphate-solubilizing bacteria lessen the demand for artificial fertilizers by converting insoluble phosphate into soluble forms that are accessible to plants (Van Der Heijden et al., 2008). Microorganisms known as plant growth-promoting microorganisms (PGPMs) include *Bacillus*, *Pseudomonas*, and mycorrhizal fungi. By promoting nutrient uptake, generating plant hormones like auxins, and enhancing resistance to infections, they promote plant growth. For example, mycorrhizal fungi grow symbiotic relationships with plant roots, lengthening the root system and enhancing nutrient and water intake, especially in soils low in phosphorus (Smith and Read, 2010).

By outcompeting pathogenic organisms for nutrients and generating their own natural antibiotics, microbiomes also aid in the suppression of plant diseases. As biocontrol agents, beneficial microorganisms like *Bacillus subtilis* and *Trichoderma* shield plants from bacterial and fungal illnesses. This encourages more sustainable farming methods and lessens the need for toxic pesticides. For example, *Trichoderma* fungi are commonly employed to shield crops from illnesses that are carried by the soil, like wilt and root rot (Harman et al., 2004). Microbes help plants adapt to abiotic conditions like drought, salinity, and heat. By generating substances like osmoprotectants and exopolysaccharides, which aid plants in surviving in the face of extreme salt concentrations and water scarcity, certain plant-associated microorganisms can cause stress resistance in plants. This is especially crucial for boosting crop resilience in locations prone to climate change-induced stress (Ngumbi and Kloepper, 2016). One approach that shows promise for sustainable agriculture is the use of microbial inoculants, often known as biofertilizers. These inoculants can improve soil fertility, lessen the need for chemical inputs, and encourage healthier crop growth because they are derived from beneficial bacteria and fungi. For instance, mycorrhizal inoculants are applied to various crops to promote nitrogen uptake and drought tolerance, whereas *Azospirillum* and *Rhizobium* are frequently employed as biofertilizers in leguminous crops (Vessey, 2003). Plant growth promotion, disease suppression, stress tolerance, and nutrient cycling are all significantly influenced by the agricultural microbiome. Sustainable agriculture approaches have the potential to decrease reliance on chemical inputs, improve soil health, and boost crop tolerance to environmental stress by harnessing the potential of beneficial bacteria. This promotes environmental sustainability in addition to increasing (Wang et al., 2024).

Microbes are critical to global nutrient cycles such as carbon (C), N, P, and sulfur. As part of the carbon cycle, organic matter is broken down by soil and marine microbiomes, which either release carbon dioxide (CO_2) back into the atmosphere or store it in the soil. Plants may utilize ammonia, which is produced when nitrogen-fixing bacteria like *Rhizobium* and *Azotobacter* transform atmospheric nitrogen. Microbial communities in the ocean, which include cyanobacteria, are important for the production of oxygen and the fixation of carbon in aquatic environments (Falkowski et al., 2008). Microbiomes play an important role in the breakdown of organic matter by dissolving and simplifying dead plants and animals. Organic matter is broken down in the soil by bacteria and fungi, which enhances soil fertility and structure while replenishing the ecosystem with essential nutrients. Microbial decomposers recycle nutrients back into the water column in aquatic ecosystems by breaking down organic detritus (Lladó et al., 2017).

Microbiomes are utilized in the bioremediation process, which uses microorganisms to break down or purify environmental toxins. Certain bacteria, such as *Pseudomonas* and *Bacillus* species, can break down hazardous chemicals like hydrocarbons, insecticides, and heavy metals. Hydrocarbon-degrading bacteria, for example, may repair oil spills in marine habitats, making them important instruments in decreasing environmental harm caused by pollution (Das and Chandran, 2011). In both terrestrial and aquatic habitats, microbial communities are crucial for controlling greenhouse gas emissions. For instance, methane, a strong greenhouse gas, is produced and consumed differently depending on the soil microbiomes. Methanotrophic bacteria eat methane produced by methanogenic archaea in anaerobic environments, reducing its atmospheric release. Similar to this, marine microbes, especially those found in deep-sea sediments, take part in carbon sequestration, which helps control the amount of carbon in the atmosphere (Singh et al., 2010).

Microbiomes contribute to ecosystem biodiversity, which is crucial for stability and resilience. Microbial diversity in soil affects nutrient availability, plant diversity, and general soil health. Microbial diversity in marine habitats influences the cycling of nutrients and ocean production. This microbial diversity contributes to the long-term sustainability of ecosystems by aiding in their recovery from shocks like pollution, drought, or invasive species (Bardgett and Van Der Putten, 2014). Extreme settings, including hot springs, acidic mines, arctic permafrost, and deep-sea hydrothermal vents, are well known microbiome habitats. These extremophiles have potential uses in biotechnology and advance our knowledge of life in harsh environments. For instance, chemosynthesis is a process used by microorganisms in deep-sea habitats to produce energy instead of sunlight. This process allows the microbes to maintain distinct ecosystems even in the absence of sunshine (Akinsemolu, 2018). Ecosystems all throughout the world depend on microbiomes to function. They promote the cycling of nutrients, break down organic matter, provide bioremediation, control greenhouse gas emissions, and support ecosystem stability and biodiversity.

5.2.1 Significance of NGS in Advancing Microbiome Research

The study of microbiomes has been transformed by NGS, which makes it possible to analyze microbial communities in a variety of settings in-depth and at high resolution. NGS enables the identification and characterization of complete microbial communities, including those that are difficult or impossible to culture in the lab, in contrast to traditional culture-based approaches, which can only detect a percentage of microbial species. Our knowledge of the composition, roles, and variety of microbiomes in many ecosystems has greatly expanded as a result of this technical breakthrough (Wensel et al., 2022). The following are some key contributions of NGS to microbiome research.

5.2.1.1 Functional Analysis of Microbial Communities

NGS, particularly metagenomic sequencing, enables researchers to investigate the functional potential of microbial communities by evaluating the genes in a given sample. This makes it possible to identify metabolic processes and roles, including the generation of antibiotics, pathogen resistance, and nutrient cycling. By sequencing RNA instead of DNA, a technique known as metatranscriptomics offers a more dynamic picture of microbial behaviors in real time by revealing the genes and processes that are actively involved in a microbiome (Franzosa et al., 2018).

5.2.1.2 Unbiased Microbial Identification and Diversity Profiling

Microbial diversity represents entire populations of microorganisms that can be sequenced at a given time point. Whole-genome sequencing (WGS) provides information about the complete genetic makeup of the microbiome, whereas methods like 16S amplicon sequencing enable researchers to distinguish between bacteria and archaea in a sample based on their genetic signatures. This makes it possible to find uncommon and unidentified microbial species, increasing our knowledge of the diversity of microbes in a variety of settings, including the human gut, soil, and seas (Quince et al., 2017).

5.2.1.3 Host–Microbiome Interactions

The intricate relationships that exist between microbiomes and their hosts have been greatly aided by NGS. For instance, NGS has shed light on how gut microbiota affects immunity, metabolism, and even mental health in humans through the gut–brain axis. Similarly, NGS has shown how plant-associated microbiomes affect nutrient uptake and disease resistance in agriculture, guiding the development of sustainable farming practices (Marchesi and Ravel, 2015).

5.2.2 Microbial Ecology and Environmental Studies

It is widely acknowledged that NGS has revolutionized microbiome research through metagenomics and the analysis of isolate genomes, facilitated by advanced computational tools and software (Das et al., 2020; Deb, Badhai, et al., 2020b, 2020a; Deb, Das, et al., 2020; Deb and Das, 2020b, 2020a; Das et al., 2021; Deb and Das, 2022; Das et al., 2022; Kujur et al., 2022; Dash et al., 2023; Debnath et al., 2023; Kumari et al., 2023; Singh et al., 2024; Deb et al., 2024). Due to its ability to examine intricate microbial ecosystems in their natural settings, NGS has significantly advanced the field of environmental microbiome research. NGS has been used to identify the roles that microbes play in pollutant degradation, carbon sequestration, and nutrient cycling in soil, ocean, and freshwater ecosystems. NGS, for instance, has demonstrated the enormous diversity of microorganisms found in marine habitats and their vital roles in the global biogeochemical cycles, particularly the nitrogen and carbon cycles (Gilbert et al., 2014).

5.2.3 Precision Medicine and Microbiome Therapies

NGS has enabled precision medical methods by offering precise insights into individual microbiomes. Understanding the structure and functioning of the gut microbiome can help guide tailored treatment plans for illnesses such as IBD, obesity, and metabolic disorders. It also allows for the development of microbiome-based therapeutics such as probiotics, prebiotics, and FMT (MetaHIT Consortium et al., 2010).

5.2.3.1 High-Throughput and Cost-Effective

NGS technologies are high-throughput, which means they can sequence millions of DNA fragments simultaneously, offering a full picture of microbial populations swiftly and affordably. This has democratized microbiome research, making it accessible to a broader range of scientific disciplines, including medical research and environmental biology (Goodwin et al., 2016).

NGS has revolutionized microbiology by enabling the direct sequencing of microbial genomes from environmental samples, bypassing the limitations of traditional culturing. This advancement has significantly expanded our understanding of microbial diversity and function. Multi-omics approaches, such as metagenomics, metatranscriptomics, and metaproteomics, offer a comprehensive knowledge of microbial communities, providing insights into their composition, function, and interactions. Metagenomics investigates the genetic potential of a population, whereas metatranscriptomics and metaproteomics focus on the active genes and proteins that are being expressed. The greater capacity, deeper sequencing, and improved accuracy of NGS have been extremely beneficial to these multi-omics approaches. By combining these techniques, a more comprehensive understanding of microbial ecosystems and their roles in various processes, such as human health and environmental bioremediation, could be achieved. Additionally, the reconstruction of metabolic networks and the identification of key regulatory pathways within microbial ecosystems could be possible. NGS-based microbiome studies could also contribute to the development of personalized medicine approaches, where treatments can be tailored to the specific microbial composition of an individual.

NGS has revolutionized microbiology by eliminating the need for traditional culturing by enabling the rapid and efficient sequencing of whole microbial genomes directly from environmental samples (Wensel et al., 2022). This discovery has led to a significant advance in our understanding of the diversity and function of microbes. This chapter highlights the importance of NGS for characterizing unculturable microbes and its advantages over traditional culture methods, and also covers the fundamentals of NGS, including metagenomic, metatranscriptomics, and metaproteomics and their applications in understanding disease pathogenesis and advancing clinical care. All these three approaches are powerful tools for studying microbial communities, providing insights into their composition, function, and interactions. Metagenomics analyzes the genetic material of all organisms in a sample, revealing the diversity of species and their potential metabolic (Riesenfeld et al., 2004). Metatranscriptomics, on the other hand, examines the active genes being expressed within the community, offering a snapshot of the metabolic processes occurring at a given moment (Tyagi and Katara, 2024). While metaproteomics focuses on the proteins being produced, providing direct evidence of the functional activities of the microbial community (Zhao et al., 2024). By combining these techniques, a comprehensive understanding of microbial ecosystems and their roles in various processes, such as human health and environmental bioremediation, can be achieved.

5.2.4 The Role of NGS in Sequencing Entire Microbial Genomes from a Sample, Bypassing the Need for Culturing

Characterizing common microorganisms suffers from multiple challenges, like identifying unknown microorganisms, their interactions, and their ambiguous functions in natural environments (Nandy and Kapley, 2024). The study of microbial communities is now possible without the limitations of conventional culturing techniques because of NGS, which has completely changed the science of microbiology (Bajaj et al., 2024). Our knowledge of microbial diversity and function has greatly increased because of NGS, which is distinguished by its high-throughput and parallel sequencing capabilities (Satam et al., 2023). Metagenomics is the study of direct sequencing of whole microbial genomes collected from environmental samples (Bragg and Tyson, 2014). A metagenomic investigation involves directly removing organisms from their native environments and analyzing their genomes (Rodrigues and De Carvalho, 2022).

5.2.5 Limitations of Culturing

Conventional laboratory approaches estimate that a great percentage of bacteria are unculturable. Our capacity to investigate their genetic make-up and ecological significance has been hampered by this constraint. The process of cultivating microorganisms can be tedious and time-consuming; it frequently takes weeks or months to produce enough biomass for sequencing.

5.3 MULTI-OMIC INSIGHTS INTO MICROBIOMES

Microbiomes, the complex communities of microorganisms inhabiting various environments, play pivotal roles in human health, agriculture, and environmental processes (Chaudhari et al., 2025). Understanding the composition, function, and interactions within these microbial ecosystems is crucial for addressing global challenges such as infectious diseases, climate change, and sustainable agriculture. Multi-omics, a comprehensive approach integrating metagenomics, metatranscriptomics, and metaproteomics, provides a powerful framework for unraveling the intricate complexities of microbiomes.

5.3.1 Metagenomics Workflow

Metagenomics, a powerful tool for studying microbial communities directly from environmental samples, involves a series of interconnected steps. First, environmental samples, such as soil, water, or human tissues, are collected, ensuring proper preservation and handling, minimizing contamination. Subsequently, the genetic material, typically DNA or RNA, is extracted from the sample using specialized techniques that disrupt cell walls and membranes while preserving the nucleic acids (Felczykowska et al., 2015). The extracted genetic material is then subjected to sequencing using high-throughput sequencing technologies, which generate millions or billions of short DNA or RNA sequences (Ambardar et al., 2016). These sequences, known as reads, are subsequently processed through a series of computational steps to analyze the microbial composition and function of the sample (Figure 5.1). Quality control measures are employed to filter out low-quality reads and ensure data accuracy (Hempel et al., 2022). Next, the reads are assembled into longer contiguous sequences, or contigs, which represent fragments of microbial genomes (Paszkiewicz and Studholme, 2010). These contigs are then binned into groups based on their taxonomic affiliation or functional characteristics, using computational methods that consider factors such as GC content, tetranucleotide frequency, and coverage depth (Saeed and Halgamuge, 2009). Once binned, the contigs are annotated to identify genes and their potential functions, often using homology-based approaches or machine learning algorithms (Maranga et al., 2023). Taxonomic profiling is performed to determine the abundance and diversity of microbial species present in the sample, leveraging databases of known microbial genomes and sequences (Ames et al., 2013). Functional profiling is conducted to analyze the metabolic pathways and functional capabilities of the microbial community, identifying genes involved in various biological processes (Zhou et al., 2022). The generated data is then interpreted and visualized to gain insights into the microbial composition, diversity, and function within the studied environment (Vanwonterghem et al., 2014).

5.3.1.1 Advantages of NGS for Microbial Genomics

NGS has revolutionized the field of microbial genomics by offering several distinct advantages. One of the most significant benefits is the ability to directly sequence DNA collected from environmental samples, such as soil, water, or human tissues. This eliminates the need for prior cultivation, allowing us to study microbial communities in their natural habitats. Moreover, NGS has enabled the emergence of metagenomics, a powerful approach for investigating the genetic material extracted directly from environmental samples (Bajaj et al., 2024). By sequencing the entire DNA content of a sample, the diversity and composition of microbial communities can be explored without the constraints of isolation and culturing. Additionally, NGS platforms are renowned for their high throughput and speed, generating massive amounts of sequence data in a short period (Schmidt and Hildebrandt, 2017). This enables the simultaneous analysis of thousands or even millions of microbial genomes, accelerating scientific research and discovery. Furthermore, NGS data can be used for de novo assembly, even for organisms without a reference genome. This is particularly valuable for studying unculturable or poorly characterized microorganisms, expanding our understanding of microbial diversity. Finally, NGS offers strain-level resolution, which could allow to differentiate between closely related strains of the same microbial species (Lefkowitz, 1975). This capability provides a more precise understanding of microbial diversity and ecology, facilitating the study of microbial evolution, adaptation, and interactions. In summary, NGS has transformed microbial genomics by providing direct sequencing capabilities, enabling metagenomics studies, offering high throughput and speed, facilitating de novo assembly, and enabling strain-level resolution. These advantages collectively empower to explore the complexities of microbial communities and their roles in various ecosystems.

5.3.2 Metatranscriptomics

Metatranscriptomics is a powerful technique that provides clarity of the active genes and functions within microbial communities. Unlike metagenomics, which analyzes the genetic potential of a community, metatranscriptomics focuses on the genes that are being expressed at a given time (Shakya et al., 2019). This allows understanding

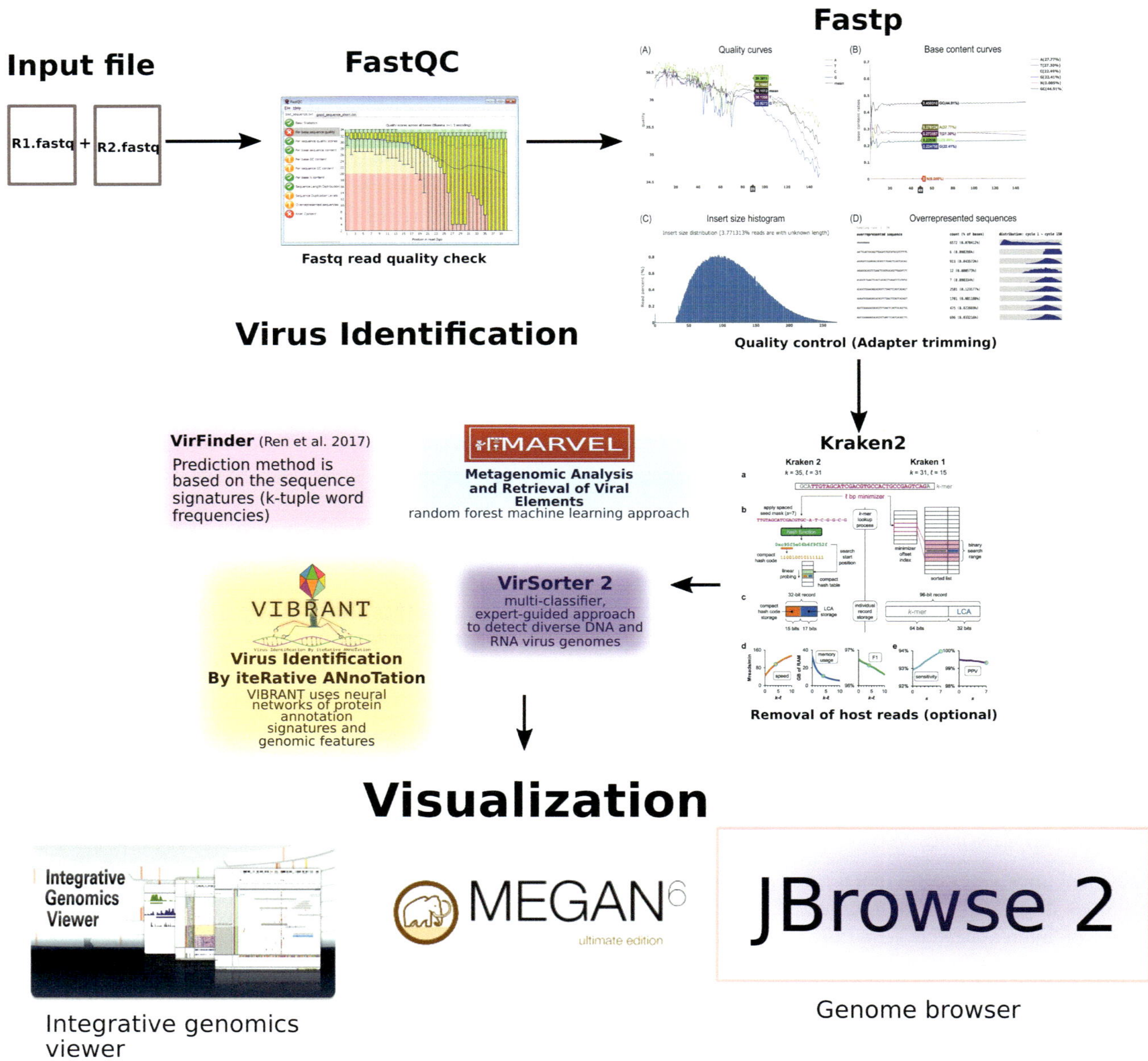

FIGURE 5.1 Schematic representation of the bioinformatics/computational pipeline for short-read metagenomic data analysis.

the metabolic processes, interactions, and responses of microbes to their environment. By analyzing the RNA transcripts present in a sample, this technique provides a better understanding of the genes that are actively being transcribed and translated into proteins. This information offers valuable insights into the metabolic pathways and functions that dominate the community. Furthermore, metatranscriptomics can help to identify the interactions between different microbial species within a community (Bikel et al., 2015). By analyzing the expression of genes involved in communication, competition, or cooperation, we can gain insights into the complex dynamics of microbial communities. Additionally, metatranscriptomics can be used to study how microbial communities respond to environmental changes. By comparing the gene expression profiles of communities under different conditions, the genes and pathways that are involved in stress responses, adaptation, and survival can be identified. This information is crucial for understanding how microbial communities adapt to changing environments and for developing strategies for managing microbial ecosystems.

5.3.2.1 Metatranscriptomics Workflow

Metatranscriptomics involves several key steps to analyze the active genes and functions within a microbial community. First, a representative sample of the microbial community of interest is collected from various environments, such as soil, water, air, or human tissues. The collected sample is then subjected to RNA extraction, which involves isolating total RNA, including messenger RNA (mRNA), ribosomal

RNA (rRNA), and transfer RNA (tRNA) (Martinez et al., 2016). To focus on mRNA, which carries the genetic information for protein synthesis, rRNA and tRNA are often depleted using specific methods. The isolated mRNA is then converted into complementary DNA (cDNA) using reverse transcriptase (Stewart et al., 2010). The cDNA library is subsequently sequenced using high-throughput sequencing technologies, generating millions or billions of short DNA sequences (reads) (Di Bella et al., 2013). Finally, the sequencing data is analyzed using bioinformatics tools to identify and quantify the RNA transcripts present in the sample (Shakya et al., 2019). This involves aligning the reads to reference genomes or transcriptomes, or assembling them de novo to identify novel transcripts (Raghavan et al., 2022). The abundance of each transcript can be used to infer the relative expression levels of the corresponding genes, providing valuable insights into the active metabolic pathways and functions within the microbial community.

5.3.2.2 Advantages of NGS for Metatranscriptomics

NGS has revolutionized the field of metatranscriptomics, offering several significant advantages over traditional sequencing methods. It has diverse applications in microbial ecology, including studying community structure, species interactions, responses to environmental changes, and molecular mechanisms of microbial pathogenesis. By analyzing gene expression, metatranscriptomics provides insights into core metabolic processes, communication, competition, cooperation, stress responses, adaptation, survival, virulence, colonization, and immune evasion, ultimately aiding in the identification of potential drug targets for infectious diseases (Edlund et al., 2018).

5.3.3 Metaproteomics

Metaproteomics is a complementary technique to metatranscriptomics that provides insights into the proteins actually being produced by a microbial community. While metatranscriptomics analyzes the RNA transcripts, metaproteomics focuses on the proteins that are translated from those transcripts. This allows to study the functional activities of microbial communities in greater detail. The wide variation in protein abundance can make it difficult to detect low-abundance proteins, necessitating advanced techniques and deep sequencing. Additionally, proteins can undergo various post-translational modifications, such as phosphorylation, acetylation, and glycosylation, which can alter their mass and identification, posing challenges for detection and quantification. Furthermore, the analysis of metaproteomics data can be computationally intensive, especially for large datasets, requiring efficient bioinformatics tools and computational resources. The accuracy of metaproteomics analysis also depends on the availability of high-quality reference proteomes for the microbial community of interest. In the absence of complete reference proteomes, de novo protein identification can be used, but it may introduce additional challenges and uncertainties.

5.3.3.1 Metaproteomics Workflow

Metaproteomics involves several key steps to analyze the proteins produced by a microbial community. First, a representative sample of the microbial community of interest is collected from various environments, such as soil, water, air, or human tissues. The collected sample is then subjected to protein extraction, which involves isolating total proteins from the cells using methods like lysis buffers, sonication, or enzymatic digestion (Kuruvilla, 2017). To analyze the extracted proteins, they are typically digested into smaller peptides using proteolytic enzymes, such as trypsin. The resulting peptide mixture is then analyzed using mass spectrometry, which measures the mass-to-charge ratio of molecules (Zhang et al., 2010). Two common types of mass spectrometry used in metaproteomics are Liquid Chromatography-Mass Spectrometry (LC-MS) and Matrix-Assisted Laser Desorption/Ionization (MALDI)-Time-of-Flight (TOF) Mass Spectrometry. The mass spectrometry data is subsequently analyzed using bioinformatics tools to identify and quantify the proteins present in the sample by comparing observed peptide masses to theoretical peptide masses derived from protein databases (Chen et al., 2017). The abundance of each protein can be inferred from the abundance of its corresponding peptides, providing valuable insights into the functional activities of the microbial community.

5.3.3.2 Advantages of NGS for Metaproteomics

Metaproteomics deals with the large-scale study of proteins that are expressed within the microbes, providing critical insights into their functionalities such as metabolism, enzyme levels, and protein–protein interactions (Wang et al., 2016; Lohmann et al., 2020). Analyzing the microbial proteome reveals the functional dynamics of ecosystems, such as nutrient cycling, secondary metabolite production, and response to various environmental stress factors. However, due to its heavy reliance on robust genomic and metagenomic data for interpretation of protein expression accurately, NGS plays a transformative role by providing comprehensive genomic and metagenomic reference databases for protein identification and integrating multi-omics analyses. In the following section, we outline the specific advantages of NGS in metaproteomics studies.

5.3.3.2.1 Reference Metagenome Generation for Protein Identification

The shotgun metagenomic sequencing platform enables whole genome sequencing from environmental samples, producing comprehensive metagenomic datasets. Such datasets serve as reference databases to be utilized in the metaproteomics analyses providing the genetic blueprints of the microbiome. Metaproteomic studies employ the identification of proteins from extracted samples through matching mass spectrometry data against predicted protein sequences derived from metagenomes. Lack of reference genomes, from unculturable microbes can severely limit protein identification, since traditional culture-based approaches cannot fully capture the diversity of microbes (Amann et al., 1995). This is mitigated by the generation of reference genomes by NGS. For instance, the human gut microbiome studies utilize the NGS-generated metagenomes allowing researchers to map peptides to particular microbial taxa, revealing their functional contributions to digestion or immune modulation (Lohmann et al., 2020).

5.3.3.2.2 Increased Detection of Low-Abundance Proteins

One of the challenges in metaproteomics studies is the detection of low-abundant proteins, which is critical in understanding specific microbial functions. NGS is important in enhancing the sensitivity of metaproteomics by producing high-coverage metagenomic data. This increases the probability of identifying rare and low-abundance microbial species and associated proteins. NGS ensures that even rare community member microbes, producing significant proteins, are represented in the reference database. This is particularly valuable in complex ecosystems such as soil, marine, and other environments, with high microbial diversity, and low-abundance species and plays important roles in processes like nutrient cycling (Wang et al., 2016).

5.3.3.2.3 Integration of Metaproteomics with Multi-Omic Approaches

Metaproteomics can be integrated with metagenomics and metatranscriptomics, which gives a holistic understanding of the microbiome functions. For example, metagenomics helps identify the genetic potential of a microbial community, metatranscriptomics reveal actively transcribed genes, while metaproteomics confirms the protein expression and its functionality. NGS facilitates this integration through high-resolution sequencing data for all three levels, enabling correlation among the presence of genes, their expression, and protein activity. Such multi-omics approach, powered by NGS, has been key in environmental microbiome research, such as those in agriculture, and oceans, where it promotes elucidating complex interactions such as plant growth and carbon sequestration (Falkowski et al., 2008; Chen et al., 2017).

5.3.3.2.4 Enhanced Resolution of Post-Translational Modifications

Post-translational modifications such as phosphorylation, acetylation, glycosylation, etc., can be detected utilizing metaproteomics analyses and are important for understanding microbial physiology and environmental responses (Wang et al., 2016). NGS data provides detailed metagenomic sequences, which permits bioinformatic tools to predict potential post-translational modification sites within proteins. For instance, integrating NGS-based metagenomic data with mass spectrometry-derived metaproteomics, researchers can identify peptide modifications and link them with specific microbial taxa or environmental cues such as stress responses in extreme environments such as deep-sea hydrothermal vents (Akinsemolu, 2018).

5.3.3.2.5 Addressing Computational Challenges with Resilient Databases

Metaproteomic-based data analyses are computationally intense, which demands extensive reference proteomes to match the mass spectrometry data accurately. Such a challenge is adequately addressed by NGS by generating large-scale, high-quality metagenomic datasets, serving as comprehensive reference proteomes. Advanced pipelines such as those using MaxQuant or MetaLab leverage NGS data to improve the accuracy and pace of protein identification. Additionally, NGS helps in reducing the dependency on generic databases, which sometimes can be incomplete, by providing sample-specific metagenomes, crucial for studying diverse and unculturable microbial communities (Van Dijk et al., 2018).

5.3.3.2.6 Tackling Cost Effectiveness and Scalability

The cost-effectiveness and scalability associated with NGS technologies enable researchers to sequence large and complex microbial communities with enhanced efficiency. This is crucial for metaproteomics analyses, which often demand large metagenomic data in order to identify proteins across diverse samples. The low cost ensures that metaproteomics studies can be carried out on a large scale, such as in the Human Microbiome Project or Earth Microbiome Project. Thousands of samples are analyzed under these projects to understand microbial functions in health and environmental contexts (Scholz et al., 2012).

5.3.3.2.7 Combining NGS with Metaproteomics for Functional Annotation, Pathway Analysis, and Reproducibility

NGS facilitates metaproteomics studies by enabling functional annotation and pathway analysis of metagenomic sequences. Mapping of proteins to annotated genes in NGS-derived metagenomes helps in identifying active metabolic pathways and biological processes. For instance, in soil microbiomes, combining NGS-based metagenomics with metaproteomics has uncovered key enzymes involved in nitrogen fixation and carbon cycling, offering insights into ecosystem-level processes (Van Der Heijden et al., 2008; Falkowski et al., 2008). Additionally, NGS addresses reproducibility and standardization challenges in metaproteomics, where variations in sample preparation, protein extraction, and data analysis can introduce inconsistencies. By providing standardized, high-quality, sample-specific metagenomic datasets, NGS minimizes biases from generic databases, improving the reliability and reproducibility of metaproteomic results. Advances in NGS protocols and bioinformatics tools further standardize data processing, ensuring consistent outcomes in metaproteomics research (Scholz et al., 2012).

5.4 QUALITY CONTROL IN MICROBIOME RESEARCH

The rapid expansion in microbiome research contributed to exponential advancements in NGS technology; however, a challenge remains in measuring variations across experiments, different laboratories, and other factors. A special project called the microbiome quality control (MBQC) has been launched to identify and quantify such variations and evaluate quality control strategies (The Microbiome Quality Control Project Consortium et al., 2017). Ensuring the quality of NGS data in microbiome research makes an important component that ensures data reliability and reproducibility of the study despite the complex intricacies of the sequencing data and analysis. Batch effects, technical variabilities, and sparsity in data are some of the challenges obscuring true biological signals (Voigt et al., 2015; Tsilimigras and Fodor, 2016; Ling et al., 2022). Therefore, quality control constitutes an important in microbiome research analysis.

Key sources of variability in microbiome studies include DNA extraction protocols, sequencing platforms, and bioinformatics pipelines (Bharti and Grimm, 2021). Differences in library preparation, sequencing depth, and sample handling can introduce biases that significantly impact results. For example, discrepancies in DNA extraction efficiency or contamination during sample handling can skew microbial composition profiles (Yuan et al., 2012). Additionally, sequencing depth variability can result in zero-inflated datasets, where rare taxa may be entirely missed in some samples (Ling et al., 2021). To address these challenges, microbiome researchers employ various QC strategies, including the use of positive and negative controls, standardized protocols, and batch effect correction methods (Kim et al., 2017; Zhou et al., 2023). Positive controls help to verify the accuracy of microbial detection, while negative controls identify potential contamination. 16S rRNA sequencing and metagenomic shotgun sequencing (MSS) are two popular NGS technologies for their use in effective taxonomic identification. Sequences are clustered into operational taxonomic units (OTUs) or amplicon sequence variants (ASVs) in order to efficiently analyze them, with tools such as DADA2, Mothur, and QIIME giving the standardized data processing pipelines (Schloss et al., 2009; Caporaso et al. 2010; Ranjan et al., 2016; Callahan et al., 2016). MSS generally sequences the total genome, providing higher resolution and identification of a broader range of microbes, such as protozoa, fungi, and viruses (Dalal et al., 2020). Factors such as zero inflation, overdispersion, high dimensionality, and sample heterogeneity makes the microbiome count data challenging to analyze. Zero inflation is the prevalence of zero counts in several samples, reflecting either true absences or technical errors, including missed low-abundance taxa (Lin and Peddada, 2020). In overdispersion, data variability exceeds tentative distributions, and higher dimensionality, with thousands of potential taxa, complicating statistical analyses (Badri et al., 2018; Jiang et al., 2021). Additionally, heterogeneity in the samples coming from varying depths of sequences creates challenges in data comparisons across samples, where larger samples can potentially generate higher artificial reads for non-differentially abundant features, resulting in spurious inference (Evans et al., 2018). Rigorous quality control measures are necessary to combat such challenges, and over the years, several normalization techniques such as ComBat-seq, RUV4, and ComBat are devised to address batch effects and minimize technical variation (Johnson et al., 2007; Leek, 2014; Hornung et al., 2016; Zhang et al., 2020). Linear mixed models, Bayesian Dirichlet-multinomial regression (BDMMA), and surrogate variable analysis (SVA) facilitate batch-specific confounders (Ritchie et al., 2015; Dai et al., 2019; Wang and LêCao, 2020). Methods such as singular value decomposition (SVD) and batch mean centering (BMC) aid in dimensionality minimization and reduce the effects of batch effect, providing more accurate interpretations of microbiome data (Alter et al., 2000; Sims et al., 2008). Proper designing of the experiments and use of appropriate statistical corrections are crucial for minimizing biases and obtaining reliable results from microbiome studies.

It is essential in microbial studies to have reproducibility for translating insights and findings into health applications. The Microbiome Quality Control (MBQC) baseline study (MBQC-base) has been established to measure variability in taxonomic profiling across 15 laboratories and 9 bioinformatics protocols using blinded specimens from human stool, chemostats, and artificial microbial communities (The Microbiome Quality Control Project Consortium et al., 2017). This project exemplifies efforts to augment quality control in microbiome research.

5.5 ASSEMBLY

NGS technology has opened avenues for metagenomics by facilitating direct analysis of genomes from environmental samples, including various microbial communities and their functional capacities (Oulas et al., 2015; Choi et al., 2017). Assembly is a crucial step in metagenomic analysis which converts sequencing reads into longer contiguous sequences for gene catalogue construction. Mainly, co-assembly and individual assembly methods are the two common approaches used for this. Co-assembly aggregate reads from multiple samples, allowing recovery of low-abundance genes by increasing coverage. But while combining diverse strains can result in fragmented assemblies because of the complex paths in de Bruijn graphs (Alneberg et al., 2018). Contrary to this, individual assembly minimizes variation at strain-level and yields more complete assemblies for abundant genomes. However, this approach can introduce redundancy, requiring clustering to create nonredundant gene sets (Steinegger, 2018). For example, the Baltic Sea gene catalogue, constructed from 2.6 billion reads, used co-assembly followed by gene calling on contigs >1000 bp. While it supports functional profiling, only ~10% of shotgun reads from typical samples map to annotated genes, potentially due to fragmentation (Alneberg et al., 2018). This highlights the need for optimized assembly strategies to improve gene catalogues. Comparisons of assembly methods, such as those for the Baltic Sea, provide valuable insights into enhancing metagenomic resources (Alneberg et al., 2018; Markussen et al., 2018; Alneberg et al., 2020).

5.6 QUALITY FILTRATION AND ERROR CORRECTION OF NGS READS

The reads generated during NGS is expressed as base call accuracy, also known as Phred quality scores, have error rates typically ranging from 0.1% to 10% (Glenn, 2011). High coverage enables error correction through de novo or reference-based assembly, while low coverage complicates distinguishing sequencing errors from true biological variation. Errors can be reduced through quality filtering where low quality reads are either discarded or truncated, paired-end reads are merged, or amplicon reads are clustered (Schloss et al., 2011; Bokulich et al., 2013). SHERA, FLASH, PANDAseq, and PEAR are examples of paired-end read merging tools that facilitate enhanced sequence prediction, though only PANDAseq integrates quality filtering (Rodrigue et al., 2010; Magoč and Salzberg, 2011; Masella et al., 2012; Zhang et al., 2014). Error correction methods such as PyroNoise, AmpliconNoise, and pre-clustering techniques address errors using abundance-based approaches, but their dependence on flowgrams limits applicability to technologies like Illumina (Huse et al., 2010; Quince et al., 2011). Neglecting error distribution tails risks inflated diversity estimates.

5.7 TAXONOMIC CLASSIFICATION: TOOLS LIKE KRAKEN, QIIME, AND METAPHLAN

Taxonomic classification of metagenomic sequencing (MGS) reads typically involves comparing unknown reads to a reference database. Several classifiers such as Kraken and MetaPhlAn are utilized to perform such applications based on sample complexity and environmental conditions. Kraken uses exact k-mer alignment with a lowest common ancestor (LCA) approach and allows for confidence threshold adjustments to refine classifications, while MetaPhlAn relies on clade-specific marker genes mapped via Bowtie 2 (Segata et al., 2012; Langmead and Salzberg, 2012; Wood and Salzberg, 2014). Braken is an add-on of Kraken that estimates abundance in taxon using the genomic information but often underutilized (Lu et al., 2017). Selecting database preferences greatly affects the accuracy of classification with larger and comprehensive databases such as NCBI RefSeq improves results but needs significant computational applications. Compromising on the number of databases used and the lack of detailed database descriptions can further complicate reproducibility (Parks et al., 2021; O'Leary et al., 2016). Confidence thresholds in Kraken significantly influence outputs, with reported values ranging from 0.05 to 0.5, though systematic evaluation of these thresholds remains limited (McIntyre et al., 2017; Walsh et al., 2018). Standardized practices and thorough benchmarking are essential for consistent metagenomic analyses (Sczyrba et al., 2017; Beghini et al., 2021).

A recent study has highlighted the variability in metagenome analysis tool performance, specifically in speed, accuracy, and functionality (Lindgreen et al., 2016). CLARK and Kraken have emerged fastest and accurate for taxonomic classification. Their results align closer to the expected distributions than others. However, there is a dearth of tools that can accurately predict bacterial phyla relative abundances. This emphasizes the need for researchers to select tools based on specific project demands and metrics. Beyond taxonomic distribution, tools capable of functional analysis showed marked differences. For example, EBI and MG-RAST successfully predicted functional shifts across three categories, whereas QIIME/PICRUSt captured two despite their taxonomic focus. MEGAN and LMAT had limited success, predicting one or no functional shifts. This variability underscores the importance of evaluating tools not only for taxonomic profiling but also for their ability to infer functional changes, a critical component in metagenomic studies. These findings serve as a guide for researchers aiming to optimize tool selection for comprehensive metagenome analysis.

5.8 FUNCTIONAL ANNOTATION

Recent advancements in NGS technology have facilitated extensive metagenomic studies, elucidating the genetic diversity within microbial consortia across diverse environments. While it offers potential to characterize uncultivated microbes through metagenomic analysis, connecting specific genes to disease phenotypes and outcomes remains challenging because of limitations in functional annotation of newly discovered microbial genes (Pasolli et al., 2019). Functional annotation is pivotal for the identification of disease-associated protein dynamics, detection of antibiotic resistance genes, and development of therapeutic strategies (Baric et al., 2016; Yan et al., 2022). But experimental approaches are laborious and find it difficult to scale with the rapid growth of uncharacterized sequences. For instance, 40–60% of human gut microbiome protein-coding genes remain functionally uncharacterized, depending on the method used, and 27.3% of Unified Human Gastrointestinal Genome genes cannot be mapped to functional databases (Vatanen et al., 2018; Pasolli et al., 2019). Conventional annotation techniques rely on homology across databases such as UniProt and RefSeq and have limited coverage. The emerging deep learning approaches predict functions at scale and are capable of overcoming dependencies on sequence homology (Sureyya Rifaioglu et al., 2019; Gligorijević et al., 2021). Metagenome annotation involves assembling reads into metagenome-assembled genomes (MAGs) or directly mapping reads to reference sequences (Kim et al., 2016; Quince et al., 2017). Tools such as HUMAnN2/3, MG-RAST, and MetaLAFFA utilize read mapping, while METABOLIC utilizes profile-oriented functional prediction by using hidden Markov models (HMMs) (Franzosa et al., 2018; Zhou et al.,

2019). MAG reconstruction promotes novel genome discovery through assembling reads into contigs along with clustering them into bins, which reveals microbial diversity and functions (Strous et al., 2012; Alneberg et al., 2014). But only 1% of UnitProtKB protein sequences have experimental functional annotations (Das and Orengo, 2016). Gene Ontology (GO) and Kyoto Encyclopedia of Genes and Genomes (KEGG) are more popularly used tools to enhance functional classifications and intermethod comparisons (Kanehisa et al., 2017; The Gene Ontology Consortium, 2019). In order to deal with the annotation gaps, a novel workflow combines the de novo reconstruction of the genome alongside taxonomic profiling and deep learning annotations via DeepFRI (Edlund et al., 2018)Validated against eggNOG, this method exhaustively annotates genes, though less specifically than homology-based tools (Huerta-Cepas et al., 2019). Using DIABIMMUNE cohort metagenomes, the workflow supports pangenome reconstruction and tracks shared genes, contributing significantly to metagenomics pipelines and understanding microbial gene functions. The integration of deep learning-based functional annotation tools such as DeepFRI with traditional homology-oriented techniques increases the scalability and accuracy of metagenomic analyses, bridging the sequence-to-function gap. Such advancements help develop comprehensive insights into microbial diversity and functional roles, further making way for improved understanding of microbial gene functions, antibiotic resistance, and novel therapeutic strategies.

5.9 INTEGRATION OF MULTI-OMICS APPROACHES

A multidimensional framework can be generated by combining genomic, transcriptomic and proteomic data, which can be pivotal in deeply exploring microbial ecosystems and their functional roles in health and disease. Genomic research is based on the foundational blueprints of microbial communities and elucidating their composition with potential capabilities. Transcriptomics complements this by quantifying active gene expression, which helps understand how microbial and host genes interact during various physiological or pathological conditions (Conesa et al., 2016). Transcriptomic data of the host further helps in the identification of active pathways during certain conditions while metatranscriptomics data provides insight into gene activities of microbes, ultimately revealing the community's functional characteristics and environmental responses (Shakya et al., 2019). Another crucial layer of information is added through proteomics capturing the gene products of these genomic as well as transcriptomics activities, which reflect the cellular functionality and post-translational modifications. Popularly, it utilizes mass spectrometry, helping to decode the dynamic microbial proteome, thus bridging the gap between genetic potential and actual biological functions (Peters et al., 2019). Integrating all the omics layers with parameters such as diet and clinical outcomes expands our contextual understanding of microbial communities in their ecosystems. Diet composition significantly shapes microbiome, and dietary interventions provide opportunities to understand microbiome-driven physiological changes (Marconi et al., 2018; Johnson et al., 2020). Clinical outcomes, supported by electronic health records, are crucial links between microbiome characteristics and disease conditions, making it possible to identify the genotype–phenotype relationships. This holistic, multi-omics approach advances our ability to elucidate the complexities of microbial ecosystems, paving the way for targeted interventions and therapeutic developments. Next, we have discussed case studies in this section which have successfully utilized the integrated omics approach.

Organohalide respiration (OHR) is crucial in the global halogen cycle and bioremediation. Challenges such as low growth yields and limited genetic accessibility in organohalide-respiring bacteria (OHRB), omics approaches—genomics, transcriptomics, and proteomics—are important for understanding their unique biology. A study utilizing *Sulfurospirillum multivorans*, *Dehalococcoides mccartyi*, and *Dehalobacter restrictus* have revealed critical pathways, such as a quinone-independent respiratory chain in *D. mccartyi*, hydrogen metabolism, and corrinoid biosynthesis (Türkowsky et al., 2018). By integrating multi-omics datasets across diverse conditions, researchers have mapped regulatory mechanisms and organohalide-respiratory chains, enhancing our understanding of OHRB's metabolic strategies. This integrative approach not only informs ecological roles but also advances bioremediation efforts, demonstrating the power of omics in tackling complex microbial processes. Another study aims to explore the local microbial landscape in non-smokers with early-stage lung adenocarcinoma, and the results reveal a significant microbial dysbiosis in comparison with the healthy controls (Sun et al., 2024). Additionally, there was a reduction in the diversity of species and potential pathogenic microbial involvement. Multi-omics analyses elucidated the host-microbiota interactions and identified a notable correlation between *Methyloversatilis discipulorum* and GOLM1, revealing prognostic significance. A diagnostic model utilizing six bacterial markers accurately differentiated ES-LUAD patients from healthy controls, offering insights into disease mechanisms and potential early diagnostic applications.

5.10 CHALLENGES IN NGS-BASED MICROBIOME RESEARCH

For the minimization of errors and obscure trends in microbiome studies, a well-designed experimental outline is important. The hypotheses should have a strong foundation through literature surveys and validation using pilot studies, reducing uncertainties and experimental failures. Studies can be geared up and improved by a rational experimental design, data processing, and reduction in confounding effects, ensuring more reliable outcomes (Honaas et al., 2016). Another important parameter is the selection of an appropriate sample size, specifically in clinical studies. Variations in microbial load among different biological replicates under similar conditions can obscure weak biological signals, specifically when effect sizes are unknown or small. Smaller sample sizes often fail to represent population-based outcomes accurately. In order to avoid such biases, a proper sample size must be ascertained depending upon the statistical principles and remain consistent during the entire study (Bhalerao and Kadam, 2010; Vandeputte et al., 2017). Setting up proper control is essential in order to differentiate true signals from spurious signals, and the experiments should contain non-manipulated samples with the targeted interventions. However, identification of suitable controls is particularly challenging, specifically in clinical trials where microbial composition is influenced by parameters such as age, gender, ethnicity, diet, and genotype. In animal studies, additional factors like housing conditions, breeding practices, and coprophagy are important confounders. Detailed documentation of these variables in metadata can help mitigate their effects during downstream analyses (Dugan et al., 2014; Goodrich et al., 2014; Laukens et al., 2016; Moore and Stanley, 2016; Martin et al., 2018). In cross-sectional experiments, where healthy subjects are compared with diseased subjects, they are easier to design and perform, but fail often due to the additive or multiplicative effects of variables. On the contrary, longitudinal studies can track similar samples under controlled conditions and give statistically robust results. However, meticulous planning is crucial in order to ensure consistent sampling times and is limited by the availability of reliable downstream analysis methods (Fritz et al., 2013; Caruana et al., 2015; Gilbert and Lynch, 2019). Metadata plays a pivotal role in microbiome research and serves as an information catalogue for all the experimental samples. They facilitate avoiding false interpretations, highlight the impact of individual parameters, and are integral to modern statistical tools. Comprehensive metadata increases the reliability of analyses and provides a base for robust interpretations of study outcomes.

5.11 ETHICAL CONSIDERATIONS IN MICROBIOME RESEARCH

In the realm of microbiome research, several ethical challenges emerge that need careful consideration. These challenges are critical to ensuring the integrity and societal trust in scientific findings.

In the case of microbiome research, transparency in sharing data and analytical workflows is essential for enhancing the integrity of research. This practice allows for independent scrutiny and validation of findings, which is vital for preventing misuse of data and ensuring responsible conduct in research (Park et al., 2022). Data sharing in microbiome research presents significant challenges that impact the field's advancement. While large-scale initiatives like the Human Microbiome Project have promoted data sharing, fundamental issues persist in metadata quality and reusability (Integrative and Research Network Consortium, 2019). A critical challenge lies in the inconsistency of terminology across research platforms, with studies revealing over 3,000 different column descriptors for similar metadata elements. Significant data sparsity contributes to heterogeneity in this context, with important variables like age and sample site location showing alarmingly high levels of incompleteness (81.6% and 94%, respectively). The research community has responded with various standardization initiatives. The Genomics Standards Consortium works to establish uniform data standards, while platforms like METAGENOTE and MGnify offer standardized analytical pipelines to enhance data accessibility and analysis (Mitchell et al., 2019; Vangay et al., 2021). These challenges reflect broader issues in modern biomedical research, including concerns about data privacy, competitive pressures in academia, and the impact of workforce turnover on data consistency. The emergence of Common Data Elements and Microbiome Metadata Standards represents ongoing efforts to address these challenges through community-driven solutions (Weber et al., 2018). The National Microbiome Data Collaboration and platforms like Nephele continue to develop tools and standards to support these initiatives (Mayer et al., 2024). However, successful implementation requires a delicate balance between standardization requirements and the practical constraints faced by researchers in the field.

The ethical implications of personalized medicine, particularly concerning microbiome data, are significant due to privacy concerns associated with the identification and re-identification of individuals. Research indicates that microbiome data can be stable over time, allowing for potential linking of personal records, which raises privacy risks in clinical settings (Cho, 2021). The increasing availability of human microbiome association studies (HMAS) data further increases these risks, as summary statistics can inadvertently reveal sensitive information about individuals (Ahmed and Hens, 2022). Ethical challenges also arise in the context of precision psychiatry, where microbiome data may inform interventions but also necessitate stringent privacy protections to prevent misuse (Coppens et al., 2022). Moreover, the need for regulatory frameworks, akin to those for genomic data, is emphasized to ensure responsible handling of microbiome information while promoting its benefits for health and well-being (Callaway, 2015).

Because microbiomes may be contaminated with human DNA, they may unintentionally disclose private information. Because microbiomes are unique, they may provide private information about a person's diet, ethnicity, or general health. The advent of publicly accessible microbiome DNA databases, which might include recognizable human DNA, heightens this worry. According to a study headed by Jonathan Allen, contamination is still an issue even after attempts to eliminate human DNA from microbiome databases (Franzosa et al., 2015). This increases the likelihood that information about a person's microbiome could be used to identify them, particularly when paired with other private data like age and health.

Researchers warn that even if it is unlikely, it is still possible to identify someone based only on their microbiota. Microbiome data management is currently "a little bit of a Wild West," according to Curtis Huttenhower, a computational biologist at the Harvard T. H. Chan School of Public Health in Boston, Massachusetts. As science develops, he highlights the importance of acknowledging that our microbiomes are extremely distinct. In the year 2013, a group of scientists demonstrated that it was feasible to identify five individuals who had participated anonymously in the international 1,000 Genomes project by employing cross-referencing techniques with a genealogy database that also encompassed ages, geographical locations, and surnames (Gymrek et al., 2013). If made public and paired with metadata like the donor's age, sex, and location, pseudonymized or anonymized data may violate the GDPR (General Data Protection Regulation) and cause privacy violations. Although technologies for cleaning human (host) DNA data have been developed to facilitate open data publications, they may inadvertently remove non-host DNA and do not completely remove human DNA. Malicious actors could use this partially cleaned data to track down sensitive medical information, including genetic mutations and health hazards, and identify donors (Irving and Clarke, 2019). In order to trace longitudinal data in the metadata fields, the first step in creating a FAIR (Findable, Accessible, Interoperable, Reusable) database for human microbiome data is to create unique identifiers, such as participant or sample collection site pseudonyms. It is impossible to eliminate these vital, unique identifiers. The general question is whether the current GDPR privacy and data laws and the recommendations for data transparency made by the FAIR principles can coexist in a world where donor privacy is protected, and scientific data reuse and transparency are fully maximized ("The National Microbiome Data Collaborative Data Portal," 2022). Ensuring GDPR compliance is crucial before implementing FAIR principles to protect the privacy of donors while promoting scientific transparency and data reuse. Transparency in study procedures and informed consent are crucial. Microbiome study participants need to understand the possible privacy hazards and the intended use of their data. To manage microbiome data ethically, consistent procedures and ethical criteria are required (Huttenhower et al., 2023).

5.12 CONCLUSION

Microbiome research has a fundamental implication in maintaining ecological balance and affecting both abiotic and biotic processes, ranging from nutrient cycling and immune modulation. Gradual advancements in the NGS technology have revolutionized our understanding of microbes, their communities, and their genetic and functional diversities. Targeted gene sequencing, such as 16S rRNA and metagenomic technologies, has made it possible to precisely identify microbial taxa and functional pathways. Despite all this, challenges remain because of the biases based on sample handling, protocols, and bioinformatics analysis. Additionally, variations in microbial profiles, inconsistencies in sample processing protocols and sample handling makes reproducibility and reliability difficult. Moreover, ethical considerations, such as safeguarding donor privacy and keeping compliance with regulations, including GDPR, are crucial in fostering responsible microbiome studies. Balancing privacy concerns with the principles of data transparency and reusability will be key to future advancements. These challenges highlight the need for a more rigorous approach to experimental design and data analysis. Standardization of laboratory procedures, sequencing technologies, and computational methods is essential for ensuring reliable, reproducible outcomes that will drive future breakthroughs in microbiome science. Addressing these challenges will unlock the true potential of microbiome research, with far-reaching implications for human health, disease prevention, and environmental sustainability.

REFERENCES

Ahmed, Eman, and Kristien Hens. 2022. "Microbiome in Precision Psychiatry: An Overview of the Ethical Challenges Regarding Microbiome Big Data and Microbiome-Based Interventions." *AJOB Neuroscience* 13 (4): 270–86. https://doi.org/10.1080/21507740.2021.1958096.

Akinsemolu, Adenike A. 2018. "The Role of Microorganisms in Achieving the Sustainable Development Goals." *Journal of Cleaner Production* 182 (May): 139–55. https://doi.org/10.1016/j.jclepro.2018.02.081.

Alneberg, Johannes, Christin Bennke, Sara Beier, et al. 2020. "Ecosystem-Wide Metagenomic Binning Enables Prediction of Ecological Niches from Genomes." *Communications Biology* 3 (1): 119. https://doi.org/10.1038/s42003-020-0856-x.

Alneberg, Johannes, Brynjar Smári Bjarnason, Ino De Bruijn, et al. 2014. "Binning Metagenomic Contigs by Coverage and Composition." *Nature Methods* 11 (11): 1144–46. https://doi.org/10.1038/nmeth.3103.

Alneberg, Johannes, John Sundh, Christin Bennke, et al. 2018. "BARM and BalticMicrobeDB, a Reference Metagenome and Interface to Meta-Omic Data for the Baltic Sea." *Scientific Data* 5 (1): 180146. https://doi.org/10.1038/sdata.2018.146.

Alter, Orly, Patrick O. Brown, and David Botstein. 2000. "Singular Value Decomposition for Genome-Wide Expression Data Processing and Modeling." *Proceedings of the National Academy of Sciences* 97 (18): 10101–6. https://doi.org/10.1073/pnas.97.18.10101.

Amann, R. I., W. Ludwig, and K. H. Schleifer. 1995. "Phylogenetic Identification and in Situ Detection of Individual Microbial Cells without Cultivation." *Microbiological Reviews* 59 (1): 143–69. https://doi.org/10.1128/mr.59.1.143-169.1995.

Ambardar, Sheetal, Rikita Gupta, Deepika Trakroo, Rup Lal, and Jyoti Vakhlu. 2016. "High Throughput Sequencing: An Overview of Sequencing Chemistry." *Indian Journal of Microbiology* 56 (4): 394–404. https://doi.org/10.1007/s12088-016-0606-4.

Ames, Sasha K., David A. Hysom, Shea N. Gardner, G. Scott Lloyd, Maya B. Gokhale, and Jonathan E. Allen. 2013. "Scalable Metagenomic Taxonomy Classification Using a Reference Genome Database." *Bioinformatics* 29 (18): 2253–60. https://doi.org/10.1093/bioinformatics/btt389.

Badri, Michelle, Zachary D. Kurtz, Richard Bonneau, and Christian L. Müller. 2018. "Shrinkage Improves Estimation of Microbial Associations under Different Normalization Methods." *Preprint, Bioinformatics*, August 31. https://doi.org/10.1101/406264.

Bajaj, Tanvi, Sayantan Bhattacharya, and Juhi Gupta. 2024. "Microbialenomics: The Changing Technological Landscape of Microbiology via NGS." In *Microbiology–2.0 Update for a Sustainable Future*, edited by Juhi Gupta and Akarsh Verma. Springer Nature Singapore. https://doi.org/10.1007/978-981-99-9617-9_14.

Bardgett, Richard D., and Wim H. Van Der Putten. 2014. "Belowground Biodiversity and Ecosystem Functioning." *Nature* 515 (7528): 505–11. https://doi.org/10.1038/nature13855.

Baric, Ralph S., Sean Crosson, Blossom Damania, Samuel I. Miller, and Eric J. Rubin. 2016. "Next-Generation High-Throughput Functional Annotation of Microbial Genomes." *mBio* 7 (5): e01245-16. https://doi.org/10.1128/mBio.01245-16.

Beghini, Francesco, Lauren J. McIver, Aitor Blanco-Míguez, et al. 2021. "Integrating Taxonomic, Functional, and Strain-Level Profiling of Diverse Microbial Communities with bioBakery 3." *eLife* 10 (May): e65088. https://doi.org/10.7554/eLife.65088.

Bhalerao, Supriya, and Prashant Kadam. 2010. "Sample Size Calculation." *International Journal of Ayurveda Research* 1 (1): 55. https://doi.org/10.4103/0974-7788.59946.

Bharti, Richa, and Dominik G. Grimm. 2021. "Current Challenges and Best-Practice Protocols for Microbiome Analysis." *Briefings in Bioinformatics* 22 (1): 178–93. https://doi.org/10.1093/bib/bbz155.

Bikel, Shirley, Alejandra Valdez-Lara, Fernanda Cornejo-Granados, et al. 2015. "Combining Metagenomics, Metatranscriptomics and Viromics to Explore Novel Microbial Interactions: Towards a Systems-Level Understanding of Human Microbiome." *Computational and Structural Biotechnology Journal* 13: 390–401. https://doi.org/10.1016/j.csbj.2015.06.001.

Bokulich, Nicholas A., Sathish Subramanian, Jeremiah J. Faith, et al. 2013. "Quality-Filtering Vastly Improves Diversity Estimates from Illumina Amplicon Sequencing." *Nature Methods* 10 (1): 57–59. https://doi.org/10.1038/nmeth.2276.

Bragg, Lauren, and Gene W. Tyson. 2014. "Metagenomics Using Next-Generation Sequencing." In *Environmental Microbiology*, edited by Ian T. Paulsen and Andrew J. Holmes, vol. 1096. Methods in Molecular Biology. Humana Press. https://doi.org/10.1007/978-1-62703-712-9_15.

Callahan, Benjamin J., Paul J. McMurdie, Michael J. Rosen, Andrew W. Han, Amy Jo A. Johnson, and Susan P. Holmes. 2016. "DADA2: High-Resolution Sample Inference from Illumina Amplicon Data." *Nature Methods* 13 (7): 581–3. https://doi.org/10.1038/nmeth.3869.

Callaway, Ewen. 2015. "Microbiomes Raise Privacy Concerns." *Nature* 521 (7551): 136. https://doi.org/10.1038/521136a.

Caporaso, J. Gregory, Justin Kuczynski, Jesse Stombaugh, et al. 2010. "QIIME Allows Analysis of High-Throughput Community Sequencing Data." *Nature Methods* 7 (5): 335–6. https://doi.org/10.1038/nmeth.f.303.

Caruana, Edward Joseph, Marius Roman, Jules Hernández-Sánchez, and Piergiorgio Solli. 2015. "Longitudinal Studies." *Journal of Thoracic Disease* 7 (11): E537.

Chaudhari, M., S. Verma, and S. Deb. 2025. "Insights into Molecular Interplay in Tuberculosis–COVID-19 Co-Infection via Integrated Multi-Omics Strategies." *Journal* 8 (4): 41. https://doi.org/10.3390/j8040041.

Chen, B., D. Zhang, X. Wang, et al. 2017. "Proteomics Progresses in Microbial Physiology and Clinical Antimicrobial Therapy." *European Journal of Clinical Microbiology & Infectious Diseases* 36 (3): 403–13. https://doi.org/10.1007/s10096-016-2816-4.

Cho, Jae-Chang. 2021. "Human Microbiome Privacy Risks Associated with Summary Statistics." *PLoS One* 16 (4): e0249528. https://doi.org/10.1371/journal.pone.0249528.

Choi, Jinlyung, Fan Yang, Ramunas Stepanauskas, et al. 2017. "Strategies to Improve Reference Databases for Soil Microbiomes." *The ISME Journal* 11 (4): 829–34. https://doi.org/10.1038/ismej.2016.168.

Conesa, Ana, Pedro Madrigal, Sonia Tarazona, et al. 2016. "A Survey of Best Practices for RNA-Seq Data Analysis." *Genome Biology* 17 (1): 13. https://doi.org/10.1186/s13059-016-0881-8.

Coppens, Frederik, Lene Lange, and Kathleen D'Hondt. 2022. "Aligning Access to Microbiome Data and Privacy Considerations for Better Solutions for Health and Wellbeing of Society and Environments." *Annales des Mines-Réalités industrielles* 2022 (3): 41–44.

Cryan, John F., and Timothy G. Dinan. 2012. "Mind-Altering Microorganisms: The Impact of the Gut Microbiota on Brain and Behaviour." *Nature Reviews Neuroscience* 13 (10): 701–12. https://doi.org/10.1038/nrn3346.

Dai, Zhenwei, Sunny H. Wong, Jun Yu, and Yingying Wei. 2019. "Batch Effects Correction for Microbiome Data with Dirichlet-Multinomial Regression." *Bioinformatics* 35 (5): 807–14. https://doi.org/10.1093/bioinformatics/bty729.

Dalal, Nishu, Rekha Jalandra, Minakshi Sharma, et al. 2020. "Omics Technologies for Improved Diagnosis and Treatment of Colorectal Cancer: Technical Advancement and Major Perspectives." *Biomedicine & Pharmacotherapy* 131 (November): 110648. https://doi.org/10.1016/j.biopha.2020.110648.

Das, Lipika, Sushanta Deb, Eiji Arakawa, Shinji Yamasaki, and Subrata K. Das. 2022. "Pufferfish (Tetraodon Cutcutia) Sampled from a Freshwater River Serves as an Intermediate Reservoir of a Sucrose Nonfermenting Variant of Vibrio Cholerae PS–4." *Microbiology Spectrum* 10 (1): e01221-21. https://doi.org/10.1128/spectrum.01221-21.

Das, Lipika, Sushanta Deb, and Subrata K. Das. 2020. "Glutamicibacter Mishrai Sp. Nov., Isolated from the Coral Favia Veroni from Andaman Sea." *Archives of Microbiology* 202 (4): 733–45. https://doi.org/10.1007/s00203-019-01783-0.

Das, Lipika, Sushanta Deb, and Subrata K. Das. 2021. "Description of Acinetobacter Kanungonis Sp. Nov., Based on Phylogenomic Analysis." *International Journal of Systematic and Evolutionar*y Microbiology 71 (6). https://doi.org/10.1099/ijsem.0.004833.

Das, Nilanjana, and Preethy Chandran. 2011. "Microbial Degradation of Petroleum Hydrocarbon Contaminants: An Overview." *Biotechnology Research Internationa*l 2011 (September): 1–13. https://doi.org/10.4061/2011/941810.

Das, Sayoni, and Christine A. Orengo. 2016. "Protein Function Annotation Using Protein Domain Family Resources." *Methods* 93 (January): 24–34. https://doi.org/10.1016/j.ymeth.2015.09.029.

Dash, Jayalaxmi, Manisha Sethi, Sushanta Deb, et al. 2023. "Biochemical, Functional and Genomic Characterization of a New Probiotic *Ligilactobacillus salivarius* F14 from the Gut of Tribes of Odisha." *World Journal of Microbiology and Biotechnology* 39 (7): 171. https://doi.org/10.1007/s11274-023-03626-z.

Deb, Sushanta, Jhasketan Badhai, and Subrata K. Das. 2020a. "Draft Genome Sequences of Two Vibrio Fortis Strains Isolated from Coral (*Fungia* Sp.) from the Andaman Sea." *Microbiology Resource Announcements* 9 (1): e01225-19. https://doi.org/10.1128/MRA.01225-19.

Deb, Sushanta, Jhasketan Badhai, and Subrata K. Das. 2020b. "Draft Genome Sequences of Vibrio Alginolyticus Strain S6–61 and Vibrio Diabolicus Strain S7–71, Isolated from Corals in the Andaman Sea." *Microbiology Resource Announcements* 9 (8): e01465-19. https://doi.org/10.1128/MRA.01465-19.

Deb, Sushanta, Jhinuk Basu, and Megha Choudhary. 2024. "An Overview of next Generation Sequencing Strategies and Genomics Tools Used for Tuberculosis Research." *Journal of Applied Microbiology* 135 (7): lxae174. https://doi.org/10.1093/jambio/lxae174.

Deb, Sushanta, Lipika Das, and Subrata K. Das. 2020. "Composition and Functional Characterization of the Gut Microbiome of Freshwater Pufferfish (Tetraodon Cutcutia)." *Archives of Microbiology* 202 (10): 2761–70. https://doi.org/10.1007/s00203-020-01997-7.

Deb, Sushanta, and Subrata K. Das. 2020a. "Draft Genome Sequence of Idiomarina Sp. Strain W-5T, Isolated from the Andaman Sea." *Microbiology Resource Announcements* 9 (5). https://doi.org/10.1128/mra.01510-19.

Deb, Sushanta, and Subrata K. Das. 2020b. "Draft Genome Sequence of *Microbacterium Oryzae* Strain MB-10, Isolated from a Rice Field in India." *Microbiology Resource Announcements* 9 (6): e01532-19. https://doi.org/10.1128/MRA.01532-19.

Deb, Sushanta, and Subrata K. Das. 2022. "Phylogenomic Analysis of Metagenome-Assembled Genomes Deciphered Novel Acetogenic Nitrogen-Fixing *Bathyarchaeota* from Hot Spring Sediments." *Microbiology Spectrum* 10 (3): e00352-22. https://doi.org/10.1128/spectrum.00352-22.

Debnath, Tanmoy, Sushanta Deb, and Subrata K. Das. 2023. "Influence of Geochemistry in the Tropical Hot Springs on Microbial Community Structure and Function." *Current Microbiology* 80 (1): 4. https://doi.org/10.1007/s00284-022-03118-7.

Di Bella, Julia M., Yige Bao, Gregory B. Gloor, Jeremy P. Burton, and Gregor Reid. 2013. "High Throughput Sequencing Methods and Analysis for Microbiome Research." *Journal of Microbiological Methods* 95 (3): 401–14. https://doi.org/10.1016/j.mimet.2013.08.011.

Dugan, Vivien G., Scott J. Emrich, Gloria I. Giraldo-Calderón, et al. 2014. "Standardized Metadata for Human Pathogen/Vector Genomic Sequences." *PLoS One* 9 (6): e99979. https://doi.org/10.1371/journal.pone.0099979.

Edlund, Anna, Youngik Yang, Shibu Yooseph, Xuesong He, Wenyuan Shi, and Jeffrey S. McLean. 2018. "Uncovering Complex Microbiome Activities via Metatranscriptomics during 24 Hours of Oral Biofilm Assembly and Maturation." *Microbiome* 6 (1): 217. https://doi.org/10.1186/s40168-018-0591-4.

Evans, Ciaran, Johanna Hardin, and Daniel M. Stoebel. 2018. "Selecting Between-Sample RNA-Seq Normalization Methods from the Perspective of Their Assumptions." *Briefings in Bioinformatics* 19 (5): 776–92. https://doi.org/10.1093/bib/bbx008.

Falkowski, Paul G., Tom Fenchel, and Edward F. Delong. 2008. "The Microbial Engines that Drive Earth's Biogeochemical Cycles." *Science* 320 (5879): 1034–9. https://doi.org/10.1126/science.1153213.

Felczykowska, Agnieszka, Anna Krajewska, Sylwia Zielińska, and Joanna Łoś. 2015. "Sampling, Metadata and DNA Extraction-Important Steps in Metagenomic Studies." *Acta Biochimica Polonica* 62 (1): 151–60.

Fierer, Noah. 2017. "Embracing the Unknown: Disentangling the Complexities of the Soil Microbiome." *Nature Reviews Microbiology* 15 (10): 579–90. https://doi.org/10.1038/nrmicro.2017.87.

Flint, Harry J., Karen P. Scott, Sylvia H. Duncan, Petra Louis, and Evelyne Forano. 2012. "Microbial Degradation of Complex Carbohydrates in the Gut." *Gut Microbes* 3 (4): 289–306. https://doi.org/10.4161/gmic.19897.

Franzosa, Eric A., Katherine Huang, James F. Meadow, et al. 2015. "Identifying Personal Microbiomes Using Metagenomic Codes." *Proceedings of the National Academy of Sciences* 112 (22): E2930–38.

Franzosa, Eric A., Lauren J. McIver, Gholamali Rahnavard, et al. 2018. "Species-Level Functional Profiling of Metagenomes and Metatranscriptomes." *Nature Methods* 15 (11): 962–8. https://doi.org/10.1038/s41592-018-0176-y.

Fritz, Joëlle V., Mahesh S. Desai, Pranjul Shah, Jochen G. Schneider, and Paul Wilmes. 2013. "From Meta-Omics to Causality: Experimental Models for Human Microbiome Research." *Microbiome* 1 (1): 14. https://doi.org/10.1186/2049-2618-1-14.

The Gene Ontology Consortium. 2019. "The Gene Ontology Resource: 20 Years and Still Going Strong." *Nucleic Acids Research* 47 (D1): D330–8. https://doi.org/10.1093/nar/gky1055.

Gilbert, Jack A., Janet K. Jansson, and Rob Knight. 2014. "The Earth Microbiome Project: Successes and Aspirations." *BMC Biology* 12 (1): 69. https://doi.org/10.1186/s12915-014-0069-1.

Gilbert, Jack A., and Susan V. Lynch. 2019. "Community Ecology as a Framework for Human Microbiome Research." *Nature Medicine* 25 (6): 884–9. https://doi.org/10.1038/s41591-019-0464-9.

Glenn, Travis C. 2011. "Field Guide to Next-Generation DNA Sequencers." *Molecular Ecology Resources* 11 (5): 759–69. https://doi.org/10.1111/j.1755-0998.2011.03024.x.

Gligorijević, Vladimir, P. Douglas Renfrew, Tomasz Kosciolek, et al. 2021. "Structure-Based Protein Function Prediction Using Graph Convolutional Networks." *Nature Communications* 12 (1): 3168. https://doi.org/10.1038/s41467-021-23303-9.

Goodrich, Julia K., Sara C. Di Rienzi, Angela C. Poole, et al. 2014. "Conducting a Microbiome Study." *Cell* 158 (2): 250–62. https://doi.org/10.1016/j.cell.2014.06.037.

Goodwin, Sara, John D. McPherson, and W. Richard McCombie. 2016. "Coming of Age: Ten Years of Next-Generation Sequencing Technologies." *Nature Reviews Genetics* 17 (6): 333–51. https://doi.org/10.1038/nrg.2016.49.

Graham, Emily B., Joseph E. Knelman, Andreas Schindlbacher, et al. 2016. "Microbes as Engines of Ecosystem Function: When Does Community Structure Enhance Predictions of Ecosystem Processes?" *Frontiers in Microbiology* 7 (February). https://doi.org/10.3389/fmicb.2016.00214.

Gymrek, Melissa, Amy L. McGuire, David Golan, Eran Halperin, and Yaniv Erlich. 2013. "Identifying Personal Genomes by Surname Inference." *Science* 339 (6117): 321–4. https://doi.org/10.1126/science.1229566.

Harman, Gary E., Charles R. Howell, Ada Viterbo, Ilan Chet, and Matteo Lorito. 2004. "*Trichoderma* Species—Opportunistic, Avirulent Plant Symbionts." *Nature Reviews Microbiology* 2 (1): 43–56. https://doi.org/10.1038/nrmicro797.

Hempel, Christopher A., Natalie Wright, Julia Harvie, Jose S. Hleap, Sarah J. Adamowicz, and Dirk Steinke. 2022. "Metagenomics versus Total RNA Sequencing: Most Accurate Data-Processing Tools, Microbial Identification Accuracy and Perspectives for Ecological Assessments." *Nucleic Acids Research* 50 (16): 9279–93. https://doi.org/10.1093/nar/gkac689.

Honaas, Loren A., Naomi S. Altman, and Martin Krzywinski. 2016. "Study Design for Sequencing Studies." In *Statistical Genomics*, edited by Ewy Mathé and Sean Davis, vol. 1418. Methods in Molecular Biology. Springer New York. https://doi.org/10.1007/978-1-4939-3578-9_3.

Hornung, Roman, Anne-Laure Boulesteix, and David Causeur. 2016. "Combining Location-and-Scale Batch Effect Adjustment with Data Cleaning by Latent Factor Adjustment." *BMC Bioinformatics* 17 (1): 27. https://doi.org/10.1186/s12859-015-0870-z.

Huerta-Cepas, Jaime, Damian Szklarczyk, Davidolubier, et al. 2019. "eggNOG 5.0: A Hierarchical, Functionally and Phylogenetically Annotated Orthology Resource Based on 5090 Organisms and 2502 Viruses." *Nucleic Acids Research* 47 (D1): D309–14. https://doi.org/10.1093/nar/gky1085.

Huse, Susan M., David Mark Welch, Hilary G. Morrison, and Mitchell L. Sogin. 2010. "Ironing out the Wrinkles in the Rare BioIere through Improved OTU Clustering." *Environmental Microbiology* 12 (7): 1889–98. https://doi.org/10.1111/j.1462-2920.2010.02193.x.

Huttenhower, Curtis, Robert D. Finn, and Alice Carolyn McHardy. 2023. "Challenges and Opportunities in Sharing Microbiome Data and Analyses." *Nature Microbiology* 8 (11): 1960–70. https://doi.org/10.1038/s41564-023-01484-x.

Integrative, HMP and Research Network Consortium. 2019. "The Integrative Human Microbiome Project." *Nature* 569 (7758): 641–8.

Irving, Rachel, and Angus John Clarke. 2019. "Ethical and Social Issues in Clinical Genetics." In *Emery and Rimoin's Principles and Practice of Medical Genetics and Genomics*. Elsevier. https://doi.org/10.1016/B978-0-12-812536-6.00013-4.

Jansson, Janet K., and Kirsten S. Hofmockel. 2018. "The Soil Microbiome—from Metagenomics to Metaphenomics." *Current Opinion in Microbiology* 43 (June): 162–8. https://doi.org/10.1016/j.mib.2018.01.013.

Jiang, Shuang, Guanghua Xiao, Andrew Y. Koh, Jiwoong Kim, Qiwei Li, and Xiaowei Zhan. 2021. "A Bayesian Zero-Inflated Negative Binomial Regression Model for the Integrative Analysis of Microbiome Data." *Biostatistics* 22 (3): 522–40. https://doi.org/10.1093/biostatistics/kxz050.

Johnson, A. J., J. J. Zheng, J. W. Kang, A. Saboe, D. Knights, and A. M. Zivkovic. 2020. "A Guide to Diet-Microbiome Study Design." *Frontiers in Nutrition* 7.

Johnson, W. Evan, Cheng Li, and Ariel Rabinovic. 2007. "Adjusting Batch Effects in Microarray Expression Data Using Empirical Bayes Methods." *Biostatistics* 8 (1): 118–27. https://doi.org/10.1093/biostatistics/kxj037.

Kanehisa, Minoru, Miho Furumichi, Mao Tanabe, Yoko Sato, and Kanae Morishima. 2017. "KEGG: New Perspectives on Genomes, Pathways, Diseases and Drugs." *Nucleic Acids Research* 45 (D1): D353–61. https://doi.org/10.1093/nar/gkw1092.

Kim, Dorothy, Casey E. Hofstaedter, Chunyu Zhao, et al. 2017. "Optimizing Methods and Dodging Pitfalls in Microbiome Research." *Microbiome* 5 (1): 52. https://doi.org/10.1186/s40168-017-0267-5.

Kim, Jiwoong, Min Soo Kim, Andrew Y. Koh, Yang Xie, and Xiaowei Zhan. 2016. "FMAP: Functional Mapping and Analysis Pipeline for Metagenomics and Metatranscriptomics Studies." *BMC Bioinformatics* 17 (1): 420. https://doi.org/10.1186/s12859-016-1278-0.

Kujur, Ritu Rani Archana, Sushanta Deb, and Subrata K. Das. 2022. "Genome Analysis of Pseudomonas Species Reveals that Pseudomonas Panacis Park et al. 2005 Is a Later Heterotypic Synonym of Pseudomonas Marginalis (Brown 1918) Stevens 1925 (Approved Lists 1980)." *International Journal of Systematic and Evolutionary Microbiology* 72 (5). https://doi.org/10.1099/ijsem.0.005354.

Kumari, Kiran, Vaishnavi Rawat, Afreen Shadan, Parva Kumar Sharma, Sushanta Deb, and Rajnish Prakash Singh. 2023. "In-Depth Genome and Pan-Genome Analysis of a Metal-Resistant Bacterium Pseudomonas Parafulva OS–1." *Frontiers in Microbiology* 14 (June): 1140249. https://doi.org/10.3389/fmicb.2023.1140249.

Kuruvilla, Jacob. 2017. *Proteomics as a Multifaceted Tool in Medicine and Environmental Assessment*. Vol. 1585. Linköping University Electronic Press.

Langmead, Ben, and Steven L. Salzberg. 2012. "Fast Gapped-Read Alignment with Bowtie 2." *Nature Methods* 9 (4): 357–9. https://doi.org/10.1038/nmeth.1923.

Laukens, Debby, Brigitta M. Brinkman, Jeroen Raes, Martine De Vos, and Peter Vandenabeele. 2016. "Heterogeneity of the Gut Microbiome in Mice: Guidelines for Optimizing Experimental Design." *FEMS Microbiology Reviews* 40 (1): 117–32. https://doi.org/10.1093/femsre/fuv036.

Leek, Jeffrey T. 2014. "Svaseq: Removing Batch Effects and Other Unwanted Noise from Sequencing Data." *Nucleic Acids Research* 42 (21): e161. https://doi.org/10.1093/nar/gku864.

Lefkowitz, R. J. 1975. "Identification of Adenylate Cyclase-Coupled Beta-Adrenergic Receptors with Radiolabeled Beta-Adrenergic Antagonists." *Biochemical Pharmacology* 24 (18): 1651–8. https://doi.org/10.1016/0006-2952(75)90001-5.

Lin, Huang, and Shyamal Das Peddada. 2020. "Analysis of Microbial Compositions: A Review of Normalization and Differential Abundance Analysis." *Npj Biofilms and Microbiomes* 6 (1): 60. https://doi.org/10.1038/s41522-020-00160-w.

Lindgreen, Stinus, Karen L. Adair, and Paul P. Gardner. 2016. "An Evaluation of the Accuracy and Speed of Metagenome Analysis Tools." *Scientific Reports* 6 (1): 19233. https://doi.org/10.1038/srep19233.

Ling, Wodan, Jiuyao Lu, Ni Zhao, et al. 2022. "Batch Effects Removal for Microbiome Data via Conditional Quantile Regression." *Nature Communications* 13 (1): 5418. https://doi.org/10.1038/s41467-022-33071-9.

Ling, Wodan, Ni Zhao, Anna M. Plantinga, et al. 2021. "Powerful and Robust Non-Parametric Association Testing for Microbiome Data via a Zero-Inflated Quantile Approach (ZINQ)." *Microbiome* 9 (1): 181. https://doi.org/10.1186/s40168-021-01129-3.

Lladó, Salvador, Rubén López-Mondéjar, and Petr Baldrian. 2017. "Forest Soil Bacteria: Diversity, Involvement in Ecosystem Processes, and Response to Global Change." *Microbiology and Molecular Biology Reviews* 81 (2): e00063-16. https://doi.org/10.1128/MMBR.00063-16.

Lohmann, Patrick, Stephanie Serena Schäpe, Sven-Bastiaan Haange, et al. 2020. "Function Is What Counts: How Microbial Community Complexity Affects Species, Proteome and Pathway Coverage in Metaproteomics." *Expert Review of Proteomics* 17 (2): 163–73. https://doi.org/10.1080/14789450.2020.1738931.

Lu, Jennifer, Florian P. Breitwieser, Peter Thielen, and Steven L. Salzberg. 2017. "Bracken: Estimating Species Abundance in Metagenomics Data." *PeerJ Computer Science* 3 (January): e104. https://doi.org/10.7717/peerj-cs.104.

Lynch, Susan V, and Oluf Pedersen. 2016. "The Human Intestinal Microbiome in Health and Disease." *New England Journal of Medicine* 375 (24): 2369–79.

Magoč, Tanja, and Steven L. Salzberg. 2011. "FLASH: Fast Length Adjustment of Short Reads to Improve Genome Assemblies." *Bioinformatics* 27 (21): 2957–63. https://doi.org/10.1093/bioinformatics/btr507.

Malla, Muneer Ahmad, Anamika Dubey, Ashwani Kumar, Shweta Yadav, Abeer Hashem, and Elsayed Fathi Abd_Allah. 2019. "Exploring the Human Microbiome: The Potential Future Role of Next-Generation Sequencing in Disease Diagnosis and Treatment." *Frontiers in Immunology* 9 (January): 2868. https://doi.org/10.3389/fimmu.2018.02868.

Maranga, Mary, Pawel Szczerbiak, Valentyn Bezshapkin, et al. 2023. "Comprehensive Functional Annotation of Metagenomes and Microbial Genomes Using a Deep Learning-Based Method." *mSystems* 8 (2): e01178-22. https://doi.org/10.1128/msystems.01178-22.

Marchesi, Julian R., and Jacques Ravel. 2015. "The Vocabulary of Microbiome Research: A Proposal." *Microbiome* 3 (1): 31. https://doi.org/10.1186/s40168-015-0094-5.

Marconi, Stefania, Alessandra Durazzo, Emanuela Camilli, et al. 2018. "Food Composition Databases: Considerations about Complex Food Matrices." *Foods* 7 (1): 2. https://doi.org/10.3390/foods7010002.

Markussen, Trine, Elisabeth M. Happel, Jonna E. Teikari, et al. 2018. "Coupling Biogeochemical Process Rates and Metagenomic Blueprints of Coastal Bacterial Assemblages in the Context of Environmental Change." *Environmental Microbiology* 20 (8): 3083–99. https://doi.org/10.1111/1462-2920.14371.

Martin, Tiphaine C., Alessia Visconti, Tim D. Spector, and Mario Falchi. 2018. "Conducting Metagenomic Studies in Microbiology and Clinical Research." *Applied Microbiology and Biotechnology* 102 (20): 8629–46. https://doi.org/10.1007/s00253-018-9209-9.

Martinez, Xavier, Marta Pozuelo, Victoria Pascal, et al. 2016. "MetaTrans: An Open-Source Pipeline for Metatranscriptomics." *Scientific Reports* 6 (1): 26447. https://doi.org/10.1038/srep26447.

Masella, Andre P., Andrea K. Bartram, Jakub M. Truszkowski, Daniel G. Brown, and Josh D. Neufeld. 2012. "PANDAseq: Paired-End Assembler for Illumina Sequences." *BMC Bioinformatics* 13 (1): 31. https://doi.org/10.1186/1471-2105-13-31.

Mayer, Rudolf, Markus Hittmeir, and Andreas Ekelhart. 2024. "Distance-Based Linkage of Personal Microbiome Records for Identification and Its Privacy Implications." *Computers & Security* 136 (January): 103538. https://doi.org/10.1016/j.cose.2023.103538.

McIntyre, Alexa B. R., Rachid Ounit, Ebrahim Afshinnekoo, et al. 2017. "Comprehensive Benchmarking and Ensemble Approaches for Metagenomic Classifiers." *Genome Biology* 18 (1): 182. https://doi.org/10.1186/s13059-017-1299-7.

MetaHIT Consortium, Junjie Qin, Ruiqiang Li, et al. 2010. "A Human Gut Microbial Gene Catalogue Established by Metagenomic Sequencing." *Nature* 464 (7285): 59–65. https://doi.org/10.1038/nature08821.

The Microbiome Quality Control Project Consortium, Rashmi Sinha, Galeb Abu-Ali, et al. 2017. "Assessment of Variation in Microbial Community Amplicon Sequencing by the Microbiome Quality Control (MBQC) Project Consortium." *Nature Biotechnology* 35 (11): 1077–86. https://doi.org/10.1038/nbt.3981.

Mitchell, Alex L., Alexandre Almeida, Martin beracochea, et al. 2019. "MGnify: The Microbiome Analysis Resource in 2020." *Nucleic Acids Research* (November 7), gkz1035. https://doi.org/10.1093/nar/gkz1035.

Moore, Robert J., and Dragana Stanley. 2016. "Experimental Design Considerations in Microbiota/Inflammation Studies." *Clinical & Translational Immunology* 5 (7): e92. https://doi.org/10.1038/cti.2016.41.

Nandy, Sampurna, and Atya Kapley. 2024. "Unraveling the Potential of Microbial Communities for Lake Bioremediation via the Metagenomics Tool: A Review." *AQUA—Water Infrastructure, Ecosystems and Society* 73 (1): 11–33. https://doi.org/10.2166/aqua.2024.154.

"The National Microbiome Data Collaborative Data Portal: An Integrated Multi-Omics Microbiome Data Resource." 2022. *Nucleic Acids Research*, ahead of print (October 30). https://doi.org/10.1093/nar/gkab990.

Ngumbi, Esther, and Joseph Kloepper. 2016. "Bacterial-Mediated Drought Tolerance: Current and Future Prospects." *Applied Soil Ecology* 105 (September): 109–25. https://doi.org/10.1016/j.apsoil.2016.04.009.

O'Leary, Nuala A., Mathew W. Wright, J. Rodney Brister, et al. 2016. "Reference Sequence (RefSeq) Database at NCBI: Current Status, Taxonomic Expansion, and Functional Annotation." *Nucleic Acids Research* 44 (D1): D733–45. https://doi.org/10.1093/nar/gkv1189.

Oulas, Anastasis, Christina Pavloudi, Paraskevi Polymenakou, et al. 2015. "Metagenomics: Tools and Insights for Analyzing Next-Generation Sequencing Data Derived from Biodiversity Studies." *Bioinformatics and Biology Insights* 9 (January): BBI.S12462. https://doi.org/10.4137/BBI.S12462.

Park, Elizabeth M., Manoj Chelvanambi, Neal Bhutiani, Guido Kroemer, Laurence Zitvogel, and Jennifer A. Wargo. 2022. "Targeting the Gut and Tumor Microbiota in Cancer." *Nature Medicine* 28 (4): 690–703. https://doi.org/10.1038/s41591-022-01779-2.

Parks, Donovan H., Fabio Rigato, Patricia Vera-Wolf, et al. 2021. "Evaluation of the Microba Community Profiler for Taxonomic Profiling of Metagenomic Datasets from the Human Gut Microbiome." *Frontiers in Microbiology* 12 (April): 643682. https://doi.org/10.3389/fmicb.2021.643682.

Pasolli, Edoardo, Francesco Asnicar, Serena Manara, et al. 2019. "Extensive Unexplored Human Microbiome Diversity Revealed by Over 150,000 Genomes from Metagenomes Spanning Age, Geography, and Lifestyle." *Cell* 176 (3): 649–662.e20. https://doi.org/10.1016/j.cell.2019.01.001.

Paszkiewicz, K., and D. J. Studholme. 2010. "De Novo Assembly of Short Sequence Reads." *Briefings in Bioinformatics* 11 (5): 457–72. https://doi.org/10.1093/bib/bbq020.

Peters, Danielle L., Wenju Wang, Xu Zhang, Zhibin Ning, Janice Mayne, and Daniel Figeys. 2019. "Metaproteomic and Metabolomic

Approaches for Characterizing the Gut Microbiome." *Proteomics* 19 (16): 1800363. https://doi.org/10.1002/pmic.201800363.

Quince, Christopher, Anders Lanzen, Russell J. Davenport, and Peter J. Turnbaugh. 2011. "Removing Noise from Pyrosequenced Amplicons." *BMC Bioinformatics* 12 (1): 38. https://doi.org/10.1186/1471-2105-12-38.

Quince, Christopher, Alan W. Walker, Jared T. Simpson, Nicholas J. Loman, and Nicola Segata. 2017. "Shotgun Metagenomics, from Sampling to Analysis." *Nature Biotechnology* 35 (9): 833–44. https://doi.org/10.1038/nbt.3935.

Raghavan, Venket, Louis Kraft, Fantin Mesny, and Linda Rigerte. 2022. "A Simple Guide to *de Novo* Transcriptome Assembly and Annotation." *Briefings in Bioinformatics* 23 (2): bbab563. https://doi.org/10.1093/bib/bbab563.

Ranjan, Ravi, Asha Rani, Ahmed Metwally, Halvor S. McGee, and David L. Perkins. 2016. "Analysis of the Microbiome: Advantages of Whole Genome Shotgun versus 16S Amplicon Sequencing." *Biochemical and Biophysical Research Communications* 469 (4): 967–77. https://doi.org/10.1016/j.bbrc.2015.12.083.

Riesenfeld, Christian S., Patrick D. Schloss, and Jo Handelsman. 2004. "Metagenomics: Genomic Analysis of Microbial Communities." *Annual Review of Genetics* 38 (1): 525–52. https://doi.org/10.1146/annurev.genet.38.072902.091216.

Ritchie, Matthew E., Belinda Phipson, Di Wu, et al. 2015. "Limma Powers Differential Expression Analyses for RNA-Sequencing and Microarray Studies." *Nucleic Acids Research* 43 (7): e47. https://doi.org/10.1093/nar/gkv007.

Rodrigue, Sébastien, Arne C. Materna, Sonia C. Timberlake, et al. 2010. "Unlocking Short Read Sequencing for Metagenomics." *PLoS One* 5 (7): e11840. https://doi.org/10.1371/journal.pone.0011840.

Rodrigues, Carlos J. C., and Carla C. C. R. De Carvalho. 2022. "Cultivating Marine Bacteria under Laboratory Conditions: Overcoming the 'Unculturable' Dogma." *Frontiers in Bioengineering and Biotechnology* 10 (August): 964589. https://doi.org/10.3389/fbioe.2022.964589.

Round, June L., and Sarkis K. Mazmanian. 2009. "The Gut Microbiota Shapes Intestinal Immune Responses during Health and Disease." *Nature Reviews Immunology* 9 (5): 313–23. https://doi.org/10.1038/nri2515.

Saeed, Isaam, and Saman K. Halgamuge. 2009. "The Oligonucleotide Frequency Derived Error Gradient and Its Application to the Binning of Metagenome Fragments." *BMC Genomics* 10 (Suppl 3): S10. https://doi.org/10.1186/1471-2164-10-S3-S10.

Satam, Heena, Kandarp Joshi, Upasana Mangrolia, et al. 2023. "Next-Generation Sequencing Technology: Current Trends and Advancements." *Biology* 12 (7): 997. https://doi.org/10.3390/biology12070997.

Schloss, Patrick D., Dirk Gevers, and Sarah L. Westcott. 2011. "Reducing the Effects of PCR Amplification and Sequencing Artifacts on 16S rRNA-Based Studies." *PLoS One* 6 (12): e27310. https://doi.org/10.1371/journal.pone.0027310.

Schloss, Patrick D., Sarah L. Westcott, Thomas Ryabin, et al. 2009. "Introducing Mothur: Open-Source, Platform-Independent, Community-Supported Software for Describing and Comparing Microbial Communities." *Applied and Environmental Microbiology* 75 (23): 7537–41. https://doi.org/10.1128/AEM.01541-09.

Schmidt, Bertil, and Andreas Hildebrandt. 2017. "Next-Generation Sequencing: Big Data Meets High Performance Computing." *Drug Discovery Today* 22 (4): 712–17. https://doi.org/10.1016/j.drudis.2017.01.014.

Scholz, Matthew B., Chien-Chi Lo, and Patrick Sg Chain. 2012. "Next Generation Sequencing and Bioinformatic Bottlenecks: The Current State of Metagenomic Data Analysis." *Current Opinion in Biotechnology* 23 (1): 9–15. https://doi.org/10.1016/j.copbio.2011.11.013.

Sczyrba, Alexander, Peter Hofmann, Peter Belmann, et al. 2017. "Critical Assessment of Metagenome Interpretation: A Benchmark of Metagenomics Software." *Nature Methods* 14 (11): 1063–71. https://doi.org/10.1038/nmeth.4458.

Segata, Nicola, Levi Waldron, Annalisa Ballarini, Vagheesh Narasimhan, Olivier Jousson, and Curtis Huttenhower. 2012. "Metagenomic Microbial Community Profiling Using Unique Clade-Specific Marker Genes." *Nature Methods* 9 (8): 811–14. https://doi.org/10.1038/nmeth.2066.

Shakya, Migun, Chien-Chi Lo, and Patrick S. G. Chain. 2019. "Advances and Challenges in Metatranscriptomic Analysis." *Frontiers in Genetics* 10 (September): 904. https://doi.org/10.3389/fgene.2019.00904.

Sims, Andrew H., Graeme J. Smethurst, Yvonne Hey, et al. 2008. "The Removal of Multiplicative, Systematic Bias Allows Integration of Breast Cancer Gene Expression Datasets—Improving Meta-Analysis and Prediction of Prognosis." *BMC Medical Genomics* 1 (1): 42. https://doi.org/10.1186/1755-8794-1-42.

Singh, Brajesh K., Richard D. Bardgett, Pete Smith, and Dave S. Reay. 2010. "Microorganisms and Climate Change: Terrestrial Feedbacks and Mitigation Options." *Nature Reviews Microbiology* 8 (11): 779–90. https://doi.org/10.1038/nrmicro2439.

Singh, Rajnish Prakash, Ayushi Sinha, Sushanta Deb, and Kiran Kumari. 2024. "First Report on In-Depth Genome and Comparative Genome Analysis of a Metal-Resistant Bacterium Acinetobacter Pittii S-30, Isolated from Environmental Sample." *Frontiers in Microbiology* 15 (April): 1351161. https://doi.org/10.3389/fmicb.2024.1351161.

Smith, Sally E., and David J. Read. 2010. *Mycorrhizal Symbiosis*. Academic Press.

Smits, Loek P., Kristien E. C. Bouter, Willem M. De Vos, Thomas J. Borody, and Max Nieuwdorp. 2013. "Therapeutic Potential of Fecal Microbiota Transplantation." *Gastroenterology* 145 (5): 946–53. https://doi.org/10.1053/j.gastro.2013.08.058.

Steinegger, M. 2018. Ultrafast and sensitive sequence search and clustering methods in the era of next generation sequencing (Doctoral dissertation, Technische Universität München).

Stewart, Frank J., Elizabeth A. Ottesen, and Edward F. DeLong. 2010. "Development and Quantitative Analyses of a Universal rRNA-Subtraction Protocol for Microbial Metatranscriptomics." *The ISME Journal* 4 (7): 896–907. https://doi.org/10.1038/ismej.2010.18.

Strous, Marc, Beate Kraft, Regina Bisdorf, and Halina E. Tegetmeyer. 2012. "The Binning of Metagenomic Contigs for Microbial Physiology of Mixed Cultures." *Frontiers in Microbiology* 3. https://doi.org/10.3389/fmicb.2012.00410.

Sun, Yaohui, Zhiming Gan, Xiaojin Wang, et al. 2024. "Integrative Metagenomic, Transcriptomic, and Proteomic Analysis Reveal the Microbiota-Host Interplay in Early-Stage Lung Adenocarcinoma among Non-Smokers." *Journal of Translational Medicine* 22 (1): 652. https://doi.org/10.1186/s12967-024-05485-0.

Sureyya Rifaioglu, Ahmet, Tunca Doğan, Maria Jesus Martin, Rengul Cetin-Atalay, and Volkan Atalay. 2019. "DEEPred: Automated Protein Function Prediction with Multi-Task Feed-Forward Deep Neural Networks." *Scientific Reports* 9 (1): 7344. https://doi.org/10.1038/s41598-019-43708-3.

Tsilimigras, Matthew C. B., and Anthony A. Fodor. 2016. "Compositional Data Analysis of the Microbiome:

Fundamentals, Tools, and Challenges." *Annals of Epidemiology* 26 (5): 330–5. https://doi.org/10.1016/j.annepidem.2016.03.002.

Türkowsky, Dominique, Nico Jehmlich, Gabriele Diekert, Lorenz Adrian, Martin Von Bergen, and Tobias Goris. 2018. "An Integrative Overview of Genomic, Transcriptomic and Proteomic Analyses in Organohalide Respiration Research." *FEMS Microbiology Ecology* 94 (3). https://doi.org/10.1093/femsec/fiy013.

Turnbaugh, Peter J., Ruth E. Ley, Michael A. Mahowald, Vincent Magrini, Elaine R. Mardis, and Jeffrey I. Gordon. 2006. "An Obesity-Associated Gut Microbiome with Increased Capacity for Energy Harvest." *Nature* 444 (7122): 1027–31. https://doi.org/10.1038/nature05414.

Tyagi, Shivani, and Pramod Katara. 2024. "Metatranscriptomics: A Tool for Clinical Metagenomics." *OMICS: A Journal of Integrative Biology* 28 (8): 394–407. https://doi.org/10.1089/omi.2024.0130.

Van Der Heijden, Marcel G. A., Richard D. Bardgett, and Nico M. Van Straalen. 2008. "The Unseen Majority: Soil Microbes as Drivers of Plant Diversity and Productivity in Terrestrial Ecosystems." *Ecology Letters* 11 (3): 296–310. https://doi.org/10.1111/j.1461-0248.2007.01139.x.

Van Dijk, Erwin L., Yan Jaszczyszyn, Delphine Naquin, and Claude Thermes. 2018. "The Third Revolution in Sequencing Technology." *Trends in Genetics* 34 (9): 666–81. https://doi.org/10.1016/j.tig.2018.05.008.

Vandeputte, Doris, Gunter Kathagen, Kevin D'hoe, et al. 2017. "Quantitative Microbiome Profiling Links Gut Community Variation to Microbial Load." *Nature* 551 (7681): 507–11. https://doi.org/10.1038/nature24460.

Vangay, Pajau, Josephine Burgin, Anjanette Johnston, et al. 2021. "Microbiome Metadata Standards: Report of the National Microbiome Data Collaborative's Workshop and Follow-On Activities." *mSystems* 6 (1): e01194-20. https://doi.org/10.1128/mSystems.01194-20.

Vanwonterghem, Inka, Paul D. Jensen, Dang P. Ho, Damien J. Batstone, and Gene W. Tyson. 2014. "Linking Microbial Community Structure, Interactions and Function in Anaerobic Digesters Using New Molecular Techniques." *Current Opinion in Biotechnology* 27 (June): 55–64. https://doi.org/10.1016/j.copbio.2013.11.004.

Vatanen, Tommi, Eric A. Franzosa, Randall Schwager, et al. 2018. "The Human Gut Microbiome in Early-Onset Type 1 Diabetes from the TEDDY Study." *Nature* 562 (7728): 589–94. https://doi.org/10.1038/s41586-018-0620-2.

Vessey, J. Kevin. 2003. "Plant Growth Promoting Rhizobacteria as Biofertilizers." *Plant and Soil* 255 (2): 571–86. https://doi.org/10.1023/A:1026037216893.

Voigt, Anita Y., Paul I. Costea, Jens Roat Kultima, et al. 2015. "Temporal and Technical Variability of Human Gut Metagenomes." *Genome Biology* 16 (1): 73. https://doi.org/10.1186/s13059-015-0639-8.

Walsh, Aaron M., Fiona Crispie, Orla O'Sullivan, Laura Finnegan, Marcus J. Claesson, and Paul D. Cotter. 2018. "Species Classifier Choice Is a Key Consideration When Analysing Low-Complexity Food Microbiome Data." *Microbiome* 6 (1): 50. https://doi.org/10.1186/s40168-018-0437-0.

Wang, Da-Zhi, Ling-Fen Kong, Yuan-Yuan Li, and Zhang-Xian Xie. 2016. "Environmental Microbial Community Proteomics: Status, Challenges and Perspectives." *International Journal of Molecular Sciences* 17 (8): 1275. https://doi.org/10.3390/ijms17081275.

Wang, Xueling, Yongkuan Chi, and Shuzhen Song. 2024. "Important Soil Microbiota's Effects on Plants and Soils: A Comprehensive 30-Year Systematic Literature Review." *Frontiers in Microbiology* 15 (March): 1347745. https://doi.org/10.3389/fmicb.2024.1347745.

Wang, Yiwen, and Kim-Anh LêCao. 2020. "Managing Batch Effects in Microbiome Data." *Briefings in Bioinformatics* 21 (6): 1954–70. https://doi.org/10.1093/bib/bbz105.

Weber, Nick, David Liou, Jennifer Dommer, et al. 2018. "Nephele: A Cloud Platform for Simplified, Standardized and Reproducible Microbiome Data Analysis." *Bioinformatics* 34 (8): 1411–13. https://doi.org/10.1093/bioinformatics/btx617.

Wensel, Caroline R., Jennifer L. Pluznick, Steven L. Salzberg, and Cynthia L. Sears. 2022. "Next-Generation Sequencing: Insights to Advance Clinical Investigations of the Microbiome." *Journal of Clinical Investigation* 132 (7): e154944. https://doi.org/10.1172/JCI154944.

Wood, Derrick E., and Steven L. Salzberg. 2014. "Kraken: Ultrafast Metagenomic Sequence Classification Using Exact Alignments." *Genome Biology* 15 (3): R46. https://doi.org/10.1186/gb-2014-15-3-r46.

Yan, Wei, A. Brantley Hall, and Xiaofang Jiang. 2022. "Bacteroidales Species in the Human Gut Are a Reservoir of Antibiotic Resistance Genes Regulated by Invertible Promoters." *Npj Biofilms and Microbiomes* 8 (1): 1. https://doi.org/10.1038/s41522-021-00260-1.

Yuan, Sanqing, Dora B. Cohen, Jacques Ravel, Zaid Abdo, and Larry J. Forney. 2012. "Evaluation of Methods for the Extraction and Purification of DNA from the Human Microbiome." *PLoS One* 7 (3): e33865. https://doi.org/10.1371/journal.pone.0033865.

Zhang, Guoan, Roland S. Annan, Steven A. Carr, and Thomas A. Neubert. 2010. "Overview of Peptide and Protein Analysis by Mass Spectrometry." *Current Protocols in Protein Science* 62 (1). https://doi.org/10.1002/0471140864.ps1601s62.

Zhang, Jiajie, Kassian Kobert, Tomáš Flouri, and Alexandros Stamatakis. 2014. "PEAR: A Fast and Accurate Illumina Paired-End Read Merger." *Bioinformatics* 30 (5): 614–20. https://doi.org/10.1093/bioinformatics/btt593.

Zhang, Yuqing, Giovanni Parmigiani, and W. Evan Johnson. 2020. "ComBat-Seq: Batch Effect Adjustment for RNA-Seq Count Data." *NAR Genomics and Bioinformatics* 2 (3): lqaa078. https://doi.org/10.1093/nargab/lqaa078.

Zhao, Zihao, Chie Amano, Thomas Reinthaler, Federico Baltar, Mónica V. Orellana, and Gerhard J. Herndl. 2024. "Metaproteomic Analysis Decodes Trophic Interactions of Microorganisms in the Dark Ocean." *Nature Communications* 15 (1): 6411. https://doi.org/10.1038/s41467-024-50867-z.

Zhou, Ruwen, Siu Kin Ng, Joseph Jao Yiu Sung, Wilson Wen Bin Goh, and Sunny Hei Wong. 2023. "Data Pre-Processing for Analyzing Microbiome Data: A Mini Review." *Computational and Structural Biotechnology Journal* 21: 4804–15. https://doi.org/10.1016/j.csbj.2023.10.001.

Zhou, Zhichao, Patricia Q. Tran, Adam M. Breister, et al. 2019. "Metabolic: High-Throughput Profiling of Microbial Genomes for Functional Traits, Biogeochemistry, and Community-Scale Metabolic Networks." *Preprint, Bioinformatics* (September 8). https://doi.org/10.1101/761643.

Zhou, Zhichao, Patricia Q. Tran, Adam M. Breister, et al. 2022. "Metabolic: High-Throughput Profiling of Microbial Genomes for Functional Traits, Metabolism, Biogeochemistry, and Community-Scale Functional Networks." *Microbiome* 10 (1): 33. https://doi.org/10.1186/s40168-021-01213-8.

6 Metagenomics Approaches for Microbiome

Marcos André Nino Rocha[1], *Karolaine Rios de Oliveira Matos*[2], *Milca da Silva Figueiredo Siqueira*[2], *Saulo Reis Nery Santos*[2], *Rodrigo Dias de Oliveira Carvalho*[2], *Vasco Ariston de Carvalho Azevedo*[1,2], *and Alessandra Estrela-Lima*[1]

[1]Escola de Medicina Veterinária e Zootecnia, Universidade Federal da Bahia, Salvador-BA, Brazil.

[2]Instituto de Biologia, Universidade Federal da Bahia, Belo Horizonte-MG, Brazil.

6.1 INTRODUCTION

Metagenomics has emerged as a powerful tool for investigating complex microbial communities. It enables the direct analysis of genetic material recovered from environmental samples without the need for prior cultivation, facilitating the identification of non-cultivable microorganisms and the study of their functional and ecological roles (Quince et al., 2017; Healey et al., 2017). Advancements in sequencing technologies and bioinformatics have revolutionized our understanding of microbiomes across diverse niches, including soil, aquatic ecosystems, and the human body (Lloyd-Price et al., 2019).

The human microbiome, composed of trillions of bacteria, viruses, fungi, and other microbes on average, performs several vital functions essential to human health, including nutrient digestion, immune system modulation, and protection against pathogens (Turnbaugh et al., 2007; Integrative HMP Research Network, 2019). Imbalances in the intestinal microbiome have been associated with various diseases, such as obesity, type 2 diabetes, inflammatory bowel diseases, and neurological disorders, including autism (Shreiner et al., 2015; Fan and Pedersen, 2021). Recent studies have shown that interventions targeting the microbiome—such as probiotics, prebiotics, and faecal transplants—can restore intestinal homeostasis and improve clinical outcomes (Zhu et al., 2020).

Beyond human health, understanding microbiomes is also essential for advancing sustainable practices in veterinary medicine and agriculture. Soil microorganisms play a critical role in plant health and productivity by promoting growth, enhancing resistance to pests and diseases, and contributing to nutrient cycling (Berendsen et al., 2012; Mendes et al., 2013). Metagenomic techniques have enabled the identification of specific microbial consortia with potential application as biofertilizers and biocontrol agents, thereby reducing the reliance on agricultural chemicals and mitigating environmental impacts (Mendes et al., 2013). In environmental sciences, metagenomics provides advanced tools for ecosystem monitoring and bioremediation. Studies on the marine microbiome have highlighted the essential roles of bacteria and archaea in regulating key biogeochemical cycles, particularly the carbon and nitrogen cycles (Sunagawa et al., 2015; Pachiadaki et al., 2017).

Moreover, gene bioprospecting in extreme environments, such as hydrothermal vents, has led to the discovery of enzymes with industrial applications, including plastic degradation and sustainable biofuel production (Joshi et al., 2014). In agriculture, a detailed understanding of microbial interactions can enhance crop productivity and promote environmental sustainability. The identification of nitrogen-fixing microorganisms in non-legume crops has the potential to transform fertilizer use, helping reduce greenhouse gas emissions (Santi et al., 2013). Additionally, manipulating the root microbiome holds great promise for enhancing plant adaptation to environmental stressors such as salinity, wind, and drought. This approach contributes to the development of more resilient and productive crops (Naylor and Coleman-Derr, 2018).

Understanding microbiomes is a key factor in addressing global challenges related to public health, food security, and environmental conservation. In human health, the microbiome has emerged as a critical component of personalized medicine, supporting the development of therapies tailored to each individual's microbial profile and clinical needs (Franzosa et al., 2019). The microbiome also plays a significant role in infectious disease management, particularly through the identification of antimicrobial resistance genes within specific microbial communities, enabling targeted therapeutic strategies (van Schaik, 2015).

The ability to monitor microbiomes in real time plays a crucial role in guiding strategies for the conservation and restoration of degraded habitats, facilitating timely interventions based on accurate data (Jansson and Hofmockel, 2020). Moreover, recent studies have underscored the role of microorganisms in mitigating environmental pollutants—such as

***Corresponding author:** vascoariston@gmail.com (Vasco Ariston Azevedo)

DOI: 10.1201/9781003544425-6

heavy metals and toxic organic compounds—through natural bioremediation processes (Mishra et al., 2021).

The impact of microbiome research extends well beyond the natural sciences, influencing national socio-economic development and public policy. It is driving innovation across various industries, particularly in the food and beverage sector, by enabling the development of personalized, health-promoting fermented products (Vinderola et al., 2022). Likewise, microbiome-based biotechnologies offer innovative solutions to industrial challenges, such as bioplastic production and wastewater treatment, thereby reducing the environmental impact of industrial waste (Wang et al., 2021).

Metagenomics represents a significant and transformative milestone in modern science, not only by expanding our understanding of microbial communities, but also by providing practical tools to enhance the quality of life and promote planetary sustainability. It offers critical insights into the interactions between microorganisms, their environments, and their ecological impacts. Integrating these advances into public policies, industrial applications, and educational strategies is essential to translate the potential of metagenomics into tangible benefits, contributing to a more sustainable and resilient future in addressing the challenges of the 21st century.

6.2 CORE CONCEPTS OF METAGENOMICS

Two decades ago, the term metagenomics emerged, referring to the analysis of genomes without the need for prior cultivation (Handelsman, 2004). Since then, sequencing technologies have become more affordable, faster, and user-friendly. The development of dedicated tools and databases has also been crucial for advancing the successful application of metagenomics (Tringe et al., 2005). This approach enables the direct analysis of deoxyribonucleic acid (DNA) from entire microbial communities. In essence, genetic material is extracted, organized into libraries, and sequenced using next-generation sequencing (NGS) platforms capable of generating either short or long reads (Trubl et al., 2020).

To perform metagenomic analysis, it is crucial to optimize several steps, including the preparation of biological material, the choice of sequencing techniques, and the execution of bioinformatic analyses (Child et al., 2024).

6.2.1 Microbial Communities DNA Isolation

For the DNA extraction stage, several approaches can be employed, beginning with microbial cell lysis. This step can involve mechanical methods, such as bead-beating, enzymatic digestion, or a combination of both. The use of lysozyme is recommended to disrupt the thick cell walls of microorganisms, particularly Gram-positive bacteria (Barea et al., 2004). Bead-beating involves agitating the sample with small solid particles to mechanically break the cells. Buffer solutions are added to prevent DNA degradation, while enzymes such as proteinase K may be included to facilitate lysis by hydrolysing membrane proteins and releasing the DNA. Mechanical disruption using solid particles is highly effective in breaking cell walls and membranes, often increasing DNA yield. However, it must be applied with caution, as it can lead to DNA shearing—an important limitation in applications requiring high-molecular-weight DNA (Burden et al., 2014). In contrast, enzymatic methods, although more costly, yield highly pure and intact DNA and are preferred when DNA integrity is critical (Greathouse et al., 2019; Child et al., 2024).

The next crucial stage involves DNA purification, which is essential for removing contaminants that could interfere with subsequent steps of metagenomic analysis. The quantity and quality of the extracted genetic material can be evaluated using spectrophotometric and electrophoretic methods to ensure its integrity for sequencing. Proper preparation at this stage requires careful assessment of parameters that influence the quality of DNA library construction. One key criterion is the amount of extracted DNA, as certain sequencing platforms recommend a minimum of 1 μg of high-quality genetic material (Simon et al., 2023; Schrader et al., 2012).

Extensive DNA fragmentation is often associated with inadequate sample storage or mechanical shearing during the purification process. Such fragmentation can result in the loss of genomic information and, consequently, reduced sequencing efficiency (Bertrand et al., 2012). Additionally, the presence of host DNA can pose a significant limitation, particularly in metagenomic approaches based on random DNA sequencing, which tend to preferentially capture the most abundant genetic material. If host-derived eukaryotic DNA predominates, it may obscure the detection of microbial DNA of interest (Nelson et al., 2019). Reagent contamination during sample preparation or sequencing steps can also compromise results (Demkina et al., 2023). To minimize this risk, it is essential to implement rigorous aseptic techniques during all stages of collection and purification, as well as to ensure proper sterilization and decontamination of the workspace where DNA extraction is performed (Davis et al., 2018).

6.2.2 Sequencing Techniques

Several sequencing technologies are currently available, including the classical Sanger method, NGS, and more advanced approaches such as single-cell sequencing. Each of these methods presents distinct advantages, applications, and limitations (Shokralla et al., 2012). Sanger sequencing, developed in the 1970s, is based on the chain termination method (Silva et al., 2022). It employs *Pyrococcus furiosus* (*Pfu*) DNA polymerase to synthesize complementary strands of the target nucleic acid, taking advantage of the enzyme's high stability and fidelity (Lundberg et al., 1991).

The methodology involves the incorporation of dideoxyribonucleotides (ddNTPs), which are structurally similar to standard nucleotides but lack a hydroxyl group (OH) at the 3′ carbon of the pentose ring. This absence prevents the formation of phosphodiester bonds, thereby terminating DNA strand elongation during synthesis. The reaction mixture contains multiple copies of the DNA template, along with DNA polymerase, oligonucleotide primers, and the four standard deoxyribonucleoside triphosphates (dATP, dTTP, dCTP, and dGTP), in addition to the fluorescently labelled ddNTPs (David et al., 2024).

Additionally, chain-terminating reagents known as ddNTPs are incorporated into the reaction, each labelled with a distinct fluorescent dye to enable detection (Silva et al., 2022). Once all components—standard and modified nucleotides, primers, and DNA polymerase—are assembled, the DNA is amplified through repeated cycles, generating fragments of varying lengths. These fragments are then separated by capillary gel electrophoresis, and the fluorescent signals are read to assemble the DNA sequence. Sanger sequencing is considered a reliable method for low-throughput applications (Kumawat et al., 2024). However, it is relatively slow and limited in throughput, typically producing reads of 700–1000 base pairs per reaction. Its primary applications include the identification of specific microorganisms, such as through 16S ribosomal RNA (rRNA) gene sequencing, and validation of NGS results for targeted genes or amplicons (David et al., 2024).

NGS, also known as massively parallel sequencing, emerged in the early 2000s and offers a high-throughput, cost-effective alternative to Sanger sequencing. NGS technologies are broadly categorized into short-read and long-read platforms. Currently, the two leading short-read platforms are Illumina and Ion Torrent (Thermo) (Ye et al., 2022). Illumina sequencing operates on the principle of sequencing by synthesis: the genetic material is fragmented, and adapters are ligated to the DNA fragments. During sequencing, fluorescent signals are emitted as nucleotides are incorporated, and these signals are detected and converted into sequence data by the platform, enabling the simultaneous reading of millions of fragments (Cook et al., 2024).

This methodology is highly precise and cost-effective on a per-read basis, resulting in a high throughput rate. However, its main limitation lies in the generation of relatively short reads—typically around 150–300 base pairs—which can make the assembly of complete genomes from complex metagenomic libraries more difficult (Goodwin et al., 2016; Quince et al., 2017). Despite this, it is widely applied in microbiome research, particularly for taxonomic classification and metagenomic sequencing. In addition to enabling structural diversity analyses to assess microbial community composition in ecological studies, it also supports the investigation of functional genetic potential (Ye et al., 2022).

Ion Torrent technology is based on sequencing by detecting hydrogen ion release during DNA synthesis. As each nucleotide is incorporated during polymerization, a proton is released, leading to a change in the pH of the solution. This pH change is detected as an electrical signal, which is then used to determine the nucleotide sequence (Liu et al., 2012). A known limitation of this method is its relatively high error rate in regions with homopolymer stretches, which can compromise accuracy in certain applications. Nevertheless, it offers the advantages of rapid sequencing and reduced costs. Similar to Illumina, Ion Torrent is widely used in metagenomics for both structural and functional analyses of microbial communities (Guo et al., 2022).

Long-read sequencing technologies are exemplified by the Pacific Biosciences (PacBio) and Oxford Nanopore platforms, both of which are highly valuable for robust gene mapping and for generating more accurate insights into the microbiome. PacBio sequencing is based on single-molecule real-time (SMRT) sequencing. In this method, nanoscale reaction chambers—known as zero-mode waveguides—are used to isolate a single DNA strand along with a DNA polymerase enzyme. The polymerase is immobilized in a matrix with the DNA template, and each nucleotide is labelled with a distinct fluorescent dye. As nucleotides are incorporated during DNA synthesis, the fluorescent signals are detected in real time, allowing the precise determination of the sequence of bases in single DNA molecules (Cook et al., 2024).

The PacBio sequencing methodology is widely employed for generating long reads, often producing sequences spanning thousands of base pairs. It offers high accuracy, particularly when using circular consensus sequencing, which enhances base-calling precision by repeatedly reading the same DNA fragment. This technology is considered highly effective for base detection, complete genome reconstruction from metagenomic samples, full-length gene analysis, and the characterization of complex microbial communities (Antil et al., 2023).

Oxford Nanopore technology utilizes nanopores—small membrane-embedded channels—through which DNA molecules are translocated linearly by an applied electrical current. As each nucleotide passes through the nanopore, it induces a characteristic change in the electrical signal, which is interpreted by dedicated software to generate the DNA sequence. This platform is known for its flexibility and affordability, and it features portable devices that are particularly advantageous for field-based analyses. Unlike other sequencing methods, it enables direct sequencing of microbial communities without extensive preparatory steps and allows real-time data acquisition. However, its precision is lower compared to Illumina platforms (Lu et al., 2016).

Finally, single-cell sequencing involves the isolation and amplification of DNA or RNA from individual cells, enabling the detailed analysis of genetic variability at the single-cell level. One of its main advantages is the ability to study rare or low-abundance microorganisms within complex communities. However, the technique is technically demanding and requires sophisticated protocols and equipment. Despite its complexity, single-cell sequencing is particularly valuable for exploring functional heterogeneity and gene expression patterns in individual cells (Liu et al., 2012).

6.2.3 Bioinformatics Approaches

Once sequencing is completed, raw data is generated and must undergo preprocessing and quality assessment to ensure data reliability and remove adapters when necessary. An initial tool commonly used for this purpose is FastQC, which evaluates the quality of DNA and RNA sequencing data produced by NGS platforms (Lokhande,

2023). FastQC is designed to identify potential issues in the sequencing output, including low read quality, the presence of adapter sequences and contaminants, GC content anomalies, and technical artifacts related to the sequencing platform (Wingett and Andrews, 2018).

Trimmomatic is a widely used tool for trimming sequencing reads, aiming to improve data quality and remove adapter sequences introduced during library preparation. It operates based on customizable filters defined by the user, with the *ILLUMINACLIP* parameter specifically designed for adapter removal (Costa, 2017). Since adapter sequences can interfere with downstream analyses, their removal is a crucial preprocessing step. Another tool commonly employed for trimming is Cutadapt, which specializes in detecting and removing adapter sequences. It is particularly useful in targeted sequencing approaches, such as 16S rRNA gene studies, polymerase chain reaction (PCR)-based assays, miRNA sequencing, and metagenomics. However, in workflows that require trimming based on both sequence quality and read length, Trimmomatic is generally considered more suitable, as Cutadapt is less effective in these scenarios (Martin, 2011).

Metagenomics can follow two main approaches: structural metagenomics, which focuses on characterizing the taxonomic composition of microbial communities to assess environmental diversity, and functional metagenomics, which investigates the biological functions encoded by the microbial genes present. Each of these approaches requires distinct analytical tools and reference databases (Zhou et al., 2022). These methodologies are commonly referred to as amplicon sequencing and shotgun metagenomic sequencing, respectively.

6.2.4 Amplicon Metagenomics

Amplicon sequencing is based on a molecular technique known as PCR (Sanschagrin et al., 2014). This method involves the amplification of a specific DNA fragment of interest to increase the number of copies in the library, thereby facilitating its detection and analysis. The resulting amplified fragments are referred to as amplicons (Klindworth et al., 2013). During PCR, primers—short nucleotide sequences complementary to target regions—are used to define the starting point for amplification. Additionally, adapters are ligated to the ends of the amplicons to enable their identification and differentiation across multiple samples during sequencing (Sanschagrin et al., 2014).

After the primers are added, the complementary DNA strand is synthesized by the action of a DNA polymerase enzyme, which also plays a central role in the natural replication of genetic material in vivo (Lorenz, 2012). The PCR process consists of several cycles, each with specific temperature requirements to ensure proper function (Garibyan et al., 2013; Green and Sambrook, 2018). The first step, denaturation, involves breaking the hydrogen bonds between the two DNA strands, typically achieved by heating the mixture to temperatures above 90°C (Green and Sambrook, 2018). Following denaturation, the temperature is lowered to around 56–60°C to allow primer annealing—where the primers hybridize to their complementary sequences on the single-stranded DNA (Lorenz, 2012; Glenn et al., 2019). Once bound, DNA synthesis begins through the action of thermostable enzymes such as Taq polymerase (*Thermus aquaticus*) or *Pfu* DNA polymerase I, which incorporate nucleotides complementary to the original template strand (Garibyan et al., 2013; Lorenz, 2012; Green and Sambrook, 2018).

These amplified regions typically correspond to conserved genes that enable the identification and differentiation of organisms, particularly microorganisms. Thus, the primary objective of amplicon sequencing is the taxonomic classification of microbial communities in various environments. Additionally, it allows for predictive ecological and functional analyses across a wide range of sample types, including soil, water, and mucosal surfaces (Notario et al., 2023). The amplified sequences are known as target gene regions—specific segments of genes that are well-characterized due to their conserved nature, which varies depending on the microbial group being studied (Notario et al., 2023). Among the most frequently analyzed targets is the 16S rRNA gene, universally present in bacteria and archaea. This gene is essential for protein synthesis, as it forms part of the 30S subunit of the prokaryotic ribosome (Klindworth et al., 2013; Nunes-Ramos et al., 2025).

The 16S rRNA gene is extensively used to determine the taxonomy of microorganisms (Matchado et al., 2024). This gene is approximately 1,500 base pairs in length and, although it contains highly conserved regions, it also includes nine hypervariable regions (V1–V9) interspersed between these conserved segments (Hassler et al., 2022; Varliero et al., 2022). During sequencing, the conserved regions serve as binding sites for primers, since they are nearly identical across bacterial and archaeal species, while the hypervariable regions enable species differentiation due to sequence variability among taxa (Hassler et al., 2022). Therefore, the primers used in PCR are designed to anneal to these conserved regions, facilitating the amplification of the 16S rRNA gene, while the variable regions allow for taxonomic resolution—typically at the genus level—based on sequence similarity (Maki et al., 2023).

To perform amplicon sequencing, it is essential to determine which gene regions will be amplified, depending on the specific research objectives. Certain regions offer higher taxonomic resolution than others, as they may allow more precise classification due to lower redundancy in base pair sequences (Maki et al., 2023; Notario et al., 2023). While the 16S rRNA gene is widely used for identifying bacteria and archaea, other target regions are employed to classify different groups of microorganisms. For instance, the internal transcribed spacer (ITS) region is commonly used for fungi due to its high sequence variability, which allows differentiation down to the species level. The ITS region is typically divided into ITS1 and ITS2 subregions, both of which are informative for fungal taxonomy (Theil et al., 2021).

Currently, amplicon sequencing is considered a relatively simple and cost-effective technique compared to approaches such as whole-genome or shotgun metagenomic sequencing. It is widely employed to characterize the microbial composition of diverse environments (Sanschagrin et al., 2014; Zhang et al., 2023). Moreover, this method enables the enrichment and detection of DNA from rare or low-abundance microorganisms, facilitating their identification within complex microbial communities across a variety of sample types (Sanschagrin et al., 2014).

However, although the hypervariable regions of marker genes assist in differentiating microbial genera, a single gene is generally insufficient for precise species-level classification. This is because the resulting amplicon may not be unique to a single species, and the PCR-based approach does not provide access to the complete genome (Hassler et al., 2022; Matchado et al., 2024). Additionally, there are limitations on the number of regions that can be analysed simultaneously, as each requires specific primers. Amplicon sequencing is, therefore, constrained primarily to taxonomic profiling. To investigate microbial metabolic functions and gain deeper insight into the functional potential of the community, complementary approaches—such as whole metagenomic sequencing—are necessary (Soriano-Lerma et al., 2020; Matchado et al., 2024).

6.2.5 Shotgun Metagenomics

Shotgun sequencing is a technique that involves the direct sequencing of all genetic material present in a sample by randomly fragmenting it into smaller segments for subsequent reading and assembly (Quince et al., 2017). Similar to amplicon sequencing, it does not require prior cultivation of the microbial community and can be applied directly to environmental samples. The random generation of DNA fragments facilitates comprehensive genetic analysis of the microbial composition in the sample of interest (Zhou et al., 2022). This approach can be employed for whole-genome sequencing of individual organisms and is particularly valuable in metagenomics, as it enables the identification and characterization of a wide range of genomes—including those of bacteria, fungi, viruses, and archaea—within a single sample (Taş et al., 2021; Roy et al., 2024).

As previously described, sample collection for shotgun sequencing can originate from various environments, including the human microbiome, soil, and marine ecosystems. After DNA extraction and fragmentation, adapters are ligated to the ends of the generated fragments to enable their subsequent sequencing on NGS platforms. The choice of sequencing platform depends on factors such as cost, intended application, advantages, and the expected size of the DNA fragments (Gawad et al., 2016). Platforms like Illumina, for instance, produce a high volume of short reads—typically around 150 base pairs each—which are later computationally assembled to reconstruct entire genomes (Mathieu et al., 2023; Mallawaarachchi et al., 2024).

Pre-processing of sequencing reads is essential to enhance the accuracy and efficiency of downstream analyses. This involves assessing read quality and removing adapter sequences that may interfere with data interpretation. Tools such as FastQC and Cutadapt are commonly used for these initial quality control and trimming steps (Martin, 2011). Additional software, including Trimmomatic (Bolger et al., 2014), Trim Galore, and Fastp, are also widely employed for filtering and trimming purposes and are compatible with multiple high-throughput NGS platforms. A significant limitation of shotgun metagenomic libraries is the presence of contaminating genetic material, particularly host-derived DNA, which can obscure microbial signals. To address this issue, one effective strategy involves aligning sequencing reads against a reference genome to identify and exclude contaminant sequences. Tools such as BWA (Li and Durbin, 2009) and Bowtie 2 (Langmead and Salzberg, 2012) are frequently used for this purpose, enabling accurate mapping of unwanted reads to the host genome and their subsequent removal from the dataset.

After the pre-processing stage, the genome assembly phase begins. This step aims to reconstruct the genomes of the organisms present in the sample, whose DNA was fragmented during sequencing (Ghurye et al., 2016; Mallawaarachchi et al., 2024). Metagenome assembly can be carried out using two main approaches: reference-based mapping or de novo assembly (Ghurye et al., 2016). In reference-based mapping, sequencing reads are aligned to existing genomic databases of known organisms. The alignment process compares the obtained reads with reference sequences and organizes them based on sequence similarity, generating a consensus sequence—a representative sequence reconstructed from the aligned reads (Ghurye et al., 2016).

De novo assembly does not rely on a reference genome; instead, it reconstructs genomes solely from the sequencing data by identifying overlapping regions among the reads. These overlapping fragments are merged into contiguous sequences known as contigs (Achudhan et al., 2024; Roy et al., 2024). When multiple contigs are generated, they can be further organized into larger sequences called scaffolds. To bridge the gaps between scaffolds or contigs, additional sequencing strategies may be employed, such as mate-pair libraries, long-read technologies, or molecular techniques like Sanger sequencing and PCR. Among the primary tools for de novo assembly are MEGAHIT and metaSPAdes, both of which utilize De Bruijn graphs to assemble the genome. These graphs are constructed based on the frequency of k-mers—short nucleotide subsequences of length k—which serve as building blocks for reconstructing the genome from complex metagenomic data (Nurk et al., 2017; Li et al., 2012, 2015).

As metagenomics involves the analysis of genomes from diverse microorganisms, a critical step in the sequencing and identification process is binning—a method used to group sequenced reads or assembled contigs into distinct clusters (*bins*), each ideally representing an individual microbial genome (Mallawaarachchi et al., 2024). Binning can be performed either *before* genome assembly, directly on the reads, or *after*, based on the assembled contigs, depending on the chosen assembly strategy. This process may rely on reference-based approaches, where sequence data is compared against known genome databases to generate a taxonomic profile based on sequence similarity (Ye et al., 2019; Wickramarachchi et al., 2020). Alternatively, reference-free (de novo) binning can be applied. In this case, grouping is based on intrinsic features of the sequences, such as k-mer composition (short nucleotide motifs), nucleotide usage patterns specific to microbial species, and sequencing coverage depth across fragments (Wickramarachchi et al., 2020; Tolstoganov et al., 2022). This allows the identification and reconstruction of genomes even from previously uncharacterized microorganisms.

Among the tools available for binning, Kraken is widely used for reference-based binning and can be applied even before genome assembly. It classifies sequences by aligning k-mers to a comprehensive reference database, enabling rapid taxonomic identification (Lamurias et al., 2022). For post-assembly binning, tools like MaxBin and MetaBAT are frequently employed. MetaBAT relies on sequence coverage and nucleotide composition to estimate the distance between contigs and cluster them accordingly (Kang et al., 2019). In contrast, MaxBin uses a probabilistic model to estimate the likelihood that specific contigs belong to the same bin, also estimating the number of genomes present in the sample (Wu et al., 2014). Another notable tool is MetaPhlAn, previously mentioned, which performs taxonomic profiling using clade-specific marker genes. Its high sensitivity allows for precise identification of microbial taxa present in a metagenomic community (Blanco-Míguez et al., 2023).

Taxonomic classification plays a fundamental role in interpreting the diversity and richness of microbial communities. Beyond identifying dominant taxa, it enables the detection of rare or fastidious microorganisms that may not be cultivable using traditional media. Among the available tools, Kraken stands out for its widespread use in metagenomic analyses. It performs taxonomic assignment by aligning k-mers from sequencing reads against customizable databases, including public repositories such as NCBI RefSeq, providing rapid and accurate classification (Jünemann et al., 2017; Wood et al., 2019).

Functional genome annotation aims to identify and characterize the biological roles of DNA elements, including genes and predicted proteins, thereby elucidating the metabolic potential of reconstructed genomes and their involvement in key ecological or biotechnological processes (Zhou et al., 2022). This analysis allows insights into microbial contributions to nutrient cycles, environmental resilience, and industrial applications. To conduct functional annotation, several widely accessible tools and databases can be employed, particularly those based on sequence orthology. Among them, Kyoto Encyclopaedia of Genes and Genomes (KEGG) and Clusters of Orthologous Groups (COG) are commonly used. KEGG emphasizes metabolic pathways and networks, enabling the identification of biosynthetic routes such as amino acid production (Kanehisa and Goto, 2000).

COG clusters orthologous genes by examining their evolutionary relationships, facilitating the identification of conserved functions across different organisms (Galperin et al., 2021). Complementing KEGG for metabolic pathway reconstruction, the KEGG Automatic Annotation Server (KAAS) enables direct pathway mapping from KEGG databases using sequence input (Moriya et al., 2007). For the annotation of specific gene categories, it is essential to employ specialized databases. One example is the Comprehensive Antibiotic Resistance Database (CARD), which is dedicated to identifying and characterizing genes associated with antibiotic resistance (Alcock et al., 2023).

Hence, the shotgun sequencing method not only enables the taxonomic classification of microorganisms but also provides insights into their environmental interactions, metabolic functions, and biological evolution (Jünemann et al., 2017). In a study by Nayfach et al. (2021), soil microbiomes were reconstructed using 10,450 metagenomes from diverse environments, highlighting the potential of this approach in uncovering genetic diversity and identifying genes and metabolic pathways with biotechnological relevance. However, a major challenge of shotgun metagenomics lies in the requirement for specialized and often costly equipment, advanced technologies, and substantial computational resources (Gilbert and Stephens, 2018).

6.3 MICROBIOME IN HEALTH AND DISEASE

Since the late 1980s, the concept of "microbiome" has been widely explored in the scientific literature. Originally, it was defined as the combination of "micro," referring to the microbial component, and "biome," encompassing the environment and its physical–chemical characteristics (Whipps et al., 1988). Today, the microbiome is recognized as a complex network of interactions among microbial communities and the ecosystems they inhabit, whether natural or host-associated (Berg et al., 2020).

Within this context, the microbiota—defined as the microbial population that inhabits or transiently resides in various environments—plays a pivotal role in host biology. Studying microbiota is essential for understanding the relationship between health and disease, as it offers valuable insights for the diagnosis, treatment, and prevention of various conditions. It directly influences host homeostasis and comprises a diverse array of microorganisms, including bacteria, fungi, viruses, and archaea, which participate in multiple biological processes (Trinh et al., 2018; Ursell et al., 2012; Colston and Jackson, 2016). Among the critical functions attributed to these microbes are the extraction of nutrients, the biosynthesis of bioactive molecules (Hou et al., 2022), the modulation of the immune, endocrine, and nervous systems, and the establishment of a protective barrier against pathogenic organisms by competing for ecological niches (Siddiqui and Khan, 2023; Filardo et al., 2024).

The diversity of functions performed by microorganisms in the microbiota is remarkable, particularly during early life. A properly established microbiota is critical during the neonatal period, as it plays a fundamental role in immune system development. Studies have shown that infants delivered vaginally tend to develop a microbiota resembling the maternal vaginal microbiota, dominated by *Lactobacillus* and *Prevotella* spp., which are essential for appropriate immune stimulation. In contrast, caesarean delivery often results in a microbiota similar to that of the skin, with a predominance of *Staphylococcus* and *Propionibacterium* spp., which has been associated with an increased risk of respiratory, dermatological, allergic, and gastrointestinal disorders later in life (Dominguez-Bello et al., 2010; Neu and Rushing, 2011; Hoffman et al., 2016). This phenomenon is not limited to humans; similar distinctions in early microbial colonization have been observed in dogs, where the mode of delivery significantly influences the microbial composition and diversity of colostrum. These differences have been linked to improved nutrient absorption, accelerated development, and enhanced immune function (Kajdič et al., 2021).

Microbial imbalance, or "dysbiosis," refers to alterations in microbial populations or reduced microbial diversity, creating conditions that favour the proliferation of pathogenic microorganisms. This disruption compromises the equilibrium of microenvironments and contributes to the development of various diseases (Hooks and O'Malley, 2017). Comparative studies between the human microbiota and those of other animal species, such as pigs and mice, are frequently conducted (Coelho et al., 2018; Xiao et al., 2015). However, among these models, the canine microbiota is considered the most similar to that of humans (Coelho et al., 2018), underscoring the value of translational research, particularly in the context of dietary interventions and metabolic disorders. Dysbiosis can affect several microenvironments across human and animal organisms—including the gastrointestinal tract, skin, oral cavity, respiratory system, and even the nervous system—and has been associated with the onset of metabolic diseases such as obesity, diabetes, cardiovascular disorders, and cancer.

6.3.1 Gut Microbiome

Diseases of the digestive tract associated with dysbiosis, particularly inflammatory bowel diseases (IBD), have been extensively investigated due to their chronic nature and substantial impact on health. Among these, Crohn's disease and ulcerative colitis stand out as significant and emerging health concerns, affecting up to 0.5% of the global population (Ng et al., 2017). Patients with IBD often exhibit distinct alterations in their intestinal microbiota when compared to healthy individuals (Oligschlaeger et al., 2019). In this context, metagenomics has emerged as a promising tool for the development of personalized therapies (Qiu et al., 2022). Notably, there are important similarities in microbiota alterations between humans and dogs with gastrointestinal diseases, particularly the increase in *Proteobacteria* and the reduction in *Bacteroidetes* populations (Frank et al., 2007; Xenoulis et al., 2008). These findings reinforce the value of translational studies, although interspecies differences in microbial composition must be considered (Axelsson et al., 2013).

6.3.2 Skin Microbiome

Cutaneous dysbiosis is primarily associated with an increase in inflammatory skin conditions, such as atopic dermatitis, which in humans is often linked to an overgrowth of *Staphylococcus aureus* (Kong et al., 2012). In animals, similar conditions are observed, including allergic dermatitis in dogs (Meason-Smith et al., 2015) and hoof diseases in ruminants (Klitgaard et al., 2008). The identification and characterization of the cutaneous microbiota have become essential tools for developing targeted therapeutic strategies, particularly those aimed at modulating microbial populations. This is especially relevant given the growing evidence that specific skin diseases may be associated with known microbial profiles (Grice, 2014; Santiago-Rodriguez et al., 2023).

6.3.3 Oral Microbiome

The balance of the oral microbiota is essential, as it contributes to multiple biological functions, including feeding, oral respiration, and interactions with surrounding skin microbiota. Disruptions in this balance can contribute to the development of local pathological conditions, such as dental caries, periodontitis (Dewhirst et al., 2010), and stomatitis (Marchini et al., 2007). Several factors may influence the composition of the human oral microbiota, including diet, alcohol and tobacco use, and oral hygiene practices (Pitiphat et al., 2003; Krishnan et al., 2017; Sedghi et al., 2021; Tribble et al., 2019; Hasan and Yang, 2019).

The phyla *Firmicutes*, *Bacteroidetes*, *Proteobacteria*, *Actinobacteria*, *Spirochaetes*, and *Fusobacteria* represent 96% of the total oral microbiota (Dewhirst et al., 2010), along with fungi such as *Candida* sp., *Cladosporium*, and *Aureobasidium* (Ghannoum et al., 2010) and viruses such as *Herpes Simplex* (Woo and Challacombe, 2007) and *Papillomavirus* (Kumaraswamy and Vidhya, 2011). Animal models, mainly using rodents, can provide important information on the dynamics of diseases in the oral cavities of both species. Although the animal model does not mimic human experience, research related to the microbiota can provide important methodologies with therapeutic potential (Hajishengallis, 2023).

6.3.4 Upper Respiratory Tract Microbiome

The respiratory tract is a complex system that extends from the nasal cavity to the lungs, harboring a diverse and region-specific microbiota. A wide variety of bacteria (Piters et al., 2020; Dickson et al., 2017), fungi—including *Ascomycetes* and *Streptomyces* species (Sharma et al., 2019)—and viruses from families such as Paramyxoviridae, Picornaviridae, and Orthomyxoviridae (Lysholm et al., 2012) inhabit this system. Alterations in the respiratory microbiota have been associated with a range of clinical conditions, from common diseases such as asthma (Chung, 2017) to complex chronic disorders such as cystic fibrosis (Lipuma, 2010). These findings underscore the importance of host–microbiota interactions in respiratory health and disease, involving mechanisms that range from relatively simple immune modulation to intricate and multifactorial pathological processes.

6.3.5 Gut–Brain Axis

Although the central nervous system (CNS) and the gastrointestinal tract are anatomically distinct, the gut–brain axis has emerged as a key area of investigation in microbiome research (Martin et al., 2018). This bidirectional communication system operates primarily through immunomodulatory pathways and the vagus nerve, which extends from the brainstem, traverses the thoracic cavity, and reaches the abdominal region. It receives input from the intestinal mucosa and systemic circulation, mediated by neuroactive molecules produced by the gut microbiota and host cells (Hajiasgharzadeh and Baradaran, 2017; Kaelberer et al., 2018).

Owing to both direct and indirect interactions along the gut–brain axis, several neurological conditions—including Alzheimer's disease, Parkinson's disease, autism spectrum disorders, depression, and anxiety—have been linked to alterations in the composition of the gut microbiota (Loh et al., 2024; Ullah et al., 2023). In particular, viral pathogens such as *Herpes Simplex Virus* type 1 and *Cytomegalovirus*, as well as bacteria like *Helicobacter* spp. and *Chlamydia pneumoniae*, have been implicated in the pathogenesis and progression of Alzheimer's disease (Bu et al., 2015). In the case of Parkinson's disease, microbial dysbiosis is often characterized by an increase in *Proteobacteria* and *Betaproteobacteria* and a reduction in *Faecalibacterium* and *Firmicutes* populations (Keshavarzian et al., 2015).

6.3.6 Metabolic Diseases

Multiple factors contribute to the development of metabolic diseases, including poor dietary habits, physical inactivity, and genetic predispositions. These elements can lead to chronic conditions such as obesity, type 2 diabetes mellitus, and cardiovascular disorders, which are widely recognized in both human and veterinary medicine (Oda, 2012; Fan and Pedersen, 2021).

Obesity, often associated with sedentary behaviour and excessive caloric intake, affects a substantial portion of the global population. In the United States, more than 40% of individuals were classified as obese in 2020 (Noor et al., 2023). This condition represents a major public health concern that is difficult to manage and also impacts various animal species, exhibiting pathophysiological dynamics similar to those observed in humans (Morelli et al., 2022). Emerging evidence suggests that alterations in the gut microbiota contribute to the development of obesity (Sonnenburg and Bäckhed, 2016; Gomes et al., 2018). Certain microbial taxa, including members of the Christensenellaceae family and genera such as *Methanobacteriales*, *Lactobacillus*, and *Bifidobacteria*, are considered beneficial due to their role in regulating energy harvest and metabolism (Depommier et al., 2019; Crovesy et al., 2017; Yin et al., 2010). In contrast, an increased abundance of *Firmicutes* has been linked to enhanced energy absorption, promoting caloric accumulation and adipose tissue deposition (Filardo et al., 2024).

Diabetes, in turn, is a metabolic disorder characterized by impaired insulin production, functionality, or receptor sensitivity, resulting in insulin resistance. Type 1 diabetes is primarily immune-mediated, whereas type 2 diabetes has a multifactorial origin, involving genetic and environmental influences (Beydag-Tasöz et al., 2023). Increasing evidence suggests that intestinal dysbiosis plays a role in the pathogenesis of type 2 diabetes by promoting systemic inflammation, enhancing intestinal permeability, and disrupting bile acid metabolism (Barlow and Mathur, 2022). These disturbances can impair pancreatic β-cell function and compromise insulin secretion (Tripathy and Chavez, 2010). Furthermore, cardiovascular disorders such as myxomatous mitral valve disease have been associated with dysbiotic changes. This condition, observed in both humans and animals, may remain asymptomatic or progress to congestive heart failure, particularly in obese or elderly individuals (Li et al., 2021).

6.3.7 Cancer

The tumour microenvironment comprises a dynamic and complex interplay among neoplastic cells, immune cells, mesenchymal and endothelial cells, the extracellular matrix—including polysaccharides, proteins, and lipids—and a variety of microorganisms that collectively constitute the tumour microbiome (Ciernikova et al., 2022; Choi et al., 2023). These components can contribute to tumour initiation, progression, and therapeutic response modulation (Banerjee et al., 2021). A growing body of evidence has linked specific microbiota profiles to the development of various cancer types, including *breast* (Urbaniak et al., 2014), lung (Huynh et al., 2023), gastric (Ferreira et al., 2018), and colorectal cancer (Burns et al., 2015). The primary mechanisms implicated in this association involve DNA damage, epigenetic modifications, disruption of genomic integrity checkpoints, modulation of cell signalling pathways, and suppression of the immune response (Zhao et al., 2023).

6.4 APPLICATIONS OF MICROBIOME TECHNOLOGIES

Studying the human microbiota through autologous models offers valuable insights into its composition and disease-associated alterations. Animal models have been widely adopted to investigate microbiota modulation, enabling the assessment of its functional impacts at therapeutic levels (Graves et al., 2008). These approaches highlight the feasibility and potential of integrating metagenomics into clinical practice.

The literature highlights rodents as effective experimental models, often offering practical advantages over other species such as dogs, pigs, and non-human primates (Hajishengallis, 2023). Among the main benefits are moderate costs, ease of handling, and the availability of numerous strains for research purposes (Webster et al., 2020). However, notable limitations include significant differences in the microbiota between rodents and other animals, as well as physiological, anatomical, and interindividual variations that can affect experimental outcomes (Hajishengallis et al., 2015).

In oncology, the study of microbiota has revealed not only its role in tumorigenesis but also its potential as a biomarker. For instance, in patients with oesophageal squamous cell carcinoma, elevated levels of *Fusobacterium nucleatum* have been associated with poorer responses to chemotherapy and a more unfavourable prognosis (Yamamura et al., 2019). Faecal microbiota transplantation (FMT) has emerged as a promising strategy for applying metagenomic insights. Despite the considerable diversity of microorganisms within the intestinal microbiota, non-culture-based identification methods have shown that only approximately 40% of these microorganisms are cultivable (Suau et al., 1999). Sequencing technologies have thus enabled a more comprehensive understanding of biological processes. FMT involves transferring faecal material from a healthy donor with a balanced intestinal microbiota to a recipient experiencing dysbiosis, and it has been applied in both human and veterinary medicine.

The feces of a healthy donor contain a balanced microbial population, along with essential vitamins and proteins that contribute to the recipient's recovery (Khoruts et al., 2010). One well-documented application of this approach is in the treatment of infections caused by *Clostridium difficile*, a common pathogen implicated in gastrointestinal disorders such as severe diarrhea, particularly in patients who have undergone prolonged antibiotic therapy (Seekatz et al., 2014). Metagenomic studies are critical for detecting such microbial variations. As microbiome research continues to evolve, its applications are becoming increasingly refined, including the ability to predict the likelihood of organ transplant acceptance or rejection. In lung transplantation, for instance, the composition of the microbiome, when analysed alongside other clinical parameters, has shown high predictive accuracy (Watzenboeck et al., 2022).

6.4.1 Animal Improvement

Research on the microbiome in farm animals, such as chickens and cattle, primarily focuses on four main objectives: improving the quantity and quality of high-demand animal products such as meat and milk; preventing diseases and promoting animal health; evaluating the impact of consuming animal-derived products on human health; and understanding how dietary changes influence animal metabolism (Peixoto et al., 2021). Optimizing the animal microbiota, particularly within the gastrointestinal tract, is crucial for enhancing productivity and overall performance. In ruminants, for instance, the microbiota plays a fundamental role in converting otherwise indigestible structural plant compounds into essential nutrients (Li et al., 2020).

In cases of dysbiosis, the disruption of microbial balance can lead to significant impairments in animal performance, particularly in ruminants. Ruminal transformation has emerged as a valuable approach for recovery. This practice involves ruminal microbiota transplantation, in which ruminal fluid from healthy donor animals is transferred to diseased recipients. The transferred fluid contains a balanced microbiota and appropriate physicochemical characteristics, facilitating the restoration of the recipient's ruminal health. This method is considered economical, rapid, and safe, as the introduced microorganisms effectively colonize the affected rumen (Steiner et al., 2020).

A study comparing the ruminal microbiota of lambs and adult sheep demonstrated that microbial composition is directly associated with the production of metabolites and short-chain volatile fatty acids, significantly influencing animal metabolism and nutritional efficiency. For instance, microorganisms such as *Clostridium* and Ruminococcaceae were found to enhance L-leucine production while reducing levels of phenylethylamine (Li et al., 2020). These findings underscore the importance of targeted research to identify microbial profiles that meet specific physiological demands and optimize animal performance.

In 2006, the European Union took a landmark step by banning the use of antimicrobial growth promoters in poultry production. This decision, driven by evidence that even subtherapeutic antibiotic doses contributed to the emergence of resistant strains (Philips et al., 2004), sparked a global movement toward safer and more sustainable practices. The search for viable alternatives has since expanded to include probiotics, prebiotics, and symbiotics, as well as organic acids, enzymes, phytogenic compounds, and metals. More recently, innovative approaches such as antibacterial peptides and bacteriophages have emerged as promising strategies (Gadde et al., 2017). This international shift highlights the scientific community's collective responsibility in combating the global threat of antibiotic resistance.

An intriguing example of indirect optimization in poultry production is the supplementation of feed with stevioside derivatives, a class of sugar substitutes. Although chickens possess relatively few receptors for sweet taste, these compounds have been shown to increase feeding frequency by modulating the intestinal microbiota. This microbial alteration appears to influence neuroactive signaling pathways within the enteric nervous system, thereby promoting enhanced feeding behaviour (Jiang et al., 2021).

Although considerable progress has been made in elucidating the relationship between soil microbiota, pastures, and animal microbiomes, many critical questions remain unanswered. The specific functions of each microbial component, their interactions, and the optimal composition of these communities still require extensive investigation. Manipulating the animal microbiome to enhance feed conversion efficiency and overall productivity holds great promise; however, further research is essential to assess the consequences of such interventions at both the individual and ecosystem levels (Attwood et al., 2019). The urgency of advancing this research lies in its potential to significantly impact animal science and sustainable agricultural practices.

6.4.2 Crop Improvement

Similar to animals, plants harbour a specific microbiota that interacts with them in complex and dynamic ways (Berendsen et al., 2012). These plant-associated microbial communities are typically categorized into three main microenvironments: the rhizosphere, which is the soil region surrounding the roots, enriched with nutrients and microbial activity (Lynch, 1990); the endosphere, referring to the internal tissues of the plant (Compant et al., 2012); and the phyllosphere, which encompasses the aerial surfaces of the plant, particularly the leaves (Vorholt, 2012). Each of these niches hosts distinct microbial signatures, shaped by both the plant's physiology and environmental conditions.

In response to climate change and the growing global population, numerous strategies are being explored to enhance agricultural productivity (Figure 6.1) (Hamilton et al., 2016; Ray et al., 2012). Among these, modulation of the microbiota associated with agronomic crops has emerged as a promising approach. These microbial communities play essential roles in plant development, including nutrient acquisition, biological nitrogen fixation,

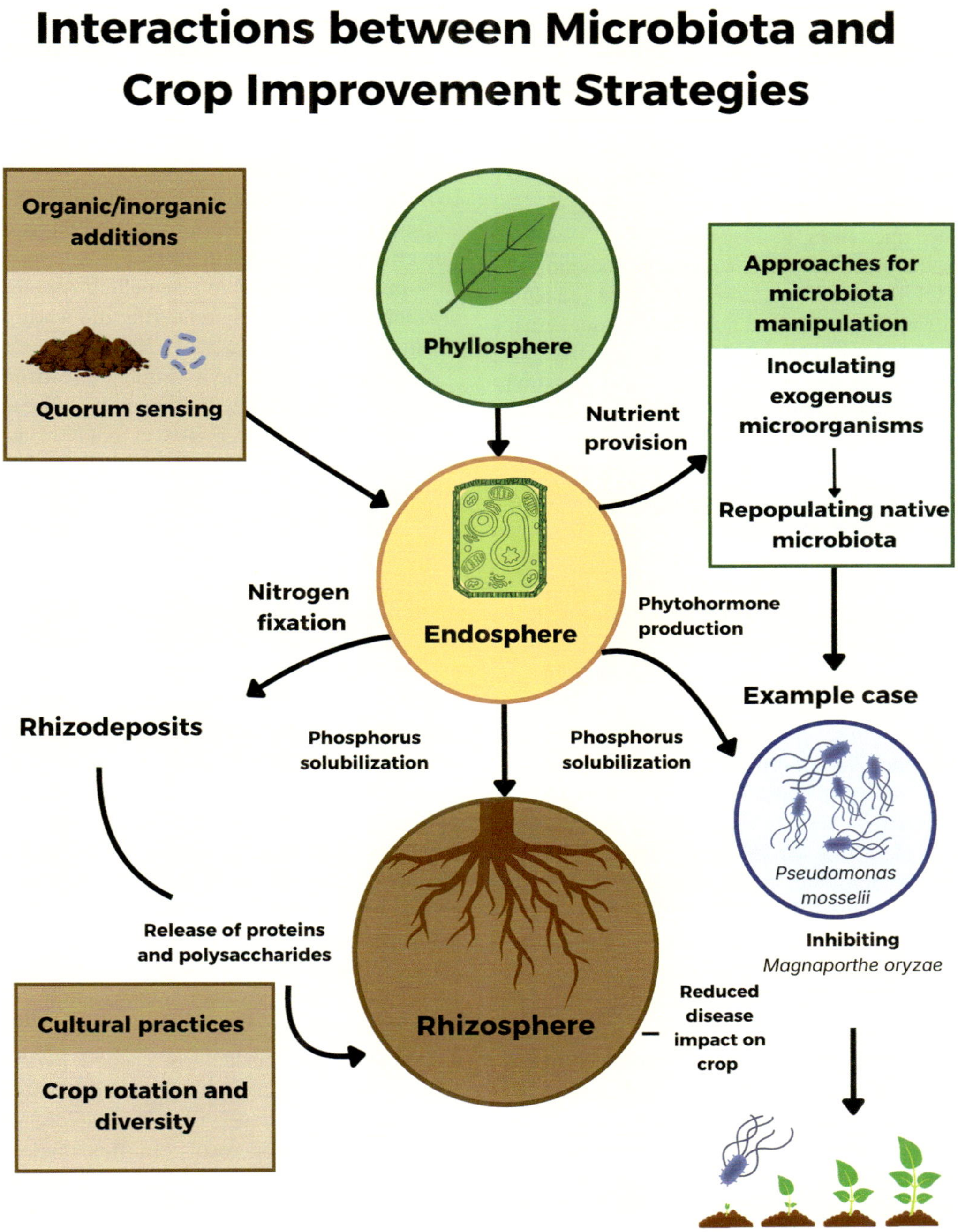

FIGURE 6.1 Interaction of microbiota with crop improvement strategies. This diagram illustrates the different regions of plant-associated microbiota, their functions, and strategies for improving crop productivity through microbiota modulation.

phytohormone production (Ahmad et al., 2011), and the solubilization of key minerals such as phosphorus, potassium, calcium, and magnesium (Altomare et al., 1999).

Soil composition can be intentionally modified to enhance microenvironmental conditions and support the development of beneficial microbiota. The incorporation of organic or inorganic components may facilitate plant–microbiota interactions (Seidt et al., 2021) and promote quorum sensing—an essential mechanism by which microorganisms regulate interactions with each other and with plant hosts. This signaling system plays a key role in symbiosis, pathogenicity, biofilm formation, and surface-associated communication (Loh et al., 2002).

During normal plant metabolism, rhizodeposits are released, consisting of macromolecules such as proteins and polysaccharides. Additionally, root exudates comprise a diverse array of soluble and insoluble low-molecular-weight organic compounds, including sugars, amino acids, and phenolic substances (Bais et al., 2006; Grayston et al., 1998; Oburger et al., 2013). These compounds play a critical role in nourishing and sustaining the associated microbiota (Ryan et al., 2009).

Since 1978, the potential of "plant growth-promoting rhizobacteria" (PGPR) has been extensively studied (Kloepper and Schroth, 1978). These bacteria colonize the root systems of plants, enhancing nutrient availability, facilitating biological nitrogen fixation, modulating phytohormone activity, and suppressing phytopathogens (Ping and Boland, 2004). Although research suggests that the rhizosphere can be manipulated to benefit various crops, the complex interactions between bacterial communities and plants remain incompletely understood. It is crucial to account for factors such as the abundance and structure of native microbial communities as well as soil physicochemical properties (Ryan et al., 2009).

Two principal strategies are employed to integrate microbiota modulation into agriculture: the inoculation of exogenous microorganisms into specific crops or the repopulation of the microenvironment with native microbial communities (Quiza et al., 2015; Ryan et al., 2009). Leveraging metagenomic approaches, these interventions aim to enhance crop gene expression, optimize enzymatic activities, and improve nutrient availability (Kaul et al., 2021; Ribeiro et al., 2020). An illustrative example of resident microbiota suppressing pathogens is *Pseudomonas mosselii*, which has demonstrated effective inhibition of the fungus *Magnaporthe oryzae*, the causative agent of rice blast disease—a condition that significantly diminishes grain yield and quality (Wu et al., 2018). This exemplifies the potential of metagenomic research in disease control and the development of targeted agricultural interventions.

A study investigating the rhizosphere microbiota in peanut plantations revealed that monoculture practices lead to microbial populations becoming increasingly specialized to the host plant (Li et al., 2019). To counteract this specialization and promote microbial diversity, agronomic techniques such as crop rotation and intercropping with different species within the same environment have been proposed as effective strategies (Suman et al., 2022).

6.4.3 Environmental Improvement

The concept of *One Health* has gained widespread acceptance by highlighting the essential balance among humans, animals, and the environment to safeguard global health (Figure 6.2) (Ma et al., 2023). Microorganisms are ubiquitously distributed across various environments and microenvironments, with their interactions exerting direct influences on both human and animal health (Trinh et al., 2018; Dominguez-Bello et al., 2010). Within this framework, genetic sequencing technologies, bioinformatics tools, and omics sciences serve as powerful instruments for characterizing the complex and diverse microbial populations inhabiting distinct ecosystems.

This analytical approach enables the detection of genetic material traces and the comprehensive characterization of the microbiota, generating extensive datasets from a single procedure (Bruno et al., 2022). Access to clean and potable water remains a critical global challenge, exacerbated by social and technological developments that often lead to surface water contamination, necessitating urgent mitigation efforts (Mahajna et al., 2022). While most microorganisms present in water are non-pathogenic, they can impact organoleptic properties, rendering the water unsuitable for consumption (Vreeburg et al., 2004).

Metagenomics has demonstrated effectiveness in assessing water quality, particularly in quantifying microbial load. For example, Calderón-Franco et al. (2023) investigated the microbiome, resistome, and mobilome of water samples collected from countries with low antibiotic usage. The findings from such studies enable a deeper understanding of microbial community structures under varying environmental conditions, facilitating the identification of genetic markers indicative of the microbiota's potential pathogenicity (Mahajna et al., 2022). Regarding airborne microbiota, culture-independent approaches, including targeted and total sequencing, provide critical insights into the interactions between environmental factors and microbial populations, thereby enabling the characterization of conditions conducive to improved microbial quality (Mat-Hussin et al., 2024).

Leung et al. (2021) analysed the airborne microbiome in public transportation systems, identifying 17 species that comprised approximately 75% of the observed microbial community, predominantly including commensal species from the human skin microbiota as well as environmental taxa. Beyond microorganisms, antimicrobial resistance genes were also detected, indicating the circulation of resistant microbes within anthropized environments. Such assessments are particularly critical given the airborne transmission of pathogens, as exemplified by the COVID-19 pandemic, during which the virus rapidly spread between individuals via bioaerosols. In this context, sequencing technologies offer valuable tools for the surveillance and study of pathogen dynamics (Naumova et al., 2022).

In soil ecosystems, microbiomes play a crucial role in the formation of biological crusts, or biocrusts, which arise from interactions among soil particles and microorganisms

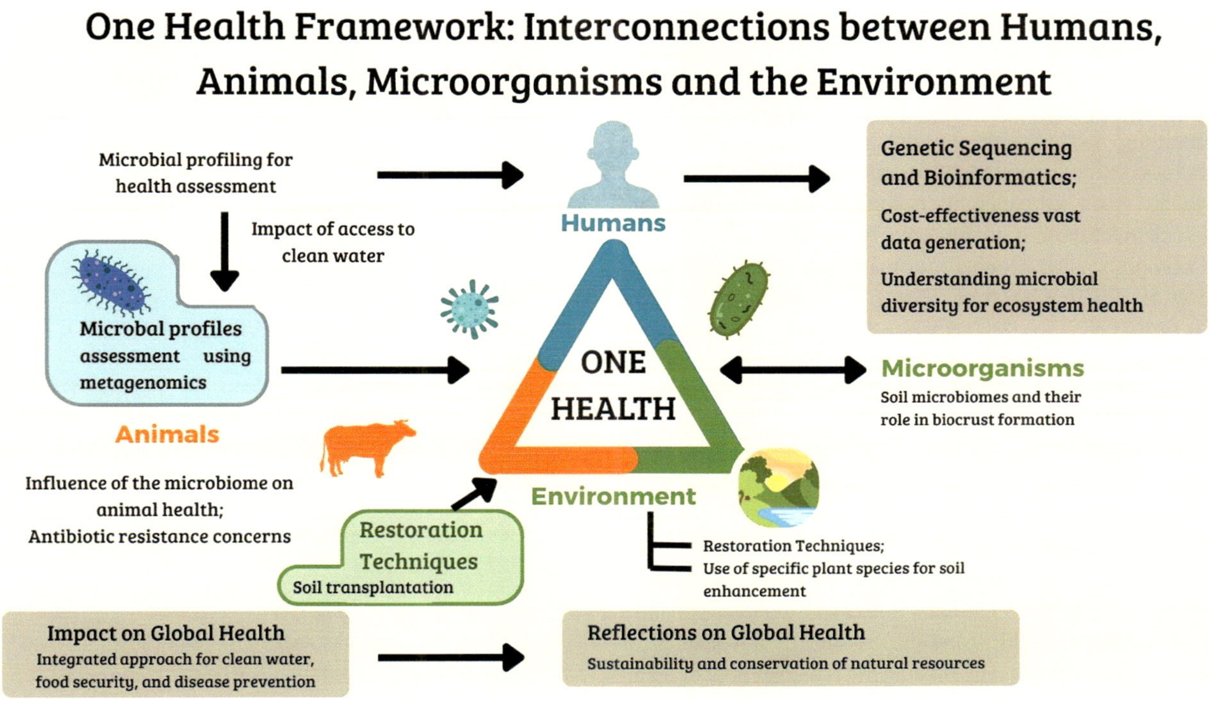

FIGURE 6.2 Metagenomics in one health. This infographic illustrates the intricate relationships and the importance of an integrated approach to health that takes into account the connections between humans, animals, and their environments.

such as algae, bacteria, and fungi. These biocrusts stabilize the soil surface, preventing erosion, facilitating carbon and nitrogen fixation, and enhancing moisture retention (Belnap and Lange, 2003). The composition and structure of these crusts vary depending on soil type (Abed et al., 2019).

Understanding the association between the microbiome and plants is fundamental for effective soil recovery, as characterizing microbial diversity is essential to identify substrate requirements and appropriate restoration strategies (Peddle et al., 2024). Approaches to restore soil microbiota include complete soil transplantation and microbial inoculation (Koziol et al., 2018); application of substrates to seeds to enhance germination and promote microbial colonization (Chua et al., 2019); use of bacterial cultures or suspensions (Vassilev et al., 2020); targeted interventions against specific agents (Thrall et al., 2001); and cultivation of particular plant species to improve soil quality for subsequent crops (Peddle et al., 2024).

The conservation and sustainable management of natural resources, including soil and water, are critical for maintaining healthy microenvironments, ensuring adequate agricultural productivity, and supporting optimal animal development. These efforts have direct and positive repercussions on human health, with genetic sequencing technologies serving as valuable tools due to their cost-effectiveness and the high quality of data they provide (Suman et al., 2022). In summary, multiple strategies exist for manipulating environmental microbiota to achieve targeted outcomes, such as selecting microorganisms to enhance flowering and plant growth (Panke-Buisse et al., 2015); boosting soil capacity for carbon sequestration through biocrust modulation (Kheirfam, 2020); and improving plant tolerance to saline soils, an area with considerable untapped potential (Mueller et al., 2021).

6.5 CHALLENGES AND FUTURE DIRECTIONS

Metagenomic studies have significantly advanced our understanding of microbiomes across diverse contexts; however, they face critical challenges that demand attention. Data analysis represents one of the foremost obstacles due to the enormous volume of sequences generated and the necessity to detect rare microorganisms, necessitating sophisticated computational approaches and high-performance computing platforms (Scholz et al., 2012; Almeida et al., 2019). Moreover, the interpretation of results is frequently constrained by the scarcity of comprehensive and up-to-date databases, resulting in incomplete or imprecise characterizations of microbial communities (Quince et al., 2017).

Ethical considerations constitute a vital component of metagenomic research. The sharing of human microbiome data, for instance, may reveal sensitive information enabling inferences about an individual's health status or identity (Olson et al., 2017). Improper handling and analysis of metagenomic datasets raise significant privacy concerns (Schmidt et al., 2019). Additionally, the application of microbiomes in diagnostics and therapeutics prompts critical questions regarding equitable access to these technologies and the potential for unintended consequences (Blaser et al., 2016). Consequently, comprehensive ethical frameworks addressing these issues must be developed and broadly endorsed within the scientific community. The adoption of stringent guidelines for informed consent and the implementation of robust data privacy protections are imperative, as advocated by the Earth Microbiome Project consortium (Gilbert and Stephens, 2018).

Emerging technologies present promising opportunities to address current challenges in metagenomics. Integrative approaches such as metatranscriptomics, metabolomics, and single-cell microbiome analysis are increasingly employed to yield comprehensive insights into microbial functions (Lloyd-Price et al., 2019; Zhu et al., 2021). Machine learning algorithms further enhance the prediction and modeling of complex microbial interactions (Segata et al., 2013). These advancements aim to broaden the applications of metagenomics across fields including personalized medicine, precision agriculture, and industrial biotechnology. Nonetheless, the standardization of protocols and the promotion of data sharing continue to pose significant challenges, as heterogeneity in data collection, processing, and analytical methodologies hampers reproducibility and comparative analyses across studies (Knight et al., 2018). Initiatives such as the Genomic Standards Consortium have been instrumental in addressing these issues by advocating for standardized practices and facilitating the establishment of open-access data repositories (Field et al., 2008; Meyer et al., 2019).

As these challenges are addressed through the implementation of ethical frameworks, best practices, and standardized protocols, metagenomics is poised to become an increasingly powerful tool for investigating microbiomes. Emerging technologies combined with collaborative initiatives hold the potential to revolutionize our understanding and practical application of microbial knowledge.

6.6 CONCLUSION

Metagenomics has become an essential tool for investigating the diversity and functionality of microbiomes across biological systems. By enabling the direct study of uncultivable microorganisms, it has driven advances in health, agriculture, and environmental sciences. Techniques such as shotgun sequencing allow not only the identification of microbial communities but also the characterization of their metabolic functions. Despite significant progress, technical, ethical, and standardization challenges still limit its full potential. The integration of multi-omics technologies and advanced computational tools is expected to overcome these limitations, establishing metagenomics as a strategic science to address global challenges in health, sustainability, and food security.

REFERENCES

Abed, R. M. M.; Tamm, A.; Hassenrück, C.; Al-Rawahi, A. N.; Rodríguez-Caballero, E.; Fiedler, S.; Maier, S.; Weber, B. Habitat-dependent composition of bacterial and fungal communities in biological soil crusts from Oman. *Scientific Reports*, v. 9, n. 1, p. 6468, Apr. 2019. https://doi.org/10.1038/s41598-019-42911-6.

Achudhan, A. B.; Kannan, P.; Gupta, A. et al. A review of web-based metagenomics platforms for analysing next-generation sequence data. *Biochemical Genetics*, v. 62, pp. 621–632, 2024. https://doi.org/10.1007/s10528-023-10467-w.

Ahmad, M.; Zahir, Z. A.; Asghar, H. N.; Asghar, M. Inducing salt tolerance in mung bean through coinoculation with rhizobia and plant-growth-promoting rhizobacteria containing 1-aminocyclopropane-1-carboxylate deaminase. *Canadian Journal of Microbiology*, v. 57, n. 7, pp. 578–589, Jul. 2011. https://doi.org/10.1139/w11-044.

Alcock, B. P. et al. CARD 2020: Antibiotic resistome surveillance with the comprehensive antibiotic resistance database. *Nucleic Acids Research*, v. 48, n. D1, pp. D517–D525, Jan. 2020. https://doi.org/10.1093/nar/gkz935. PMID: 31665441; PMCID: PMC7145624.

Almeida, A.; Mitchell, A. L.; Boland, M.; Forster, S. C.; Gloor, G. B.; Tarkowska, A. et al. A new genomic blueprint of the human gut microbiota. *Nature*, v. 568, n. 7753, pp. 499–504, 2019.

Altomare, C.; Norvell, W. A.; Bjorkman, T.; Harman, G. E. Solubilization of phosphates and micronutrients by the plant-growth-promoting and biocontrol fungus *Trichoderma harzianum rifai* 1295-22. *Applied and Environmental Microbiology*, v. 65, n. 7, pp. 2926–2933, Jul. 1999. https://doi.org/10.1128/AEM.65.7.2926-2933.1999.

Antil, S. et al. DNA barcoding, an effective tool for species identification: A review. *Molecular Biology Reports*, v. 50, n. 1, pp. 761–775, Jan. 2023. https://doi.org/10.1007/s11033-022-08015-7. Epub 2022 Oct 29. PMID: 36308581.

Attwood, G. T.; Wakelin, S. A.; Leahy, S. C.; Rowe, S.; Clarke, S.; Chapman, D. F.; Muirhead, R.; Jacobs, J. M. E. Applications of the soil, plant, and rumen microbiomes in pastoral agriculture. *Frontiers in Nutrition*, v. 6, p. 107, Jul. 2019. https://doi.org/10.3389/fnut.2019.00107.

Axelsson, E.; Ratnakumar, A.; Arendt, M. L. et al. The genomic signature of dog domestication reveals adaptation to a starch-rich diet. *Nature*, v. 495, pp. 360–364, 2013. https://doi.org/10.1038/nature11837.

Bais, H. P.; Weir, T. L.; Perry, L. G.; Gilroy, S.; Vivanco, J. M. The role of root exudates in rhizosphere interactions with plants and other organisms. *Annual Review of Plant Biology*, v. 57, pp. 233–266, 2006. https://doi.org/10.1146/annurev.arplant.57.032905.105159.

Banerjee, S.; Wei, Z.; Tian, T.; Bose, D.; Shih, N. N. C.; Feldman, M. D.; Khoury, T.; De Michele, A.; Robertson, E. S. Prognostic correlations with the microbiome of breast cancer subtypes. *Cell Death & Disease*, v. 12, n. 9, p. 831, Sep. 2021. https://doi.org/10.1038/s41419-021-04092-x.

Barea, J. A.; Pardini, M. I. M. C.; Gushiken, T. Extração de DNAolubteriais de arquivo e fontes escassas para utilização em reação de polimerização em cadeia (PCR). *Revista Brasileira de Hematologia e Hemoterapia*, v. 26, n. 4, pp. 274–281, 2004. https://doi.org/10.1590/S1516-84842004000400008.

Barlow, G. M.; Mathur, R. Type 2 diabetes and the microbiome. *Journal of the Endocrine Society*, v. 7, n. 2, p. bvac184, Nov. 2022. https://doi.org/10.1210/jendso/bvac184.

Belnap, J.; Lange, O. L. Biological soil crusts: Structure, function, and management. In: Baldwin, I. T.; Caldwell, M. M.; Heldmaier, G.; Lange, O. L.; Mooney, H. A.; Schulze, E.-D.; Sommer, U. (eds.). *Ecological Studies Series 150*. Berlin: Springer-Verlag, 2003, p. 503.

Berendsen, R. L.; Pieterse, C. M.; Bakker, P. A. The rhizosphere microbiome and plant health. *Trends in Plant Science*, v.

17, n. 8, pp. 478–486, Aug. 2012. https://doi.org/10.1016/j.tplants.2012.04.001.

Berg, G.; Rybakova, D.; Fischer, D.; Cernava, T.; Vergès, M. C.; Charles, T.; Chen, X.; Cocolin, L.; Eversole, K.; Corral, G. H.; Kazou, M.; Kinkel, L.; Lange, L.; Lima, N.; Loy, A.; Macklin, J. A.; Maguin, E.; Mauchline, T.; McClure, R.; Mitter, B.; Ryan, M.; Sarand, I.; Smidt, H.; Schelkle, B.; Roume, H.; Kiran, G. S.; Selvin, J.; Souza, R. S. C.; Van Overbeek, L.; Singh, B. K.; Wagner, M.; Walsh, A.; Sessitsch, A.; Schloter, M. Microbiome definition re-visited: Old concepts and new challenges. *Microbiome*, v. 8, n. 1, p. 103, Jun. 2020. https://doi.org/10.1186/s40168-020-00875-0. Erratum in: *Microbiome*, v. 8, n. 1, p. 119, Aug. 2020. https://doi.org/10.1186/s40168-020-00905-x.

Bertrand, D. et al. Hybrid metagenomic assembly enables high-resolution analysis of resistance determinants and mobile elements in human microbiomes. *Nature Biotechnology*, v. 37, pp. 937–944, 2019. https://doi.org/10.1038/s41587-019-0191-2.

Beydag-Tasöz, B. S.; Yennek, S.; Grapin-Botton, A. Towards a better understanding of diabetes mellitus using organoid models. *Nature Reviews Endocrinology*, v. 19, pp. 232–248, 2023. https://doi.org/10.1038/s41574-022-00797-x.

Blanco-Míguez, A. et al. Extending and improving metagenomic taxonomic profiling with uncharacterized species using MetaPhlAn 4. *Nature Biotechnology*, v. 41, pp. 1633–1644, 2023. https://doi.org/10.1038/s41587-023-01688-w.

Blaser, M. J.; Bork, P.; Fraser, C.; Knight, R.; Wang, J. The microbiome explored: Recent insights and future challenges. *Nature Reviews Microbiology*, v. 14, n. 12, pp. 708–713, 2016.

Bolger, A. M.; Lohse, M.; Usadel, B. Trimmomatic: A flexible trimmer for Illumina sequence data. *Bioinformatics*, v. 30, n. 15, pp. 2114–2120, Aug. 2014. https://doi.org/10.1093/bioinformatics/btu170.

Bruno, A.; Agostinetto, G.; Fumagalli, S.; Ghisleni, G.; Sandionigi, A. It's a long way to the tap: Microbiome and DNA-based omics at the core of drinking water quality. *International Journal of Environmental Research and Public Health*, v. 19, n. 7940, 2022. https://doi.org/10.3390/ijerph19137940.

Bu, X. L.; Yao, X. Q.; Jiao, S. S.; Zeng, F.; Liu, Y. H.; Xiang, Y.; Liang, C. R.; Wang, Q. H.; Wang, X.; Cao, H. Y.; Yi, X.; Deng, B.; Liu, C. H.; Xu, J.; Zhang, L. L.; Gao, C. Y.; Xu, Z. Q.; Zhang, M.; Wang, L.; Tan, X. L.; Xu, X.; Zhou, H. D.; Wang, Y. J. A study on the association between infectious burden and Alzheimer's disease. *European Journal of Neurology*, v. 22, n. 12, pp. 1519–1525, Des. 2015. https://doi.org/10.1111/ene.12477.

Burden, D.; Brangs, H.; Gibbons, L. Bead beating: A primer. *ResearchGate*, 2014. https://doi.org/10.13140/RG.2.1.4223.7207.

Burns, M. B.; Lynch, J.; Starr, T. K.; Knights, D.; Blekhman, R. Virulence genes are a signature of the microbiome in the colorectal tumor microenvironment. *Genome Medicine*, v. 7, n. 1, p. 55, Jun. 2015. https://doi.org/10.1186/s13073-015-0177-8.

Calderón-Franco, D.; Corbera-Rubio, F.; Cuesta-Sanz, M.; Pieterse, B.; De Ridder, D.; Van Loosdrecht, M. C. M.; Van Halem, D.; Laureni, M.; Weissbrodt, D. G. Microbiome, resistome and mobilome of chlorine-free drinking water treatment systems. *Water Research*, v. 235, p. 119905, May 2023. https://doi.org/10.1016/j.watres.2023.119905.

Child, A. et al. Comparative evaluation of soil DNA extraction kits for long read metagenomic sequencing. *Access Microbiology*, v. 6, n. 9, p. 000868.v3, Sep. 2024. https://doi.org/10.1099/acmi.0.000868.v3.

Choi, E.; Murray, B.; Choi, S. Biofilm and cancer: Interactions and future directions for cancer therapy. *International Journal of Molecular Sciences*, v. 24, n. 16, p. 12836, Aug. 2023. https://doi.org/10.3390/ijms241612836.

Chua, M.; Erickson, T. E.; Merritt, D. J.; Chilton, A. M.; Ooi, M. K. J.; Muñoz-Rojas, M. Bio-priming seeds with cyanobacteria: effects on native plant growth and soil properties. *Restoration Ecology*, v. 28, suppl. 1, pp. S168–S176, 2020. https://doi.org/10.1111/rec.13040

Chung, K. F. Airway microbial dysbiosis in asthmatic patients: A target for prevention and treatment? *Journal of Allergy and Clinical Immunology*, v. 139, n. 4, pp. 1071–1081, Apr. 2017. https://doi.org/10.1016/j.jaci.2017.02.004.

Ciernikova, S.; Sevcikova, A.; Stevurkova, V.; Mego, M. Tumor microbiome—an integral part of the tumor microenvironment. *Frontiers in Oncology*, v. 12, p. 1063100, Nov. 2022. https://doi.org/10.3389/fonc.2022.1063100.

Coelho, L. P.; Kultima, J. R.; Costea, P. I. et al. Similarity of the dog and human gut microbiomes in gene content and response to diet. *Microbiome*, v. 6, p. 72, 2018. https://doi.org/10.1186/s40168-018-0450-3.

Colston, T. J.; Jackson, C. R. Microbiome evolution along divergent branches of the vertebrate tree of life: What is known and unknown. *Molecular Ecology*, v. 25, n. 16, pp. 3776–3800, Aug. 2016. https://doi.org/10.1111/mec.13730.

Compant, S.; Clément, C.; Sessitsch, A. Plant growth-promoting bacteria in the rhizo- and endosphere of plants: Their role in plant growth and adaptation to stress conditions. *Microbiological Research*, v. 167, n. 4, pp. 169–180, 2012. https://doi.org/10.1016/j.soilbio.2009.11.024.

Cook, R. et al. The long and short of it: Benchmarking viromics using Illumina, Nanopore and PacBio sequencing technologies. *Microbial Genomics*, v. 10, n. 2, p. 001198, Feb. 2024. https://doi.org/.1099/mgen.0.001198. PMID: 38376377; PMCID: PMC10926689.

Costa, J. C. S. Development of an automated pipeline for meta-omics data analysis. Dissertação de Mestrado (Mestrado em Ciência da Computação)—Departamento de Informática, Universidade do Minho, Braga, Portugal, Oct. 2017.

Crovesy, L.; Ostrowski, M.; Ferreira, D. M. T. P.; Rosado, E. L.; Soares-Mota, M. Effect of *Lactobacillus* on body weight and body fat in overweight subjects: A systematic review of randomized controlled clinical trials. *International Journal of Obesity (London)*, v. 41, n. 11, pp. 1607–1614, Nov. 2017. https://doi.org/10.1038/ijo.2017.161.

David, A.; Priya, J. D. A.; Gautam, A. DNA sequencing technologies and DNA barcoding. *Methods in Molecular Biology*, v. 2744, pp. 139–154, 2024. https://doi.org/10.1007/978-1-0716-3581-0_8. PMID: 38683316.

Davis, N. M. et al. Simple statistical identification and removal of contaminant sequences in marker-gene and metagenomics data. *Microbiome*, v. 6, n. 1, p. 226, Dec. 2018. https://doi.org/10.1186/s40168-018-0605-2. PMID: 30558668; PMCID: PMC6298009.

Demkina, A. et al. Benchmarking DNA isolation methods for marine metagenomics. *Scientific Reports*, v. 13, n. 1, p. 22138, Dec. 2023. https://doi.org/10.1038/s41598-023-48804-z. PMID: 38092853; PMCID: PMC10719357.

Depommier, C.; Everard, A.; Druart, C.; Plovier, H.; Van Hul, M.; Vieira-Silva, S.; Falony, G.; Raes, J.; Maiter,

D.; Delzenne, N. M.; De Barsy, M.; Loumaye, A.; Hermans, M. P.; Thissen, J. P.; De Vos, W. M.; Cani, P. D. Supplementation with *Akkermansia muciniphila* in overweight and obese human volunteers: A proof-of-concept exploratory study. *Nature Medicine*, v. 25, n. 7, pp. 1096–1103, Jul. 2019. https://doi.org/10.1038/s41591-019-0495-2.

Dewhirst, F. E.; Chen, T.; Izard, J.; Paster, B. J.; Tanner, A. C.; Yu, W. H.; Lakshmanan, A.; Wade, W. G. The human oral microbiome. *Journal of Bacteriology*, v. 192, n. 19, pp. 5002–5017, Oct. 2010. https://doi.org/10.1128/JB.00542-10.

Dickson, R. P.; Erb-Downward, J. R.; Freeman, C. M.; McCloskey, L.; Falkowski, N. R.; Huffnagle, G. B.; Curtis, J. L. Bacterial topography of the healthy human lower respiratory tract. *mBio*, v. 8, 2017. https://doi.org/10.1128/mbio.02287-16.

Dominguez-Bello, M. G.; Costello, E. K.; Contreras, M.; Magris, M.; Hidalgo, G.; Fierer, N.; Knight, R. Delivery mode shapes the acquisition and structure of the initial microbiota across multiple body habitats in newborns. *Proceedings of the National Academy of Sciences of the United States of America*, v. 107, n. 26, pp. 11971–11975, Jun. 2010. https://doi.org/10.1073/pnas.1002601107.

Fan, Y.; Pedersen, O. Gut microbiota in human metabolic health and disease. *Nature Reviews Microbiology*, v. 19, pp. 55–71, 2021. https://doi.org/10.1038/s41579-020-0433-9.

Ferreira, R. M.; Pereira-Marques, J.; Pinto-Ribeiro, I.; Costa, J. L.; Carneiro, F.; Machado, J. C.; Figueiredo, C. Gastric microbial community profiling reveals a dysbiotic cancer-associated microbiota. *Gut*, v. 67, n. 2, pp. 226–236, Feb. 2018. https://doi.org/10.1136/gutjnl-2017-314205.

Field, D.; Garrity, G.; Gray, T.; Morrison, N.; Selengut, J.; Sterk, P.; Womack, M. The minimum information about a genome sequence (MIGS) specification. *Nature Biotechnology*, v. 26, n. 5, pp. 541–547, 2008.

Filardo, S.; Di Pietro, M.; Sessa, R. Current progresses and challenges for microbiome research in human health: A perspective. *Frontiers in Cellular and Infection Microbiology*, v. 14, p. 1377012, Apr. 2024. https://doi.org/10.3389/fcimb.2024.1377012.

Frank, D. N.; St Amand, A. L.; Feldman, R. A.; Boedeker, E. C.; Harpaz, N.; Pace, N. R. Molecular-phylogenetic characterization of microbial community imbalances in human inflammatory bowel diseases. *Proceedings of the National Academy of Sciences of the United States of America*, v. 104, n. 34, pp. 13780–13785, Aug. 2007. https://doi.org/10.1073/pnas.0706625104.

Franzosa, E. A. et al. Gut microbiome structure and metabolic activity in inflammatory bowel disease. *Nature Microbiology*, v. 4, n. 2, pp. 293–305, 2019.

Gadde, U.; Kim, W. H.; Oh, S. T.; Lillehoj, H. S. Alternatives to antibiotics for maximizing growth performance and feed efficiency in poultry: A review. *Animal Health Research Reviews*, v. 18, n. 1, pp. 26–45, Jun. 2017. https://doi.org/10.1017/S1466252316000207.

Galperin, M. Y.; Wolf, Y. I.; Makarova, K. S.; Vera Alvarez, R.; Landsman, D.; Koonin, E. V. COG database update: Focus on microbial diversity, model organisms, and widespread pathogens. *Nucleic Acids Research*, v. 49, n. D1, pp. D274–D281, Jan. 2021. https://doi.org/10.1093/nar/gkaa1018.

Garibyan, L.; Avashia, N. Polymerase chain reaction. *Journal of Investigative Dermatology*, v. 133, n. 3, pp. 1–4, 2013. https://doi.org/10.1038/jid.2013.1.

Gawad, C.; Koh, W.; Quake, S. Single-cell genome sequencing: Current state of the science. *Nature Reviews Genetics*, v. 17, pp. 175–188, 2016. https://doi.org/10.1038/nrg.2015.16.

Ghannoum, M. A.; Jurevic, R. J.; Mukherjee, P. K.; Cui, F.; Sikaroodi, M.; Naqvi, A.; Gillevet, P. M. Characterization of the oral fungal microbiome (mycobiome) in healthy individuals. *PLoS Pathogens*, v. 6, n. 1, p. e1000713, Jan. 2010. https://doi.org/10.1371/journal.ppat.1000713.

Ghurye, J. S.; Cepeda-Espinoza, V.; Pop, M. Metagenomic assembly: Overview, challenges and applications. *Yale Journal of Biology and Medicine*, v. 89, pp. 353–362, 2016.

Gilbert, J. A.; Stephens, B. Microbiology of the built environment. *Nature Reviews Microbiology*, v. 16, pp. 661–670, 2018. https://doi.org/10.1038/s41579-018-0065-5.

Glenn, T. C.; Pierson, T. W.; Bayona-Vásquez, N. J.; Kieran, T. J.; Hoffberg, S. L.; Thomas, J. C., IV; Lefever, D. E.; Finger, J. W.; Gao, B.; Bian, X.; Louha, S.; Kolli, R. T.; Bentley, K. E.; Rushmore, J.; Wong, K.; Shaw, T. I.; Rothrock, M. J., Jr.; McKee, A. M.; Guo, T. L.; Mauricio, R.; Molina, M.; Cummings, B. S.; Lash, L. H.; Lu, K.; Gilbert, G. S.; Hubbell, S. P.; Faircloth, B. C. Adapterama II: Universal amplicon sequencing on Illumina platforms (TaggiMatrix). *PeerJ*, v. 7, p. e7786, 2019. https://doi.org/10.7717/peerj.7786.

Gomes, A. C.; Hoffmann, C.; Mota, J. F. The human gut microbiota: Metabolism and perspective in obesity. *Gut Microbes*, v. 9, n. 4, pp. 308–325, Jul. 2018. https://doi.org/10.1080/19490976.2018.1465157.

Goodwin, S.; McPherson, J. D.; McCombie, W. R. Coming of age: Ten years of next-generation sequencing technologies. *Nature Reviews Genetics*, v. 17, n. 6, pp. 333–351, Jun. 2016. https://doi.org/10.1038/nrg.2016.49.

Graves, D. T.; Fine, D.; Teng, Y. T.; Van Dyke, T. E.; Hajishengallis, G. The use of rodent models to investigate host-bacteria interactions related to periodontal diseases. *Journal of Clinical Periodontology*, v. 35, n. 2, pp. 89–105, Feb. 2008. https://doi.org/10.1111/j.1600-051X.2007.01172.x.

Grayston, S. J.; Wang, S.; Campbell, C. D.; Edwards, A. C. Selective influence of plant species on microbial diversity in the rhizosphere. *Soil Biology and Biochemistry*, v. 30, n. 3, pp. 369–378, 1998. https://doi.org/10.1016/S0038-0717(97)00124-7.

Greathouse, K. L.; Sinha, R.; Vogtmann, E. DNA extraction forIan microbiome studies: The issue of standardization. *Genome Biology*, v. 20, n. 1, p. 212, Oct. 2019. https://doi.org/10.1186/s13059-019-1843-8. PMID: 31639026; PMCID: PMC6802309.

Green, M. R.; Sambrook, J. The basic Polymerase Chain Reaction (PCR). *Cold Spring Harbor Protocols*, 2018. https://doi.org/10.1101/pdb.prot095117.

Grice, E. A. The skin microbiome: Potential for novel diagnostic and therapeutic approaches to cutaneous disease. *Seminars in Cutaneous Medicine and Surgery*, v. 33, n. 2, pp. 98–103, Jun. 2014. https://doi.org/10.12788/j.sder.0087.

Guo, F. et al. Ion torrent ™ genexus ™ integrated sequencer and ForeNGS analysis software: An automatic NGS-STR workflow from DNA to profile for forensic science. *Forensic Science International: Genetics*, v. 61, p. 102753, Nov. 2022. https://doi.org/10.1016/j.fsigen.2022.102753. PMID: 35931007.

Hajiasgharzadeh, K.; Baradaran, B. Cholinergic anti-inflammatory pathway and the liver. *Advanced Pharmaceutical Bulletin*, v. 7, n. 4, pp. 507–513, Dec. 2017. https://doi.org/10.15171/apb.2017.063.

Hajishengallis, G. Illuminating the oral microbiome and its host interactions: Animal models of disease. *FEMS Microbiology Reviews*, v. 47, n. 3, p. fuad018, May 2023. https://doi.org/10.1093/femsre/fuad018.

Hajishengallis, G.; Lamont, R. J.; Graves, D. T. The enduring importance of animal models in understanding periodontal disease. *Virulence*, v. 6, n. 3, pp. 229–235, 2015. https://doi.org/10.4161/21505594.2014.990806.

Hamilton, C. E.; Bever, J. D.; Labb, J.; Yang, X.; Yin, H. Mitigating climate change through managing constructed-microbial communities in agriculture. *Agriculture, Ecosystems & Environment*, v. 216, pp. 304–308, 2016. https://doi.org/10.1016/j.agee.2015.10.006.

Handelsman, J. Metagenomics: Application of genomics to uncultured microorganisms. *Microbiology and Molecular Biology Reviews*, v. 68, n. 4, pp. 669–685, Dec. 2004. https://doi.org/10.1128/MMBR.68.4.669-685.2004. PMID: 15590779; PMCID: PMC539003.

Hasan, N.; Yang, H. Factors affecting the composition of the gut microbiota, and its modulation. *PeerJ*, v. 7, p. e7502, Aug. 2019. https://doi.org/10.7717/peerj.7502.

Hassler, H. B.; Probert, B.; Moore, C. et al. Phylogenies of the 16S rRNA gene and its hypervariable regions lack concordance with core genome phylogenies. *Microbiome*, v. 10, p. 104, 2022. https://doi.org/10.1186/s40168-022-01295-y.

Healey, G. R.; Murphy, R.; Brough, L.; Butts, C. A.; Coad, J. Interindividual variability in gut microbiota and host response to dietary interventions. *Nutrition Reviews*, v. 75, n. 12, pp. 1059–1080, Dec. 2017. https://doi.org/10.1093/nutrit/nux062. PMID: 29190368.

Hoffmann, A. R.; Proctor, L. M.; Surette, M. G.; Suchodolski, J. S. The microbiome: The trillions of microorganisms that maintain health and cause disease in humans and companion animals. *Veterinary Pathology*, v. 53, n. 1, pp. 10–21, 2016. https://doi.org/10.1177/0300985815595517.

Hooks, K. B.; O'Malley, M. A. Dysbiosis and its discontents. *mBio*, v. 8, n. 5, p. e01492-17, Oct. 2017. https://doi.org/10.1128/mBio.01492-17.

Hou, K.; Wu, Z. X.; Chen, X. Y. et al. Microbiota in health and diseases. *Signal Transduction and Targeted Therapy*, v. 7, p. 135, 2022. https://doi.org/10.1038/s41392-022-00974-4.

Huynh, M.; Crane, M. J.; Jamieson, A. M. The lung, the niche, and the microbe: Exploring the lung microbiome in cancer and immunity. *Frontiers in Immunology*, v. 13, p. 1094110, Jan. 2023. https://doi.org/10.3389/fimmu.2022.1094110

Integrative HMP Research Network. The integrative human microbiome project. *Nature*, v. 569, n. 7758, pp. 641–648, 2019.

Jansson, J. K.; Hofmockel, K. S. Soil microbiomes and climate change. *Nature Reviews Microbiology*, v. 18, n. 1, pp. 35–46, 2020.

Jiang, J.; Qi, L.; Lv, Z.; Wei, Q.; Shi, F. Dietary stevioside supplementation increases feed intake by altering the hypothalamic transcriptome profile and gut microbiota in broiler chickens. *Journal of the Science of Food and Agriculture*, v. 101, n. 5, pp. 2156–2167, Mar. 2021. https://doi.org/10.1002/jsfa.10838.

Joshi, G. K.; Jugran, J.; Bhatt, J. P. Metagenomics: The exploration of unculturable microbial world. In: Ravi, I.; Baunthiyal, M.; Saxena, J. (eds.). *Advances in Biotechnology*. New Delhi: Springer, 2014. https://doi.org/10.1007/978-81-322-1554-7_7.

Jünemann, S.; Kleinbölting, N.; Jaenicke, S.; Henke, C.; Hassa, J.; Nelkner, J.; Stolze, Y.; Albaum, S. P.; Schlüter, A.; Goesmann, A.; Sczyrba, A.; Stoye, J. Bioinformatics for NGS-based metagenomics and the application to biogas research. *Journal of Biotechnology*, v. 261, pp. 10–23, 2017. https://doi.org/10.1016/j.jbiotec.2017.08.012.

Kaelberer, M. M.; Buchanan, K. L.; Klein, M. E.; Barth, B. B.; Montoya, M. M.; Shen, X.; Bohórquez, D. V. A gut-brain neural circuit for nutrient sensory transduction. *Science*, v. 361, n. 6408, p. eaat5236, Sep. 2018. https://doi.org/10.1126/science.aat5236.

Kajdič, L.; Plavec, T.; Zdovc, I.; Kalin, A.; Zakošek Pipan, M. Impact of type of parturition on colostrum microbiota composition and puppy survival. *Animals (Basel)*, v. 11, n. 7, p. 1897, Jun. 2021. https://doi.org/10.3390/ani11071897.

Kanehisa, M.; Goto, S. KEGG: Kyoto encyclopedia of genes and genomes. *Nucleic Acids Research*, v. 28, n. 1, pp. 27–30, Jan. 2000. https://doi.org/10.1093/nar/28.1.27. PMID: 10592173; PMCID: PMC102409.

Kang, D. D.; Li, F.; Kirton, E.; Thomas, A.; Egan, R.; An, H.; Wang, Z. MetaBAT 2: An adaptive binning algorithm for robust and efficient genome reconstruction from metagenome assemblies. *PeerJ*, v. 7, p. e7359, 2019. https://doi.org/10.7717/peerj.7359.

Kaul, S.; Choudhary, M.; Gupta, S.; Dhar, M. K. Engineering host microbiome for crop improvement and sustainable agriculture. *Frontiers in Microbiology*, v. 12, p. 635917, May 2021. https://doi.org/10.3389/fmicb.2021.635917.

Keshavarzian, A.; Green, S. J.; Engen, P. A.; Voigt, R. M.; Naqib, A.; Forsyth, C. B.; Mutlu, E.; Shannon, K. M. Colonic bacterial composition in Parkinson's disease. *Movement Disorders*, v. 30, n. 10, pp. 1351–1360, Sep. 2015. https://doi.org/10.1002/mds.26307.

Kheirfam, H. Increasing soil potential for carbon sequestration using microbes from biological soil crusts. *Journal of Arid Environments*, v. 172, p. 104022, 2020. https://doi.org/10.1016/j.jaridenv.2019.104022.

Khoruts, A.; Dicksved, J.; Jansson, J. K.; Sadowsky, M. J. Changes in the composition of the human fecal microbiome after bacteriotherapy for recurrent *Clostridium difficile*-associated diarrhea. *Journal of Clinical Gastroenterology*, v.–4, n. 5, pp. 354–360, May–Jun. 2010. https://doi.org/10.1097/MCG.0b013e3181c87e02.

Klindworth, A.; Pruesse, E.; Schweer, T.; Peplies, J.; Quast, C.; Horn, M.; Glöckner, F. O. Evaluation of general 16S ribosomal RNA gene PCR primers for classical and next-generation sequencing-based diversity studies. *Nucleic Acids Research*, v. 41, n. 1, p. e1, Jan. 2013. https://doi.org/10.1093/nar/gks808.

Klitgaard, K.; Boye, M.; Capion, N.; Jensen, T. K. Evidence of multiple *Treponema* phylotypes involved in bovine digital dermatitis as shown by 16S rRNA gene analysis and fluorescence in situ hybridization. *Journal of Clinical Microbiology*, v. 46, n. 9, pp. 3012–3020, Sep. 2008. https://doi.org/10.1128/JCM.00670-08.

Kloepper, J.; Schroth, M. N. Plant growth-promoting rhizobacteria on radishes. *IV International Conference on Plant Pathogenic Bacteria*, França, v. 2, pp. 879–882, 1978.

Knight, R.; Vrbanac, A.; Taylor, B. C.; Aksenov, A.; Callewaert, C.; Debelius, J.; Gilbert, J. A. Best practices for analysing microbiomes. *Nature Reviews Microbiology*, v. 16, n. 7, pp. 410–422, 2018.

Kong, H. H.; Oh, J.; Deming, C.; Conlan, S.; Grice, E. A.; Beatson, M. A.; Nomicos, E.; Polley, E. C.; Komarow, H. D.; NISC Comparative Sequence Program; Murray, P. R.; Turner,

M. L.; Segre, J. A. Temporal shifts in the skin microbiome associated with disease flares and treatment in children with atopic dermatitis. *Genome Research*, v. 22, n. 5, pp. 850–859, May 2012. https://doi.org/10.1101/gr.131029.111.

Koziol, L.; Schultz, P. A.; House, G. L.; Bauer, J. T.; Middleton, E. L.; Bever, J. D. The plant microbiome and native plant restoration: The example of native mycorrhizal fungi. *BioScience*, v. 68, n. 12, pp. 996–1006, Dec. 2018. https://doi.org/10.1093/biosci/biy125.

Krishnan, K.; Chen, T.; Paster, B. J. A practical guide to the oral microbiome and its relation to health and disease. *Oral Diseases*, v. 23, n. 3, pp. 276–286, Apr. 2017. https://doi.org/10.1111/odi.12509.

Kumaraswamy, K. L.; Vidhya, M. Human papilloma virus and oral infections: An update. *Journal of Cancer Research and Therapeutics*, v. 7, n. 2, pp. 120–127, Apr–Jun. 2011. https://doi.org/10.4103/0973-1482.82915.

Kumawat, R. L. et al. Advancement of next-generation DNA sequencing through ionic blockade and transverse tunneling current methods. *Small*, v. 20, n. 36, p. e2401112, Sep. 2024. https://doi.org/10.1002/smll.202401112. Epub May 2024. PMID: 38716623.

Lamurias, A.; Sereika, M.; Albertsen, M.; Hose, K.; Nielsen, T. D. Metagenomic binning with assembly graph embeddings. *Bioinformatics*, v. 38, n. 19, pp. 4481–4487, Oct. 2022. https://doi.org/10.1093/bioinformatics/btac557.

Langmead, B.; Salzberg, S. Fast gapped-read alignment with Bowtie 2. *Nature Methods*, v. 9, pp. 357–359, 2012. https://doi.org/10.1038/nmeth.1923.

Leung, M. H. Y.; Tong, X.; BØIfot, K. O.; Bezdan, D.; Butler, D. J.; Danko, D. C.; Gohli, J.; Green, D. C.; Hernandez, M. T.; Kelly, F. J.; Levy, S.; Mason-Buck, G.; Nieto-Caballero, M.; Syndercombe-Court, D.; Udekwu, K.; Young, B. G.; Mason, C. E.; Dybwad, M.; Lee, P. K. H. Characterization of the public transit air microbiome and resistome reveals geographical specificity. *Microbiome*, v. 9, n. 1, p. 112, May 2021. https://doi.org/10.1186/s40168-021-01044-7.

Li, D.; Liu, C.-M.; Luo, R.; Sadakane, K.; Lam, T.-W. MEGAHIT: An ultra-fast single-node solution for large and complex metagenomics assembly via succinct de Bruijn graph. *Bioinformatics*, v. 31, n. 10, pp. 1674–1676, May 2015. https://doi.org/10.1093/bioinformatics/btv033.

Li, H.; Durbin, R. Fast and accurate short read alignment with Burrows-Wheeler transform. *Bioinformatics*, v. 25, n. 14, pp. 1754–1760, Jul. 2009. https://doi.org/10.1093/bioinformatics/btp324.

Li, H.; Yu, Q.; Li, T.; Shao, L.; Su, M.; Zhou, H.; Qu, J. Rumen microbiome and metabolome of tibetan sheep (*Ovis aries*) reflect animal age and nutritional requirement. *Frontiers in Veterinary Science*, v. 7, p. 609, Sep. 2020. https://doi.org/10.3389/fvets.2020.00609.

Li, Q.; Larouche-Lebel, É.; Loughran, K. A.; Huh, T. P.; Suchodolski, J. S.; Oyama, M. A. Metabolomics analysis reveals deranged energy metabolism and amino acid metabolic reprogramming in dogs with myxomatous mitral valve disease. *Journal of the American Heart Association*, v. 10, n. 9, p. e018923, May 2021. https://doi.org/10.1161/JAHA.120.018923.

Li, X.; Jousset, A.; De Boer, W.; Carrión, V. J.; Zhang, T.; Wang, X.; Kuramae, E. E. Legacy of land use history determines reprogramming of plant physiology by soil microbiome. *ISME Journal*, v. 13, n. 3, pp. 738–751, Mar. 2019. https://doi.org/10.1038/s41396-018-0300-0.

Li, Z. et al. Comparison of the two major classes of assembly algorithms: Overlap–layout–consensus and de-bruijn-graph. *Briefings in Functional Genomics*, v. 11, n. 1, pp. 25–37, Jan. 2012. https://doi.org/10.1093/bfgp/elr035.

Lipuma, J. J. The changing microbial epidemiology in cystic fibrosis. *Clinical Microbiology Reviews*, v. 23, n. 2, pp. 299–323, Apr. 2010. https://doi.org/10.1128/CMR.00068-09.

Liu, L. et al. Comparison of next-generation sequencing systems. *Journal of Biomedical Biotechnology*, v. 2012, p. 251364, Jul. 2012. https://doi.org/10.1155/2012/251364. PMID: 22829749; PMCID: PMC3398667.

Lloyd-Price, J. et al. Multi-omics of the gut microbial ecosystem in inflammatory bowel diseases. *Nature*, v. 569, n. 7758, pp. 655–662, 2019.

Loh, J. S.; Mak, W. Q.; Tan, L. K. S. et al. Microbiota–gut–brain axis and its therapeutic applications in neurodegenerative diseases. *Signal Transduction and Targeted Therapy*, v. 9, p. 37, 2024. https://doi.org/10.1038/s41392-024-01743-1.

Loh, J. S.; Pierson, E. A.; Pierson, L. S., 3rd; Stacey, G.; Chatterjee, A. Quorum sensing in plant-associated bacteria. *Current Opinion in Plant Biology*, v. 5, n. 4, pp. 285–290, Aug. 2002. https://doi.org/10.1016/s1369-5266(02)00274-1.

Lokhande, H. A. Bioinformatics analysis of miRNA sequencing data. *Methods in Molecular Biology*, v. 2595, pp. 225–237, 2023. https://doi.org/10.1007/978-1-0716-2823-2_16. PMID: 36441466.

Lorenz, T. C. Polymerase chain reaction: Basic protocol plus troubleshooting and optimization strategies. *Journal of Visualized Experiments*, n. 63, p. e3998, 2012. https://doi.org/10.3791/3998.

Lu, H.; Giordano, F.; Ning, Z. Oxford nanopore MinION sequencing and genome assembly. *Genomics, Proteomics & Bioinformatics*, v. 14, n. 5, pp. 265–279, Oct. 2016. https://doi.org/10.1016/j.gpb.2016.05.004. PMID: 27646134; PMCID: PMC5093776.

Lundberg, K. S.; Shoemaker, D. D.; Adams, M. W.; Short, J. M.; Sorge, J. A.; Mathur, E. J. High-fidelity amplification using a thermostable DNA polymerase isolated from Pyrococcus furiosus. *Gene*, v. 108, n. 1, pp. 1–6, Dec. 1991. https://doi.org/10.1016/0378-1119(91)90480-Y.

Lynch, J. M. *The Rhizosphere*. Chichester: John Wiley & Sons, 1990.

Lysholm, F.; Wetterbom, A.; Lindau, C.; Darban, H.; Bjerkner, A.; Fahlander, K.; Lindberg, A. M.; Persson, B.; Allander, T.; Andersson, B. Characterization of the viral microbiome in patients with severe lower respiratory tract infections, using metagenomic sequencing. *PLoS One*, v. 7, n. 2, p. e30875, 2012. https://doi.org/10.1371/journal.pone.0030875.

Ma, L. C.; Zhao, H. Q.; Wu, L. B.; Cheng, Z. L.; Liu, C. Impact of the microbiome on human, animal, and environmental health from a One Health perspective. *Science of One Health*, v. 2, p. 100037, Aug. 2023. https://doi.org/10.1016/j.soh.2023.100037.

Mahajna, A.; Dinkla, I. J. T.; Euverink, G. J. W.; Keesman, K. J.; Jayawardhana, B. Clean and safe drinking water systems via metagenomics data and artificial intelligence: State-of-the-art and future perspective. *Frontiers in Microbiology*, v. 13, p. 832452, May 2022. https://doi.org/10.3389/fmicb.2022.832452.

Maki, K. A.; Wolff, B.; Varuzza, L.; Green, S. J.; Barb, J. J. Multi-amplicon microbiome data analysis pipelines for mixed orientation sequences using QIIME2: Assessing reference database, variable region and pre-processing bias in classification of mock bacterial community samples. *PLoS One*, v. 18, n. 1, p. e0280293, 2023. https://doi.org/10.1371/journal.pone.0280293.

Mallawaarachchi, V.; Wickramarachchi, A.; Xue, H.; Papudeshi, B.; Grigson, S. R.; Bouras, G.; Prahl, R. E.; Kaphle, A.; Verich, A.; Talamantes-Becerra, B.; Dinsdale, E. A.; Edwards, R. A. Solving genomic puzzles: Computational methods for metagenomic binning. *Briefings in Bioinformatics*, v. 25, n. 5, bbae372, Sep. 2024. https://doi.org/10.1093/bib/bbae372.

Marchini, L.; Campos, M. S.; Silva, A. M.; Paulino, L. C.; Nobrega, F. G. Bacterial diversity in aphthous ulcers. *Oral Microbiology and Immunology*, v. 22, n. 4, pp. 225–231, Aug. 2007. https://doi.org/10.1111/j.1399-302X.2006.00345.x.

Martin, C. R.; Osadchiy, V.; Kalani, A.; Mayer, E. A. The brain-gut-microbiome axis. *Cellular and Molecular Gastroenterology and Hepatology*, v. 6, n. 2, pp. 133–148, Apr. 2018. https://doi.org/10.1016/j.jcmgh.2018.04.003.

Martin, M. Cutadapt removes adapter sequences from high-throughput sequencing reads. *EMBnet Journal*, v. 17, n. 1, pp. 10–12, 2011. https://doi.org/10.14806/ej.17.1.200.

Matchado et al. Microbial genomics, v. 10, p. 001203, 2024. https://doi.org/10.1099/mgen.0.001203.

Mathieu, E.; LÉJard, V.; Ezzine, C.; Govindin, P.; Morat, A.; Giat, M.; Lapaque, N.; Doré, J.; Blottière, H. M. An insight into functional metagenomics: A high-throughput approach to decipher food–microbiota–host interactions in the human gut. *International Journal of Molecular Sciences*, v. 24, p. 17630, 2023. https://doi.org/10.3390/ijms242417630.

Mat-Hussin, N. H.; Siew, S. W.; Maghpor, M. N.; Gan, H. M.; Ahmad, H. F. Method for detection of pathogenic bacteria from indoor air microbiome samples using high-throughput amplicon sequencing. *MethodsX*, v. 12, p. 102636, Feb. 2024. https://doi.org/10.1016/j.mex.2024.102636.

Meason-Smith, C.; Diesel, A.; Patterson, A. P.; Older, C. E.; Mansell, J. M.; Suchodolski, J. S.; Rodrigues Hoffmann, A. What is living on your dog's skin? Characterization of the canine cutaneous mycobiota and fungal dysbiosis in canine allergic dermatitis. *FEMS Microbiology Ecology*, v. 91, n. 12, p. fiv139, Dec. 2015. https://doi.org/10.1093/femsec/fiv139.

Mendes, R. et al. The rhizosphere microbiome: Significance of plant beneficial, plant pathogenic, and human pathogenic microorganisms. *FEMS Microbiology Reviews*, v. 37, n. 5, pp. 634–663, 2013.

Meyer, F.; Fritz, A.; Deng, Z. L.; Koslicki, D.; Lesker, T. R.; Gurevich, A.; McHardy, A. C. Critical assessment of metagenome interpretation: A benchmark of computational metagenomics software. *Nature Methods*, v. 16, n. 11, pp. 1063–1071, 2019.

Mishra, B.; Varjani, S.; Kumar, G.; Awasthi, M. K.; Awasthi, S. K.; Sindhu, R.; Binod, P.; Rene, E. R.; Zhang, Z. Microbial approaches for remediation of pollutants: Innovations, future outlook, and challenges. *Energy & Environment*, v. 32, n. 6, pp. 1029–1058, 2021.

Morelli, G.; Patuzzi, I.; Losasso, C.; Ricci, A.; Contiero, B.; Andrighetto, I.; Ricci, R. Characterization of intestinal microbiota in normal weight and overweight Border Collie and Labrador Retriever dogs. *Scientific Reports*, v. 12, n. 1, p. 9199, Jun. 2022. https://doi.org/10.1038/s41598-022-13270-6.

Moriya, Y.; Itoh, M.; Okuda, S.; Yoshizawa, A.; Kanehisa, M. KAAS: An automatic genome annotation and pathway reconstruction server. *Nucleic Acids Research*, v. 35, pp. W182–W185, 2007.

Mueller, U. G.; Juenger, T. E.; Kardish, M. R.; Carlson, A. L.; Burns, K. M.; Edwards, J. A.; Smith, C. C.; Fang, C. C.; Des Marais, D. L. Artificial selection on microbiomes to breed microbiomes that confer salt tolerance to plants. *mSystems*, v. 6, n. 6, p. e0112521, Dec. 2021. https://doi.org/10.1128/mSystems.01125-21.

Naumova, N. B.; Kabilov, M. R. About the biodiversity of the air microbiome. *Acta Naturae*, – 14, n. 4, pp. 50–56, Oct.–Dec. 2022. https://doi.org/10.32607/actanaturae.11671.

Nayfach, S.; Roux, S.; Seshadri, R. et al. A genomic catalog of Earth's microbiomes. *Nature Biotechnology*, v. 39, pp. 499–509, 2021. https://doi.org/10.1038/s41587-020-0718-6.

Naylor, D.; Coleman-Derr, D. Drought stress and root-associated bacterial communities. *Frontiers in Plant Science*, v. 8, p. 2223, 2018.

Nelson, M. T. et al. Human and extracellular DNA depletion for metagenomic analysis of complex clinical infection samples yields optimized viable microbiome profiles. *Cell Reports*, v. 26, n. 8, pp. 2227–2240.e5, Feb. 2019. https://doi.org/10.1016/j.celrep.2019.01.091. PMID: 30784601; PMCID: PMC6435281.

Neu, J.; Rushing, J. Cesarean versus vaginal delivery: Long-term infant outcomes and the hygiene hypothesis. *Clinical Perinatology*, v. 38, n. 2, pp. 321–331, Jun. 2011. https://doi.org/10.1016/j.clp.2011.03.008.

Ng, S. C.; Shi, H. Y.; Hamidi, N.; Underwood, F. E.; Tang, W.; Benchimol, E. I.; Panaccione, R.; Ghosh, S.; Wu, J. C. Y.; Chan, F. K. L.; Sung, J. J. Y.; Kaplan, G. G. Worldwide incidence and prevalence of infla[mm]atory bowel disease in the 21st century: A systematic review of population-based studies. *Lancet*, v. 390, n. 10114, pp. 2769–2778, Dec. 2017. https://doi.org/10.1016/S0140-6736(17)32448-0.

Noor, J.; Chaudhry, A.; Batool, S.; Noor, R.; Fatima, G. Exploring the impact of the gut microbiome on obesity and weight loss: A review article. *Cureus*, v. 15, n. 6, p. e40948, Jun. 2023. https://doi.org/10.7759/cureus.40948.

Notario, E.; Visci, G.; Fosso, B.; Gissi, C.; Tanaskovic, N.; Rescigno, M.; Marzano, M.; Pesole, G. Amplicon-based microbiome profiling: From second- to third-generation sequencing for higher taxonomic resolution. *Genes*, v. 14, p. 1567, 2023. https://doi.org/10.3390/genes14081567.

Nunes-Ramos, J.; Veloso-da-Costa, L.; Viana-Vieira, V.; Lima-Brandão, M. L. Challenges in the identification of environmental bacterial isolates from a pharmaceutical industry facility by 16S rRNA gene sequences. *DNA*, v. 5, n. 3, p. 33, 2025. https://doi.org/10.3390/dna5030033.

Nurk, S.; Meleshko, D.; Korobeynikov, A.; Pevzner, P. A. MetaSPAdes: A new versatile metagenomic assembler. *Genome Research*, v. 27, n. 5, pp. 824–834, May 2017. https://doi.org/10.1101/gr.213959.116.

Oburger, E.; Dell'Mour, M.; Hann, S.; Wieshammer, G.; Puschenreiter, M.; Wenzel, W. W. Evaluation of a novel tool for sampling root exudates from soil-grown plants compared to conventional techniques. *Environmental and Experimental Botany*, v. 87, pp. 235–247, 2013. https://doi.org/10.1016/j.envexpbot.2012.11.007.

Oda, E. Metabolic syndrome: Its history, mechanisms, and limitations. *Acta Diabetologica*, v. 49, n. 2, pp. 89–95, Apr. 2012. https://doi.org/10.1007/s00592-011-0309-6.

Oligschlaeger, Y.; Yadati, T.; Houben, T.; Condello Oliván, C. M.; Shiri-Sverdlov, R. Inflammatory bowel disease: A stressed "gut/feeling". *Cells*, v. 8, n. 7, p. 659, Jun. 2019. https://doi.org/10.3390/cells8070659.

Olson, N. D.; Treangen, T. J.; Hill, C. M.; Cepeda-Espinoza, V.; Ghurye, J.; Koren, S.; Phillippy, A. M. Metagenomic assembly through the lens of validation: Recent advances in assessing and improving the quality of genomes assembled from metagenomes. *Briefings in Bioinformatics*, v. 20, n. 4, pp. 1070–1083, 2017.

Pachiadaki, M. G. et al. Major role of nitrite-oxidizing bacteria in dark ocean carbon fixation. *Science*, v. 358, n. 6366, pp. 1046–1051, 2017.

Panke-Buisse, K.; Poole, A. C.; Goodrich, J. K.; Ley, R. E.; Kao-Kniffin, J. Selection on soil microbiomes reveals reproducible impacts on plant function. *ISME Journal*, v. 9, n. 4, pp. 980–989, Mar. 2015. https://doi.org/10.1038/ismej.2014.196.

Peddle, S. D.; Hodgson, R. J.; Borrett, R. J.; Brachmann, S.; Davies, T. C.; Erickson, T. E.; Liddicoat, C.; Muñoz-Rojas, M.; Robinson, J. M.; Watson, C. D.; Krauss, S. L.; Breed, M. F. Practical applications of soil microbiota to improve ecosystem restoration: Current knowledge and future directions. *Biological Reviews*, 2024. https://doi.org/10.1111/brv.13124.

Peixoto, R. S.; Harkins, D. M.; Nelson, K. E. Advances in microbiome research for animal health. *Annual Review of Animal Biosciences*, v. 9, pp. 289–311, Feb. 2021. https://doi.org/10.1146/annurev-animal-091020-075907.

Phillips, I.; Casewell, M.; Cox, T.; De Groot, B.; Friis, C.; Jones, R.; Nightingale, C.; Preston, R.; Waddell, J. Does the use of antibiotics in food animals pose a risk to human health? A critical review of published data. *Journal of Antimicrobial Chemotherapy*, v. 53, n. 1, pp. 28–52, Jan. 2004. https://doi.org/10.1093/jac/dkg483.

Ping, L.; Boland, W. Signals from the underground: Bacterial volatiles promote growth in *Arabidopsis*. *Trends in Plant Science*, v. 9, n. 6, pp. 263–266, Jun. 2004. https://doi.org/10.1016/j.tplants.2004.04.008.

Piters, W. A. A. S.; Binkowska, J.; Bogaert, D. Early life microbiota and respiratory tract infections. *Cell Host & Microbe*, v. 28, pp. 223–232, 2020. https://doi.org/10.1016/j.chom.2020.07.004.

Pitiphat, W.; Merchant, A. T.; Rimm, E. B.; Joshipura, K. J. Alcohol consumption increases periodontitis risk. *Journal of Dental Research*, v. 82, n. 7, pp. 509–513, Jul. 2003. https://doi.org/10.1177/154405910308200704.

Qiu, P.; Ishimoto, T.; Fu, L.; Zhang, J.; Zhang, Z.; Liu, Y. The gut microbiota in inflammatory bowel disease. *Frontiers in Cellular and Infection Microbiology*, v. 12, p. 733992, Feb. 2022. https://doi.org/10.3389/fcimb.2022.733992.

Quince, C.; Walker, A. W.; Simpson, J. T.; Loman, N. J.; Segata, N. Shotgun metagenomics, from sampling to analysis. *Nature Biotechnology*, v. 35, n. 9, pp. 833–844, 2017.

Quiza, L.; St-Arnaud, M.; Yergeau, E. Harnessing phytomicrobiome signaling for rhizosphere microbiome engineering. *Frontiers in Plant Science*, v. 6, p. 507, Jul. 2015. https://doi.org/10.3389/fpls.2015.00507.

Ray, D. K.; Ramankutty, N.; Mueller, N. D.; West, P. C.; Foley, J. A. Recent patterns of crop yield growth and stagnation. *Nature Communications*, v. 3, p. 1293, 2012. https://doi.org/10.1038/ncomms2296.

Ribeiro, I. D. A.; Volpiano, C. G.; Vargas, L. K.; Granada, C. E.; Lisboa, B. B.; Passaglia, L. M. P. Use of mineral weathering bacteria to enhance nutrient availability in crops: A review. *Frontiers in Plant Science*, v. 11, p. 590774, Dec. 2020. https://doi.org/10.3389/fpls.2020.590774.

Roy, G.; Prifti, E.; Belda, E.; Zucker, J. D. Deep learning methods in metagenomics: A review. *Microbial Genomics*, v. 10, p. 001231, 2024. https://doi.org/10.1099/mgen.0.001231.

Ryan, P. R.; Dessaux, Y.; Thomashow, L. S. et al. Rhizosphere engineering and management for sustainable agriculture. *Plant and Soil*, v. 321, pp. 363–383, 2009. https://doi.org/10.1007/s11104-009-0001-6.

Sanschagrin, S.; Yergeau, E. Next-generation sequencing of 16S ribosomal RNA gene amplicons. *Journal of Visualized Experiments*, n. 90, p. e51709, 2014. https://doi.org/10.3791/51709.

Santi, C. et al. Biological nitrogen fixation in non-legume plants. *Annals of Botany*, v. 111, n. 5, pp. 743–767, 2013.

Santiago-Rodriguez, T. M.; Le François, B.; MacKlaim, J. M.; Doukhanine, E.; Hollister, E. B. The skin microbiome: Current techniques, challenges, and future directions. *Microorganisms*, v. 11, p. 1222, 2023. https://doi.org/10.3390/microorganisms11051222.

Schmidt, T. S. B.; Raes, J.; Bork, P. The human gut microbiome: From association to modulation. *Cell*, v. 172, n. 6, pp. 1198–1215, 2019.

Scholz, M.; Ward, D. V.; Pasolli, E.; Tolio, T.; Zolfo, M.; Asnicar, F.; Segata, N. Strain-level microbial epidemiology and population genomics from shotgun metagenomics. *Nature Methods*, v. 19, n. 5, pp. 429–437, 2012.

Schrader, C. et al. PCR inhibitors—occurrence, properties and removal. *Journal of Applied Microbiology*, v. 113, n. 5, pp. 1014–1026, Nov. 2012. https://doi.org/10.1111/j.1365-2672.2012.05384.x. PMID: 22747964.

Sedghi, L.; Dimassa, V.; Harrington, A.; Lynch, S. V.; Kapila, Y. L. The oral microbiome: Role of key organisms and complex networks in oral health and disease. *Periodontology 2000*, v. 87, n. 1, pp. 107–131, Oct. 2021. https://doi.org/10.1111/prd.12393.

Seekatz, A. M.; Aas, J.; Gessert, C. E.; Rubin, T. A.; Saman, D. M.; Bakken, J. S.; Young, V. B. Recovery of the gut microbiome following fecal microbiota transplantation. *mBio*, v. 5, p. e00893-14, 2014. https://doi.org/10.1128/mbio.00893-14.

Segata, N.; Izard, J.; Waldron, L.; Gevers, D.; Miropolsky, L.; Garrett, W. S.; Huttenhower, C. Metagenomic biomarker discovery and explanation. *Genome Biology*, v. 12, n. 6, p. R60, 2013.

Sharma, A.; Laxman, B.; Naureckas, E. T.; Hogarth, D. K.; Sperling, A. I.; Solway, J.; Ober, C.; Gilbert, J. A.; White, S. R. Associations between fungal and bacterial microbiota of airways and asthma endotypes. *Journal of Allergy and Clinical Immunology*, v. 144, n. 5, pp. 1214–1227.e7, Nov. 2019. https://doi.org/10.1016/j.jaci.2019.06.025.

Shokralla, S. et al. Next-generation sequencing technologies for environmental DNA research. *Molecular Ecology*, v. 21, n. 8, pp. 1794–1805, Apr. 2012. https://doi.org/10.1111/j.1365-294X.2012.05538.x. PMID: 22486820.

Shreiner, A. B. et al. The gut microbiome in health and disease. *Current Opinion in Gastroenterology*, v. 31, n. 1, pp. 69–75, 2015.

Siddiqui, R.; Khan, N. A. Microbiome and One Health: Potential of novel metabolites from the gut microbiome of unique species for human health. *Microorganisms*, v. 11, n. 2, p. 481, Feb. 2023. https://doi.org/10.3390/microorganisms 11020481.

Siedt, M.; Schäffer, A.; Smith, K. E. C.; Nabel, M.; Roß-Nickoll, M.; Van Dongen, J. T. Comparing straw, compost, and biochar regarding their suitability as agricultural soil amendments to affect soil structure, nutrient leaching, microbial communities, and the fate of pesticides. *Science of the Total Environment*, v. 751, p. 141607, Jan. 2021. https://doi.org/10.1016/j.scitotenv.2020.141607.

Silva, R. C.; Lima, A.; Souza, L. C. S. Principais métodos de sequenciamento de DNA. *Scientific Electronic Archives*, v. 15, n. 10, 2022. https://doi.org/10.36560/15820221603.

Simon, S. A. et al. Dancing the Nanopore limbo—Nanopore metagenomics from small DNA quantities for bacterial genome reconstruction. *BMC Genomics*, v. 24, p. 727, 2023. https://doi.org/.1186/s12864-023-09853-w.

Sonnenburg, J. L.; Bäckhed, F. Diet-microbiota interactions as moderators of human metabolism. *Nature*, v. 535, n. 7610, pp. 56–64, Jul. 2016. https://doi.org/10.1038/nature18846.

Soriano-Lerma, A.; Pérez-Carrasco, V.; Sánchez-Marañón, M. et al. Influence of 16S rRNA target region on the outcome of microbiome studies in soil and saliva samples. *Scientific Reports*, v. 10, p. 13637, 2020. https://doi.org/10.1038/s41598-020-70141-8.

Steiner, S.; Linhart, N.; Neidl, A.; Baumgartner, W.; Tichy, A.; Wittek, T. Evaluation of the therapeutic efficacy of rumen transfaunation. *Journal of Animal Physiology and Animal Nutrition (Berl)*, v. 104, n. 1, pp. 56–63, Jan. 2020. https://doi.org/10.1111/jpn.13232.

Suau, A.; Bonnet, R.; Sutren, M.; Godon, J. J.; Gibson, G. R.; Collins, M. D.; Doré, J. Direct analysis of genes encoding 16S rRNA from complex communities reveals many novel molecular species within the human gut. *Applied and Environmental Microbiology*, v. 65, n. 11, pp. 4799–4807, Nov. 1999. https://doi.org/10.1128/AEM.65.11.4799-4807.1999.

Suman, J.; Rakshit, A.; Ogireddy, S. D.; Singh, S.; Gupta, C.; Chandrakala, J. Microbiome as a key player in sustainable agriculture and human health. *Frontiers in Soil Science*, v. 2, p. 821589, 2022. https://doi.org/10.3389/fsoil.2022.821589.

Sunagawa, S. et al. Structure and function of the global ocean microbiome. *Science*, v. 348, n. 6237, p. 1261359, 2015.

Taş, N.; De Jong, A. E. E.; Li, Y.; Trubl, G.; Xue, Y.; Dove, N. C. Metagenomic tools in microbial ecology research. *Current Opinion in Biotechnology*, v. 67, pp. 184–191, 2021. https://doi.org/10.1016/j.copbio.2021.01.019.

Theil, S.; Rifa, E. Ranomaly: Amplicon workflow for microbial community Analysis [version 1; peer review: 2 approved]. *F1000Research*, v. 10, p. 7, 2021. https://doi.org/10.12688/f1000research.27268.1.

Thrall, P.; Murray, B. R.; Watkin, E.; Woods, M. J.; Baker, K.; Burdon, J.; Brockwell, J. Bacterial partnerships enhance the value of native legumes in rehabilitation of degraded agricultural lands. *Ecological Management and Restoration*, v. 2, pp. 233–235, 2001.

Tolstoganov, I. et al. BinSPreader: Refine binning results for fuller MAG reconstruction. *iScience*, v. 25, n. 8, p. 104770, 2022. https://doi.org/10.1016/j.isci.2022.104770.

Tribble, G. D.; Angelov, N.; Weltman, R.; Wang, B. Y.; Eswaran, S. V.; Gay, I. C.; Parthasarathy, K.; Dao, D. V.; Richardson, K. N.; Ismail, N. M.; Sharina, I. G.; Hyde, E. R.; Ajami, N. J.; Petrosino, J. F.; Bryan, N. S. Frequency of tongue cleaning impacts the human tongue microbiome composition and enterosalivary circulation of nitrate. *Frontiers in Cellular and Infection Microbiology*, v. 9, p. 39, Mar. 2019. https://doi.org/10.3389/fcimb.2019.00039.

Tringe, S. G. et al. Comparative metagenomics of microbial communities. *Science*, v. 308, pp. 554–557, 2005. https://doi.org/10.1126/science.1107851.

Trinh, P.; Zaneveld, J. R.; Safranek, S.; Rabinowitz, P. M. One Health relationships between human, animal, and environmental microbiomes: A mini-review. *Frontiers in Public Health*, v. 6, p. 235, Aug. 2018. https://doi.org/10.3389/fpubh.2018.00235.

Tripathy, D.; Chavez, A. O. Defects in insulin secretion and action in the pathogenesis of type 2 diabetes mellitus. *Current Diabetes Reports*, v. 10, n. 3, pp. 184–191, Jun. 2010. https://doi.org/10.1007/s11892-010-0115-5.

Trubl, G. et al. Coming-of-age characterization of soil viruses: A user's guide to virus isolation, detection within metagenomes, and viromics. *Soil Systems*, v. 4, p. 23, 2020.

Turnbaugh, P. J. et al. The human microbiome project. *Nature*, v. 449, n. 7164, pp. 804–810, 2007.

Ullah, H.; Arbab, S.; Tian, Y.; Liu, C. Q.; Chen, Y.; Li, Q.; Khan, M. I. U.; Hassan, I. U.; Li, K. The gut microbiota-brain axis in neurological disorder. *Frontiers in Neuroscience*, v. 17, p. 1225875, Aug. 2023. https://doi.org/10.3389/fnins.2023.1225875.

Urbaniak, C.; Cummins, J.; Brackstone, M.; MacKlaim, J. M.; Gloor, G. B.; Baban, C. K.; Scott, L.; O'Hanlon, D. M.; Burton, J. P.; Francis, K. P.; Tangney, M.; Reid, G. Microbiota of human breast tissue. *Applied and Environmental Microbiology*, v. 80, n. 10, pp. 3007–3014, May 2014. https://doi.org/10.1128/AEM.00242-14.

Ursell, L. K.; Metcalf, J. L.; Parfrey, L. W.; Knight, R. Defining the human microbiome. *Nutrition Reviews*, v. 70, supl. 1, pp. S38–44, Aug. 2012. https://doi.org/10.1111/j.1753-4887.2012.00493.x.

Van Schaik, W. The human gut resistome. *Philosophical Transactions of the Royal Society B: Biological Sciences*, v. 370, n. 1670, p. 20140087, 2015.

Varliero, G.; Lebre, P. H.; Stevens, M. I.; Czechowski, P.; Makhalanyane, T.; Cowan, D. A. The use of different 16S rRNA gene variable regions in biogeographical studies. *Environmental Microbiology Reports*, v. 15, pp. 216–228, 2023. https://doi.org/10.1111/1758-2229.13145.

Vassilev, N.; Vassileva, M.; Martos, V.; Garcia Del Moral, L. F.; Kowalska, J.; Tylkowski, B.; Malusá, E. Formulation of microbial inoculants by encapsulation in natural polysaccharides: Focus on beneficial properties of carrier additives and derivatives. *Frontiers in Plant Science*, v. 11, p. 270, Mar. 2020. https://doi.org/10.3389/fpls.2020.00270.

Vinderola, G.; Sanders, M. E.; Salminen, S. The concept of postbiotics. *Foods*, v. 11, n. 8, p. 1077, Apr. 2022. https://doi.org/10.3390/foods11081077. PMID: 35454664; PMCID: PMC9027423.

Vorholt, J. Microbial life in the phyllosphere. *Nature Reviews Microbiology*, v. 10, pp. 828–840, 2012. https://doi.org/10.1038/nrmicro2910.

Vreeburg, J. H. G.; Schaap, P. G.; Van Dijk, J. C. Particles in the drinking water system: From source to discolouration. *Water Science and Technology: Water Supply*, v. 4, pp. 431–438, 2004. https://doi.org/10.2166/ws.2004.0135.

Wang, Y. et al. Applications of synthetic biology in wastewater treatment. *Nature Reviews Microbiology*, v. 19, n. 5, pp. 305–318, 2021.

Watzenboeck, M. L.; Gorki, A. D.; Quattrone, F.; Gawish, R.; Schwarz, S.; Lambers, C.; Jaksch, P.; Lakovits, K.; Zahalka, S.; Rahimi, N.; Starkl, P.; Symmank, D.; Artner, T.; Pattaroni, C.; Fortelny, N.; Klavins, K.; Frommlet, F.; Marsland, B. J.; Hoetzenecker, K.; Widder, S.; Knapp, S. Multi-omics profiling predicts allograft function after lung transplantation. *European Respiratory Journal*, v. 59, n. 2, p. 2003292, Feb. 2022. https://doi.org/10.1183/13993003.03292-2020.

Webster, J. D.; Santagostino, S. F.; Foreman, O. Applications and considerations for the use of genetically engineered mouse models in drug development. *Cell and Tissue Research*, v. 380, pp. 325–340, 2020. https://doi.org/10.1007/s00441-019-03101-y.

Whipps, J. M.; Lewis, K.; Cooke, R. C. Mycoparasitism and plant disease control. In: *Fungi in Biological Control Systems*. Manchester, UK: Manchester University Press; New York, NY: St. Martin's Press, 1988, pp. 161–187.

Wickramarachchi, A.; Mallawaarachchi, V.; Rajan, V.; Lin, Y. MetaBCC-LR: Metagenomics binning by coverage and composition for long reads. *Bioinformatics*, v. 36, supl. 1, pp. i3–i11, Jul. 2020. https://doi.org/10.1093/bioinformatics/btaa441.

Wingett, S. W.; Andrews, S. FastQ screen: A tool for multi-genome mapping and quality control. *F1000Research*, v. 7, p. 1338, Aug. 2018. https://doi.org/10.12688/f1000research.15931.2. PMID: 30254741; PMCID: PMC6124377.

Woo, S. B.; Challacombe, S. J. Management of recurrent oral herpes simplex infections. *Oral Surgery, Oral Medicine, Oral Pathology, Oral Radiology, and Endodontology*, v. 103, supl., pp. S12.e1–18, Mar. 2007. https://doi.org/10.1016/j.tripleo.2006.11.004.

Wood, D. E.; Lu, J.; Langmead, B. Improved metagenomic analysis with Kraken 2. *Genome Biology*, v. 20, p. 257, 2019. https://doi.org/10.1186/s13059-019-1891-0.

Wood, D. E.; Salzberg, S. L. Kraken: Ultrafast metagenomic sequence classification using exact alignments. *Genome Biology*, v. 15, p. R46, 2014. https://doi.org/10.1186/gb-2014-15-3-r46.

Wu, L.; Xiao, W.; Chen, G.; Song, D.; Khaskheli, M. A.; Li, P.; Zhang, S.; Feng, G. Identification of Pseudomonas mosselii BS011 gene clusters required for suppression of Rice Blast Fungus *Magnaporthe oryzae*. *Journal of Biotechnology*, v. 282, pp. 1–9, Sep. 2018. https://doi.org/10.1016/j.jbiotec.2018.04.016.

Wu, Y. W. et al. MaxBin: An automated binning method to recover individual genomes from metagenomes using an expectation-maximization algorithm. *Microbiome*, v. 2, p. 26, 2014. https://doi.org/10.1186/2049-2618-2-26.

Xenoulis, P. G.; Palculict, B.; Allenspach, K.; Steiner, J. M.; Van House, A. M.; Suchodolski, J. S. Molecular-phylogenetic characterization of microbial communities imbalances in the small intestine of dogs with inflammatory bowel disease. *FEMS Microbiology Ecology*, v. 66, n. 3, pp. 579–589, Dec. 2008. https://doi.org/10.1111/j.1574-6941.2008.00556.x.

Xiao, L.; Feng, Q.; Liang, S. et al. A catalog of the mouse gut metagenome. *Nature Biotechnology*, v. 33, pp. 1103–1108, 2015. https://doi.org/10.1038/nbt.3353.

Yamamura, K.; Izumi, D.; Kandimalla, R.; Sonohara, F.; Baba, Y.; Yoshida, N.; Kodera, Y.; Baba, H.; Goel, A. Intratumoral *Fusobacterium nucleatum* levels predict therapeutic response to neoadjuvant chemotherapy in esophageal squamous cell carcinoma. *Clinical Cancer Research*, v. 25, n. 20, pp. 6170–6179, Oct. 2019. https://doi.org/10.1158/1078-0432.CCR-19-0318.

Ye, L. et al. High-resolution metagenomics of human gut microbiota generated by nanopore and illumina hybrid metagenome assembly. *Frontiers in Microbiology*, v. 13, 2022. https://doi.org/10.3389/fmicb.2022.801587.

Ye, S. H. et al. Benchmarking metagenomics tools for taxonomic classification. *Cell*, v. 178, n. 4, pp. 779–794, 2019. https://doi.org/10.1016/j.cell.2019.07.010.

Yin, Y. N.; Yu, Q. F.; Fu, N.; Liu, X. W.; Lu, F. G. Effects of four *Bifidobacteria* on obesity in high-fat diet-induced rats. *World Journal of Gastroenterology*, v. 16, n. 27, pp. 3394–3401, Jul. 2010. https://doi.org/10.3748/wjg.v16.i27.3394.

Zhang, W.; Fan, X.; Shi, H.; Li, J.; Zhang, M.; Zhao, J.; Su, X. Comprehensive assessment of 16S rRNA gene amplicon sequencing for microbiome profiling across multiple habitats. *Microbiology Spectrum*, v. 11, p. e00563-23, 2023. https://doi.org/10.1128/spectrum.00563-23.

Zhao, L. Y.; Mei, J. X.; Yu, G. et al. Role of the gut microbiota in anticancer therapy: From molecular mechanisms to clinical applications. *Signal Transduction and Targeted Therapy*, v. 8, p. 201, 2023. https://doi.org/10.1038/s41392-023-01406-7.

Zhou, Y.; Liu, M.; Yang, J. Recovering metagenome-assembled genomes from shotgun metagenomic sequencing data: Methods, applications, challenges, and opportunities. *Microbiological Research*, v. 260, p. 127023, 2022. https://doi.org/10.1016/j.micres.2022.127023.

Zhu, A.; Sunagawa, S.; Mende, D. R.; Bork, P. Inter-individual differences in the gene content of human gut bacterial species. *Nature*, v. 582, n. 7812, pp. 107–111, 2021.

Zhu, S. et al. Gut microbiota in atherosclerosis: Focus on trimethylamine N-oxide. *Atherosclerosis*, v. 293, pp. 97–104, 2020.

7 Metagenomics in Microbiome Consortium

Mona M. Ali[1], *Hanan M. K. Abdel-Fatah*[2], *Mennatullah A. Turky*[1], *and Tarek A. A. Moussa*[2]*
[1]Biotechnology Department, Faculty of Science, Cairo University, Giza 12613, Egypt
[2]Botany and Microbiology Department, Faculty of Science, Cairo University, Giza 12613, Egypt

7.1 INTRODUCTION

A consortium (pl. consortia) is two or more organisms that share everyday activity or use resources to attain a common objective. A microbial consortium is a group of two or more microbial organisms living in an endosymbiotic, ectosymbiotic, or both relationships. The concept of microbial consortium was first proposed in 1872. It reflects the real world of microbial life rather than the isolation of a single colony. Recent advances in metagenomics, proteomics, metabolomics, and computational modeling have shed light on the numerous applications of microbial consortia in biotechnological applications.

Using a microbial consortium has several advantages because complex carbon sources can be used in mixed cultures, production is higher than using a single organism as multistep reactions are performed faster, certain uncultivable microorganisms may be utilized in a consortium, and stability is achieved in the consortium because undesired microbes are not able to grow (Mishra et al., 2021). Anyway, the application of microbial consortia still has several disadvantages and drawbacks. Among these defects is that the consortium cannot detect microbial contamination. Another problem is the domination of certain microorganisms over others, which causes productivity issues. The lack of knowledge about metabolite profiling, functioning, and nutritional requirements of the microbial consortium members influences the construction of a practical consortium. The viability of the microbes in the consortium is also a limiting factor. The choice of microbes for constructing consortia is a crucial factor in their success. Interactions and cross-talk between microbes, residual metabolic product accumulation, and competition for resources build obstacles in optimizing consortia utilization for boosting productivity (Mishra et al., 2021).

7.1.1 Classification of Microbial Consortia Based on Their Origin and Human Conciliation

The primary way to classify microbial consortia is based on the origin of microbial members (one source or several sources), the pre-existence of interactions between microorganisms in the source material, human intervention in their composition (including genetic manipulation), and the method used to obtain them. They are classified on that basis into four categories: natural microbial consortia (NMCs), artificial microbial consortia, synthetic microbial consortia, and cross-kingdom consortia (Bernstein and Carlson, 2012).

7.1.1.1 Natural Microbial Consortia

NMCs are composed of microorganisms from a single source that may interact naturally or be obtained through successive enrichment and selection cultures. They are often derived from nature. Although enrichment cultures alter the original proportions of microorganisms, the resulting consortium is still considered natural because the interactions between the consortium members are still present. The possibility exists that the conditions of the enrichment culture induce new abilities and associations. These abilities and associations would differ from those active in the original community (Joergensen and Wichern 2018; Spini et al., 2018). Microbes in this microbial consortium have several physiological activities regarding bioremediation and wastewater treatment (Marchevsky et al., 2015; Massot et al., 2022)

So generally, NMCs are characterized by microbial members originating from the same sample source and low human intervention. Their function, rather than their members, defines them, and previous interactions between the consortium members can exist. They may contain unculturable microbes (Villaverde et al., 2018; Massot et al., 2022).

7.1.1.2 Artificial Microbial Consortia

In artificial microbial consortia, their members originate from different sources and combine (including natural and culture collections); microorganisms are previously isolated. They are characterized by high human conciliation. They are defined by their members rather than their function. There is no previous interaction between the consortia members in nature (Shankar et al., 2014; Mnif et al., 2015; Patowary et al., 2016).

***Corresponding author:** tarekmoussa@cu.edu.eg (Tarek A. A. Moussa)

DOI: 10.1201/9781003544425-7

7.1.1.3 Synthetic Microbial Consortia

Synthetic microbial consortia are a class of artificial microbial consortia in which microbial members and their proportions are chosen to achieve specific desired functions. In this type of consortia, microorganisms originate from different sources, and the highest human conciliation characterizes them. They could include a genetically modified microorganism to optimize a specific function (sequential biocatalysis), but it is not necessary (Gilbert et al., 2003; Alnahhas et al., 2020). If the consortia members are a mixture of wild-type and engineered microbes, they are considered a semi-synthetic consortium.

7.1.1.4 Cross-Kingdom Consortia

NMCs, artificial microbial consortia, and synthetic microbial consortia can consist of members from the same kingdom (single-kingdom consortia: bacteria, archaea, or eukarya) or consist of members from different kingdoms, resulting in cross-kingdom consortia. Microbes in cross-kingdom consortia can come from a single source or from various sources (Seong et al., 2018; Yuan et al., 2018). Cross-kingdom consortia have several biotechnological applications, including fungal, bacterial, and archaeal-bacterial consortia, especially in bioremediation. In addition, microalgae–bacteria consortia are promising for the bioremediation of wastewater (Chan et al., 2022). Finally, the existence of gray zones in the limits between each proposed microbial consortia category is addressed.

7.1.2 Classification of Microbial Consortium Based on the Functions

Another way to classify microbial consortia is based on the functions of the microbial consortium members. On that basis, they are classified into four categories: environmental maintenance consortium, nutrient exchange consortium, substrate facilitator consortium, and signal exchange consortium (Mishra et al., 2021).

7.1.2.1 Environmental Maintenance Consortium

An environmental maintenance consortium is established to enhance the growth of microbial consortium members by creating suitable conditions. A clear example is an aerobic/anaerobic consortium comprising two types of microbes: aerobic and anaerobic. The aerobic microbes remove oxygen to create an oxygen-free environment, enhancing and allowing anaerobic microbes to grow (Wu et al., 2016).

7.1.2.2 Nutrient Exchange Consortium

A nutrient exchange consortium is a type in which one group of microbes provides the other members in the consortium with the essential supplements and requirements for their growth. An illustration of that is the auxotroph growth (nutritional mutant), which is provoked as a result of the presence of a microorganism that supplies its development, providing nutrients that the auxotroph cannot produce by itself (Shou et al., 2007).

7.1.2.3 Substrate Facilitator Consortium

In the substrate facilitator consortium, one of its microbial members provides the other members with free carbon substrates that they can use. This is critical because all the available kinds of complex carbohydrates (carbon sources) are needed and used by various microorganisms (Dwidar et al., 2013).

7.1.2.4 Signal Exchange Consortium

The microorganisms of this consortium produce signaling molecules. The exchange of these signals results in the production of antibiotics and other metabolites due to the expression of specific cryptic genes by other microbes (Pérez et al., 2011).

7.2 METAGENOMIC METHODS FOR ANALYZING MICROBIAL CONSORTIA

Exploring the microbial communities within specific environmental niches has garnered considerable interest due to the crucial functions these microbial communities perform. Different techniques have been developed over the years to identify the taxonomic composition and functional potential of microbial consortia in agriculture. Due to the advantages that microbial consortia provide over single-strain usage, it has become increasingly popular to recruit microbial consortia to enhance the growth of economically essential cultivars, increase yield, or even maintain soil quality (McGuire and Treseder, 2010). Meanwhile, numerous methods relying on the amplification of a single gene have been established to reveal the composition of the microbiome. Ranging from fingerprinting-based techniques such as T-RFLP, DGGE, and TGGE, which mainly adopt molecular biology protocols, to sequencing-based technologies, Metagenomics, which includes two types: 16s rRNA sequencing and DNA shotgun technology (Sterflinger and Piñar, 2021).

Although fingerprinting-based techniques and their synonyms are cost-effective, they do not provide enough physiological information about the microbiota found in each specimen, nor fully identify the unculturable microbes. Thus, sequencing-based methods have prevailed in microbiome identification due to their proven capability to identify unculturable microbes and reveal their biological activity by studying the microbial genetic composition in nature (Kumari and Rai, 2020). This subchapter will elucidate and comparatively analyze genomics and metagenomics methodologies employed for mixed microbial culture and community analysis. Moreover, we will pinpoint how these methods are transformed into tools applicable in environmental biotechnology, broadly, and microbiology, specifically.

7.2.1 Mixed Microbial Consortium's Comparative Metagenomic Analysis

Our understanding of the mixed microbial consortium's distinctive symbiotic interaction in natural habitats, the mechanism behind the network formation between the microbes, or how these consortia bring about the synergistic effect that enhances crops and soil quality altogether, remains unclear (Anderson and Cairney, 2004). Therefore, we need to compare metagenomics data from different consortia to answer these two questions: What commonalities and differences do each consortium exhibit, and how do microbial consortia contribute to the ecosystem's distinguished quality upon their recruitment? Hence, comparative genomics is the process of choice by which we can better understand how mixed microbial consortia behave in nature.

Comparative Metagenomics is a computational process used to compare various metagenomes from different samples or datasets, allowing for the analysis of the commonalities and differences between datasets (Maltez Thomas et al., 2018). Multiple comparative analysis methods could be used according to the researcher's preference, research question, and available datasets. Briefly, these comparison methods could be one of the following: Analysis taking account of the ecological measurements perspective that includes evaluating the alpha diversity (richness, evenness, diversity), the beta diversity (samples set variation in their communities structure), or generating microbial association network for the comparison groups; Weighing up the abundance of either essential genes' copy number, relative abundance, absolute abundance, high-level phenotype abundance of particular microbial strain that can further integrate differential analysis of microbial pathways, or selecting a feature taxa representing the comparison groups; Identifying specific microbial taxa's codon bias which is more common with metagenomics data generated from shotgun sequencing; last but not least, differential analysis of comparison microbial groups from the statistical point of view. It is worth noting that statistical comparisons between available metagenomics datasets are not feasible using parametric tests due to the sparsity and high dimensionality of metagenomics data. Critically, the filtration of datasets for comparative metagenomics analysis must be done with caution, as the experimental protocol has a significant impact on the analysis results (Nayfach and Pollard, 2016). Therefore, the experimental protocol for the data sets must be scrutinized according to multiple quality control parameters established by the International Human Microbiome Standards (IHMS) project or the Microbiome Quality Control (MBQC) project (Voigt et al., 2015; Sinha et al., 2015).

Recently, comparative metagenomics analysis methodology has gained wide popularity for examining and comparing mixed microbial cultures on numerous aspects, including taxonomic, phenotypic, functional, and strain-level comparisons. Additionally, the analysis of metagenomics assemblies from mixed microbial consortia is pioneering. For example, a recent study has made a comparative metagenomics analysis for the polycyclic aromatic hydrocarbon (PAH)-polluted soil; the research identified the taxonomic and functional differences caused by corn-stover-stimulated native microbiota or a bio-augmented fungal-bacterial consortium (Zafra et al., 2016). The study procedures also included witnessing the effect of some environmental changes on the consortium's functional gene composition internally and the soil-acquired capability for PAH degradation externally. On the other hand, the metagenomic assemblies of the methanogenic, anaerobic, benzene-degrading cultures were constructed in a study aiming to learn more about the genomic sequences of notorious microbial consortia made for benzene degradation (Toth et al., 2023), named DGG-B and OR. The results revealed the presence of low-abundance microbial strains within the examined consortia. They recovered the relative abundance of some metagenome-assembled genomes (MAGs) in the DGG-B consortium, thus revealing the complex structural composition and the functional role played by the consortia in question in the benzene degradation process. Statistical-based structural composition analysis is one last example of the methods commonly used to compare mixed microbial culture metagenomics. Hug et al. (2012) compared three *Dehalococcoides*-containing enriched communities to recognize standard features of these consortia, including the metabolic profile and the statistically significant structural variance between communities, as well.

Comparative metagenomics still holds considerable potential for discovery in the era of artificial intelligence. Hence, more research can exploit standard identification techniques using AI-assisted models. Although the analysis of comparative metagenomics has contributed to the identification of differential unculturable microbial strains, much more advancement is anticipated with the substantial reduction in the cost of high-throughput sequencing technologies, allowing for the prompt and massive generation of samples. Moreover, consolidating well-trained machine learning and deep learning models with comparative metagenomics can foster the analysis of microbial consortia.

7.2.2 Genomic Analysis of Microbial Communities

Genomic analysis involves capturing the genome characteristics of a given biological sample, whether it is a single nucleotide polymorphism (SNP), insertion, duplication, deletion, or translocation of sequences from one portion of the genome to another, examining the whole genome sequence, generally, or the functional and regulatory elements, precisely (Fasman and Salzberg, 1998). Furthermore, comparative and evolutionary genomic analysis studies have revolutionized this field, focusing on comparing genomes that originate from the same ancestor and identifying novel structural or functional features on the DNA level (Kawashima, 2019). Meanwhile, spectacular sequencing technologies have emerged, favoring the advancement of genomic analysis, with the short-read-generating technologies being the most common, producing reads ranging from 150 to 400 bp (Mohamed and Syed, 2013). While third-generation sequencing allowed for the sequencing of much longer fragments that can reach 10 kb (Berger and Yu, 2023). Furthermore, specialized technologies were created to identify the three-dimensional (3D) chromosomal structure for Hi-C technology (Belton et al., 2012).

Microbial communities serve as the unknown soldiers that configure the balance in the ecosystem; however, these communities exhibit high levels of complexity. However, unraveling the genetic composition of microbial communities significantly enhances our understanding of the mechanisms by which microorganisms interact with their natural habitat. Before the era of next-generation sequencing (NGS), identifying the genomes of microbial communities was nearly impossible because only ~1% of the microbes are culturable (Williamson and Yooseph, 2012). However, metagenomics has refrained researchers from the need for tedious culturing and isolation processes to obtain access to the microbiota genome. Metagenomics studies are analyses conducted to identify the function and structure of microbes isolated from specific habitats. Hence, it is the total genome analysis of microorganisms. The process is as follows: Nucleic acid is extracted in high quality directly from the natural habitat, followed by either sequencing of the 16S rRNA or shotgun sequencing of fragmented DNA, without any isolation or culturing process (Wang et al., 2016). Although Metagenomics counteracts the pitfalls of fingerprinting-based techniques, special attention should be given to the environmental habitat where the sample is collected, as metagenomic analysis is greatly affected by DNA quality (Ghurye et al., 2016).

The genomic analysis of microbial communities mainly revolves around two questions: Who is there? What are they doing when revealing the composition of the environmental sample? Identifying the role these communities exhibit in nature. Notably, the shotgun methodology has been more implemented in taxonomic composition identification as it counteracts the bias introduced by the 16S rRNA techniques in the phylogenetic tree construction (von Mering et al., 2007). The first question is the sequence-driven analysis approach, which inspects the reads based on various factors, the most important of which is distinguishing between microbial community members and the genetic constituents that can interact with the surrounding environment. The sequence-driven genomic analysis of shotgun sequencing data comprises three steps following the generation of sequencing reads: genome assembly, binning, and annotation (Wang et al., 2016). However, selecting assembly and binning algorithms can substantially shift the analysis results. The genome assembly step aims to construct either the genomic contigs if the study uses shotgun sequencing of the community's complete DNA, or the assembly of the marker gene fragments, 16s rRNA, which is most commonly used (Ghurye et al., 2016). It is advisable to navigate multiple studies that benchmarked the performance of different de novo assemblers and whether the assembly step affects the taxonomic and functional annotation to scrutinize the assembly method according to the aim of the study (Sutton et al., 2019; Tamames et al., 2019). Meanwhile, the binning step involves clustering assembled contigs to form discrete groups of microorganisms' marker genes or genome bins. The clustering process differs according to the algorithm used for benchmarking the widely available binning tools, and precise selection for the algorithm is critical to perform accurate genomic analysis and capture important genetic features such as virulence factors and antimicrobial genes (Maguire et al., 2020; Yue et al., 2020).

7.2.3 Applied Microbiology and Biotechnology Tools

The recent advancements in sequencing technologies, cost reduction, and high accessibility in nearly every research laboratory have led to a vast generation of metagenomic data, unraveling the secrets of microbial communities' system biology. Therefore, there are widespread applications for exploiting comparative and metagenomics analysis. Out of the immense applications of metagenomics, we will focus on the implementation of metagenomics in microbiological, environmental, and biotechnological applications. Assessing functional and microbial diversity is one of the primary applications that advanced the identification of microbial communities and assisted in developing microbial consortia. In contrast, some studies have explored the taxonomic composition of microbial systems found in environmental habitats (Hugenholtz et al., 2016). The target of such studies varied considerably according to the study preference. It may open the doors for exploring microbial evolution in the ecosystem through one of the following: the phylogenetic tree analysis (Czech et al., 2022), studying strain-level genomic diversity (Nayfach et al., 2016), or gene abundance by calculating the copy number of the genes of interest found per taxa (Greenblum et al., 2015). Additionally, it can infer the differential abundance of microbes across different time points and fluctuations in biotic or abiotic factors, particularly in natural habitats (Finucane et al., 2014).

Recently, the entire gene set of a particular species has become a topic of interest, a field known as pan-genome analysis (Tettelin et al., 2005). For the past decade, the pangenome discovery workflow has widely utilized metagenome-assembled genomes (MAGs) with certain restrictions (Bowers et al., 2017). MAGs are genomes assembled from shotgun sequencing, filtered, assembled, binned, and then assigned their taxonomy (Hugerth et al., 2015). Therefore, pangenome analysis has evolved since the rise of the metagenomics technique, and the revolution of the species-specific pangenome of bacteria 62,63, fungi 64, and economically significant plants (Song et al., 2020; J. Wang et al., 2023) is ongoing with trials to assess pangenome analysis methods to bring about more precise and efficient results (Zhang et al., 2024). Metagenomics has established a new gateway to bioprospecting through the discovery of novel biomolecules, including genes, enzymes, and metabolites, from diverse microbial habitats, including hard-to-explore extreme environments (Sambles et al., 2017). Metagenomics-assisted bioprospecting involves identifying gene clusters that confer functional roles. For example, gene clusters orchestrate the production of antibiotics and antitumor agents from certain microbes (Yue et al., 2020); enzymes are utilized in biodegradation, the biotechnological industry, or for identifying industrially important enzymes with distinctive biochemical characteristics (Pabbathi et al., 2023). On the other hand, metagenomics emerged as a hot topic in agriculture in the identification of microbial community mechanisms contributing to the elevation of crops' tolerance to biotic and abiotic stressors (Z. Wang et al., 2023; Wu et al., 2024), thus leading to the consolidation of tolerant cultivars breeding programs and the exploitation of either the tolerant microbial communities or their byproducts in enhancing crops production. Furthermore, metagenomics has a vast application in bioremediation as bacterial communities identified genomes can be further used in degrading hydrocarbons polluting the soil and improving soil quality (Naghavi et al., 2024) as a biosurfactant (Thies et al., 2016), or to reveal candidate communities that could be utilized to remove heavy metal contamination (Li et al., 2023).

7.3 METAGENOME AND STUDYING MICROBIAL CONSORTIA FOR THE AGRICULTURAL FUTURE PERSPECTIVE

The study of microbial consortia through metagenomics has revolutionized our understanding of the complex interactions between microorganisms and their environments (Handelsman, 2004). Researchers can develop new strategies to improve soil health, reduce fertilizer use, and promote sustainable agriculture. Metagenomics involves the direct sequencing of DNA from environmental samples, allowing researchers to access the genetic information of entire microbial communities (Moussa et al., 2017). For example, novel genes and enzymes involved in nitrogen fixation, phosphate solubilization, and other vital processes for plant growth can be identified (Timofeeva et al., 2023). This approach has been instrumental in agriculture, where microbial consortia are crucial in soil fertility, plant nutrition, and disease suppression (Nannipieri et al., 2017). Additionally, metagenomics has been used to study the diversity and composition of microbial communities in different agricultural soils, providing insights into the factors that influence soil fertility and plant productivity (Timofeeva et al., 2023).

7.3.1 Microbial Consortium as Promising Biostimulants for Plant Health

Microbial consortia, composed of multiple microorganisms, have emerged as promising biostimulants for plant health (Berg, 2009). These consortia can promote plant growth, enhance nutrient uptake, and induce systemic resistance against pathogens (Greff et al., 2022). Using microbial consortia as biostimulants offers several advantages over traditional chemical fertilizers and pesticides, including improved soil health, reduced environmental pollution, and increased crop yields (Vessey, 2003).

Microbial consortia can be designed to provide specific benefits to plants, such as nitrogen fixation, phosphate solubilization, and production of plant growth-promoting substances (Verma et al., 2022). For example, a consortium of nitrogen-fixing bacteria, such as Rhizobia, and phosphate-solubilizing bacteria, such as Pseudomonas, can promote plant growth and enhance nutrient uptake (Kuhl-Nagel et al., 2022). Additionally, microbial consortia can induce systemic resistance against pathogens, reducing the need for chemical pesticides (Pieterse et al., 2014).

The use of microbial consortia as biostimulants has been demonstrated to offer several benefits for plant health, including improved soil structure, increased nutrient availability, and enhanced disease resistance (Berg et al., 2020). A microbial consortium composed of bacteria and fungi enhanced soil structure and promoted plant growth in degraded soil (Greff et al., 2022; Kumari et al., 2023). Another study found that a microbial consortium induced systemic resistance against a fungal pathogen in tomato plants (Saeed et al., 2021).

Despite the potential benefits of microbial consortia as biostimulants, several challenges need to be addressed, including selecting compatible microorganisms, developing effective delivery systems, and optimizing application rates (Aloo et al., 2022). Further research is needed on how microbial consortia interact with plants and the environment (Kuhl-Nagel et al., 2022).

7.3.2 Plant Growth Stimulation and Enhancing Nutrient Availability

A microbial consortium refers to the community of microorganisms that coexist and interact with each other in a specific environment. The microbial consortium plays a crucial role in promoting plant health and stimulating plant growth. Studies have shown that microbial consortia can enhance plant growth by producing plant growth-promoting substances, such as auxins, cytokinins, and gibberellins (Santoyo et al., 2021; Nunes et al., 2024).

The microbial consortium can also enhance plant nutrient availability by solubilizing phosphorus, producing siderophores, and fixing nitrogen. For example, certain bacteria in the microbial consortium can solubilize phosphorus from insoluble sources, making it available for plant uptake (Kaur et al., 2022; Luo et al., 2024). Similarly, microbial consortia can produce siderophores, which are iron-chelating compounds that enhance plant iron availability (Lurthy et al., 2020; Moussa et al., 2023; Deb and Tatung, 2024).

In addition to enhancing nutrient availability, the microbial consortium can promote plant growth by producing antibiotics and antifungal compounds that can suppress plant pathogens. Certain bacteria in the microbial consortium can produce antibiotics that can inhibit the growth of fungal pathogens (Maciag et al., 2023; Wang et al., 2024). This can improve plant health and reduce the incidence of disease.

Microbial consortia can also interact with plants to promote plant growth and development. For example, certain microorganisms in the microbial consortium can produce signaling molecules that stimulate plant growth and development (Santoyo et al., 2021; Maciag et al., 2023). This can lead to improved plant yields and enhanced plant productivity.

Microbial consortia play a crucial role in enhancing nutrient availability and promoting plant growth. They can improve plant health and productivity by producing chemicals that stimulate plant development, solubilize phosphorus, produce siderophores, and suppress plant diseases. The potential of microbial consortia in plant production and sustainable agriculture needs more investigation (Negi et al., 2024; Nunes et al., 2024).

7.3.3 Enhancing Nitrogen Fixation

Nitrogen is an essential nutrient for plant growth, but it is often limited in soil ecosystems. The microbial consortium can play a crucial role in enhancing nitrogen fixation, the process by which atmospheric nitrogen is converted into a form that plants can use. Nitrogen-fixing microorganisms, such as rhizobia and agrobacteria, can form symbiotic relationships with plants and convert atmospheric nitrogen into ammonia, which the plant can then use (Aczel 2019; Aasfar et al., 2021).

The microbial consortium can enhance nitrogen fixation by providing a diverse array of microorganisms that work together to fix nitrogen. For example, certain microorganisms in the consortium can produce plant growth-promoting substances that can stimulate plant growth and increase the availability of carbohydrates for nitrogen-fixing microorganisms (Timofeeva et al., 2023; Nunes et al., 2024). The microbial consortium can also solubilize phosphorus, an essential nutrient for nitrogen-fixing microorganisms (Kaur et al., 2022; Luo et al., 2024).

Microbial consortia can also enhance nitrogen fixation by reducing the competition for nitrogen between microorganisms and plants. For example, certain microorganisms in the consortium can produce antibiotics that can suppress the growth of competing microorganisms, allowing nitrogen-fixing microorganisms to dominate the community (Di Benedetto et al., 2017; Samantaray et al., 2024). Additionally, the microbial consortium can also produce siderophores, which are iron-chelating compounds that can enhance iron availability for nitrogen-fixing microorganisms (Timofeeva et al., 2023; Zhang et al., 2025).

The use of microbial consortia to enhance nitrogen fixation is effective in various agricultural systems. For example, the use of microbial consortia has been shown to increase crop yields and reduce the need for synthetic fertilizers in rice and wheat production systems (Kaur et al., 2022; Aasfar et al., 2024). Additionally, the use of microbial consortia has also been shown to enhance nitrogen fixation in legume-based cropping systems (Berza Beyene et al., 2022; Chamkhi et al., 2022).

Ultimately, microbial consortia have the potential to significantly enhance nitrogen fixation, a crucial process for plant growth and agricultural productivity. A microbial consortium can boost crop yields and lessen the need for synthetic fertilizers by offering a variety of microorganisms that can cooperate to fix nitrogen, decreasing competition for nitrogen, and improving iron availability (Mahmud et al., 2021; Nunes et al., 2024).

7.3.4 Biological Control Approach, Disease Suppression, and Plant Defense Mechanisms Amelioration

Microbial consortia have been increasingly recognized as a promising approach for the biological control of plant diseases. Applying a consortium of microorganisms rather than a single strain makes it possible to achieve more effective and consistent disease suppression. This is because different organisms can work together to exploit different mechanisms of disease suppression, such as antibiosis, competition for resources, and induction of plant defense mechanisms (Maciag et al., 2023; Nunes et al., 2024).

One of the key benefits of using microbial consortia for biological control is that it can lead to more durable and sustainable disease suppression. This is because the consortium can adapt to changing environmental conditions and evolve to overcome the development of resistance in the pathogen population (Haq et al., 2024; Nunes et al., 2024). The microbial consortium can also promote plant growth and health, further enhancing disease suppression (Khan et al., 2022).

The microbial consortium can also ameliorate plant defense mechanisms, making plants more disease-resistant. For example, certain microorganisms in the consortium can produce elicitors that stimulate plant defense responses, such as the production of phytoalexins and the activation of defense-related genes (Meena et al., 2022; Hassan et al., 2023). Additionally, microbial consortia can produce plant growth-promoting substances, such as auxins and cytokinins, enhancing plant growth and development (Singh et al., 2023).

The use of microbial consortia for the biological control of plant diseases is effective in various crops, including tomatoes, peppers, and cucumbers (Lahlali et al., 2022). The microbial consortium has also been used to control soil-borne diseases, such as root rot and damping off, which can be difficult to control using traditional chemical fungicides (Berg et al., 2021).

To summarize, microbial consortia offer a viable method for biologically controlling plant diseases, providing a more sustainable and efficient alternative to conventional chemical fungicides. Microbial consortia can offer long-lasting and reliable disease suppression by utilizing several disease suppression mechanisms while fostering plant development and health, making them a desirable choice for integrated pest management plans (Collinge et al., 2022; Fenta and Mekonnen, 2024).

7.3.5 Arbuscular Mycorrhizal Fungi

Arbuscular mycorrhizal fungi (AMF) form symbiotic relationships with plant roots, providing essential nutrients such as phosphorus, nitrogen, and water in exchange for carbohydrates. A microbial consortium comprising diverse microorganisms can interact with AMF to enhance plant growth, health, and nutrient uptake. Studies have shown that microbial consortia can enhance AMF colonization, improve nutrient exchange, and promote plant growth (Abdelmoneim et al., 2013; Etesami et al., 2021; Bhantana et al., 2021).

The microbial consortium can also influence AMF community structure and diversity, impacting plant performance. For example, certain microorganisms in the consortium can produce compounds that stimulate AMF spore germination and hyphal growth, leading to increased colonization and nutrient uptake (Bortolot et al., 2024; Ghorui et al., 2024). Additionally, microbial consortia can produce antibiotics and antifungal compounds that can suppress plant pathogens, reducing competition for resources and promoting plant health (Maciag et al., 2023).

The microbial consortium and AMF interaction can also improve soil structure and fertility. For example, AMF can produce glomalin, a glycoprotein that can act as a soil glue, improving soil aggregation and water-holding capacity (Wahab et al., 2023). Microbial consortia can also produce enzymes that break down organic matter, releasing nutrients that plants can absorb (Upadhayay and Upadhayay, 2023; Negi et al., 2024).

Microbial consortium and AMF can also interact to promote plant tolerance to abiotic stresses such as drought and salinity. For example, AMF can produce compounds that stimulate plant antioxidant defenses, reduce oxidative stress, and improve plant growth under stress conditions (Ruiz-Lozano et al., 2016). Microbial consortia can also produce compounds that stimulate plant hormone production, promoting plant growth and development under stress conditions (Koza et al., 2022).

7.3.6 Phytohormone Modulation

The microbial consortia comprising diverse microorganisms can interact with plants to modulate phytohormone production, improving plant growth, health, and productivity. Phytohormones, such as auxins, cytokinins, gibberellins, and ethylene, play crucial roles in plant development, stress response, and defense against pathogens (S. Ali et al., 2024; Shafqat et al., 2024). Microbial consortia can produce phytohormones or modulate plant phytohormone production, leading to changes in plant morphology, physiology, and biochemistry changes (Bilal et al., 2018).

The microbial consortium can modulate auxin production, resulting in alterations in plant root architecture, shoot growth, and leaf development. For example, certain microorganisms in the consortium can produce indole-3-acetic acid (IAA), a major auxin phytohormone that stimulates plant cell elongation and division (Koza et al., 2022; Ratnaningsih et al., 2023). Additionally, the microbial consortium can also modulate cytokinin production, leading to changes in plant cell division, differentiation, and senescence (Akhtar et al., 2019).

Microbial consortium can also modulate gibberellin production, leading to changes in plant stem elongation, leaf growth, and flowering. For example, certain microorganisms in the consortium can produce gibberellins, which can stimulate plant stem elongation and leaf growth (Timofeeva et al., 2023). The microbial consortium can also modulate ethylene production, leading to changes in plant stress response, senescence, and fruit ripening (Ravanbakhsh et al., 2018; Tipu and Sherif, 2024).

The modulation of phytohormone production by microbial consortia can also improve plant tolerance to abiotic stresses, such as drought, salinity, and temperature extremes. For example, certain microorganisms in the consortium can produce phytohormones that stimulate plant antioxidant defenses, reducing oxidative stress and improving plant growth under stress conditions (Shaffique et al., 2023; Nivedita et al., 2024).

7.3.7 Microbial Biopesticides

The microbial consortium, comprising a diverse array of microorganisms, can be utilized as a biopesticide to control plant pathogens and pests. Microbial biopesticides are a promising alternative to chemical pesticides, offering a more environmentally friendly and sustainable approach to pest management (Ayilara et al., 2023; Chaudhary et al., 2024). Microbial consortia can be formulated into various products, such as powders, liquids, and granules, making them easy to apply to crops.

Microbial consortia can control plant pathogens through various mechanisms, including antibiosis, competition for resources, and induction of plant defense responses (Maciag et al., 2023; Nunes et al., 2024). For example, certain microorganisms in the consortium can produce antibiotics that can inhibit the growth of plant pathogens, while others can produce enzymes that can degrade pathogen cell walls (Maciag et al., 2023; M.A. Ali et al., 2024). Additionally, the microbial consortium can also induce plant defense responses, making plants more resistant to pathogen attack (Kumar et al., 2023).

Microbial consortia can also control insect pests through various mechanisms, including the production of insecticidal compounds, competition for resources, and alteration of insect behavior (Spescha et al., 2023). For example, certain microorganisms in the consortium can produce insecticidal compounds, such as *Bacillus thuringiensis* (*Bt*) toxins, which are effective in killing insect pests. Additionally, microbial consortia can also compete with insect pests for resources, such as nutrients and water (Wu et al., 2023; Ragasruthi et al., 2024).

Using microbial consortia as biopesticides has several advantages over chemical pesticides, including reduced environmental impact, improved human safety, and increased crop yields. Microbial consortia can also be tailored to specific crops and pests, making them more effective and efficient. However, the development and commercialization of microbial biopesticides require careful consideration of formulation, application, and regulation factors (Nunes et al., 2024; Tadesse Mawcha et al., 2024).

7.3.7.1 Microbes in Biological Pest Control

The microbial consortium, comprising a diverse array of microorganisms, can be utilized as a biological pest control agent to manage insect pests and plant diseases. Microbes, such as bacteria, fungi, and viruses, can be used as biopesticides to control a wide range of pests, including insects, mites, and plant pathogens. Microbial consortia can be formulated into various products, such as powders, liquids, and granules, making them easy to apply to crops (Chaudhary et al., 2024; Galli et al., 2024; Vermelho et al., 2024).

Microbes can also control plant diseases through various mechanisms, including antibiosis, competition for resources, and induction of plant defense responses (Bonaterra et al., 2022). For example, certain bacteria, such as *Pseudomonas fluorescens*, can produce antibiotics that can inhibit the growth of plant pathogens. Additionally, microbes can also induce plant defense responses, making plants more resistant to disease (Mehmood et al., 2023; Yang et al., 2025).

The use of microbial consortia as a biological pest control agent has several advantages over chemical pesticides, including reduced environmental impact, improved human safety, and increased crop yields. Microbial consortia can also be tailored to specific pests and crops, making them more effective and efficient. However, the development and commercialization of microbial biopesticides require careful consideration of factors such as formulation, application, and regulation (Chaudhary and Shukla, 2019; Nunes et al., 2024).

7.3.7.2 Pesticide Degradation by the Insect Gut Microbiota

A microbial consortium, comprising diverse microorganisms, can play a crucial role in degrading pesticides in the environment. Insect gut microbiota are capable of degrading a wide range of pesticides, including organophosphates, carbamates, and pyrethroids. Pesticides can be degraded by insect gut microbiota through various mechanisms, including hydrolysis, oxidation, and reduction (Bernal et al., 2017).

The insect gut microbiota is a complex community of microorganisms that live within the insect's gut and play a crucial role in the insect's nutrition, health, and development. The microbiota can also influence the insect's ability to degrade pesticides, as certain microorganisms produce enzymes that break down pesticide molecules (Siddiqui et al., 2022; Swathy et al., 2024). For example, certain bacteria in the insect gut microbiota can produce enzymes that hydrolyze organophosphate pesticides, thereby reducing their toxicity to the insect (Jaffar et al., 2022; Yasika and Shivakumar, 2025).

Microbial consortia can also be used to enhance the degradation of pesticides by insect gut microbiota. By introducing specific microorganisms into the insect gut, it is possible to improve the degradation of pesticides and reduce their toxicity to the insect (Bernal et al., 2017; Siddiqui et al., 2022). For example, certain microorganisms can be introduced into the insect gut to enhance the degradation of pyrethroid pesticides, which can be toxic to certain insects (Galadima et al., 2021).

The degradation of pesticides by insect gut microbiota can also impact the development of pesticide resistance in insect populations. By degrading pesticides, the insect gut microbiota can reduce the selection pressure on the insect population, making it less likely for pesticide-resistant individuals to emerge (Galadima et al., 2021; Han et al., 2024). This can help to slow the development of pesticide resistance and extend the useful life of pesticides.

7.4 IDENTIFICATION AND RESOLUTION OF MICRODIVERSITY THROUGH METAGENOMIC SEQUENCING OF PARALLEL CONSORTIA

The microbial consortium, comprising a diverse array of microorganisms, plays a crucial role in various ecological and biotechnological processes. However, the complexity of microbial communities can make it challenging to understand the interactions and dynamics within the consortium. Recent advances in metagenomic sequencing have enabled the identification and resolution of microdiversity within microbial consortia (Kang et al., 2019; Sharon et al., 2019). By analyzing the metagenomic data from parallel consortia, researchers can gain insights into the genetic and functional diversity of microorganisms.

Metagenomic sequencing of parallel consortia involves analyzing DNA sequences from multiple samples of the same consortium. This approach allows researchers to identify and quantify microdiversity within the consortium, including the presence of rare or low-abundance microorganisms. By comparing the metagenomic data from different samples, researchers can also identify patterns of microdiversity that are associated with specific environmental or operational conditions (Nelson et al., 2016; Navgire et al., 2022).

The identification and resolution of microdiversity within microbial consortia can have significant implications for our understanding of ecosystem function and biotechnological processes. For example, the presence of rare microorganisms can influence the overall function and stability of the consortium (Pascoal et al., 2021). By understanding the microdiversity within microbial consortia, researchers can develop more effective strategies for manipulating and optimizing these communities for specific applications (Duncker et al., 2021; Contreras-Salgado et al., 2024).

The use of metagenomic sequencing to analyze microbial consortia has several advantages over traditional approaches, including the ability to analyze complex communities and identify rare or low-abundance microorganisms. However, the analysis of metagenomic data also presents several challenges, including the need for sophisticated bioinformatic tools and the potential for bias in the data (Zhou et al., 2015; Yen and Johnson, 2021).

Metagenomic sequencing of parallel consortia is a powerful method for identifying and resolving microdiversity within microbial consortia, which is essential for understanding the complexity of these communities. By examining the metagenomic data from parallel consortia, researchers can learn more about the genetic and functional variety of microorganisms and create more efficient methods for modifying and tailoring these communities for use (Quince et al., 2021; Nam et al., 2023).

7.5 BIOREMEDIATION

Bioremediation utilizes biological systems to eliminate toxic and hazardous compounds, which can be applied to the environment and the ecosystem to remove or stabilize contaminants (Cabral et al., 2022). It is an advantageous method as it is cost-effective and eco-friendly. The concept of applying various microbial consortia for pollutant biodegradation and bioremediation represents a very promising perspective (Massot et al., 2022). Bioremediation can be utilized to eliminate both organic and inorganic pollutants (Verma and Kuila, 2019; Madadi and Bester, 2021). For the bioremediation of organic compounds, two approaches can be employed: biostimulation and bioaugmentation (Gutiérrez et al., 2020; Massot et al., 2022). Biostimulation is the process of adding nutrients and/or oxygen to encourage the biodegradation of pollutants by indigenous microbes that possess the complete catalytic ability to degrade the pollutant (Della-Flora et al., 2022). On the other hand, bioaugmentation is used when the indigenous microbes lack the catabolic capability to degrade the pollutant, and pollutant-degrading microbes are inoculated. A critical approach to microbial associations, also known as a consortium, involves using inoculants that can degrade pollutants instead of relying on a single microorganism. Microbial consortia are more auspicious in this field than using single microbes for several reasons. Microbial consortia can utilize complex carbon substrates, resulting in higher production compared to using a single microbe. This is achieved through multistep reactions that occur quickly. Additionally, uncultivable microorganisms can be utilized in a consortium, and stability is maintained because unwanted microbes are unable to grow (Mishra et al., 2021). With recent advancements in tools for gene editing and synthetic biology, the efficiency and stability of artificial microbial consortia systems are expected to increase. A microbial consortium can be established by leveraging the relationships and interactions among microbes. Each member in the microbial consortia has a definite job, and microbes within the consortia can exchange substances and communicate using signaling molecules and complex metabolic regulation networks. The consortia are a strong approach to estimating the potential of viable but non-culturable bacteria and for understanding the function of several unknown genes. Additionally, studying the ecology of microbial consortia is crucial for understanding their roles and niches in bioremediation (Li et al., 2022). Thus, microbial consortia can perform complex tasks that a single strain alone cannot accomplish (Mee et al., 2014). The division of labor in microbial consortia and the complementary degradation are regarded as an auspicious approach (Zhou et al., 2011; Tsoi et al., 2018).

Microbial consortia have proved to be very encouraging in the biodegradation of PAHs (Aruotu et al., 2023), several xenobiotics bioremediation (Li et al., 2022), wastewater treatment (Cheng et al., 2021); detoxification of heavy metals (Sharma et al., 2024); biodegradation of complex recalcitrant materials such as lignocellulosic wastes, oil-based plastics, and azo-dyes (Cao et al., 2022) and sustainable mitigation of saline soils (Kapadia et al., 2021).

7.5.1 Bioremediation of PAHs

PAHs are compounds that are composed of two or more aromatic rings. They are very toxic compounds and represent one of the most severe hazardous contaminants to the environment. They are mutagens, teratogenic, and carcinogenic (Okere and Semple, 2012). They do not degrade quickly in nature (Muthukumar et al., 2022). The PAH environment can undergo volatilization, bioaccumulation, or adsorption as well as photo-oxidation and chemical oxidation in the environment (Kumar et al., 2021). PAHs, such as naphthalene and phenanthrene, are found in insecticides, fungicides, detergents, dyes, and mothballs (Vacca et al., 2005). The presence of these pollutants affects the biogeochemical cycle and microbial biodiversity, harming both the environment and the ecosystem (Raj and Maiti, 2020). Naphthalene, with the molecular weight among PAHs, represents a serious pollutant according to the Environmental Protection Agency (EPA) in the US (Yan et al., 2004). This is due to low solubility in water and a high solid–liquid distribution ratio, which makes its biodegradation by microbes hard, resulting in its accumulation in the environment (Maki et al., 2001). The metabolism mechanisms, aerobic and anaerobic, have been studied extensively (Okere and Semple, 2012). Fungi and bacteria are both capable of biodegrading naphthalene in soil (Okoh, 2006). Soil microbes biodegrade PAHs with varying efficiency compared to liquid media because the sorption of highly lipophilic PAHs to soils reduces their bioavailability to microbial action, thereby inhibiting bioremediation (Oudot et al., 1998). Degrading microbes must form sufficient biomass and grow on PAHs as carbon and energy sources in polluted environments to remove them (Pelletier et al., 2004). Enrichment methods for naphthalene biodegradation by microbes have revealed the presence of *Pseudomonas* sp., *Geobacillus* sp., and *Bacillus cereus* (Bubinas et al., 2008; Pathak et al., 2009; Rehman et al., 2021). Microbes in mixed culture were found to have a faster degradation rate (Musat et al., 2009; Kafilzadeh and Pour, 2012; Aruotu et al., 2023).

7.6 MONITORING THE IMPACT OF XENOBIOTICS ON THE ECOSYSTEM

Herbicides and pesticides are phytotoxic, causing a decrease in crop yield and quality. They alter the soil structure and reduce the microbial flora's enzyme activities. They enter the aquatic ecosystem directly or indirectly, resulting in water pollution and promoting the growth of undesired aquatic organisms. Microbes can be utilized to biodegrade herbicide residues environmentally, making this process cost-effective, eco-friendly, and safe. No secondary pollution results from environmental restoration by microbes. On the other hand, it has several disadvantages, including low efficiency, incomplete biodegradation, instability, and repellence by indigenous microbes during the bioremediation of *in situ* soil. Complex pollutants in the ecosystem cannot be biodegraded by a single microbial strain, but rather by the combined metabolism and syntrophism of consortia. Several bacterial consortia have been able to degrade different phytotoxic herbicides, and consortia were better than a single microorganism in xenobiotic biodegradation (Li et al., 2022). Herbicide biodegradation by the NMC has not been investigated sufficiently, and studies have primarily focused on degradation efficiency. Bioaugmentation by adding two different bacterial consortia and biostimulation to biodegrade pentachlorophenol (an organochlorine pesticide and disinfectant) resulted in a 71% reduction of pentachlorophenol. This indicates that combining both biostimulation and bioaugmentation represents a great approach to the remediation of pesticides (Ammeri et al., 2022)

7.6.1 Bioaugmentation and Biostimulation for Wastewater Treatment

The treatment of wastewater has become a serious economic and environmental challenge, as the water resources in the plant are limited, and there is a pressing need to address this issue. Textile industry wastewater contains a significant number of contaminants, including sizing agents, surfactants, fiber impurities, carboxylic acids, thickeners, complexing agents, salts, and dyes (EU report 2003). Textile wastewater treatment can employ physical, chemical, or biological methods; the latter is preferred as it is a cheap and efficient method (Bilińska et al., 2016). Biological wastewater treatment usually occurs via microbial degradation by cells or microbial enzymes. There are different approaches to enhance the microbial biodegradation process of wastewater: (1) biostimulation of aboriginal microbiomes by the addition of co-substrates. The scarcity of nutrients can limit the catabolic activity of the indigenous microbial community and the bioremediation process (Sarkar et al., 2016). Added substrate could be carbon or nitrogen sources (Khan et al. 2013), (2) Bioaugmentation (co-culture). The aboriginal microflora can be stimulated by adding one or more microbes that support aboriginal microflora growth (Forss et al., 2017), (3) Co-metabolism, which stimulates aboriginal microbial community functions. The microbiome of wastewater influences the bioremediation process (Carvalho et al., 2020), while the substrates fed affect microbial growth, structure, dynamics, and the bioremediation process through the metabolites produced (Chen and Wu, 2022). These metabolites can act as redox mediators, playing a role in the breakdown of dye molecules and other pollutants (Watanabe et al., 2009). Redox mediators were found to be easily used by other microbes rather than their own (Rosenbaum et al. 2011). Riboflavin was found to play a role in dye degradation (Gomaa et al., 2019). Additionally, the addition of cobalamin to the medium has been reported to influence the shape of microbial community and to direct its biodegradation (Chen and Wu, 2022).

Bioaugmentation of textile wastewater with *Bacillus spizizenii* DN and the addition of its metabolites to the wastewater to use the three approaches: co-substrate addition, co-culture, and co-metabolism was found to be very promising in decolorization and bioremediation of wastewater (Gomaa et al., 2023). Bioaugmentation and biosimulation can improve the quality of effluent in both lab-scale experiments and industrial-scale processes. The bioaugmentation of beneficial microbes has been used in wastewater treatment (Cheng et al., 2021). Kim et al. (Kim et al., 2013; Kim et al., 2014) reported that removal rates of chemical oxygen demand (COD) and total nitrogen (TN) reached 91% and 79%, respectively, in lab-scale tannery wastewater treatments. The removal rates of COD and TN were estimated to be 98.3% and 88.5%, respectively, under continuous inoculation with beneficial microorganisms at an industrial-scale tannery wastewater. Components of tannery wastewater that are resistant to biodegradation can inhibit biological treatment, reducing biotreatment efficiencies (Lofrano et al., 2013). Hence, an effective solution must be developed to address these complex issues. Chromium ions have been removed by microbial consortium addition to the tannery wastewater (Fernandez et al., 2019; Elkarrach et al. 2020). Research on utilizing beneficial microbes to treat industrial-scale tannery wastewater is limited and requires further evaluation. Moreover, the functional genes of beneficial microbes associated with the removal of organic material from wastewater were also identified (Cheng et al., 2021).

7.6.2 Detoxification of Heavy Metals

Heavy metal pollution and accumulation can alter the quality of soil. They accumulate in the ecosystem as a result of industrial practices, mining, intense use of pesticides and fertilizers, fossil fuels, sewage, and unsuitable waste management (Ali et al., 2019). Heavy metals are not biodegradable, and if they are broken down, they are not degraded into harmless products and remain in the ecosystem, causing severe environmental damage (Anusha et al., 2021). Heavy metals can penetrate the plant parts, animals, and humans and accumulate there, causing several diseases and disorders in animals, humans, and plants (Deng et al., 2015). Restoration of polluted ecosystems using biological systems is highly promising for detoxifying various heavy metals, which are among the most prevalent pollutants in the soil worldwide.

Microbes, organisms, and plants require certain heavy metals as essential micronutrients, including zinc (Zn), nickel (Ni), copper (Cu), and iron (Fe). Other heavy metals such as mercury (Hg), cadmium (Cd), arsenic (As), and lead (Pb) are not essential for microbes and other organisms. Each metal can be harmful if it is present in amounts higher than natural levels and if it is exposed for an extended period. The most frequent heavy metals present in the polluted soils are Cd, Cr, Cu, Hg, Zn, Pb, Ni, and As (Ali et al., 2019).

The application of the bioaugmentation (co-culture) method to construct microbial consortia for detoxifying heavy metals has been extensively investigated in several studies (Sharma et al., 2024). A consortium of *Bacillus cereus and Bacillus pumilus* was able to detoxify Pb^{2+} and Ni^{2+} by nearly 95% sorption efficiency (Sharma et al., 2021). Cu, Zn, and Pb were detoxified with sorption efficiencies of 92.7%, 90.3%, and 86%, respectively, by a microbial consortium of *Bacillus thuringiensis* A1–3, *Pseudomonas aeruginosa* A-33, *Bacillus cereus* A1–5, and *Bacillus anthracis* A1–7 (Anusha et al., 2021). A microbial consortium of *Acinetobacter* sp. IrC1, *Acinetobacter* sp. IrC2, and *Cupriavidus* sp. IrC4 was able to detoxify Cu with a 33% sorption efficiency (Irawati et al., 2018). A consortium of *Bacillus licheniformis* and *C. parapsilosis* was able to detoxify Hg with an 85% sorption efficiency (Muneer et al., 2013). Similarly, a consortium of *Chlorella vulgaris* and *Enterobacter* sp. MN17 detoxified Cd^{2+} by 93% sorption efficiency (Mubashar et al., 2020). Sulfate-reducing bacteria can effectively remove metals such as Pb, Ni, Fe, and Zn by sulfide precipitation (Sharma et al., 2024). Microbial consortia combat arsenic poisoning in soil by facilitating mechanisms such as oxidation–reduction, methylation–demethylation, volatilization, immobilization, and arsenic mobilization (Kaya et al., 2024). Fungal consortia also play a significant role in the bioremediation and detoxification of heavy metals, as individual fungal species contribute substantially to this process. Mishra and Malik (2014) reported that *Aspergillus lentutlus* FJ172995 bioaccumulated Cu (56%) and Pb (100%) from the waste of electroplating industries. A consortium of *Aspergillus lentulus*, *Aspergillus terreus*, and *Rhizopus oryzae* detoxified Cr^{6+} and Cu^{2+}. The *Aspergillus flavus* (FS4) and *Aspergillus fumigatus* (FS6) consortium was able to detoxify Cr^{6+} and Cd^{2+} with sorption efficiencies 33%, 70%, and 74%, respectively (Talukdar et al., 2020). A consortium of *Aspergillus fumigatus*, *Aspergillus terreus*, and *Paenibacillus dendritiformis* detoxified Cd^{2+} by nearly 91% sorption efficiency (Zango et al. 2018).

Various mechanisms have been proposed by microbes in consortia to cope with heavy metal stress. These mechanisms involve biofilm formation, Siderophore production, biosurfactant production, biosorption, extracellular polymeric substances production, and the use of the exopolymer binding mechanism (Sharma et al., 2024). Biofilm formation could be used in heavy metal bioremediation by accumulating toxic metal ions or modifying them biochemically. Siderophores may form stable complexes with heavy metals, solubilizing unavailable forms into more soluble ones. The biosurfactant of *Candida sphaerica* UCP0995 could remediate Zn, Pb, and Fe from the soil of battery industries (Luna et al., 2016). Biosurfactants increase metal bioavailability and solubility. Microbes produce several mucopolysaccharides and extracellular polymeric substances, which can form complexes with heavy metals and decrease their movement in the environment, reducing their bioavailability.

7.6.3 Biodegradation of Complex Recalcitrant Materials

For recalcitrant materials that are resistant to decomposition, microbes were found to be very promising for application in this process (Shong et al., 2012). A vast number of microbes were isolated that were capable of decomposing recalcitrant materials such as lignocellulose and oil-based plastics (Singh et al., 2014). Microbial consortia are superior to single strains (Mikesková et al., 2012). Complex compounds are often difficult to degrade by natural microbes due to their complex structures. Microbial consortia are promising in this situation. Microbial consortia degrade complex compounds more effectively than isolated bacteria due to the reasons mentioned in Section 6. Additionally, with advancements in tools for gene editing and synthetic biology, artificial microbial consortia systems could be reconstructed to be more effective and stable. Microbial consortia have proven to be favorable for biodegrading several complex, recalcitrant compounds and achieving various biotechnological perspectives. They could degrade lignocellulosic materials and wastes to produce biofuel. They also have promising potential in degrading petroleum and plastic compounds to valorize them into highly valuable compounds. They can be beneficial in the bioremediation of non-biodegradable hazardous compounds in soil and sewage wastewater (Cao et al., 2022).

The lignocellulose decomposition remains mysterious, as the interaction between different microbial members in the community and the dynamics of enzyme complements remain poorly understood. Few studies on the community structure and enzyme systems of a lignocellulolytic microbial consortium by using metagenome sequencing and metaproteomics were performed (Zhu et al., 2016). Using a mixed culture for the decomposition of lignocellulose is preferred because the removal of lignin by a single species is not natural. The lignocellulose is biodegraded by multiple microbial species under aerobic or anaerobic conditions (Sharma et al., 2023). The artificial consortium could be utilized to degrade recalcitrant lignocellulosic waste and use it as a substrate to obtain highly valuable products. In the artificial synthetic consortium, one strain would degrade complex cellulose into more minor compounds, and the other microbial strain would use those more minor compounds to give the desired products (Wen et al., 2020).

A synthetic consortium of *Citrobacter cellulovorans* and *Citrobacter* beijerinckii was able to produce butanol five times higher than the control (Wen et al., 2020). Also, a consortium of *Thermoanaerobacterium thermosaccharolyticum* and *Clostridium acetobutylicum* produces butanol from non-treated corncob (Jiang et al., 2020). Bioethanol was obtained from lignocellulosic wastes of cotton by the action of a consortium consisting of *Saccharomyces cerevisiae* and *Pachysolen tannophilus* (Malik et al. 2021).

Microbial consortia were vigorously investigated for degrading synthetic petroleum-based plastics such as polyethylene terephthalate (PET), polyurethane PU, polyethylene PE, polystyrene PS, and polyethylene PE (Cao et al., 2022). The accumulation of these types of plastics, resulting from extensive production compared to the amount being recycled, represents a severe environmental and ecological problem. Additionally, petroleum-based plastics are not easily degraded due to their complex nature. Specific isolated NMCs were able to biodegrade these kinds of synthetic plastics. Studying interactions between strains of NMCs is an important demand for constructing them. Therefore, microorganisms can be added to an isolated microbial consortium according to the interactions and relationships between its members to enhance their efficiency and gain a deeper understanding of degradability, gene regulation, or enzymatic functions. A consortium of *Bacillus* sp. and *Pseudomonas* sp. was constructed to biodegrade polystyrene, resulting in a 12.4% weight reduction in 30 days (Ho et al., 2018). A consortium of Rhodobacterales, Rhizobiales, Burkholderiales, Actinomycetales, and Sphingobacteriales was able to break down polyurethane by 50.3% of proprietary aromatic PE-PU-A copolymer in 25 days (Gaytán et al., 2020). Low-density polyethylene was biodegraded with an 81% reduction in weight in 120 days by a consortium consisting of *Enterobacter* sp. bengaluru-btdsce01, *Enterobacter* sp. bengaluru-btdsce02, and engineered *Bacillus subtilis* species (Qi, Ma, et al., 2021; Qi, Yan, et al., 2021). Thermophilic Consortia of *Brevibacillus* sp. and *Aneurinibacillus* sp. have been isolated from the environment to enhance polymer degradation (Skariyachan et al., 2018).

Nearly 90% of n-alkanes and aromatic hydrocarbons could be biodegraded by microbial consortia. *Pseudomonas stutzeri* and *Dietzia* sp. consortium were able to synergistically biodegrade 90% of n-alkane in diesel oil by nearly 85.5% (Hu et al., 2020). For polycyclic aromatic hydrocarbons, a consortium of *Sphingomonas* sp., *Sphingobium* sp., *Pseudomonas* sp., *Dokdonella* sp., and *Luteimonas* sp. was able to biodegrade fluorene and phenanthrene after five days, nearly completely (Bacosa and Inoue, 2015). Microbial consortia are also capable of biodegrading recalcitrant azo dyes. A halophilic bacterial consortium isolated from textile wastewater decolorized azo dye by 98%. (Shi et al., 2021). Microbial consortia whose members are *Vanrija humicola*, *Meyerozyma caribbica, Debaryomyces hanenii, and Meyerozyma guilliemondii* were able to decolorize all Orange II (250 mg) (S.S. Ali et al., 2022).

7.6.4 Sustainable Mitigation of Saline Soil

Soil salinity represents a serious problem. Nearly one billion hectares of the world's agricultural soil are influenced by salinity. Saline soil increases by 10% every year. Soil salinity replaces Ca^{2+} and Mg^{2+} ions, which are then absorbed onto the soil aggregates and particle surfaces (Orhan, 2016). Organic manures and gypsum are used to mitigate the salinity's influence on crops, but they also disrupt the diversity of soil microbes and the biogeochemical reactions performed by these microbes (Shrivastava and Kumar, 2015; Schirawski and Perlin, 2018). The physical and chemical methods to solve the salinity problem are not compelling enough, are not eco-friendly, and cannot be applied on a large scale. Plant growth-promoting rhizobacteria (PGPR) are regarded as a promising solution for reclaiming saline soils. Genera such as *Rhizobium*, *Bacillus*, *Klebsiella*, *Enterobacter*, Azotobacter, Pseudomonas, and Streptomyces have been found to improve crop yield in salinity stress (Abbas et al., 2019). PGPR microbes provide several antioxidant enzymes and osmolytes, and using different microbial genera can significantly enhance plant growth in saline soil and mitigate salinity (79, 242, 243). Microbial consortia can play a significant role in ameliorating saline soil by producing plant growth promoters, antioxidants, biocontrol agents, and osmoprotectants (Ilangumaran and Smith, 2017). Research has demonstrated that a microbial consortium is more encouraging than the individual strains (Nuti and Giovannetti, 2015; Sekar et al., 2016). Using a microbiome consortium composed of halophilic or halotolerant bacterial species such as *Delftia* sp., *Enterobacter* sp., *Bacillus* sp., and *Achromobacter* sp. Enhanced the growth and mineral uptake of tomatoes (Solanum lycopersicum L.) under saline conditions (Kapadia et al., 2021).

7.7 METAGENOMICS AND MICROBIOME ENGINEERING (SUSTAINABLE AGRICULTURE THROUGH MICROBIOME ENGINEERING)

The microbiome plays a fundamental role in regulating the crops' sustainability, owing to the direct interaction between the microbiome and the plant, or the microbiome members' interaction (Nadarajah and Abdul Rahman, 2023). Since different microbiome compositions make plants acquire varied properties, special attention must be given to the microbiome of the economically important crops and plants impacted by climate change (Singh et al., 2020). However, identifying microbiome composition would not be possible without recent advancements in metagenomics techniques, which enable the identification of metagenome-assembled genomes (MAGs) and the gapless circularized MAGs (cMAGs) for agricultural specimens (Xue et al., 2024). On the other hand, environmental stress, which is considerably associated with climate change, has a disastrous impact on the health and microbiome composition of crops. Customizing microbiome composition, also known as microbiome engineering, holds great potential in enabling crops to tolerate various environmental stresses; it can also confer additional desired characteristics, such as increasing crop yield and providing plants with essential nutrients.

7.7.1 Microbiome Engineering in the Alleviation of Drought as a Stress Condition

Engineering the microbiome of soil primarily involves manipulating the constituents, genetic composition, or functional abundance of the microbiota members. Through one of the following methods: introducing metabolite amendments to the soil itself; these metabolites interact with the soil microbiota to form a metabolite-microbiome-host network of interactions, causing the formation of novel pathways in the crops and their surroundings (Kaul et al., 2021). Another approach to customize the soil microbiome is the establishment of synthetic microbial consortia; we will discuss the approaches and principles upon which the synthetic microbial consortia are constructed in the next part of the chapter. In brief, the synthesis process of man-made microbial consortia involves one or multiple regulatory genetic constructs, known as biobricks, into the microorganisms' genome to introduce new features to the consortia according to the experimental purpose (Ke et al., 2021). Such an approach surpasses traditional methods for changing the microbiome composition of the soil, as it can engineer a dynamic genetic network (Han and Yoshikuni 2022). Hence, consortia persist in the environment. Another method to generate a man-made microbiome is the host-mediated microbiome engineering (HMME) strategy, which integrates molecular biology and biochemical approaches to yield an engineered microbiome that is artificially selected to match the host genotype (Mueller et al., 2021; Han and Yoshikuni 2022).

Among the various stresses affecting crops, drought is a leading one. Drought has become a significant abiotic stress conferring the sustainability of manifold crops due to the global water scarcity problem provoked by climate change (Salam et al., 2022). Therefore, multiple attempts are ongoing to mitigate the drought crisis that undermines agricultural sustainability, among which microbiome engineering is frequently employed. Host-mediated microbiome engineering was done on water-deficient wheat seedlings. Moreover, the engineered rhizosphere application from the sixth round of the artificial selection increased the biomass, dry weight, root length, and plant surface area. In addition, different dilutions of the rhizosphere have proven their ability to tolerate water stress (Jochum et al., 2019). Another study illustrated the competency of iterative artificial selection on drought tolerance in a rice japonica cultivar named "super dwarf" (Styer et al., 2024). However, a study constructed a synthetic consortium using a strategy other than HMME. Molina-Romero et al. (2017) combined four strains with varying drought tolerance capacities and observed the effect the synthetic consortium had on maize. The study manifested a significant surge in roots and shoots' dry weight, in addition to an increase in plant height, as well (Molina-Romero et al., 2017). Moreover, chickpea-drought-sensitive and tolerant seeds were inoculated with PGPR and PGRs, and various enhancements in metabolite accumulation were observed in the drought-sensitive plants grown compared to the control (Khan et al., 2019). The previous findings emphasize the prosperous ability of engineered microbiomes to confer the mounting abiotic stress, not only on the molecular and metabolic levels but also at the physicochemical and morphological characteristics, consequently leading to crops' enhanced yield (S. Ali et al., 2022). Although the research is ongoing, the epic advancement in genetic engineering technologies and their application in plant microbiome synthesis (Afridi et al., 2022; Berruto and Demirer 2024) opens the door to conquering environmental stressors and paving the way to improving agricultural sustainability.

7.7.2 Metagenome Contribution for Harnessing Endophytic Microbial Consortia of Medicinal Plants

Endophytes are microorganisms that reside inside the host plant. They mainly do not harm their host. These microbes can be found intercellularly in various plant parts, including roots, shoots, leaves, and flowers. Endophytes can survive environmental stress within the ecosystem due to their presence inside plants. Moreover, they live with their host in either commensal or mutualistic relationship in which endophytes may provide some essential elements and nutrients to the host plant, support their host by performing multiple biological processes inducing metabolites production by plants, sequester harmful compounds produced from biological processes such as Reactive Oxygen Species (ROS), or protect the plant against biotic or abiotic stress (Semenzato and Fani 2024). Meanwhile, endophytes play a key role in plants' sustainability due to their ability to synthesize novel secondary metabolites that act as the host's defense mechanism against pathogens' attacks. However, the endophytes of medicinal plants have captured special attention from the scientific community due to their rich production of novel bioactive compounds, such as antibiotics, antivirals, anticancer agents, antioxidants, and/or immunosuppressants (Narayanan and Glick 2022). Since medicinal plants are dispersed across various geographical regions, the endophytic composition differs accordingly (Wu et al., 2021). Independent of the cultivability of the microorganism, metagenomics analysis techniques have been used to identify the metagenome of endophytes in medicinal plants, providing insights into their taxonomic composition, function, biodiversity, and the mechanisms behind their symbiotic association with these plants (Adeleke et al., 2023).

For example, one study focused on identifying the microorganisms abundant in the root zone of *Datora anoxia* using two metagenomics techniques: 16S rRNA and Whole Genome Sequencing (WGS) for the roots of *Datora* sp. The study defined the taxonomic abundance of microorganisms and predicted the function of which to find out that the rootzone microbes are involved in producing precursors essential for the flavonoid biosynthesis pathway (Senn et al., 2022). Meanwhile, flavonoids are notoriously known for boosting plants' defense against biotic and abiotic stress (Pandey et al., 2022), and they are considered a powerful source of antioxidants, anticancer, antiviral, and antimutagenic agents (Shomali et al., 2022). Another study employed metagenomic analysis of grapevines to determine the relative abundance of two *Bacillus* species using 16S rRNA, followed by shotgun sequencing. The research results were lab-validated using an anti-microbial test, illustrating that endophytes in grapevine vegetative tissues possess antimicrobial activity (Mardanov et al., 2019).

Furthermore, an antitumor natural product called Astin was found to originate from an endophytic fungus in *Aster tataricus* (Schafhauser et al., 2019). The study used the *Cyanodermella asteris* genome that was previously sequenced and pinpointed gene clusters hypothesized to produce Astins (Jahn et al., 2017). Since metagenomic analysis paved the way to understand how medicinal plants can acquire such beneficial characteristics, researchers started to draft whole genome sequences of endophytes to explore their genetic composition further, contributing to bioactive compound production (Passari et al., 2019).

7.7.3 Role of Microbial Consortia Metagenome in Exploring Microbial Dark Matter

Since the discovery of metagenomics and single cell metagenomic techniques, the mystery surrounding unculturable microbes has been unveiled. Not only has high-throughput sequencing uncovered the genomic sequence of microbial communities, but it has also enabled the discovery of gene clusters concerned with the crux of biological pathways of distinguished biomes, conquering environmental stress, counteracting the pathogens' attack on their host organism, serving as diagnostic biomarkers, or playing a culprit in disease development (Iqbal et al., 2025). Besides, metagenome-assembled genomes (MAGs) and single-cell-assembled genomes (SAGs) created a paramount turn, whether in the discovery of novel enzymes from extreme habitats serving the biotechnology industry, taking Taq polymerases as an example, which have been used ever since in daily molecular biology (Ishino and Ishino 2014), or unmasking microbial dark matter (MDM) functional genetic mechanisms like the CRISPR–CAS system that revolutionized genetic engineering (Burstein et al., 2017).

Additionally, the metagenome discovery of microbial consortia has evolved the pharmaceutical field due to the exploration of MDM possessing an antitumor activity (Maglangit et al., 2020). Also, microbial communities have been suggested as a therapeutic strategy to divert the immunity of cancer patients (Park et al., 2022). More importantly, microbial communities' metagenome added to agricultural sustainability by guiding beneficial microbial communities' usage in the soil, consequently enhancing crop production (Iquebal et al., 2022). Alongside its wide application in various fields, the microbial consortia metagenome serves as a building block in constructing a range of online resources for microbial annotation, microbial data integration for functional and structural analysis, and inferring novel species and gene clusters. It is worth noting that multiple studies now recommend integrating available MAGs and SAGs online with modern AI models. After contamination removal, the resources are integrated using the appropriate model, predicting the communities' dynamic patterns and behavior in the ecosystem, so that it can reveal the hidden function of unculturable strains of the consortia (Pavlopoulos et al., 2023).

7.8 DIFFERENT APPROACHES FOR SYNTHETIC MICROBIAL CONSORTIA CONSTRUCTION (SYNCOMS)

The construction of synthetic microbial consortia was first inspired by synthetic biology, a science that branches from systems biology. Synthetic Biology is the construction of living systems using engineering-driven concepts; thus, it is an applied intersection between genetics and engineering disciplines. The aim behind introducing synthesized microbial communities is due to the proven advantages of multiple-strain consortia over a single organism. This advantage was evident in multiple aspects, ranging from enhancing the functional performance of such a biological community to the traits acquired by the consortia due to communication between the consortia members. Building on the natural merits of microbial consortia, applying engineering concepts to tailor the consortia members and incorporating the synthetic biology learning cycle (Design-Build-Test-Learn) would significantly enhance their applications in numerous fields. Hence, the idea of synthesizing microbial consortia was raised on the horizon. This part of the chapter is dedicated to discussing three key points: synthetic-biology-inspired strategies for constructing synthetic microbial communities (hereafter referred to as SynComs), various principles behind designing SynComs, and, finally, examples of some synthetic biology tools used in building SynComs.

7.8.1 How the Essence of Synthetic Biology Was Added to Construct SynComs

Initially, SynComs' building strategies revolved around top-down and bottom-up concepts, which were both motivated by the engineering principles of synthetic biology. The bottom-up strategy implies selecting several single strains to be grown together. Then comes the long-term optimization process for the newly formed consortium. The optimization requires multiple trials and errors to get the best condition-fitting member strains in the SynComs. Although the bottom-up strategy gives more prosperous results than the single-strain culture, the strategy bears mainly two pitfalls: Only one member becomes dominant after long-term culturing, and the unsuitability of multiple strains for culturing (Xie et al., 2021).

On the other hand, the top–down method relies on the sequential selection of the desired consortium members through continuous culture enrichment or serial dilution (Puentes-Téllez and Falcao Salles 2018). This method primarily relies on the filtration of naturally occurring microbial communities, rather than co-culturing single strains, as in the bottom–up approach. The top-down strategy offers several advantages, counteracting some of the bottom–up limitations by providing more stable consortia with additional functional redundancy. However, the top-down concept relies heavily on the randomness of the selection, so directing the evolution of the community can substantially promote the selection of stable and functional consortia. Therefore, a multi-strategy has evolved as a trial to improve the previous methods, which includes assembling the desired consortia members based on their metabolic network and then fluctuating certain environmental factors to modulate the SynCom competition (Liang et al., 2022).

7.8.2 Common SynComs Constructing Principles

Researchers classify the principles upon which the consortia are built into two broad branches: principles based on the species composition, whether the consortia contain one, two, or multiple species, and the construction principles branch, concerning nutritional, signaling, phenotypic, and spatial-based design. Since the latter branch has the lion's share nowadays, we shall explain each principle more by giving examples. The nutritional-based principle for building SynComs is an approach where all or some consortia members play roles to feed each other either by the metabolism of certain compounds, exchanging amino acids, or producing essential metabolites for members; all the previous could be put under the cross-feeding umbrella (Kõljalg et al., 2005; Liang et al., 2022).

Cross-feeding inside SynComs could be unidirectional, bidirectional, or multidirectional; also, it often involves metabolites, mainly or less commonly, extracellular enzymes (Nadell and Foster, 2012). A single-species consortia was once constructed that contained isoleucine and leucine auxotrophs; both strains overproduced the amino acid needed by their partner strain upon establishing the consortia, forming a mutualistic relationship (Hosoda et al., 2011). Another example of metabolite exchange in SynComs is the cooperator–cheater consortia, where the cooperators produce essential metabolites that the cheaters use without reciprocating. Against all odds, this consortium reached a steady state after passing the time of overproducing essential metabolites to the cheaters; this could be explained by the genetic variability occurring inside the cooperators upon producing an excess amount of metabolites, urged by the cheater strain (Gore et al., 2009). This excess serves as a signal to the cooperator strain to stop production when the cheaters compete for survival in the community. Hence, both strains coexisted (Nadell and Foster 2012). However, the investigation of the behavior of cheaters when they are superior in all nutrients produced by the mutualist species and supplied eternally is not yet understood. It was found that the cheater persisted in the culture, leading to the extinction of one or both mutualists (Sun et al., 2019). Although the cross-feeding principle is conventional when building SynComs, the concepts bear some drawbacks. Ranging from a lack of genetic stability across generations of the culture, the possibility of gene mutation that may drive the system to be broken, and the environmental selection caused metabolite scarcity in the cross-feeding system (Liang et al., 2022). Thus, it is preferable to understand the precise network between SynComs members when constructing a community that adopts cross-feeding relationships.

Moreover, signaling-based construction plays a fundamental role in the evolution of SynComs. Signaling between microorganisms often depends on cell–cell communication, a process known as Quorum Sensing (QS). QS involves the transduction of the signal carrier (autoinducer) from cells to the inter-populations' extracellular environment, forming an organized genetic fluctuation in the members of the community that, in turn, generates specific activity (Papenfort and Bassler 2016). QS signal transduction can exhibit multiple manifestations, including spatio-temporal, pulse-generator, or pattern-forming systems (Jia et al., 2016). The spatiotemporal system usually generates changes in the spatial position of the population in response to changes in cell density. An example of spatio-temporal-based QS SynComs is the LasR/LasI system, which serves as a switch that controls the motility of the *E. coli* population in response to a threshold of cell density (Liu et al., 2011). Meanwhile, it is recommended that the complexity of the QS systems of SynComs originating from microbes other than E. coli and S. cerevisiae be studied further. To overcome the pitfalls of QS and cross-feeding-based systems, Phenotype-based SynComs could be established, as they consider the functional trait proposed to be present in the community, rather than the mechanism behind it. Additionally, a trait-based model for SynComs was used to inform the environmental-driven changes in the consortia (Le Roux et al., 2016). One last principle that has emerged due to advancements in microfluidic technology is the spatial structure, which utilizes the customized separation and assembly of consortia members as an indication of the interactions between them (Sanati Nezhad 2014; Michielsen et al. 2024). Nevertheless, the spatially structured consortia synthesis is still in its primary stages, which require integrations of other principles, like cross-feeding or QS, to optimize the synthesis of microbial consortia (Ben Said et al., 2020).

Computational model integration offers a promising solution for designing, predicting, or enhancing the efficiency of SynComs. Dynamic, stoichiometric, or agent-based models are examples of models used in SynComs construction (McCarty and Ledesma-Amaro 2019). Therefore, consolidating manifold principles of construction and the advanced computational models with multi-omics analysis of the consortia proposes a promising direction for SynComs construction.

7.9 CONCLUSION

Metagenomics has revolutionized the study of microbiome consortia, providing unprecedented insights into the complex interactions and dynamics of microbial communities. By leveraging metagenomic approaches, researchers can uncover novel microbial interactions, metabolic pathways, and ecosystem services, transforming our understanding of microbiome function and its impact on human health, environmental sustainability, and biotechnology. The integration of metagenomics with other omics techniques has elucidated the functional roles of microbiome members, revealing intricate networks of cooperation and competition. These insights have far-reaching implications for disease prevention, ecosystem restoration, and bio-based production of valuable compounds. As metagenomics continues to evolve, it is poised to transform our understanding of microbial ecosystems and unlock the full potential of microbiome consortia. Future research directions include the development of novel metagenomic tools, the integration of metagenomics with other disciplines, and the translation of metagenomic discoveries into practical applications. Ultimately, metagenomics will remain a powerful tool for deciphering the complexities of microbiome consortia and harnessing their potential to improve human and environmental health.

REFERENCES

Aasfar A, Bargaz A, Yaakoubi K, Hilali A, Bennis I, Zeroual Y, Meftah Kadmiri I. 2021. Nitrogen fixing azotobacter species as potential soil biological enhancers for crop nutrition and yield stability. Front Microbiol. 12:628379.

Aasfar A, Meftah Kadmiri I, Azaroual SE, Lemriss S, Mernissi N El, Bargaz A, Zeroual Y, Hilali A. 2024. Agronomic advantage of bacterial biological nitrogen fixation on wheat plant growth under contrasting nitrogen and phosphorus regimes. Front Plant Sci. 15:1388775.

Abbas R, Rasul S, Aslam K, Baber M, Shahid M, Mubeen F, Naqqash T. 2019. Halotolerant PGPR: A hope for cultivation of saline soils. J King Saud Univ. 31(4):1195–1201.

Abdelmoneim TS, Moussa TAA, Almaghrabi OA, Alzahrani HS, Abdelbagi I. 2013. Increasing plant tolerance to drought stress by inoculation with arbuscular mycorrhizal fungi. Life Sci J. 10(4):3273–3280.

Aczel MR. 2019. What is the nitrogen cycle and why is it key to life? Front Young Minds. 7(March):1–9.

Adeleke BS, Muller D, Babalola OO. 2023. A metagenomic lens into endosphere microbial communities, promises, and discoveries. Lett Appl Microbiol. 76(2):ovac030.

Afridi MS, Javed MA, Ali S, De Medeiros FHV, Ali B, Salam A, Sumaira, Marc RA, Alkhalifah DHM, Selim S. 2022. New opportunities in plant microbiome engineering for increasing agricultural sustainability under stressful conditions. Front Plant Sci. 13:899464.

Akhtar SS, Mekureyaw MF, Pandey C, Roitsch T. 2019. Role of cytokinins for interactions of plants with microbial pathogens and pest insects. Front Plant Sci. 10:1777.

Ali H, Khan E, Ilahi I. 2019. Environmental chemistry and ecotoxicology of hazardous heavy metals: Environmental persistence, toxicity, and bioaccumulation. J Chem. 2019(1):6730305.

Ali MA, Ahmed T, Ibrahim E, Rizwan M, Chong KP, Yong JWH. 2024. A review on mechanisms and prospects of endophytic bacteria in biocontrol of plant pathogenic fungi and their plant growth-promoting activities. Heliyon 10(11):e31573.

Ali S, Akhtar MS, Siraj M, Zaman W. 2024. Molecular communication of microbial plant biostimulants in the rhizosphere under abiotic stress conditions. Int J Mol Sci. 25(22).

Ali S, Tyagi A, Park S, Mir RA, Mushtaq M, Bhat B, Mahmoudi H, Bae H. 2022. Deciphering the plant microbiome to improve drought tolerance: Mechanisms and perspectives. Environ Exp Bot. 201:104933.

Ali SS, Al-Tohamy R, Khalil MA, Ho S-H, Fu Y, Sun J. 2022. Exploring the potential of a newly constructed manganese peroxidase-producing yeast consortium for tolerating lignin degradation inhibitors while simultaneously decolorizing and detoxifying textile azo dye wastewater. Bioresour Technol. 351:126861.

Alnahhas RN, Sadeghpour M, Chen Y, Frey AA, Ott W, Josić K, Bennett MR. 2020. Majority sensing in synthetic microbial consortia. Nat Commun. 11(1):3659.

Aloo BN, Tripathi V, Makumba BA, Mbega ER. 2022. Plant growth-promoting rhizobacterial biofertilizers for crop production: The past, present, and future. Front Plant Sci. 13:1002448.

Ammeri RW, Simeone GDR, Hidri Y, Abassi MS, Mehri I, Costa S, Hassen A, Rao MA. 2022. Combined bioaugmentation and biostimulation techniques in bioremediation of pentachlorophenol contaminated forest soil. Chemosphere. 290:133359.

Anderson IC, Cairney JWG. 2004. Diversity and ecology of soil fungal communities: Increased understanding through the application of molecular techniques. Environ Microbiol. 6(8):769–779.

Anusha P, Natarajan D, Narayanan M. 2021. Heavy metal removal competence of individual and bacterial consortium, evolved from metal contaminated soil. In: Proceedings of the International Conference on Emerging Trends in Materials Science, Technology and Engineering (ICMSTE2K21).

Aruotu JO, Chikere CB, Okafor CP, Edamkue I. 2023. Microbial consortium for polycyclic aromatic hydrocarbons degradation from petroleum hydrocarbon polluted soils in rivers state, Nigeria. Appl Sci. 13(16):9335.

Ayilara MS, Adeleke BS, Akinola SA, Fayose CA, Adeyemi UT, Gbadegesin LA, Omole RK, Johnson RM, Uthman QO, Babalola OO. 2023. Biopesticides as a promising alternative to synthetic pesticides: A case for microbial pesticides, phytopesticides, and nanobiopesticides. Front Microbiol. 14:1040901.

Bacosa HP, Inoue C. 2015. Polycyclic aromatic hydrocarbons (PAHs) biodegradation potential and diversity of microbial consortia enriched from tsunami sediments in Miyagi, Japan. J Hazard Mater. 283:689–697.

Belton J-M, McCord RP, Gibcus JH, Naumova N, Zhan Y, Dekker J. 2012. Hi–C: A comprehensive technique to capture the conformation of genomes. Methods. 58(3):268–276.

Ben Said S, Tecon R, Borer B, Or D. 2020. The engineering of spatially linked microbial consortia—potential and perspectives. Curr Opin Biotechnol. 62:137–145.

Berg G. 2009. Plant-microbe interactions promoting plant growth and health: Perspectives for controlled use of microorganisms in agriculture. Appl Microbiol Biotechnol. 84(1):11–18.

Berg G, Kusstatscher P, Abdelfattah A, Cernava T, Smalla K. 2021. Microbiome modulation-toward a better understanding of plant microbiome response to microbial inoculants. Front Microbiol. 12:650610.

Berg G, Rybakova D, Fischer D, Cernava T, Vergès M-CC, Charles T, Chen X, Cocolin L, Eversole K, Corral GH. 2020. Microbiome definition re-visited: Old concepts and new challenges. Microbiome. 8:1–22.

Berger B, Yu YW. 2023. Navigating bottlenecks and trade-offs in genomic data analysis. Nat Rev Genet. 24(4):235–250.

Bernal P, Allsopp LP, Filloux A, Llamas MA. 2017. The Pseudomonas putida T6SS is a plant warden against phytopathogens. ISME J. 11(4):972–987.

Bernstein HC, Carlson RP. 2012. Microbial consortia engineering for cellular factories: In vitro to in silico systems. Comput Struct Biotechnol J. 3(4):e201210017.

Berruto CA, Demirer GS. 2024. Engineering agricultural soil microbiomes and predicting plant phenotypes. Trends Microbiol. 32(9):858–873.

Berza Beyene B, Pagano MC, Vaiyapuri R P, Assefa Tuji F. 2022. Microbial consortia inoculation of woody legume Erythrina brucei increases nodulation and shoot nitrogen and phosphorus under greenhouse conditions. Biotechnol Reports. 33:e00707.

Bhantana P, Rana MS, Sun X, Moussa MG, Saleem MH, Syaifudin M, Shah A, Poudel A, Pun AB, Bhat MA, et al., 2021. Arbuscular mycorrhizal fungi and its major role in plant growth, zinc nutrition, phosphorous regulation and phytoremediation. Symbiosis. 84(1):19–37.

Bilal S, Shahzad R, Khan AL, Kang SM, Imran QM, Al-Harrasi A, Yun BW, Lee IJ. 2018. Endophytic microbial consortia of phytohormones-producing fungus *Paecilomyces formosus* lhl10 and bacteria *Sphingomonas* sp. lk11 to glycine max l. regulates physio-hormonal changes to attenuate aluminum and zinc stresses. Front Plant Sci. 9(September):1–18.

Bilińska L, Gmurek M, Ledakowicz S. 2016. Comparison between industrial and simulated textile wastewater treatment by AOPs—Biodegradability, toxicity and cost assessment. Chem Eng J. 306:550–559.

Bonaterra A, Badosa E, Daranas N, Francés J, Roselló G, Montesinos E. 2022. Bacteria as biological control agents of plant diseases. Microorganisms. 10(9):1759.

Bortolot M, Buffoni B, Mazzarino S, Hoff G, Martino E, Fiorilli V, Salvioli Di Fossalunga A. 2024. The importance of mycorrhizal fungi and their associated bacteria in promoting crops' performance: An applicative perspective. Horticulturae. 10(12):1326.

Bowers RM, Kyrpides NC, Stepanauskas R, Harmon-Smith M, Doud D, Reddy TBK, Schulz F, Jarett J, Rivers AR, Eloe-Fadrosh EA, et al. 2017. Minimum information about a single amplified genome (MISAG) and a metagenome-assembled genome (MIMAG) of bacteria and archaea. Nat Biotechnol. 35(8):725–731.

Bubinas A, Giedraitytė G, Kalėdienė L, Nivinskiene O, Butkiene R. 2008. Degradation of naphthalene by thermophilic bacteria via a pathway, through protocatechuic acid. Open Life Sci. 3(1):61–68.

Burstein D, Harrington LB, Strutt SC, Probst AJ, Anantharaman K, Thomas BC, Doudna JA, Banfield JF. 2017. New CRISPR–Cas systems from uncultivated microbes. Nature. 542(7640):237–241.

Cabral L, Giovanella P, Pellizzer EP, Teramoto EH, Kiang CH, Sette LD. 2022. Microbial communities in petroleum-contaminated sites: Structure and metabolisms. Chemosphere. 286:131752.

Cao Z, Yan W, Ding M, Yuan Y. 2022. Construction of microbial consortia for microbial degradation of complex compounds. Front Bioeng Biotechnol. 10:1051233.

Carvalho JRS, Amaral FM, Florencio L, Kato MT, Delforno TP, Gavazza S. 2020. Microaerated UASB reactor treating textile wastewater: The core microbiome and removal of azo dye Direct Black 22. Chemosphere. 242:125157.

Chamkhi I, Cheto S, Geistlinger J, Zeroual Y, Kouisni L, Bargaz A, Ghoulam C. 2022. Legume-based intercropping systems promote beneficial rhizobacterial community and crop yield under stressing conditions. Ind Crops Prod [Internet]. 183:114958.

Chan SS, Khoo KS, Chew KW, Ling TC, Show PL. 2022. Recent advances biodegradation and biosorption of organic compounds from wastewater: Microalgae-bacteria consortium: A review. Bioresour Technol. 344:126159.

Chaudhary R, Nawaz A, Khattak Z, Butt MA, Fouillaud M, Dufossé L, Munir M, Haq I ul, Mukhtar H. 2024. Microbial bio-control agents: A comprehensive analysis on sustainable pest management in agriculture. J Agric Food Res [Internet]. 18:101421.

Chaudhary T, Shukla P. 2019. Bioinoculants for bioremediation applications and disease resistance: Innovative perspectives. Indian J Microbiol. 59:129–136.

Chen W-Y, Wu J-H. 2022. Microbiome composition resulting from different substrates influences trichloroethene dechlorination performance. J Environ Manage. 303:114145.

Cheng Y, Chon K, Ren X, Kou Y, Hwang M-H, Chae K-J. 2021. Bioaugmentation treatment of a novel microbial consortium for degradation of organic pollutants in tannery wastewater under a full-scale oxic process. Biochem Eng J. 175:108131.

Collinge DB, Jensen DF, Rabiey M, Sarrocco S, Shaw MW, Shaw RH. 2022. Biological control of plant diseases: What has been achieved and what is the direction? Plant Pathol. 71(5):1024–1047.

Contreras-Salgado EA, Sánchez-Morán AG, Rodríguez-Preciado SY, Sifuentes-Franco S, Rodríguez-Rodríguez R, Macías-Barragán J, Díaz-Zaragoza M. 2024. Multifaceted applications of synthetic microbial communities: Advances in biomedicine, bioremediation, and industry. Microbiol Res (Pavia). 15(3):1709–1727.

Czech L, Stamatakis A, Dunthorn M, Barbera P. 2022. Metagenomic analysis using phylogenetic placement: A review of the first decade. Front Bioinforma. 2:871393.

Deb CR, Tatung M. 2024. Siderophore producing bacteria as biocontrol agent against phytopathogens for a better environment: A review. South African J Bot. 165:153–162.

Della-Flora IK, Clerici NJ, Dupont GK, Serafini CG, Daroit DJ. 2022. Remediation of soil contaminated with a commercial diesel-biodiesel blend (B12): A microcosm evaluation on the effects of (in) organic amendments. Chemosphere. 287:132059.

Deng L, Zeng G, Fan C, Lu L, Chen X, Chen M, Wu H, He X, He Y. 2015. Response of rhizosphere microbial community structure and diversity to heavy metal co-pollution in arable soil. Appl Microbiol Biotechnol. 99:8259–8269.

Di Benedetto NA, Corbo MR, Campaniello D, Cataldi MP, Bevilacqua A, Sinigaglia M, Flagella Z. 2017. The role of Plant Growth Promoting Bacteria in improving nitrogen use efficiency for sustainable crop production: A focus on wheat. AIMS Microbiol. 3(3):413–434.

Duncker KE, Holmes ZA, You L. 2021. Engineered microbial consortia: Strategies and applications. Microb Cell Fact. 20(1):211.

Dwidar M, Kim S, Jeon BS, Um Y, Mitchell RJ, Sang B-I. 2013. Co-culturing a novel *Bacillus* strain with *Clostridium tyrobutyricum* ATCC 25755 to produce butyric acid from sucrose. Biotechnol Biofuels. 6:1–10.

Elkarrach K, Merzouki M, Biyada S, Benlemlih M. 2020. Bioaugmentation process for the treatment of tannery effluents in Fez, Morocco: An eco-friendly treatment using novel chromate bacteria. J Water Process Eng. 38:101589.

Etesami H, Jeong BR, Glick BR. 2021. Contribution of arbuscular mycorrhizal fungi, phosphate-solubilizing bacteria, and silicon to P uptake by plant. Front Plant Sci. 12:699618.

Fasman KH, Salzberg SL. 1998. An introduction to biological sequence analysis. In: Computational Methods in Molecular Biology, volume 32. Elsevier B.V; pp. 29–42.

Fenta L, Mekonnen H. 2024. Microbial biofungicides as a substitute for chemical fungicides in the control of phytopathogens: Current perspectives and research directions. Scientifica (Cairo). 2024:5322696.

Fernandez M, Paisio CE, Perotti R, Pereira PP, Agostini E, González PS. 2019. Laboratory and field microcosms as useful experimental systems to study the bioaugmentation treatment of tannery effluents. J Environ Manage. 234:503–511.

Finucane MM, Sharpton TJ, Laurent TJ, Pollard KS. 2014. A taxonomic signature of obesity in the microbiome? Getting to the guts of the matter. PLoS One. 9(1):e84689.

Forss J, Lindh MV, Pinhassi J, Welander U. 2017. Microbial biotreatment of actual textile wastewater in a continuous sequential rice husk biofilter and the microbial community involved. PLoS One. 12(1):e0170562.

Galadima M, Singh S, Pawar A, Khasnabis S, Dhanjal DS, Anil AG, Rai P, Ramamurthy PC, Singh J. 2021. Toxicity, microbial degradation and analytical detection of pyrethroids: A review. Environ Adv. 5:100105.

Galli M, Feldmann F, Vogler UK, Kogel K-H. 2024. Can biocontrol be the game-changer in integrated pest management? A review of definitions, methods and strategies. J Plant Dis Prot. 131(2):265–291.

Gaytán I, Sánchez-Reyes A, Burelo M, Vargas-Suárez M, Liachko I, Press M, Sullivan S, Cruz-Gómez MJ, Loza-Tavera H. 2020. Degradation of recalcitrant polyurethane and xenobiotic additives by a selected landfill microbial community and its biodegradative potential revealed by proximity ligation-based metagenomic analysis. Front Microbiol. 10:2986.

Ghorui M, Chowdhury S, Balu P, Burla S. 2024. Arbuscular Mycorrhizal inoculants and its regulatory landscape. Heliyon [Internet]. 10(9):e30359.

Ghurye JS, Cepeda-Espinoza V, Pop M. 2016. Metagenomic assembly: Overview, challenges and applications. Yale J Biol Med. 89(3):353–362.

Gilbert ES, Walker AW, Keasling JD. 2003. A constructed microbial consortium for biodegradation of the organophosphorus insecticide parathion. Appl Microbiol Biotechnol. 61:77–81.

Gomaa OM, Fapetu S, Kyazze G, Keshavarz T. 2019. Probing the mechanism of simultaneous bioenergy production and biodegradation process of Congo red in microbial fuel cells. J Chem Technol Biotechnol. 94(7):2092–2097.

Gomaa OM, Ibrahim SAEM, Mansour NM. 2023. B acillus spizizenii DN and microbial consortia biostimulation followed by gamma irradiation for efficient textile wastewater treatment. Environ Sci Pollut Res. 30(12):33907–33916.

Gore J, Youk H, van Oudenaarden A. 2009. Snowdrift game dynamics and facultative cheating in yeast. Nature. 459(7244):253–256.

Greenblum S, Carr R, Borenstein E. 2015. Extensive strain-level copy-number variation across human gut microbiome species. Cell. 160(4):583–594.

Greff B, Szigeti J, Nagy Á, Lakatos E, Varga L. 2022. Influence of microbial inoculants on co-composting of lignocellulosic crop residues with farm animal manure: A review. J Environ Manage. 302:114088.

Gutiérrez EJ, Abraham M del R, Baltazar JC, Vázquez G, Delgadillo E, Tirado D. 2020. Pseudomonas fluorescens: A bioaugmentation strategy for oil-contaminated and nutrient-poor soil. Int J Environ Res Public Health. 17(19):6959.

Han S, Akhtar MR, Xia X. 2024. Functions and regulations of insect gut bacteria. Pest Manag Sci. 80(10):4828–4840.

Han S-W, Yoshikuni Y. 2022. Microbiome engineering for sustainable agriculture: Using synthetic biology to enhance nitrogen metabolism in plant-associated microbes. Curr Opin Microbiol. 68:102172.

Handelsman J. 2004. Metagenomics: Application of genomics to uncultured microorganisms. Microbiol Mol Biol Rev. 68(4):669–685.

Haq IU, Rahim K, Yahya G, Ijaz B, Maryam S, Paker NP. 2024. Eco-smart biocontrol strategies utilizing potent microbes for sustainable management of phytopathogenic diseases. Biotechnol Reports. 44:e00859.

Hassan A, Akram W, Rizwana H, Aftab Z-E-H, Hanif S, Anjum T, Alwahibi MS. 2023. The imperative use of *Bacillus consortium* and quercetin contributes to suppress fusarium wilt disease by direct antagonism and induced resistance. Microorganisms. 11(10):2603.

Ho BT, Roberts TK, Lucas S. 2018. An overview on biodegradation of polystyl and modified polystyrene: The microbial approach. Crit Rev Biotechnol. 38(2):308–320.

Hosoda K, Suzuki S, Yamauchi Y, Shiroguchi Y, Kashiwagi A, Ono N, Mori K, Yomo T. 2011. Cooperative adaptation to establishment of a synthetic bacterial mutualism. PLoS One. 6(2):e17105.

Hu B, Wang M, Geng S, Wen L, Wu M, Nie Y, Tang Y-Q, Wu X-L. 2020. Metabolic exchange with non-alkane-consuming *Pseudomonas stutzeri* SLG510A3-8 improves n-alkane biodegradation By the alkane degrader *Dietzia* sp. strain DQ12-45-1b. Appl Environ Microbiol. 86(8):e02931-19.

Hug LA, Beiko RG, Rowe AR, Richardson RE, Edwards EA. 2012. Comparative metagenomics of three *Dehalococcoides* containing enrichment cultures: The role of the non-dechlorinating community. BMC Genomics. 13:327.

Hugenholtz P, Skarshewski A, Parks DH. 2016. Genome-based microbial taxonomy coming of age. Cold Spring Harb Perspect Biol. 8:a018085.

Hugerth LW, Larsson J, Alneberg J, Lindh M V., Legrand C, Pinhassi J, Andersson AF. 2015. Metagenome-assembled genomes uncover a global brackish microbiome. Genome Biol. 16(1):279.

Ilangumaran G, Smith DL. 2017. Plant growth promoting rhizobacteria in amelioration of salinity stress: A systems biology perspective. Front Plant Sci. 8:1768.

Iqbal S, Begum F, Ullah I, Jalal N, Shaw P. 2025. Peeling off the layers from microbial dark matter (MDM): Recent advances, future challenges, and opportunities. Crit Rev Microbiol. 51(1):1–21.

Iquebal MA, Jagannadham J, Jaiswal S, Prabha R, Rai A, Kumar D. 2022. Potential use of microbial community genomes in various dimensions of agriculture productivity and its management: A review. Front Microbiol. 13:708335.

Irawati W, Ompusunggu NP, Yuwono T. 2018. Influence of bacterial consortium for copper biosorption and accumulation. In: Inventing Prosperous Future Through Biological Research and Tropical Biodiversity Management: Proceedings of the 5th International Conference on Biological Science. AIP Conf Proc. Vol. 2002. AIP Publishing; p. 020072.

Ishino S, Ishino Y. 2014. DNA polymerases as useful reagents for biotechnology–the history of developmental research in the field. Front Microbiol. 5:465.

Jaffar S, Ahmad S, Lu Y. 2022. Contribution of insect gut microbiota and their associated enzymes in insect physiology and biodegradation of pesticides. Front Microbiol. 13:979383.

Jahn L, Schafhauser T, Wibberg D, Rueckert C, Winkler A, Kulik A, Weber T, Flor L, van Pee K-H, Kalinowski J. 2017. Linking secondary metabolites to biosynthesis genes in the fungal endophyte *Cyanodermella asteris*: The anti-cancer bisanthraquinone skyrin. J Biotechnol. 257:233–239.

Jia X, Liu C, Song H, Ding M, Du J, Ma Q, Yuan Y. 2016. Design, analysis and application of synthetic microbial consortia. Synth Syst Biotechnol. 1(2):109–117.

Jiang Y, Lv Y, Wu R, Lu J, Dong W, Zhou J, Zhang W, Xin F, Jiang M. 2020. Consolidated bioprocessing performance of a two-species microbial consortium for butanol production from lignocellulosic biomass. Biotechnol Bioeng. 117(10):2985–2995.

Jochum MD, McWilliams KL, Pierson EA, Jo Y-K. 2019. Host-mediated microbiome engineering (HMME) of drought tolerance in the wheat rhizosphere. PLoS One. 14(12):e0225933.

Joergensen RG, Wichern F. 2018. Alive and kicking: Why dormant soil microorganisms matter. Soil Biol Biochem. 116:419–430.

Kafilzadeh F, Pour FH. 2012. Degradation of naphthalene, phenanthrene and pyrene by *Pseudomonas* sp. and *Corynebacterium* sp. in the landfills. Int J Biosci. 2(9):77–84.

Kang D-W, Adams JB, Coleman DM, Pollard EL, Maldonado J, McDonough-Means S, Caporaso JG, Krajmalnik-Brown R. 2019. Long-term benefit of Microbiota Transfer Therapy on autism symptoms and gut microbiota. Sci Rep. 9(1):5821.

Kapadia C, Sayyed RZ, El Enshasy HA, Vaidya H, Sharma D, Patel N, Malek RA, Syed A, Elgorban AM, Ahmad K. 2021. Halotolerant microbial consortia for sustainable mitigation of salinity stress, growth promotion, and mineral uptake in tomato plants and soil nutrient enrichment. Sustainability. 13(15):8369.

Kaul S, Choudhary M, Gupta S, Dhar MK. 2021. Engineering host microbiome for crop improvement and sustainable agriculture. Front Microbiol. 12:635917.

Kaur T, Devi R, Kumar S, Sheikh I, Kour D, Yadav AN. 2022. Microbial consortium with nitrogen fixing and mineral solubilizing attributes for growth of barley (Hordeum vulgare L.). Heliyon. 8(4):e09326.

Kawashima T. 2019. Comparative and evolutionary genomics. In: Encyclopedia of Bioinformatics and Computational Biology. Elsevier; pp. 257–267.

Kaya C, Uğurlar F, Ashraf M, Hou D, Kirkham MB, Bolan N. 2024. Microbial consortia-mediated arsenic bioremediation in agricultural soils: Current status, challenges, and solutions. Sci Total Environ. 917:170297.

Ke J, Wang B, Yoshikuni Y. 2021. Microbiome engineering: Synthetic biology of plant-associated microbiomes in sustainable agriculture. Trends Biotechnol. 39(3):244–261.

Khan MY, Nadeem SM, Sohaib M, Waqas MR, Alotaibi F, Ali L, Zahir ZA, Al-Barakah FNI. 2022. Potential of plant growth promoting bacterial consortium for improving the growth and yield of wheat under saline conditions. Front Microbiol. 13:958522. https://doi.org/10.3389/fmicb.2022.958522

Khan N, Bano A, Rahman MA, Guo J, Kang Z, Babar MA. 2019. Comparative physiological and metabolic analysis reveals a complex mechanism involved in drought tolerance in chickpea (Cicer arietinum L.) induced by PGPR and PGRs. Sci Rep. 9(1):2097.

Khan R, Bhawana P, Fulekar MH. 2013. Microbial decolorization and degradation of synthetic dyes: A review. Rev Environ Sci Bio/Technology. 12:75–97.

Kim I-S, Ekpeghere KI, Ha S-Y, Kim B-S, Song B, Kim J-T, Kim H-G, Koh S-C. 2014. Full-scale biological treatment of tannery wastewater using the novel microbial consortium BM-S–1. J Environ Sci Heal Part A. 49(3):355–364.

Kim I-S, Ekpeghere KI, Ha S-Y, Kim S-H, Kim B-S, Song B, Chun J, Chang J-S, Kim H-G, Koh S-C. 2013. An eco-friendly treatment of tannery wastewater using bioaugmentation with a novel microbial consortium. J Environ Sci Heal Part A. 48(13):1732–1739.

Kõljalg U, Larsson K-H, Abarenkov K, Nilsson RH, Alexander IJ, Eberhardt U, Erland S, Høiland K, Kjøller R, Larsson E, et al. 2005. UNITE: A database providing web-based methods for the molecular identification of ectomycorrhizal fungi. New Phytol. 166(3):1063–1068.

Koza NA, Adedayo AA, Babalola OO, Kappo AP. 2022. Microorganisms in plant growth and development: Roles in abiotic stress tolerance and secondary metabolites secretion. Microorganisms. 10(8).

Kuhl-Nagel T, Rodriguez PA, Gantner I, Chowdhury SP, Schwehn P, Rosenkranz M, Weber B, Schnitzler J-P, Kublik S, Schloter M, et al. 2022. Novel pseudomonas sp. SCA7 promotes plant growth in two plant families and induces systemic resistance in *Arabidopsis thaliana*. Front Microbiol. 13:923515.

Kumar S, Chandra R, Behera L, Sudhir I, Meena M, Singh S, Keswani C. 2023. Microbial consortium mediated acceleration of the defense response in potato against Alternaria solani through prodigious inflation in phenylpropanoid derivatives and redox homeostasis. Heliyon [Internet]. 9(11):e22148.

Kumar V, Shahi SK, Ferreira LFR, Bilal M, Biswas JK, Bulgariu L. 2021. Detection and characterization of refractory organic and inorganic pollutants discharged in biomethanated distillery effluent and their phytotoxicity, cytotoxicity, and genotoxicity assessment using Phaseolus aureus L. and Allium cepa L. Environ Res. 201:111551.

Kumari M, Swarupa P, Kesari KK, Kumar A. 2023. Microbial inoculants as plant biostimulants: A review on risk status. Life. 13(1):12.
Kumari N, Rai LC. 2020. Cyanobacterial diversity: Molecular insights under multifarious environmental conditions. In: Advances in Cyanobacterial Biology. Elsevier; pp. 17–33.
Lahlali R, Ezrari S, Radouane N, Kenfaoui J, Esmaeel Q, El Hamss H, Belabess Z, Barka EA. 2022. Biological control of plant pathogens: A global perspective. Microorganisms. 10(3):596.
Le Roux X, Bouskill NJ, Niboyet A, Barthes L, Dijkstra P, Field CB, Hungate BA, Lerondelle C, Pommier T, Tang J, et al. 2016. Predicting the responses of soil nitrite-oxidizers to multi-factorial global change: A trait-based approach. Front Microbiol. 7:628.
Li L, Meng D, Yin H, Zhang T, Liu Y. 2023. Genome-resolved metagenomics provides insights into the ecological roles of the keystone taxa in heavy-metal-contaminated soils. Front Microbiol. 14:1203164.
Li Xiang, Lu C, Dai Y, Yu Z, Gu W, Li T, Li Xinyu, Li Xu, Wang X, Su Z. 2022. Characterizing the microbial consortium L1 capable of efficiently degrading chlorimuron-ethyl via metagenome combining 16S rDNA sequencing. Front Microbiol. 13:912312.
Liang Y, Ma A, Zhuang G. 2022. Construction of environmental synthetic microbial consortia: Based on engineering and ecological principles. Front Microbiol. 13:829717.
Liu C, Fu X, Liu L, Ren X, Chau CKL, Li S, Xiang L, Zeng H, Chen G, Tang L-H, et al. 2011. Sequential establishment of stripe patterns in an expanding cell population. Science 334:238–241.
Lofrano G, Meriç S, Zengin GE, Orhon D. 2013. Chemical and biological treatment technologies for leather tannery chemicals and wastewaters: A review. Sci Total Environ. 461:265–281.
Luna JM, Rufino RD, Sarubbo LA. 2016. Biosurfactant from Candida sphaerica UCP0995 exhibiting heavy metal remediation properties. Process Saf Environ Prot. 102:558–566.
Luo D, Shi J, Li M, Chen J, Wang T, Zhang Q, Yang L, Zhu N, Wang Y. 2024. Consortium of phosphorus-solubilizing bacteria promotes maize growth and changes the microbial community composition of rhizosphere soil. Agronomy. 14(7):1535.
Lurthy T, Cantat C, Jeudy C, Declerck P, Gallardo K, Barraud C, Leroy F, Ourry A, Lemanceau P, Salon C, Mazurier S. 2020. Impact of bacterial siderophores on iron status and ionome in pea. Front Plant Sci. 11:730.
Maciag T, Kozieł E, Rusin P, Otulak-Kozieł K, Jafra S, Czajkowski R. 2023. Microbial consortia for plant protection against diseases: More than the sum of its parts. Int J Mol Sci. 24(15):12227.
Madadi R, Bester K. 2021. Fungi and biochar applications in bioremediation of organic micropollutants from aquatic media. Mar Pollut Bull. 166:112247.
Maglangit F, Fang Q, Kyeremeh K, Sternberg JM, Ebel R, Deng H. 2020. A co-culturing approach enables discovery and biosynthesis of a bioactive indole alkaloid metabolite. Molecules. 25(2):256.
Maguire F, Jia B, Gray KL, Lau WYV, Beiko RG, Brinkman FSL. 2020. Metagenome-assembled genome binning methods with short reads disproportionately fail for plasmids and genomic Islands. Microb Genomics. 6(10):mgen000436.
Mahmud K, Missaoui A, Lee K, Ghimire B, Presley HW, Makaju S. 2021. Rhizosphere microbiome manipulation for sustainable crop production. Curr Plant Biol. 27:100210.
Maki H, Sasaki T, Harayama S. 2001. Photo-oxidation of biodegraded crude oil and toxicity of the photo-oxidized products. Chemosphere. 44(5):1145–1151.
Malik K, Salama E-S, El-Dalatony MM, Jalalah M, Harraz FA, Al-Assiri MS, Zheng Y, Sharma P, Li X. 2021. Co-fermentation of immobilized yeasts boosted bioethanol production from pretreated cotton stalk lignocellulosic biomass: Long-term investigation. Ind Crops Prod. 159:113122.
Thomas AM, Lima FP, Moura LMS, da Silva AM, Dias-Neto E, Setubal JC. 2018. Comparative metagenomics. In: Setubal J, Stoye J, Stadler P, editors. Methods in Molecular Biology, vol 1704. New York, NY: Humana Press; pp. 243–260.
Marchevsky N, Urbieta MS, Bernardelli C, Mas M, Donati ER. 2015. Zinc recovery during refractory ore biooxidation by an indigenous consortium. Int J Miner Process. 138:30–37.
Mardanov AV, Chizhevskaya EP, Lazarev AM, Rakitin AL, Beletsky AV, Chebotar VK, Ravin NV. 2019. Complete genome sequences of endophytic Bacilli isolated from Grapevine plants. Microbiol Resour Announc. 8(48):10–1128.
Massot F, Bernard N, Alvarez LMM, Martorell MM, MacCormack WP, Ruberto LAM. 2022. Microbial associations for bioremediation: What does "microbial consortia" mean? Appl Microbiol Biotechnol. 106(7):2283–2297.
McCarty NS, Ledesma-Amaro R. 2019. Synthetic biology tools to engineer microbial communities for biotechnology. Trends Biotechnol. 37(2):181–197.
McGuire KL, Treseder KK. 2010. Microbial communities and their relevance for ecosystem models: Decomposition as a case study. Soil Biol Biochem. 42(4):529–535.
Mee MT, Collins JJ, Church GM, Wang HH. 2014. Syntrophic exchange in synthetic microbial communities. Proc Natl Acad Sci. 111(20):E2149–E2156.
Meena M, Yadav G, Sonigra P, Nagda A, Mehta T, Swapnil P, Harish, Marwal A. 2022. Role of elicitors to initiate the induction of systemic resistance in plants to biotic stress. Plant Stress. 5:100103.
Mehmood N, Saeed M, Zafarullah S, Hyder S, Rizvi ZF, Gondal AS, Jamil N, Iqbal R, Ali B, Ercisli S, Kupe M. 2023. Multifaceted impacts of plant-beneficial pseudomonas spp. in managing various plant diseases and crop yield improvement. ACS Omega. 8(25):22296–22315.
Michielsen S, Vercelli GT, Cordero OX, Bachmann H. 2024. Spatially structured microbial consortia and their role in food fermentations. Curr Opin Biotechnol. 87:103102.
Mikesková H, Novotný Č, Svobodová K. 2012. Interspecific interactions in mixed microbial cultures in a biodegradation perspective. Appl Microbiol Biotechnol. 95:861–870.
Mishra A, Malik A. 2014. Metal and dye removal using fungal consortium from mixed waste stream: Optimization and validation. Ecol Eng. 69:226–231.
Mishra M, Singh SK, Kumar A. 2021. Microbial consortia: Approaches in crop production and yield enhancement. In: Microbiome Stimulants for Crops (Mechanisms and Applications). Elsevier; pp. 293–303.
Mnif I, Mnif S, Sahnoun R, Maktouf S, Ayedi Y, Ellouze-Chaabouni S, Ghribi D. 2015. Biodegradation of diesel oil by a novel microbial consortium: Comparison between co-inoculation with biosurfactant-producing strain and

exogenously added biosurfactants. Environ Sci Pollut Res. 22:14852–14861.
Mohamed S, Syed BA. 2013. Commercial prospects for genomic sequencing technologies. Nat Rev Drug Discov. 12(5):341–342.
Molina-Romero D, Baez A, Quintero-Hernández V, Castañeda-Lucio M, Fuentes-Ramírez LE, Bustillos-Cristales M del R, Rodríguez-Andrade O, Morales-García YE, Munive A, Muñoz-Rojas J. 2017. Compatible bacterial mixture, tolerant to desiccation, improves maize plant growth. PLoS One. 12(11):e0187913.
Moussa TAA, Al-Qaysi SAS, Thabit ZA, Kadhem SB. 2017. Microbial levan from *Brachybacterium phenoliresistens*: Characterization and enhancement of production. Process Biochem. 57:9-15.
Moussa TAA, Rashad YM, Baka ZAM. 2023. New perspectives on fungal siderophores. In: Rashad YM, Baka ZAM, Moussa TAA, editors. Plant Mycobiome Divers Interact Uses. Cham: Springer; pp. 225–239.
Mubashar M, Naveed M, Mustafa A, Ashraf S, Shehzad Baig K, Alamri S, Siddiqui MH, Zabochnicka-Świątek M, Szota M, Kalaji HM. 2020. Experimental investigation of Chlorella vulgaris and Enterobacter sp. MN17 for decolorization and removal of heavy metals from textile wastewater. Water. 12(11):3034.
Mueller UG, Juenger TE, Kardish MR, Carlson AL, Burns KM, Edwards JA, Smith CC, Fang C-C, Des Marais DL. 2021. Artificial selection on microbiomes to breed microbiomes that confer salt tolerance to plants. MSystems. 6(6):e01125-21.
Muneer B, Iqbal MJ, Shakoori FR, Shakoori AR. 2013. Tolerance and biosorption of mercury by microbial consortia: Potential use in bioremediation of wastewater. Pak J Zool. 45(1):247–254.
Musat F, Galushko A, Jacob J, Widdel F, Kube M, Reinhardt R, Wilkes H, Schink B, Rabus R. 2009. Anaerobic degradation of naphthalene and 2-methylnaphthalene by strains of marine sulfate-reducing bacteria. Environ Microbiol. 11(1):209–219.
Muthukumar B, Al Salhi MS, Narenkumar J, Devanesan S, Rao TN, Kim W, Rajasekar A. 2022. Characterization of two novel strains of Pseudomonas aeruginosa on biodegradation of crude oil and its enzyme activities. Environ Pollut. 304:119223.
Nadarajah K, Abdul Rahman NSN. 2023. The microbial connection to sustainable agriculture. Plants. 12(12):2307.
Nadell CD, Foster KR. 2012. Mutually helping microbes can evolve by hitchhiking. Proc Natl Acad Sci. 109(47):19037–19038.
Naghavi M, Vollset SE, Ikuta KS, Swetschinski LR, Gray AP, Wool EE, Robles Aguilar G, Mestrovic T, Smith G, Han C, et al. 2024. Global burden of bacterial antimicrobial resistance 1990–2021: A systematic analysis with forecasts to 2050. Lancet. 404(10459):1199–1226.
Nam NN, Do HD, Loan Trinh KT, Lee NY. 2023. Metagenomics: An effective approach for exploring microbial diversity and functions. Foods. 12(11):2140.
Nannipieri P, Ascher J, Ceccherini MT, Landi L, Pietramellara G, Renella G. 2017. Microbial diversity and soil functions. Eur J Soil Sci [Internet]. 68(1):12–26.
Narayanan Z, Glick BR. 2022. Secondary metabolites produced by plant growth-promoting bacterial endophytes. Microorganisms. 10(10):2008.
Navgire GS, Goel N, Sawhney G, Sharma M, Kaushik P, Mohanta YK, Mohanta TK, Al-Harrasi A. 2022. Analysis and interpretation of metagenomics data: An approach. Biol Proced Online. 24(1):18.
Nayfach S, Pollard KS. 2016. Toward accurate and quantitative comparative metagenomics. Cell. 166(5):1103–1116.
Nayfach S, Rodriguez-Mueller B, Garud N, Pollard KS. 2016. An integrated metagenomics pipeline for strain profiling reveals novel patterns of bacterial transmission and biogeography. Genome Res. 26(11):1612–1625.
Negi R, Sharma B, Jan T, Kaur T, Chowdhury S, Kapoor M, Singh S, Kumar A, Rai AK, Rustagi S, et al. 2024. Microbial consortia: Promising tool as plant bioinoculants for agricultural sustainability. Curr Microbiol. 81(8):222.
Nelson WC, Maezato Y, Wu Y-W, Romine MF, Lindemann SR. 2016. Identification and resolution of microdiversity through metagenomic sequencing of parallel consortia. Appl Environ Microbiol. 82(1):255–267.
Nivedita S, Behera SS, Behera PK, Parwez Z, Giri S, Behera HT, Ray L. 2024. Salt-resistant Streptomyces consortia promote growth of rice (Oryza sativa var. Swarna) alleviating salinity and drought stress tolerance by enhancing photosynthesis, antioxidant function, and proline content. The Microbe. 4:100124.
Nunes PSO, Lacerda-Junior GV, Mascarin GM, Guimarães RA, Medeiros FHV, Arthurs S, Bettiol W. 2024. Microbial consortia of biological products: Do they have a future? Biol Control. 188:105439.
Nuti M, Giovannetti G. 2015. Borderline products between biofertilizerIo-effectors and plant protectants: The role of microbial consortia. J Agric Sci Technol A. 5:305–315.
Okere U, Semple K. 2012. Biodegradation of PAHs in 'pristine' soils from different climatic regions. J Bioremed Biodegrad S. 1(2):1–11.
Okoh AI. 2006. Biodegradation alternative in the cleanup of petroleum hydrocarbon pollutants. Biotechnol Mol Biol Rev. 1(2):38–50.
Orhan F. 2016. Alleviation of salt stress by halotolerant and halophilic plant growth-promoting bacteria in wheat (Triticum aestivum). Brazilian J Microbiol. 47(3):621–627.
Oudot J, Merlin FX, Pinvidic P. 1998. Weathering rates of oil components in a bioremediation experiment in estuarine sediments. Mar Environ Res. 45(2):113–125.
Pabbathi NPP, Velidandi A, Tavarna T, Gupta S, Raj RS, Gandam PK, Baadhe RR. 2023. Role of metagenomics in prospecting novel endoglucanases, accentuating functional metagenomics approach in second-generation biofuel production: A review. Biomass Convers Biorefinery. 13(2):1371–1398.
Pandey SS, Jain R, Bhardwaj P, Thakur A, Kumari M, Bhushan S, Kumar S. 2022. Plant probiotics–endophytes pivotal to plant health. Microbiol Res. 263:127148.
Papenfort K, Bassler BL. 2016. Quorum sensing signal–response systems in Gram-negative bacteria. Nat Rev Microbiol. 14(9):576–588.
Park EM, Chelvanambi M, Bhutiani N, Kroemer G, Zitvogel L, Wargo JA. 2022. Targeting the gut and tumor microbiota in cancer. Nat Med. 28(4):690–703.
Pascoal F, Costa R, Magalhães C. 2021. The microbial rare biosphere: Current concepts, methods and ecological principles. FEMS Microbiol Ecol [Internet]. 97(1):fiaa227.
Passari AK, Rajput V, Zothanpuia, Priya LPM, Dharne M, Dastager S, Mathew OK, Hashem A, Abd_ Allah EF,

Singh BP. 2019. Draft genome sequence of plant growth-promoting endophytic Microbacterium hydrothermale BPSAC84, isolated from the medicinal plant Mirabilis jalapa. Microbiol Resour Announc. 8(22):10–1128.

Pathak H, Kantharia D, Malpani A, Madamwar D. 2009. Naphthalene degradation by Pseudomonas sp. HOB1: In vitro studies and assessment of naphthalene degradation efficiency in simulated microcosms. J Hazard Mater. 166(2–3):1466–1473.

Patowary K, Patowary R, Kalita MC, Deka S. 2016. Development of an efficient bacterial consortium for the potential remediation of hydrocarbons from contaminated sites. Front Microbiol. 7:1092.

Pavlopoulos GA, Baltoumas FA, Liu S, Selvitopi O, Camargo AP, Nayfach S, Azad A, Roux S, Call L, Ivanova NN. 2023. Unraveling the functional dark matter through global metagenomics. Nature. 622(7983):594–602.

Pelletier E, Delille D, Delille B. 2004. Crude oil bioremediation in sub-Antarctic intertidal sediments: Chemistry and toxicity of oiled residues. Mar Environ Res. 57(4):311–327.

Pérez J, Muñoz-Dorado J, Braña AF, Shimkets LJ, Sevillano L, Santamaría RI. 2011. Myxococcus xanthus induces actinorhodin overproduction and aerial mycelium formation by Streptomyces coelicolor. Microb Biotechnol. 4(2):175–183.

Pieterse CMJ, Zamioudis C, Berendsen RL, Weller DM, Van Wees SCM, Bakker PAHM. 2014. Induced systemic resistance by beneficial microbes. Annu Rev Phytopathol. 52:347–375.

Puentes-Téllez PE, Falcao Salles J. 2018. Construction of effective minimal active microbial consortia for lignocellulose degradation. Microb Ecol. 76(2):419–429.

Qi X, Ma Y, Chang H, Li B, Ding M, Yuan Y. 2021. Evaluation of PET degradation using artificial microbial consortia. Front Microbiol. 12:778828.

Qi X, Yan W, Cao Z, Ding M, Yuan Y. 2021. Current advances in the biodegradation and bioconversion of polyethylene terephthalate. Microorganisms. 10(1):39.

Quince C, Nurk S, Raguideau S, James R, Soyer OS, Summers JK, Limasset A, Eren AM, Chikhi R, Darling AE. 2021. Strong: Metagenomics strain resolution on assembly graphs. Genome Biol. 22(1):214.

Ragasruthi M, Balakrishnan N, Murugan M, Swarnakumari N, Harish S, Sharmila DJS. 2024. *Bacillus thuringiensis* (Bt)-based biopesticide: Navigating success, challenges, and future horizons in sustainable pest control. Sci Total Environ. 954:176594.

Raj D, Maiti SK. 2020. Risk assessment of potentially toxic elements in soils and vegetables around coal-fired thermal power plant: A case study of Dhanbad, India. Environ Monit Assess. 192(11):699.

Ratnaningsih HR, Noviana Z, Dewi TK, Loekito S, Wiyono S, Gafur A, Antonius S. 2023. IAA and ACC deaminase producing-bacteria isolated from the rhizosphere of pineapple plants grown under different abiotic and biotic stresses. Heliyon. 9(6):e16306.

Ravanbakhsh M, Sasidharan R, Voesenek LACJ, Kowalchuk GA, Jousset A. 2018. Microbial modulation of plant ethylene signaling: Ecological and evolutionary consequences. Microbiome [Internet]. 6(1):52.

Rehman R, Ali MI, Ali N, Badshah M, Iqbal M, Jamal A, Huang Z. 2021. Crude oil biodegradation potential of biosurfactant-producing Pseudomonas aeruginosa and Meyerozyma sp. J Hazard Mater. 418:126276.

Rosenbaum M, Aulenta F, Villano M, Angenent LT. 2011. Cathodes as electron donors for microbial metabolism: Which extracellular electron transfer mechanisms are involved? Bioresour Technol. 102(1):324–333.

Ruiz-Lozano JM, Aroca R, Zamarreño ÁM, Molina S, Andreo-Jiménez B, Porcel R, García-Mina JM, Ruyter-Spira C, López-Ráez JA. 2016. Arbuscular mycorrhizal symbiosis induces strigolactone biosynthesis under drought and improves drought tolerance in lettuce and tomato. Plant Cell Environ. 39(2):441–452.

Saeed Q, Xiukang W, Haider FU, Kučerik J, Mumtaz MZ, Holatko J, Naseem M, Kintl A, Ejaz M, Naveed M, et al. 2021. Rhizosphere bacteria in plant growth promotion, biocontrol, and bioremediation of contaminated sites: A comprehensive review of effects and mechanisms. Int J Mol Sci. 22(19).

Salam A, Ali A, Afridi MS, Ali S, Ullah Z. 2022. Agrobiodiversity: Effect of drought stress on the eco-physiology and morphology of wheat. In: Öztürk M, Khan SM, Altay V, Efe R, Egamberdieva D, Khassanov FO, editors. Biodiversity, Conservation and Sustainability in Asia. Cham: Springer; pp. 597–618.

Samantaray A, Chattaraj S, Mitra D, Ganguly A, Kumar R, Gaur A, Mohapatra PKD, Santos-Villalobos S de los, Rani A, Thatoi H. 2024. Advances in microbial based bio-inoculum for amelioration of soil health and sustainable crop production. Curr Res Microb Sci. 7:100251.

Sambles C, Moore K, Lux TM, Jones K, Littlejohn GR, Gouveia JD, Aves SJ, Studholme DJ, Lee R, Love J. 2017. Metagenomic analysis of the complex microbial consortium associated with cultures of the oil-rich alga *Bot*ryococcus braunii. Microbiologyopen. 6(4).

Sanati NA. 2014. Microfluidic platforms for plant cells studies. Lab Chip. 14(17):3262–3274.

Santoyo G, Guzmán-Guzmán P, Parra-Cota FI, Santos-Villalobos SD, Orozco-Mosqueda MD, Glick BR. 2021. Plant growth stimulation by microbial consortia. Agronomy. 11(2):219.

Sarkar J, Kazy SK, Gupta A, Dutta A, Mohapatra B, Roy A, Bera P, Mitra A, Sar P. 2016. Biostimulation of indigenous microbial community for bioremediation of petroleum refinery sludge. Front Microbiol. 7:1407.

Schafhauser T, Jahn L, Kirchner N, Kulik A, Flor L, Lang A, Caradec T, Fewer DP, Sivonen K, van Berkel WJH. 2019. Antitumor astins originate from the fungal endophyte *Cyanodermella asteris* living within the medicinal plant *Aster tataricus*. Proc Natl Acad Sci. 116(52):26909–26917.

Schirawski J, Perlin MH. 2018. Plant–microbe interaction 2017—the good, the bad and the diverse. Int J Mol Sci. 19(5):1374.

Sekar J, Raj R, Prabavathy VR. 2016. Microbial consortial products for sustainable agriculture: Commercialization and regulatory issues in India. In: Agriculturaly Important Microorganisms. pp. 107–132. Springer Science+Business Media Singapore.

Semenzato G, Fani R. 2024. Endophytic bacteria: A sustainable strategy for enhancing medicinal plant cultivation and preserving microbial diversity. Front Microbiol. 15:1477465.

Senn S, Pangell K, Bowerman AL. 2022. Metagenomic insights into the composition and function of microbes associated with the rootzone of datura inoxia. BioTech. 11(1):1.

Seong CN, Kang JW, Lee JH, Seo SY, Woo JJ, Park C, Bae KS, Kim MS. 2018. Taxonomic hierarchy of the phylum

Firmicutes and novel Firmicutes species originated from various environments in Korea. J Microbiol [Internet]. 56(1):1–10.

Shaffique S, Hussain S, Kang SM, Imran M, Injamum-Ul-Hoque M, Khan MA, Lee IJ. 2023. Phytohormonal modulation of the drought stress in soybean: Outlook, research progress, and cross-talk. Front Plant Sci. 14:1237295.

Shafqat A, Abbas S, Ambreen M, Siddiqa Bhatti A, Kausar H, Gull T. 2024. Exploring the vital role of phytohormones and plant growth regulators in orchestrating plant immunity. Physiol Mol Plant Pathol. 133:102359.

Shankar S, Kansrajh C, Dinesh MG, Satyan RS, Kiruthika S, Tharanipriya A. 2014. Application of indigenous microbial consortia in bioremediation of oil-contaminated soils. Int J Environ Sci Technol. 11:367–376.

Sharma N, Yadav G, Varma A, Koul M, Mishra A. 2024. Heavy metal remediation from rhizospheric soil by using microbial consortium. In: Gupta J, Verma A, editors. Microbiology-2.0 Update for a Sustainable. Springer; pp. 57–76.

Sharma R, Jasrotia T, Kumar Rajeev, Kumar Raman, Umar A, Alharthi FA, Alghamdi AA, Al-Zaqri N. 2021. An insight into the mechanism of 'symbiotic-bioremoval' of heavy metal ions from synthetic and industrial samples using bacterial consortium. Environ Technol Innov. 21:101302.

Sharma V, Bhise K, Bhagwat P, Pillai S, Ranveer R. 2023. Waste-based bioethanol production by using food, fruit, and grain wastes. In: Value-Addition Agri-food Ind Waste Through Enzym Technol. [place unknown]: Elsevier; pp. 281–298.

Sharon G, Cruz NJ, Kang D-W, Gandal MJ, Wang B, Kim Y-M, Zink EM, Casey CP, Taylor BC, Lane CJ, et al. 2019. Human gut microbiota from autism spectrum disorder promote behavioral symptoms in mice. Cell. 177(6):1600–1618.e17.

Shi Y, Yang Z, Xing L, Zhang X, Li X, Zhang D. 2021. Recent advances in the biodegradation of azo dyes. World J Microbiol Biotechnol. 37:1–18.

Shomali A, Das S, Arif N, Sarraf M, Zahra N, Yadav V, Aliniaeifard S, Chauhan DK, Hasanuzzaman M. 2022. Diverse physiological roles of flavonoids in plant environmental stress responses and tolerance. Plants. 11(22):3158.

Shong J, Diaz MRJ, Collins CH. 2012. Towards synthetic microbial consortia for bioprocessing. Curr Opin Biotechnol. 23(5):798–802.

Shou W, Ram S, Vilar JMG. 2007. Synthetic cooperation in engineered yeast populations. Proc Natl Acad Sci. 104(6):1877–1882.

Shrivastava P, Kumar R. 2015. Soil salinity: A serious environmental issue and plant growth promoting bacteria as one of the tools for its alleviation. Saudi J Biol Sci. 22(2):123–131.

Siddiqui JA, Khan MM, Bamisile BS, Hafeez M, Qasim M, Rasheed MT, Rasheed MA, Ahmad S, Shahid MI, Xu Y. 2022. Role of insect gut microbiota in pesticide degradation: A review. Front Microbiol. 13:870462.

Singh A, Yadav VK, Chundawat RS, Soltane R, Awwad NS, Ibrahium HA, Yadav KK, Vicas SI. 2023. Enhancing plant growth promoting rhizobacterial activities through consortium exposure: A review. Front Bioeng Biotechnol. 11:1099999.

Singh BK, Trivedi P, Egidi E, Macdonald CA, Delgado-Baquerizo M. 2020. Crop microbiome and sustainable agriculture. Nat Rev Microbiol. 18(11):601–602.

Singh N, Saha P, Rajkumar K, Abraham J. 2014. Biosynthesis of silver and selenium nanoparticles by *Bacillus* sp. JAPSK2 and evaluation of antimicrobial activity. Der Pharm Lett. 6(1):175–181.

Sinha R, Abnet CC, White O, Knight R, Huttenhower C. 2015. The microbiome quality control project: Baseline study design and future directions. Genome Biol. 16(1):276.

Skariyachan S, Patil AA, Shankar A, Manjunath M, Bachappanavar N, Kiran S. 2018. Enhanced polymer degradation of polyethylene and polypropylene by novel thermophilic consortia of *Brevibacillus* sps. and *Aneurinibacillus* sp. screened from waste management landfills and sewage treatment plants. Polym Degrad Stable 149:52–68.

Song J-M, Guan Z, Hu J, Guo C, Yang Z, Wang S, Liu D, Wang B, Lu S, Zhou R, et al. 2020. Eight high-quality genomes reveal pan-genome architecture and ecotype differentiation of Brassica napus. Nat Plants. 6(1):34–45.

Spescha A, Zwyssig M, Hess Hermida M, Moix A, Bruno P, Enkerli J, Campos-Herrera R, Grabenweger G, Maurhofer M. 2023. When competitors join forces: Consortia of entomopathogenic microorganisms increase killing speed and mortality in leaf- and root-feeding insect hosts. Microb Ecol [Internet]. 86(3):1947–1960.

Spini G, Spina F, Poli A, Blieux A-L, Regnier T, Gramellini C, Varese GC, Puglisi E. 2018. Molecular and microbiological insights on the enrichment procedures for the isolation of petroleum degrading bacteria and fungi. Front Microbiol. 9:2543.

Sterflinger K, Piñar G. 2021. Molecular-based techniques for the study of microbial communities in artworks. In: Microorganisms in the Deterioration and Preservation of Cultural Heritage. Cham: Springer; pp. 59–77.

Styer A, Pettinga D, Caddell D, Coleman-Derr D. 2024. Improving rice drought tolerance through host-mediated microbiome selection. *eLife*. 13:RP97015.

Sun Z, Koffel T, Stump SM, Grimaud GM, Klausmeier CA. 2019. Microbial cross-feeding promotes multiple stable states and species coexistence, but also susceptibility to cheaters. J Theor Biol. 465:63–77.

Sutton TDS, Clooney AG, Ryan FJ, Ross RP, Hill C. 2019. Choice of assembly software has a critical impact on virome characterisation. Microbiome. 7(1):12.

Swathy K, Vivekanandhan P, Yuvaraj A, Sarayut P, Kim JS, Krutmuang P. 2024. Biodegradation of pesticide in agricultural soil employing entomopathogenic fungi: Current state of the art and future perspectives. Heliyon. 10(1):e23406.

Tadesse Mawcha K, Malinga L, Muir D, Ge J, Ndolo D. 2024. Recent advances in biopesticide research and development with a focus on microbials. F1000Research. 13:1071.

Talukdar D, Jasrotia T, Sharma R, Jaglan S, Kumar Rajeev, Vats R, Kumar Raman, Mahnashi MH, Umar A. 2020. Evaluation of novel indigenous fungal consortium for enhanced bioremediation of heavy metals from contaminated sites. Environ Technol Innov. 20:101050.

Tamames J, Cobo-Simón M, Puente-Sánchez F. 2019. Assessing the performance of different approaches for functional and taxonomic annotation of metagenomes. BMC Genomics. 20(1):960.

Tettelin H, Masignani V, Cieslewicz MJ, Donati C, Medini D, Ward NL, Angiuoli SV, Crabtree J, Jones AL, Durkin AS, et al. 2005. Genome analysis of multiple pathogenic isolates of *Streptococcus agalactiae*: Implications for the microbial "pan-genome." Proc Natl Acad Sci. 102(39):13950–13955.

Thies S, Rausch SC, Kovacic F, Schmidt-Thaler A, Wilhelm S, Rosenau F, Daniel R, Streit W, Pietruszka J, Jaeger K-E. 2016. Metagenomic discovery of novel enzymes and biosurfactants in a slaughterhouse biofilm microbial community. Sci Rep. 6(1):27035.

Timofeeva AM, Galyamova MR, Sedykh SE. 2023. Plant growth-promoting bacteria of soil: Designing of consortia beneficial for crop production. Microorganisms. 11(12).

Tipu MMH, Sherif SM. 2024. Ethylene and its crosstalk with hormonal pathways in fruit ripening: Mechanisms, modulation, and commercial exploitation. Front Plant Sci. 15:1–19.

Toth CRA, Molenda O, Nesbø C, Luo F, Devine C, Guo S, Chen X, Edwards EA. 2023. Metagenomic and genomic sequences from a methanogenic benzene-degrading consortium. Microbiol Resour Announc. 12(5):e0134222.

Tsoi R, Wu F, Zhang C, Bewick S, Karig D, You L. 2018. Metabolic division of labor in microbial systems. Proc Natl Acad Sci. 115(10):2526–2531.

Upadhayay VK, Upadhayay K. 2023. Microbial consortium as an important bioproduct for sustainable agriculture. Vigyan Varta an International E-Magazine for Science Enthusiasts 4:5–8.

Vacca DJ, Bleam WF, Hickey W. 2005. Isolation of soil bacteria adapted to degrade humic acid-sorbed phenanthrene. Appl Environ Microbiol. 71(7):3797–3805.

Verma H, Kant C, Singh SK, White JF, Kumar A, Droby S. 2022. Biocontrol potential of microbial consortia: Approaches in food security and disease management. In: Kumar A, editor. Microbial Biocontrol: Sustainable Agriculture and Phytopathogen Management. Cham: Springer; pp. 187–203.

Verma S, Kuila A. 2019. Bioremediation of heavy metals by microbial process. Environ Technol Innov. 14:100369.

Vermelho AB, Moreira JV, Akamine IT, Cardoso VS, Mansoldo FRP. 2024. Agricultural pest management: The role of microorganisms in biopesticides and soil bioremediation. Plants (Basel, Switzerland). 13(19):2762.

Vessey JK. 2003. Plant growth promoting rhizobacteria as biofertilizers. Plant Soil [Internet]. 255(2):571–586.

Villaverde J, Rubio-Bellido M, Lara-Moreno A, Merchan F, Morillo E. 2018. Combined use of microbial consortia isolated from different agricultural soils and cyclodextrin as a bioremediation technique for herbicide contaminated soils. Chemosphere. 193:118–125.

Voigt AY, Costea PI, Kultima JR, Li SS, Zeller G, Sunagawa S, Bork P. 2015. Temporal and technical variability of human gut metagenomes. Genome Biol. 16(1):73.

von Mering C, Hugenholtz P, Raes J, Tringe SG, Doerks T, Jensen LJ, Ward N, Bork P. 2007. Quantitative phylogenetic assessment of microbial communities in diverse environments. Science (80-). 315(5815):1126–1130.

Wahab A, Muhammad M, Munir A, Abdi G, Zaman W, Ayaz A, Khizar C, Reddy SPP. 2023. Role of arbuscular mycorrhizal fungi in regulating growth, enhancing productivity, and potentially influencing ecosystems under abiotic and biotic stresses. Plants. 12(17):3102.

Wang J, Yang W, Zhang S, Hu H, Yuan Y, Dong J, Chen L, Ma Y, Yang T, Zhou L, et al. 2023. A pangenome analysis pipeline provides insights into functional gene identification in rice. Genome Biol. 24(1):19.

Wang R, Zhou Y, Cao S, Wang Y, Zhang J, Deng H-W. 2016. Metagenomic profiling, interaction of genomics with metagenomics. In: Wang X, Baumgartner C, Shields D, Deng HW, Beckmann J, editors. Application of Clinical Bioinformatics. Translational Bioinformatics, vol 11. Dordrecht: Springer; pp. 241–267.

Wang Y, Pei Y, Wang X, Dai X, Zhu M. 2024. Antimicrobial metabolites produced by the plant growth-promoting rhizobacteria (PGPR): *Bacillus* and *Pseudomonas*. Adv Agrochem [Internet]. 3(3):206–221.

Wang Z, Peng D, Fu C, Luo X, Guo S, Li L, Yin H. 2023. Pan-metagenome reveals the abiotic stress resistome of cigar tobacco phyllosphere microbiome. Front Plant Sci. 14:1248476.

Watanabe K, Manefield M, Lee M, Kouzuma A. 2009. Electron shuttles in biotechnology. Curr Opin Biotechnol. 20(6):633–641.

Wen Z, Ledesma-Amaro R, Lu M, Jiang Y, Gao S, Jin M, Yang S. 2020. Combined evolutionary engineering and genetic manipulation improve low pH tolerance and butanol production in a synthetic microbial Clostridium community. Biotechnol Bioeng. 117(7):2008–2022.

Williamson SJ, Yooseph S. 2012. From bacterial to microbial ecosystems (Metagenomics). In: van Helden J, Toussaint A, Thieffry D, editors. Bacterial Molecular Networks. Methods in Molecular Biology, vol 804. New York, NY: Springer; pp. 35–55.

Wu H, Cui H, Fu C, Li R, Qi F, Liu Z, Yang G, Xiao K, Qiao M. 2024. Unveiling the crucial role of soil microorganisms in carbon cycling: A review. Sci Total Environ. 909:168627.

Wu P, Wang Genyu, Wang Gehua, Børresen BT, Liu H, Zhang J. 2016. Butanol production under microaerobic conditions with a symbiotic system of *Clostridium acetobutylicum* and *Bacillus cereus*. Microb Cell Fact. 15:1–11.

Wu S, Zhong J, Lei Q, Song H, Chen S-F, Wahla AQ, Bhatt K, Chen S. 2023. New roles for *Bacillus thuringiensis* in the removal of environmental pollutants. Environ Res. 236:116699.

Wu W, Chen W, Liu S, Wu J, Zhu Y, Qin L, Zhu B. 2021. Beneficial relationships between endophytic bacteria and medicinal plants. Front Plant Sci. 12:646146.

Xie B-B, Li M, Anantharaman K, Ravin NV. 2021. Editorial: The uncultured microorganisms: Novel technologies and applications. Front Microbiol. 12:756287.

Xue S, Kui L, Sharifi R, Chen J. 2024. The role of microbiome in sustainable agriculture. Front Microbiol. 15:1388926.

Yan J, Wang L, Fu PP, Yu H. 2004. Photomutagenicity of 16 polycyclic aromatic hydrocarbons from the US EPA priority pollutant list. Mutat Res Toxicol Environ Mutagen. 557(1):99–108.

Yang R, Du X, Khojasteh M, Ali Shah SM, Peng Y, Zhu Z, Xu Z, Chen G. 2025. Green guardians: The biocontrol potential of Pseudomonas-derived metabolites for sustainable agriculture. Biol Control [Internet]. 201:105699.

Yasika Y, Shivakumar MS. 2025. A comprehensive account of functional role of insect gut microbiome in insect orders. J Nat Pestic Res. 11:100110.

Yen S, Johnson JS. 2021. Metagenomics: A path to understanding the gut microbiome. Mamm Genome. 32(4):282–296.

Yuan X, Zhang X, Chen X, Kong D, Liu X, Shen S. 2018. Synergistic degradation of crude oil by indigenous bacterial consortium and exogenous fungus *Scedosporium boydii*. Bioresour Technol. 264:190–197.

Yue Y, Huang H, Qi Z, Dou H-M, Liu X-Y, Han T-F, Chen Y, Song X-J, Zhang Y-H, Tu J. 2020. Evaluating metagenomics tools for genome binning with real metagenomic datasets and CAMI datasets. BMC Bioinformatics. 21(1):334.

Zafra G, Taylor TD, Absalón AE, Cortés-Espinosa DV. 2016. Comparative metagenomic analysis of PAH degradation in soil by a mixed microbial consortium. J Hazard Mater. 318:702–710.

Zango UU, Ahluwalia SS, Sharma AK. 2018. Microbial consortium of *Aspergillus fumigatus*, *Aspergillus terreus* and *Paenibacillus dendritiformis* in the bioremoval of cadmium. Int J Pharm Res. 10(3):230–238.

Zhang Y, Wang E, Feng B, Xu L, Xue Y, Chen Y. 2025. Siderophore production capability of Nitrogen-Fixing Bacterium (NFB) GXGL-4A regulates cucumber rhizosphere soil microecology. Microorganisms. 13(2):346.

Zhang Y, Yang Z, He Y, Liu D, Liu Y, Liang C, Xie M, Jia Y, Ke Q, Zhou Y, et al. 2024. Structural variation reshapes population gene expression and trait variation in 2,105 Brassica napus accessions. Nat Genet. 56(11):2538–2550.

Zhou J, He Q, Hemme CL, Mukhopadhyay A, Hillesland K, Zhou A, He Z, Van Nostrand JD, Hazen TC, Stahl DA. 2011. How sulphate-reducing microorganisms cope with stress: Lessons from systems biology. Nat Rev Microbiol. 9(6):452–466.

Zhou J, He Z, Yang Y, Deng Y, Tringe SG, Alvarez-Cohen L. 2015. High-throughput metagenomic technologies for complex microbial community analysis: Open and closed formats. MBio. 6(1). https://doi.org/10.1128/mbio.02288-14.

Zhu N, Yang J, Ji L, Liu J, Yang Y, Yuan H. 2016. Metagenomic and metaproteomic analyses of a corn stover-adapted microbial consortium EMSD5 reveal its taxonomic and enzymatic basis for degrading lignocellulose. Bolubiol Biofuels. 9:1–23.

8 Metagenomics in Prospecting for Genes of Agricultural Importance

*Soumya Chatterjee[1,2], Anasuya Karjee[1], Partha Barman[1], Chandan Barman[2], Sukanta Majumdar[3], and Ranadhir Chakraborty[1]**

[1]OMICS Laboratory, Department of Biotechnology, University of North Bengal, Siliguri 734013, West Bengal, India

[2]Reproductive Ecology of Angiosperms Laboratory, Department of Botany, University of Gour Banga, P.O. Mokdumpur, Malda, 732103, West Bengal, India

[3]Microbiology and Microbial Biotechnology Laboratory, Department of Botany, University of Gour Banga, P.O. Mokdumpur, Malda, 732103, West Bengal, India

8.1 INTRODUCTION

Metagenomics, encompassing the genomics of environment and communities, is recognised as an essential requirement in microbial analysis. It enables the examination of community population dynamics with respect to heterogeneity and complexity over time. The objective of any metagenomic study is to thoroughly characterize a microbial community by addressing the questions: "Who is present?" and "What are their functions?" This approach helps elucidate the microbiome composition, functional roles of each microbial constituent, and intra-species variability data (Scholz et al., 2012).

Advances in sequencing and computational tools have revolutionized metagenomic analysis. The reduced cost of next-generation sequencing (NGS), coupled with increased analytical applications, has significantly transformed microbiology. Microbial community characterization now employs diverse sequencing technologies, including Sanger, ABI SOLiD, pyrosequencing, Illumina, and advanced assembly approaches. Metagenomics involves extracting and analyzing DNA from a wide variety of microbes. Extensive research has shown that microbes constitute a substantial population present in nearly all environments on Earth, particularly due to the use of 16S rRNA sequencing, which facilitates the identification of numerous novel microbial lineages (Iquebal et al., 2022). While 16S rRNA transformed microbial community identification, metagenomics has enabled in-depth exploration of environmental microorganisms and their ecological and physiological roles.

Metagenomics also plays a critical role in understanding the rhizosphere, the zone of soil directly influenced by plant roots. The rhizospheric microbiome is essential for plant health as it supports nutrient uptake, pathogen defense, and plant growth promotion through various mechanisms (Mendes et al., 2013). Metagenomic investigations of the rhizosphere have identified specific microbial taxa and functions that are linked to enhancing plant growth, disease suppression, and nutrient acquisition. By uncovering the genetic basis of these functions, metagenomics provides insights that can be harnessed to develop microbial inoculants or to breed crops that better utilize their associated microbiomes. This chapter highlights the wide-ranging applications of metagenomics in agriculture, including livestock health, crop productivity, and responses to environmental factors such as abiotic stress.

8.2 METAGENOMIC STRATEGIES FOR GENE PROSPECTING

8.2.1 Approaches for Metagenomic Gene Discovery

Metagenomics has brought about a groundbreaking change in gene discovery by enabling the finding of new genes and enzymes from a wide range of microorganisms that are difficult or impossible to culture in the laboratory setup. Gene prospecting refers to the search for genes with useful properties, such as those that encode enzymes for industrial processes, pharmaceuticals, or agricultural applications. Traditional gene discovery methods rely on the cultivation of microorganisms, which severely limits microbial exploration, as it is estimated that fewer than 1% of microbes can be cultured using conventional methods (Amann et al., 1995). Metagenomic approaches, by contrast, allow researchers to gain direct access to the genetic content of complete microbial populations, opening up new avenues for gene discovery across diverse environments, including soils, oceans, and extreme habitats (Riesenfeld et al., 2004).

There are two primary approaches to metagenomic gene discovery: functional metagenomics and sequence-based metagenomics. Each approach offers distinct advantages and challenges, and the choice between them depends on the specific goals of the study, the nature of the microbial community being studied, and the resources available.

***Corresponding author:** rcnbusiliguri@gmail.com (Ranadhir Chakraborty)

DOI: 10.1201/9781003544425-8

8.2.1.1 Functional Metagenomics

Functional metagenomics is a strategy that involves the direct screening of metagenomic libraries for specific activities or phenotypes of interest. In this approach, DNA is extracted from an environmental sample and fragmented into smaller pieces, which are then cloned into suitable vectors and introduced into a host organism, typically *Escherichia coli* or another model bacterium. The generated metagenomic library is then evaluated for the presence of targeted characteristics, such as the synthesis of enzymes that degrade specific substrates, confer antibiotic resistance, or produce bioactive compounds (Handelsman, 2004). Functional metagenomics is particularly powerful for discovering novel enzymes and other gene products that may not be predicted based on sequence homology alone.

One of the main advantages of functional metagenomics is its ability to discover genes based on actual function rather than sequence similarity to known genes. This is especially important for identifying novel genes with no known homologs, as well as for discovering enzymes with unique properties, such as stability under extreme conditions or activity on unusual substrates (Aminov, 2011). The application of functional metagenomics has led to the successful discovery of new antibiotics, enzymes for biomass degradation, and other valuable biocatalysts (Nwachukwu & Babalola, 2022).

However, functional metagenomics also has limitations. The expression of metagenomic DNA in a heterologous host can be challenging, particularly when the genes of interest require specific cofactors, chaperones, or other regulatory elements that are not present in the host organism. Additionally, the screening process can be labor-intensive and may miss genes that are expressed at low levels or that require specific environmental conditions for activity (Simon & Daniel, 2011). Despite these challenges, functional metagenomics remains a powerful tool for gene discovery, especially when integrated with large-scale screening methods and advanced bioinformatics tools.

8.2.1.2 Sequence-Based Metagenomics

Sequence-based metagenomics, also known as shotgun metagenomics, entails the immediate sequencing of genetic material obtained from environmental sources without requiring prior cultivation or cloning. This approach produces large volumes of sequence data that can be assembled into contigs and analyzed to identify genes, pathways, and other functional elements within the metagenome (Quince et al., 2017). Sequence-based metagenomics is especially useful for examining genetic variation and metabolic capacity in microbial communities, as well as for discovering novel genes based on sequence similarity to known genes.

In sequence-based metagenomics, genomic DNA is isolated from an environmental source and subjected to high-throughput sequencing using different platforms, such as Illumina, PacBio, or Oxford Nanopore. The resulting reads are assembled into contigs, which can then be annotated using bioinformatics tools to identify coding regions, predict protein functions, and assign taxonomic classifications (Meyer et al., 2008). Sequence-based metagenomics allows for the comprehensive analysis of microbial communities, providing insights into the abundance, diversity, and functional potential of different taxa.

A major benefit of sequence-based metagenomics is its ability to generate a complete picture of the genetic landscape within a microbial community, including both known and unknown genes. This approach is particularly valuable for exploring environments with high microbial diversity or for studying microbial communities that are poorly understood. Moreover, sequence-based metagenomics can be used to reconstruct metagenome-assembled genomes (MAGs) from individual species within a community, providing detailed information about the metabolic pathways and ecological roles of specific microbes (Parks et al., 2015).

However, sequence-based metagenomics also has its limitations. The vast amount of information obtained by shotgun sequencing requires substantial computational resources for assembly, annotation, and analysis. Additionally, the identification of novel genes often relies on sequence homology to known genes, which can limit the discovery of truly novel functions. Despite these challenges, sequence-based metagenomics is a powerful tool for gene prospecting, particularly when combined with advanced bioinformatics pipelines and functional validation experiments.

8.2.2 Bioinformatics Pipelines for Gene Annotation and Function Prediction

The successful application of metagenomic strategies for gene prospecting relies heavily on bioinformatics tools and pipelines for data analysis, gene annotation, and function prediction. These pipelines are essential for processing the vast amounts of sequence data generated by metagenomic studies, identifying genes of interest, and predicting their functions based on sequence similarity and structural features.

8.2.2.1 Data Preprocessing and Quality Control

The first step in a metagenomic analysis is data preprocessing, which involves filtering out low-quality reads, removing contaminants, and trimming adapter sequences. Tools such as FastQC and Trimmomatic are commonly used for these tasks (Andrews, 2010; Bolger et al., 2014). High-quality reads are essential for accurate assembly and annotation, as errors or contaminants can lead to incorrect predictions of gene function.

8.2.2.2 Sequence Assembly

Following preprocessing, the high-quality reads are assembled into contigs using tools such as SPAdes or MEGAHIT (Bankevich et al., 2012; Li et al., 2015). Assembly is particularly important in shotgun metagenomics, as it allows for the reconstruction of longer sequences that can be analyzed for gene content. In some cases, de novo assembly is performed, while in others, reference-based assembly is used to align reads to known genomes.

8.2.2.3 Gene Annotation

Once contigs are assembled, the next step is gene annotation, which involves identifying coding regions and predicting the functions of the encoded proteins. Annotation tools, namely Prokka (rapid prokaryotic genome annotation) and Rapid Annotation using Subsystem Technology (RAST) (Aziz et al., 2008; Seemann, 2014) are commonly used for this purpose. These tools compare metagenomic sequences to databases of known genes, such as the Kyoto Encyclopedia of Genes and Genomes (KEGG) database or the NCBI RefSeq database, to predict gene functions.

8.2.2.4 Functional Prediction

Beyond sequence-based annotation, functional prediction tools are used to infer the biochemical activities and ecological roles of the identified genes. For example, the tools Pfam and InterProScan (Finn et al., 2016; Jones et al., 2014) use protein domain analysis to forecast the gene products' functions. Additionally, metabolic pathway reconstruction tools such as MetaCyc (Caspi et al., 2014) and KEGG Mapper (Kanehisa et al., 2016) allow researchers to place identified genes within broader metabolic networks, offering an understanding of their ecological function and possible applications.

8.2.2.5 Comparative Analysis and Data Visualization

Comparative metagenomics involves the analysis of multiple metagenomes to identify differences in gene content, abundance, and functional potential across different environments or treatment conditions. Tools such as Anvi'o and MG-RAST provide comprehensive platforms for comparative analysis and data visualization, allowing researchers to explore the relationships between microbial communities, gene functions, and ecological conditions (Eren et al., 2015; Meyer et al., 2008).

8.3 AGRICULTURAL IMPORTANCE OF SOIL MICROBIOMES

8.3.1 Soil Microbiome Composition and Diversity

The assembly of microbial populations within the soil and the factors influencing their activity at relevant spatial scales (i.e., soil composition and physical properties) remain poorly understood (Wilpiszeski et al., 2019). Soils consist of mineral particles and biological materials compacted into aggregates of varying sizes, arranged in three-dimensional porous frameworks.

The main components of soil are mineral particles categorized by size into clay particles (less than 2 μm), silt particles (approx. 2–63 μm), and sand particles (approx. 63–2,000 μm) (Blott & Pye, 2012). Soil particles combine with carbonates, biotic components, metal oxides, organic matter, and organic macromolecules (such as mucilage, lipids, polysaccharides, extracellular polymeric substances, and proteins) to create organo-mineral complexes, resulting in silt-sized aggregates (<50 μm) or microaggregates (50–250 μm) (Totsche et al., 2018); the forces binding particles within aggregates exceed those between neighboring particles (Martin et al., 1955). Microaggregates and silt-sized aggregates are organized into larger, more fragile macroaggregates (typically >250 μm), which are mostly kept together through fungal hyphae (Chotte, 2005) and root complexes. Macroaggregates initially form and serve as a nucleus for the development of microaggregates within their centre (Oades, 1984; Six et al., 2000). Decomposed plant and fungal material, in conjunction with organic particles from roots, gradually become enveloped by clay particles and microbial byproducts, leading to the formation of the core of durable microaggregates and silt-sized clumps (Six et al., 2004). When organic material in the center is exhausted, microbial processes and metabolite synthesis come to a halt, leading to a decline in aggregate cohesion and the eventual disintegration of the aggregate.

Microbial diversity in soil is typically characterized by alpha diversity (the diversity within an individual sample) and beta diversity (the diversity among different samples). High microbial diversity is generally associated with greater soil health and resilience, as a diverse community can perform a wide range of ecological functions and adapt to environmental changes. For example, diverse microbial communities are better equipped to degrade a variety of organic substrates, fix nitrogen, and suppress soil-borne pathogens (Mendes et al., 2013). The spatial and temporal distribution of soil microbes is also highly heterogeneous, with different microbial taxa occupying distinct ecological niches within the soil matrix (Table 8.1). Soil microbiomes are composed of a vast array of taxa, with bacterial communities being particularly abundant. Dominant bacterial phyla typically include Proteobacteria, Actinobacteria, Acidobacteria, and Firmicutes, while fungi are represented primarily by Ascomycota and Basidiomycota (Fierer, 2017). This heterogeneity is influenced by soil structure, nutrient availability, and the presence of plant roots, which create microhabitats with varying physical and chemical conditions (Nannipieri et al., 2003).

Consequently, soil aggregates' pore spaces constitute a critical microhabitat for soil microorganisms (Table 8.1). The majority of soil microbes exist as interconnected communities linked to these particles, so making soil structure a determinant of their resources via water movement, oxygen diffusion, nutrient availability, and organic matter accessibility (Schlüter et al., 2022; Wilpiszeski et al., 2019). Therefore, comprehending the impact of agricultural management on the relationship between soil composition and microbial communities is crucial for predicting alterations in soil microbial populations and their ecosystem services across various agricultural management practices.

TABLE 8.1
Connections between the Microbiome and Soil Structure

Soil Composition	Characteristics	Microbial Diversity and Function of Soil	References
Particle size fractions	1. Clay particle, silt particle, sand particle. 2. Differ in chemical composition, organic matter decomposability, cation exchange capacity, surface reactivity, and sorption characteristics.	1. Smaller fractions (e.g., clay) host larger (~10×) and more diverse (~48–52%) bacterial and archaeal taxa. 2. Microenvironment fosters changes in enzymatic activity and nutrient conversion (carbon, nitrogen, phosphorus).	Acosta et al. (2011), Briar et al. (2011), Christensen (1987), Christensen and Olesen (1998), Hemkemeyer et al. (2015, 2018, 2019), Luo et al. (2017), Sessitsch et al. (2001), and Zhang et al. (2016)
Soil aggregates	1. Microaggregates (50–250 μm) are stable and contain recalcitrant carbon. 2. Macroaggregates (>250 μm) are less stable, contain labile carbon.	1. Approx. 90% of soil bacteria associate with aggregates. 2. Microaggregates promote slow-growing oligotrophs. 3. Macroaggregates support fast-growing copiotrophs adapted to labile carbon.	Mummey et al. (2006), Ranjard et al. (2000), Raynaud and Nunan (2014), Rillig et al. (2017), and Trivedi et al. (2017)
Soil pores	1. Macropores (>75 μm): Habitat for microarthropods. 2. Pores between aggregates: Protists, nematodes, fungi. 3. Pores within microaggregates: Bacteria and archaea. 4. Sharp oxygen gradients.	1. Aerobic bacteria in larger pores; anaerobic niches in microaggregate pores. 2. Microaggregate pores serve as hotspots for carbon stabilization and N_2O emissions. 3. Anaerobic microbes like denitrifiers thrive.	Bhattacharyya et al. (2022), Borer et al. (2018), Kong et al. (2010), Oades (1984), Sessitsch et al. (2001), Six et al. (2004), and Zhang et al. (2021)
Surface reactivity and sorption	1. Variations depend on mineralogy. 2. Determines microbial survival and enzymatic activity	1. Influences carbon (C), nitrogen (N), and phosphorus (P) transformation. 2. Affects pollutant degradation.	Acosta et al. (2011), Hemkemeyer et al. (2015), Sessitsch et al. (2001), and Zhang et al. (2016)
Degraded soil structure	1. Collapse of aggregates and pore networks reduces water and oxygen infiltration. 2. Favors anaerobic conditions.	1. Reduced microbial diversity and activity. 2. Promotes anaerobic processes leading to CH_4 and N_2O production. 3. Negatively impacts nutrient cycling, plant growth, and pollutant degradation.	Bhattacharyya et al. (2022), Kong et al. (2010), Sessitsch et al. (2001), and Zhang et al. (2021)

8.3.2 The Influence of Microbial Communities in the Phyllosphere and Rhizosphere on Plant Growth Promotion

The aerial leaf surfaces, or phyllosphere, host a community of anoxygenic phototrophic bacteria that exhibit various metabolic characteristics (Atamna-Ismaeel et al., 2012; Bakshi et al., 2020) (Figure 8.1). The occurrence of bacteria on the surface of leaves affects the production of compound secretions or surface-active agents from leaves, improving water retention and moisture levels in the leaves by regulating the stomatal aperture (Hardoim et al., 2008; Mohanty et al., 2016). The toxins produced by these bacteria alter the leaf cuticle, leading to modifications in the ion movement across the cellular membranes (Schreiber et al., 2005). The alliance of *Rumohra adiantiformis* as an endophyte on the leaf surfaces of leather ferns results in leaf abnormalities (Kloepper et al., 2013). Bacterial strains obtained from the leaf surface of various crops, including potato, mungbean, cotton, wheat, and pearl millet, were analyzed for a range of features that promote plant growth, including secretion of ammonia, auxin production, solubilization of phosphate, and assimilation of nitrate. The utilization of these isolated strains by foliar spray promotes the growth of plants, particularly regarding height and productivity, in potatoes (Kumar et al., 2018a). The interactions between plants and rhizospheric bacteria are essential for developing eco-friendly farming practices and the generation of advantageous bioactive chemicals or biofertilizers (Figure 8.1). Plant roots release exudates high in organic carbon or nitrogen molecules into the soil. These secretions promote interactions between plants and microbes, perhaps providing nutrients to the plants, hence improving development and resistance against diseases (Huang et al., 2014). The presence of bacteria in plant roots can enhance the overall health and vigor of flora and is designated as Plant Growth Promoting Rhizobacteria (PGPR) (Ahemad & Kibret, 2014; Glick, 2014). Various energy conversion systems enable bacteria to compete with others and form associations with beneficial partners (Gupta et al., 2018). The alliance between microbes of soil and the roots of plants is necessary for carbon sequestration and recycling of nutrients (Abhilash et al., 2012). For instance, rhizobacteria in the Indo-Gangetic region facilitate the solubilization of tricalcium phosphates through mineral solubilization processes in the soil (Bahadur et al., 2016). In addition to serving as bioindicators, the diversity of soil microbes in the rhizosphere influences plant development, disease responses, soil fertility, and soil structure (Bramhachari et al., 2017). Microorganisms linked to

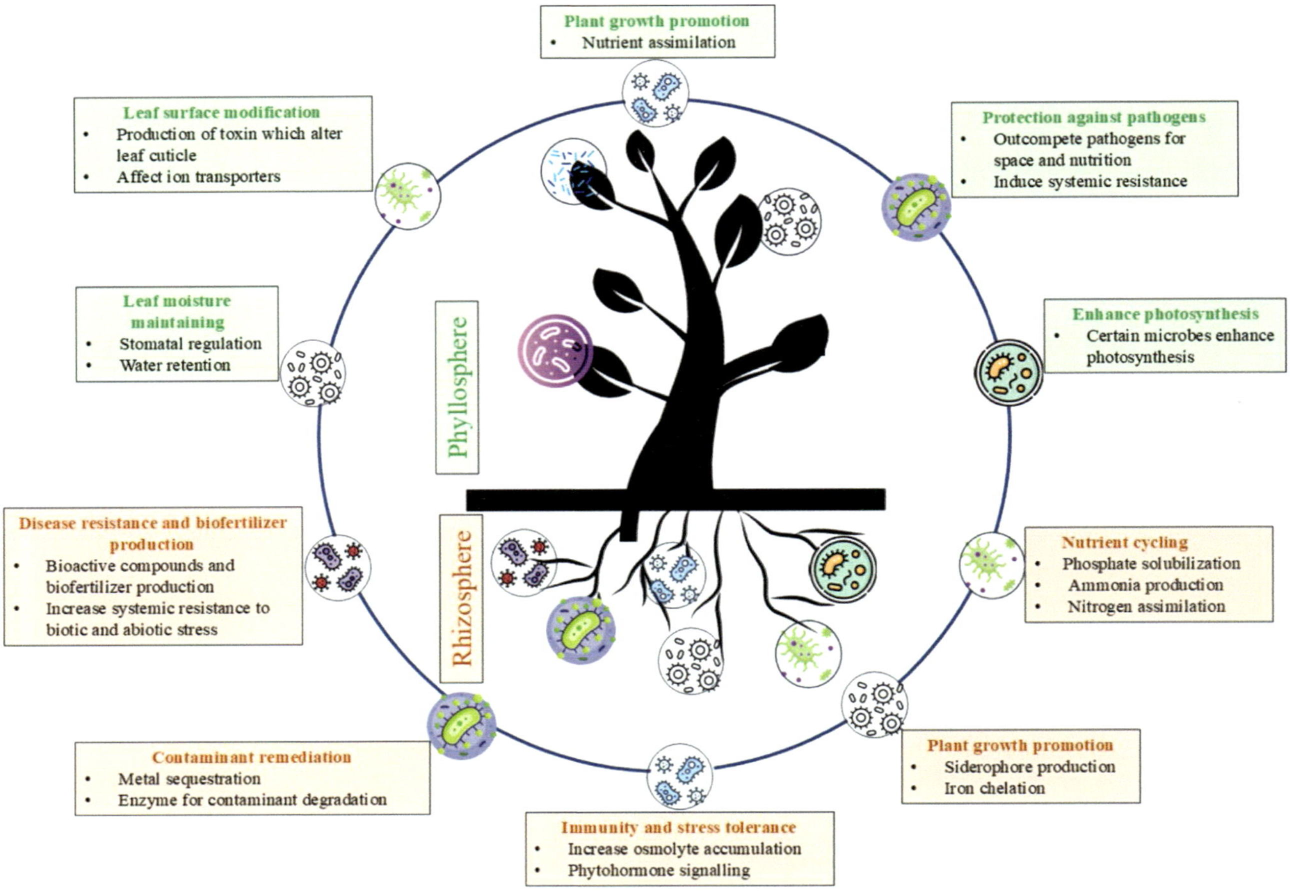

FIGURE 8.1 A schematic representation of the multifaceted roles of microbial communities in the phyllosphere and rhizosphere in promoting plant growth and sustainable agriculture. In the phyllosphere, microbes contribute to leaf surface modification, moisture maintenance, pathogen protection, nutrient assimilation, and enhanced photosynthesis. In the rhizosphere, microbes facilitate nutrient cycling (phosphate solubilization, nitrogen assimilation, ammonia production), promote plant growth (via siderophore production and iron chelation), confer immunity and stress tolerance through osmolyte accumulation and phytohormone signaling, aid in disease resistance and biofertilizer production, and participate in contaminant remediation. These interactions, revealed and better understood through metagenomic technologies, highlight the critical importance of unculturable plant growth promoting rhizobacteria (PGPR) and other microbial taxa in supporting sustainable and climate-resilient agriculture.

plant roots can affect significant plant characteristics, including blooming time, overall yield, and resistance to biotic as well as abiotic stresses (Marasco et al., 2012; Panke-Buisse et al., 2015; Van Wees et al., 2008). Metagenomics, or the application of molecular techniques, is an innovative methodology for evaluating microbial populations, encompassing individual cells to entire microbial communities inside the rhizosphere (Bloem et al., 2005; Sørensen et al., 2009). Recent developments in metagenomics, encompassing comprehensive genome sequencing and computational analysis, have enabled the discovery of unique microorganisms (Hoff et al., 2008) and have optimized the detection of genetic elements and biomolecules that demonstrate PGPR properties in crops. Through functional metagenomics, the PGPR populations as endophytes have been found to be important in the roots of rice plants (Sessitsch et al., 2012) and identified the existence of bacteria responsible for the oxidation of ammonia and/or archaea in soil (Schauss et al., 2009). Certain species of bacteria, such as *Rhizobium*, *Azospirillum*, *Frankia*, *Agrobacterium*, *Thiobacillus*, *Burkholderia*, *Bacillus*, and *Pseudomonas,* are identified as PGPRs having the properties related to plant growth promotion and other biocontrol activities (Ramesh et al., 2009; Vessey, 2003). Bacteria in the rhizosphere possess evolved features that assist plants in coping with both biological and environmental stressors, perhaps enhancing plant immunity by triggering resistance mechanisms throughout the plant (Yang et al., 2009). PGPR enhances the production of osmolytes and the regulation of phytohormones in response to environmental stress. Transgenic *Arabidopsis thaliana* consisting of the *Bacillus subtilis proBA* gene shows enhanced synthesis of proline, leading to greater salt tolerance (Chen et al., 2007). Usage of both traditional methods and hydroponic setups in inoculating rice with the SN23 isolate and maize with the SQR9 isolate of *B. amyloliquefaciens* improves salt stress tolerance, while concurrently activating the genes linked to osmotic and ionic stresses (Chen et al., 2016; Nautiyal et al., 2013).

Agricultural practices, such as tillage, significantly bring structural changes and influence the diversity within the soil microbiome (Carbonetto et al., 2014). Treating wheat seedlings with rhizosphere-associated bacteria enhanced drought stress tolerance and increased overall growth (Timmusk et al., 2014). Association of *B. cereus* AR156, *B. subtilis* SM21, and *Serratia* sp. XY21 exhibited protective effects and resistance to drought in cucumber plants (Wang et al., 2012). *Trichoderma harzianum* induces an antioxidant defense system in tomato seedlings, promoting tolerance to drought through improved scavenging of reactive oxygen species (ROS) and increased ascorbate and glutathione oxidation (Mastouri et al., 2012). PGPR also mitigated salt stress in tomato plants and helped in phosphate and potassium solubilization (Nath et al., 2017; Tank & Saraf, 2010). Rhizobacteria help in solubilizing ferric ions by efficiently synthesizing siderophores, a low molecular weight iron-chelating compound (Pahari & Mishra, 2017). Moreover, the exploration of the metagenomic study of the root microbiome in grass family crops, including sorghum, rice, and maize, is essential for developing methods to enhance yield attributes and stress resistance in plants without any modification at the gene level. The obtainment of microbiota possessing metal-binding or resistance genes, as well as genes or enzymes involved in the breakdown of contaminants, is potentially beneficial for the transformation and remediation of polluted soils with less toxicity.

8.3.3 Impact of Farming Practices on Soil Microbial Populations

Farming practices profoundly influence the soil microbiome, influencing its structure, variety, and functional roles. Methods such as tilling, crop rotation, fertilization, pesticide application, and irrigation can alter the organization and behavior of soil microbiomes, significantly affecting soil health and long-term agricultural sustainability.

8.3.3.1 Tillage

Tillage is a key agricultural practice affecting the soil microbiome. Conventional tillage disrupts the soil structure, breaking up microbial habitats and exposing organic matter to rapid decomposition. This disturbance often results in a reduction of microbial population and the elimination of advantageous microbes, particularly those involved in nutrient cycling and disease control (Francioli et al., 2016). No-till or reduced-tillage systems, on the other hand, have been shown to preserve soil structure, maintain higher microbial population, and promote the richness of advantageous microbes, like mycorrhizal fungi and nitrogen-fixing bacteria (Schmidt et al., 2018). These preservation tillage techniques also enhance soil organic matter content, which serves as a food source for microbes, further enhancing microbial activity and soil health.

8.3.3.2 Crop Rotation and Cover Cropping

Practices like crop rotation and cover cropping can positively influence soil microbial communities by increasing plant diversity and providing a continuous supply of organic matter for the soil. Diverse cropping systems create a variety of root secretions, which act as substrates for different microbial taxa, thereby enhancing microbial diversity and functional redundancy (Abdalla et al., 2019). For example, legumes in crop rotations can increase the abundance of nitrogen-fixing bacteria, while cover crops like clover and rye can promote beneficial mycorrhizal associations and improve soil structure (Bowles et al., 2017). These practices also help break the life cycles of soil-borne pathogens by disrupting their host plants, thereby reducing disease incidence.

8.3.3.3 Fertilization

The use of chemical fertilizers directly affects soil microbial communities, frequently resulting in changes to microbial composition and activity. While fertilizers can enhance crop yields by providing readily available nutrients, they can also disrupt natural nutrient cycles and reduce the richness of microbial communities (Jiang et al., 2019). For instance, the extensive application of nitrogen fertilizers can lead to a decline in the population of nitrogen-fixing bacteria and other microbes involved in natural nitrogen cycling, as these processes become less critical in the presence of abundant synthetic nitrogen. Additionally, chemical fertilizers can result in soil acidification and nutrient imbalances, further altering microbial communities and reducing soil health (Geisseler & Scow, 2014). In contrast, the use of organic inputs, including manure or compost, can enhance microbial diversity and activity through the gradual availability of nutrients and organic material that supports a wide range of microbial functions (Zhen et al., 2014).

8.3.3.4 Pesticide Application

Pesticides, including herbicides, insecticides, and fungicides, can significantly impact non-target soil microbes, leading to a reduction in the variety of microbial populations and disruptions in key ecological processes (Gomez et al., 2001). For instance, certain pesticides can inhibit the proliferation of beneficial microbes, like nitrogen-fixing bacteria as well as mycorrhizal fungi, thereby impairing nutrient cycling and plant–microbe interactions. Pesticides can also contribute to the selection of resistant microbial strains, reducing the effectiveness of biocontrol organisms and leading to the persistence of pathogenic microbes (Kalia & Gosal, 2011). Integrated pest management practices that minimize pesticide use and promote biological control are essential for maintaining healthy soil microbiomes and sustainable agricultural systems.

8.3.3.5 Irrigation

Irrigation practices can influence soil microbial communities by altering soil moisture levels, which are a critical determinant of microbial activity and community structure. Excessive irrigation can lead to waterlogging, creating anaerobic conditions that favor the growth of facultative and obligate anaerobes, such as denitrifying bacteria and methanogens (Maestre et al., 2015). These shifts can result in increased emissions of greenhouse gases and the loss of essential nutrients through leaching or volatilization. Conversely, water scarcity can limit microbial activity and reduce the efficiency of nutrient cycling, resulting in nutrient deficiencies and decreased crop yield (Hartmann et al., 2015). Optimizing irrigation practices to maintain proper soil hydration is essential for supporting a healthy and active soil microbiome.

8.4 KEY GENES OF AGRICULTURAL IMPORTANCE IDENTIFIED THROUGH METAGENOMICS

Metagenomic approaches have been increasingly applied to agriculture to identify genes capable of boosting crop yield, enhancing soil health, and reducing the need for chemical inputs. Genes with agricultural relevance include those involved in nutrient cycling, plant development, disease resistance, and stress resilience (Table 8.2). The screening and identification of such genes from metagenomic libraries can lead to the development of biofertilizers, biopesticides, and other bioproducts that support sustainable agriculture.

8.4.1 Nutrient Cycling

Soil metagenomes are rich sources of genes responsible for the cycling of vital nutrients, including sulfur, nitrogen,

TABLE 8.2
Agronomically Important Genes/Enzymes Obtained from Various Natural Sources

Genes/Enzymes	Function	Nature of Sample	Applications in Agriculture	Reference
Nitrogenase (Nif)	Nitrogen fixation	Agricultural soil	Biofertilizers to enhance soil nitrogen availability	Wolińska et al. (2017)
Polysaccharide-degrading enzymes	Cellulose and hemicellulose degradation	Gut of wood feeding higher termite	These enzymes are essential in numerous agricultural processes because of their capacity to decompose complex polysaccharides into simpler sugars and play a significant part in enhancing soil vigor, boosting crop production, and promoting environmental sustainability.	Hegyi et al. (2021)
Cellulase	Convert cellulose into simple sugars	Soil	Utilized in biofuel production through the transformation of plant biomass into fermentable sugars, it also improves soil fertility by decomposing organic material.	Prayogo et al. (2020)
Proteases	Catalyze the breakdown of proteins into amino acids and peptides	Mangrove sediment	Utilized in animal nutrition to enhance protein digestibility, it breaks down organic matter, contributing to soil fertility and nutrient cycling.	Prayogo et al. (2020)
Genes help in sulfur metabolism	Regulate sulfur uptake, assimilation, and conversion into essential compounds like amino acids.	Acidic peatland	Enhance sulfur utilization in crops, particularly in sulfur-deficient soils, and bolster crop resilience against environmental stresses, namely drought or pollution.	Hausmann et al. (2018)
PhdB (toluene producing enzyme)	Catalyzes the conversion of phenolic compounds into toluene and other derivatives.	Sediments of anoxic lake	Utilized in bioremediation processes to decompose environmental contaminants such as toluene, there is potential for genetically engineered plants to neutralize aromatic pollutants present in soil and water.	Beller et al. (2018)
IAA producing gene	Produces plant hormone IAA (auxin) which regulates growth and development.	Rhizobacteria from fly ash dumps	Utilized in substances that control plant growth to boost root development, foster plant development, and boost crop production.	Malhotra et al. (2017)
Lipase	Catalyzes the hydrolysis of fats and lipids into fatty acids and glycerol.	Soil	Utilized in the processing of vegetable oils and enhancing soil aeration through the decomposition of organic matter, with the potential to boost crop health by increasing nutrient availability from organic matter.	Prayogo et al. (2020)
Endoglucanase	Breaks down cellulose into smaller oligosaccharides.	Hot spring soil	Facilitates the breakdown of plant biomass and optimizes the cycling of soil organic matter.	Prayogo et al. (2020)
Chitinase	Degrades chitin.	Wetland environment	Employed in the realm of biological pest management through the targeting of fungal pathogens and insect pests.	Prayogo et al. (2020)
α-Amylase	Catalyzes the breakdown of starch into sugars.	Rumen of black goat	Utilized in agriculture to promote plant development by decomposing starch reserves in seeds, thereby enhancing germination.	Prayogo et al. (2020)
Genes for hydrocarbon degradation	Break down hydrocarbons, often from petroleum or plant-derived sources.	Samples from the cold costal environment at high latitudes	Utilized in bioremediation for the cleanup of oil spills and various environmental contaminants.	Espínola et al. (2018)

(Continued)

TABLE 8.2 ***(Continued)***
Agronomically Important Genes/Enzymes Obtained from Various Natural Sources

Genes/Enzymes	Function	Nature of Sample	Applications in Agriculture	Reference
Genes or enzymes associated with carbon, sulfur, nitrogen, and phosphorus cycle	Regulate the cycling of carbon, sulfur, nitrogen, and phosphorus within ecosystems.	Soil deficient in phosphorus and soil rich in phosphorus	Utilized to enhance soil health through the improvement of nutrient availability and the stimulation of microbial activity within the soil.	Yao et al. (2018)
Genes for secondary metabolite biosynthesis	Impart synthesis of secondary metabolites like alkaloids, terpenes, and phenolics.	Grassland soil of Northern California	Employed in the synthesis of important botanical substances for medicinal applications, pest management, and flavor enhancement.	Crits-Christoph et al. (2018)
Genes for salt resistance	Enable plants to tolerate high salinity by regulating ion balance and stress responses.	Rhizosphere with brines and moderate salinity	Essential for advancing the cultivation of salt-resistant plants in saline environments or coastal regions.	Mirete et al. (2015)
Genes for arsenic resistance	Aid in detoxifying and compartmentalizing arsenic in plant cells.	Drainage from an acid mine	Employed in the process of phytoremediation to eliminate arsenic contamination from soil and water.	Morgante et al. (2015)
Genes conferring resistance to nickel	Help plants tolerate and accumulate nickel in soils with high concentrations of this metal.	Rhizospheric soil	Valuable for the restoration of soils polluted with nickel.	Mirete et al. (2007)
Phosphate-solubilizing genes	Break down insoluble phosphates into forms accessible to plants.	Soil of Barley rhizosphere	Utilized to enhance phosphorus accessibility in soils lacking sufficient phosphate.	Chhabra et al. (2013)
Catechol 2,3-dioxygenases and phenol hydroxylases	Impart a breakdown of aromatic compounds, breaking down pollutants.	Wastewater treatment plant	Utilized in bioremediation to break down aromatic contaminants in soil and water.	Silva et al. (2013)
Oxygenases	Catalyze the incorporation of oxygen into organic molecules, often involved in biodegradation.Soil/water artificially polluted, along with contaminated oil	Soil/water artificially polluted, along with contaminated oil	Employed in the breakdown of environmental contaminants and enhancing plant development through alterations in plant metabolism.	Nagayama et al. (2015)
Novel phytase genes	Hydrolyze phytate, releasing available phosphorus for plants.	Agricultural soil	Enhance the accessibility of phosphorus for crops, especially in soils that lack sufficient phosphorus.	Tan et al. (2014)
Sulfate or sulfite-reducing genes	Involved in reducing sulfate or sulfite to sulfur, which can then be incorporated into amino acids.	Terrestrial and marine ecosystem	Utilized to enhance sulfur cycling in soils and boost sulfur nutrition for crops.	Anantharaman et al. (2018)
Genes conferring copper resistance	Aid in detoxifying or compartmentalizing excess copper in plants.	Agricultural soils polluted with copper	Utilized to create crops capable of withstanding copper contamination in polluted soils.	Altimira et al. (2012)
Naphthalene dioxygenase	Catalyzes the degradation of naphthalene and similar compounds.	Oil-contaminated soil	Utilized in bioremediation to decompose naphthalene and various polycyclic aromatic hydrocarbons (PAHs).	Ono et al. (2007)
ACC (1-aminocyclopropane-1-carboxylate) *deaminase*	Breaks down ACC, a precursor of ethylene, reducing plant stress.	Wheat rhizosphere	Utilized to improve plant development in challenging environments like drought and heavy metal exposure through the adjustment of ethylene concentrations.	Govindasamy et al. (2008)
Extradiodioxygenases	Catalyze the breakdown of aromatic compounds through an extradiol cleavage mechanism.	Activated sludge	Utilized in bioremediation for the breakdown of environmental contaminants, including aromatic compounds.	Suenaga et al. (2007)
ShcA (a secretion-related gene)	Involved in the secretion of metabolites and proteins, often associated with bacterial virulence.	Tobacco leaves inoculated with *Pseudomonas syringae*	Possible applications in agriculture for enhancing beneficial microbial activity or suppressing plant pathogens.	Dijk et al. (2002)

and phosphorus. For instance, genes encoding nitrogenase enzymes, which catalyze the conversion of atmospheric nitrogen into ammonia, have been identified in metagenomes from agricultural soils (Bahram et al., 2018) (Table 8.2). Similarly, genes involved in phosphate solubilization and sulfur oxidation have been discovered through metagenomic studies, providing potential targets for the development of biofertilizers that improve nutrient availability to crops (Richardson & Simpson, 2011).

8.4.2 Plant Growth Promotion

Metagenomic screening has also identified genes encoding plant growth-enhancing traits, such as the synthesis of phytohormones, siderophores, and antifungal compounds (Table 8.2). For instance, metagenomic analysis of rhizosphere soils has revealed genes responsible for the synthesis of indole-3-acetic acid (IAA), a key phytohormone that induces root growth and development (Lugtenberg & Kamilova, 2009). Other metagenomic studies have identified genes associated with siderophore biosynthesis, which enhance iron absorption by plants and restrict the growth of phytopathogens (Saha et al., 2016).

8.4.3 Disease Resistance

Functional metagenomics has been used to discover genes that confer resistance to phytopathogens, resulting in the development of biocontrol agents. For example, genes encoding lytic enzymes, such as chitinases and glucanases, have been identified in metagenomic libraries and shown to have activity against fungal pathogens that cause crop diseases (Banerjee et al., 2022) (Table 8.2). These enzymes degrade the cell walls of pathogenic fungi, providing a natural means of disease control.

8.4.4 Stress Tolerance

Genes that enhance plant tolerance to abiotic stresses, such as salinity, drought, and heavy metal contamination, have also been identified through metagenomic approaches. For example, genes encoding trehalose biosynthesis enzymes, which protect plants from osmotic stress, have been discovered in metagenomic libraries from saline soils (McBride & Ensign, 1987). Discovering and characterizing such genes may lead to the establishment of stress-resistant crop varieties or microbial inoculants that enhance plant resilience under adverse conditions.

8.5 CASE STUDIES AND APPLICATIONS

8.5.1 Successful Examples of Gene Prospecting using Metagenomics

8.5.1.1 Discovery of Novel Nitrogen-Fixing Bacteria

Nitrogen is critical for vegetative growth and maturation, as it plays a key role in numerous metabolic activities, structural components, and biochemical reactions in plants. Plants cannot convert atmospheric nitrogen into ammonia on their own and depend on biological agents such as nitrogen-fixing bacteria and Cyanobacteria (blue-green algae), or on industrial processes like the Haber–Bosch method, which synthesizes ammonia from nitrogen and hydrogen under high pressure and temperature.

Plants primarily depend on bacteria possessing the nitrogenase enzyme, which converts atmospheric nitrogen into ammonia (NH_3^+) and/or nitrate (NO_3^-) via biological nitrogen fixation (BNF)—a process especially beneficial in nitrogen-deficient soil (Franche et al., 2009).

The nitrogenase enzyme consists of two main components: the dinitrogenase subunit, encoded by *nifD* and *nifK,* responsible for catalytic activity; and the dinitrogenase reductase subunit, encoded by *nif*H, which donates electrons for the reduction of nitrogen into ammonia (Inoue et al., 2015; Pi et al., 2022, 2024). Additional genes, such as *nifE*, and *nifN,* are homologous to *nifD* and *nifK* (Inoue et al., 2015), *nifU, nifS*, and *nifV* play vital roles in the synthesis of the nitrogenase complex. The *nifE* gene is associated with the maturation of the molybdenum cofactor, which is vital for the activity of the nitrogenase enzyme complex, *nifN* gene encodes a protein that incorporates molybdenum cofactor into the nitrogenase enzyme complex, and *nifU* is involved in Fe-S cluster synthesis, essential for nitrogenase function; *nifS* encodes a protein responsible for cysteine synthesis, and *nifV* gene is required for MoFe cofactor biosynthesis (Dos Santos et al., 2012). Some genes, such as *fixA*, *fixB*, and *fixC*, play regulatory roles in nitrogen fixation (Fani et al., 2000).

A detailed study of nitrogen-fixation genes has paved the way for the discovery of numerous bacterial strains that significantly enhance soil quality for sustainable agriculture (Table 8.3).

8.5.1.2 Identification of Phosphate-Solubilizing Microbes

Phosphate solubilization refers to the process of converting insoluble forms of phosphorus into soluble forms, which can be absorbed by plants, as it is essential for plant metabolism. Certain microbes, like fungi and bacteria, convert this insoluble phosphate into its soluble and absorbable form for the plants. The conversion of insoluble phosphates into their bioavailable forms facilitates easier nutrient availability for plants.

Microbes solubilize the inorganic phosphate and mineralize organic phosphate by dropping the pH of the surrounding soil and by the action of certain enzymes, respectively. Microbes like *Bacillus*, *Pseudomonas*, *Acinetobacter*, etc., consist of the *gcd* gene that encodes glucose dehydrogenase, responsible for producing gluconic acid, which in turn lowers the pH of the surrounding soil (Pedrinho et al., 2023; Wu et al., 2022). Alkaline phosphatases encoded by *phoA* and *phoD,* and acid phosphatases encoded by *phoN* and *olpA,* mineralize phosphorus from organic phosphate, exopolyphosphatase like *ppx* degrades inorganic polyphosphate to phosphate, and the release of phosphorus in soil takes place. Carbon–phosphorus (C–P) lyase, encoded by the *phn*

TABLE 8.3
Novel Nitrogen-Fixing Bacteria Discovered over the Years and Their Respective Sources

Year	Name	Source	Reference
1970	*Desulfovibrio vulgaris*, *D. gigas*	Soil (sulfate-reducing bacteria)	Riederer-Henderson and Wilson (1970)
1983	Cellulolytic nitrogen-fixing bacterium	Deshayes gland in shipworms	Waterbury et al. (1983)
1983	*Campylobacter nitrofigilis* sp. nov.	Associated with the roots of Spartina alterniflora	McClung et al. (1983)
1989	*Azospirillum irakense* sp. nov.	Rhizosphere soil of rice	Khammas et al. (1989)
1993	*Azoarcus* spp.	Kallar grass roots	Reinhold-Hurek et al. (1993)
2001	*Azospirillum doebereinerae* sp. nov.	Roots of Miscanthus	Eckert et al. (2001)
2001	*Burkholderia* spp., *Herbaspirillum* spp,	Banana and pineapple roots	Magalhães Cruz et al. (2001)
2004	*Burkholderia tropica* sp. nov.	Rhizosphere and endophytes of sugarcane	Reis et al. (2004)
2005	*Pseudomonas azotifigens*	Compost pile	Hatayama et al. (2005)
2006	*Acetobacter nitrogenifigens* sp. nov.	Kombucha tea	Dutta and Gachhui (2006)
2008	*Pseudacidovorax intermedius* sp. nov.	Soil	Kampfer et al. (2008)
2008	*Gluconacetobacter* spp., *Swaminathania* spp.	Sugarcane, coffee, rice plants	Pedraza (2008)
2009	*Rhizobium* spp., *Frankia* spp.	Leguminous plants, non-leguminous plants	Franche et al. (2009)
2014	*Mangrovibacterium diazotrophicum*	Mangrove sediment	Huang et al. (2014)
2017	*Paenibacillus azotifigens* sp. nov.	Paddy soil	Siddiqi et al. (2017)
2021	*Azotobacter* spp.	Soil	Aasfar et al. (2021)
2024	Advances in Diazotroph diversity	Agricultural soil and legumes	de Lima et al. (2024)

gene also helps in cleaving C–P bonds and thus, the release of phosphorus from phosphonates takes place (Liao et al., 2023; Wu et al., 2022). In a low-phosphate environment or high-phosphate environment transporter proteins like phosphate-specific transporters (*pst*) and low-affinity inorganic phosphate transporters (*pit*) are associated with the assimilation of optimum phosphorus content (Wu et al., 2022). Similarly, genes *phoB* and *phoR* regulate the expression of phosphate solubilization and transport genes and enable them to function properly in low-phosphorus environments (Liao et al., 2023). Recent metagenomics studies have identified novel phosphate-solubilizing bacteria carrying *gcd* genes in their genomes, a very important biomarker, which had not previously reported as belonging to various phyla, Bacteroidetes, Acidobacteria, Planctomycetes, and Gemmatimonadota. Actinobacteria, Firmicutes, and Proteobacteria were the only three phyla previously reported to contain phosphate-solubilizing bacteria (Liang et al., 2020). These findings contribute to a deeper understanding of microbial diversity involved in solubilizing phosphate and also the genes involved in this process, which may enhance phosphorus availability to crops while preserving soil health and microbial diversity without the use of chemical fertilizers.

8.5.1.3 Isolation of Disease-Suppressive Soil Microbiomes

The soil microbiome plays a vital role in suppressing diseases in crops by enhancing plant health, inducing defense mechanisms in plants, antibiotic production, metabolites production, and competitive inhibition against pathogens. Such microbial communities responsible for the suppression of diseases in plants are defined as disease-suppressive soil (Jayaraman et al., 2021). They are classified into two categories: (1) general soil suppression—in which microbial biomass is high, but protection against specific pathogens is low, and (2) Specific soil suppression—in which the soil is rich in specific microbial species, and thus provides resistance against specific pathogens (Spooren et al., 2024).

Metagenomic approaches are essential for understanding the genetic composition and functional potential of soil microbial communities. For instance, several biosynthetic gene clusters (BGCs) responsible for siderophore production have been identified in disease-suppressive soils, particularly those effective against the pathogen *Fusarium culmorum*. Novel *non-ribosomal peptide synthetase* (*NRPS*) genes involved in the synthesis of antifungal bioactive compounds have also been detected (Tracanna et al., 2021). Additionally, genes encoding antimicrobial peptides, such as defensins and lantibiotics like grisin produced by *Streptomycetes* against *F. oxysporum*, have been reported (Sagova-Mareckova et al., 2023). Key genes involved in the biosynthesis of non-ribosomal peptides and polyketides—major classes of bioactive metabolites—play a critical role in disease suppression. Moreover, genes responsible for the production of antimicrobial compounds like 2,4-diacetylphloroglucinol (*phlD*) and hydrogen cyanide (*hcnAB*) have been identified in suppressive soils, highlighting the immense potential of metagenomics in uncovering microbial-mediated disease resistance mechanisms.

8.5.2 Translation of Metagenomic Findings to Agricultural Practice

8.5.2.1 Development of Biofertilizers and Biopesticides

Advancements in metagenomics have significantly accelerated the development of biofertilizer and biopesticide innovation by enabling the identification and analysis of beneficial bacteria from various environmental samples. Metagenomic methods have enabled the identification of new strains of bacteria, fungi, and viruses with significant plant growth-promoting and pest-controlling capabilities. Researchers can isolate bacteria that improve soil fertility, facilitate nutrient absorption, and offer natural pest control by utilizing the genetic diversity of microbial populations. The application of these metagenomic discoveries in developing biofertilizer and biopesticide formulations is facilitating the agricultural sector's transition to more sustainable methods. These bioproducts provide environmentally sustainable alternatives to chemical fertilizers and pesticides, enhancing ecosystem health and increasing crop yields.

8.5.2.1.1 Biofertilizer

Biofertilizers, a crucial element of sustainable farming, are formulations consisting of live or dormant cells of advantageous microbes (Table 8.4). These are subjected to different seeds, soil, or compost to enhance the abundance of such microbes and accelerate microbial activities that improve the accessibility of nutrients readily absorbed by plants. They lack intrinsic nutrients but facilitate plant nutrition through nitrogen fixation, phosphorus and zinc solubilization, potassium mobilization, and the mobilization of phosphates and micronutrients, among other processes (Kumar et al., 2017; Meena et al., 2015a; Priyadharsini & Muthukumar, 2016) (Table 8.4).

Biofertilizers additionally enhance plant development by many indirect pathways (Gopalakrishnan et al., 2015; Zahedi, 2016; Meena et al., 2015b). Biofertilizers significantly enhance soil fertility by converting atmospheric nitrogen into a bioavailable form, as well as breaking down insoluble phosphates and other essential nutrients. Besides the above-listed functions, specific bacteria and fungi stimulate plant growth by excreting hormones such as IAA, abscisic acid, gibberellins, auxins, and cytokinins, which directly facilitate plant development. The primary kinds of advantageous microbes employed as biofertilizers comprise fungi, bacteria, and cyanobacteria, most of which form symbiotic associations with plants. Currently, numerous commercial biofertilizer products are accessible in India. Various microbes and their relationships with agricultural plants are being utilized in the manufacture of biofertilizers. They can be categorised in several ways according to their characteristics. A list of several biofertilizers, together with their functions on target crops, is provided below (Table 8.4).

8.5.2.1.2 Biopesticides

Biopesticides originate from natural sources, including animals, plants, microorganisms, and specific minerals, and are extensively utilized for the management of insects, pests, and various pathogens (Table 8.5). By early 2013, approximately 400 active ingredients were registered in biopesticides, with over 1,250 biopesticide products actively registered (Raja, 2013). To quantitatively and qualitatively increase agricultural yields, sustainable organic farming methods require balanced nutritional supplements along with appropriate environment-friendly pest and disease management techniques (Raja & Masresha, 2015). Generally, biopesticides are insecticides derived from microbes or natural ingredients that control different pests using non-toxic methods. For the management of various pests, biopesticides and their derivatives are extensively employed. They are employed to safeguard crops from a range of insects, pests, as well as bacterial, viral, and fungal illnesses (Table 8.5). The most prevalent biopesticides produced and utilized in India include *Bacillus Thuringiensis* (*Bt*), *Trichoderma*, NPV (nuclear polyhedrosis virus), and neem-based pesticides. Numerous bacterial species, including *Pseudomonas* sp., *Bacillus* sp., *Burkholderia* sp., *Xanthomonas* sp., *Streptomyces* sp., *Enterobacter* sp., and *Serratia* sp., as well as fungal species such as *Beauveria bassiana*, *Trichoderma* sp., *Verticillium lecanii*, *Metarhizium* sp., *Paecilomyces fumosoroseus*, *Pythium* sp., *Fusarium* sp., and *Penicillium* sp., along with viruses like *Neodiprion sertifer* and *Lymantria dispar*, possess the ability to safeguard crops against various diseases (Ali et al., 2002; Wahla et al., 2012). Commonly, biopesticides are classified into three categories: (1) microbial biopesticides, (2) plant-incorporated protectants, and (3) biochemical biopesticides. The quantity of microorganisms, plant derivatives, and insect components utilized as active ingredients for the control of diseases, pests, and insects, along with their beneficial effects, has been documented globally in sustainable crop production.

8.5.3 Genetic Engineering of Crops with Beneficial Microbial Genes

8.5.3.1 Genetic Elements and Microbial Genes Utilized for the Transformation of Plants

Bioengineering enables the alteration of plant genomes through the manipulation or introduction of particular traits. A widely used method for plant modification involves the bacterium *Agrobacterium tumefaciens*, a gram-negative soil microorganism that naturally transfers DNA (Riva et al., 1998). Several bacterial virulence (*vir*) genes located on the Ti plasmid, but outside the T-DNA region, facilitate the transfer of T-DNA to the host plant cell (Christie, 1997; Yadav et al., 1982). The gene transfer process and the successful incorporation of genes into host plants necessitate a selection system to identify transformed cells from non-transformed cells. A significant variety of selection marker genes used in plant genetic modification have originated from microbial sources. These genes can be classified into three distinct categories according to their selective traits: antimetabolite marker genes, genes for antibiotic resistance, and genes for herbicide resistance. The *npII* gene (neomycin phosphotransferase II), derived from the *E. coli* Tn5 transposon, remains one of the most commonly used antibiotic selection markers in commercially released modified crops (Fraley et al., 1983; Yenofsky et al., 1990).

TABLE 8.4
Categories, Functions, and Commercial Availabilities of Biofertilizers in India

Biofertilizer Category	Type of Microorganisms	Representative Species	Functional Role	Beneficial for Crops	Name of Biofertilizer Product	References
Nitrogen fixing	Symbiotic bacteria	*Rhizobium* spp., *Frankia* spp.	Enhance nitrogen fixation and siderophore production	Leguminous crops	Rhizobium (symbiotic) (Kisaan Rhizobium culture), Nitrogen Biofertilizer,	Parewa et al. (2021) and Ramasamy et al. (2020)
	Free-living bacteria	*Azotobacter* spp.	Nitrogen fixation, vitamins, and plant hormones production	Non-leguminous plants	Azetobacter (non-symbiotic) (Kisaan Azotobacter culture), nitrogen bio-fertilizer	Singh et al. (2014)
	Associative symbiotic	*Azospirillum* sp.	Nitrogen fixation, siderophore and IAA production, phosphate solubilization	Non-leguminous crops and barley, maize, oats, sugarcane, sorghum, millets	Reap®-N3, Bio NPK™, Sardar Liquid Biofertilizers (culture of *Azopirillum*), Biospirillum, and Green Plus	Sharma et al. (2012)
Phosphate solubilizing	Bacteria	*Pseudomonas* sp., *Bacillus megaterium*, *Burkholderia* sp.	Phosphate solubilization, Synthesis of plant growth hormones (cytokinins, IAA, GAs, and spermidines)	For all kinds of crops like wheat, rice, cucumber, tomato, mustard, etc.	Phosphorus Biofertilizer (Kisaan P.S.B. culture), Reap®-P, Phospho-109	Ramasamy et al. (2020), Singh et al. (2014), and Swain and Ray (2009)
	Fungi	Mycorrhiza (Arbuscular mycorrhiza, orchid mycorrhiza, etc.)	Enhance nutrients and water uptake	For all kinds of crops	R-Jaal, Symbion-VAM	Mittal et al. (2008), Ramasamy et al. (2020), and Sharma et al. (2012)
Sulfur solubilizing	Sulfur-oxidizing or solubilizing bacteria	*Thiobacillus* spp., *Thermothrix thiopara*, etc.	Oxidize sulfur compounds into sulfate	Oilseed crops	Siron, S-Sol B®, MicroS-109	Ehrlich (2002) and Sharma et al. (2012)
Potassium solubilizing	Bacteria	*Paenibacillus glucanolyticus*, *Bacillus mucilaginosus*, *B. edaphicus*	Solubilize potash, helps in nutrient uptake, and gain biomass	Root crops and cereals	Reap®-K, Premium potash active, Potash-109	Sheng (2005) and Zarjani et al. (2013)
Plant growth promoters	Bacteria producing phytohormones	*Rhizobium* spp., *Pseudomonas fluorescens*, *Bacillus* spp., *Bradyrhizobium* sp., *Alcaligenes* sp., etc.	Stimulates root growth, helps in nutrient uptake, disease resistance, and gains biomass	All kinds of crops	Esvin Pseudo, Trichoguard-L, P-Suraksha, WP	Ramasamy et al. (2020), Sharma et al. (2012), Singh et al. (2014), and Vessey (2003)
Micronutrient solubilizing	Zinc solubilizing bacteria	*Thiobacillus thioxidans*, *Bacillus subtilis*, *T. ferrooxidans*, etc.	Solubilize zinc into bioavailable forms	Cereals and vegetables	Reap®-Z, Premium zinc activator, Zinc Extra, MicroZ-109	Ullah et al. (2015) and Saravanan et al. (2007)
	Silicate-solubilizing microbes	*Pseudomonas* spp., *Bacillus caldolyticus*, *Proteus mirabilis*	Solubilizes silicate and helps in nutrient uptake	Various crops	Silica-109	Meena et al. (2015a, 2015c, 2014)

The frequently utilized screenable markers derived from microbes include luciferase, β-glucuronidase (GUS), chloramphenicol acetyl transferase (CAT), and β-galactosidase. The *E. coli* derived *uidA* gene, which synthesizes β-glucuronidase (GUS), has been globally utilized as a reporter gene in numerous transgenic studies (Joersbo & Okkels, 1996). Table 8.6 represents a selected list of commonly used microbial marker genes. It is essential to recognize that marker genes generally do not offer agronomic or economic advantages to plants, and their incorporation into plant genomes has been driven solely by practical considerations in genetic modification techniques.

Additionally, numerous microbial genes are utilized as regulatory components, including promoters, terminators,

TABLE 8.5
An Overview of the Several Types of Biopesticides, Their Target Organisms, and Host Crops

Biopesticide Category	Organism	Target Pests	Host Crops	References
Bacterial	*Bacillus* sp.	*Alternaria* sp., *Botrytis* sp., *Ascochyta* sp., *Erwinia* sp., *Cercospora* sp., *Colletotrichum* sp., *Erysiphe polygoni*, *Fusarium* sp., *Macrophomina* sp., *Myrothecium* sp., *Phytophthora* sp., *Pythium* sp., *Ramularia* sp., *Rhizoctonia* sp., *Sclerotinia* sp., *Verticillium* sp., *Xanthomonas* sp.	Chili, soybean, tomato, wheat, fava bean, black gram, grapevine	Akhtar et al. (2012), Belete et al. (2015), Khan et al. (2001), and Zalila-Kolsi et al. (2016)
	B. thuringiensis	Bugs, caterpillars, leaf-feeding insects, leafhoppers, weevils.	Vegetables, fruits, cereals, ornamentals	Bravo et al. (2007), Koul (2012), and Saberi et al., 2020)
	Pseudomonas sp.	*Alternaria alternata*, *Aspergillus aculatus*, *Rhizoctonia solani*, *Botryodiplodia theobromae*, *Dreshlera biseptata*, *Fusarium oxysporum*, *Sclerotium rolfsii*, *Vibrio harveyi*	Mango, rice, eggplant, wheat, tomato, grapevine	Khokhar et al. (2011), Maurya et al. (2014), Prasad et al. (2015), Priyaja et al. (2014), and Sivanantham et al. (2013)
	Agrobacterium radiobacter strain 84	*Agrobacterium tumefaciens*	Fruits, vegetables, nuts, ornamentals	Kumar (2018)
	Alcaligenes sp.	*Fusarium* sp., *Aspergillus* sp., *Alternaria* sp.	Banana, tomato, groundnuts, maize, wheat	Kumar (2018)
Fungal	*Beauveria bassiana*	*Cyclocephala lurida*, *Dendrolimus tabulaeformis*, *Frankliniella occidentalis*, *Hylobius abietis*, *Polyphylla fullo*, *Spodoptera litura*, *Tuta absoluta*, *Uvarovistia zebra*, *Oryzaephilus surinamensis*	Oats, cabbage, kidney beans, white pine, wheat	Erler and Ates (2015) and Fan et al. (2013)
	Trichoderma sp.	*Alternaria* sp., *Ascochyta* sp., *Cercospora* sp., *Colletotrichum* sp., *Fusarium* sp., *Macrophomina* sp., *Myrothecium* sp., *Phytophthora* sp., *Pythium* sp., *Ralstonia* sp., *Rhizoctonia* sp., *Sclerotinia* sp.	Tomato, potato, pea, lentil, soybean, sunflower, cucumber, alfalfa, beans, eggplant, rice, sugar beet	Kumar (2018)
	Verticillium lecanii	Aphids, mealybugs, thrips, whiteflies	*Kidney bean*	Bouhous and Larous (2012) and Xu et al. (2009)
	Paecilomyces spp.	*Phytophthora Palmivora*, *Meloidogyne javanica*	Tomato, cocoa, black gram	Adebola and Amadi (2010), Bhat et al. (2011), Ganate and Khan (2010), Oclarit and Cumagun (2009), and Siddiqui and Akhtar (2008)
	Gliocladium sp.	*Penicillium* sp., *Chaetomium* sp., *Alternaria* sp., *Rhizopus* sp., *Fusarium* sp., *Aspergillus* sp.	Tomato, potato, cabbage, cauliflower, Groundnut, maize, Cereals, legumes, Banana, Citrus, apple	Kumar (2018)
	Metarhizium sp.	*Callosobruchus maculatus*, *Ceratitis capitata*, *Maruca vitrata*, *Musca domestica*, *Polyphylla fullo*, *Spodoptera litura*, *Tetranychus urticae*, *Tuta absoluta*	Sweet orange, kidney bean, tomato	Bugeme et al. (2015), Contreras et al. (2014), Erler and Ates (2015), Lozano-Tovar et al. (2015), Mehinto et al. (2014), Nabaei et al. (2012), Sharififard et al. (2011), and Sowjanya Sree et al. (2008)
Viral	*Granulosis* virus and NPV	Alfalfa looper, cotton leafworm, almond moth, armyworms, cabbage looper, corn earworm, Douglas-fir tussock moth, European corn borer, gypsy moth, imported cabbageworm, pine sawfly, cotton bollworm, spruce budworm, tobacco budworm	Cotton, maize, cabbage, rice, tomato, chili, etc.	Kumar (2018)

enhancers, and transit peptides in plant genetic modification frameworks. The CaMV 35S (cauliflower mosaic virus 35S) promoter, a widely used constitutive promoter derived from a microorganism, is frequently employed in the transformation of crops (Kumar et al., 2018b). In addition to CaMV 35S, the promoter of nos (nopaline synthase) derived from *Agrobacterium* stands as the second most commonly employed microbial origin promoter for the creation of transgenic plants. In a similar manner, CaMV 35S and nos terminators serve as gene termination signals in plant transformation constructs, whereas the 35S enhancer is employed to boost the expression of foreign genes in transgenic plants.

8.5.3.2 Insect-Resistant Microbial Genes

In agriculture, insects pose a significant threat. Diverse insect pests infest agricultural plants, causing substantial losses.

Microbes are essential in agricultural ecosystems and have considerable potential to mitigate plant damage by insect pests. Various approaches to the bio-control of insects have been documented, employing an array of microorganisms including bacteria, viruses, fungi, and nematodes (Kumar et al., 2018b). Approximately two decades ago, the first transgenic plant with insect resistance, which incorporated a microbially derived gene, was developed. Since that time, numerous new resistance genes originating from various microorganisms have been identified and introduced into plant genetic modification. Different *cry* genes isolated from *Bacillus thuringiensis* have been extensively used to enhance resistance against insect pests in agricultural crops. In the mid-1990s, cotton and later maize containing *cry* genes were introduced for commercial farming. There are already about 350 known endotoxins from *B. thuringiensis* (Crickmore, 2006). During sporulation, *B. thuringiensis* generates insecticidal protein crystals (Kumar et al., 2018b). Their mode of action involves a sequence of events that leads to insect mortality within hours of ingestion (Höfte & Whiteley, 1989; Knowles, 1994). A substantial family of delta-endotoxins is categorized as Cry I, II, III, IV, V, etc., based on molecular affinity and efficacy toward insect larvae. The insect larvae consume leaves, ingesting the toxin, which subsequently binds to precise receptors in the stomach. The toxin is solubilized and activated by proteases within the insect's middle intestine epithelium. The active toxins cause rupture of the middle intestine epithelial barrier, resulting in cell lysis and subsequent larval mortality (Lightwood et al., 2000). Table 8.6 presents a compilation of commercially cultivated crops that have been modified with *cry* insecticidal genes derived from various microorganisms. In addition to Bt, there are other insecticidal genes classified as vegetative insecticidal proteins (vip), which also come from *Bacillus* spp., specifically *B. thuringiensis* (Fang et al., 2007). Vip proteins demonstrate a mechanism distinct from Bt endotoxins, rendering them an attractive option for the enhancement of insect-resistance characteristics (Llewellyn et al., 2007). The *vip3Aa20* and *vip3A(a)* genes were effectively incorporated into maize and cotton cultivars, respectively.

8.5.3.3 Abiotic Stress Tolerance Genes from Microbes

Crop plants frequently encounter detrimental environmental conditions, such as drought, which significantly reduce crop output (Boyer, 1982). A new maize variety that is resistant to drought has been developed, featuring the *cspB* (cold-shock protein) gene sourced from *B. subtilis*. The *cspB* gene is pivotal in helping plants cope with drought stress through the stabilization of RNA molecules (Castiglioni et al., 2008). Compared to other drought-resistant hybrids lacking this microbial trait, field testing showed an average yield increase of five bushels per acre under drought conditions (Waltz, 2014). In addition to maize, a drought-resistant sugarcane cultivar incorporating the microbial gene *betA* has been introduced (Waltz, 2014). Two variants of the *betA* gene, encoding choline dehydrogenase, have been extracted from *Rhizobium meliloti* or *E. coli*. The enzyme, choline dehydrogenase, converts choline to betaine aldehyde, which is then transformed into the osmoprotectant glycine betaine by betaine aldehyde dehydrogenase. Glycine betaine, an osmoprotectant, helps plants resist abiotic stressors such as salinity, drought, and cold by preserving macromolecular structures and maintaining plasma membrane stability. This variety of sugarcane produces higher sugar yields (approx. 20–30%) under drought conditions compared to its traditional variety (Waltz, 2014). Table 8.6 provides an overview of genes obtained from microbes employed in creating drought-resistant crop plants.

8.5.3.4 Microbe-Derived Pathogen Resistance Genes

Genes obtained from diverse microbes have been used to combat microbial pathogens responsible for major plant diseases. Currently, resistance based on microbial genes has been successfully implemented only against viral pathogens. Research and development efforts are ongoing for numerous other pathogens, including bacteria, fungi, and nematodes. Transgenic crops resistant to viruses were mainly developed using gene silencing techniques, such as RNA interference (RNAi) and antisense RNA targeting viral sequences (Ramesh et al., 2007; Yan, 2007). Approaches include introducing viral coat protein genes, modifying replicase genes, synthesizing antiviral proteins, or expressing antibodies that target viral proteins. Various viral genes have been used to create crops resistant to viral diseases (Tricoll et al., 1995). A well-known example is the papaya ringspot virus (PRSV)-resistant papaya developed using a viral coat protein gene (Gonsalves et al., 2004).

8.6 CHALLENGES AND FUTURE DIRECTIONS

8.6.1 Technical and Methodological Challenges in Metagenomic Gene Prospecting

Although many studies have effectively utilized metagenomic sequence data to assist in the isolation and cultivation of microbes, several challenges and limitations remain that must be addressed to achieve broader success.

Antibody engineering guided by genomic data opens promising avenues for cultivating microbes with precision, yet several obstacles remain. One of the biggest challenges is finding appropriate surface-exposed epitopes due to limited structural data from homologous proteins in related organisms,

TABLE 8.6
Detailed Information Regarding Genetic Elements and Microbial Genes Utilized for Plant Transformation and Stress Tolerance

Parameter	Obtained from	Gene	Product of a Gene	Function of Gene	Crops
Microbial genes utilized as a transformation marker to develop commercial transgenic crop	*Streptomyces hygroscopicus*	*bar*	Phosphinothricin acetyl transferase	Inhibits the herbicidal efficacy of phosphinothricin herbicides through acetylation.	Rice, soybean, cotton, chicory, maize
	E. coli Tn5 transposon	*nptII*	Neomycin phosphotransferase II enzyme	Enables genetically modified plants to metabolize antibiotics (kanamycin and neomycin) during selection.	Alfaalfa, chicory, apple, cotton, eggplant, potato, plum, maize, melon, papaya, sugarcane, tomato, tobacco, sugar beet, squash
	*S. viridochromo*genes	*pat*	Phosphinothricin *N*-acetyltransferase (PAT) enzyme	Neutralizes the herbicidal efficacy of phosphinothricin herbicides through acetylation.	Cotton, soybean, maize, sugar beet
	E. coli	*aad*	3″(9)-O-aminoglycoside adenylyltransferase enzyme	Facilitates screening for resistance to aminoglycoside class of antibiotics.	Potato, cotton, eggplant
	E. coli	*aph4* (*hpt*)	Hygromycin B phosphotransferase (hph) enzyme	Facilitates screening for resistance to the hygromycin B antibiotic.	Rice, sugarcane, cotton
	E. coli	*bla*	β-lactamase enzyme	Neutralize the β-lactam class of Antibiotics, namely, ampicillin.	Soybean, maize
	A. tumefaciens strain CP4	*nos*	Nos enzyme	Facilitates the production of nopaline that enables the detection of modified plant embryos.	Flax
	E. coli	*spc*	Spectinomycin adenyltransferase enzyme (not expressed in plant tissues)	Imparts tolerance to streptomycin and spectinomycin antibiotics.	Flax
	E. coli	*uidA*	β-D-glucuronidase (GUS) enzyme	Generates a blue colour on modified tissue, facilitating the selection visually.	Soybean, cotton, papaya, palm
	E. coli	*pmi*	Phosphomannose isomerase (PMI) enzyme	Facilitates the metabolism of sugar mannose and enables favorable selection for the regeneration of modified plants.	Maize
Microbial genes utilized as a transformation marker to develop a commercial insect-resistant crop	*B. thuringiensis* subsp. *kurstaki*	*cry1AB*	Cry1Ab delta-endotoxin	Imparts protection against winged insects by specifically injuring middle intestine epithelium.	Rice, maize, cotton
	B. thuringiensis subsp. kumamotoensis	*cry1A.105*	Cry1A.105 protein that comprises the Cry1Ab, Cry1F, and Cry1Ac proteins	Imparts protection against winged insects by specifically injuring middle intestine epithelium.	Soybean, maize
	B. thuringiensis	*cry1A*	Delta-endotoxin of the Cry1A group	Imparts protection against winged insects by specifically injuring middle intestine epithelium.	Maize, cotton
	B.thuringiensis subsp. *kurstaki* strain HD73	*cry1Ac*	Cry1Ac delta-endotoxin	Imparts protection against winged insects by specifically injuring middle intestine epithelium.	Cotton, eggplant, maize, tomato, rice, soybean
	Synthetic fusion gene derived from *B. thuringiensis*	*cry1Ab-Ac*	Cry1Ab-Ac deltaendotoxin (fusion protein)	Imparts protection against winged insects by specifically injuring middle intestine epithelium.	Cotton
	Synthetic Cry1ab gene obtained from *B. thuringiensis* subsp. *kumamotoensis*	*cry1Ab* (truncated)	Cry1Ab delta-endotoxin	Imparts protection against winged insects by specifically injuring middle intestine epithelium.	Rice, maize

	Synthetic *cry1F* gene obtained from *B. thuringiensis* var. aizawai	*cry1Fa2*	Modified Cry1F protein	Imparts protection against winged insects by specifically injuring middle intestine epithelium.	Maize
	B. thuringiensis var. aizawai	*cry1F*	Cry1F delta-endotoxin	Imparts protection against winged insects by specifically injuring middle intestine epithelium.	Soybean, cotton, maize
	Synthetic gene obtained from *B. thuringiensis*	*cry1C*	Cry1C delta-endotoxin	Imparts protection against winged insects, specifically *Spodoptera*	Cotton
	B. thuringiensis strain PS149B1	*cry34Ab1*	Cry34Ab1 delta-endotoxin	Imparts protection against beetles, by selectively destroying middle intestine epithelium.	Maize
	B.thuringiensis subsp. *tolworthi* strain BTS02618A	*cry9C*	Cry9C delta-endotoxin	Imparts protection against winged insects by specifically injuring middle intestine epithelium.	Maize
	B. thuringiensis strain AB88	*vip3A(a)*	Vip3A vegetative insecticidal protein	Imparts protection against herbivory damage caused by winged insects by selectively injuring middle intestine lining	Cotton
	B. thuringiensis subsp. *kumamotoensis*	*cry2Ab2*	Cry2Ab delta-endotoxin	Imparts protection against winged insects by specifically injuring middle intestine epithelium.	Soybean, cotton, maize
Microbial genes utilized as transformation marker to develop commercial drought resistant crop	*Rhizobium meliloti*	*RmBetA*	Choline dehydrogenase	Facilitates the formation of the glycine betaine, hence enhancing hydric stress tolerance.	Sugarcane
	B. subtilis	*cspB*	Cold-shock-protein B	Preserves RNA stability and translation to sustain normal cellular processes during water-stress circumstances.	Maize
	E. coli	*EcBetA*	Choline dehydrogenase	Facilitates the formation of the glycine betaine, hence enhancing hydric tolerance.	Sugarcane
Microbial genes utilized as transformation marker to develop commercial pathogen resistant crop	ZYMV (Zucchini-yellow-mosaic-potyvirus)	*zymv*_cp	ZYMV's Coat protein	Imparts protection against zucchini yellow mosaic potyvirus via a pathogen-derived resistance mechanism.	Squash
	BGMV (Bean-golden-mosaic-virus)	*ac1* (sense as well as antisense)	Sense and antisense RNA of viral replication protein (Rep); any protein that helps in functional replication is not synthesized	The synthesis of the protein from BGMV, which helps in viral replication, is restricted, providing tolerance to BGMV	Beans
	PPV (Plum-pox-virus)	*ppv*_cp	PPV's coat protein	Imparts protection against PPV through a resistance mechanism obtained from the pathogen	Plum
	Potato leafroll virus (PLRV)	*plrv*_ORF1 and ORF2	ORF1 encodes a putative replicase domain, whereas ORF2 encodes a putative helicase in PLRV	Imparts protection against PLRV	Potato
	PRSV	*prsv*_cp	PRSV's coat protein	Imparts protection against PRSV	Papaya
	PVY (Potato-virus-Y)	*pvy*_cp	PVY's coat protein	Imparts protection against PVY	Potato
	PRSV	*prsv*_rep	Replicase domain of the PRSV	Imparts protection against PRSV via a gene silencing mechanism	Papaya
	WMV2 (Watermelon-mosaic-potyvirus 2)	*wmv*_cp	WMV2's coat protein	Imparts protection against WMV2	Squash
	CMV (Cucumber-mosaic-virus)	*cmv*_cp	CMV's coat protein	Imparts protection against CMV	Tomato, sweet pepper, squash

According to the GM Approval Database by ISAAA (http://www.isaaa.org/) and Kumar et al. (2018b).

making accurate prediction difficult. Furthermore, reduced in situ expression levels of the target membrane protein may hinder effective antibody binding. Nonetheless, the accessibility of gene expression or proteomics information may facilitate the identification of membrane proteins with increased abundance. Moreover, modifications such as glycosylation may obstruct antibody recognition of specific regions (Zhao et al., 2020). To alleviate these restrictions, using numerous epitopes or employing complete protein domains as antigens may prove effective, but at the expense of specificity.

Isolating anaerobic microorganisms presents further technological challenges for specific approaches, particularly those utilizing cell sorters (Hailemariam et al., 2020). The dimensions of cell sorters frequently limit their application within standard anaerobic chambers. Creating fluorescence-based cell sorting technology suitable for anaerobic conditions would significantly mitigate this constraint.

To sustain the continuous proliferation of isolated target organisms, it is essential to select an appropriate growth medium and optimize physicochemical parameters. Genome sequences often fail to provide sufficient information to determine all the requirements for the successful proliferation of certain microorganisms, as the chemical makeup of their native habitats is frequently unknown. Lavy et al. (2014), developed a growing medium informed by genetic data for *Candidatus Poribacteria* sp. WGA-4E. Despite genomic data indicating metabolic characteristics, including potential urea utilization as a nitrogen source and assimilatory sulfate reduction, the formulated medium failed to facilitate the development of *Poribacteria* under the tested conditions. Medium component concentrations may have been suboptimal, or that the addition of signaling molecules was necessary (Lavy et al., 2014). Integrating metagenomic strategies with empirical methods like culturomics (Amrane et al., 2019), in situ cultivation (Ben-Dov et al., 2009; Kaeberlein et al., 2002), single-cell techniques (Gross et al., 2015; Jiang et al., 2016), and systematic trial-and-error approaches may enhance the likelihood of successful cultivation. The acquisition of genomic data from target species to guide cultivation practices is mostly contingent upon their representation and the efficacy of DNA extraction methods. Prevalent problems encompass improper cell extraction, inadequate lysis, or DNA degradation (Wesolowska-Andersen et al., 2014), which may influence the representation of particular species in metagenomic datasets. Single-cell genomics provides a method for reconstructing draft genomes of specific microorganisms (Rinke et al., 2013; Wurch et al., 2016), albeit it requires specialized technical expertise.

8.6.1.1 Complexity of Soil Samples

The composition and activity of the soil microbiome rely on the soil's physicochemical properties (Liu et al., 2019). Nonetheless, a significant number of studies on soil microbiomes either overlook the inclusion of physicochemical properties of soil or limit their focus only to parameters such as acidity and alkalinity of the soil, soil organic carbon, total carbon and nitrogen content, moisture, and/or temperature. These limited measurements fail to capture the intricate chemical matrix of soil and its diverse constituent elements, including nitrogen, carbon, potassium, phosphorus, zinc, iron, calcium, manganese, and magnesium. Soil characteristics change with depth, as evidenced by the variations in soluble organic carbon, total nitrogen, and total phosphorus. Furthermore, soil characteristics, especially pH, are subject to variability and can change in response to shifts in climate (Shi et al., 2020), environmental factors (Bahram et al., 2018), and the aboveground biomass (Zhao et al., 2022).

A significant limitation in soil microbiome research is the inadequate evaluation of the study of soil's physicochemical properties that are essential for agricultural practices (Liu et al., 2019), hence impacting nutrient availability and cycling (Bossolani et al., 2021). These alterations elucidate the transformations noted in root-soil interactions, soil microbial populations (Bulgarelli et al., 2022), pathogen suppression (Su et al., 2022), and microbial activities (Reichenbach et al., 2021). The assessment of these qualities, if not done before the start of an experiment, which is supposed to be a long period, then the control will not be useful to study the changes in the soil's physicochemical properties, as these properties alter with time. The treatment's direct impact will be inseparable from the treatment's indirect effect due to the change in the parameters of the soil's physicochemical properties.

In the metagenomic investigation of the soil, the incorporation of the soil physicochemical parameters further facilitates the study comparisons and enhances understanding of the system. Further elucidation of the soil microbiome will be achieved by associating individual bacteria or their roles with particular chemical alterations (Saghaï et al., 2021). The responses of the soil microbes depend upon varying environmental aspects, namely, soil disturbance (Seitz et al., 2022), climate, and nutrient availability (She et al., 2018). The bacteria that favor root growth are influenced by the prevalence of nutrient availability, which may elucidate variations in plant productivity. The microbial populations and their interactions are also affected by these settings (Zhou et al., 2021). Excluding physicochemical parameters from a soil microbial community study results in a substantial gap in the understanding of the microbial population responses of the soil.

Moreover, considering soil features might enhance comprehension of why some microorganisms exhibit heightened sensitivity to specific abiotic alterations and the probable ramifications of this susceptibility on functionality of the ecosystem (Leite et al., 2021), recently been demonstrated that the efficacy of microorganisms favoring plant growth is contingent upon nitrogen availability. A comprehensive understanding of the influence of physicochemical parameters of soil on its microbiome will elucidate the formation of these communities and their roles in soil vigor and crop yield.

8.6.1.2 Limitations of Current Sequencing Technologies

During metagenomic analysis, the vast microbial diversity in soil presents considerable obstacles. Bacterial diversity is estimated to vary between 10^3 and 10^7 species per sample

(Lombard et al., 2011). In contrast, archaeal diversity shows even greater variation depending on soil type (Fierer et al., 2007). Constructing either a comprehensive or generalized representation of these communities requires customized strategies for sampling and DNA extraction.

Fingerprinting methods relying on PCR, which are frequently used for analyzing microbial communities, often face challenges such as the selective amplification of specific gene segments (Wintzingerode et al., 1997). Humic acids can adversely affect PCR performance (Wilson, 1997), leading to finding a balance between DNA purity and the preservation of genetic data. Methods such as T-RFLP and microarrays have been utilized to analyze bacterial communities in rhizospheres and soil aggregates (Mummey & Stahl, 2004; Sanguin et al., 2006). Comprehensive DNA recovery is frequently unnecessary in such comparisons; nonetheless, the employed procedure can be pivotal. Martin-Laurent et al. (2001), revealed that the employed approach must possess sufficient sensitivity to identify significant changes in bacterial populations relevant to the experiment.

DNA is incorporated into vectors to construct metagenomic libraries, which offer useful functional and phylogenetic information about microbial populations (van Elsas et al., 2008). The libraries display susceptibility to biases, including DNA degradation during extraction, perhaps giving rise to the under-representation of specific species (Temperton et al., 2009). The dimensions of DNA fragments influence their incorporation into vectors, possibly omitting smaller bits. Adverse gene expression in host organisms may result in the loss of genetic information (Feingersch & Béjà, 2009). To address these difficulties, dual-size libraries have been suggested (Daniel, 2005).

Millions of clones must be produced to create a complete metagenomic library because of the variation in the microbiome and genome sizes (Ginolhac et al., 2004). Moreover, populations inside the rare biosphere, crucial for ecosystem dynamics and industrial processes, are sometimes difficult to detect using conventional approaches (Brazelton et al., 2010; Pester et al., 2010). Enrichment strategies improve detection by reducing library size and focusing on the screening of active targets.

Recent developments in NGS have reduced expenses and upgraded the efficacy of metagenomic research (Margulies et al., 2005; Rothberg & Leamon, 2008). Nonetheless, other challenges persist, including limited read lengths, sequencing errors, and substantial processing requirements. Specialized assemblers such as Newbler and Euler-SR are engineered to handle error-prone data; nonetheless, they lack the capability to reconstruct whole genomes (Niu et al., 2010; Wooley & Ye, 2010). Managing the extensive datasets generated necessitates considerable computational power, given that conventional assembly techniques face challenges with chimeric contigs and elevated error rates (Dohm et al., 2008).

Technologies in bioinformatics, such as megablast, gpu-blast, cd-hit, and uclust, significantly improve similarity searching and clustering techniques (Edgar, 2010; W. Li & Godzik, 2006). Notwithstanding these gains, 20–70% of sequences fail to correspond with any entries in reference databases. Gene prediction methods such as MetaGeneMark and MetaGeneAnnotator enhance precision, even for brief data (Hoff, 2009). Software systems like QIIME, RAMMCAP, and web-based applications such as MG-RAST and CAMERA facilitate data analysis, although they necessitate specialized computational proficiency (Caporaso et al., 2010; Markowitz et al., 2006).

The advancements in sequencing and analytical technologies have facilitated the formulation of highly ambitious projects, including the analysis of an entire soil microbiome (Vogel et al., 2009). The TerraGenome sequencing initiative will initially begin with a single reference soil metagenome and subsequently expand to additional soil metagenomes. Vogel et al. (2009), want to employ "omics" methodologies, specifically transcriptomics, proteomics, and metabolomics, to investigate the gene expression, activity, and functionality of the soil microbiome, beyond merely doing a sequencing study.

8.6.1.3 Data Analysis and Interpretation Hurdles

The metagenomic investigations generated data that pose two main setbacks: storage of the generated data and the requirement for significant computational resources. The simpler tasks, like taxonomic profiling utilizing marker genes, can be performed with the help of available reference metagenome, and more intricate processes, including assembly of metagenome, functional characterization, and gene discovery, necessitate substantially enhanced computational capabilities. Cloud computing offers a cost-effective solution; numerous server-based versions of pipelines of metagenomics exist (Lee et al., 2021; Mitchell et al., 2018; Raknes & Bongo, 2018), and additional packages can be accessed as installable images on cloud computing platforms (McIver et al., 2018). The complexities of microbial communities in soils and plants pose considerable hurdles in agricultural metagenomics, shaped by many habitats and environmental conditions. Identification of bacteria of agricultural value may be hindered by the limitations of gene prediction methods and gaps in reference databases, which complicate the interpretation of metagenomic data. This issue has prompted the investigation of safe computation for metagenomic analysis (Wagner et al., 2016).

8.6.2 Future Trends in Metagenomics for Agriculture

The traits of microorganisms associated with plants provide vital nutrients, promoting healthy growth and enhancing sustainable and efficient agricultural productivity. Crop improvement depends on many factors, including the identification of bioactive compounds of agronomic importance, such as biocontrol agents, stress-responsive compounds, and secondary metabolites (Müller et al., 2016). A thorough examination of various microbial communities or isolated strains has led to the discovery of compounds that control weeds (Kao-Kniffin et al., 2013).

Additionally, establishing fast and cost-effective approaches for acquiring comprehensive genetic data of microbial communities is crucial for metagenomic research. The utilization of metagenomics for identifying bioactive compounds in sustainable agriculture relies on an efficient vector-host expression system. To overcome this problem, it is essential to devise more advanced techniques for analyzing a large number of clones. Metagenomics can reveal the genetic data of microbial communities residing under severe conditions, including those exposed to extreme temperatures, salinity, and acidity (Guazzaroni et al., 2013; Schröder et al., 2014). The function of those genes, which confer resistance to harsh environmental challenges, might be utilized in plants and applied in agricultural yield enhancement initiatives. Notwithstanding its advantages and constraints, metagenomics has unequivocally yielded insights into unique microbial genes and pathways that facilitate their survival in diverse contexts. Molecular metagenomics, in conjunction with sequence-based methodologies, facilitates the comprehension of microbial community characteristics and elucidates their fundamental processes or pathways. Despite tremendous research conducted over the past few decades, there is a pressing demand for breakthrough computational software for processing and analyzing large metagenomic databases. The screening of metagenomic clones based on function is a successful approach that relies on foreign gene expression inside the host-expression system. Therefore, developing more precise vectors for discovering new genes and metabolic processes is essential. The management and preservation of huge metagenomic data are important considerations, and tackling this issue is critical. There is a need for the development of a storage system designed for extensive data volume. The detection of an individual genome through the incorporation of a labeled compound can assist in the discovery of non-cultivated microorganisms with agricultural significance. The DNA-stable isotope probing (SIP) approach is used to identify active bacteria in various ecosystems. These microorganisms are detected on the basis of their capacity to uptake or incorporate various isotope-labeled compounds (Malmstrom & Eloe-Fadrosh, 2019). In recent years, metagenomics has significantly contributed to the progress of clustered regularly interspaced short palindromic repeats (CRISPR)/Cas9 technology (Barrangou et al., 2007; Barrangou & van der Oost, 2013). This method enables accurate genome editing in various eukaryotic organisms and has emerged as an innovative tool for genome modification in agriculture. CRISPR and Cas9, RNA-guided endonucleases, can target precise genomic sites using a short RNA guide (Barrangou et al., 2007). DpbCasX, a recently identified CasX protein, was discovered in Deltaproteobacteria from groundwater samples through metagenomic techniques (Song, 2019). DpbCasX's distinct domains for binding the non-target strand and loading the target strand are crucial for distorting an RNA–DNA complex and facilitating the cutting of a specific DNA strand (Song, 2019). CRISPR/Cas system holds the potential to improve breeding procedures in sustainable farming by incorporating desirable traits into superior crop varieties.

8.7 CONCLUSION

The sophisticated instruments of metagenomics provide numerous pathways to gain an expanded understanding of the agricultural soil microbiome and its specific activities. The quantification of plant microbiome impacts has significantly enhanced our conceptual framework in plant–microbe metagenomics, leading to practical applications that could boost crop production through improved plant trait expression and the better use of previously underutilized habitats. Recent advancements in metagenomics have improved our comprehension of the relationships between plants and bacteria, as well as among microbes within the rhizosphere, while simultaneously presenting numerous inquiries for further investigation. The domain of metagenomics, especially regarding soil microbial communities, possesses considerable promise that it remains inadequately investigated and examined. The identification of novel PGPR traits, through either activity-based screening or genomic sequencing, would significantly enhance our comprehension of the mechanistic variations in PGPR phenotypes. Moreover, metagenomics can clarify the gene clusters linked to biosynthetic genes and the production of antimicrobial compounds. Anticipating the continued progress of metagenomics to facilitate effective innovation, the application of PGPR will undoubtedly become a reality, contributing to essential physio-biochemical processes that maintain the sustainability and efficiency of farming, thereby guiding toward an optimal sustainable agricultural setup. The integration of metagenomics and traditional or innovative molecular techniques will undoubtedly facilitate the identification of new mechanisms underlying PGPR activity, novel PGPR identities, and a fresh perspective on the biology and practical applications of PGPR.

REFERENCES

Aasfar, A., Bargaz, A., Yaakoubi, K., Hilali, A., Bennis, I., Zeroual, Y., & Meftah Kadmiri, I. (2021). Nitrogen fixing azotobacter species as potential soil biological enhancers for crop nutrition and yield stability. *Frontiers in Microbiology*, *12*. https://doi.org/10.3389/fmicb.2021.628379

Abdalla, M., Hastings, A., Cheng, K., Yue, Q., Chadwick, D., Espenberg, M., Truu, J., Rees, R. M., & Smith, P. (2019). A critical review of the impacts of cover crops on nitrogen leaching, net greenhouse gas balance and crop productivity. *Global Change Biology*, *25*(8), 2530–2543. https://doi.org/10.1111/gcb.14644

Abhilash, P. C., Powell, J. R., Singh, H. B., & Singh, B. K. (2012). Plant–microbe interactions: Novel applications for exploitation in multipurpose remediation technologies. *Trends in Biotechnology*, *30*(8), 416–420. https://doi.org/10.1016/j.tibtech.2012.04.004

Acosta, J. A., Martínez-Martínez, S., Faz, A., & Arocena, J. (2011). Accumulations of major and trace elements in particle size fractions of soils on eight different parent materials. *Geoderma*, *161*(1–2), 30–42. https://doi.org/10.1016/j.geoderma.2010.12.001

Adebola, M. O., & Amadi, J. E. (2010). Antagonistic activities of Paecilomyces and Rhizopus species against the cocoa black pod pathogen (Phytophthora palmivora). *African Scientist*, *4*, 235–239.

Ahemad, M., & Kibret, M. (2014). Mechanisms and applications of plant growth promoting rhizobacteria: Current perspective. *Journal of King Saud University—Science*, *26*(1), 1–20. https://doi.org/10.1016/j.jksus.2013.05.001

Akhtar, A., Abbasi, H., & Sharf, R. (2012). Antagonistic effects of *Pseudomonas fluorescens* and *Bacillus subtilis* on *Meloidogyne incognita* infecting *Vigna mungo* L. *International Journal of Plant, Animal and Environmental Sciences*, *1*, 55–63.

Ali, N. I., Siddiqui, I. A., Shahid Shaukat, S., & Zaki, M. J. (2002). Nematicidal activity of some strains of Pseudomonas spp. *Soi*l Biology and Biochemistry, 34(8), 1051–1058. https://doi.org/10.1016/S0038-0717(02)00029-9

Altimira, F., Yáñez, C., Bravo, G., González, M., Rojas, L. A., & Seeger, M. (2012). Characterization of copper-resistant bacteria and bacterial communities from copper-polluted agricultural soils of central Chile. BMC Microbiology, 12(1), 193. https://doi.org/10.1186/1471-2180-12-193

Amann, R. I., Ludwig, W., & Schleifer, K. H. (1995). Phylogenetic identification and in situ detection of individual microbial cells without cultivation. *Microbiological Reviews*, *59*(1), 143–169. https://doi.org/10.1128/mr.59.1.143-169.1995

Aminov, R. I. (2011). Horizontal gene exchange in environmental microbiota. *Frontiers in Microbiology*, *2*. https://doi.org/10.3389/fmicb.2011.00158

Amrane, S., Hocquart, M., Afouda, P., Kuete, E., Pham, T.-P.-T., Dione, N., Ngom, I. I., Valles, C., Bachar, D., Raoult, D., & Lagier, J. C. (2019). Metagenomic and culturomic analysis of gut microbiota dysbiosis during Clostridium difficile infection. *Scientific Reports*, *9*(1), 12807. https://doi.org/10.1038/s41598-019-49189-8

Anantharaman, K., Hausmann, B., Jungbluth, S. P., Kantor, R. S., Lavy, A., Warren, L. A., Rappé, M. S., Pester, M., Loy, A., Thomas, B. C., & Banfield, J. F. (2018). Expanded diversity of microbial groups that shape the dissimilatory sulfur cycle. *The ISME Journal*, *12*(7), 1715–1728. https://doi.org/10.1038/s41396-018-0078-0

Andrews, S. (2010). FastQC: A quality control tool for high throughput sequence data. Http://Www.Bioinformatics.Babraham.Ac.Uk/Projects/Fastqc.

Atamna-Ismaeel, N., Finkel, O., Glaser, F., von Mering, C., Vorholt, J. A., Koblížek, M., Belkin, S., & Béjà, O. (2012). Bacterial anoxygenic photosynthesis on plant leaf surfaces. *Environmental Microbiology Reports*, *4*(2), 209–216. https://doi.org/10.1111/j.1758-2229.2011.00323.x

Aziz, R. K., Bartels, D., Best, A. A., DeJongh, M., Disz, T., Edwards, R. A., Formsma, K., Gerdes, S., Glass, E. M., Kubal, M., Meyer, F., Olsen, G. J., Olson, R., Osterman, A. L., Overbeek, R. A., McNeil, L. K., Paarmann, D., Paczian, T., Parrello, B., . . . Zagnitko, O. (2008). The RAST server: Rapid annotations using subsystems technology. *BMC Genomics*, *9*(1), 75. https://doi.org/10.1186/1471-2164-9-75

Bahadur, I., Maurya, B. R., Meena, V. S., Saha, M., Kumar, A., & Aeron, A. (2016). Mineral release dynamics of tricalcium phosphate and waste muscovite by mineral-solubilizing rhizobacteria isolated from indo-gangetic plain of India. *Geomicrobiology Journal*, 1–13. https://doi.org/10.1080/01490451.2016.1219431

Bahram, M., Hildebrand, F., Forslund, S. K., Anderson, J. L., Soudzilovskaia, N. A., Bodegom, P. M., Bengtsson-Palme, J., Anslan, S., Coelho, L. P., Harend, H., Huerta-Cepas, J., Medema, M. H., Maltz, M. R., Mundra, S., Olsson, P. A., Pent, M., Põlme, S., Sunagawa, S., Ryberg, M., . . . Bork, P. (2018). Structure and function of the global topsoil microbiome. *Nature*, *560*(7717), 233–237. https://doi.org/10.1038/s41586-018-0386-6

Bakshi, A., Moin, M., & Madhav, M. S. (2020). Metagenomics in agriculture: State-of-the-art. In *Metagenomics: Techniques, Applications, Challenges and Opportunities* (pp. 167–187). Springer Singapore. https://doi.org/10.1007/978-981-15-6529-8_11

Banerjee, S., Maiti, T. K., & Roy, R. N. (2022). Enzyme producing insect gut microbes: An unexplored biotechnological aspect. *Critical Reviews in Biotechnology*, *42*(3), 384–402. https://doi.org/10.1080/07388551.2021.1942777

Bankevich, A., Nurk, S., Antipov, D., Gurevich, A. A., Dvorkin, M., Kulikov, A. S., Lesin, V. M., Nikolenko, S. I., Pham, S., Prjibelski, A. D., Pyshkin, A. V., Sirotkin, A. V., Vyahhi, N., Tesler, G., Alekseyev, M. A., & Pevzner, P. A. (2012). SPAdes: A new genome assembly algorithm and its applications to single-cell sequencing. *Journal of Computational Biology*, *19*(5), 455–477. https://doi.org/10.1089/cmb.2012.0021

Barrangou, R., Fremaux, C., Deveau, H., Richards, M., Boyaval, P., Moineau, S., Romero, D. A., & Horvath, P. (2007). CRISPR provides acquired resistance against viruses in prokaryotes. *Science*, *315*(5819), 1709–1712. https://doi.org/10.1126/science.1138140

Barrangou, R., & van der Oost, J. (2013). *CRISPR-Cas Systems: RNA-Mediated Adaptive Immunity in Bacteria and Archaea* (R. Barrangou & J. van der Oost, Eds.). Springer Berlin Heidelberg. https://doi.org/10.1007/978-3-642-34657-6

Belete, E., Ayalew, A., & Ahmed, S. (2015). Antagonistic effect of native *Bacillus* isolates against black root rot of faba bean. *African Crop Science Journal*, *23*(3), 249–259.

Beller, H. R., Rodrigues, A. V., Zargar, K., Wu, Y.-W., Saini, A. K., Saville, R. M., Pereira, J. H., Adams, P. D., Tringe, S. G., Petzold, C. J., & Keasling, J. D. (2018). Discovery of enzymes for toluene synthesis from anoxic microbial communities. *Nature Chemical Biology*, *14*(5), 451–457. https://doi.org/10.1038/s41589-018-0017-4

Ben-Dov, E., Kramarsky-Winter, E., & Kushmaro, A. (2009). An in situ method for cultivating microorganisms using a double encapsulation technique. *FEMS Microbiology Ecology*, *68*(3), 363–371. https://doi.org/10.1111/j.1574-6941.2009.00682.x

Bhat, M. Y., wani, A. H., & Fazal, M. (2011). Effect of Paecilomyces Lilacinus and Plant Growth Promoting Rhizobacteria on Meloidogyne Incognita Inoculated Black Gram, Vigna Mungo Plants. *Journal of Biopesticides*, *5*(1), 36–43. https://doi.org/10.57182/jbiopestic.5.1.36-43

Bhattacharyya, S. S., Leite, F. F. G. D., France, C. L., Adekoya, A. O., Ros, G. H., de Vries, W., Melchor-Martínez, E. M., Iqbal, H. M. N., & Parra-Saldívar, R. (2022). Soil carbon sequestration, greenhouse gas emissions, and water pollution under different tillage practices. *Science of The Total Environment*, *826*, 154161. https://doi.org/10.1016/j.scitotenv.2022.154161

Bloem, J., Hopkins, D. W., & Benedetti, A. (2005). *Microbiological Methods for Assessing Soil Quality*. Cabi Publishing.

Blott, S. J., & Pye, K. (2012). Particle size scales and classification of sediment types based on particle size distributions: Review and recommended procedures. *Sedimentology*, *59*(7), 2071–2096. https://doi.org/10.1111/j.1365-3091.2012.01335.x

Bolger, A. M., Lohse, M., & Usadel, B. (2014). Trimmomatic: A flexible trimmer for Illumina sequence data. *Bioinformatics*, *30*(15), 2114–2120. https://doi.org/10.1093/bioinformatics/btu170

Borer, B., Tecon, R., & Or, D. (2018). Spatial organization of bacterial populations in response to oxygen and carbon counter-gradients in pore networks. *Nature Communications*, *9*(1), 769. https://doi.org/10.1038/s41467-018-03187-y

Bossolani, J. W., Crusciol, C. A. C., Leite, M. F. A., Merloti, L. F., Moretti, L. G., Pascoaloto, I. M., & Kuramae, E. E. (2021). Modulation of the soil microbiome by long-term Ca-based soil amendments boosts soil organic carbon and physicochemical quality in a tropical no-till crop rotation system. *Soil Biology and Biochemistry*, *156*, 108188. https://doi.org/10.1016/j.soilbio.2021.108188

Bouhous, M., & Larous, L. (2012). Efficiency of the entomopathogenic fungus Verticillium lecanii in the biological control of Trialeurodes vaporariorum, (Homoptera: Aleyrodidae), a greenhouse culture pest. *African Journal of Microbiology Research*, *6*(10), 2435–2442.

Bowles, T. M., Jackson, L. E., Loeher, M., & Cavagnaro, T. R. (2017). Ecological intensification and arbuscular mycorrhizas: A meta-analysis of tillage and cover crop effects. *Journal of Applied Ecology*, *54*(6), 1785–1793. https://doi.org/10.1111/1365-2664.12815

Boyer, J. S. (1982). Plant productivity and environment. *Science*, *218*(4571), 443–448. https://doi.org/10.1126/science.218.4571.443

Bramhachari, P. V., Nagaraju, G. P., & Kariali, E. (2017). Metagenomic approaches in understanding the mechanism and function of PGPRs: Perspectives for sustainable agriculture. In *Agriculturally Important Microbes for Sustainable Agriculture* (pp. 163–182). Springer Singapore. https://doi.org/10.1007/978-981-10-5589-8_8

Bravo, A., Gill, S. S., & Soberón, M. (2007). Mode of action of *Bacillus thuringiensis* Cry and Cyt toxins and their potential for insect control. *Toxicon*, *49*(4), 423–435. https://doi.org/10.1016/j.toxicon.2006.11.022

Brazelton, W. J., Ludwig, K. A., Sogin, M. L., Andreishcheva, E. N., Kelley, D. S., Shen, C.-C., Edwards, R. L., & Baross, J. A. (2010). Archaea and bacteria with surprising microdiversity show shifts in dominance over 1,000-year time scales in hydrothermal chimneys. *Proceedings of the National Academy of Sciences*, *107*(4), 1612–1617. https://doi.org/10.1073/pnas.0905369107

Briar, S. S., Fonte, S. J., Park, I., Six, J., Scow, K., & Ferris, H. (2011). The distribution of nematodes and soil microbial communities across soil aggregate fractions and farm management systems. *Soil Biology and Biochemistry*, *43*(5), 905–914. https://doi.org/10.1016/j.soilbio.2010.12.017

Bugeme, D. M., Knapp, M., Ekesi, S., Chabi-Olaye, A., Boga, H. I., & Maniania, N. K. (2015). Efficacy of *Metarhizium anisopliae* in controlling the two-spotted spider mite *Tetranychus urticae* on common bean in screenhouse and field experiments. *Insect Science*, *22*(1), 121–128. https://doi.org/10.1111/1744-7917.12111

Bulgarelli, R. G., Leite, M. F. A., de Hollander, M., Mazzafera, P., Andrade, S. A. L., & Kuramae, E. E. (2022). Eucalypt species drive rhizosphere bacterial and fungal community assembly but soil phosphorus availability rearranges the microbiome. *Science of The Total Environment*, *836*, 155667. https://doi.org/10.1016/j.scitotenv.2022.155667

Caporaso, J. G., Kuczynski, J., Stombaugh, J., Bittinger, K., Bushman, F. D., Costello, E. K., Fierer, N., Peña, A. G., Goodrich, J. K., Gordon, J. I., Huttley, G. A., Kelley, S. T., Knights, D., Koenig, J. E., Ley, R. E., Lozupone, C. A., McDonald, D., Muegge, B. D., Pirrung, M., . . . Knight, R. (2010). QIIME allows analysis of high-throughput community sequencing data. *Nature Methods*, *7*(5), 335–336. https://doi.org/10.1038/nmeth.f.303

Carbonetto, B., Rascovan, N., Álvarez, R., Mentaberry, A., & Vázquez, M. P. (2014). Structure, composition and metagenomic profile of soil microbiomes associated to agricultural land use and tillage systems in argentine pampas. *PLoS One*, *9*(6), e99949. https://doi.org/10.1371/journal.pone.0099949

Caspi, R., Altman, T., Billington, R., Dreher, K., Foerster, H., Fulcher, C. A., Holland, T. A., Keseler, I. M., Kothari, A., Kubo, A., Krummenacker, M., Latendresse, M., Mueller, L. A., Ong, Q., Paley, S., Subhraveti, P., Weaver, D. S., Weerasinghe, D., Zhang, P., & Karp, P. D. (2014). The MetaCyc database of metabolic pathways and enzymes and the BioCyc collection of Pathway/Genome Databases. *Nucleic Acids Research*, *42*(D1), D459–D471. https://doi.org/10.1093/nar/gkt1103

Castiglioni, P., Warner, D., Bensen, R. J., Anstrom, D. C., Harrison, J., Stoecker, M., Abad, M., Kumar, G., Salvador, S., D'Ordine, R., Navarro, S., Back, S., Fernandes, M., Targolli, J., Dasgupta, S., Bonin, C., Luethy, M. H., & Heard, J. E. (2008). Bacterial RNA chaperones confer abiotic stress tolerance in plants and improved grain yield in maize under water-limited conditions. *Plant Physiology*, *147*(2), 446–455. https://doi.org/10.1104/pp.108.118828

Chen, L., Liu, Y., Wu, G., Veronican Njeri, K., Shen, Q., Zhang, N., & Zhang, R. (2016). Induced maize salt tolerance by rhizosphere inoculation of *Bacillus amyloliquefaciens* SQR9. *Physiologia Plantarum*, *158*(1), 34–44. https://doi.org/10.1111/ppl.12441

Chen, M., Wei, H., Cao, J., Liu, R., Wang, Y., & Zheng, C. (2007). Expression of bacillus subtilis proBA genes and reduction of feedback inhibition of proline synthesis increases proline production and confers osmotolerance in transgenic *Arabidopsis*. *BMB Reports*, *40*(3), 396–403. https://doi.org/10.5483/BMBRep.2007.40.3.396

Chhabra, S., Brazil, D., Morrissey, J., Burke, J. I., O'Gara, F., & N. Dowling, D. (2013). Characterization of mineral phosphate solubilization traits from a barley rhizosphere soil functional metagenome. *MicrobiologyOpen*, *2*(5), 717–724. https://doi.org/10.1002/mbo3.110

Chotte, J.-L. (2005). Importance of microorganisms for soil aggregation. In *Microorganisms in Soils: Roles in Genesis and Functions* (pp. 107–119). Springer-Verlag. https://doi.org/10.1007/3-540-26609-7_5

Christensen, B. T. (1987). Decomposability of organic matter in particle size fractions from field soils with straw incorporation. *Soil Biology and Biochemistry*, *19*(4), 429–435. https://doi.org/10.1016/0038-0717(87)90034-4

Christensen, B. T., & Olesen, J. E. (1998). Nitrogen mineralization potential of organomineral size separates from soils with annual straw incorporation. *European*

Journal of Soil Science, *49*(1), 25–36. https://doi.org/10.1046/j.1365-2389.1998.00130.x

Christie, P. J. (1997). Agrobacterium tumefaciens T-complex transport apparatus: A paradigm for a new family of multifunctional transporters in eubacteria. *Journal of Bacteriology*, *179*(10), 3085–3094. https://doi.org/10.1128/jb.179.10.3085-3094.1997

Contreras, J., Mendoza, J. E., Martínez-Aguirre, M. R., García-Vidal, L., Izquierdo, J., & Bielza, P. (2014). Efficacy of enthomopathogenic Fungus *Metarhizium anisopliae* against *Tuta absoluta* (Lepidoptera: Gelechiidae). *Journal of Economic Entomology*, *107*(1), 121–124. https://doi.org/10.1603/EC13404

Crickmore, N. (2006). Beyond the spore—past and future developments of Bacillus thuringiensis as a biopesticide. *Journal of Applied Microbiology*, *101*(3), 616–619. https://doi.org/10.1111/j.1365-2672.2006.02936.x

Crits-Christoph, A., Diamond, S., Butterfield, C. N., Thomas, B. C., & Banfield, J. F. (2018). Novel soil bacteria possess diverse genes for secondary metabolite biosynthesis. *Nature*, *558*(7710), 440–444. https://doi.org/10.1038/s41586-018-0207-y

Daniel, R. (2005). The metagenomics of soil. *Nature Reviews Microbiology*, *3*(6), 470–478. https://doi.org/10.1038/nrmicro1160

de Lima, J. D., de Souza, A. J., Nunes, A. L. P., Rivadavea, W. R., Zaro, G. C., & da Silva, G. J. (2024). Expanding agricultural potential through biological nitrogen fixation: Recent advances and diversity of diazotrophic bacteria. *Australian Journal of Crop Science*, *18*(6), 324–333. https://doi.org/10.21475/ajcs.24.18.06.p4104

Dijk, K. van, Tam, V. C., Records, A. R., Petnicki-Ocwieja, T., & Alfano, J. R. (2002). The ShcA protein is a molecular chaperone that assists in the secretion of the HopPsyA effector from the type III (Hrp) protein secretion system of *Pseudomonas syringae*. *Molecular Microbiology*, *44*(6), 1469–1481. https://doi.org/10.1046/j.1365-2958.2002.02979.x

Dohm, J. C., Lottaz, C., Borodina, T., & Himmelbauer, H. (2008). Substantial biases in ultra-short read data sets from high-throughput DNA sequencing. *Nucleic Acids Research*, *36*(16). https://doi.org/10.1093/nar/gkn425

Dos Santos, P. C., Fang, Z., Mason, S. W., Setubal, J. C., & Dixon, R. (2012). Distribution of nitrogen fixation and nitrogenase-like sequences amongst microbial genomes. *BMC Genomics*, *13*(1), 162. https://doi.org/10.1186/1471-2164-13-162

Dutta, D., & Gachhui, R. (2006). Novel nitrogen-fixing Acetobacter nitrogenifigens sp. nov., isolated from Kombucha tea. *International Journal of Systematic and Evolutionary Microbiology*, *56*(8), 1899–1903. https://doi.org/10.1099/ijs.0.64101-0

Eckert, B., Weber, O. B., Kirchhof, G., Halbritter, A., Stoffels, M., & Hartmann, A. (2001). *Azospirillum doebereinerae* sp. nov., a nitrogen-fixing bacterium associated with the C4-grass Miscanthus. *International Journal of Systematic and Evolutionary Microbiology*, *51*(1), 17–26. https://doi.org/10.1099/00207713-51-1-17

Edgar, R. C. (2010). Search and clustering orders of magnitude faster than BLAST. *Bioinformatics*, *26*(19), 2460–2461. https://doi.org/10.1093/bioinformatics/btq461

Ehrlich, H. L. (2002). *Geomicrobiology* (H. L. Ehrlich, Ed.). CRC Press. https://doi.org/10.1201/9780824744458

Eren, A. M., Esen, Ö. C., Quince, C., Vineis, J. H., Morrison, H. G., Sogin, M. L., & Delmont, T. O. (2015). Anvi'o: An advanced analysis and visualizatio*r*Jn platform for 'omics data'. *Pee*, *3*, e1319. https://doi.org/10.7717/peerj.1319

Erler, F., & Ates, A. O. (2015). Potential of two entomopathogenic fungi, Beauveria bassiana and Metarhizium anisopliae (Coleoptera: Scarabaeidae), as biological control agents against the June beetle. *Journal of Insect Science*, *15*(1), 44–44. https://doi.org/10.1093/jisesa/iev029

Espínola, F., Dionisi, H. M., Borglin, S., Brislawn, C. J., Jansson, J. K., Mac Cormack, W. P., Carroll, J., Sjöling, S., & Lozada, M. (2018). Metagenomic analysis of subtidal sediments from polar and subpolar coastal environments highlights the relevance of anaerobic hydrocarbon degradation processes. *Microbial Ecology*, *75*(1), 123–139. https://doi.org/10.1007/s00248-017-1028-5

Fan, J., Xie, Y., Xue, J., & Liu, R. (2013). The effect of *Beauveria brongniartii* and its secondary metabolites on the detoxification enzymes of the pine caterpillar, *Dendrolimus tabulaeformis*. *Journal of Insect Science*, *13*(44), 1–13. https://doi.org/10.1673/031.013.4401

Fang, J., Xu, X., Wang, P., Zhao, J.-Z., Shelton, A. M., Cheng, J., Feng, M.-G., & Shen, Z. (2007). Characterization of chimeric *Bacillus thuringiensis* Vip3 Toxins. *Applied and Environmental Microbiology*, *73*(3), 956–961. https://doi.org/10.1128/AEM.02079-06

Fani, R., Gallo, R., & Liò, P. (2000). Molecular evolution of nitrogen fixation:
olubevolutionary history of the nifD, nifK, nifE, and nifN genes. *Journal of Molecular Evolution*, *51*(1), 1–11. https://doi.org/10.1007/s002390010061

Feingersch, R., & Béjà, O. (2009). Bias in assessments of marine SAR11 biodiversity in environmental fosmid and BAC libraries? *The ISME Journal*, *3*(10), 1117–1119. https://doi.org/10.1038/ismej.2009.80

Fierer, N. (2017). Embracing the unknown: Disentangling the complexities of the soil microbiome. *Nature Reviews Microbiology*, *15*(10), 579–590. https://doi.org/10.1038/nrmicro.2017.87

Fierer, N., Breitbart, M., Nulton, J., Salamon, P., Lozupone, C., Jones, R., Robeson, M., Edwards, R. A., Felts, B., Rayhawk, S., Knight, R., Rohwer, F., & Jackson, R. B. (2007). Metagenomic and small-subunit rRNA analyses reveal the genetic diversity of bacteria, archaea, fungi, and viruses in soil. *Applied and Environmental Microbiology*, *73*(21), 7059–7066. https://doi.org/10.1128/AEM.00358-07

Finn, R. D., Coggill, P., Eberhardt, R. Y., Eddy, S. R., Mistry, J., Mitchell, A. L., Potter, S. C., Punta, M., Qureshi, M., Sangrador-Vegas, A., Salazar, G. A., Tate, J., & Bateman, A. (2016). The Pfam protein families database: Towards a more sustainable future. *Nucleic Acids Research*, *44*(D1), D279–D285. https://doi.org/10.1093/nar/gkv1344

Fraley, R. T., Rogers, S. G., Horsch, R. B., Sanders, P. R., Flick, J. S., Adams, S. P., Bittner, M. L., Brand, L. A., Fink, C. L., Fry, J. S., Galluppi, G. R., Goldberg, S. B., Hoffmann, N. L., & Woo, S. C. (1983). Expression of bacterial genes in plant cells. *Proceedings of the National Academy of Sciences*, *80*(15), 4803–4807. https://doi.org/10.1073/pnas.80.15.4803

Franche, C., Lindström, K., & Elmerich, C. (2009). Nitrogen-fixing bacteria associated with leguminous and non-leguminous plants. *Plant and Soil*, *321*(1–2), 35–59. https://doi.org/10.1007/s11104-008-9833-8

Francioli, D., Schulz, E., Lentendu, G., Wubet, T., Buscot, F., & Reitz, T. (2016). Mineral vs. organic amendments: Microbial community structure, activity and abundance of agriculturally relevant microbes are driven by long-term fertilization strategies. *Frontiers in Microbiology*, *7*. https://doi.org/10.3389/fmicb.2016.01446

Ganate, M. A., & Khan, T. A. (2010). Biological potential of paecilomyces lilacinus on pathogenesis of meloidodyne javanica infecting tomato plant. *European Journal of Applied Sciences*, *2*, 80–84.

Geisseler, D., & Scow, K. M. (2014). Long-term effects of mineral fertilizers on soil microorganisms: A review. *Soil Biology and Biochemistry*, *75*, 54–63. https://doi.org/10.1016/j.soilbio.2014.03.023

Ginolhac, A., Jarrin, C., Gillet, B., Robe, P., Pujic, P., Tuphile, K., Bertrand, H., Vogel, T. M., Perrière, G., Simonet, P., & Nalin, R. (2004). Phylogenetic analysis of polyketide synthase I domains from soil metagenomic libraries allows selection of promising clones. *Applied and Environmental Microbiology*, *70*(9), 5522–5527. https://doi.org/10.1128/AEM.70.9.5522-5527.2004

Glick, B. R. (2014). Bacteria with ACC deaminase can promote plant growth and help to feed the world. *Microbiological Research*, *169*(1), 30–39. https://doi.org/10.1016/j.micres.2013.09.009

Gomez, E., Ferreras, L., Toresani, S., Ausilio, A., & Bisaro, V. (2001). Changes in some soil properties in a Vertic Argiudoll under short-term conservation tillage. *Soil and Tillage Research*, *61*(3–4), 179–186. https://doi.org/10.1016/S0167-1987(01)00193-3

Gonsalves, D., Gonsalves, C., Ferreira, C., & Fitch, M. (2004). Transgenic virus-resistant papaya: From hope to reality in controlling papaya ringspot virus in Hawaii. *APSnet Feature Articles*. https://doi.org/10.1094/APSnetFeature-2004-0704

Gopalakrishnan, S., Sathya, A., Vijayabharathi, R., Varshney, R. K., Gowda, C. L. L., & Krishnamurthy, L. (2015). Plant growth promoting rhizobia: Challenges and opportunities. *3 Biotech*, *5*(4), 355–377. https://doi.org/10.1007/s13205-014-0241-x

Govindasamy, V., Senthilkumar, M., Gaikwad, K., & Annapurna, K. (2008). Isolation and characterization of ACC deaminase gene from two plant growth-promoting rhizobacteria. *Current Microbiology*, *57*(4), 312–317. https://doi.org/10.1007/s00284-008-9195-8

Gross, A., Schoendube, J., Zimmermann, S., Steeb, M., Zengerle, R., & Koltay, P. (2015). Technologies for single-cell isolation. *International Journal of Molecular Sciences*, *16*(8), 16897–16919. https://doi.org/10.3390/ijms160816897

Guazzaroni, M., Morgante, V., Mirete, S., & González-Pastor, J. E. (2013). Novel acid resistance genes from the metagenome of the T into R iver, an extremely acidic environment. *Environmental Microbiology*, *15*(4), 1088–1102. https://doi.org/10.1111/1462-2920.12021

Gupta, N., Vats, S., & Bhargava, P. (2018). Sustainable agriculture: Role of metagenomics and metabolomics in exploring the soil microbiota. In *In Silico Approach for Sustainable Agriculture* (pp. 183–199). Springer Singapore. https://doi.org/10.1007/978-981-13-0347-0_11

Hailemariam, S., Zhao, S., & Wang, J. (2020). Complete genome sequencing and transcriptome analysis of nitrogen metabolism of succinivibrio dextrinosolvens strain Z6 isolated from dairy cow rumen. *Frontiers in Microbiology*, *11*. https://doi.org/10.3389/fmicb.2020.01826

Handelsman, J. (2004). Metagenomics: Application of genomics to uncultured microorganisms. *Microbiology and Molecular Biology Reviews*, *68*(4), 669–685. https://doi.org/10.1128/MMBR.68.4.669-685.2004

Hardoim, P. R., van Overbeek, L. S., & Elsas, J. D. van. (2008). Properties of bacterial endophytes and their proposed role in plant growth. *Trends in Microbiology*, *16*(10), 463–471. https://doi.org/10.1016/j.tim.2008.07.008

Hartmann, M., Frey, B., Mayer, J., Mäder, P., & Widmer, F. (2015). Distinct soil microbial diversity under long-term organic and conventional farming. *The ISME Journal*, *9*(5), 1177–1194. https://doi.org/10.1038/ismej.2014.210

Hatayama, K., Kawai, S., Shoun, H., Ueda, Y., & Nakamura, A. (2005). Pseudomonas azotifigens sp. nov., a novel nitrogen-fixing bacterium isolated from a compost pile. *International Journal of Systematic and Evolutionary Microbiology*, *55*(4), 1539–1544. https://doi.org/10.1099/ijs.0.63586-0

Hausmann, B., Pelikan, C., Herbold, C. W., Köstlbacher, S., Albertsen, M., Eichorst, S. A., Glavina del Rio, T., Huemer, M., Nielsen, P. H., Rattei, T., Stingl, U., Tringe, S. G., Trojan, D., Wentrup, C., Woebken, D., Pester, M., & Loy, A. (2018). Peatland *Acidobacteria* with a dissimilatory sulfur metabolism. *The ISME Journal*, *12*(7), 1729–1742. https://doi.org/10.1038/s41396-018-0077-1

Hegyi, A., Nguyen, T. B. K., & Posta, K. (2021). Metagenomic analysis of bacterial communities in agricultural soils from Vietnam with special attention to phosphate solubilizing bacteria. *Microorganisms*, *9*(9), 1796. https://doi.org/10.3390/microorganisms9091796

Hemkemeyer, M., Christensen, B. T., Martens, R., & Tebbolubi. (2015). Soil particle size fractions harbour distinct microbial communities and differ in potential for microbial mineralisation of organic pollutants. *Soil Biology and Biochemistry*, *90*, 255–265. https://doi.org/10.1016/j.soilbio.2015.08.018

Hemkemeyer, M., Christensen, B. T., Tebbe, C. C., & Hartmann, M. (2019). Taxon-specific fungal preference for distinct soil particle size fractions. *European Journal of Soil Biology*, *94*, 103103. https://doi.org/10.1016/j.ejsobi.2019.103103

Hemkemeyer, M., Dohrmann, A. B., Christensen, B. T., & Tebbe, C. C. (2018). Bacterial preferences for specific soil particle size fractions revealed by community analyses. *Frontiers in Microbiology*, *9*. https://doi.org/10.3389/fmicb.2018.00149

Hoff, K. J. (2009). The effect of sequencing errors on metagenomic gene prediction. *BMC Genomics*, *10*(1), 520. https://doi.org/10.1186/1471-2164-10-520

Hoff, K. J., Tech, M., Lingner, T., Daniel, R., Morgenstern, B., & Meinicke, P. (2008). Gene prediction in metagenomic fragments: A large scale machine learning approach. *BMC Bioinformatics*, *9*(1), 217. https://doi.org/10.1186/1471-2105-9-217

Höfte, H., & Whiteley, H. R. (1989). Insecticidal crystal proteins of Bacillus thuringiensis. *Microbiological Reviews*, *53*(2), 242–255. https://doi.org/10.1128/mr.53.2.242-255.1989

Huang, X.-F., Chaparro, J. M., Reardon, K. F., Zhang, R., Shen, Q., & Vivanco, J. M. (2014). Rhizosphere interactions: Root exudates, microbes, and microbial communities. *Botany*, *92*(4), 267–275. https://doi.org/10.1139/cjb-2013-0225

Inoue, J., Oshima, K., Suda, W., Sakamoto, M., Iino, T., Noda, S., Hongoh, Y., Hattori, M., & Ohkuma, M. (2015). Distribution and evolution of nitrogen fixation genes in the phylum *Bacteroidetes*. *Microbes and Environments*, *30*(1), 44–50. https://doi.org/10.1264/jsme2.ME14142

Iquebal, M. A., Jagannadham, J., Jaiswal, S., Prabha, R., Rai, A., & Kumar, D. (2022). Potential use of microbial community genomes in various dimensions of agriculture productivity and its management: A review. *Frontiers in Microbiology, 13*. https://doi.org/10.3389/fmicb.2022.708335

Jayaraman, S., Naorem, A. K., Lal, R., Dalal, R. C., Sinha, N. K., Patra, A. K., & Chaudhari, S. K. (2021). Disease-suppressive soils—beyond food production: A critical review. *Journal of Soil Science and Plant Nutrition, 21*(2), 1437–1465. https://doi.org/10.1007/s42729-021-00451-x

Jiang, C.-Y., Dong, L., Zhao, J.-K., Hu, X., Shen, C., Qiao, Y., Zhang, X., Wang, Y., Ismagilov, R. F., Liu, S.-J., & Du, W. (2016). High-throughput single-cell cultivation on microfluidic streak plates. *Applied and Environmental Microbiology, 82*(7), 2210–2218. https://doi.org/10.1128/AEM.03588-15

Jiang, S.-Q., Yu, Y.-N., Gao, R.-W., Wang, H., Zhang, J., Li, R., Long, X.-H., Shen, Q.-R., Chen, W., & Cai, F. (2019). High-throughput absolute quantification sequencing reveals the effect of different fertilizer applications on bacterial community in a tomato cultivated coastal saline soil. *Science of The Total Environment, 687*, 601–609. https://doi.org/10.1016/j.scitotenv.2019.06.105

Joersbo, M., & Okkels, F. T. (1996). A novel principle for selection of transgenic plant cells: Positive selection. *Plant Cell Reports, 16*(3–4), 219–221. https://doi.org/10.1007/BF01890871

Jones, P., Binns, D., Chang, H.-Y., Fraser, M., Li, W., McAnulla, C., McWilliam, H., Maslen, J., Mitchell, A., Nuka, G., Pesseat, S., Quinn, A. F., Sangrador-Vegas, A., Scheremetjew, M., Yong, S.-Y., Lopez, R., & Hunter, S. (2014). InterProScan 5: Genome-scale protein function classification. *Bioinformatics, 30*(9), 1236–1240. https://doi.org/10.1093/bioinformatics/btu031

Kaeberlein, T., Lewis, K., & Epstein, S. S. (2002). Isolating "uncultivable" microorganisms in pure culture in a simulated natural environment. *Science, 296*(5570), 1127–1129. https://doi.org/10.1126/science.1070633

Kalia, A., & Gosal, S. K. (2011). Effect of pesticide application on soil microorganisms. *Archives of Agronomy and Soil Science, 57*(6), 569–596. https://doi.org/10.1080/03650341003787582

Kampfer, P., Thummes, K., Chu, H.-I., Tan, C.-C., Arun, A. B., Chen, W.-M., Lai, W.-A., Shen, F.-T., Rekha, P. D., & Young, C.-C. (2008). Pseudacidovorax intermedius gen. nov., sp. nov., a novel nitrogen-fixing betaproteobacterium isolated from soil. *International Journal of Systematic and Evolutionary Microbiology, 58*(2), 491–495. https://doi.org/10.1099/ijs.0.65175-0

Kanehisa, M., Sato, Y., Kawashima, M., Furumichi, M., & Tanabe, M. (2016). KEGG as a reference resource for gene and protein annotation. *Nucleic Acids Research, 44*(D1), D457–D462. https://doi.org/10.1093/nar/gkv1070

Kao-Kniffin, J., Carver, S. M., & DiTommaso, A. (2013). Advancing weed management strategies using metagenomic techniques. *Weed Science, 61*(2), 171–184. https://doi.org/10.1614/WS-D-12-00114.1

Khammas, K. M., Ageron, E., Grimont, P. A. D., & Kaiser, P. (1989). *Azospirillum irakense* sp. nov., a nitrogen-fixing bacterium associated with rice roots and rhizosphere soil. *Research in Microbiology, 140*(9), 679–693. https://doi.org/10.1016/0923-2508(89)90199-X

Khan, H. U., Ahmad, R., Ahmed, W., Khan, S. M., & Khan, M. A. (2001). Evaluation of the combined effects of *Paecilomyces lilacinus* and *Trichoderma harzianum* against root-knot disease of tomato. *Journal of Biological Sciences, 1*(3), 139–142. https://doi.org/10.3923/jbs.2001.139.142

Khokhar, I., Haider, S. M., Ali, A., Mukhtar, I., & Mushtaq, S. (2011). Evaluation of antagonistic activity of soil bacteria against plant pathogenic fungi. *Pakistan Journal of Phytopathology, 2*, 166–169.

Kloepper, J. W., McInroy, J. A., Liu, K., & Hu, C.-H. (2013). Symptoms of fern distortion syndrome resulting from inoculation with opportunistic endophytic fluorescent pseudomonas spp. *PLoS One, 8*(3), e58531. https://doi.org/10.1371/journal.pone.0058531

Knowles, B. H. (1994). *Mechanism of Action of Bacillus Thuringiensis Insecticidal δ-Endotoxins* (pp. 275–308). https://doi.org/10.1016/S0065-2806(08)60085-5

Kong, A. Y. Y., Hristova, K., Scow, K. M., & Six, J. (2010). Impacts of different N management regimes on nitrifier and denitrifier communities and N cycling in soil microenvironments. *Soil Biology and Biochemistry, 42*(9), 1523–1533. https://doi.org/10.1016/j.soilbio.2010.05.021

Koul, O. (2012). Microbial biopesticides: Opportunities and challenges. *CABI Reviews*, 1–26. https://doi.org/10.1079/PAVSNNR20116056

Kumar, A., Maurya, B. R., Raghuwanshi, R., Meena, V. S., & Tofazzal Islam, M. (2017). Co-inoculation with enterobacter and rhizobacteria on yield and nutrient uptake by wheat (Triticum aestivum L.) in the alluvial soil under indo-gangetic plain of India. *Journal of Plant Growth Regulation, 36*(3), 608–617. https://doi.org/10.1007/s00344-016-9663-5

Kumar, K., Aggarwal, C., Sapna, Singh, I., & Yadava, P. (2018b). Microbial genes in crop improvement. In *Crop Improvement Through Microbial Biotechnology* (pp. 39–56). Elsevier. https://doi.org/10.1016/B978-0-olu63987-5.00003-7

Kumar, S., Chaudhary, D., shmi, R., Jangra, R., Kumari, A., & Kumar, R. (2018a). Exploring phyllosphere bacteria for growth promotion and yield of potato (Solanum tuberosum L.). *International Journal of Current Microbiology and Applied Sciences, 7*(4), 1065–1071. https://doi.org/10.20546/ijcmas.2018.704.117

Kumar, V. V. (2018). Biofertilizers and biopesticides in sustainable agriculture. In *Role of Rhizospheric Microbes in Soil* (pp. 377–398). Springer Singapore. https://doi.org/10.1007/978-981-10-8402-7_14

Lavy, A., Keren, R., Haber, M., Schwartz, I., & Ilan, M. (2014). Implementing sponge physiological and genomic information to enhance the diversity of its culturable associated bacteria. *FEMS Microbiology Ecology, 87*(2), 486–502. https://doi.org/10.1111/1574-6941.12240

Lee, B., Min, H., & Yoon, S. (2021). MUGAN: Multi-GPU accelerated AmpliconNoise server for rapid microbial diversity assessment. *Bioinformatics, 37*(11), 1562–1570. https://doi.org/10.1093/bioinformatics/bty096

Leite, M. F. A., Dimitrov, M. R., Freitas-Iório, R. P., de Hollander, M., Cipriano, M. A. P., Andrade, S. A. L., da Silveira, A. P. D., & Kuramae, E. E. (2021). Rearranging the sugarcane holobiont via plant growth-promIg bacteria and nitrogen input. *Science of the Total Environment, 800*, 149493. https://doi.org/10.1016/j.scitotenv.2021.149493

Li, D., Liu, C.-M., Luo, R., Sadakane, K., & Lam, T.-W. (2015). MEGAHIT: An ultra-fast single-node solution for large and complex metagenomics assembly via succinct *de Bruijn* graph. *Bioinformatics, 31*(10), 1674–1676. https://doi.org/10.1093/bioinformatics/btv033

Li, W., & Godzik, A. (2006). Cd-hit: A fast program for clustering and comparing large sets of protein or nucleotide sequences. *Bioinformatics*, *22*(13), 1658–1659. https://doi.org/10.1093/bioinformatics/btl158

Liang, J.-L., Liu, J., Jia, P., Yang, T., Zeng, Q., Zhang, S., Liao, B., Shu, W., & Li, J. (2020). Novel phosphate-solubilizing bacteria enhance soil phosphorus cycling following ecological restoration of land degraded by mining. *The ISME Journal*, *14*(6), 1600–1613. https://doi.org/10.1038/s41396-020-0632-4

Liao, X., Zhao, J., Yi, Q., Li, J., Li, Z., Wu, S., Zhang, W., & Wang, K. (2023). Metagenomic insights into the effects of organic and inorganic agricultural managements on soil phosphorus cycling. *Agriculture, Ecosystems & Environment*, *343*, 108281. https://doi.org/10.1016/j.agee.2022.108281

Lightwood, D. J., Ellar, D. J., & Jarrett, P. (2000). Role of proteolysis in determining potency of *Bacillus thuringiensis* Cry1Ac δ-endotoxin. *Applied and Environmental Microbiology*, *66*(12), 5174–5181. https://doi.org/10.1128/AEM.66.12.5174-5181.2000

Liu, C., Jin, Y., Hu, Y., Tang, J., Xiong, Q., Xu, M., Bibi, F., & Beng, K. C. (2019). Drivers of soil bacterial community structure and diversity in tropical agroforestry systems. *Agriculture, Ecosystems & Environment*, *278*, 24–34. https://doi.org/10.1016/j.agee.2019.03.015

Llewellyn, D. J., Mares, C. L., & Fitt, G. P. (2007). Field performance and seasonal changes in the efficacy against *Helicoverpa armigera* (Hübner) of transgenic cotton expressing the insecticidal protein vip3A. *Agricultural and Forest Entomology*, *9*(2), 93–101. https://doi.org/10.1111/j.1461-9563.2007.00332.x

Lombard, N., Prestat, E., van Elsas, J. D., & Simonet, P. (2011). Soil-specific limitations for access and analysis of soil microbial communities by metagenomics. *FEMS Microbiology Ecology*, *78*(1), 31–49. https://doi.org/10.1111/j.1574-6941.2011.01140.x

Lozano-Tovar, M. D., Garrido-Jurado, I., Lafont, F., & Quesada-Moraga, E. (2015). Insecticidal activity of a destruxin-containing extract of metarhizium brunneum against ceratitis capitata (Diptera: Tephritidae). *Journal of Economic Entomology*, *108*(2), 462–472. https://doi.org/10.1093/jee/tov041

Lugtenberg, B., & Kamilova, F. (2009). Plant-growth-promoting rhizobacteria. *Annual Review of Microbiology*, *63*(1), 541–556. https://doi.org/10.1146/annurev.micro.62.081307.162918

Luo, G., Ling, N., Nannipieri, P., Chen, H., Raza, W., Wang, M., Guo, S., & Shen, Q. (2017). Long-term fertilisation regimes affect the composition of the alkaline phosphomonoesterase encoding microbial community of a vertisol and its derivative soil fractions. *Biology and Fertility of Soils*, *53*(4), 375–388. https://doi.org/10.1007/s00374-017-1183-3

Maestre, F. T., Delgado-Baquerizo, M., Jeffries, T. C., Eldridge, D. J., Ochoa, V., Gozalo, B., Quero, J. L., García-Gómez, M., Gallardo, A., Ulrich, W., Bowker, M. A., Arredondo, T., Barraza-Zepeda, C., Bran, D., Florentino, A., Gaitán, J., Gutiérrez, J. R., Huber-Sannwald, E., Jankju, M., . . . Singh, B. K. (2015). Increasing aridity reduces soil microbial diversity and abundance in global drylands. *Proceedings of the National Academy of Sciences*, *112*(51), 15684–15689. https://doi.org/10.1073/pnas.1516684112

Magalhães Cruz, L., Maltempi de Souza, E., Weber, O. B., Baldani, J. I., Döbereiner, J., & de Oliveira Pedrosa, F. (2001). 16S ribosomal DNA characterization of nitrogen-fixing bacteria isolated from banana (*Musa* spp.) and pineapple (*Ananas comosus* (L.) Merril). *Applied and Environmental Microbiology*, *67*(5), 2375–2379. https://doi.org/10.1128/AEM.67.5.2375-2379.2001

Malhotra, S., Mishra, V., Karmakar, S., & Sharma, R. S. (2017). Environmental predictors of indole acetic acid producing rhizobacteria at fly ash dumps: Nature-based solution for sustainable restoration. *Frontiers in Environmental Science*, *5*. https://doi.org/10.3389/fenvs.2017.00059

Malmstrom, R. R., & Eloe-Fadrosh, E. A. (2019). Advancing genome-resolved metagenomics beyond the shotgun. *MSystems*, *4*(3). https://doi.org/10.1128/mSystems.00118-19

Marasco, R., Rolli, E., Ettoumi, B., Vigani, G., Mapelli, F., Borin, S., Abou-Hadid, A. F., El-Behairy, U. A., Sorlini, C., Cherif, A., Zocchi, G., & Daffonchio, D. (2012). A drought resistance-promoting microbiome is selected by root system under desert farming. *PLoS One*, *7*(10), e48479. https://doi.org/10.1371/journal.pone.0048479

Margulies, M., Egholm, M., Altman, W. E., Attiya, S., Bader, J. S., Bemben, L. A., Berka, J., Braverman, M. S., Chen, Y.-J., Chen, Z., Dewell, S. B., Du, L., Fierro, J. M., Gomes, X. V., Godwin, B. C., He, W., Helgesen, S., Ho, C. H., Irzyk, G. P., . . . Rothberg, J. M. (2005). Genome sequencing in microfabricated high-density picolitre reactors. *Nature*, *437*(7057), 376–380. https://doi.org/10.1038/nature03959

Markowitz, V. M., Ivanova, N., Palaniappan, K., Szeto, E., Korzeniewski, F., Lykidis, A., Anderson, I., Mavrommatis, K., Kunin, V., Garcia Martin, H., Dubchak, I., Hugenholtz, P., & Kyrpides, N. C. (2006). An experimental metagenome data management and analysis system. *Bioinformatics*, *22*(14), e359–e367. https://doi.org/10.1093/bioinformatics/btl217

Martin, J. P., Martin, W. P., Page, J. B., Raney, W. A., & de Ment, J. D. (1955). Soil aggregation. In *Advances in Agronomy* (Vol. 7, pp. 1–37). https://doi.org/10.1016/S0065-2113(08)60333-8

Martin-Laurent, F., Philippot, L., Hallet, S., Chaussod, R., Germon, J. C., Soulas, G., & Catroux, G. (2001). DNA extraction from soils: Old bias for new microbial diversity analysis methods. *Applied and Environmental Microbiology*, *67*(5), 2354–2359. https://doi.org/10.1128/AEM.67.5.2354-2359.2001

Mastouri, F., Björkman, T., & Harman, G. E. (2012). *Trichoderma harzianum* enhances antioxidant defense of tomato seedlings and resistance to water deficit. *Molecular Plant-Microbe Interactions®*, *25*(9), 1264–1271. https://doi.org/10.1094/MPMI-09-11-0240

Maurya, M. K., Singh, R., & Tomer, A. (2014). In vitro evaluation of antagonistic activity of pseudomonas fluorescensagainst fungal pathogen. *Journal of Biopesticides*, *7*(1), 43–46. https://doi.org/10.57182/jbiopestic.7.1.43-46

McBride, M. J., & Ensign, J. C. (1987). Metabolism of endogenous trehalose by Streptomyces griseus spores and by spores or cells of other actinomycetes. *Journal of Bacteriology*, *169*(11), 5002–5007. https://doi.org/10.1128/jb.169.11.5002-5007.1987

McClung, C. R., Patriquin, D. G., & Davis, R. E. (1983). Campylobacter nitrofigilis sp. nov., a nitrogen-fixing bacterium associated with roots of spartina alterniflora loisel. *International Journal of Systematic Bacteriology*, *33*(3), 605–612. https://doi.org/10.1099/00207713-33-3-605

McIver, L. J., Abu-Ali, G., Franzosa, E. A., Schwager, R., Morgan, X. C., Waldron, L., Segata, N., & Huttenhower, C. (2018). bioBakery: A meta'omic analysis environment. *Bioinformatics*, *34*(7), 1235–1237. https://doi.org/10.1093/bioinformatics/btx754

Meena, R. S., Meena, V. S., Meena, S. K., & Verma, J. P. (2015a). The needs of healthy soils for a healthy world. *Journal of Cleaner Production*, *102*, 560–561. https://doi.org/10.1016/j.jclepro.2015.04.045

Meena, R. S., Meena, V. S., Meena, S. K., & Verma, J. P. (2015b). Towards the plant stress mitigate the agricultural productivity: A book review. *Journal of Cleaner Production*, *102*, 552–553. https://doi.org/10.1016/j.jclepro.2015.04.047

Meena, V. S., Maurya, B. R., & Bahadur, I. (2015c). Potassium solubilization by bacterial strain in waste mica. *Bangladesh Journal of Botany*, *43*(2), 235–237. https://doi.org/10.3329/bjb.v43i2.21680

Meena, V. S., Maurya, B. R., & Verma, J. P. (2014). Does a rhizospheric microorganism enhance K+ availability in agricultural soils? *Microbiological Research*, *169*(5–6), 337–347. https://doi.org/10.1016/j.micres.2013.09.003

Mehinto, J. T., Atachi, P., Douro Kpindou, O., Dannon, E., & Tamò, M. (2014). Mortality of maruca vitrata (Lepidoptera: Crambidae) larval stages induced by different doses of the entomopathogenic fungi metarhizium anisopliae and Beauveria bassiana. *International Journal of Advanced Research*, *4*, 273–285.

Mendes, R., Garbeva, P., & Raaijmakers, J. M. (2013). The rhizosphere microbiome: Significance of plant beneficial, plant pathogenic, and human pathogenic microorganisms. *FEMS Microbiology Reviews*, *37*(5), 634–663. https://doi.org/10.1111/1574-6976.12028

Meyer, F., Paarmann, D., D'Souza, M., Olson, R., Glass, E., Kubal, M., Paczian, T., Rodriguez, A., Stevens, R., Wilke, A., Wilkening, J., & Edwards, R. (2008). The metagenomics RAST server: A public resource for the automatic phylogenetic and functional analysis of metagenomes. *BMC Bioinformatics*, *9*(1), 386. https://doi.org/10.1186/1471-2105-9-386

Mirete, S., de Figueras, C. G., & González-Pastor, J. E. (2007). Novel nickel resistance genes from the rhizosphere metagenome of plants adapted to acid mine drainage. *Applied and Environmental Microbiology*, *73*(19), 6001–6011. https://doi.org/10.1128/AEM.00048-07

Mirete, S., Mora-Ruiz, M. R., Lamprecht-Grandío, M., de Figueras, C. G., Rosselló-Móra, R., & González-Pastor, J. E. (2015). Salt resistance genes revealed by functional metagenomics from brines and moderate-salinity rhizosphere within a hypersaline environment. *Frontiers in Microbiology*, *6*. https://doi.org/10.3389/fmicb.2015.01121

Mitchell, A. L., Scheremetjew, M., Denise, H., Potter, S., Tarkowska, A., Qureshi, M., Salazar, G. A., Pesseat, S., Boland, M. A., Hunter, F. M. I., ten Hoopen, P., Alako, B., Amid, C., Wilkinson, D. J., Curtis, T. P., Cochrane, G., & Finn, R. D. (2018). EBI Metagenomics in 2017: Enriching the analysis of microbial communities, from sequence reads to assemblies. *Nucleic Acids Research*, *46*(D1), D726–D735. https://doi.org/10.1093/nar/gkx967

Mittal, V., Singh, O., Nayyar, H., Kaur, J., & Tewari, R. (2008). Stimulatory effect of phosphate-solubilizing fungal strains (Aspergillus awamori and Penicillium citrinum) on the yield of chickpea (Cicer arietinum L. cv. GPF2). *Soil Biology and Biochemistry*, *40*(3), 718–727. https://doi.org/10.1016/j.soilbio.2007.10.008

Mohanty, S. R., Dubey, G., Ahirwar, U., Patra, A. K., & Kollah, B. (2016). Prospect of phyllosphere microbiota: A case study on bioenergy crop jatropha curcas. In *Plant-Microbe Interaction: An Approach to Sustainable Agriculture* (pp. 453–462). Springer Singapore. https://doi.org/10.1007/978-981-10-2854-0_20

Morgante, V., Mirete, S., de Figueras, C. G., Postigo Cacho, M., & González-Pastor, J. E. (2015). Exploring the diversity of arsenic resistance genes from acid mine drainage microorganisms. *Environmental Microbiology*, *17*(6), 1910–1925. https://doi.org/10.1111/1462-2920.12505

Müller, C. A., Obermeier, M. M., & Berg, G. (2016). Bioprospecting plant-associated microbiomes. *Journal of Biotechnology*, *235*, 171–180. https://doi.org/10.1016/j.jbiotec.2016.03.033

Mummey, D. L., Holben, W., Six, J., & Stahl, P. (2006). Spatial stratification of soil bacterial populations in aggregates of diverse soils. *Microbial Ecology*, *51*(3), 404–411. https://doi.org/10.1007/s00248-006-9020-5

Mummey, D. L., & Stahl, P. D. (2004). Analysis of soil whole- and inner-microaggregate bacterial communities. *Microbial Ecology*, *48*(1), 41–50. https://doi.org/10.1007/s00248-003-1000-4

Nabaei, N., Mehrvar, A., & Saber, B. M. (2012). Efficacy of entomopathogenic fungi in combination with diatomaceous earth against Callosobruchus maculatus (Coleoptera: Bruchidae). *Acta Entomologica Sinica*, *55*, 1282–1288.

Nagayama, H., Sugawara, T., Endo, R., Ono, A., Kato, H., Ohtsubo, Y., Nagata, Y., & Tsuda, M. (2015). Isolation of oxygenase genes for indigo-forming activity from an artificially polluted soil metagenome by functional screening using Pseudomonas putida strains as hosts. *Applied Microbiology and Biotechnology*, *99*(10), 4453–4470. https://doi.org/10.1007/s00253-014-6322-2

Nannipieri, P., Ascher, J., Ceccherini, M. T., Landi, L., Pietramellara, G., & Renella, G. (2003). Microbial diversity and soil functions. *European Journal of Soil Science*, *54*(4), 655–670. https://doi.org/10.1046/j.1351-0754.2003.0556.x

Nath, D., Maurya, B. R., & Meena, V. S. (2017). Documentation of five potassium- and phosphorus-solubilizing bacteria for their K and P-solubilization ability from various minerals. *Biocatalysis and Agricultural Biotechnology*, *10*, 174–181. https://doi.org/10.1016/j.bcab.2017.03.007

Nautiyal, C. S., Srivastava, S., Chauhan, P. S., Seem, K., Mishra, A., & Sopory, S. K. (2013). Plant growth-promoting bacteria Bacillus amyloliquefaciens NBRISN13 modulates gene expression profile of leaf and rhizosphere community in rice during salt stress. *Plant Physiology and Biochemistry*, *66*, 1–9. https://doi.org/10.1016/j.plaphy.2013.01.020

Niu, B., Fu, L., Sun, S., & Li, W. (2010). Artificial and natural duplicates in pyrosequencing reads of metagenomic data. *BMC Bioinformatics*, *11*(1), 187. https://doi.org/10.1186/1471-2105-11-187

Nwachukwu, B. C., & Babalola, O. O. (2022). Metagenomics: A tool for exploring key microbiome with the potentials for improving sustainable agriculture. *Frontiers in Sustainable Food Systems*, *6*. https://doi.org/10.3389/fsufs.2022.886987

Oades, J. M. (1984). Soil organic matter and structural stability: Mechanisms and implications for management. *Plant and Soil*, *76*(1–3), 319–337. https://doi.org/10.1007/BF02205590

Oclarit, E., & Cumagun, C. (2009). Evaluation of efficacy of paecilomyces lilacinus as biological control agent of meloidogyne incognita attacking tomato. *Journal of Plant Protection Research*, *49*(4), 337–340. https://doi.org/10.2478/v10045-009-0053-x

Ono, A., Miyazaki, R., Sota, M., Ohtsubo, Y., Nagata, Y., & Tsuda, M. (2007). Isolation and characterization of naphthalene-catabolic genes and plasmids from oil-contaminated soil by using two cultivation-independent approaches. *Applied Microbiology and Biotechnology*, *74*(2), 501–510. https://doi.org/10.1007/s00253-006-0671-4

Pahari, A., & Mishra, B. B. (2017). Characterization of siderophore producing rhizobacteria and its effect on growth performance of different vegetables. *International Journal of Current Microbiology and Applied Sciences*, *6*(5), 1398–1405. https://doi.org/10.20546/ijcmas.2017.605.152

Panke-Buisse, K., Poole, A. C., Goodrich, J. K., Ley, R. E., & Kao-Kniffin, J. (2015). Selection on soil microbiomes reveals reproducible impacts on plant function. *The ISME Journal*, *9*(4), 980–989. https://doi.org/10.1038/ismej.2014.196

Parewa, H. P., Joshi, N., Meena, V. S., Joshi, S., Choudhary, A., Ram, M., Meena, S. C., & Jain, L. K. (2021). Role of biofertilizers and biopesticides in organic farming. In *Advances in Organic Farming* (pp. 133–159). Elsevier. https://doi.org/10.1016/B978-0-12-822358-1.00009-2

Parks, D. H., Imelfort, M., Skennerton, C. T., Hugenholtz, P., & Tyson, G. W. (2015). CheckM: Assessing the quality of microbial genomes recovered from isolates, single cells, and metagenomes. *Genome Research*, *25*(7), 1043–1055. https://doi.org/10.1101/gr.186072.114

Pedraza, R. O. (2008). Recent advances in nitrogen-fixing acetic acid bacteria. *International Journal of Food Microbiology*, *125*(1), 25–35. https://doi.org/10.1016/j.ijfoodmicro.2007.11.079

Pedrinho, A., Mendes, L. W., do Rêgo Barros, F. M., Merloti, L. F., Martins e Martins, M., Cotta, S. R., Andreote, F. D., & Tsai, S. M. (2023). Impacts of deforestation and forest regeneration on soil bacterial communities associated with phosphorus transformation processes in the Brazilian Amazon region. *Ecological Indicators*, *146*, 109779. https://doi.org/10.1016/j.ecolind.2022.109779

Pester, M., Bittner, N., Deevong, P., Wagner, M., & Loy, A. (2010). A 'rare biosphere' microorganism contributes to sulfate reduction in a peatland. *The ISME Journal*, *4*(12), 1591–1602. https://doi.org/10.1038/ismej.2010.75

Pi, H.-W., Chiang, Y.-R., & Li, W.-H. (2024). Mapping geological events and nitrogen fixation evolution onto the timetree of the evolution of nitrogen-fixation genes. *Molecular Biology and Evolution*, *41*(2). https://doi.org/10.1093/molbev/msae023

Pi, H.-W., Lin, J.-J., Chen, C.-A., Wang, P.-H., Chiang, Y.-R., Huang, C.-C., Young, C.-C., & Li, W.-H. (2022). Origin and evolution of nitrogen fixation in prokaryotes. *Molecular Biology and Evolution*, *39*(9). https://doi.org/10.1093/molbev/msac181

Prasad, R., Chandra, H., Sinha, B. K., & Srivastava, J. (2015). Antagonistic effect of Pseudomonas fluorescens isolated from soil of Doon Valley (Dehradun-India)on certain Phyo-Pathogenic Fungi. *Octa Journal of Biosciences*, *2*, 92–95.

Prayogo, F. A., Budiharjo, A., Kusumaningrum, H. P., Wijanarka, W., Suprihadi, A., & Nurhayati, N. (2020). Metagenomic applications in exploration and development of novel enzymes from nature: A review. *Journal of Genetic Engineering and Biotechnology*, *18*(1), 39. https://doi.org/10.1186/s43141-020-00043-9

Priyadharsini, P., & Muthukumar, T. (2016). Interactions between arbuscular mycorrhizal fungi and potassium-solubilizing microorganisms on agricultural productivity. In *Potassium Solubilizing Microorganisms for Sustainable Agriculture* (pp. 111–125). Springer India. https://doi.org/10.1007/978-81-322-2776-2_8

Priyaja, P., Jayesh, P., Correya, N. S., Sreelakshmi, B., Sudheer, N. S., Philip, R., & Singh, I. S. B. (2014). Antagonistic effect of pseudomonas aeruginosa isolates from various ecological niches on vibrio species pathogenic to crustaceans. *Journal of Coastal Life Medicine*, *1*, 76–84.

Quince, C., Walker, A. W., Simpson, J. T., Loman, N. J., & Segata, N. (2017). Shotgun metagenomics, from sampling to analysis. *Nature Biotechnology*, *35*(9), 833–844. https://doi.org/10.1038/nbt.3935

Raja, N. (2013). Biopesticides and biofertilizers: Ecofriendly sources for sustainable agriculture. *Journal of Fertilizers & Pesticides*, *4*(1). https://doi.org/10.4172/2155-6202.1000e112

Raja, N., & Masresha, G. (2015). Plant based biopesticides: Safer alternative for organic food production. *Journal of Fertilizers & Pesticides*, *6*(2). https://doi.org/10.4172/2471-2728.1000e128

Raknes, I. A., & Bongo, L. A. (2018). META-pipe authorization service. *F1000Research*, *7*, 32. https://doi.org/10.12688/f1000research.13256.2

Ramasamy, M., Geetha, T., & Yuvaraj, M. (2020). Role of biofertilizers in plant growth and soil health. In *Nitrogen Fixation*. IntechOpen. https://doi.org/10.5772/intechopen.87429

Ramesh, R., Joshi, A. A., & Ghanekar, M. P. (2009). Pseudomonads: Major antagonistic endophytic bacteria to suppress bacterial wilt pathogen, Ralstonia solanacearum in the eggplant (Solanum melongena L.). *World Journal of Microbiology and Biotechnology*, *25*(1), 47–55. https://doi.org/10.1007/s11274-008-9859-3

Ramesh, S. V., Mishra, A. K., & Praveen, S. (2007). Hairpin RNA-mediated strategies for silencing of *Tomato Leaf Curl Virus* AC1 and AC4 genes for effective resistance in plants. *Oligonucleotides*, *17*(2), 251–257. https://doi.org/10.1089/oli.2006.0063

Ranjard, L., Poly, F., Combrisson, J., Richaume, A., Gourbière, F., Thioulouse, J., & Nazaret, S. (2000). Heterogeneous cell density and genetic structure of bacterial pools associated with various soil microenvironments as determined by enumeration and DNA fingerprinting approach (RISA). *Microbial Ecology*, *39*(4), 263–272.

Raynaud, X., & Nunan, N. (2014). Spatial ecology of bacteria at the microscale in soil. *PLoS One*, *9*(1), e87217. https://doi.org/10.1371/journal.pone.0087217

Reichenbach, M., Fiener, P., Garland, G., Griepentrog, M., Six, J., & Doetterl, S. (2021). The role of geochemistry in organic carbon stabilization against microbial decomposition in tropical rainforest soils. *Soil*, *7*(2), 453–475. https://doi.org/10.5194/soil-7-453-2021

Reinhold-Hurek, B., Hurek, T., Claeyssens, M., & van Montagu, M. (1993). Cloning, expression in Escherichia coli, and characterization of cellulolytic enzymes of Azoarcus sp., a root-invading diazotroph. *Journal of Bacteriology*, *175*(21), 7056–7065. https://doi.org/10.1128/jb.175.21.7056-7065.1993

Reis, V. M., Santos, P. E., Tenorio-Salgado, S., Vogel, J., Stoffels, M., Guyon, S., Mavingui, P., Baldani, V. L. D., Schmid, M.,

Baldani, J. I., Balandreau, J., Hartmann, A., & Caballero-Mellado, J. (2004). Burkholderia tropica sp. nov., a novel nitrogen-fixing, plant-associated bacterium. *International Journal of Systematic and Evolutionary Microbiology*, *54*(6), 2155–2162. https://doi.org/10.1099/ijs.0.02879-0

Richardson, A. E., & Simpson, R. J. (2011). Soil microorganisms mediating phosphorus availability update on microbial phosphorus. *Plant Physiology*, *156*(3), 989–996. https://doi.org/10.1104/pp.111.175448

Riederer-Henderson, M.-A., & Wilson, P. W. (1970). Nitrogen fixation by sulphate-reducing bacteria. *Journal of General Microbiology*, *61*(1), 27–31. https://doi.org/10.1099/00221287-61-1-27

Riesenfeld, C. S., Schloss, P. D., & Handelsman, J. (2004). Metagenomics: Genomic analysis of microbial communities. *Annual Review of Genetics*, *38*(1), 525–552. https://doi.org/10.1146/annurev.genet.38.072902.091216

Rillig, M. C., Muller, L. A. H., & Lehmann, A. (2017). Soil aggregates as massively concurrent evolutionary incubators. *The ISME Journal*, *11*(9), 1943–1948. https://doi.org/10.1038/ismej.2017.56

Rinke, C., Schwientek, P., Sczyrba, A., Ivanova, N. N., Anderson, I. J., Cheng, J.-F., Darling, A., Malfatti, S., Swan, B. K., Gies, E. A., Dodsworth, J. A., Hedlund, B. P., Tsiamis, G., Sievert, S. M., Liu, W.-T., Eisen, J. A., Hallam, S. J., Kyrpides, N. C., Stepanauskas, R., . . . Woyke, T. (2013). Insights into the phylogeny and coding potential of microbial dark matter. *Nature*, *499*(7459), 431–437. https://doi.org/10.1038/nature12352

Riva, D. L., González-Cabrera, G. A., Vázquez-Padrón, R., & Ayra-Pardo, C. (1998). Agrobacterium tumefaciens: A natural tool for plant transformation. *Electronic Journal of Biotechnology*, *1*(3), 24–25.

Rothberg, J. M., & Leamon, J. H. (2008). The development and impact of 454 sequencing. *Nature Biotechnology*, *26*(10), 1117–1124. https://doi.org/10.1038/nbt1485

Saberi, F., Marzban, R., Ardjmand, M., Pajoum Shariati, F., & Tavakoli, O. (2020). Optimization of culture media to enhance the ability of local Bacillus thuringiensis var. tenebrionis. *Journal of the Saudi Society of Agricultural Sciences*, *19*(7), 468–475. https://doi.org/10.1016/j.jssas.2020.08.004

Saghaï, A., Banerjee, S., Degrune, F., Edlinger, A., García-Palacios, P., Garland, G., Heijden, M. G. van der, Herzog, C., Maestre, F. T., Pescador, D. S., Philippot, L., Rillig, M., Romdhane, S., & Hallin, S. (2021). Diversity of archaea and niche preferences among putative ammonia-oxidizing Nitrososphaeria dominating across European arable soils. https://doi.org/10.21203/rs.3.rs-429240/v1

Sagova-Mareckova, M., Omelka, M., & Kopecky, J. (2023). The golden goal of soil management: Disease-suppressive soils. *Phytopathology®*, *113*(4), 741–752. https://doi.org/10.1094/PHYTO-09-22-0324-KD

Saha, M., Sarkar, S., Sarkar, B., Sharma, B. K., Bhattacharjee, S., & Tribedi, P. (2016). Microbial siderophores and their potential applications: A review. *Environmental Science and Pollution Research*, *23*(5), 3984–3999. https://doi.org/10.1007/s11356-015-4294-0

Sanguin, H., Remenant, B., Dechesne, A., Thioulouse, J., Vogel, T. M., Nesme, X., Moënne-Loccoz, Y., & Grundmann, G. L. (2006). Potential of a 16S rRNA-based taxonomic microarray for analyzing the rhizosphere effects of maize on *Agrobacterium* spp. and bacterial communities. *Applied and Environmental Microbiology*, *72*(6), 4302–4312. https://doi.org/10.1128/AEM.02686-05

Saravanan, V. S., Kalaiarasan, P., Madhaiyan, M., & Thangaraju, M. (2007). Solubilization of insoluble zinc compounds by Gluconacetobacter diazotrophicus and the detrimental action of zinc ion (Zn 2+) and zinc chelates on root knot nematode Meloidogyne incognita. *Letters in Applied Microbiology*, *44*(3), 235–241. https://doi.org/10.1111/j.1472-765X.2006.02079.x

Schauss, K., Focks, A., Leininger, S., Kotzerke, A., Heuer, H., Thiele-Bruhn, S., Sharma, S., Wilke, B., Matthies, M., Smalla, K., Munch, J. C., Amelung, W., Kaupenjohann, M., Schloter, M., & Schleper, C. (2009). Dynamics and functional relevance of ammonia-oxidizing archaea in two agricultural soils. *Environmental Microbiology*, *11*(2), 446–456. https://doi.org/10.1111/j.1462-2920.2008.01783.x

Schlüter, S., Leuther, F., Albrecht, L., Hoeschen, C., Kilian, R., Surey, R., Mikutta, R., Kaiser, K., Mueller, C. W., & Vogel, H.-J. (2022). Microscale carbon distribution around pores and particulate organic matter varies with soil moisture regime. *Nature Communications*, *13*(1), 2098. https://doi.org/10.1038/s41467-022-29605-w

Schmidt, R., Gravuer, K., Bossange, A. V., Mitchell, J., & Scow, K. (2018). Long-term use of cover crops and no-till shift soil microbial community life strategies in agricultural soil. *PLoS One*, *13*(2), e0192953. https://doi.org/10.1371/journal.pone.0192953

Scholz, M. B., Lo, C.-C., & Chain, P. S. (2012). Next generation sequencing and bioinformatic bottlenecks: The current state of metagenomic data analysis. *Current Opinion in Biotechnology*, *23*(1), 9–15. https://doi.org/10.1016/j.copbio.2011.11.013

Schreiber, L., Krimm, U., Knoll, D., Sayed, M., Auling, G., & Kroppenstedt, R. M. (2005). Plant–microbe interactions: Identification of epiphytic bacteria and their ability to alter leaf surface permeability. *New Phytologist*, *166*(2), 589–594. https://doi.org/10.1111/j.1469-8137.2005.01343.x

Schröder, C., Elleuche, S., Blank, S., & Antranikian, G. (2014). Characterization of a heat-active archaeal β-glucosidase from a hydrothermal spring metagenome. *Enzyme and Microbial Technology*, *57*, 48–54. https://doi.org/10.1016/j.enzmictec.2014.01.010

Seemann, T. (2014). Prokka: Rapid prokaryotic genome annotation. *Bioinformatics*, *30*(14), 2068–2069. https://doi.org/10.1093/bioinformatics/btu153

Seitz, T. J., Schütte, U. M. E., & Drown, D. M. (2022). Unearthing shifts in microbial communities across a soil disturbance gradient. *Frontiers in Microbiology*, *13*. https://doi.org/10.3389/fmicb.2022.781051

Sessitsch, A., Hardoim, P., Döring, J., Weilharter, A., Krause, A., Woyke, T., Mitter, B., Hauberg-Lotte, L., Friedrich, F., Rahalkar, M., Hurek, T., Sarkar, A., Bodrossy, L., van Overbeek, L., Brar, D., van Elsas, J. D., & Reinhold-Hurek, B. (2012). Functional characteristics of an endophyte community colonizing rice roots as revealed by metagenomic analysis. *Molecular Plant-Microbe Interactions®*, *25*(1), 28–36. https://doi.org/10.1094/MPMI-08-11-0204

Sessitsch, A., Weilharter, A., Gerzabek, M. H., Kirchmann, H., & Kandeler, E. (2001). Microbial population structures in soil particle size fractions of a long-term fertilizer field experiment. *Applied and Environmental*

Microbiology, *67*(9), 4215–4224. https://doi.org/10.1128/AEM.67.9.4215-4224.2001

Sharififard, M., Mossadegh, M., Vazirianzadeh, B., & Zarei-Mahmoudabadi, A. (2011). Interactions between entomopathogenic fungus, metarhizium anisopliae and sublethal doses of spinosad for control of house fly, musca domestica. *Iranian Journal of Arthropod-Borne Diseases*, *5*(1), 28–36.

Sharma, P., Kaur, N., & Garg, K. (2012). Revitalizing soil health with bio-fertilizers. In K. Chandra (Ed.), *Biofertiliser News Letter: Vol. 20(2)* (pp. 5–10). National Centre of Organic Farming, National Project on Organic Farming, Ministry of Agriculture, Government of India.

She, W., Bai, Y., Zhang, Y., Qin, S., Feng, W., Sun, Y., Zheng, J., & Wu, B. (2018). Resource availability drives responses of soil microbial communities to short-term precipitation and nitrogen addition in a desert shrubland. *Frontiers in Microbiology*, *9*. https://doi.org/10.3389/fmicb.2018.00186

Sheng, X. F. (2005). Growth promotion and increased potassium uptake of cotton and rape by a potassium releasing strain of Bacillus edaphicus. *Soil Biology and Biochemistry*, *37*(10), 1918–1922. https://doi.org/10.1016/j.soilbio.2005.02.026

Shi, Y., Su, C., Wang, M., Liu, X., Liang, C., Zhao, L., Zhang, X., Minggagud, H., Feng, G., & Ma, W. (2020). Modern climate and soil properties explain functional structure better than phylogenetic structure of plant communities in Northern China. *Frontiers in Ecology and Evolution*, *8*. https://doi.org/10.3389/fevo.2020.531947

Siddiqi, M. Z., Choi, G.-M., Choi, K. D., & Im, W.-T. (2017). Paenibacillus azotifigens sp. nov., a novel nitrogen-fixing bacterium isolated from paddy soil. *International Journal of Systematic and Evolutionary Microbiology*, *67*(12), 4917–4922. https://doi.org/10.1099/ijsem.0.002269

Siddiqui, Z. A., & Akhtar, M. S. (2008). Synergistic effects of antagonistic fungi and a plant growth promoting rhizobacterium, an arbuscular mycorrhizal fungus, or composted cow manure on populations of *Meloidogyne incognita* and growth of tomato. *Biocontrol Science and Technology*, *18*(3), 279–290. https://doi.org/10.1080/09583150801896043

Silva, C. C., Hayden, H., Sawbridge, T., Mele, P., De Paula, S. O., Silva, L. C. F., Vidigal, P. M. P., Vicentini, R., Sousa, M. P., Torres, A. P. R., Santiago, V. M. J., & Oliveira, V. M. (2013). Identification of genes and pathways related to phenol degradation in metagenomic libraries from petroleum refinery wastewater. *PLoS One*, *8*(4), e61811. https://doi.org/10.1371/journal.pone.0061811

Simon, C., & Daniel, R. (2011). Metagenomic analyses: Past and future trends. *Applied and Environmental Microbiology*, *77*(4), 1153–1161. https://doi.org/10.1128/AEM.02345-10

Singh, S., Singh, B. K., Yadav, S. M., & Gupta, A. K. (2014). Potential of biofertilizers in crop production in Indian agriculture. *American Journal of Plant Nutrition and Fertilization Technology*, *4*(2), 33–40. https://doi.org/10.3923/ajpnft.2014.33.40

Sivanantham, T., Rasaiyah, V., Satkunanathan, N., & Thavaranjit, A. C. (2013). In vitro screening of antagonistic effect of soil borne bacteria on some selected phytopathogenic fungi. *Archives of Applied Science Research*, *1*, 1–4.

Six, J., Bossuyt, H., Degryze, S., & Denef, K. (2004). A history of research on the link between (micro)aggregates, soil biota, and soil organic matter dynamics. *Soil and Tillage Research*, *79*(1), 7–31. https://doi.org/10.1016/j.still.2004.03.008

Six, J., Elliott, E. T., & Paustian, K. (2000). Soil macroaggregate turnover and microaggregate formation: A mechanism for C sequestration under no-tillage agriculture. *Soil Biology and Biochemistry*, *32*(14), 2099–2103. https://doi.org/10.1016/S0038-0717(00)00179-6

Song, Y. (2019). A new CRISPR scissor. *Nature Chemical Biology*, *15*(4), 315–315. https://doi.org/10.1038/s41589-019-0261-2

Sørensen, J., Haubjerg Nicolaisen, M., Ron, E., & Simonet, P. (2009). Molecular tools in rhizosphere microbiology—from single-cell to whole-community analysis. *Plant and Soil*, *321*(1–2), 483–512. https://doi.org/10.1007/s11104-009-9946-8

Sowjanya Sree, K., Padmaja, V., & Murthy, Y. L. (2008). Insecticidal activity of destruxin, a mycotoxin from *Metarhizium anisopliae* (Hypocreales), against *Spodoptera litura* (Lepidoptera: Noctuidae) larval stages. *Pest Management Science*, *64*(2), 119–125. https://doi.org/10.1002/ps.1480

Spooren, J., van Bentum, S., Thomashow, L. S., Pieterse, C. M. J., Weller, D. M., & Berendsen, R. L. (2024). Plant-driven assembly of disease-suppressive soil microbiomes. *Annual Review of Phytopathology*, *62*(1), 1–30. https://doi.org/10.1146/annurev-phyto-021622-100127

Streit, W. R., & Schmitz, R. A. (2004). Metagenomics—the key to the uncultured microbes. *Current Opinion in Microbiology*, *7*(5), 492–498. https://doi.org/10.1016/j.mib.2004.08.002

Su, L., Feng, H., Mo, X., Sun, J., Qiu, P., Liu, Y., Zhang, R., Kuramae, E. E., Shen, B., & Shen, Q. (2022). Potassium phosphite enhanced the suppressive capacity of the soil microbiome against the tomato pathogen Ralstonia solanacearum. *Biology and Fertility of Soils*, *58*(5), 553–563. https://doi.org/10.1007/s00374-022-01634-z

Suenaga, H., Ohnuki, T., & Miyazaki, K. (2007). Functional screening of a metagenomic library for genes involved in microbial degradation of aromatic compounds. *Environmental Microbiology*, *9*(9), 2289–2297. https://doi.org/10.1111/j.1462-2920.2007.01342.x

Swain, M. R., & Ray, R. C. (2009). Biocontrol and other beneficial activities of Bacillus subtilis isolated from cowdung microflora. *Microbiological Research*, *164*(2), 121–130. https://doi.org/10.1016/j.micres.2006.10.009

Tan, H., Mooij, M. J., Barret, M., Hegarty, P. M., Harington, C., Dobson, A. D. W., & O'Gara, F. (2014). Identification of novel phytase genes from an agricultural soil-derived metagenome. *Journal of Microbiology and Biotechnology*, *24*(1), 113–118. https://doi.org/10.4014/jmb.1307.07007

Tank, N., & Saraf, M. (2010). Salinity-resistant plant growth promoting rhizobacteria ameliorates sodium chloride stress on tomato plants. *Journal of Plant Interactions*, *5*(1), 51–58. https://doi.org/10.1080/17429140903125848

Temperton, B., Field, D., Oliver, A., Tiwari, B., Mühling, M., Joint, I., & Gilbert, J. A. (2009). Bias in assessments of marine microbial biodiversity in fosmid libraries as evaluated by pyrosequencing. *The ISME Journal*, *3*(7), 792–796. https://doi.org/10.1038/ismej.2009.32

Timmusk, S., Abd El-Daim, I. A., Copolovici, L., Tanilas, T., Kännaste, A., Behers, L., Nevo, E., Seisenbaeva, G., Stenström, E., & Niinemets, Ü. (2014). Drought-tolerance of wheat improved by rhizosphere bacteria from harsh environments: Enhanced biomass production and reduced emissions of stress volatiles. *PLoS One*, *9*(5), e96086. https://doi.org/10.1371/journal.pone.0096086

Totsche, K. U., Amelung, W., Gerzabek, M. H., Guggenberger, G., Klumpp, E., Knief, C., Lehndorff, E., Mikutta, R., Peth, S., Prechtel, A., Ray, N., & Kögel-Knabner, I. (2018).

Microaggregates in soils. *Journal of Plant Nutrition and Soil Science*, *181*(1), 104–136. https://doi.org/10.1002/jpln.201600451

Tracanna, V., Ossowicki, A., Petrus, M. L. C., Overduin, S., Terlouw, B. R., Lund, G., Robinson, S. L., Warris, S., Schijlen, E. G. W. M., van Wezel, G. P., Raaijmakers, J. M., Garbeva, P., & Medema, M. H. (2021). Dissecting disease-suppressive rhizosphere microbiomes by functional amplicon sequencing and 10× metagenomics. *MSystems*, *6*(3). https://doi.org/10.1128/mSystems.01116-20

Tricoll, D. M., Carney, K. J., Russell, P. F., McMaster, J. R., Groff, D. W., Hadden, K. C., Himmel, P. T., Hubbard, J. P., Boeshore, M. L., & Quemada, H. D. (1995). Field evaluation of transgenic squash containing single or multiple virus coat protein gene constructs for resistance to cucumber mosaic virus, watermelon mosaic virus 2 and zucchini yellow mosaic virus. *Nature Biotechnology*, *13*(12), 1458–1465. https://doi.org/10.1038/nbt1295-1458

Trivedi, P., Delgado-Baquerizo, M., Jeffries, T. C., Trivedi, C., Anderson, I. C., Lai, K., McNee, M., Flower, K., Pal Singh, B., Minkey, D., & Singh, B. K. (2017). Soil aggregation and associated microbial communities modify the impact of agricultural management on carbon content. *Environmental Microbiology*, *19*(8), 3070–3086. https://doi.org/10.1111/1462-2920.13779

Ullah, A., Heng, S., Munis, M. F. H., Fahad, S., & Yang, X. (2015). Phytoremediation of heavy metals assisted by plant growth promoting (PGP) bacteria: A review. *Environmental and Experimental Botany*, *117*, 28–40. https://doi.org/10.1016/j.envexpbot.2015.05.001

van Elsas, J. D., Costa, R., Jansson, J., Sjöling, S., Bailey, M., Nalin, R., Vogel, T. M., & van Overbeek, L. (2008). The metagenomics of disease-suppressive soils—experiences from the METACONTROL project. *Trends in Biotechnology*, *26*(11), 591–601. https://doi.org/10.1016/j.tibtech.2008.07.004

Van Wees, S. C., Van der Ent, S., & Pieterse, C. M. (2008). Plant immune responses triggered by beneficial microbes. *Current Opinion in Plant Biology*, *11*(4), 443–448. https://doi.org/10.1016/j.pbi.2008.05.005

Vessey, J. K. (2003). Plant growth promoting rhizobacteria as biofertilizers. *Plant and Soil*, *255*(2), 571–586. https://doi.org/10.1023/A:1026037216893

Vogel, T. M., Simonet, P., Jansson, J. K., Hirsch, P. R., Tiedje, J. M., van Elsas, J. D., Bailey, M. J., Nalin, R., & Philippot, L. (2009). TerraGenome: A consortium for the sequencing of a soil metagenome. *Nature Reviews Microbiology*, *7*(4), 252–252. https://doi.org/10.1038/nrmicro2119

Wagner, J., Paulson, J. N., Wang, X., Bhattacharjee, B., & Corrada Bravo, H. (2016). Privacy-preserving microbiome analysis using secure computation. *Bioinformatics*, *32*(12), 1873–1879. https://doi.org/10.1093/bioinformatics/btw073

Wahla, V., Maheshwari, D. K., & Bajpai, V. K. (2012). Nematicidal fluorescent pseudomonads for the *in vitro* and *in vivo* suppression of root knot (*Meloidogyne incognita*) of *Capsicum annuum* L. *Pest Management Science*, *68*(8), 1148–1155. https://doi.org/10.1002/ps.3276

Waltz, E. (2014). Beating the heat. *Nature Biotechnology*, *32*(7), 610–613. https://doi.org/10.1038/nbt.2948

Wang, C.-J., Yang, W., Wang, C., Gu, C., Niu, D.-D., Liu, H.-X., Wang, Y.-P., & Guo, J.-H. (2012). Induction of drought tolerance in cucumber plants by a consortium of three plant growth-promoting rhizobacterium strains. *PLoS One*, *7*(12), e52565. https://doi.org/10.1371/journal.pone.0052565

Waterbury, J. B., Calloway, C. B., & Turner, R. D. (1983). A cellulolytic nitrogen-fixing bacterium cultured from the gland of deshayes in shipworms (Bivalvia: Teredinidae). *Science*, *221*(4618), 1401–1403. https://doi.org/10.1126/science.221.4618.1401

Wesolowska-Andersen, A., Bahl, M. I., Carvalho, V., Kristiansen, K., Sicheritz-Ponten, T., Gupta, R., & Licht, T. R. (2014). Choice of bacterial DNA extraction method from fecal material influences community structure as evaluated by metagenomic analysis. *Microbiome*, *2*(1), 19. https://doi.org/10.1186/2049-2618-2-19

Wilpiszeski, R. L., Aufrecht, J. A., Retterer, S. T., Sullivan, M. B., Graham, D. E., Pierce, E. M., Zablocki, O. D., Palumbo, A. V., & Elias, D. A. (2019). Soil aggregate microbial communities: Towards understanding microbiome interactions at biologically relevant scales. *Applied and Environmental Microbiology*, *85*(14). https://doi.org/10.1128/AEM.00324-19

Wilson, I. G. (1997). Inhibition and facilitation of nucleic acid amplification. *Applied and Environmental Microbiology*, *63*(10), 3741–3751. https://doi.org/10.1128/aem.63.10.3741-3751.1997

Wintzingerode, F. V., Göbel, U. B., & Stackebrandt, E. (1997). Determination of microbial diversity in environmental samples: Pitfalls of PCR-based rRNA analysis. *FEMS Microbiology Reviews*, *21*(3), 213–229. https://doi.org/10.1111/j.1574-6976.1997.tb00351.x

Wolińska, A., Kuźniar, A., Zielenkiewicz, U., Banach, A., Izak, D., Stępniewska, Z., & Błaszczyk, M. (2017). Metagenomic analysis of some potential nitrogen-fixing bacteria in arable soils at different formation processes. *Microbial Ecology*, *73*(1), 162–176. https://doi.org/10.1007/s00248-016-0837-2

Wooley, J. C., & Ye, Y. (2010). Metagenomics: Facts and artifacts, and computational challenges. *Journal of Computer Science and Technology*, *25*(1), 71–81. https://doi.org/10.1007/s11390-010-9306-4

Wu, X., Rensing, C., Han, D., Xiao, K.-Q., Dai, Y., Tang, Z., Liesack, W., Peng, J., Cui, Z., & Zhang, F. (2022). Genome-resolved metagenomics reveals distinct phosphorus acquisition strategies between soil microbiomes. *MSystems*, *7*(1). https://doi.org/10.1128/msystems.01107-21

Wurch, L., Giannone, R. J., Belisle, B. S., Swift, C., Utturkar, S., Hettich, R. L., Reysenbach, A.-L., & Podar, M. (2016). Genomics-informed isolation and characterization of a symbiotic Nanoarchaeota system from a terrestrial geothermal environment. *Nature Communications*, *7*(1), 12115. https://doi.org/10.1038/ncomms12115

Xu, D., Ali, S., Huang, Z., Zhou, F. C., Afzal, M., & Bashir, M. H. (2009). Influence of the entomopathogenic fungus, verticillium lecanii on the whitefly predator, axinoscymnus cardilobus (Coleoptera: Coccinellidae) under laboratory conditions. *Pakistan Journal of Zoology*, *4*, 289–295.

Yadav, N. S., Vanderleyden, J., Bennett, D. R., Barnes, W. M., & Chilton, M.-D. (1982). Short direct repeats flank the T-DNA on a nopaline Ti plasmid. *Proceedings of the National Academy of Sciences*, *79*(20), 6322–6326. https://doi.org/10.1073/pnas.79.20.6322

Yan, F. (2007). Transgenic wheat expressing virus-derived hairpin RNA is resistant to Barley yellow dwarf virus. *Hereditas*, *29*(1), 97. https://doi.org/10.1360/yc-007-0097

Yang, J., Kloepper, J. W., & Ryu, C.-M. (2009). Rhizosphere bacteria help plants tolerate abiotic stress. *Trends in Plant Science*, *14*(1), 1–4. https://doi.org/10.1016/j.tplants.2008.10.004

Yao, Q., Li, Z., Song, Y., Wright, S. J., Guo, X., Tringe, S. G., Tfaily, M. M., Paša-Tolić, L., Hazen, T. C., Turner, B. L., Mayes, M. A., & Pan, C. (2018). Community proteogenomics reveals the systemic impact of phosphorus availability on microbial functions in tropical soil. *Nature Ecology & Evolution*, *2*(3), 499–509. https://doi.org/10.1038/s41559-017-0463-5

Yenofsky, R. L., Fine, M., & Pellow, J. W. (1990). A mutant neomycin phosphotransferase II gene reduces the resistance of transformants to antibiotic selection pressure. *Proceedings of the National Academy of Sciences*, *87*(9), 3435–3439. https://doi.org/10.1073/pnas.87.9.3435

Zahedi, H. (2016). Growth-promoting effect of potassium-solubilizing microorganisms on some crop species. In *Potassium Solubilizing Microorganisms for Sustainable Agriculture* (pp. 31–42). Springer India. https://doi.org/10.1007/978-81-322-2776-2_3

Zalila-Kolsi, I., Ben Mahmoud, A., Ali, H., Sellami, S., Nasfi, Z., Tounsi, S., & Jamoussi, K. (2016). Antagonist effects of Bacillus spp. strains against Fusarium graminearum for protection of durum wheat (Triticum turgidum L. subsp. durum). *Microbiological Research*, *192*, 148–158. https://doi.org/10.1016/j.micres.2016.06.012

Zarjani, J. K., Aliasgharzad, N., Oustan, S., Emadi, M., & Ahmadi, A. (2013). Isolation and characterization of potassium solubilizing bacteria in some Iranian soils. *Archives of Agronomy and Soil Science*, *59*(12), 1713–1723. https://doi.org/10.1080/03650340.2012.756977

Zhang, Q., Liang, G., Zhou, W., Sun, J., Wang, X., & He, P. (2016). Fatty-acid profiles and enzyme activities in soil particle-size fractions under long-term fertilization. *Soil Science Society of America Journal*, *80*(1), 97–111. https://doi.org/10.2136/sssaj2015.07.0255

Zhang, W., Zhou, H., Sheng, R., Qin, H., Hou, H., Liu, Y., Chen, A., Chen, C., & Wei, W. (2021). Differences in the nitrous oxide emission and the nitrifier and denitrifier communities among varying aggregate sizes of an arable soil in China. *Geoderma*, *389*, 114970. https://doi.org/10.1016/j.geoderma.2021.114970

Zhao, M., Sun, M., Xiong, T., Tian, S., & Liu, S. (2022). On the link between tree size and ecosystem carbon sequestration capacity across continental forests. *Ecosphere*, *13*(6). https://doi.org/10.1002/ecs2.4079

Zhao, P., Praissman, J. L., Grant, O. C., Cai, Y., Xiao, T., Rosenbalm, K. E., Aoki, K., Kellman, B. P., Bridger, R., Barouch, D. H., Brindley, M. A., Lewis, N. E., Tiemeyer, M., Chen, B., Woods, R. J., & Wells, L. (2020). Virus-receptor interactions of glycosylated SARS-CoV-2 spike and human ACE2 receptor. *Cell Host & Microbe*, *28*(4), 586–601.e6. https://doi.org/10.1016/j.chom.2020.08.004

Zhen, Z., Liu, H., Wang, N., Guo, L., Meng, J., Ding, N., Wu, G., & Jiang, G. (2014). Effects of manure compost application on soil microbial community diversity and soil microenvironments in a temperate Cropland in China. *PLoS One*, *9*(10), e108555. https://doi.org/10.1371/journal.pone.0108555

Zhou, L., Guan, D., Yuan, X., Zhang, M., & Gao, W. (2021). Quantifying the spatiotemporal characteristics of ecosystem services and livelihoods in China's poverty-stricken counties. *Frontiers of Earth Science*, *15*(3), 553–579. https://doi.org/10.1007/s11707-020-0832-2

9 Uses of Microbiome for Improvements of Plant Ability to Utilize Nutrients

Harshita Sharma[1], *Sakshi Singla*[1], *Aastha Dua*[1], *and Indra Mani*[2]*
[1]J.C. Bose University of Science and Technology, Faridabad, Haryana, India
[2]Department of Microbiology, Gargi College, University of Delhi, New Delhi, India

9.1 INTRODUCTION

Concerns regarding the long-term economical and environmental effects of agriculture's conventional dependency on chemical products like fertilizers to increase the availability of nutrients have grown, as it can cause issues like resource depletion, soil erosion, and climate change. This has increased the need for the introduction of sustainable methods (Bhattacharyya and Jha, 2012). Given their diverse plant growth-promoting (PGP) traits, plant-associated microorganisms have a lot of potential to help with these problems. As a result, the plant microbiome has become a viable substitute, providing a biological method of improving plant productivity and health (Kaur and Karnwal, 2023). These microbes have proven to be useful assets for sustainable agriculture because of their ability to increase plant growth, strengthen resistance to diseases, and most importantly, improve uptake of nutrients (Mitter et al., 2016).

Plant microbiota can be split into two groups: core microbiota and satellite taxa. The core plant microbiota is made up of microbes that are persistently linked to the particular genotypes or species of plants, regardless of changes in the environment (Toju et al., 2018). Because of their consistent existence, these microbes are regarded as functionally significant. They possess vital genes that help enhance plant fitness as a result of evolutionary selection. Potato-associated microbiome comprising *Bradyrhizobium*, *Microvirga*, and *Sphingobium* (Pfeiffer et al., 2017), and Hyphomicrobiaceae, Pseudomonadaceae, and Micrococcaceae in grapevines are noteworthy examples (Zarraonaindia et al., 2015). These microbes play an important part in preserving ecosystem stability and boosting plant health by the processes that increase growth, control pathogens, and cycle nutrients (Lemanceau et al., 2017). However, satellite taxa are not as prevalent and only appear briefly in particular environments. Such microbes might not be as ubiquitous, but they contribute significantly to the ecosystem, especially in terms of improving its stability and resilience. They perform vital roles in processes like nutrient cycling and disease control (Hanski, 1982; Gera Hol et al., 2015; Jousset et al., 2017). Exemplary satellite microbiota that participate in biological regulation and development of plants, as well as decreasing aluminum toxicity, consist of microbes like *Cladorrhinum* sp. and *Lipomyces* sp. (Kunito et al., 2012; Tayyab et al., 2021).

The carposphere, rhizosphere, phyllosphere, spermosphere, and anthosphere are some of the environmental reservoirs from which plants obtain their microbiome (Hardoim et al., 2015). One of the most complex ecosystems in all is the rhizosphere, which is a thin film made up of soil and has direct contact with roots. It is thought to be a hotspot of microbial activity and a major source for the acquisition of microorganisms as it provides an organic nutrient rich environment because of the root exudates (Raaijmakers et al., 2009). Microbes thriving here can have direct association with the plants in the form of endophytes or epiphytes, or they can be free-living. Microorganisms can be transferred horizontally or vertically. The root microbiome mostly originates from the soil environment and thus is transferred horizontally (Shade et al., 2017). However, microorganisms can spread via seeds vertically as well (Fierer, 2017).

These advantageous microorganisms offer a sustainable method of enhancing plant nutrition. These microorganisms can improve the acquisition of nutrients by plants through various methods, such as nutrient mobilization, transcriptional regulation, and conversion of nutrients from an available form to one that can be absorbed by the plants (Figure 9.1). So far, many types of microorganisms have been identified that can improve the uptake of nutrients by plants, providing environmentally safe ways to satisfy plant nutrition needs (Singh et al., 2022). To sustain proper development and growth, terrestrial plants typically need an assortment of important nutrients found in the soil, aside from the C (carbon), H (hydrogen), and O (oxygen), which can be obtained via carbon dioxide and water. These vital nutrients fall into two categories according to their requirements. Macronutrients are those that are used by the plants in higher quantities, like potassium (K), nitrogen (N), calcium (Ca), sulfur (S), magnesium (Mg), and phosphorus (P). On the other hand, micronutrients are required in lesser amounts by the plants and include nutrients like iron (Fe), zinc (Zn), Copper (Cu), manganese (Mn), molybdenum (Mo), copper (Cu), boron (B), chlorine (Cl), and nickel (Ni). At the molecular and phenotypic levels, deficiencies in any of these nutrients can cause various problems (Andresen et al., 2018; Gong et al., 2020).

***Corresponding author:** indra.mani@gargi.du.ac.in, indramanibhu@gmail.com (Dr. Indra Mani)

DOI: 10.1201/9781003544425-9

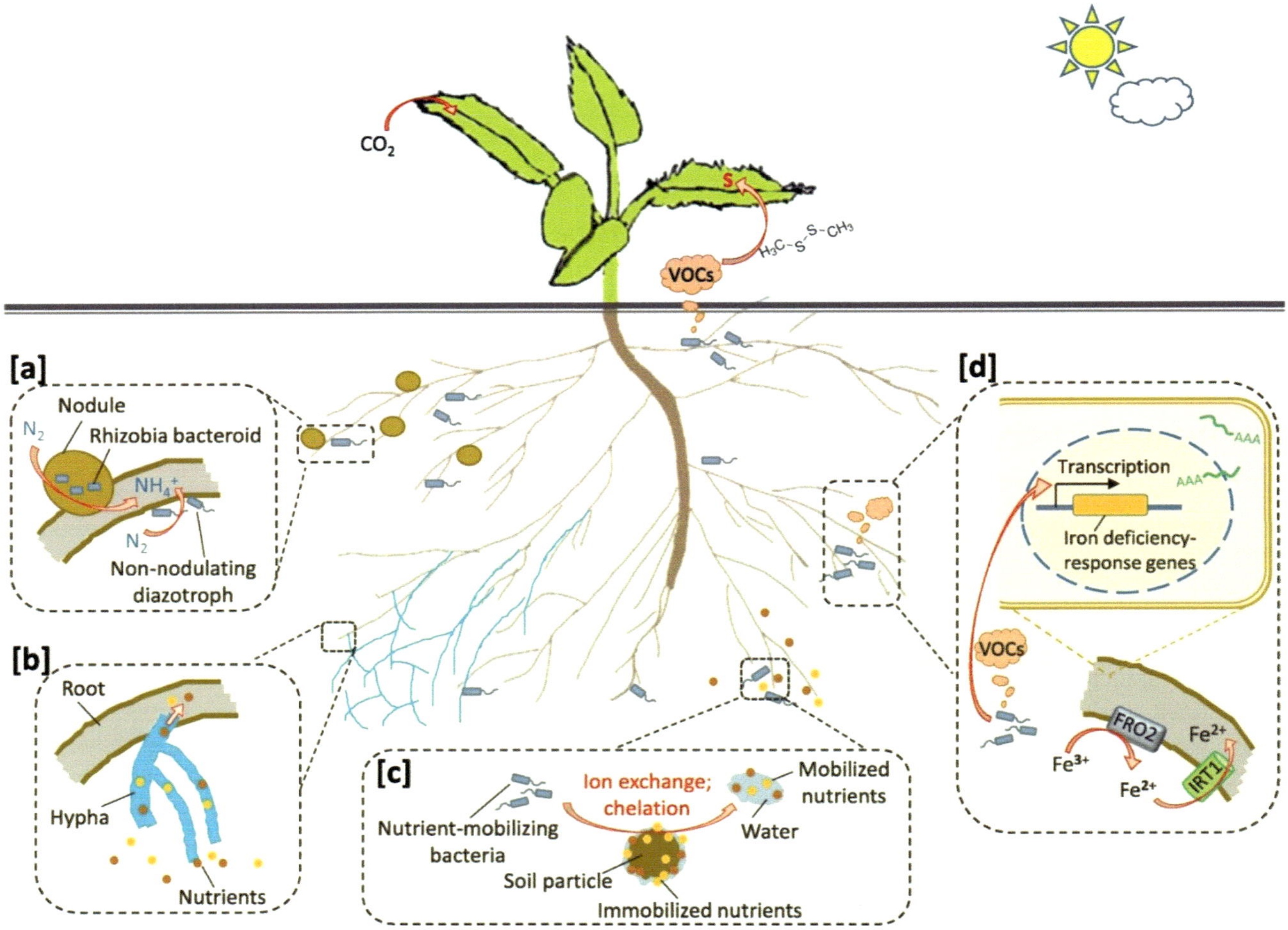

FIGURE 9.1 Microorganism-based improvement of plant ability to acquire nutrients. Microorganisms can enhance the uptake of various macronutrients and micronutrients by the plants through different mechanisms like (a) nitrogen fixation, (b) uptake and transport of nutrients via mycorrhizal mycelium that extend beyond the root and can reach far away soil, (c) nutrient mobilization with the help of siderophores and organic acids released by microorganisms, and (d) regulation of transcription of the genes associated with uptake of the nutrients by microbes (Singh et al., 2022).

Plants coexist with a diverse and intricate community of microorganisms, particularly concentrated within the rhizosphere—the narrow zone of soil surrounding plant roots. These plant–microbe interactions span a spectrum from commensal and beneficial to pathogenic, significantly shaping plant health, development, and productivity. Notable advantages of these associations include microbial synthesis of phytohormones, promotion of biological nitrogen fixation (BNF), and activation of biostimulants that enhance plant growth and tolerance to environmental stressors. Beneficial microbes are also integral to nutrient cycling, aiding in the solubilization and mobilization of key macronutrients such as phosphorus, potassium, and iron, thus boosting nutrient uptake and improving plant nutrient-use efficiency.

The rhizosphere microbiome—a dynamic consortium comprising bacteria, fungi, archaea, viruses, and various eukaryotic microorganisms—forms a complex ecological network. Within this network, microorganisms interact not only with the plant host but also with one another, often through synergistic or competitive relationships. These intricate microbial functions present promising opportunities to advance sustainable agriculture by fostering plant growth, enhancing stress resilience, and decreasing reliance on chemical fertilizers and pesticides.

Although extensive research has highlighted the rhizosphere microbiome's potential to suppress pathogens and promote crop productivity, translating these insights into effective field applications remains a significant challenge. Therefore, a deeper understanding of the multifaceted interactions among plants, rhizospheric microbes, and the surrounding environment is essential. This chapter delves into these microbial dynamics, emphasizing their ecological relevance and potential to revolutionize sustainable agricultural practices (Zhang et al., 2021).

9.2 MICROBIOME FOR IMPROVEMENT OF PLANT MACRONUTRIENT ACQUISITION

Plant macronutrients are known to be crucial for robust development and growth because they serve a variety of purposes, including being redox-sensitive agents and structural components. These are required in higher quantities (more than 0.1% dry mass) by the plant and include nutrients like K (potassium), S (sulfur), and N (nitrogen), P (phosphorus), Mg (magnesium), and Ca (calcium). These aforementioned nutrients are necessary for a plant to complete the course of its existence. Typically, root systems absorb these nutrients from the soil in their ionic state. S and P are present as oxyanions, sulfate (SO_4^{2-}), and phosphate (PO_4^{3-}), respectively, in the soil. N can be found in the soil as NO_3^- (anionic nitrate) or NH_4^+ (cation ammonium), and the metals Ca^{2+}, K^+, and Mg^{2+} are available as free cations. A number of variables, including precipitation, temperature, wind, soil pH, and type, can cause significant variations in soil's nutrient availability over time and place. Since plants are sessile creatures, they have evolved with flexible and adaptive nutrient-acquisition mechanisms, which are mechanistically similar for all macronutrients (Maathuis, 2009). The most beneficial and effective ones are based on the association of plants with the microorganisms to increase the availability and uptake of these nutrients.

9.2.1 Nitrogen

Nitrogen makes up about 80% of our atmospheric air. However, plants do not have access to this very stable form of nitrogen (N_2). Plant species and soil conditions determine the most common form of N uptake. Generally, plants that have acclimated to soil that have low pH and reducing conditions are more likely to absorb NH_4^+. In more aerobic soils with higher pH values, NO_3^- predominates. In the soil, NH_4^+ and NO_3^- are very mobile (Marschner, 1995; Miller and Cramer, 2008). Microbes have a major impact on how well plants uptake these. Nitrogenase, which is present in nitrogen-fixing archaea and bacteria such as diazotrophs, catalyzes the transformation of atmospheric nitrogen into NH_3, and it is subsequently ionized to NH_4^+, promoting low ambient pH. The breakdown of organic waste by bacteria or fungi also produces ammonium, which nitrifying bacteria can oxidize to NH_4^+ and NO_3^-. While some N_2-fixing bacteria live freely in soil, others are associated with plants or even need to work in symbiosis with them in order to fix N_2. Nitrogen is symbiotically fixed in nodules found in roots. The unique environment there guarantees that the N_2 is fixed, and in exchange, bacteroids get the organic nutrients derived from plants (Roy et al., 2020; Mus et al., 2016). Most plants that develop nodules are members of the Leguminosae family. Rhizobia that belong to the Rhizobiaceae family, form species-specific symbiotic relationship with these leguminous plants (Andrews and Andrews, 2017). Along with the symbiotic relationship between rhizobia and legumes, actinorhizal plants that harbor the actinomycete *Frankia* and *Parasponia* species also exhibit N_2-fixing nodulation (Svistoonoff et al., 2014; van Velzen et al., 2018). Other diazotrophs that do not form nodules also help in nitrogen acquisition. So, the microbes can be used to improve the uptake of nutrients. For example, inoculation of *Nostoc punctiforme* in the roots of rice plant and *Azospirillum brasilense* on root surface of wheat results in detectable nitrogen fixation (Álvarez et al., 2020; Santos et al., 2017). In addition to this, mycorrhizal fungi can improve nutrient uptake by plants, because their mycelia can deliver the absorbed nitrogen through a mutualistic interaction with roots as it can reach extra soil present farther away from the root for additional nutrients (Govindarajulu et al., 2005). Additionally, symbiotic interactions of fungi like *Rhizophagus irregularis* with plants can upregulate nitrogen acquisition by plant transporters like *OsNPF4.5* (Wang et al., 2020).

9.2.2 Phosphorus

The majority of the phosphorus found in soil naturally is insoluble, making it unavailable to plants (Richardson and Simpson, 2011). Plants can absorb the inorganic P (Pi). But because of how slowly its release occurs, its deficiency is common. Although the form of Pi in the soil varies with pH, at normal pH, Pi is almost entirely present as $H_2PO_4^-$, which is the form that plants can absorb. Phosphorus is supplemented in farming practices using enormous quantities of fertilizer made from a limited resource like phosphate rock (Maathuis, 2009). But plants rely on their microbiome to acquire sufficient amounts of this macronutrient. Numerous types of fungi and bacteria can mineralize organic phosphorus and/or solubilize inorganic phosphorus, yielding bioavailable nutrients that plants can easily use. *Rhizobium, Pseudomonas*, and *Bacillus* are some of the strongest bacteria that solubilize phosphorus (Rodríguez and Fraga, 1999). In soil, the most prevalent organic phosphorus molecule is phytate (Lim et al., 2007). It is hydrolyzed by phytases to form myo-inositol, which is a part of root exudates excreted by various plant species (Mitchell et al., 1997; Vílchez et al., 2020). Myo-inositol recruits *Bacillus megaterium* species, which are recognized for their potent phosphorus-solubilizing properties (Vílchez et al., 2020). Microbes, especially bacteria, can also improve phosphorus uptake by controlling phosphorus transporters. For example, the tomato plant's *LePT4* phosphate transporter's transcription level was markedly elevated by AM fungal invasion, and phosphate uptake correspondingly doubled (Xu et al., 2007). Since fungi produce higher levels of organic acids, can access additional resources, and have symbiotic relationships with many plant cells, they may be regarded as more useful agents than bacteria to boost plant phosphorus acquisition (Sharma et al., 2013). But both fungi and bacteria together can further improve the uptake of this nutrient. For example, when AM fungi and rhizobacteria were inoculated together in durum wheat, the transcript levels of *Pht1* and *PT2-1* phosphorus transporters were upregulated, and the phosphorus contents above the ground also improved. However, when fungi were used alone, only the gene expression of Pht2 transporter was induced, and extra organic nitrogen supply was needed to increase plant phosphorus contents in a similar manner (Saia et al., 2015).

9.2.3 Sulfur

Both organic and inorganic forms of sulfur are found in soils. Inorganic sulfur predominates in sodic and saline soils. Sulfate (SO_4^{2-}) is the primary inorganic form of sulfur found in aerobic environments, and plants absorb the majority of sulfur in this form (Maathuis, 2009). However, over 95% sulfur in soils related to agriculture is in the form of sulfur bonded to carbon or sulfate esters like amino acid sulfur and sulfonates (Zhao et al., 2006; Kertesz and Mirleau, 2004). Plants can collect S from the atmosphere as well, where it exists as SO_4^{2} and H_2S (Maathuis, 2009).

Various fungal and bacterial species can degrade sulfate esters to produce sulfur. This mineralization can be facilitated by sulfatases, such as arylsulfatase, which cleaves the O–S bond and alkylsulfatase, which breaks the bond between C–O of aliphatic and aromatic esters, respectively. But only bacteria can release sulfur from sulfonates as it can encode a gene cluster, *ssu*, to form a mono-oxygenase enzyme complex that catalyzes the reaction (Eichhorn et al., 1999). While fungi cannot break sulfonates, some saprotrophic fungal species can help bacteria in this mineralization by breaking down polymeric units into oligomeric and monomeric units of sulfonates (Kertesz et al., 2007; Gahan and Schmalenberger, 2015). Along with boosting the SO_4^{2-} supply, through transcriptional modulation of the plant sulfate absorption pathway, the plant microbiome can improve sulfur nutrition. Transcriptional levels of multiple genes implicated in the assimilation pathway of the sulfate, including ATPS (ATP sulfurylase) and APR (adenosine 5′-phosphosulfate reductase), were up-regulated in *Arabidopsis thaliana* subjected to the volatile organic compounds (VOCs) excreted by *Bacillus amyloliquefaciens* GB03; concurrently, acquisition of sulfate was noted with radio-labeled isotope of sulfur in 30 minutes (Aziz et al., 2016). Uptake of sulfur from air can also be increased using a specific microbe, *Bacillus* sp. B55 (Meldau et al., 2013).

9.2.4 Potassium

About 2.6% of the surface of the Earth is made of potassium. Most of the K^+ in soils is dehydrated and bonded to oxygen molecules that plants cannot access. The soil typically contains 0.1–1 mM concentration of K^+. Although deficiency of K^+ is uncommon, potash fertilization is widely used in many crop-producing regions, as any extra K^+ supplies typically promote plant growth (Maathuis, 2009). Potassium uptake can also be improved with the help of microbiome. Numerous strains of fungi and bacteria are capable of liberating potassium from silicate minerals like orthoclase, biotite, and mica (Masood and Bano, 2016; Meena et al., 2014). Microorganisms can increase in-vitro weathering of these minerals and can enhance solubilization of potassium in soil. It transpires due to the secretion of organic acids by microbes that result in the processes like chelation, exchange reactions, and acidolysis (Masood and Bano, 2016; Meena et al., 2014). Additionally, Arbuscular mycorrhizal fungal invasion can improve potassium absorption. in various plant parts like *Zea mays* root steles, *Pelargonium peltatum* shoots, and *Lactuca sativa* leaves (Perner et al., 2007; Baslam et al., 2013). Saprophytic fungi are also crucial for the solubilization of potassium (Meena et al., 2014). For example, inoculation of *Liquidambar styraciflua* (American sweetgum) with *Clitopilus hobsonii* can enhance growth and K acquisition of plants (Peng et al., 2021). If minerals containing potassium are also inoculated along with the microbes, it can enhance nutrient uptake. It was studied in perennial ryegrass, which was inoculated with *A. aculeatus* along with feldspar comprising potassium (Li et al., 2021).

9.2.5 Magnesium

The word magnesium originates from the Greek word "Magnesia," which refers to a mining area where talc was extracted. The typical concentration of soil magnesium ranges from 0.05% to 0.5%. Because of its thin hydration shell, Mg^{2+} has a limited adsorption to soil particles, leading to significant rates of leaching and a typical Mg^{2+} deficit. Plants grown in sandy and acidic soils frequently suffer from magnesium deficits as a result of leaching and competition from other cations (Hermans et al., 2010). As compared to other macronutrients, only a few instances have been recorded for microbe-mediated improvement of magnesium uptake. The colonization of roots with *Piriformospora indica*, an endophytic fungus that can encode Mg transporter PiMgT1, improved the plant's Mg nutrition in *A. thaliana*, especially while it was under magnesium-deficient conditions (Prasad et al., 2019). Increased growth and suppression of magnesium deficiency symptoms were also studied in *Poncirus trifoliata* seedlings inoculated with the AM fungi, *Funneliformis mosseae*; the fungus-induced increases in superoxide dimutase and root catalase activities along with increased levels of soluble protein, may be responsible for the decrease in magnesium deficiency symptoms (Zhang et al., 2015).

9.3 MICROBIOME FOR IMPROVEMENT OF PLANT MICRONUTRIENT ACQUISITION

The processes of photosynthesis, respiration, biomolecule synthesis, redox homeostasis, cell growth, and immune system function are all critically regulated by micronutrients like iron (Fe), manganese (Mn), copper (Cu), zinc (Zn) (Sherman, 1992; Welch and Shuman, 1995; Fischer et al., 2015). Even though organisms only need a minimal amount of micronutrients, their absence severely restricts biological processes and organism growth (Dai et al., 2023). The function and structure of the soil microbiome are thought to be primarily predicted by the pH and macronutrients of the soil (Rousk et al., 2010; Liu et al., 2015).

Micronutrients serve as fundamental pillars in orchestrating the structure and function of soil microbiomes for three primary reasons. To begin with, trace elements like iron (Fe) and manganese (Mn) are indispensable for microbial respiration and energy metabolism, powering vital life processes. Additionally, micronutrients such as copper (Cu) and zinc (Zn) fortify microbial immunocompetence, enhancing their ability to respond to environmental stressors and pathogenic threats. Moreover, elements like Fe, molybdenum (Mo), and nickel (Ni) are central to complex enzymatic reactions, particularly in nitrogen fixation—a cornerstone of soil fertility. Together, these micronutrients not only sustain microbial survival but also drive the intricate biochemical symphony essential for robust and dynamic soil ecosystems (Dai et al., 2023). Second, micronutrient elements that can both give and absorb electrons, such as Fe and Mn, are linked to a number of soil C, N, and S redox processes (Dubinsky et al., 2010; Whalen et.al., 2018; Feng et al., 2019). Third, contrasting habitats created by soils derived from secondary minerals or parent materials that have unique micronutrients would favor some bacteria but not others (Moreno-Jiménez et al., 2023).

Additionally, microorganisms and plant roots develop enduring mutualistic interactions that support both nutrient intake and plant immunity (Duchene et al., 2017).

9.3.1 Iron

To improve iron mobility in the rhizosphere, plants employ a variety of techniques. Graminaceous monocots secrete siderophores to chelate ferric iron (Fe^{3+}) before its root uptake, whereas non-graminaceous monocots and dicots employ a combination strategy that includes rhizosphere acidification, reduction of Fe^{3+} to ferrous Fe^{2+} plasma membrane ferric reductase, and transporter-mediated Fe^{2+} import into roots (Curie and Briat, 2003).

Plants can directly absorb Fe-bound forms of microbial siderophores. When iron-siderophore binding occurs, the two primary functional groups involved are hydroxymates and catechols (Rajkumar et al., 2010).

It is possible to distinguish different microorganisms by their structurally distinct siderophores. *Rhizobium meliloti* and *Bradyrhizobium japonicum*, for instance, release siderophores in the form of citrate, which is neither a hydroxamate nor a catechol, and rhizobactin, which contains catechol (Smith et al., 1985; Guerinot et al., 1990). Since acidity increases Fe^{3+} mobility, rhizosphere acidification that is either directly caused by bacteria or promoted by them helps plants acquire iron. Extracellular metabolites of microorganisms frequently contain organic acids.

For example, glyoxylic acid, 3-methyl-butanoic acid, and diethyl acetic acid are present in the volatile organic compounds (VOCs) derived from *B. amyloliquefaciens* GB03 (Farag et al., 2006), which directly contributes to the acidity of the rhizosphere (Zhang et al., 2009). VOC-exposed *Arabidopsis* that were moved to a new growth medium devoid of VOCs demonstrated that additionally, GB03 VOCs increase the rhizosphere's plant acidity (Zhang et al., 2009). Importantly, in contrast to the plant's stimulation of rhizosphere acidification, *Arabidopsis* responded to GB03 VOCs under Fe-sufficient conditions with systemic Fe-deficiency responses, including increased FRO2 enzyme activity and transcriptional induction of the Fe^{3+} reductase FRO2 and the Fe^{2+} transporter IRT1 (Zhang et al., 2009). Consequently, the VOC-exposed plants had higher Fe levels (Zhang et al., 2009), in encouraging plants to produce more Fe-containing photosynthetic equipment (Zhang et al., 2008; Zhang et al., 2009). The mechanism or phenotype of the GB03 VOC-enhanced plant numerous subsequent research, such as those using different plant species, VOCs from other microorganisms, and GB03 or other inoculation of the soil advantageous microbes, also reported comparable findings for Fe acquisition (Castulo-Rubio et al., 2015; Del Carmen O-MM et al., 2013a, 2013b; Freitas et al., 2015; Hernández-Calderón et al., 2018; Kong et al., 2021; Martínez-Medina et al., 2017; Montejano-Ramírez et al., 2020; Zamioudis et al., 2015; Zhou et al., 2016). Thus, a key factor in microbe-enhanced plant Fe uptake is the microbial control of plant Fe-deficiency responses.

9.3.2 Copper

AM fungus can improve plant fitness in toxic and Cu-deficient environments (Ferrol et al., 2016). Extraradical mycelia extending from the roots into the surrounding environment to seek additional nutrients were responsible for enhancing copper nutrition in white clover (*Trifolium repens* L.) plants infected with *Glomus mosseae* (Li et al., 1991). Similarly, because of extraradical mycelia-dependent nutrient uptake, *G. mosseae* colonization raised the Cu contents in cucumber (*Cucumis sativus*) shoots (Lee and George, 2005). The fungal contribution to plant Cu uptake in mycorrhizal white clover was unaffected by the availability of phosphorus; however, increases in external phosphorus sources from soil altered the distribution patterns of the Cu contents contributed by *G. mosseae*, causing the roots and shoots to decrease and increase, respectively(Li et al., 1991). Likewise, reduced root Cu concentrations were the outcome of increased P supply to hyphae in mycorrhizal cucumber plants (Lee and George, 2005), hence highlighting the interactions between several nutritional enhancements in mycorrhizal plants. Microbial siderophores or the alteration of specific plant genes has been linked to bacterial enhancements of plant Cu nutrition. Two strains of *Rhizobium leguminosarum* and *P. fluorescens* bv phaseoli bacteria that produce siderophores improved the absorption of Fe and Cu in alfalfa (*Medicago sativa*) seedlings (Carrillo-Castañeda et al., 2002). The plant-beneficial bacteria *A. brasilense* reduced the symptoms of plant stress in cucumber plants suffering from Cu deficit or dual deficiency of Fe and Cu, and improved root growth and nutrient uptake (Marastoni et al., 2019).

9.3.3 Manganese

Plants only absorb the divalent form of magnesium (Mn^{2+}). In well-aerated calcareous soils, Mn is not easily absorbed by plants because oxidative and alkaline conditions encourage the formation of water-insoluble Mn oxides. However, due to the rapid oxidation of Mn^{2+}, Mn^{2+} soil dressing is not very effective in reducing deficiencies (Andresen et al., 2018). Since soil pH affects Mn availability, microbe-mediated rhizosphere acidification would increase plant Mn intake. Bacterial mobilization of Mn is exemplified by certain acidophilic bacterial strains, such as *Acinetobacter* sp. and *Lysinibacillus* sp., which were used for Mn bioleaching from ores or mining wastewater (Sanket et al., 2017; Ghosh et al., 2018). *Aspergillus terreus* and *Penicillium daleae* are two examples of Mn-solubilizing fungal strains that have been isolated from low-grade Mn mine tailings. The synthesis of organic acids such as gluconic acid, citric acid, oxalic acid, and maleic acid by the mycelia has been connected to their ability to solubilize Mn (Mohanty et al., 2017).

The Fe^{2+} importer IRT1 is also one of the Mn transporters in plants; however, its selectivity is limited (Barberon et al., 2014). Therefore, microorganisms that can cause plant iron deficiency responses may improve plant manganese uptake. Populations of Fe^{3+}-reducing and Mn^{4+}-reducing bacteria were found in the rhizosphere microbiome of maize (*Z. mays*) in a calcareous soil; the latter was one magnitude larger than the former (Kothari et al., 1991). Plants growing in natural soil acquired more Mn, which is in line with this Mn^{4+}-reducing capacity, compared to plants cultivated on sterilized soil. Interestingly, the populations of Mn^{4+}-reducing and Fe^{3+}-reducing microorganisms were almost completely eradicated when the AM fungus *G. mosseae* was added to the non-sterilized soil, although the overall microbial community remained largely constant in terms of number. While the mycorrhizal plants continued to show modest increases in plant Mn levels, most likely as a result of the fungus-assisted uptake, the fungus also considerably decreased the native microbiome-dependent enhancement of plant Mn accumulation (Kothari et al., 1991). These results demonstrated how plants' uptake of Mn might be enhanced by the rhizosphere microbiome.

9.3.4 Zinc

The productivity and quality of staple food crops are often negatively impacted by zinc deficiency. Roughly half of rice-growing soils are thought to be zinc-deficient, and fertilizing with soluble $ZnSO_4$ usually results in very low fertilizer usage efficiency (1–5%) because of Zn immobilization in the soil (Krithika and Balachandar, 2016). Alternative methods to improve plant zinc acquisition are provided by Zn-mobilizing microorganisms. $ZnSO_4$ fertilization and the Zinc accumulation in rice (*O. sativa*) seedlings were similarly affected by the administration of insoluble zinc oxide bound with zinc-mobilizing bacteria (Krithika and Balachandar, 2016). Applying Zn-mobilizing bacteria, at a rate of 2.5 mg Zn as $ZnSO_4.7H_2O$ per kilogram soil, had a bigger effect on plant zinc levels than Zn fertilization in a different experiment utilizing naturally existing soil that was zinc-deficient (Vaid et al., 2014). Higher zinc contents are known to promote plant growth in a number of bacteria, including *Pseudomonas* and *Bacillus* strains.

The microbial mobilization of zinc has been ascribed to various mechanisms, including zinc sequestering from organic acids by siderophores or anions, oxido-reductive mechanisms on cell membranes, and rhizosphere acidification (Kamran et al., 2017). It is also clear that fungi help plants acquire zinc.

In mycorrhizal bread wheat (*Triticum aestivum*) and barley (*Hordeum vulgare*), the fungal contribution to plant zinc uptake was shown to be as high as 24.3% and 12.7% of total above-ground zinc in wheat and barley, respectively (Coccina et al., 2019). With greater nutrient uptake, mycorrhizal maize enhanced plant physiological functions like photosynthesis in Zn-deficient soils (Saboor et al., 2021). Apart from supplying plants with mycorrhizal zinc, bacteria and fungus can also control plant zinc transporters.

In addition to enhanced grain and straw Zn contents, inoculation of barley with the AM fungus *R. irregularis* led to a considerable up-regulation of *HvZIP13*, a Zn transporter gene activated by low Zn availability (Watts-Williams and Cavagnaro, 2018). Similarly, inoculating *M. truncatula* with *R. irregularis* enhanced plant resistance to soil zinc deficit; concurrently, the fungus stimulated the expression of the *MtZIP5* gene, which is triggered by zinc deficiency (Nguyen et al., 2019). In rice plants inoculated with the Zn-mobilizing bacteria *Enterobacter cloacae*, gene expression of *OsZIP1* and *OsZIP5* was induced when Zn oxide, the insoluble source of zinc in the growth medium, was absent; in contrast, the bacteria suppressed the expression of *OsZIP4* and *OsZIP1* when Zn oxide was present (Krithika and Balachandar, 2016). The microbial suppression of these *ZIP* genes may be a feedback control by the microbe-dependent Zn sufficiency, as was the case with *R. irregularis*-inoculated barley and Medicago (Watts-Williams and Cavagnaro, 2018; Nguyen et al., 2019). Crucially, whereas AM fungi can enhance plant zinc feeding in low-Zn soil, they have also been shown to protect plants against excessive zinc buildup in plants with high Zn availability, including tomato, Medicago, and red clover (*Trifolium pretense*) (Li and Christie, 2001; Watts-Williams et al., 2013; Nguyen et al., 2019). AM fungi are excellent possibilities as microbial tools for maximizing crop zinc intake because of their two plant-beneficial impacts.

9.4 NUTRITION-DEPENDENT PLANT–MICROBIOME ASSOCIATION

As established earlier, the acquisition of macronutrients and micronutrients by plants is significantly facilitated through interactions with various microbial communities. Accordingly, a fundamental prerequisite is the comprehensive study of plant–microbe associations that confer mutual benefits, particularly those that enhance nutrient uptake and overall plant vitality. These interactions, such as secretion of biostimulants by microbial community, nitrogen fixation biologically, or microbial biosynthesis of hormones and many more that have a vital role in plant development.

Microbial communities comprise a diverse array of organisms, each fulfilling specific functional roles that work synergistically to maintain soil health and boost plant growth and productivity. These communities embody intricate and ever-evolving networks, shaped by a variety of interactions, including resource competition, metabolic interdependencies, including nutrient sharing mechanisms like cross-feeding, spatial organization through mechanisms like horizontal gene exchange, biofilm development, cell-to-cell communication, co-adaptation, and viral interactions (Geller and Levy, 2023).

Maintaining the operation of soil ecosystem services is influenced by the richness and variety of microbial species present in the soil. Microbes having the ability to protect and promote plant growth are crucial for sustainable farming methods. An improved comprehension of plant–microbe interactions within agroecosystems and natural habitats will provide us with a biotechnological approach for controlling fertility of the soil and creating a system of enhanced agricultural output (Suman et al., 2022). Thus, microbial interactions serve as a foundational component in facilitating plant nutrient acquisition, highlighting their critical contribution to maintaining plant development and ensuring agricultural productivity.

9.4.1 Bioactive Microbes for Agriculture

In nature, microbes and plants coexist, but they can be free-living, attached, or share a mutually beneficial relationship with their host plants. Over the span of plant evolution, they have developed a range of ecological interactions with microorganisms, including mutualism, commensalism, and parasitism. As plants evolved, they engaged with a variety of rhizosphere bacteria that stimulate plant growth, known as plant growth-promoting rhizobacteria (PGPR). Bacterial rhizobiome mapping is advancing quickly due to novel identification methods, huge genome-sequencing strategies, and current developments in metagenomics. These discoveries identified new bacterial species and their biocontrol and plant growth-promoting processes (Khan, 2019; Mani, 2020b).

In particular, by fixing atmospheric nitrogen, the production of siderophores, phytohormones (such as auxins, cytokinins, and gibberellins [GAs]), volatile organic compounds, and by solubilizing essential nutrients and minerals like phosphorus, potassium, and zinc, soil microorganisms can indirectly promote plant growth. Moreover, certain PGPR species have an involvement in the resistance to stress in plants. They can enhance the plant's capacity to uptake heavy metals or other harmful substances and assist the host plant in overcoming saline conditions and drought-related stress (Li et al., 2014; Aira et al., 2010; Husna et al., 2023).

Plant growth-enhancing rhizobacteria are essential bacteria for agriculture because they increase biomass, germination rate, nitrogen content, and chlorophyll—all of which promote plant growth. These could improve foliar expansion, shoot and root elongation, water transport efficiency, and agricultural output. Moreover, *PGPRS* give plants resistance against biotic stressors (such as pathogenic bacteria, fungi, yeasts, pests, and insects) and abiotic challenges (including drought, flooding, heavy metal toxicity, salinity, and temperature extremes) (Yadav et al., 2018).

9.4.2 Nitrogen-Fixing Biological Processes

Nitrogen fixation anchors the entire planetary nitrogen cycle, making it its paramount process. Chemically triple-bonded N is how di-nitrogen enters the biosphere. When BNF occurs, nitrogenase activity quickly transforms the NH_3 into NH_4^+ (ammonium), which is then used for the biochemical formation of proteins and amino acids, and is temporarily retained within microbial or plant biomass (Pate, 1980). Upon senescence and decomposition, these organic nitrogen sources contribute to the soil organic nitrogen (SON) reservoir. Subsequently, a portion of this SON is mineralized by soil microorganisms, transforming it into plant-available forms of nitrogen—such as nitrate (NO_3^-)—which can then be taken up by plant roots and other soil microbial inhabitants (Ladha et al., 2022).

Contrary to other elements, organisms do not recycle nitrogen entirely. Rather, it is continuously lost through a variety of loss pathways, such as: (1) microbial-driven denitrification of nitrate (NO_3^-) in oxygen-limited conditions which releases strong greenhouse gases like nitrous oxide (N_2O) from the soil; (2) ammonia (NH_3) loss through volatilization from compost and high-pH (alkaline) soils and animal urine patches; (3) NO_3^- leaching to groundwater beyond the root zone of crops; and (4) loss of nitrogen, in both organic and inorganic forms, through surface runoff and erosion (Stevenson, 1982; Fillery, 2001; Ledgard, 2001; Peoples et al., 2009; Galloway et al., 2008; Timilsina et al., 2020).

Therefore, fixing nitrogen biologically is the key mechanism for restocking the organic N. Soil is in dynamic flux and functions as both a source of N (loss) and a sink (gain). The equilibrium between net additions and net losses determines a soil system's capacity to hold nitrogen. Therefore, the fundamental mechanisms behind the N-cycle can be regarded as BNF and the numerous N-loss pathways (Stevenson, 1982).

BNF occurs through the enzymatic action of nitrogenase in nitrogen-fixing bacteria, which convert atmospheric nitrogen into ammonia. This process is carried out by three main types of bacteria: free-living, associative, and symbiotic, all of which possess the capability to facilitate BNF. Free-living N_2-fixing bacteria are found in several genera, including *Gluconacetobacter*, *Azospirillum*, and *Azotobacter* spp.; however, their share in the BNF is quite small. Rhizobia are nitrogen-fixing bacteria that form symbiotic relationships with host plants that account for the largest percentage of BNF. Furthermore, diazotrophs—certain other PGPRs equipped with nitrogenase enzymes—are capable of fixing atmospheric nitrogen

(N_2) in association with non-leguminous plants, such as grains. The nitrogenase complex functions as a metal-containing enzyme (metalloenzyme) that consists of two components. The initial component is an iron-containing homodimeric structure protein known as dinitrogenase reductase, which serves as the reducing agent in the nitrogenase complex. They are in charge of supplying electrons and have a high reducing capacity. The second component is dinitrogenase, a molybdenum-iron (Mo-Fe) protein with a heterotetrameric structure that converts N_2 to NH_3 by using the available electrons (Ladha et al., 2022).

Leguminous plants and *rhizobium* bacteria are symbiotic partners. Cyanobacteria, also known as blue-green algae—including genera such as *Anabaena* and *Nostoc*—are capable of nitrogen fixation, and other species from various genera, including *Beijerinckia*, *Azotobacter*, and *Clostridium*, are the nonsymbiotic or free-living type N_2-fixing bacteria. Associative nitrogen-fixing microbes, including *Alcaligenes* sp., *Klebsiella* sp., *Azotobacter* sp., and *Azospirillum* sp. (maize, rice, and wheat), reside in the rhizosphere surrounding roots and are responsible for supplying the plant with fixed nitrogen.

Cereals, grasses, and sugarcane are associated with endophytic nitrogen-fixing microbes, including *Gluconacetobacter diazotrophicus* (sorghum and sugarcane), *Burkholderia* sp. (rice), *Herbaspirillum* sp. (rice, sorghum, and sugarcane), and *Azoarcus* sp. (Kallar grass, sorghum, and rice) (Laslo and Mara, 2019; Turan et al., 2016).

9.4.3 Synthesis of Plant Regulatory Compounds

Organic substances known as phytohormones have an impact on plant physiological functions even at extremely low concentrations. There is a possibility that phytohormones are produced by the soil bacteria. GAs, ethylene, cytokinins, abscisic acids (ABAs), and auxins (indoleacetic acid (IAA) are examples of phytohormones, produced by microorganisms. Nutrients and metabolic energy are needed in significant quantities for synthesis (Ladha et al., 2022; de Bruijn and Hungria, 2022). A core signaling backbone that synchronizes defense responses against phytopathogens is formed by the stress-related regulators of plant immunity, ethylene, jasmonic acid, and salicylic acid (Matilla and Krell, 2018).

Phytohormones are essential for plant growth, particularly during key developmental stages such as cell division and expansion, seed germination, root formation, and stem elongation. Certain bacteria synthesize and secrete these growth-regulating compounds into the plant system, thereby enhancing overall plant development. Research has shown that some bacterial strains can produce plant growth regulators at levels up to 60 times greater than those synthesized by the plants themselves. Under natural conditions, the genera *Bradyrhizobium, Rhizobium, Azospirillum, Bacillus, Sinorhizobium, Paenibacillus,* and *Pseudomonas* can create phytohormones (auxins, ABA, cytokinins, and GAs) that enhance plant growth and production (Geller and Levy, 2023; Ladha et al., 2022).

Auxins are produced by all plant-associated microorganisms; however, GA is not produced by all PGP bacteria. This ability has to do with bacteria that are connected to roots. 80% of rhizosphere bacteria produce auxins, primarily indoleacetic acid. The amino acid tryptophan is the primary precursor in bacteria for the production of IAA. It is possible for bacteria that support IAA production to absorb tryptophan from root exudates. In *Azospirillum brasilense*, for example, five separate pathways exist, including both tryptophan-dependent and tryptophan-independent routes, and it is uncertain what the biochemical intermediates are (Ladha et al., 2022).

ABA is a pivotal stress-responsive phytohormone that governs photoperiodic flowering and is crucial for plant growth and resilience to environmental challenges like low temperatures, high salinity, and drought. Certain plant-associated bacteria are capable of producing ABA themselves, thereby elevating the overall phytohormone levels within the plant system and influencing plant physiology. Notably, ABA also modulates plant immune responses; plants with impaired ABA biosynthesis or signaling often exhibit enhanced resistance to pathogens. Production of ABA is done by various endophytic bacteria such as *Achromobacter xylosoxidans*, *Bacillus licheniformis*, *Bacillus pumilus*, *Brevibacterium halotolerans*, and *Lysinibacillus fusiformis* (Salazar-Cerezo et al., 2018).

GAs are a group of diterpenoid phytohormones essential for controlling multiple facets of plant growth, such as leaf enlargement and stem elongation. The external application of GAs has been demonstrated to trigger parthenocarpic fruit development, stimulate bolting, break tuber dormancy, and enhance both fruit size and bud proliferation. Several soil-dwelling microorganisms are capable of synthesizing GAs, thereby influencing plant growth and symbiotic interactions. These microbial-derived GAs can either promote nodule organogenesis or suppress nodulation during the early stages of rhizobial infection, depending on the context. The first bacterium identified to produce GAs was *Rhizobium phaseoli*, which synthesizes GA1 and GA4. Additionally, other endophytic and rhizospheric bacteria such as *Azospirillum lipoferum*, *Acetobacter diazotrophicus*, and *Herbaspirillum seropedicae* are documented to produce biologically active GAs, including GA1 and GA3 (Salazar-Cerezo et al., 2018).

Cytokinins (CKs) are a crucial group of phytohormones that play a central role in orchestrating key developmental processes in plants. They actively drive cell division, initiate root and root hair development, awaken dormant axillary buds, and trigger seed germination, making them indispensable for early growth and morphogenesis. In addition to modulating apical dominance, cytokinins play a crucial role in the modulation of nodule formation and symbiotic nitrogen assimilation. Both supportive and detrimental microbial species have been reported to biosynthesize cytokinins. Among these, PGPR from the *Pseudomonas* and *Bacillus* genera are notable producers, particularly of the active cytokinin zeatin. *Pseudomonas fluorescens* and species of *Rhizobium* have been widely researched for their ability to produce cytokinins and their significant contribution to promoting plant growth and development (Yadav et al., 2018; Ladha et al., 2022).

In summary, diverse groups of plant-linked microbes are key players in influencing plant physiology by synthesizing and releasing key phytohormones such as cytokinins, GAs, and ABA. Through these biochemical interactions, microbes influence critical developmental and stress-responsive pathways in plants, highlighting their integral contribution to plant growth, resilience, and symbiotic efficiency.

In addition to phytohormone production and many more, beneficial plant-associated microbes engage in a variety of interactions that significantly enhance plant health and production. Many rhizosphere-associated growth-enhancing bacteria exhibit biocontrol activity by producing bioactive compounds that restrict the growth of disease-causing microbes in plants. These microbes can release antibiotics and cell wall-degrading enzymes, including chitinases and glucanases, which target pathogenic fungi and bacteria, thereby protecting plants from disease. Moreover, PGPRs contribute to nutrient availability by solubilizing essential macronutrients; for instance, phosphorus, these bacteria break down insoluble phosphate into bioavailable forms for plant uptake improving plant uptake. Similarly, certain microbes facilitate potassium mobilization and enhance iron acquisition through the excretion of siderophores, potent iron-chelating agents that make iron more accessible to plants under deficient conditions. Collectively, these microbial functions underscore their critical role in long-term agricultural sustainability by supporting plant growth, improving nutrient use efficiency, and reducing the reliance on chemical fertilizers and pesticides (Backer et al., 2018).

9.5 PHYTO-MICROBIOME ENGINEERING TO IMPROVE NUTRIENT UPTAKE

The deliberate modification of microbial populations associated with plants, to improve its growth and resilience, is known as phyto-microbiome engineering. Identifying advantageous microorganisms, maximizing their use in fields related to agriculture, and encouraging symbiotic partnerships that improve plant development are all part of this engineering strategy. Scientists can create microbial consortia particularly suited to certain crops and environmental circumstances by comprehending the molecular principles underlying these interactions. In addition to increasing agricultural yields and the ability to withstand both biotic and abiotic stresses, this could result in a decreased reliance on chemical fertilizers for nutrition (Urvasha et al., 2025).

New advances in computational and meta-omics approaches have substantially improved our knowledge of the genes involved in distributions that are specific to the niche, the host plant functions, as well as diversity of microbiomes. Nowadays, one of the most popular techniques for thoroughly investigating and discovering new gene functions in microbiomes is genome-wide functional genomics (Ke et al., 2021; Mani, 2020a). This helps in the discovery of the genes that are required in various pathways involved in nutrient uptake. This knowledge can be further used with genetic engineering to improve nutrient acquisition. For example, *nif* gene cluster that codes for the nitrogenase enzyme complex required during nitrogen fixation was remodeled and synthesized to express it in *Escherichia coli* (Smanski et al., 2014). It is used to create nitrogen-fixing strains of *Herbaspirillum seropedicae* and *Azospirillum brasilense* that naturally colonize grasses including wheat. These strains were used to increase nitrogen acquisition in wheat which only occurs in leguminous crops (Fukamachi et al., 2019).

Pseudomonas fluorescens strains were also engineered to overexpress the genes involved in phosphate solubilization using CRISPR technology. Because phosphorus is trapped in insoluble forms, rice plants frequently experience its deficiency in the soil (Stringlis et al., 2019). A microbial consortium, a microbial group having different properties, can also be used to overcome different problems at the same time. For example, to increase lettuce growth, a microbial consortium made up of microbes with various beneficial characteristics was created. It consists of *Trichoderma*, a biocontrol agent against root diseases, *Bacillus* to increase solubilization of phosphate, and *Azospirillum*, which helps in fixing nitrogen (Pascale et al., 2020). By meticulously mapping plant–microbe interactions, phyto-microbiome engineering has enormous potential to improve crop performance. It can be used to encourage advantageous interactions that enhance plant growth and nutrient uptake by comprehending and adjusting the intricate links that exist between microbiota and plants (Urvasha et al., 2025).

9.6 CONCLUSION AND FUTURE PERSPECTIVE

Harnessing the power of plant microbiomes presents exciting and transformative opportunities for enhancing the acquisition of plant nutrients and promoting ultimate agricultural sustainability. The uptake of important micronutrients such as iron, manganese, and zinc, along with macronutrients like potassium, phosphorus, and nitrogen, is greatly aided by useful microbes, especially plant growth-promoting rhizobacteria (PGPRs). These microbes enhance nutrient availability and uptake using a variety of methods, including solubilization of phosphate, nitrogen fixation, and the secretion of siderophores that chelate and mobilize otherwise inaccessible micronutrients in the soil.

Additionally, the synthesis of phytohormones by microbes, including GAs, cytokinins, and auxins significantly influences root architecture and growth dynamics, further improving the plant's nutrition absorption capacity efficiently. The complex and dynamic relationship of plants and their related microbial communities indicates the requirement for a holistic approach of managing plant health and productivity. Advances in phyto-microbiome engineering are now enabling the development of customized microbial consortia adapted to certain soil types, crops, and environmental circumstances, thereby optimizing nutrient uptake efficiency and decreasing reliance on synthetic fertilizers.

As we continue to learn more about plant–microbe symbioses, the integration of microbial technologies into modern crop management strategies offers immense potential to cultivate more resilient, productive, and environmentally sustainable agricultural systems in the upcoming years.

REFERENCES

Aira M, Gómez-Brandón M, Lazcano C, Bååth E (2010) Plant genotype strongly modifies the structure and growth of maize rhizosphere microbial communities. Soil Biol Biochem 42:2276–2281.

Álvarez C, Navarro JA, Molina-Heredia FP, Mariscal V (2020) Endophytic colonization of rice (Oryza sativa L.) by the symbiotic strain Nostoc punctiforme PCC 73102. Mol Plant Microbe Interact 33(8):1040–1045. https://doi.org/10.1094/MPMI-01-20-0015-SC

Andresen E, Peiter E, Kupper H (2018) Trace metal metabolism in plants. J Exp Bot 69(5):909–954. https://doi.org/10.1093/jxb/erx465

Andrews M, Andrews ME (2017) Specificity in legume-rhizobia symbioses. Int J Mol Sci 18(4):705. https://doi.org/10.3390/ijms18040705

Aziz M, Nadipalli RK, Xie X, Sun Y, Surowiec K, Zhang JL, Paré PW (2016) Augmenting sulfur metabolism and herbivore defense in *Arabidopsis* by bacterial volatile signaling. Front Plant Sci 7(–):458. https://doi.org/10.3389/fpls.2016.00458

Backer R, Rokem JS, Ilangumaran G, Lamont J, Praslickova D, Ricci E, Subramanian S, Smith DL (2018) Plant growth-promoting rhizobacteria: Context, mechanisms of action, and roadmap to commercialization of biostimulants for sustainable agriculture. Front Plant Sci 9:1473. https://doi.org/10.3389/fpls.2018.01473

Barberon M, Dubeaux G, Kolb C, Isono E, Zelazny E, Vert G (2014) Polarization of Iron-Regulated Transporter 1 (IRT1) to the plant-soil interface plays crucial role in metal homeostasis. PNAS USA 111(22):8293–8298. https://doi.org/10.1073/pnas.1402262111

Baslam M, Garmendia I, Goicoechea N (2013) The arbuscular mycorrhizal symbiosis can overcome reductions in yield and nutritional quality in greenhouse-lettuces cultivated at inappropriate growing seasons. Sci Hortic 164:145–154. https://doi.org/10.1016/j.scienta.2013.09.021

Bhattacharyya PN, Jha DK (2012) Plant growth-promoting rhizobacteria (PGPR): Emergence in agriculture. World J Microbiol Biotechnol 28:1327–1350. https://doi.org/10.1007/s11274-011-0979-9

Carrillo-Castañeda G, Muños JJ, Peralta-Videa J, Gomez E, Tiemannb K, Duarte-Gardea M et al (2002) Alfalfa growth promotion by bacteria grown under iron limiting conditions. Adv Environ Res 6(3):391–399. https://doi.org/10.1016/S1093-0191(02)00054-0

Castulo-Rubio DY, Alejandre-Ramírez NA, Del Carmen O-MM, Santoyo G, Macías-Rodríguez LI, Valencia-Cantero E (2015) Volatile organic compounds produced by the rhizobacterium *Arthrobacter agilis* UMCV2 modulate *Sorghum bicolor* (strategy II plant) morphogenesis and SbFRO1 transcription in vitro. J Plant Growth Regul 34(3):611–623. https://doi.org/10.1007/s00344-015-9495-8

Coccina A, Cavagnaro TR, Pellegrino E, Ercoli L, Mclaughlin MJ, Watts-Williams SJ (2019) The mycorrhizal pathway of zinc uptake contributes to zinc accumulation in barley and wheat grain. BMC Plant Biol 19(1):1–14. https://doi.org/10.1186/s12870-019-1741-y

Curie C, Briat J-F (2003) Iron transport and signaling in plants. Annu Rev Plant Biol 54(1):183–206. https://doi.org/10.1146/annurev.arplant.54.031902.135018

Dai Z, Guo X, Lin J, Wang X, He D, Zeng R, Meng J, Luo J, Delgado-Baquerizo M, Moreno-Jiménez E, Brookes PC (2023) Metallic micronutrients are associated with the structure and function of the soil microbiome. Nat Commun 14(1):8456.

de Bruijn FJ, Hungria M (2022) Biological nitrogen fixation. In Good Microbes in Medicine, Food Production, Biotechnology, Bioremediation, and Agriculture; John Wiley & Sons, Ltd.: Hoboken, NJ, USA; pp. 466–475, ISBN 978-1-119-76262-1.

Del Carmen O-MM, Macías-Rodríguez LI, Santoyo G, Farías-Rodríguez R, Valencia-Cantero E (2013a) Medicago truncatula increases its iron-uptake mechanisms in response to volatile organic compounds produced by Sinorhizobium meliloti. Folia Microbiol (Praha) 58(6):579–585. https://doi.org/10.1007/s12223-013-0243-9

Del Carmen O-MM, Velázquez-Becerra C, Macías-Rodríguez LI, Santoyo G, Flores-Cortez I, Alfaro-Cuevas R et al (2013b) Arthrobacter agilis UMCV2 induces iron acquisition in Medicago truncatula (strategy I plant) in vitro via dimethylhexadecylamine emission. Plant Soil 362(1):51–66. https://doi.org/10.1007/s11104-012-1263-y

Dubinsky EA, Silver WL, Firestone MK (2010) Tropical forest soil microbial communities couple iron and carbon biogeochemistry. Ecology 91:2604–2612.

Duchene O, Vian JF, Celette F (2017) Intercropping with legume for agroecological cropping systems: Complementarity and facilitation processes and the importance of soil microorganisms: A review. Agr Ecosyst Environ 240:148–161.

Eichhorn E, Van Der Ploeg JR, Leisinger T (1999) Characterization of a two-component alkanesulfonate monooxygenase from Escherichia coli. J Biol Chem 274(38):26639–26646. https://doi.org/10.1074/jbc.274.38.26639

Farag MA, Ryu C-M, Sumner LW, Paré PW (2006) GC–MS SPME profiling of rhizobacterial volatiles reveals prospective inducers of growth promotion and induced systemic resistance in plants. Phytochemistry 67(20):2262–2268. https://doi.org/10.1016/j.phytochem.2006.07.021

Feng J et al (2019) Coupling and decoupling of soil carbon and nutrient cycles across an aridity gradient in the drylands of northern china: Evidence from ecoenzymatic stoichiometry. Glob Biogeochem Cy 33:559–569.

Ferrol N, Tamayo E, Vargas P (2016) The heavy metal paradox in arbuscular mycorrhizas: From mechanisms to biotechnological applications. J Exp Bot 67(22):erw403. https://doi.org/10.1093/jxb/erw403

Fierer N (2017) Embracing the unknown: Disentangling the complexities of the soil microbiome. Nat Rev Microbiol 15:579–590. https://doi.org/10.1038/nrmicro.2017.87

Fillery IRP (2001) The fate of biologically fixed nitrogen in legume-based dryland farming systems: A review. Aust J Exp Agric 41(3):361–381. https://doi.org/10.1071/EA99141

Fischer WW, Hemp J, Johnson JE (2015) Manganese and the evolution of photosynthesis. Orig Life Evol B 45:351–357.

Freitas MA, Medeiros FH, Carvalho SP, Guilherme LR, Teixeira WD, Zhang H, Paré PW (2015) Augmenting iron accumulation in cassava by the beneficial soil bacterium *Bacillus subtilis* (GBO3). Front Plant Sci 6:596. https://doi.org/10.3389/fpls.2015.00596

Fukamachi K, Konishi Y, Nomura T (2019) Disease control of Phytophthora infestans using cyazofamid encapsulated in poly lactic-co-glycolic acid (PLGA) nanoparticles. Colloid Surf A: Physicochem Eng Asp 577:315–322.

Gahan J, Schmalenberger A (2015) Arbuscular mycorrhizal hyphae in grassland select for a diverse and abundant hyphospheric bacterial community involved in sulfonate desulfurization. Appl Soil Ecol 89:113–121. https://doi.org/10.1016/j.apsoil.2014.12.008

Galloway JN, Townsend AR, Erisman JW, Bekunda M, Cai Z, Freney JR, Martinelli LA, Seitzinger SP, Sutton MA (2008) Transformation of the nitrogen cycle: Recent trends, questions, and potential solutions. Sci 320(5878):889–892. https://doi.org/10.1126/science.1136674

Geller AM, Levy A (2023) "What I cannot create, I do not understand": Elucidating microbe-microbe interactions to facilitate plant microbiome engineering. Curr Opin Microbiol 72:102283.

Gera Hol WH, de Boer W, de Hollander M, Kuramae EE, Meisner A, van der Putten WH (2015) Context dependency and saturating effects of loss of rare soil microbes on plant productivity. Front Plant Sci 6:485. https://doi.org/10.3389/fpls.2015.00485

Ghosh S, Bal B, Das A (2018) Enhancing manganese recovery from low-grade ores by using mixed culture of indigenously isolated bacterial strains. Geomicrobiol J 35(3):242–246. https://doi.org/10.1080/01490451.2017.1362080

Gong Z, Xiong L, Shi H, Yang S, Herrera-Estrella LR, Xu G, Chao DY, Li J, Wang PY, Qin F, Li J, Ding Y, Shi Y, Wang Y, Yang Y, Guo Y, Zhu JK. (2020) Plant abiotic stress response and nutrient use efficiency. Sci China Life Sci 63(5):635–674. https://doi.org/10.1007/s11427-020-1683-x

Govindarajulu M, Pfeffer PE, Jin H, Abubaker J, Douds DD, Allen JW, Bücking H, Lammers PJ, Shachar-Hill Y (2005) Nitrogen transfer in the arbuscular mycorrhizal symbiosis. Nature 435(7043):819–823. https://doi.org/10.1038/nature03610

Guerinot ML, Meidl EJ, Plessner O (1990) Citrate as a siderophore in *Bradyrhizobium japonicum*. J Bacteriol 172(6):3298–3303. https://doi.org/10.1128/jb.172.6.3298-3303.1990Iski I (1982) Dynamics of regional distribution: The Core and satellite species hypothesis. Oikos 38:210. https://doi.org/10.2307/3544021

Hardoim PR, van Overbeek LS, Berg G, Pirttilä AM, Compant S, Campisano A et al. (2015) The hidden world within plants: Ecological and evolutionary considerations for defining functioning of microbial endophytes. Microbiol Mol Biol Rev 79:293–320. https://doi.org/10.1128/mmbr.00050-14

Hermans C, Vuylsteke M, Coppens F, Cristescu SM, Harren FJ, Inzé D et al (2010) Systems analysis of the responses to long-term magnesium deficiency and restoration in *Arabidopsis thaliana*. New Phytol 187(1):132–144. https://doi.org/10.1111/j.1469-8137.2010.03257

Hernández-Calderón E, Aviles-Garcia ME, Castulo-Rubio DY, Macías-Rodríguez L, Ramírez VM, Santoyo G, López-Bucio J, Valencia-Cantero E (2018) Volatile compounds from beneficial or pathogenic bacteria differentially regulate root exudation, transcription of iron transporters, and defense signaling pathways in *Sorghum bicolor*. Plant Mol Biol 96(3):291–304. https://doi.org/10.1007/s11103-017-0694-5

Husna, Hussain A, Shah M, Hamayun M, Iqbal A, Qadir M, Alataway A, Dewidar AZ, Elansary HO, Lee I-J (2023) Phytohormones producing rhizobacteria alleviate heavy metals stress in soybean through multilayered response. Microbiol Res 266:127237.

Jousset A, Bienhold C, Chatzinotas A, Gallien L, Gobet A, Kurm V et al (2017) Where less may be more: How the rare biosphere pulls ecosystems strings. ISME J 11:853–862. https://doi.org/10.1038/ismej.2016.174

Kamran S, Shahid I, Baig DN, Rizwan M, Malik KA, Mehnaz S (2017) Contribution of zinc solubilizing bacteria in growth promotion and zinc content of wheat. Front Microbiol 8:2593. https://doi.org/10.3389/fmicb.2017.02593

Kaur M, Karnwal A (2023) Screening of endophytic bacteria from stress-tolerating plants for abiotic stress tolerance and plant growth-promoting properties: Identification of potential strains for bioremediation and crop enhancement. J Agric Food Res 14:Article 100723.

Ke J, Wang B, Yoshikuni Y (2021) Microbiome engineering: Synthetic biology of plant-associated microbiomes in sustainable agriculture. Trends Biotechnol 39(3):244–261.

Kertesz MA, Fellows E, Schmalenberger A (2007) Rhizobacteria and plant sulfur supply. Adv Appl Microbiol 62:235–268. https://doi.org/10.1016/S0065-2164(07)62008-5.

Kertesz MA, Mirleau P (2004) The role of soil microbes in plant Sulphur nutrition. J Exp Bot 55(404):1939–1945. https://doi.org/10.1093/jxb/erh176

Khan AAH (2019) Plant-bacterial association and their role as growth promoters and biocontrol agents. In Plant Growth Promoting Rhizobacteria for Sustainable Stress Management; Springer: Cham, Switzerland, 2019; pp. 389–419, ISBN 9789811369858.

Kong W-L, Wang Y-H, Wu X-Q (2021) Enhanced iron uptake in plants by volatile emissions of Rahnella aquatilis JZ-GX1. Front Plant Sci 12:704000. https://doi.org/10.3389/fpls.2021.704000

Kothari S, Marschner H, Römheld V (1991) Effect of a vesicular–arbuscular mycorrhizal fungus and rhizosphere microorganisms on manganese reduction in the rhizosphere and manganese concentrations in maize (*Zea mays* L.). New Phytol 117(4):649–655. https://doi.org/10.1111/j.1469-8137.1991.tb00969.x

Krithika S, Balachandar D (2016) Expression of zinc transporter genes in rice as influenced by zinc-solubilizing Enterobacter cloacae strain ZSB14. Front Plant Sci 7:446. https://doi.org/10.3389/Fpls.2016.00446

Kunito T, Owaki M, Ihyo Y, Sumi H, Toda H, Fukuda D et al (2012) Genera Burkholderia and Lipomyces are predominant aluminum-resistant microorganisms isolated from acidic forest soils using cycloheximide-amended growth media. Ann Microbiol 62:1339–1344.

Ladha JK, Peoples MB, Reddy PM, Biswas JC, Bennett A, Jat ML, Krupnik TJ (2022) Biological nitrogen fixation and prospects for ecological intensification in cereal-based cropping systems. Field Crops Res 283:108541. [CrossRef] [PubMed].

Laslo É, Mara G (2019) Is PGPR an alternative for npk fertilizers in sustainable agriculture? In Singh D, Gupta V,

Prabha R. (eds) Microbial Interventions in Agriculture and Environment; Springer: Singapore. https://doi.org/10.1007/978-981-13-8391-5_3
Ledgard SF (2001) Nitrogen cycling in low input legume-based agriculture, with emphasis on legume/grass pastures. Plant and Soil, 228(1):43–59.
Lee Y-J, George E (2005) Contribution of mycorrhizal hyphae to the uptake of metal cations by cucumber plants at two levels of phosphorus supply. Plant Soil 278(1):361–370. https://doi.org/10.1007/s11104-005-0373-1
Lemanceau P, Blouin M, Muller D, Moënne-Loccoz Y (2017) Let the Core microbiota be functional. Trends Plant Sci 22:583–595. https://doi.org/10.1016/j.tplants.2017.04.008
Li X, Christie P (2001) Changes in soil solution Zn and pH and uptake of Zn by arbuscular mycorrhizal red clover in Zn-contaminated soil. Chemosphere 42(2):201–207. https://doi.org/10.1016/s0045-6535(00)00126-0
Li X, Rui J, Mao Y, Yannarell A, Mackie R (2014) Dynamics of the bacterial community structure in the rhizosphere of a maize cultivar. Soil Biol Biochem 68:392–401.
Li X, Yin Y, Fan S, Xu X, Amombo E, Xie Y, Fu J (2021) Aspergillus aculeatus enhances potassium uptake and photosynthetic characteristics in perennial ryegrass by increasing potassium availability. J Appl Microbiol. https://doi.org/10.1111/jam.15186
Li X-L, Marschner H, George E (1991) Acquisition of phosphorus and copper by VA-mycorrhizal hyphae and root-to-shoot transport in white clover. Plant Soil 136(1):49–57. https://doi.org/10.1007/Bf02465219
Lim BL, Yeung P, Cheng C, Hill JE (2007) Distribution and diversity of phytate-mineralizing bacteria. ISME J 1(4):321–330. https://doi.org/10.1038/ismej.2007.40
Liu JJ et al (2015) Soil carbon content drives the biogeographical distribution of fungal communities in the black soil zone of northeast China. Soil Biol Biochem 83:29–39.
Maathuis FJ (2009) Physiological functions of mineral macronutrients. Curr Opin Plant Biol 12(3):250–258. https://doi.org/10.1016/j.pbi.2009.04.003. Epub 2009 May 25. PMID: 19473870.
Mani I (2020a) Current status and challenges of DNA sequencing. In Singh V (eds) Advances in Synthetic Biology; Springer: Singapore. https://doi.org/10.1007/978-981-15-0081-7_5
Mani I (2020b) Metagenomics Approach for Bioremediation: Challenges and Perspectives, Bioremediation of Pollutants; Elsevier; pp. 275–285, ISBN 9780128190258. https://doi.org/10.1016/B978-0-12-819025-8.00012-0
Marastoni L, Pii Y, Maver M, Valentinuzzi F, Cesco S, Mimmo T (2019) Role of *Azospirillum brasilense* in triggering different Fe chelate reductase enzymes in cucumber plants subjected to both nutrient deficiency and toxicity. Plant Physiol Biochem 136:118–126. https://doi.org/10.1016/j.plaphy.2019.01.013
Marschner H (1995) Mineral Nutrition in Higher Plants; Academic Press: London.
Martínez-Medina A, Van Wees SC, Pieterse CM (2017) Airborne signals from *Trichoderma* fungi stimulate iron uptake responses in roots resulting in priming of jasmonic acid-dependent defences in shoots of *Arabidopsis thaliana* and *Solanum lycopersicum*. Plant Cell Environ 40(11):2691–2705. https://doi.org/10.1111/pce.13016
Masood S, Bano A (2016) Mechanism of potassium solubilization in the agricultural soils by the help of soil microorganisms. In Potassium Solubilizing Microorganisms for Sustainable Agriculture; Springer; pp 137–147. https://doi.org/10.1007/978-81-322-2776-2_10
Matilla M, Krell T (2018) Plant growth promotion and biocontrol mediated by plant-associated bacteria. In Plant Microbiome: Stress Response; Springer: Cham, Switzerland; pp. 45–80, ISBN 978-981-10-5513-3.
Meena V, Maurya B, Bahadur I (2014) Potassium solubilization by bacterial strain in waste mica. Bangladesh J Bot 43(2):235–237. https://doi.org/10.3329/bjb.v43i2.21680
Meldau DG, Meldau S, Hoang LH, Underberg S, Wünsche H, Baldwin IT (2013) Dimethyl disulfide produced by the naturally associated bacterium Bacillus sp B55 promotes Nicotiana attenuata growth by enhancing sulfur nutrition. Plant Cell 25(7):2731–2747. https://doi.org/10.1105/tpc.113.114744
Miller AJ, Cramer MD (2008) Root nitrogen acquisition and assimilation. Plant Soil 274:1–36.
Mitchell DB, Vogel K, Weimann BJ, Pasamontes L, Van Loon AP (1997) The phytase subfamily of histidine acid phosphatases: Isolation of genes for two novel phytases from the fungi aspergillus terreus and *Myceliophthora thermophila*. Microbiology 143(1):245–252. https://doi.org/10.1099/00221287-143-1-245
Mitter B, Pfaffenbichler N, Sessitsch A (2016) Plant–microbe partnerships in 2020. Microb Biotechnol 9:635–640. https://doi.org/10.1111/1751-7915.12382
Mohanty S, Ghosh S, Nayak S, Das A (2017) Isolation, identification and screening of manganese solubilizing fungi from low-grade manganese ore deposits. Geomicrobiol J 34(4):309–316. https://doi.org/10.1080/01490451.2016.1189016
Montejano-Ramírez V, García-Pineda E, Valencia-Cantero E (2020) Bacterial compound N, N-dimethylhexadecylamine modulates expression of iron deficiency and defense response genes in *Medicago truncatula* independently of the jasmonic acid pathway. Plants 9(5):624. https://doi.org/10.3390/plants9050624
Moreno-Jiménez E, Maestre FT, Flagmeier M, Guirado E, Berdugo M, Bastida F, Dacal M, Díaz-Martínez P, Ochoa-Hueso R, Plaza C, Rillig MC (2023) Soils in warmer and less developed countries have less micronutrients globally. Glob Change Biol 29(2):522–532.
Mus F, Crook MB, Garcia K, Garcia Costas A, Geddes BA, Kouri ED, Paramasivan P, Ryu MH, Oldroyd GED, Poole PS, Udvardi MK, Voigt CA, Ané JM, Peters JW (2016) Symbiotic nitrogen fixation and the challenges to its extension to nonlegumes. Appl Environ Microbiol 82(13):3698–3710. https://doi.org/10.1128/AEM.01055-16
Nguyen TD, Cavagnaro TR, Watts-Williams SJ (2019) The effects of soil phosphorus and zinc availability on plant responses to mycorrhizal fungi: A physiological and molecular assessment. Sci Rep 9(1):1–13. https://doi.org/10.1038/s41598-019-51369-5
Pascale A, Proietti S, Pantelides IS, Stringlis IA (2020) Modulation of the root microbiome by plant molecules: The basis for targeted disease suppression and plant growth promotion. Front Plant Sci 10:1741.
Pate JS (1980) Transport and partitioning of nitrogenous solutes. Ann Rev Plant Biol 31:313–340. https://doi.org/10.1146/annurev.pp.31.060180.001525
Peng L, Shan X, Yang Y, Wang Y, Druzhinina IS, Pan X, Jin W, He X, Wang X, Zhang X, Martin FM, Yuan Z (2021) Facultative

symbiosis with a saprotrophic soil fungus promotes potassium uptake in American sweetgum trees. Plant Cell Environ 44(8):2793–2809. https://doi.org/10.1111/pce.14053

Peoples MB, Brockwell J, Herridge DF, Rochester IJ, Alves BJR, Urquiaga S, Boddey RM, Dakora FD, Bhattarai S, Maskey SL, Sampet C, Rerkasem B, Khan DF, Hauggaard-Nielsen H, Jensen ES (2009) The contributions of nitrogen-fixing crop legumes to the productivity of agricultural systems. Symbiosis 48(1–3):1–17. https://doi.org/10.1007/BF03179980

Perner H, Schwarz D, Bruns C, Mäder P, George E (2007) Effect of arbuscular mycorrhizal colonization and two levels of compost supply on nutrient uptake and flowering of pelargonium plants. Mycorrhiza 17(5):469–474. https://doi.org/10.1007/s00572-007-0116-7

Pfeiffer S, Mitter B, Oswald A, Schloter-Hai B, Schloter M, Declerck S et al (2017) Rhizosphere microbiomes of potato cultivated in the high Andes show stable and dynamic core microbiomes with different responses to plant development. FEMS Microbiol Ecol 93:fiw242. https://doi.org/10.1093/femsec/fiw242

Prasad D, Verma N, Bakshi M, Narayan OP, Singh AK, Dua M, Johri AK (2019) Functional characterization of a magnesium transporter of root endophytic fungus Piriformospora indica. Front Microbiol 9:3231. https://doi.org/10.3389/fmicb.2018.03231

Raaijmakers JM, Paulitz TC, Steinberg C, Alabouvette C, Moënne-Loccoz Y (2009) The rhizosphere: A playground and battlefield for soilborne pathogens and beneficial microorganisms. Plant Soil 321:341–361. https://doi.org/10.1007/s11104-008-9568-6

Rajkumar M, Ae N, Prasad MNV, Freitas H (2010) Potential of siderophore-producing bacteria for improving heavy metal phytoextraction. Trends Biotechnol 28(3):142–149. https://doi.org/10.1016/j.tibtech.2009.12.002

Richardson AE, Simpson RJ (2011) Soil microorganisms mediating phosphorus availability update on microbial phosphorus. Plant Physiol 156(3):989–996. https://doi.org/10.1104/pp.111.175448

Rodrı́Guez H, Fraga R (1999) Phosphate solubilizing bacteria and their role in plant growth promotion. Biotechnol Adv 17(4–5):319–339. https://doi.org/10.1016/s0734-9750(99)00014-2

Rousk J et al (2010) Soil bacterial and fungal communities across a pH gradient in an arable soil. ISME J 4:1340–1351.

Roy S, Liu W, Nandety RS, Crook A, Mysore KS, Pislariu CI, Frugoli J, Dickstein R, Udvardi MK (2020) Celebrating 20 years of genetic discoveries in legume nodulation and symbiotic nitrogen fixation. Plant Cell 32(1):15–41. https://doi.org/10.1105/tpc.19.00279

Saboor A, Ali MA, Danish S, Ahmed N, Fahad S, Datta R, Ansari MJ, Nasif O, Rahman MH, Glick BR (2021) Effect of arbuscular mycorrhizal fungi on the physiological functioning of maize under zinc-deficient soils. Sci Rep 11(1):1–11. https://doi.org/10.1038/s41598-021-97742-1

Saia S, Rappa V, Ruisi P, Abenavoli MR, Sunseri F, Giambalvo D, Frenda AS, Martinelli F (2015) Soil inoculation with symbiotic microorganisms promotes plant growth and nutrient transporter genes expression in durum wheat. Front. Plant Sci 6:815. https://doi.org/10.3389/fpls.2015.00815

Salazar-Cerezo S, Martínez-Montiel N, García-Sánchez J, Pérez-y-Terrón R, Martínez-Contreras RD (2018) Gibberellin biosynthesis and metabolism: A convergent route for plants, fungi and bacteria. Microbiol Res 208:85–98. https://doi.org/10.1016/j.micres.2018.01.010

Sanket A, Ghosh S, Sahoo R, Nayak S, Das A (2017) Molecular identification of acidophilic manganese (Mn)-solubilizing bacteria from mining effluents and their application in mineral beneficiation. Geomicrobiol J 34(1):71–80. https://doi.org/10.1080/01490451.2016.1141340

Santos K, Moure V, Hauer V, Santos AS, Donatti L, Galvão C et al (2017) Wheat colonization by an *Azospirillum brasilense* ammonium-excreting strain reveals upregulation of nitrogenase and superior plant growth promotion. Plant Soil 415(1):245–255. https://doi.org/10.1007/s11104-016-3140-6

Shade A, Jacques MA, Barret M (2017) Ecological patterns of seed microbiome diversity, transmission, and assembly. Curr Opin Microbiol 37:15–22. https://doi.org/10.1016/j.mib.2017.03.010

Sharma SB, Sayyed RZ, Trivedi MH, Gobi TA (2013) Phosphate solubilizing microbes: Sustainable approach for managing phosphorus deficiency in agricultural soils. SpringerPlus 2(1):1–14. https://doi.org/10.1186/2193-1801-2-587

Sherman AR (1992) Zinc, copper, and iron nutriture and immunity. J Nutr 122:604–609.

Singh SK, Wu X, Shao C, Zhang H (2022) Microbial enhancement of plant nutrient acquisition. Stress Biol Jan 10;2(1):3. https://doi.org/10.1007/s44154-021-00027-w

Smanski M, Bhatia S, Zhao D. *et al.* (2014) Functional optimization of gene clusters by combinatorial design and assembly. Nat Biotechnol 32:1241–1249. https://doi.org/10.1038/nbt.3063

Smith MJ, Shoolery J, Schwyn B, Holden I, Neilands J (1985) Rhizobactin, a structurally novel siderophore from *Rhizobium meliloti*. J Am Chem Soc 107(6):1739–1743. https://doi.org/10.1021/Ja00292a047

Stevenson FJ (ed) (1982) Nitrogen in agricultural soils. American Society of Agronomy. ISBN:9780891180708. https://doi.org/10.2134/agronmonogr22

Stringlis IA, De Jonge R, Pieterse CM (2019) The age of coumarins in plant–microbe interactions. Plant and Cell Physiology 60(7):1405–1419.

Suman J, Rakshit A, Ogireddy SD, Singh S, Gupta C, Chandrakala J (2022) Microbiome as a key player in sustainable agriculture and human health. Front Soil Sci 2:821589.

Svistoonoff S, Hocher V, Gherbi H (2014) Actinorhizal root nodule symbioses: What is signalling telling on the origins of nodulation. Curr Opin Plant Biol 20:11–18. https://doi.org/10.1016/j.pbi.2014.03.001

Tayyab M, Yang Z, Zhang C, Islam W, Lin W, Zhang H (2021) Sugarcane monoculture drives microbial community composition, activity and abundance of agricultural-related microorganisms. Environ Sci Pollut Res 28:48080–48096. https://doi.org/10.1007/s11356-021-14033-y

Timilsina A, Yadav RK, Pandey D (2020) Potential pathways of nitrous oxide formation in plants. Front Plant Sci 11:1177. https://doi.org/10.3389/fpls.2020.01177

Toju H, Peay KG, Yamamichi M, Narisawa K, Hiruma K, Naito K et al (2018) Core microbiomes for sustainable agroecosystems. Nat Plants 4:247–257.

Turan M, Topcuo˘glu B, Kıtır N, Alkaya Ü, Erçelik F, Nikerel E, Güne¸s A, Turan M, Topcuo˘glu B, Kıtır N et al (2016) Plant Growth Promoting Rhizobacteria's (PGPRS) enzyme dynamics in soil remediation. In Soil Contamination—Current

Consequences and Further Solutions; IntechOpen: London, UK, ISBN 978-953-51-2816-8.

Urvasha P, Bala R, Kaur M, Faizan M, Alam P (2025) Phytomicrobiome engineering: Designing plant-microbe interactions for improved crop performance. The Microbe 6:100272, ISSN 2950-1946. https://doi.org/10.1016/j.microb.2025.100272

Vaid SK, Kumar B, Sharma A, Shukla A, Srivastava P (2014) Effect of Zn solubilizing bacteria on growth promotion and Zn nutrition of rice. J Soil Sci Plant Nutr 14(4):889–910. https://doi.org/10.4067/S0718-95162014005000071

van Velzen R, Holmer R, Bu F, Rutten L, van Zeijl A, Liu W, Santuari L, Cao Q, Sharma T, Shen D, Roswanjaya Y, Wardhani TAK, Kalhor MS, Jansen J, van den Hoogen J, Güngör B, Hartog M, Hontelez J, Verver J, Yang WC, Schijlen E, Repin R, Schilthuizen M, Schranz ME, Heidstra R, Miyata K, Fedorova E, Kohlen W, Bisseling T, Smit S, Geurts R (2018) Comparative genomics of the nonlegume *Parasponia* reveals insights into evolution of nitrogen-fixing rhizobium symbioses. Proc Natl Acad Sci U S A. 15;115(20):E4700–E4709. https://doi: 10.1073/pnas.1721395115. Epub 2018 May 1. PMID: 29717040; PMCID: PMC5960304.

Vílchez JI, Yang Y, He D, Zi H, Peng L, Lv S, Kaushal R, Wang W, Huang W, Liu R, Lang Z, Miki D, Tang K, Paré PW, Song CP, Zhu JK, Zhang H (2020) DNA demethylases are required for myo-inositol-mediated mutualism between plants and beneficial rhizobacteria. Nat Plants 6(8):983–995. https://doi.org/10.1038/s41477-020-0707-2

Wang S, Chen A, Xie K, Yang X, Luo Z, Chen J, Zeng D, Ren Y, Yang C, Wang L, Feng H, López-Arredondo DL, Herrera-Estrella LR, Xu G (2020) Functional analysis of the OsNPF4. 5 nitrate transporter reveals a conserved mycorrhizal pathway of nitrogen acquisition in plants. PNAS USA 117(28):16649–16659. https://doi.org/10.1073/pnas.2000926117

Watts-Williams SJ, Cavagnaro TR (2018) Arbuscular mycorrhizal fungi increase grain zinc concentration and modify the expression of root ZIP transporter genes in a modern barley (Hordeum vulgare) cultivar. Plant Sci 274:163–170. https://doi.org/10.1016/j.plantsci.2018.05.015

Watts-Williams SJ, Patti AF, Cavagnaro TR (2013) Arbuscular mycorrhizas are beneficial under both deficient and toxic soil zinc conditions. Plant Soil 371(1):299–312. https://doi.org/10.1007/s11104-013-1670-8

Welch RM, Shuman L (1995) Micronutrient nutrition of plants. Crit Rev Plant Sci 14:49–82.

Whalen ED, Smith RG, Grandy AS, Frey SD (2018) Manganese limitation as a mechanism for reduced decomposition in soils under atmospheric nitrogen deposition. Soil Biol Biochem 127:252–263.

Xu G-H, Chague V, Melamed-Bessudo C, Kapulnik Y, Jain A, Raghothama KG, Levy AA, Silber A (2007) Functional characterization of LePT4: A phosphate transporter in tomato with mycorrhiza-enhanced expression. J Exp Bot 58(10):2491–2501. https://doi.org/10.1093/jxb/erm096

Yadav AN, Kumar V, Dhaliwal H, Prasad R, Saxena A (2018) Microbiome in crops: Diversity, distribution, and potential role in crop improvement. In New and Future Developments in Microbial Biotechnology and Bioengineering: Crop Improvement through Microbial Biotechnology; Elsevier: Amsterdam, The Netherlands, 2018; pp. 305–332, ISBN 978-0-444-63987-5.

Zamioudis C, Korteland J, Van Pelt JA, Van Hamersveld M, Dombrowski N, Bai Y et al (2015) Rhizobacterial volatiles and photosynthesis-related signals coordinate MYB 72 expression in *Arabidopsis* roots during onset of induced systemic resistance and iron-deficiency responses. Plant J 84(2):309–322. https://doi.org/10.1111/tpj.12995

Zarraonaindia I, Owens SM, Weisenhorn P, West K, Hampton-Marcell J, Lax S et al (2015) The soil microbiome influences grapevine-associated microbiota. MBio 6:10–1128. https://doi.org/10.1128/mBio.02527-14

Zhang F, Du P, Song C-X, Wu Q-S (2015) Alleviation of mycorrhiza to magnesium deficiency in trifoliate orange: Changes in physiological activity. Emir J Food Agric 27:763–769.

Zhang H, Sun Y, Xie X, Kim MS, Dowd SE, Paré PW (2009) A soil bacterium regulates plant acquisition of iron via deficiency-inducible mechanisms. Plant J 58(4):568–577. https://doi.org/10.1111/j.1365-313X.2009.03803.x

Zhang H, Xie X, Kim MS, Kornyeyev DA, Holaday S, Paré PW (2008) Soil bacteria augment *Arabidopsis* photosynthesis by decreasing glucose sensing and abscisic acid levels in planta. Plant J 56(2):264–273. https://doi.org/10.1111/j.1365-313X.2008.03593.x

Zhang J, Cook J, Nearing JT, Zhang J, Raudonis R, Glick BR, Langille MGI, Cheng Z (2021) Harnessing the plant microbiome to promote the growth of agricultural crops. Microbiol Res 245:126690. https://doi.org/10.1016/j.micres.2020.126690. Epub 2021 Jan 6. PMID: 33460987.

Zhao F, Lehmann J, Solomon D, Fox M, Mcgrath S (2006) Sulphur speciation and turnover in soils: Evidence from Sulphur K-edge XANES spectroscopy and isotope dilution studies. Soil Biol Biochem 38(5):1000–1007. https://doi.org/10.1016/j.soilbio.2005.08.013

Zhou C, Guo J, Zhu L, Xiao X, Xie Y, Zhu J et al (2016) Paenibacillus polymyxa BFKC01 enhances plant iron absorption via improved root systems and activated iron acquisition mechanisms. Plant Physiol Biochem 105:162–173. https://doi.org/10.1016/j.plaphy.2016.04.025

10 Microbiome in Improvements of Crop Yield to Utilize Biofertilizers

Arpan Mukherjee[1,2], *Bansh Narayan Singh*[1], *and Jay Prakash Verma*[1*]
[1]Plant-Microbe Interaction Lab, Institute of Environment and Sustainable Development, Banaras Hindu University, Varanasi, Uttar Pradesh, India
[2]Department of Basic Sciences, School of Sciences and Humanities, SR University, Warangal, Telangana, India

10.1 INTRODUCTION

The increasing worldwide need for food production, along with the negative environmental impacts of traditional farming methods, has driven considerable attention toward sustainable agricultural alternatives (Rai et al., 2023). In modern agriculture, biofertilizers have become a key focus due to their potential to improve soil quality and increase crop yields. Innovative farming techniques that utilize plant–microbe relationships have proven particularly effective in pest management, soil enrichment, and productivity enhancement (Mukherjee et al., 2022b). These methods often involve traps and repellent crops to control pests while simultaneously enhancing soil microbial activity. Although these systems are already beneficial, their efficiency can be further improved by incorporating biofertilizers, which enhance beneficial microbial populations in the plant–soil environment (Chouhan et al., 2021). To optimize the use of biofertilizers in sustainable farming, it is crucial to investigate how they interact with crops and soil microbiomes (Kumar et al., 2024).

Biofertilizers, such as phosphate-solubilizing bacteria (PSB), potassium-solubilizing bacteria (KSB), and nitrogen-fixing bacteria (NFB), are essential for improving nutrient availability and supporting beneficial microbial activity (Nabati et al., 2025). Studies demonstrate that biofertilizers enhance soil health and plant growth by promoting microbial communities involved in nutrient cycling, disease resistance, and stress adaptation (Singh et al., 2023). Their ability to boost plant growth and defense mechanisms aligns well with sustainable farming goals, leading to higher productivity and reduced dependence on chemical fertilizers. By harnessing these microbial processes, agricultural systems can achieve greater resilience and efficiency while lowering environmental harm (Mukherjee et al., 2020).

However, effectively integrating biofertilizers into farming systems requires a deep understanding of the complex interactions between crops, microbes, and environmental conditions. Research has advanced in identifying how biofertilizers contribute to agricultural productivity (Verma et al., 2017). Key processes include plant growth stimulation through phytohormones, increased nutrient accessibility via nitrogen fixation and phosphate solubilization, and pathogen suppression through antimicrobial compounds and induced resistance (Meena et al., 2023). For example, PSB enhance phosphorus uptake, a crucial nutrient for plant development, while KSB improve potassium availability, strengthening crop resilience. Similarly, NFB increase soil nitrogen levels, supporting healthier plant growth and higher yields (Mukherjee et al., 2021b). These findings highlight how biofertilizers can enhance sustainable agriculture, offering a pathway to more eco-friendly and productive farming.

The strategic use of biofertilizers in sustainable agriculture can improve system performance while maintaining ecological balance. This approach reduces reliance on synthetic fertilizers and enhances soil quality (Pahalvi et al., 2021). As a result, biofertilizers represent a viable solution for developing more sustainable and adaptive farming systems. Yet a deeper exploration of how biofertilizers interact with plant microbiomes is needed to maximize their benefits. Additionally, biofertilizers have been found to improve soil moisture retention and drought resistance, which are increasingly important in climate-variable conditions (Mikiciuk et al., 2024). Recent studies suggest that combining biochar with microbial inoculants can enhance water retention and crop output, further advancing sustainable agriculture (Zhang et al., 2023). Despite substantial evidence supporting biofertilizers in sustainable farming, further research is necessary to understand microbial behavior within plant–soil systems. Challenges such as inconsistent performance, complex soil–plant-microbe interactions, and the lack of standardized production and application methods hinder widespread adoption.

10.1.1 The Plant Microbiome

The plant microbiome encompasses a rich and varied community of microorganisms that includes bacteria, fungi, archaea, viruses, etc. that inhabit or surround plant tissues like roots, stems, leaves, flowers, and seeds, forming a complex and ever-evolving system (Rout, 2014; Mukherjee et al.,

*Corresponding author: verma_bhu@yahoo.co.in (Dr Jay Prakash Verma)

DOI: 10.1201/9781003544425-10

2024). These microbes engage with their host plants across distinct ecological zones, including the rhizosphere (soil near roots), phyllosphere (leaf surfaces), and endosphere (internal plant tissues), significantly influencing plant vitality, growth, and adaptability (Berendsen et al., 2012; Compant et al., 2019). They contribute to nutrient uptake, bolster resilience against stress, and shield plants from pathogens. Among the key players are beneficial symbionts like NFB, which transform atmospheric nitrogen into a form plants can use in the rhizosphere, and mycorrhizal fungi, which enhance phosphorus absorption, alongside pathogens and neutral microbes that also affect plant well-being (Dellagi et al., 2020) (Figure 10.1). The composition and effectiveness of this microbiome are driven by factors such as plant species, genotype, soil characteristics, and environmental conditions, highlighting its importance in maintaining ecosystem balance and advancing sustainable farming practices, with cutting-edge metagenomics shedding light on opportunities for microbiome manipulation (Mukherjee et al., 2024).

10.1.2 Microbial Community

The plant microbiome comprises distinct microbial communities that colonize specific plant-associated habitats: the rhizosphere, phyllosphere, and endosphere. The rhizosphere refers to the soil region immediately surrounding plant roots, teeming with bacteria, fungi, and other microbes attracted by root exudates, which provide nutrients and shape microbial diversity (Vives-Peris et al., 2020). In contrast, the phyllosphere encompasses the aerial surfaces of plants, primarily leaves, where microorganisms like fungi and bacteria adapt to fluctuating conditions such as ultraviolet (UV) radiation and moisture scarcity (Sohrabi et al., 2023). The endosphere, meanwhile, consists of microbes living within plant tissues, often entering through roots or wounds, and includes endophytic bacteria and fungi that may form symbiotic relationships with the host (Hardoim et al., 2015). These spatially distinct communities collectively contribute to plant health and adaptation.

10.1.3 Functions of the Microbiome in Nutrient Cycling, Disease Suppression, and Stress Tolerance

The plant microbiome plays a pivotal role in supporting plant vitality through nutrient cycling, disease suppression, and stress tolerance. In nutrient cycling, microorganisms such as rhizobial bacteria fix atmospheric nitrogen, while PSB and mycorrhizal fungi enhance the availability of phosphorus and other essential nutrients, boosting plant growth (Yadav et al., 2021). Disease suppression is mediated by beneficial microbes, including *Pseudomonas* and *Trichoderma* species, which produce antimicrobial compounds or outcompete pathogens, reducing the incidence of diseases like root rot (Meena et al., 2017). Additionally, microbiome enhances stress tolerance by synthesizing hormones like auxin or inducing systemic resistance, helping plants withstand drought, salinity, and temperature extremes (Mukherjee et al., 2021a). These functions highlight the microbiome's integral role in sustainable plant ecosystems.

10.1.4 Factors Influencing Microbiome Composition

The composition of the plant microbiome is shaped by a variety of interacting factors, including soil type, plant genotype, and environmental conditions. Soil type influences microbial diversity by determining nutrient availability, pH, and texture, with sandy soils fostering different communities than clay-rich soils (Zhu et al., 2021). Plant genotype

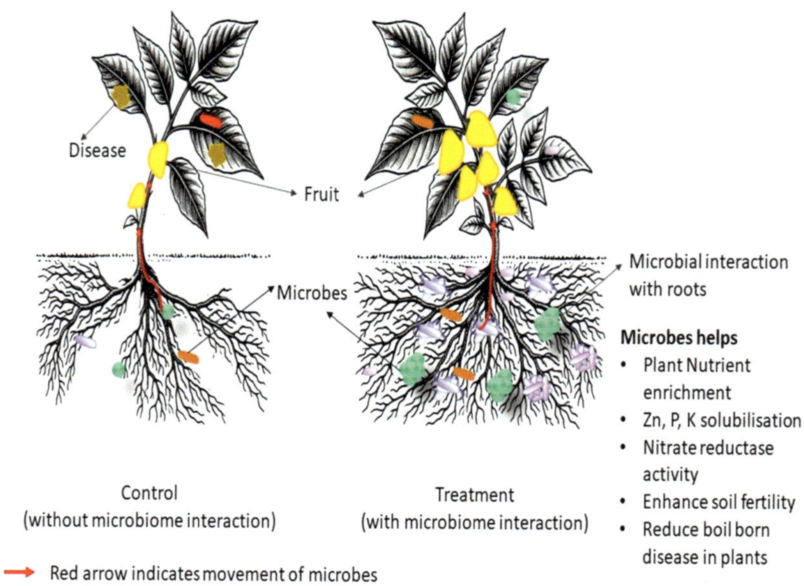

FIGURE 10.1 Effect of microbiome on the plant growth attributes.

exerts a selective effect, as different species or cultivars release unique root exudates that attract specific microbes, a phenomenon known as the "host effect" (Bulgarelli et al., 2013). Environmental conditions, such as temperature, moisture, and seasonality, further modulate microbiome structure by favoring the survival of adapted microbial taxa (Coleine et al., 2024). These factors create a dynamic interplay that governs the assembly and functionality of microbial communities associated with plants.

10.2 BIOFERTILIZERS

Biofertilizers are living microbial inoculants that enhance plant growth by improving nutrient availability and uptake in the soil, offering an eco-friendly alternative to chemical fertilizers. They consist of specific microorganisms that colonize plant roots or soil, facilitating the cycling of essential nutrients like nitrogen, phosphorus, and potassium (Mukherjee et al., 2021c). Biofertilizers are classified based on their functional roles into categories such as nitrogen-fixing, phosphate-solubilizing, potassium-mobilizing, and PGPR (Nosheen et al., 2021). These microbial agents, including bacteria, fungi, and cyanobacteria, work symbiotically with plants to boost productivity and soil health (Mahanty et al., 2017). PGPR are a diverse group of beneficial bacteria, including *Pseudomonas*, *Bacillus*, and *Azospirillum*, that colonize the rhizosphere and promote plant growth through multiple mechanisms. Their ability to fix nitrogen, solubilize phosphates, and secrete siderophores (iron-chelating compounds) further aids plant nutrition (Glick, 2012). PGPR are widely applied in agriculture to enhance crop resilience and yield sustainably. Their application supports sustainable agriculture by reducing reliance on synthetic inputs (Table 10.1).

10.2.1 Nitrogen Fixing Biofertilizers

Nitrogen-fixing biofertilizers comprise microorganisms capable of converting atmospheric nitrogen (N_2) into ammonia, a form plants can assimilate, thus enriching soil fertility. *Rhizobium*, a symbiotic bacterium, forms nodules on legume roots, where it fixes nitrogen in exchange for nutrients from the host plant (Kumar et al., 2020). In contrast, free-living bacteria like *Azotobacter* thrive in the rhizosphere, fixing nitrogen independently while also producing growth-promoting substances (Jehani et al., 2023). The biofertilizers are vital for crops in nitrogen-deficient soils, reducing the need for chemical nitrogen fertilizers (Chaudhary et al., 2022). Their efficacy depends on soil conditions and plant compatibility.

10.2.1.1 Phosphate-Solubilizing Biofertilizers

Phosphate-solubilizing biofertilizers include microorganisms like *Pseudomonas* and *Bacillus* that solubilize insoluble phosphates in the soil, making phosphorus available to plants (Abawari et al., 2020). These bacteria secrete organic acids, such as citric and gluconic acid, which lower soil pH and dissolve bound phosphates, enhancing nutrient uptake (Rawat et al., 2021). Fungi like *Aspergillus* and *Penicillium* also contribute by producing enzymes like phosphatases (Sharma et al., 2013). Widely used in phosphorus-deficient soils, these biofertilizers improve root development and crop yield, offering a sustainable solution to phosphorus management in agriculture.

10.2.1.2 Potassium-Mobilizing Biofertilizers

Potassium-mobilizing biofertilizers consist of microbes that release potassium from insoluble soil minerals, such as silicates, making it accessible to plants (Mukherjee et al., 2019). Bacteria like *Bacillus mucilaginosus* and

TABLE 10.1
Commonly Used Microbes for Preparation of Biofertiliser

Microbes Type	Examples	Working Mechanism	Recent Reference
Nitrogen-fixing bacteria	*Rhizobium*, *Azotobacter*	Convert atmospheric nitrogen (N_2) into ammonia via nitrogenase enzyme, enhancing plant nitrogen supply.	Singh et al. (2022)
Phosphate-solubilizing bacteria	*Pseudomonas*, *Bacillus*	Produce organic acids (e.g., gluconic acid) to solubilize insoluble phosphates, increasing P availability.	Sharma et al. (2023)
Potassium-mobilizing bacteria	*Bacillus mucilaginosus*	Release potassium from silicates through acidolysis and chelation, improving K uptake by plants.	Meena et al. (2021)
Plant growth-promoting rhizobacteria (PGPR)	*Azospirillum*, *Pseudomonas*	Synthesize phytohormones (e.g., auxin, gibberellins) and induce systemic resistance against pathogens.	Glick and Gamalero (2021)
Mycorrhizal fungi	*Glomus*, *Gigaspora*	Form symbiotic networks with roots, enhancing nutrient (P, Zn) and water uptake via hyphal extension.	Zhang et al. (2022)
Cyanobacteria	*Anabaena*, *Nostoc*	Fix nitrogen photosynthetically and enrich soil with organic matter, aiding nutrient cycling.	Prasanna et al. (2021)

Frateuria aurantia produce organic acids and polysaccharides that break down potassium-bearing compounds, enhancing soil fertility (Soumare et al., 2023). Though less studied than nitrogen or phosphate biofertilizers, their role is critical in potassium-deficient regions, supporting plant functions like water regulation and enzyme activation (Meena et al., 2014). These biofertilizers contribute to balanced nutrient management and sustainable crop production.

10.2.2 Mechanisms of Action

10.2.2.1 Nutrient Solubilization and Fixation

Biofertilizers enhance nutrient availability through solubilization and fixation processes. Nitrogen-fixing bacteria, such as *Rhizobium* and *Azotobacter*, convert atmospheric N_2 into ammonia via the enzyme nitrogenase, directly supplying plants with nitrogen (Vats et al., 2021). Phosphate-solubilizing microbes, like *Pseudomonas*, release organic acids to solubilize insoluble phosphates, while potassium-mobilizing bacteria degrade silicates to free potassium (Bhardwaj et al., 2014). These mechanisms replenish soil nutrient pools, reducing dependency on chemical fertilizers and supporting sustainable agriculture.

10.2.2.2 Production of Phytohormones

Biofertilizers produce phytohormones such as auxins, gibberellins, and cytokinins, which regulate plant growth and development (Aloo et al., 2022). For instance, *Azospirillum* secretes indole-3-acetic acid (IAA), a type of auxin that stimulates root elongation and branching, improving nutrient and water uptake (Pham et al., 2022). These hormones enhance cell division, seed germination, and overall plant vigor, particularly under stress conditions (Vessey, 2003). This mechanism underscores the role of biofertilizers in boosting crop productivity naturally.

10.2.2.3 Induction of Systemic Resistance in Plants

Biofertilizers induce systemic resistance (ISR) in plants, enhancing their defense against pathogens and environmental stress (Purwaningsih et al., 2023). PGPRs, such as *Bacillus subtilis*, trigger ISR by producing signaling molecules, such as salicylic acid or jasmonates, which activate plant immune responses (Jalmi & Sinha, 2022). This primes plants to resist diseases like fungal infections or insect attacks without direct antagonism (Pieterse et al., 2014). ISR complements other biofertilizer benefits, promoting resilient and healthy crops.

10.2.2.4 Enhancement of Soil Structure and Fertility

Biofertilizers improve soil structure and fertility by producing extracellular polysaccharides and other organic compounds that bind soil particles, enhancing aggregation and water retention (Dar et al., 2021). Microbes like *Rhizobium* and mycorrhizal fungi contribute to soil organic matter through biomass turnover, while phosphate solubilizers increase nutrient availability (Kour et al., 2020). These processes mitigate soil degradation, promote long-term fertility, and support sustainable farming systems.

10.3 ROLE OF THE MICROBIOME IN ENHANCING CROP YIELD

10.3.1 Microbiome-Mediated Nutrient Acquisition and Utilization

The plant microbiome significantly enhances crop yield by improving nutrient acquisition and utilization through symbiotic microbial interactions (Mukherjee et al., 2021b). Nitrogen-fixing bacteria, such as *Rhizobium* in legumes, convert atmospheric nitrogen into plant-usable forms, while phosphate-solubilizing bacteria like *Pseudomonas* release bound phosphorus via organic acid production (Chen et al., 2025). Mycorrhizal fungi extend root surface area, facilitating the uptake of water and micronutrients like zinc and copper (Jacoby et al., 2017). These processes optimize nutrient efficiency, reducing the need for chemical fertilizers and supporting higher yields in nutrient-poor soils.

10.3.2 Improvement of Plant Growth and Development through Microbial Interactions

Microbial interactions within the plant microbiome promote growth and development by producing phytohormones and enhancing root architecture. PGPR, such as *Bacillus* and *Azospirillum*, synthesize auxins, gibberellins, and cytokinins, which stimulate cell division, root elongation, and shoot growth (Raza, 2023; Mukherjee et al., 2021b). These microbes also improve nutrient uptake by increasing root branching, as seen with *Azotobacter* inoculation in cereals (Compant et al., 2010). Such interactions lead to vigorous plants with higher biomass and yield potential, critical for agricultural productivity.

10.3.3 Microbiome Contributions to Abiotic Stress Tolerance

The plant microbiome enhances crop resilience to abiotic stresses like drought and salinity, thereby sustaining yields under adverse conditions. Microbes such as *Pseudomonas fluorescens* produce exopolysaccharides that improve soil moisture retention during drought, while others, like *Bacillus subtilis*, synthesize osmoprotectants such as proline to stabilize plant cells under salinity stress (Naseem et al., 2024). Additionally, mycorrhizal fungi alleviate water stress by improving hydraulic conductivity in roots (Santos et al., 2021). These adaptations enable crops to maintain growth and productivity despite environmental challenges.

10.3.4 Microbiome-Mediated Biotic Stress Resistance

The microbiome bolsters crop yield by conferring resistance to biotic stresses, including pathogens and pests, through direct and indirect mechanisms (Mukherjee Et al., 2020). Beneficial microbes like *Trichoderma* spp. antagonize soil-borne pathogens by producing antifungal compounds, while PGPR such as *Pseudomonas syringae* ISR, priming plants against fungal and bacterial diseases (Natsiopoulos et al., 2022). Certain rhizobacteria also deter pests by synthesizing volatile organic compounds that repel herbivores (Berendsen et al., 2012). This natural biocontrol reduces crop losses and enhances yield stability.

Microbiome-based biofertilizers have demonstrated tangible success in boosting crop yields across diverse agricultural systems. In India, *Rhizobium* inoculation in chickpea fields increased yields by up to 30% by enhancing nitrogen fixation, as reported in a 2022 study (Singh et al., 2022). Similarly, in Brazil, the application of *Azospirillum brasilense* in maize improved grain yield by 15–20% under drought conditions through phytohormone production and stress mitigation (Galindo et al., 2022). Phosphate-solubilizing *Bacillus megaterium* boosted wheat yields in phosphorus-deficient soils (Iqbal et al., 2024).

10.4 CHALLENGES AND LIMITATIONS

Despite the promising benefits of the plant microbiome and biofertilizers in enhancing crop yield, several challenges and limitations hinder their widespread adoption (Table 10.2). Variability in microbial efficacy due to soil type, climate, and plant genotype often leads to inconsistent results, complicating their reliability across diverse agricultural systems. Additionally, the production, storage, and application of biofertilizers face logistical hurdles, including limited shelf life and sensitivity to environmental conditions like temperature and humidity, which can reduce microbial viability. Regulatory frameworks for biofertilizer approval and commercialization also lag, slowing their integration into mainstream farming practices (Timmusk et al., 2017). Furthermore, a lack of farmer awareness and technical knowledge about microbiome-based solutions poses a barrier to their effective use, necessitating improved education and infrastructure.

10.5 CONCLUSION AND FUTURE PERSPECTIVES

The future of leveraging the plant microbiome for sustainable agriculture holds immense potential through advances in microbiome engineering, synthetic biology, and integrative technologies. Innovations in microbiome engineering and synthetic biology promise to design microbial consortia with enhanced capabilities, such as minerals solubilization, optimized nitrogen fixation, or stress tolerance, tailored to specific agricultural needs. The integration of omics technologies genomics, metagenomics, and transcriptomics will deepen our understanding of microbial interactions and functions, enabling precise manipulation of the microbiome for improved crop performance. This knowledge will facilitate the development of tailored biofertilizers customized for specific crops and environments, ensuring efficacy across diverse agroecosystems. Additionally, policy and economic incentives, such as subsidies or certification programs, could accelerate biofertilizer adoption by addressing cost barriers and encouraging farmer uptake, paving the way for a greener agricultural future (Trivedi et al., 2020).

The plant microbiome and biofertilizers represent transformative tools for sustainable agriculture, enhancing crop yield through improved nutrient acquisition, growth promotion, and resilience to biotic and abiotic stresses. By harnessing microbial communities in the rhizosphere, phyllosphere, and endosphere, biofertilizers like nitrogen-fixing *Rhizobium*, phosphate-solubilizing *Pseudomonas*, and plant growth-promoting rhizobacteria offer eco-friendly alternatives to chemical inputs, as demonstrated in

TABLE 10.2
Impact of Biofertilizers (Microbes Based) on Crop Yields

Biofertilizer Type	Crop(s)	Crops Yield Improvement	Reference
Bacillus spp., AMF, *Rhizobium* spp., *Trichoderma* spp.	Different types of vegetables, legumes, and millet	Improve productivity of millet up to 65%, up to 35% in Cabbage, and 54% in kidney beans. The key factor of biofertilizer is that it improves soil enzyme activity and microbial biomass.	Pei et al. (2025)
AMF, N fixers, and P solubilizers	Cereals, fruits, and herbs	Increased productivity of approximately up to 30–55%, by improvingsoilqualityandnutrientsuptakeandotherPGPactivity–	Schütz et al. (2018)
Bacillus sp. and *Trichoderma* sp.-based biofertilizers	Maize	Overall yield increased up to approximately 29% in *Bacillus/Trichoderma* combinations by improving soil quality.	Gan et al. (2023)
Bacillus and *Trichoderma-based* biofertilizers	Corn and soybean	Overall yield increased approximately up to 16% in Corn and 17% in soybeans. The biofertilizer directly improves P and NH_4^+, N, and reshapes microbial community networks.	de Lima Gonilha et al. (2024)

successful case studies worldwide. However, challenges such as inconsistent efficacy, logistical constraints, and limited adoption highlight the need for further innovation and support. Looking ahead, advances in microbiome engineering, omics technologies, tailored biofertilizer development, and supportive policies promise to overcome these hurdles, unlocking the full potential of microbial solutions. As agriculture faces growing pressures from climate change and population demands, integrating microbiome-based strategies will be crucial for ensuring food security and environmental sustainability.

REFERENCES

Abawari, R. A., Tuji, F. A., & Yadete, D. M. (2020). Phosphate solubilizing bio-fertilizers and their role in bio-available P nutrient: An overview. *International Journal of Applied Agriculture Sciences*, *6*(6), 162.

Aloo, B. N., Tripathi, V., Makumba, B. A., & Mbega, E. R. (2022). Plant growth-promoting rhizobacterial biofertilizers for crop production: The past, present, and future. *Frontiers in Plant Science*, *13*, 1002448.

Berendsen, R. L., Pieterse, C. M. J., & Bakker, P. A. H. M. (2012). The rhizosphere microbiome and plant health. *Trends in Plant Science*, *17*(8), 478–486.

Bhardwaj, D., Ansari, M. W., Sahoo, R. K., & Tuteja, N. (2014). Biofertilizers function as key player in sustainable agriculture by improving soil fertility, plant tolerance and crop productivity. *Microbial Cell Factories*, *13*, 66.

Bulgarelli, D., Schlaeppi, K., Spaepen, S., Ver Loren van Themaat, E., & Schulze-Lefert, P. (2013). Structure and functions of the bacterial microbiota of plants. *Annual Review of Plant Biology*, *64*, 807–838.

Chaudhary, P., Singh, S., Chaudhary, A., Sharma, A., & Kumar, G. (2022). Overview of biofertilizers in crop production and stress management for sustainable agriculture. *Frontiers in Plant Science*, *13*, 930340.

Chen, W., Li, Y., Shi, G., Fan, G., Tong, F., Liu, L., . . . & Gao, Y. (2025). The role of symbiotic nitrogen-fixing bacteria, Rhizobium and Sinorhizobium, as "bridges" in the rhizosphere of legumes after fomesafen application. *Applied Soil Ecology*, *209*, 106013.

Chouhan, G. K., Verma, J. P., Jaiswal, D. K., Mukherjee, A., Singh, S., de Araujo Pereira, A. P., . . . & Singh, B. K. (2021). Phytomicrobiome for promoting sustainable agriculture and food security: Opportunities, challenges, and solutions. *Microbiological Research*, *248*, 126763.

Coleine, C., Delgado-Baquerizo, M., DiRuggiero, J., Guirado, E., Harfouche, A. L., Perez-Fernandez, C., . . . & Egidi, E. (2024). Dryland microbiomes reveal community adaptations to desertification and climate change. *The ISME Journal*, *18*(1), wrae056.

Compant, S., Clément, C., & Sessitsch, A. (2010). Plant growth-promoting bacteria in the rhizo- and endosphere of plants: Their role, colonization, mechanisms involved and prospects for utilization. *Soil Biology and Biochemistry*, *42*(5), 669–678.

Compant, S., Samad, A., Faist, H., & Sessitsch, A. (2019). A review on the plant microbiome: Ecology, functions, and emerging trends in microbial application. *Journal of Advanced Research*, *19*, 29–37.

Dar, A., Zahir, Z. A., Iqbal, M., Mehmood, A., Javed, A., Hussain, A., . . . & Ahmad, M. (2021). Efficacy of rhizobacterial exopolysaccharides in improving plant growth, physiology, and soil properties. *Environmental Monitoring and Assessment*, *193*, 1–15.

de Lima Gonilha, D. B., Santos, C. H. B., Frezarin, E. T., Siqueira, J. S., & Rigobelo, E. C. (2024). Biological strategies to minimize fertilizer use in maize: Efficacy of *Trichoderma harzianum* and *Bacillus subtilis*. *Microbiology Research*, *15*(4), 2261–2273.

Dellagi, A., Quillere, I., & Hirel, B. (2020). Beneficial soil-borne bacteria and fungi: A promising way to improve plant nitrogen acquisition. *Journal of Experimental Botany*, *71*(15), 4469–4479.

Galindo, F. S., Rodrigues, W. L., Fernandes, G. C., Boleta, E. H. M., Jalal, A., Rosa, P. A. L., . . . & Teixeira Filho, M. C. M. (2022). Enhancing agronomic efficiency and maize grain yield with *Azospirillum brasilense* inoculation under Brazilian savannah conditions. *European Journal of Agronomy*, *134*, 126471.

Gan, Y. W., Feng, X. J., Li, Y. H., Yang, L., & Ning, P. (2023). Effects of microbial fertilizer application on maize yield in China: A meta-analysis. *Journal of Plant Nutrition and Fertilizers*, *29*, 2247–2257.

Glick, B. R. (2012). Plant growth-promoting bacteria: Mechanisms and applications. *Scientifica*, *2012*, 963401.

Glick, B. R., & Gamalero, E. (2021). Recent developments in the study of plant microbiomes. *Microorganisms*, *9*(5), 999.

Hardoim, P. R., van Overbeek, L. S., & van Elsas, J. D. (2015). Properties of bacterial endophytes and their proposed role in plant growth. *Trends in Microbiology*, *23*(10), 622–632.

Iqbal, Z., Ahmad, M., Raza, M. A., Hilger, T., & Rasche, F. (2024). Phosphate-solubilizing Bacillus sp. modulate soil exoenzyme activities and improve wheat growth. *Microbial Ecology*, *87*(1), 31.

Jacoby, R., Peukert, M., Succurro, A., Koprivova, A., & Kopriva, S. (2017). The role of soil microorganisms in plant mineral nutrition—current knowledge and future directions. *Frontiers in Plant Science*, *8*, 1617.

Jalmi, S. K., & Sinha, A. K. (2022). Ambiguities of PGPR-induced plant signaling and stress management. *Frontiers in Microbiology*, *13*, 899563.

Jehani, M. D., Singh, S., Kumar, D., & Kumar, G. (2023). *Azospirillum*: A free-living nitrogen-fixing bacterium. In *Rhizobiome* (pp. 285–308). Academic Press.

Kour, D., Rana, K. L., Yadav, A. N., Yadav, N., Kumar, M., Kumar, V., . . . & Saxena, A. K. (2020). Microbial biofertilizers: Bioresources and eco-friendly technologies for agricultural and environmental sustainability. *Biocatalysis and Agricultural Biotechnology*, *23*, 101487.

Kumar, G. A., Kumar, S., Bhardwaj, R., Swapnil, P., Meena, M., Seth, C. S., & Yadav, A. (2024). Recent advancements in multifaceted roles of flavonoids in plant–rhizomicrobiome interactions. *Frontiers in Plant Science*, *14*, 1297706.

Kumar, N., Srivastava, P., Vishwakarma, K., Kumar, R., Kuppala, H., Maheshwari, S. K., & Vats, S. (2020). The rhizobium–plant symbiosis: State of the art. *Plant Microbe Symbiosis*, 1–20.

Mahanty, T., Bhattacharjee, S., Goswami, M., Bhattacharyya, P., Das, B., Ghosh, A., & Tribedi, P. (2017). Biofertilizers: A potential approach for sustainable agriculture development.

Environmental Science and Pollution Research, *24*(4), 3315–3335.

Meena, M., Mehta, T., Nagda, A., Yadav, G., & Sonigra, P. (2023). PGPR-mediated synthesis and alteration of different secondary metabolites during plant–microbe interactions. In *Plant-Microbe Interaction-Recent Advances in Molecular and Biochemical Approaches* (pp. 229–255). Academic Press.

Meena, M., Swapnil, P., Zehra, A., Aamir, M., Dubey, M. K., Goutam, J., & Upadhyay, R. S. (2017). Beneficial microbes for disease suppression and plant growth promotion. *Plant-Microbe Interactions in Agro-Ecological Perspectives: Volume 2: Microbial Interactions and Agro-Ecological Impacts*, 395–432.

Meena, V. S., Bahadur, I., & Saha, P. (2021). Potassium-solubilizing bacteria: Mechanisms and their role in sustainable agriculture. *Microbiological Research*, *252*, 126865.

Meena, V. S., Maurya, B. R., & Verma, J. P. (2014). Does a rhizospheric microorganism enhance K^+ availability in agricultural soils? *Microbiological Research*, *169*(5–6), 337–347.

Mikiciuk, G., Miller, T., Kisiel, A., Cembrowska-Lech, D., Mikiciuk, M., Łobodzińska, A., & Bokszczanin, K. (2024). Harnessing beneficial microbes for drought tolerance: A review of ecological and agricultural innovations. *Agriculture*, *14*(12), 2228.

Mukherjee, A., Bhowmick, S., Yadav, S., Rashid, M. M., Chouhan, G. K., Vaishya, J. K., & Verma, J. P. (2021a). Re-vitalizing of endophytic microbes for soil health management and plant protection. *3 Biotech*, *11*, 1–17.

Mukherjee, A., Chouhan, G. K., Gaurav, A. K., Jaiswal, D. K., & Verma, J. P. (2021b). Development of indigenous microbial consortium for biocontrol management. In *New and Future Developments in Microbial Biotechnology and Bioengineering* (pp. 91–104). Elsevier.

Mukherjee, A., Gaurav, A. K., Patel, A. K., Singh, S., Chouhan, G. K., Lepcha, A., . . . & Verma, J. P. (2021c). Unlocking the potential plant growth-promoting properties of chickpea (Cicer arietinum L.) seed endophytes bio-inoculants for improving soil health and crop production. *Land Degradation & Development*, *32*(15), 4362–4374.

Mukherjee, A., Gaurav, A. K., Singh, S., Chouhan, G. K., Kumar, A., & Das, S. (2019). Role of potassium (K) solubilising microbes (KSM) in growth and induction of resistance against biotic and abiotic stress in plant. *Climate Change Environmental Sustainability*, *7*, 212–214.

Mukherjee, A., Gaurav, A. K., Singh, S., Yadav, S., Bhowmick, S., Abeysinghe, S., & Verma, J. P. (2022a). The bioactive potential of phytohormones: A review. *Biotechnology Reports*, *35*, e00748.

Mukherjee, A., Singh, B. K., & Verma, J. P. (2020). Harnessing chickpea (Cicer arietinum L.) seed endophytes for enhancing plant growth attributes and bio-controlling against Fusarium sp. *Microbiological Research*, *237*, 126469.

Mukherjee, A., Singh, B. N., Kaur, S., Sharma, M., de Araújo, A. S. F., de Araujo Pereira, A. P., . . . & Verma, J. P. (2024). Unearthing the power of microbes as plant microbiome for sustainable agriculture. *Microbiological Research*, *286*, 127780.

Mukherjee, A., Singh, S., Gaurav, A. K., Chouhan, G. K., Jaiswal, D. K., de Araujo Pereira, A. P., . . . & Verma, J. P. (2022b). Harnessing of phytomicrobiome for developing potential biostimulant consortium for enhancing the productivity of chickpea and soil health under sustainable agriculture. *Science of the Total Environment*, *836*, 155550.

Nabati, J., Nezami, A., Yousefi, A., Oskoueian, E., Oskoueian, A., & Ahmadi-Lahijani, M. J. (2025). Biofertilizers containing plant growth promoting rhizobacteria enhance nutrient uptake and improve the growth and yield of chickpea plants in an arid environment. *Scientific Reports*, *15*(1), 8331.

Naseem, M., Chaudhry, A. N., Jilani, G., Alam, T., Naz, F., Ullah, R., . . . & Zaman, S. (2024). Exopolysaccharide-producing bacterial cultures of Bacillus cereus and Pseudomonas aeruginosa in soil augment water retention and maize growth. *Heliyon*, *10*(4).

Natsiopoulos, D., Tziolias, A., Lagogiannis, I., Mantzoukas, S., & Eliopoulos, P. A. (2022). Growth-promoting and protective effect of *Trichoderma atrobrunneum* and *T. simmonsii* on tomato against soil-borne fungal pathogens. *Crops*, *2*(3), 202–217.

Nosheen, S., Ajmal, I., & Song, Y. (2021). Microbes as biofertilizers, a potential approach for sustainable crop production. *Sustainability*, *13*(4), 1868.

Pahalvi, H. N., Rafiya, L., Rashid, S., Nisar, B., & Kamili, A. N. (2021). Chemical fertilizers and their impact on soil health. *Microbiota and Biofertilizers, Vol 2: Ecofriendly Tools for Reclamation of Degraded Soil Environs*, 1–20.

Pei, B., Liu, T., Xue, Z., Cao, J., Zhang, Y., Yu, M., . . . & Zhang, Z. (2025). Effects of biofertilizer on yield and quality of crops and properties of soil under field conditions in China: A meta-analysis. *Agriculture*, *15*(10), 1066.

Pham, T. M., Bui, X. D., Le, T. M., Nguyen, M. L., Trinh, D. M., Phuong, N. T. D., . . . & Show, P. L. (2022). Isolation of indole-3-acetic acid-producing *Azospirillum brasilense* from Vietnamese wet rice: Co-immobilization of isolate and microalgae as a sustainable biorefinery. *Journal of Biotechnology*, *349*, 12–20.

Pieterse, C. M. J., Zamioudis, C., Berendsen, R. L., Weller, D. M., Van Wees, S. C. M., & Bakker, P. A. H. M. (2014). Induced systemic resistance by beneficial microbes. *Annual Review of Phytopathology*, *52*, 347–375.

Prasanna, R., Pattnaik, S., & Kanwat, M. (2021). Cyanobacterial biofertilizers: A sustainable approach for rice cultivation. *Bioresource Technology*, *325*, 124703.

Purwaningsih, P., Hadijah, S., Budi, S., & Rahayu, S. (2023). Phosphate solubilizing bacteria inducing systemic resistance with a potential for use as biofertilizer for rice. *JPBIO (Jurnal Pendidikan Biologi)*, *8*(1), 93–105.

Rai, A. K., Bana, S. R., Sachan, D. S., & Singh, B. (2023). Advancing sustainable agriculture: A comprehensive review for optimizing food production and environmental conservation. *International Journal of Plant Soil & Science*, *35*(16), 417–425.

Rawat, P., Das, S., Shankhdhar, D., & Shankhdhar, S. C. (2021). Phosphate-solubilizing microorganisms: Mechanism and their role in phosphate solubilization and uptake. *Journal of Soil Science and Plant Nutrition*, *21*(1), 49–68.

Raza, A. A. (2023). Microbial capacity for auxin biosynthesis: A comparative exploration. *Indus Journal of Science*, *1*(02), 65–70.

Rout, M. E. (2014). The plant microbiome. In *Advances in Botanical Research* (Vol. 69, pp. 279–309). Academic Press.

Santos, I. S., Almeida, F. A., da Silva, T. B., & Melo, Y. L. (2021). Plant microbiome: A reservoir of novel genes and metabolites for abiotic stress tolerance. *Frontiers in Microbiology*, *12*, 765518.

Schütz, L., Gattinger, A., Meier, M., Müller, A., Boller, T., Mäder, P., & Mathimaran, N. (2018). Improving crop yield and nutrient use efficiency via biofertilization—A global meta-analysis. *Frontiers in Plant Science*, *8*, 2204.

Sharma, S. B., Chen, Q., & Kumar, S. (2023). Phosphate solubilizing bacteria: Mechanisms and applications in agriculture. *Frontiers in Microbiology*, *14*, 1094567.

Sharma, S. B., Sayyed, R. Z., Trivedi, M. H., & Gobi, T. A. (2013). Phosphate solubilizing microbes: Sustainable approach for managing phosphorus deficiency in agricultural soils. *SpringerPlus*, *2*, 587.

Singh, A., Mazahar, S., Chapadgaonkar, S. S., Giri, P., & Shourie, A. (2023). Phyto-microbiome to mitigate abiotic stress in crop plants. *Frontiers in Microbiology*, *14*, 1210890.

Singh, R. P., Pandey, D. K., & Jha, P. N. (2022). Harnessing plant-microbe interactions for sustainable agriculture: A case study on biofertilizers in chickpea production. *Journal of Applied Microbiology*, *133*(4), 1987–1998.

Sohrabi, R., Paasch, B. C., Liber, J. A., & He, S. Y. (2023). Phyllosphere microbiome. *Annual Review of Plant Biology*, *74*(1), 539–568.

Soumare, A., Djibril, S. A. R. R., & Diédhiou, A. G. (2023). Potassium sources, microorganisms and plant nutrition: Challenges and future research directions. *Pedosphere*, *33*(1), 105–115.

Timmusk, S., Behers, L., Hett, J., Kask, K., & Niinemets, Ü. (2017). Challenges and opportunities for microbial inoculants in agriculture: A review. *Frontiers in Plant Science*, *8*, 1965.

Trivedi, P., Leach, J. E., Tringe, S. G., Sa, T., & Singh, B. K. (2020). Plant–microbiome interactions: From community ecology to plant health. *Nature Reviews Microbiology*, *18*(11), 607–621.

Vats, S., Srivastava, P., Saxena, S., Mudgil, B., & Kumar, N. (2021). Beneficial effects of nitrogen-fixing bacteria for agriculture of the future. *In Soil Nitrogen Ecology* (pp. 305–325). Cham: Springer International Publishing.

Verma, J. P., Jaiswal, D. K., Singh, S., Kumar, A., Prakash, S., & Curá, J. A. (2017) Consequence of phosphate solubilising microbes in sustainable agriculture as efficient microbial consortium: A review. *Climate Change and Environmental Sustainability*, *5*(1), 1–19.

Vessey, J. K. (2003). Plant growth promoting rhizobacteria as biofertilizers. *Plant and Soil*, *255*(2), 571–586.

Vives-Peris, V., De Ollas, C., Gómez-Cadenas, A., & Pérez-Clemente, R. M. (2020). Root exudates: From plant to rhizosphere and beyond. *Plant Cell Reports*, *39*(1), 3–17.

Yadav, A. N., Kour, D., Kaur, T., Devi, R., Yadav, A., Dikilitas, M., . . . & Saxena, A. K. (2021). Biodiversity, and biotechnological contribution of beneficial soil microbiomes for nutrient cycling, plant growth improvement and nutrient uptake. *Biocatalysis and Agricultural Biotechnology*, *33*, 102009.

Zhang, F., Wang, P., & Zou, Y. N. (2022). Mycorrhizal fungi and their role in sustainable crop production: A review. *Applied Soil Ecology*, *171*, 104324.

Zhang, M., Liu, Y., Wei, Q., Gou, J., Liu, L., Gu, X., & Wang, M. (2023). The co-application of PGPR and biochar enhances the production capacity of continuous cropping peppers in the Karst yellow soil region of Southwest China. *Horticulturae*, *9*(10), 1104.

Zhu, Y., Guo, B., Liu, C., Lin, Y., Fu, Q., Li, N., & Li, H. (2021). Soil fertility, enzyme activity, and microbial community structure diversity among different soil textures under different land use types in coastal saline soil. *Journal of Soils and Sediments*, *21*, 2240–2252.

11 Rhizosphere Microbiome in Improvements of Nutrient Uptake of Plants

Rehab M. Hafez[1]*, *Aya A. Mostafa*[2], *Rana Y. Maher*[1], *Mona M. Soliman*[1], *and Mohamed S. Zaki*[1]

[1]Botany and Microbiology Department, Faculty of Science, Cairo University, Giza 12613, Egypt

[2]Biotechnology Department, Faculty of Science, Cairo University, Giza 12613, Egypt

11.1 INTRODUCTION

The rhizosphere is the thin strip of soil immediately surrounding roots that holds a very diverse and dynamic microbiota whose collective gene potential is often more important than that of the plant and which plays a vital job in the acquisition of plant nutrients, control of growth, and tolerance of stress (Berendsen et al., 2012; Mendes et al., 2013). Recent syntheses emphasize that rhizosphere communities are structured by a combination of host genotype, soil physicochemical characteristics, and environmental variables, and that their functions (N cycling, P solubilization, siderophore-dependent Fe mobilization, phytohormone biosynthesis, disease suppression) are central to sustainable crop growth and minimized fertilizer dependence (Pantigoso et al., 2022). Advances in high-throughput "omics" platforms, marker-gene analyses, shotgun metagenomics, metatranscriptomics, metaproteomics, and metabolomics, and stable-isotope probing, now make it possible for scientists to relate taxonomy to in-situ function and to identify the active taxa and biochemical pathways that underlie nutrient conversions within the rhizosphere (Allen et al., 2017). These strategies have demonstrated great taxonomic richness (Proteobacteria, Actinobacteria, Acidobacteria, Bacteroidetes, Ascomycota, and Basidiomycota among the better-represented phyla) and functional redundancy, maintaining ecosystem stability and nutrient cycling. Root exudation is a key point of regulation by which plants organize rhizosphere communities: exudates (sugars, amino acids, organic acids, flavonoids, strigolactones, and mucilage) are chemoattractants or repellents, draw mutualists (rhizobia, AMF, phosphate-solubilizing bacteria), govern microbial quorum sensing, and can alter nutrient availability through rhizosphere priming effects (Sasse et al., 2018; Canarini et al., 2019). Environmental stresses (drought, salinity, and heat) and developmental stage change exudation profiles and thus remould microbial assemblages and their functional outputs, with immediate implications for stress tolerance and nutrient acquisition. Translating this knowledge into agronomic application involves two complementary strategies: (1) immediate microbiome manipulation through targeted bioinoculants or designed microbial consortia to enhance N fixation, P solubilization, and mobilization of micronutrients; and (2) host-directed approaches (breeding and genetic improvement) to facilitate selection of plant characters (root architecture, exudate composition) that are favorable for attracting beneficial taxa. Combined with multi-omics monitoring and gene editing technologies, these methods have the potential to reduce chemical application and enhance crop resilience, while ecological safety, persistence, and field efficacy remain active research concerns that must be tackled (Park et al., 2023).

11.2 RHIZOSPHERE MICROBIOME

11.2.1 Definition

Rhizosphere microbiomes are the collection of microbial genomes that are closely associated with rhizosphere soil particles and play a vital job for kipping plant health and nutrient supply (Wang et al., 2020). Their combined genome is larger than the plant's, leading to its designation as "the second genome of a plant" (Berendsen et al., 2012). The rhizosphere is colonized by a diversity of microorganisms, including algae, archaea, arthropods, microbes, nematodes, protozoa, and viruses, which have positive, negative, or neutral impacts on the growth, adaptation, and development of plants (Wu et al., 2023). Most of these microorganisms are members of a complex food chain that employs a high level of nutrients exudated by the plant (Mendes et al., 2013). Plants liberate organic carbon into the soil through their roots, which attracts and influences the activity of rhizosphere microorganisms. These interactions can affect the absorption of water and nutrients, growth promotion, hormonal levels, and soilborne pathogen defense (De Faria et al., 2021). The soil rhizosphere microbiome is complex

***Corresponding author:** rehabhafez@sci.cu.edu.eg (Rehab M. Hafez)

DOI: 10.1201/9781003544425-11

and divergent, comprising archaea (10^7–10^8 g^{-1}), bacteria (10^8–10^9 g^{-1}), and fungi (10^5–10^6 g^{-1}) (Afridi et al., 2022).

The rhizosphere microbiome is comprised of various functional groups, including rhizobia, mycorrhizal fungi, nitrogen-fixing bacteria, biocontrol agents, and growth-promoting rhizobacteria (PGPR), which sustain homeostasis in the soil ecosystem. However, it also includes pathogenic microorganisms that can harm plant growth and health (Mendes et al., 2013; J. Li et al., 2021; Tedersoo et al., 2024). The prokaryotic phyla in the rhizosphere are Acidobacteria, Actinobacteria, Bacteroidetes, Firmicutes, and Proteobacteria, whereas the fungal phyla are Ascomycota and Basidiomycota (Pantigoso et al., 2022). *Acidobacteria, Actinobacteria, Chloroflexi, Gemmatimonadetes, Planctomycetes, Proteobacteria*, and *Verrucomicrobia* were the most potent bacterial phyla in the rhizosphere soil, while *Bacteroidetes* were highly active in roots. The most prevalent fungal species were more differentiated. *Flavobacterium, Rhizobium*, and *Streptomyces* were major in roots, while *Rhodoplanes* and *Sphingomonas* (*Kaistobacter*) were superior in rhizosphere soil (Gkarmiri et al., 2017).

11.2.2 Omics Technologies to Identify the Rhizosphere Microbiome Profiles

Understanding the taxonomic, genetic, and functional elements of the rhizosphere microbiome is beneficial in their characterization to determine the assembly of the microbiome. It is essential to regulate them for effective crop yield (Xu et al., 2018). Several technologies are available to study the microbiome (Galloway-Peña and Hanson, 2020). The development of next-generation sequencing (NGS), conventional culture-dependent methods, and meta-analyses technologies has become essential for characterizing microbial assemblages (J. Li et al., 2021).

New developments in molecular biology have revealed that studying the microbial composition of specific environments is possible without using artificial media. There are various culture-independent approaches for evaluating the microbial community's biomass, diversity, and catabolic activity to realize the structure and function of organisms that were not previously discovered or cultivated. In the new era of sequencing and genomic sciences, new generations of life scientists can develop methods for producing and utilizing knowledge (Bhimani et al., 2024). The meta-omics methods, like genomics, transcriptomics, and metabolomics, aim to detect microorganisms, gene transcripts, proteins, variants, metabolic pathways, and ecological functions that can represent microbial community in the sample and have proven to be extremely helpful in studying the biochemical, molecular, and physiological aspects of plant–microbe interactions under different status (Afridi et al., 2022; Tiwari et al., 2024) (Figure 11.1). Technological developments in omics, including genomics and gene manipulation, have incredible capability for increasing crop production (Jain et al., 2024).

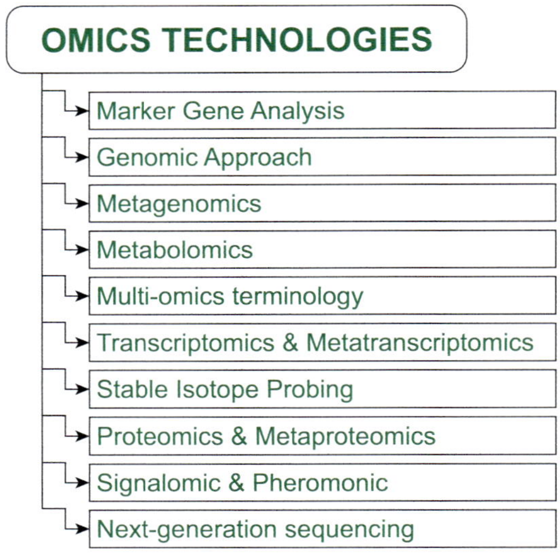

FIGURE 11.1 Some omics approaches are used to identify rhizosphere microbiome profiles.

11.2.2.1 Marker Gene Analysis

The most common sequencing approaches involve sequencing specific genes, such as the internal transcribed spacer (ITS) region for fungi and the 16S ribosomal RNA gene for bacteria. These genes have created a distinct genetic signature that can be categorized according to specific taxonomies and used to estimate their prevalence within the microbial community (Galloway-Peña and Hanson 2020). The microbial species (more than 2000 species per gram of rhizosphere soil) in *Arabidopsis thaliana* rhizosphere can be seen using 16S rRNA PCR amplicon sequencing (Bulgarelli et al., 2012). Analysis of the cowpea rhizosphere by 16S rRNA gene sequencing found that *Achromobacter, Bradyrhizobium, Burkholderia*, and *Rhizobium* species could form nodules on cowpea plants and effectively fix atmospheric nitrogen (Azarias Guimarães et al., 2012).

One challenge that arises in marker gene analysis is finding a distinct sequence. Furthermore, individual taxa (*Escherichia* and *Shigella*) may have identical gene sequences, confounding the decision of a unique sequence and differentiating individual taxa (Větrovský and Baldrian 2013). PCR errors could affect the identification of a distinct sequence (Xue et al., 2018). Additionally, the presence of humic acids, polyphenols, and other degraded macromolecules can interfere with PCR by inhibiting amplification and hindering the extraction of biological molecules like DNA, RNA, proteins, and metabolites. For high-resolution omics results, it is critical to separate these biological molecules from interfering substances (Jain et al., 2024). Amplicon-based community profiling fails to yield genetic or functional information regarding the microbiome (Xu et al., 2018).

11.2.2.2 Genomic Approach

The genome is the whole genetic material of an organism, contained within DNA or RNA sequences in the nucleus or other organelles of the cell. Individual genome sequencing is a modern and powerful method that enables direct and complete sequencing of genomes. Genomic technologies are also used to investigate the effect of an individual's microbiome on health and disease, which is a collection of microbes found in the rhizosphere (Micheel et al., 2012).

11.2.2.3 Metagenomics

Metagenomics implies a whole-community sampling of all the genetic information encoded by the DNA of all microorganisms present within the sample, where DNA is randomly fragmented and directly sequenced to gain insight into the function, genomics, and taxonomy of microorganisms (Allen et al., 2017). Metagenomics offers a more comprehensive understanding of microbial communities by analyzing their functional genes compared to phylogenetic surveys, which primarily focus on a single gene's diversity. Metagenomics gives insights into the evolutionary history of microbial communities, identifies new enzymes or biocatalysts, and reveals the relationship between genetic information and function in uncultured organisms (Thomas et al., 2012). Metagenomic analyses are also used to dictate protein-encoding gene fragments that can speculate key microbe survival traits and colonization mechanisms (Pantigoso et al., 2022). The analysis of genetic information from a microbiome sample, including bacteria, fungi, DNA viruses, and other microbes, depends on existing reference genomes and scientific knowledge. This information can be applied to identify the types of microorganisms present, their relative abundance, their gene coding sequences, and their functional ability (Galloway-Peña and Hanson, 2020).

Metagenomics used 16S rRNA and random shotgun sequencing to recognize novel genes in entire microbial communities and complex soil communities (Xu et al., 2018; Regalado et al., 2020). Metagenomics emerged as a genomic tool by integrating various techniques to examine the whole genetic material of microbial communities using the microbial genome as a template. Bioinformatics tools are then used to extract valuable information from the resulting data (Priya et al., 2021).

De Novo Assembly of metagenomic data facilitates error correction of reads, links protein-coding gene fragments, and enhances the accuracy of functional and taxonomic assignments through increased alignment length to reference databases. Illumina Moleculo, Oxford, and Nanopore PacBio are the best sequencing platforms for long reads for high-complexity microbial communities. Metagenomics cannot directly identify the metabolic processes that are occurring within a microbial community at the time of sampling (Allen et al., 2017). QIIME, UPARSE, and MOTHUR are valuable programs for analyzing bioinformatics data and generating operational taxonomic units (OTUs) in metagenomic studies, whereas MG-RAST and MetaPhlAn2 enable phylogenetic and functional analyses of metagenomic data (Jain et al., 2024).

Metagenomic analysis of the rhizosphere microbiome declared a diverse community of organisms from five kingdoms, 51 phyla, 94 classes, 207 orders, 415 families, 1,220 genera, and 3,356 species. Proteobacteria is the most dominant phylum, accounting for 47.5% of the microbial community, followed by Thaumarchaeota, Gemmatimonadetes, Nitrospirae, and Actinobacteria. At the genus level, *Nitrososphaera*, *Nitrospira*, *Gemmatirosa*, *Pseudomonas*, and *Burkholderia* were the most abundant genera, accounting for 2.5%, 2.4%, 2.2%, 1.0%, and 0.8%, respectively (Jia et al., 2022).

Metagenomics has been used to discover new nitrogen-fixing genes in the symbiotic relationship between *Rhizobium* bacteria and legumes. Additionally, it helped to detect genes involved in mycotoxin production in *Alternaria alternata* and the corresponding host response in apples (Regalado et al., 2020). Diazotrophs, free-living anaerobic bacteria, like *Azotobacter vinelandii*, *Klebsiella pneumoniae*, *Rhodobacter capsulatus,* and *Rhodospirillum rubrum* have Nif genes for nitrogen fixation (Usman and Wali, 2024). Nitrogen-fixing bacteria like *Rhizobia* and *Mesorhizobium* use specific genes (NifA2 gene and s54 sigma factor) to regulate acdS (1-aminocyclopropane-1-carboxylic acid (ACC) deaminase) gene expression (Larekeng et al., 2024). *Rhizobium* and *Hydrogenophaga* species have genes that enhance plant growth through metabolic functions and compound synthesis (Ilangumaran et al., 2024). *Pseudomonas sivasensis* enhances canola growth through genes conducted for ACC deaminase, indole-3-acetic acid production, and siderophore formation (Świątczak et al., 2024).

11.2.2.4 Epigenomics

Epigenomics is a scientific approach that investigates how environmental elements can influence gene expression without changing the underlying DNA sequence (Samantara et al., 2021). Epigenomics involves studying the divergences in DNA methylation, histone modifications, and non-coding RNAs. It also involves studying chromatin remodeling and understanding how these changes enable microbes to adapt to different environments and pass on these adaptations to their offspring (M.H. Ali et al., 2023). Epi-effectors (zinc finger, transcription activator-like effector, or modified CRISPR/Cas9 complexes) can be utilized to precisely edit the epigenome by targeting specific DNA sequences and modifying epigenetic marks (Doddavarapu et al., 2024). The proline–alanine–valine gene in fungi may be linked to specific genetic and epigenetic traits, providing potential predictive insights about how these fungal pathogens adapt to their hosts and helping in the development of more potent disease prevention strategies (Joubert and Krasileva 2024). This approach recognizes key microbial species, their metabolic pathways, and epigenetic factors that influence host–microbe interactions, and it can lead to novel insights into agriculture applications by combining different approaches (Masenya et al., 2024).

11.2.2.5 Transcriptomics and Metatranscriptomics

Transcriptomic analysis uses NGS to identify and measure RNA molecules in biological samples and to understand how gene activity changes under specific conditions (Chen et al., 2022). Transcriptomics can be utilized to investigate regulatory alterations, mutations, sequence variations, gene expression differences, and alternative splicing (Weidemüller et al., 2021). Metatranscriptomics explores gene expression (total mRNA) across the microbial community, revealing dynamic functions and giving insights into complex microbial physiology (Dubey et al., 2020). It is a valuable method for studying functional activities, understanding the structure of microbial communities and their interaction with plants, and measuring their functional diversity (Gamalero et al., 2022).

Transcriptomics uses microarray analysis, SOLiD-SAGE, and RNA sequencing (RNA-seq) methods for transcriptional profiling (Chen et al., 2022). Metatranscriptomic techniques mainly counter upon Illumina sequencing methodologies (Illumina HiSeq) due to the high throughput and low cost per base. Metatranscriptomic analysis methods likeHUMAnN2, MetaTrans, and SAMSA provide pathways for gene expression analysis (Liu et al., 2021). Metatranscriptomics can also be performed on longer-read technologies, and sequencing can be done directly on Oxford Nanopore (Allen et al., 2017). Bioinformatics programs such as SOAPdenovo can be applied to analyze and assemble metatranscriptomic data (Galloway-Peña and Hanson 2020).

The transcriptomic studies examine gene expression that diverges in the host (rice, wheat, sugarcane, and tobacco) during interactions with PGPB, including *Azospirillum brasilense*, *Herbaspirillum seropedicae*, *Bacillus subtilis*, *Burkholderia anthina*, and *Paraburkholderia kururiensis*. Alterations in gene expression have been seen in host plants concerning defense responses, hormone signaling, and nutrient transport during interactions with PGPB (Bhimani et al., 2024). Transcriptomics is used to identify the infection induced by *Penicillium expansum* and mycotoxin production. *Delftia acidovorans* RAY209 was clarified by RNA-seq transcriptomic analysis, with a particular emphasis on the colonization approach (Gamalero et al., 2022). Dubey et al. (2020) used metatranscriptomics to investigate changes in gene expression caused by *Xylella fastidiosa*, which causes Pierce's disease in grapevines.

11.2.2.6 Metabolomics

Metabolomics techniques focus on profiling the metabolites formed by microbiota and their biochemical interaction with both microbiota and host (Bhimani et al., 2024). Metabolites are the molecules produced or consumed by cells. These biochemical intermediates or end products are important for various cellular functions. The complete collection of metabolites is known as the metabolome, which comprises endogenous and exogenous metabolites. Endogenous metabolites are divided into primary metabolites (amino acids, lipids, and carbohydrates), which are crucial for fundamental cellular functions like growth, development, and reproduction, as well as secondary metabolites (flavonoids, phenolics, terpenoids, and alkaloids), which are not directly involved in growth but have specialized functions like defense, signaling, and pigmentation (Tanna and Mishra 2018).

Metabolomics is an important tool for understanding the specific metabolic pathways participating in the phenotypic responses of organisms (Sardans et al., 2011). These methods often quantify small molecules like antibiotics and intermediates of host and microbiome (Galloway-Peña and Hanson, 2020). Recognizing bioactive compounds related to disease resistance, stress responses, and growth promotion is vital in improving crop health and yield amidst alternating environmental conditions (Olanrewaju et al., 2024). There are two types of metabolomics: highly specific, which concentrates on the identification and measurement of metabolites chosen in advance based on the scientific question (Jones et al., 2014), and unspecific, which seeks to concurrently analyze as many metabolites as possible in a single analysis, producing hundreds of metabolites across numerous samples (Rivas-Ubach et al., 2013).

Metabolomics often utilizes liquid chromatography–mass spectrometry (LC-MS), gas chromatography–mass spectrometry (GC-MS), and nuclear magnetic resonance spectroscopy (NMR) techniques to determine a wide range of metabolites in a single run (Alseekh et al., 2021). Without requiring significant sample preparation or fractionation, NMR can identify the most prevalent metabolites in samples with excellent accuracy, precision, and consistency. NMR can reliably measure short-chain fatty acids (SCFAs) like acetate, propionate, and butyrate, as well as small molecular weight volatile organic molecules as alcohols. It can also identify metabolites in salty samples (Dalcin Martins et al., 2017). Mass spectrometers coupled to LC or GC are becoming more powerful and sensitive techniques for identifying metabolites and proteins of plant exudates and microbes in the rhizosphere and soil samples. LC-MS and GC-MS are more sensitive methods than NMR spectroscopy (Rivas-Ubach et al., 2012). GC-MS is a high-resolution chromatographic separation method, and it has been highly employed to estimate volatile and semi-volatile compounds (van Dam and Bouwmeester 2016). GC-MS has also proven to be ideal for identifying metabolites from primary metabolism, like fatty acids, carbohydrates, and organic acids (Gullberg et al., 2004). LC-MS has been the most common technique to resolve a large diversity of non-volatile compounds, primary metabolites (acids, carbohydrates, and organic acids), and metabolites (flavonoids, alkaloids, phenolic acids, and saponins) (De Vos et al., 2007). However, ion-mobility spectrometry mass spectrometry (IMS-MS) allows the detection and characterization of hundreds of metabolites and lipidomics in a single run (El-ghfar et al., 2016). Metabolomic data analysis is facilitated by methods such as Xcms, MZMine, and Metabolome MetaboAnalyst, which provide functionalities for pathway mapping and analysis (Jain et al., 2024).

Metabolomics profiles the metabolite variations related to mycorrhizal connections for nutrient uptake in *Glomus intraradices* and tomatoes (Chen et al., 2022). Metabolomics has recognized lipid metabolites from *Plasmopara viticola* inoculated in grapevines, whereas a modification of maize plant metabolites was identified as a result of bacterial nitrogen fixation (Diwan et al., 2022). Detailed metabolomic analysis showed that *Pseudomonas syringae* infection led to hyperaccumulation of dihydrocamalexic acid (DHCA) in plants, whereas DHCA helps to restrict *P. syringae* and restore plant resistance (Singh et al., 2023). Metabolomic profiling of *Dendrobium nobile* when cultured with *Trichoderma longibrachiatum* disclosed an increase in the level of dendrobium (Sarsaiya et al., 2024). In maize, the resistance to *Fusarium graminearum* is associated with the production of smiglaside and smilaside, while resistance to southern corn leaf blight is related to the production of polyphenols, lignin, and flavonoids via metabolite profiling (Manickam et al., 2023).

11.2.2.7 Proteomics and Metaproteomics

Proteomic investigation has examined microbial proteins and their metabolic function in various environments. Proteomic analysis gives valuable insights into the plant–microbe interactions (Jain et al., 2024). Proteins are essential biomolecules that play a major role in different biological processes, like metabolism, signaling pathways, cellular function, cellular homeostasis, defense responses, and organismal systems. They provide valuable insights into complex mechanisms linked to health and diseases (Birhanu 2023). Proteomics relies heavily on the identification of proteins and their divergence during infection. These methods include mass spectrometry-based sample analysis, protein extraction, fractionation, and isolation, along with further testing using a proteome database. Proteomics provides information on the exact number of active pathways in the sample as well as the components of a functional protein that the cell creates (Sharma et al., 2020). Signal transduction pathways can be identified thanks to proteomics, which monitors phosphorylation changes in proteins that are essential to protein function (J. Li et al., 2022). Metaproteomics focuses on recognizing and quantifying the protein present within a microbiome (Galloway-Peña and Hanson 2020).

Proteomic analysis techniques, such as Metaproteome Analyzer, SECIMTools, MetaLab, and Galaxy-P, offer automated processes to analyze proteomic data analysis. Galaxy-P provides a comprehensive solution, including database generation from sequencing data, iterative database searches, and taxonomic and functional analyses via methods like Unipept and MEGAN. Gel-based resolution tools, like two-dimensional gel electrophoresis and differential gel electrophoresis combined with mass spectrometry, can be used to identify and quantify proteins (Jain et al., 2024). Mass spectrometric data analysis techniques like PeptIdent, MultiIdent, MASCOT, SEQUEST, and Sherpa enable researchers to recognize and investigate protein interactions and functions. In addition, programs like ProFound, MS-Fit, MOWSE, PepSea, PepFrag, and MS-Tag enhance the depth of proteomic investigation (Chandok et al., 2022). Bottom-up proteomics is the most frequently used strategy for metaproteomics. The bottom-up approach needed that proteins within the sample be digested into smaller peptides using a protease, separated by several multidimensional LC strategies, and analyzed using MS/MS. One of the greatest techniques currently employed for microbial community study is bottom-up metaproteomics, which gives high proteome coverage (Szabo and Janaky, 2015). Proteins, polar metabolites, and lipids may all be extracted from rhizospheric soil at the same time using a 270-novel metaproteomic technique (Nakayasu et al., 2016). Perhaps the most widely used analytical tools for protein identification and profiling are still 2-DE and MS (Bhimani et al., 2024).

Metaproteogenomic approaches study complex relationships between plants and rhizosphere microorganisms in rhizosphere systems. Matrix-assisted laser desorption ionization (MALDI) TOF MS recognized 29 proteins of fungi and 43 proteins of bacteria from rhizosphere samples. Most of these bacterial proteins belonged to the Proteobacteria and Actinobacteria (Wang et al., 2011). Wiggins et al. (2022) conducted a detailed proteomic analysis that revealed the interaction between *Oryza sativa* and the plant growth-promoting bacteria (PGPB), *Azospirillum brasilense* and *Herbaspirillum seropedicae*.

In response to root exudates from various host plants, the beneficial symbiont Sinorhizobium meliloti and the phytopathogen *Pseudomonas syringae* release different proteins, according to comparative proteomic investigations. Additionally, secreted proteins associated with fire blight disease in apples caused by *Erwinia amylovora* have been identified using proteomics (Jain et al., 2021). Proteomics was utilized to identify the pathogen effectors and host proteins involved in *Phytophthora infestans*-induced disease in potatoes (Khatabi et al., 2019). The proteomes secreted by *Botrytis cinerea* are serine proteases, metalloproteases, threonine proteases, and pectinases that are conducted for cell wall degradation (Sarim et al., 2020). Rane et al. (2022) employed metaproteomics to recognize proteins in *Trichoderma harzianum* that contribute to its biocontrol ability, whereas Priya et al. (2021) used metaproteomics to recognize proteins participating in the pathogenicity of *Fusarium oxysporum*, which causes vascular wilt and the host defense mechanism in watermelon.

11.2.2.8 Signalomic and Pheromonic Approach

Rhizosphere microorganisms communicate with the plant, and each other uses chemical signals. Plants release various metabolites in response to alterations in their gene expression, influencing their interactions with the surrounding microbial community. These interactions play an essential role in root-to-root interactions, nutrient availability, microbe accumulation, biofilm formation, and the suppression of soil-borne diseases developed by such communication (Mhlongo et al., 2018). Effective communication by signaling between rhizobacteria, other soil microbes, and their plant hosts is essential for a healthy symbiotic relationship. Many signaling molecules liberated in the rhizospheric zone influence the type and quantities of rhizobacteria and the soil's structural and physical heterogeneity. Rhizospheric signaling systems can be classified into three major groups: (1) plant-to-microorganisms signaling, where plants release low molecular weight chemicals that affect microbial behavior; (2) microorganism-to-microorganism signaling, where microorganisms communicate with each other using quorum-sensing molecules; and (3) microorganism-to-plant signaling, where microorganisms form compounds that can affect plant growth and development (Singh et al., 2022). In response to microbial signals, plants release chemical substances called root exudates. These comprise low molecular weight compounds like sugar, organic acids, flavonoids, fatty acids, amino acids, and secondary metabolites, as well as big molecular weight molecules such as proteins and mucilage. Volatile organic compounds (VOCs), which are liberated into the atmosphere by plant parts like leaves, flowers, fruits, and roots, make up 1% of secondary metabolites. VOCs are highly permeable, attracting pathogens and restricting growth. Rhizobia and legumes have a symbiotic relationship by secretion of phytochemicals from the plant roots that attract rhizobia and activate their nodulation genes, forming lipochitooligosaccharides (LCOs). LCOs are essential signaling molecules that stimulate the formation of nodules. Likely, the formation of symbiosis between plants and arbuscular mycorrhizal fungi (AMF) is influenced by the release of mycorrhizal LCOs by the AMF in response to signals from strigolactones present in plant secretions (Jamil et al., 2022). Microorganisms use quorum sensing (QS) to build interactions with their hosts. This involves the production and detection of signal molecules, such as *N*-acyl homoserine lactones (AHLs), which are associated with biofilm formation and pathogenicity in bacterial populations (Abdul Hamid and Nadarajah 2022).

11.2.2.9 Multi-Omics Terminology

Multi-omics investigations utilized an integrated tool to study many molecular levels within organisms. Researchers can thoroughly comprehend intricate biological systems by integrating various "-omics" fields like proteomics, metabolomics, transcriptomics, and genomics. Multi-omics methods allow for a comprehensive analysis of biological systems by examining multiple types of biological information, including genes, DNA, mRNA, proteins, metabolites, and other biomolecules. This integrated approach provides unprecedented insights into complex biological processes and phenomena (Bhimani et al., 2024). Individual species (archaea, bacteria, fungi, picoeukaryotes, plants, mammals, and viruses), multi-organism communities, and consortia can all produce multi-omics data. Data that has been directly measured, examined, and combined using multivariate statistics is referred to as multi-omics (Allen et al., 2017). Multi-omics techniques interpret the functional and structural properties of genes, provide insights into the complete microbial community, evaluate transcript sequences, identify protein-protein interaction networks, and develop models of metabolic networks (Mishra et al., 2022; Zulfiqar et al., 2024).

Multi-omics approaches permit the investigation of diverse interactions, ranging from obligate symbiotic plants and arbuscular mycorrhizal fungi interaction to interactions with non-mycorrhizal microorganisms as biofilms (Mishra et al., 2022). Technological progress enables the examination of trace chemical compounds and supports the study of allelochemical biosynthesis and responses of plants, which enhances the understanding of rhizosphere chemistry (Weidenhamer et al., 2023). The presence of carbohydrate-active enzymes indicates possible symbiotic interactions between plants, bacteria, and fungi in the rhizosphere, thereby guiding the application of omics tools to develop methods for promoting plant resilience to environmental stress (Alshareef, 2024).

11.2.2.10 Next-Generation Sequencing

NGS platforms are powerful tools applied for direct measurements of whole communities' metagenomes and metatranscriptomes (White III et al., 2016). NGS technology advancements have lowered sequencing prices while also speeding up access to de novo assemblies, whole-genome sequences, and resequencing of various strains within a species (Crandall et al., 2020). NGS has been utilized to study various aspects of the rhizospheric microbial community, including microbial community structure (Bulgarelli et al., 2012); the core microbial community (Lundberg et al., 2012); microbial metabolic potential (Mendes et al., 2014); plant–microbe interactions (Turner et al., 2013), and microbial community expression under pesticide treatment (i.e., glyphosate) (Newman et al., 2016).

Illumina, Pacific Biosciences (PacBio), and Oxford Nanopore Technologies are the leading manufacturers of NGS platforms. Among these, Illumina is the most mature maker of NGS platforms and offers a wide range of sequencing capabilities (Bentley et al., 2008). PacBio and Oxford nanopore technologies generally offer lower-to-medium throughput than Illumina platforms. Both PacBio and Oxford nanopore could be techniques that access whole community modification profiles of nucleic acids (White III et al., 2016). Despite its advancements, NGS still faces several challenges when applied to rhizosphere studies. These challenges include data analysis, data storage, nucleic acids extraction, the presence of free and high molecular weight nucleic acids, and nucleic acid stability (Allen et al., 2017). Since no single extraction method can extract all members of a microbial community, more than one extraction technique, as well as various techniques of lysis, must be used to get more complete coverage (Kajiura et al., 2015).

11.2.2.11 Stable Isotope Probing

Stable isotope probing (SIP) is a potential tool for recognizing microorganisms that consume recently fixed plant carbon. Researchers can track which microbes actively use plant-derived carbon by incorporating a stable isotope into a labeled substrate. This method is particularly valuable for studying the plants and microbes interactions in the rhizosphere (Lugtenberg and Kamilova, 2009). Gkarmiri et al. (2017) used RNA-SIPm to analyze the structures and compositions of bacterial and fungal communities in the rhizosphere soil of *Brassica napus* plants. *Streptomyces*, *Rhizobium*, *Clonostachys*, and *Fusarium* were shown to be effective at using carbon generated from plants, indicating that their potential as inoculants to enhance oilseed crop health and productivity could be investigated to be efficient at utilizing plant-derived carbon, suggesting their potential as inoculants to improve the productivity and health of oilseed crops could be explored (Gkarmiri et al., 2017).

11.2.3 Biodiversity

Soil microbial communities are incredibly diverse, representing one of Earth's largest reservoirs of biological diversity. The rhizosphere can hold up to 1011 microbial cells per gram of root and more than 30,000 prokaryotic species (Berendsen et al., 2012). This complex microbial community in the rhizosphere is significantly influenced by the various organic compounds secreted by plant roots. These rhizodeposits (exudates, border cells, mucilage) play a crucial role in regulating the diversity and activity of microorganisms around plant roots by activating or repressing members of the rhizosphere community (Mendes et al., 2013). Various factors, consisting of the plant genotype and developmental stage, the soil type, nutrition, carbon level, soil pH, biogeography, and microbial interaction, influence the composition and quantity of root exudates. These factors also contribute to the diversity and abundance of bacteria in the rhizosphere (De Faria et al., 2021). Numerous microorganisms, including microbes, oomycetes, viruses, protists, and archaea, thrive in the varied soil texture. These microbes interact with one another in intricate food webs. Microorganisms in the rhizosphere can affect the host plant's health, composition, and productivity both directly and indirectly (J. Li et al., 2021).

11.2.3.1 Beneficial Rhizospheric Microbiome for Plants

Beneficial rhizosphere organisms that positively impact plant growth and health consist of nitrogen-fixing bacteria, mycorrhizal fungi, plant growth-promoting rhizobacteria (PGPR), biocontrol microorganisms, mycoparasitic fungi, and protozoa (Mendes et al., 2013). Microbes form signaling molecules that react to signals from plants, control plant hormone pathways, create specialized nutrient exchange structures, boost nutrient availability, stop pathogen colonization, and improve resistance to biotic and abiotic stressors. These advantageous rhizospheric microbiomes promote nutrient cycling, soil fertility, and the resilience and sustainability of entire ecosystems (Jain et al., 2024). These beneficial microbiomes, including rhizobacteria (*Proteobacteria* and *Firmicutes*), *Pseudomonas*, *Bacillus*, mycorrhizas, and Deuteromycetes (*Trichoderma* and *Gliocladium*), consist of biofertilization, enhancement of root growth, rhizoremediation, control of abiotic stress, and disease control (J. Li et al., 2021).

11.2.3.1.1 Nutrient Acquisition

Rhizosphere microorganisms play a crucial role in nutrient cycling, which can diminish the uptake of toxic elements by plants and release the nutritive cations from soil minerals needed for plant growth (Mendes et al., 2013). Microbes develop enzymes and metabolites that break down complex molecules and increase nutrient availability (Jain et al., 2024). *Fusarium oxysporum* produces extracellular enzymes that facilitate nutrient acquisition (Niazi et al., 2023). Rhizosphere bacteria, symbiotic rhizobia, and endophytes can promote the breakdown of 1-aminocyclopropane-1-carboxylate (ACC) into a-ketobutyrate and ammonia. This breakdown helps to regulate plant nutrition and growth by reducing the ethylene concentration (Kumari et al., 2023). Phytohormones, such as auxins and cytokinins, formed by *Rhizobium leguminosarum*, enhance plant growth and nutrient uptake by signaling through plant receptors (Meena et al., 2023).

Rhizobacteria (Nitrogen-fixing bacteria), like *Rhizobium leguminosarum*, have nitrogen fixation capacity through the nitrogenase enzymes by transforming atmospheric nitrogen (N_2) into ammonia (NH_4^+) or by increasing the efficiency of N absorption by plants, thereby stimulating plant growth and soil fertility (Goyal and Habtewold 2023). The enriched taxa in the rhizosphere soils from mining areas are *Chloroflexi*, *Proteobacteria*, *Actinobacteria*, and *Acidobacteria*, which contribute to enhancing soil nutrient contents such as phosphorus (P), sulfur (S), and molybdenum (Mo). Mo is a crucial element for the biosynthesis of metalloenzyme cofactors, which are essential for microbial metabolism, particularly N-fixation (Lan et al., 2024). Furthermore, *Azospirillum*, *Burkholderia*, *Stenotrophomonas*, and *Rhizobium* can improve nutrient acquisition for plants by fixing nitrogen and solubilizing phosphorus and iron (De Faria et al., 2021). Rhizobacteria-induced root modifications can lead to an extensive root system, using a large volume of soil, increasing nutrient accumulation and use, making the root system more efficient, and increasing yield as well as reducing chemical fertilizers and production costs (Khan et al., 2021). Rhizobacteria produce indole and organic acid compounds that acidify the root medium, generating a proton gradient in the cell, which can activate the ATP hydrolysis proton pump, facilitating the movement of molecules and the flow of nutrients and water in the apoplast (Guan and Liu 2020). Rhizobacteria produce various organic acids (fumaric, oxalic, and tartaric) and enzymes, which chelate and mobilize potassium that improves nutrient absorption (Valente Lima et al., 2021).

Phosphate-solubilizing microbes make organic acids like acetic, butyric, citric, lactic, and malic or enzymes (acid phosphatases), which can break the mineral and enhance the availability of phosphate ions in the soil (Silva et al., 2023). P is one of the major factors influencing the structure and assembly of microbial communities in As/Sb-contaminated rhizosphere soil (H. Yu et al., 2021). Phosphate-solubilizing bacteria (*Klebsiella* sp. and *Kluyvera* sp.) can enhance the availability of phosphorus in the rhizosphere soil, thereby decreasing the uptake of Cd by plants (Qin et al., 2023). In addition to phosphorus solubilization, the acid phosphatases function as chelators of many minerals, like Fe^{2+}, Ca^{2+}, and Zn^{2+}, which can increase plant uptake of these minerals. Additionally, siderophores produced by bacteria can chelate iron and make it more available in soluble form for plant uptake (Valente Lima et al., 2021). *Pseudomonas fluorescens* and *Rhizopus arrhizus* can promote iron nutrition by producing siderophores. *Bacillus subtilis* can stimulate the plant's iron acquisition machinery (Mendes et al., 2013).

Plants are mainly dependent on Arbuscular mycorrhiza fungi to collect and transfer phosphorus or essential nutrients, limiting plant growth (Afridi et al., 2022). *Penicillium* and *Aspergillus* liberate phosphorus and potassium during the breakdown of minerals, benefiting plant growth and defense (Paul et al., 2024). *Salvia miltiorrhiza* enhances plant stress resistance by signals transmitted through arbuscular mycorrhiza hyphal networks against *Fusarium solani* (Han et al., 2023). The genera *Enterococcus*, *Stenotrophomonas*, *Serratia*, and *Bifidobacterium* are crucial for detoxifying harmful substances and increasing nutrient availability in the rhizosphere soils of mining regions by encouraging P solubilization (Abbaszadeh-Dahaji et al., 2019). *S. marcescens* exhibits resistance to vanadium and facilitates plant absorption (Wang et al., 2024). *Bifidobacterium* can bind and absorb more metals through hydroxyl, phosphate, and carboxyl molecules (Arun et al., 2021). These microbes enhance nutrient bioavailability by immobilizing toxic elements via chelation and by stimulating nutrient uptake by pioneer plants in tungsten mining areas (Lan et al., 2024). The microbiome assists in weathering minerals and degrading recalcitrant organic matter and assisting plants in the uptake of phosphorus and nitrogen (Berendsen et al., 2012).

Sulfur-oxidizing bacteria in the rhizosphere can enhance the bioavailability of sulfur, which indirectly reduces the uptake of toxic elements by plants (Wu et al., 2021). In soils contaminated with cadmium, sulfur controls the distribution and metabolism of soil microorganisms (Sun et al., 2021); *Thiobacillus* and *Rhizobium* are the significant populations that faster the oxidation of As^{3+} and S^{2-} (Sun et al., 2024).

11.2.3.1.2 Plant Growth Promotion

Rhizospheric microorganisms can form or deteriorate the plant hormones ethylene, auxin, cytokinin, gibberellin, or indole-3-acetic acid (IAA), which is connected directly to plant growth. The root growth may be connected to the indole-3-acetic acid biosynthesis. Auxin synthesis from the rhizosphere microorganisms is involved in root growth and development. Plant growth-promoting rhizobacteria (PGPR) can trigger hormonal action that increases root volume, morphoanatomical changes, and root hair development. These changes offer numerous benefits to plants, including promoting plant growth, improving nutrient absorption, ensuring robustness, precocity, and adaptation of the seedling (Valente Lima et al., 2021). The best-known PGPR colonizing in the rhizosphere strains include *Bacillus, Rhizobium*, *Acinetobacter*, *Alcaligenes*, *Azotobacter*, *Arthrobacter*, *Enterobacter*, *Pseudomonas*, *Serratia*, and *Burkholderia* (Afridi et al., 2022). VOCs emitted by rhizosphere microbes can stimulate plant growth, influence systemic resistance in plants, and influence the uptake of specific nutrients (iron and sulfur) (Pantigoso et al., 2022). Various bacterial species, consisting of *Stenotrophomonas maltophilia*, *Serratia plymuthica*, *Pseudomonas trivialis*, *P. fluorescens*, *B. subtilis*, and *Burkholderia cepacia*, form VOCs to inhibit mycelial growth of fungal plant pathogens (Mendes et al., 2013).

11.2.3.1.3 Disease Suppression

Beneficial rhizosphere microbes can directly protect plants against pathogens via various mechanisms including antagonism, competition for nutrients and space, parasitism, antibiosis through biosynthesis of microbial compounds consisting of volatile compounds, iron-chelating siderophores, hydrogen cyanide (HCN), lytic enzymes (chitinases and β-1,3-glucanase), and antibiotics (bacteriocins, polyketide and non-ribosomal) that inhibit pathogens, interference with quorum sensing impact on virulence, and induced systemic resistance (ISR) (De Faria et al., 2021; J. Li et al., 2021; Afridi et al., 2022). Bacterial volatiles such as dimethylsulfide can interfere with quorum sensing of phylogenetically various bacteria by suppressing the transcription of the N-acyl-homoserine lactone synthase genes (Chernin et al., 2011). Fungi communicate with each other by using pheromones, which produce secondary metabolites, change structure, and secrete enzymes (Bhimani et al., 2024).

Furthermore, beneficial rhizosphere bacteria, including Chitinophaga, Chryseobacterium, Flavobacterium, and Pseudomonas, a group of PGPR, and nitrogen-fixing microorganisms (Pedobacter, Rhizobium, and Variovorax) are exhibiting antagonistic activities to the soil-borne pathogens *Rhizoctonia solani*. Additionally, Janthinobacterium showed broad antagonism against *Pythium ultimum*, *R. solani*, and *R. oryzae* (Yin et al., 2021). Comamonadaceae and Xanthomonadaceae, Pseudomonas, and Stenotrophomonas control *Fusarium* wilt disease by secreting more organic acids (Wen et al., 2020). Transgenic expression of antimicrobial proteins from *Bacillus thuringiensis* into plants can enhance host resistance against *Fusarium* ear rot (Adams et al., 2024). The antagonistic fluorescent *Pseudomonas* spp. produces the antifungal compound 2,4-diacetyl phloroglucinol against G. graminis var. tritici that causes the root disease of wheat (Berendsen et al., 2012). *Trichoderma* is the most frequently commercialized biocontrol agent production of antimicrobial compounds that evade various plant diseases (Roberts et al., 2010) and secretes plant growth promoters (Contreras-Cornejo et al., 2016). Many beneficial fungi as mycorrhizal fungi are well-known hunters of plant parasitic nematodes (Haraguchi and Yoshiga 2020). In addition to producing vitamins, mycorrhizal fungus also provide antioxidant protection. By creating a protective layer around plant roots and acidifying the soil, ectomycorrhiza (EcM) fungus demonstrate a high level of defense against soil-borne diseases (Tedersoo et al., 2020).

11.2.3.1.4 Plant Defense Activation

Beneficial rhizosphere microorganisms, such as mycorrhiza, nitrogen-fixing rhizobia, and PGPR, can indirectly induce plant defense mechanisms by improving host-ISR, producing signaling molecules, and modifying plant volatiles (Kong et al., 2021). Also, rhizosphere microbiomes can modulate and enhance the host plant's immune system, physiology, and metabolism. In addition, quorum-sensing molecules from rhizobacteria can trigger a variety of plant responses, as the activation of various defense-related genes (Valente Lima et al., 2021).

Rhizosphere microbes activate plant response through microbial elicitors so-called microbe-associated molecular patterns (MAMPs), including lipopolysaccharide, peptidoglycans, flagellin, chitin, and AHL-type QS molecules, which trigger systemic defense responses (ISR) by rhizosphere-beneficial bacteria or systemic acquired resistance (SAR) by pathogens, reduce stress responses by controlling plant nutritional and hormonal balance, and promote plant growth (Pantigoso et al., 2022). Specific pathogens can establish specific effector proteins into plant cells and cause effector-triggered immunity (ETI) (Bhimani et al., 2024). Microbial species, including *R. solanacearum*, *E. carotovora*, *D. solani*, *E. amylovora*, *Pseudomonas carotovorum*, *Bacillus*, *Serratia*, and *Trichoderma* successfully stimulate disease resistance against bacterial plant pathogens (De Faria et al., 2021; Afridi et al., 2022).

Plant phytohormones including auxin, cytokinin, indole acetic acid, abscisic acid, GA, salicylic acid, auxins, and VOCs are formed by rhizosphere microbes. Plants may uptake VOCs produced by bacteria for growth and development (Pantigoso et al., 2022). *Trichoderma harzianum* produces VOCs that trigger the systemic response of plants (Contreras-Cornejo et al., 2024). Stimulation of ISR in *Arabidopsis thaliana* by *Pseudomonas fluorescens* can suppress flagellin-triggered immune responses through apoplastic secretion of low-molecular-weight molecules. The ISR primes for accelerated defense response, including increased gene expression related to defense response and increased callose deposition at pathogen entry sites. Also, mycorrhizal fungi, *Trichoderma* spp., have been found to stimulate ISR in plants (Berendsen et al., 2012). Effectors from *Pseudomonas syringae* manipulate host cellular processes to modulate plant immunity (Kalita et al., 2024). *Agrobacterium tumefaciens* can trigger systemic plant responses by producing quorum-sensing molecules, while *Pseudomonas aeruginosa* induces plant defense mechanisms by producing lipopolysaccharides (Jain et al., 2024). *Pseudomonas putida* stimulates systemic resistance in maize against *Colletotrichum graminicola* by triggering the release of VOCs (J. Li et al., 2021).

11.2.3.2 Harmful Microbiome for Plants

The harmful rhizosphere microbes, including pathogenic microbes, oomycetes, and nematodes, can significantly reduce plant growth and lead to yield losses of food, feed, fiber, and fuel crops (Yin et al., 2021). Microbes employ various strategies to colonize plants, subverting plant defense responses through exopolysaccharide production, adhesin secretion, pili/fimbriae binding, enzymatic degradation of plant cell wall components, mimicking plant hormones, interfering with signaling pathways, and suppressing immune surveillance to promote colonization and infection. These mechanisms demonstrate the capability of microorganisms to overcome plant immune defenses (Niazi et al., 2023; Jain et al., 2024). *Agrobacterium tumefaciens*, *Dickeya dadanthi*, *D. solani*, *Pectobacterium atrosepticum*, *P. carotovorum*, and *Ralstonia solanacearum* can cause substantial economic damage in some crops (Mansfield et al., 2012). Plant pathogens that are soilborne are activated and drawn to root exudates. To start an infection, oomycete diseases create motile zoospores that swim in the direction of the plant root (Hua et al., 2008). Contaminated horticulture items, such as strawberries, lettuce, cabbage, and other raw salad vegetables, are increasingly being associated with salmonellosis. *Escherichia coli* has also been isolated from fresh food products more frequently (Mendes et al., 2013). *Fusarium oxysporum* generates fusaric acid, which inhibits the formation of the antibiotic molecule DAPG responsible for the antagonistic activity of *P. fluorescens* against this pathogen (Berendsen et al., 2012).

11.3 PLANT–RHIZOSPHERE INTERACTIONS

The rhizosphere surrounding plant roots is a hotspot for interactions between plants and diverse microbes, such as microbes, nematodes, and oomycetes. This zone is vital for nutrient cycling, organic matter decomposition, and maintaining soil health (Chamkhi et al., 2022). The rhizosphere is a highly dynamic zone where plant roots interact with a complex community of microbes, consisting of microbes, archaea, and protists. These microbes are crucial in nutrient cycling, plant growth, and soil health. Microbial interactions in this region influence plant health, resistance to diseases, and environmental stress tolerance, thereby contributing to plant productivity and ecosystem sustainability (Pantigoso et al., 2022).

The diversity of microbial communities in the rhizosphere is directly related to soil fertility. Different microbial populations support various functions, from nutrient cycling to decomposition of organic matter. This diversity contributes to soil structure, improving water retention and reducing soil erosion (Raimi et al., 2023).

Microbes in the rhizosphere play a crucial job in nutrient acquisition for plants by solubilizing phosphorus, fixing nitrogen, and cycling essential nutrients like sulfur and potassium. This decreases the reliance on chemical fertilizers in agriculture (Thepbandit and Athinuwat 2024). The rhizosphere provides a buffering zone where plants can better withstand environmental stress, like drought and salinity. Microbial interactions in this zone help plants adapt by stimulating nutrient uptake and decreasing oxidative damage (Chandra et al., 2022).

11.3.1 Role of Root Exudates in Microbial Interactions

The rhizosphere priming impact describes how root exudates stimulate microbial decomposition of soil organic matter, affecting carbon and nutrient cycling (Dijkstra et al., 2013). Roots excrete various organic compounds, like sugars, amino acids, and organic acids, which are used as food for rhizosphere microbes. These exudates nourish beneficial microbes and influence microbial community composition by promoting the growth of certain species over others (M. Li et al., 2022). Rhizosphere microbes can influence root architecture, with beneficial microbes promoting root elongation and branching. These alterations in root structure allow plants to access more nutrients and water, thereby stimulating overall growth (Gonin et al., 2023).

Root exudates play an essential role in affecting plant–microbe interactions and shaping the soil ecosystem. These exudates are composed of a variety of compounds, such as sugars, amino acids, organic acids, secondary metabolites, and signaling molecules. They act as mediators of communication between plant roots and microorganisms, regulating microbial activity, composition, and behavior within the rhizosphere, the soil area are directly affected by root secretions.

Root exudates produce essential carbon and energy sources for soil microbes, promoting microbial growth and activity. For example, simple sugars like glucose and fructose can enhance the development of specific microbial groups in the rhizosphere (Bais et al., 2006). Exudates help recruit beneficial microbes and suppress harmful pathogens. For instance, legumes release flavonoids that attract nitrogen-fixing *Rhizobium* bacteria, while antimicrobial substances such as phenolics can inhibit the growth of pathogens (Badri and Vivanco 2009).

Certain exudates, including strigolactones, are vital for initiating symbiosis. These compounds encourage arbuscular mycorrhizal fungal spores to germinate and guide their hyphae toward plant roots, facilitating nutrient exchange (Sasse et al., 2018). Molecules within root exudates can regulate microbial communication, including interfering with bacterial quorum sensing, which affects activities like biofilm formation and pathogenicity (Bais et al., 2006).

Root exudates affect microbial community structures, which in turn play a role in critical nutrient cycling processes such as nitrogen fixation, phosphate solubilization, and organic matter decomposition (Canarini et al., 2019).

Drought stress alters rhizosphere dynamics by changing root exudation patterns, promoting beneficial microbial populations, and increasing production of osmolytes that enhance plant water use efficiency (Kruthika et al., 2024). Root exudates and microbial activity in the rhizosphere can alter soil pH, affecting nutrient solubility and microbial community structure (Huang et al., 2014).

Exudates and microbial metabolites promote soil aggregation by binding soil particles, enhancing soil structure, and stimulating water retention (Chenu and Cosentino, 2011). Microbes in the rhizosphere produce VOCs that influence plant growth, root development, and defense responses, acting as an additional communication layer (Ali et al., 2025). Metagenomic analyses have provided insights into the functional diversity of rhizosphere microbial communities, revealing genes involved in nutrient cycling and stress tolerance (Liu et al., 2023).

11.3.2 Microbial Pathogens and Disease Suppression

Some rhizosphere microbes, like certain bacteria and fungi, are noted to suppress plant diseases by inhibiting the growth of pathogens. These beneficial microbes use several mechanisms, such as forming antimicrobial compounds, competing for space and nutrients, and inducing plant immune responses (Berendsen et al., 2012; Hacquard et al., 2017).

Microbial pathogens and their interactions with plants are central to understanding agricultural productivity and plant health. Pathogens can infect plants by secreting virulence factors, forming biofilms, or directly invading plant tissues, leading to diseases that reduce yield. However, plants and soil ecosystems have evolved multiple defense strategies to suppress these diseases, often mediated by beneficial microbes. The interplay between pathogens and disease suppression highlights the complexity and importance of microbial communities in the soil.

Pathogens such as *Phytophthora infestans* and *Xanthomonas* species use sophisticated mechanisms to invade plants. These include the secretion of effector proteins that suppress plant immune responses and enzymes such as cellulases and pectinases that degrade plant cell walls. Additionally, pathogens like *Ralstonia solanacearum* can form biofilms to enhance survival and virulence in the rhizosphere (Zhang et al., 2022).

Beneficial microorganisms in the rhizosphere often outcompete pathogens for nutrients and space. For example, siderophore-producing bacteria such as *Pseudomonas fluorescens* can sequester iron, making it unavailable to pathogens. This resource competition limits pathogen growth and reduces disease prevalence (Berendsen et al., 2012).

Certain microbes suppress pathogens by producing antifungal, antibacterial, or antiviral metabolites. For instance, *Streptomyces* species produces streptomycin and other antibiotics, which inhibit a broad range of soil pathogens. Similarly, VOCs formed by beneficial microbes can act as airborne suppressants, affecting pathogen viability in the surrounding soil (Andargie et al., 2023).

Plants can employ specific useful microbes by modifying their root exudates in response to pathogen attacks. For example, under pathogen stress, plants may release phenolic compounds or flavonoids that attract beneficial microbes capable of suppressing the invading pathogen. This phenomenon has been observed in the inhibition of *Fusarium* wilt in crops like bananas and tomatoes (Park et al., 2023).

Suppressive soils are naturally occurring ecosystems where pathogens fail to establish or cause significant damage despite their presence. These soils are often enriched with specific microbial communities, such as *Trichoderma* spp. and *Bacillus subtilis*, that inhibit pathogen activity. Studies on these soils have revealed the value of microbial diversity and synergy in suppressing diseases like *Rhizoctonia* damping-off and *Pythium* root rot (Jayaraman et al., 2021).

Plants rely on both ISR and SAR mechanisms to defend against pathogens. ISR is influenced by beneficial microorganisms, such as rhizobacteria, which activate plant defense genes via jasmonic acid and ethylene signaling pathways. However, SAR involves salicylic acid-dependent pathways and is often induced by prior pathogen infection or specific biocontrol agents. These mechanisms allow plants to respond faster and more robustly to subsequent pathogen attacks (Wilson et al., 2023).

High microbial diversity in the rhizosphere is a vital factor in suppressing pathogens. Diverse microbial communities create a competitive and dynamic environment, limiting pathogen establishment. Additionally, microbial diversity supports functional redundancy, ensuring that key processes like nutrient cycling and pathogen suppression remain active even when specific microbial populations fluctuate (Berendsen et al., 2012).

Understanding microbial interactions and disease suppression has significant implications for sustainable agriculture. Biocontrol agents like *Bacillus*, *Pseudomonas*, and *Streptomyces* are being developed as alternatives to chemical pesticides. These microbes not only inhibit pathogens but also stimulate plant growth and soil fertility, enhancing more sustainable and eco-friendly farming practices (Jayaraman et al., 2021).

Under salinity stress, rhizosphere microbes help maintain ionic balance, reduce sodium uptake, and increase antioxidant activity in plants, promoting salt tolerance (Liu et al., 2022). Enzymatic activities, like phosphatase and dehydrogenase production, in the rhizosphere, contribute to organic matter decomposition and nutrient mineralization, enhancing soil fertility (Yang et al., 2012).

Plants contribute to soil carbon sequestration through rhizosphere interactions by stabilizing organic matter and stimulating microbial activities that fix carbon in the soil (Bhaduri et al., 2015). Invasive plant species often modify rhizosphere interactions to outcompete native species by altering microbial communities or producing allelopathic compounds (Sanon et al., 2009).

11.3.2.1 Factor Promoting Microbiome Assembly

The whole structure of the soil microbiome, whether rhizosphere or rhizoplane microbiota, is intensively linked and functionally correlated. Microorganisms and other living organisms in a definite habitat connect and interact with each other differently, whether directly or indirectly, so any evolved alteration in ecosystem constituents or function will lead to a positive or negative influence on the microbiota composition and function (Hartmann and Six 2023). Significant variations may be noticed in how various species react to their surroundings (Jiao and Lu 2020). This range of different microbial species reactions and responses influences the shape and performance of the microbiome(Nemergut et al., 2013). Many studies differentiate between core and spontaneous random factors that influence microbial structure assembly in specific environments; the core factors contain biotic and abiotic drivers, while the spontaneous random ones include certain species mortality, migration, and/or propagation. Any alteration in microbiota environmental conditions leads to a change in microbial community assembly (Jiao and Lu 2020; Trivedi et al., 2020).

11.3.3 Biotic Parameters that Promote Rhizosphere Microbiome

11.3.3.1 Interaction between Microbes

The rhizosphere is a reservoir of many living organisms like fungi, oomycetes, bacteria, nematodes, and even viruses, which coexist in a dynamic state of continuous interaction, whether in a mutualistic or even antagonistic relationship. Out of sharing the same nutrient resources, whether from root exudates or the original organic material in the soil itself, different species may exert antagonistic effects on each other, competing for survival throughout, secreting many exoenzymes, mycotoxins, antibiotics, flavonoids, alkaloids, terpenoids, and many natural products, which potentially would lead to shaping the rhizosphere mycobiome community (Yang et al., 2017).

Moreover, rhizospheres are specific microbial spots where many mold species share the same nutrient source and substrate. Some fungal species, like *Trichoderma*, which are considered among the commercial bio-fungicides (Harman et al., 2004), and biofertilizers (Vinale et al., 2008) may act as microbial parasites that exert antagonistic effects over other fungal species, which lead to the exclusion or suppression of some taxa and change the community assembly. On the contrary, the rhizosphere may witness suppression and elimination of some pathogenic and parasitic organisms and pull others of symbiotic and beneficial ones, owing to microbial-plant and inter-microbial interactions that altogether formulate a dynamic battle zone (Ehrmann and Ritz 2014).

Although many studies have been conducted to figure out the incidences and prevalence of microbial interaction within the microbiome structure in a definite environment, the repercussions of microbial interaction in the same environment on the microbiome assembly (whether it is exploitation relations by predatory behavior of some parasitic microbes on others competing for nutrients or even collaboration to share the same resource in mutualism) are still totally unclear and subjected to more study and research (Spragge et al., 2023).

Foster and Bell (2012) tested numerous culturable bacterial strains from plant holes filled with raindrops and ended up that those bacterial strains developed and proliferated better when singly grown than in competition with other species (Foster and Bell, 2012). On the other side, when Kehe et al. (2021) examined the combination of many bacterial strains of 20 soil samples applying the high-throughput (HTP) technique, it turned out that around 40%of more than 180 thousand bacterial strains in the samples examined showed better growth in combination with others than being grown separately, considering the majority of these interactions were parasitic (Kehe et al., 2021). After compiling data from several research on microbiomes, Palmer and Foster (2022) recently concluded that "negative relationships among the microbial species dominate, while beneficial collaboration between species is typically rare"(Palmer and Foster 2022).

11.3.3.2 Plant–Microbiota Interaction Influences Rhizosphere Microbiome Assembly

Every plant species possesses a significant impact on the rhizosphere microbiome structure (Fierer 2017). This effect may be directly through the root secretions, which are thought to be a particularly alluring spring for the attracting of microorganisms; these secretions contain organic matter needed for rhizosphere microbiome growth and development, or indirectly via the potential of the plant-microbiota network, which has the power to either attract or repel other microbes (Baetz and Martinoia 2014). It is assumed that rhizosphere microbiome composition is a plant species-dependent structure (Sui et al., 2017).

There is a microbial network evolved between plant microbiota; this network combines the endophytic and rhizospheric microbiomes that surround the plant in a dynamic relation, so many researchers recognized the whole plant microbiome as endophytic in addition to the rhizospheric microbial constituents, considering the effect of plant microbiome on the surrounding soil microbes (Berendsen et al., 2012). An example of a plant possessing influence on the rhizosphere microbiome is noticed in the secretion of some definite phytohormones that impose a selective pressure force on the rhizosphere microbiome that reassembles its structure (Liu et al., 2020). For instance, abscisic acid, which plants have secreted, was demonstrated to influence the attraction of some rhizobacteria like *Rhodococcus* and *Novoshingobium*, where it is considered their carbon source (Eichmann et al., 2021). Some Streptomycetaceae species can also utilize plant auxins as a carbon source for growth and development, so these plant auxins and phytohormones influence alteration in metabolic pathways within the rhizosphere microbiome (Eichmann et al., 2021). The potentiality of some other phytohormones like cytokinins, gibberellic acid, and brassinosteroids has been recognized in selective pressure toward some rhizobia and AMF in the rhizosphere (Yu and Hochholdinger 2018; Carvalhais et al., 2019; Liu et al., 2020).

Other strategies plants may apply and lead to reshaping the microbiome assembly, which is the alteration in the regulation of some genome related to specific signaling pathways throughout plant lifespan, which will lead to significant impact in related microbiome structure, like moderation in the regulation of genes responsible for the salicylic acid pathway in young *Ginkgo biloba* and adjustment of regulation of other genes that are concerned of the jasmonic acid pathway in adult ones (S. Wang et al., 2022). The transcriptomic analysis projected some microbial metabolic pathway alteration during the associated plant growth, reporting a significant shift in the metabolic pathway of some bacteria that are related to newly cultivated plants, from just consuming carbon to contributing to cycles of N_2 and S in aging plants (Qu et al., 2020). This alteration in plant nutrition pattern during its lifecycle led to changes in associated microbiota metabolic pathways as well, and these were reported significantly in *Pinus koraiensis*, *Picea koraiensis*, and their associated microbiome (Ren et al., 2023). Hart et al. (2014) recognized that the abundance of arbuscular mycorrhizal fungal composition associated with some forest plants, like *Artocarpus altilis*, changes during the plant life cycle, where it witnessed more prevalence in the aging stage than the newly grown stage (Hart et al., 2014).

11.3.4 Biotic Parameters that Promote Rhizosphere Microbiome

11.3.4.1 Environmental Conditions

Other environmental factors are recognized to possess significant influence on the rhizosphere microbiome assembly as well (Vetrovský et al., 2019). High-throughput sequencing (HTS) gave a significant understanding of the diversity and functions of microbiome composition. Much conducted research demonstrated that applying HTS to rhizosphere samples revealed an abundance of some fungal taxa and suppression of others, which was attributed to climatic factors' influence and these taxa's adaptivity. Some of the fungal taxa that prevailed were Basidiomycota and Ascomycota, so HTS data analysis proved the strong relation between definite taxa and specific environmental circumstances (Větrovský et al., 2019).

A big project was organized for studying the metaproteome of the world's soil by Fernandes et al. (2022), which resulted in the finding that rhizosphere microbiome assembly showed a strong linkage to environmental factors like temperature, pH, and moisture, so that any potential climatic alteration would come to the diversification of rhizosphere microbiome assembly (Fernandes et al., 2022).

Soil bacterial species are considered crucial parameters in sustaining the C cycle's balance and stability, besides promoting plant proliferation (Song et al., 2024). It is predicted that global warming will have a fatal effect on earth's C composition because disturbances in the photosynthesis ratio, in addition to the heat condition, will impact the balance of C concentration in soil and atmosphere and thus will melt out, forcing alteration in the microbiome metabolism, functions, and assembly (Castañeda-Gómez et al., 2020).

Some abiotic stressors, like salinity, greatly affect the development and structure of microbiome constituents in both organic and cultivated environments, as they exert pressure on attracting microbial species with a genetic tendency to accommodate salty environments. Rath et al. (2019) recognized that in salty soil, there is a prevalence of bacterial species that is more tolerable to the salty conditions than in soil of low salt concentration (Rath et al., 2019). Also, he mentioned that water content in the salty-soil media, besides the accessible organic materials needed for nutrition, both as environmental factors, impose a significant effect on microbiome development (Etesami and Beattie, 2018; H. Li et al., 2021). Another environmental factor that promotes the rhizosphere microbiome assembly is scarcity of water, as in extreme drought conditions, plants discharge distinct bioactive substances into the rhizosphere, therefore, modifying the microbiome population inhabiting the root's rhizosphere, which does not happen in normal conditions (Ait-El-Mokhtar et al., 2023).

11.3.4.2 Soil Type Promotes the Microbiome Structure

Soil is a key constituent of an ecosystem with a significant profile of varied chemical and physical structure. Every soil profile induces the development of microbial species with similar characteristics, features, and living patterns and shares the same responses to environmental factors. On the other hand, each soil profile may suppress or inhibit the growth of various microbial species; this pattern melts out the formation of a soil-dependent microbiome structure. It has been demonstrated that alterations in the physical and chemical composition of the rhizosphere cause shifts in the structure of the rhizosphere microbiome (Yan et al., 2024). Many investigations highlighted the correlation between the microbiome and habitat. R. Wang et al. (2023) demonstrated that bacterial species of halophytic plants' rhizospheres, which inhabit coastal regions, have shown differences according to soil types, whether it is a temperate or subtropical coastal wetland. Deforestation and afforestation activities exert a significant role in promoting microbiome diversity and reshaping.

11.4 ROLE OF PLANTS IN RESHAPING THE RHIZOSPHERE MICROBIOME

The rhizosphere microbiome is a significantly dynamic ecosystem because of many abiotic and biotic factors (Figure 11.2). The abiotic factors are the non-living effectors of the ecosystem that can be categorized into atmospheric and terrestrial levels. For instance, the atmospheric effectors are UV irradiation, drastic fluctuations in temperature, rainfall, and wind velocity (Schlechter et al., 2019). The terrestrial effectors are related to the soil, such as the physicochemical characteristics of the soil, including the soil type, temperature, pH, moisture/water content, availability of nutrients and oxygen, and presence of organic and/or inorganic pollutants and contaminants (Jain et al., 2020). Contrary to the abiotic effectors, the biotic ones are the results of interactions between the living organisms composing an ecosystem, such as plants, microbiota, and animals. Here, we introduce one of the biotic effectors: the

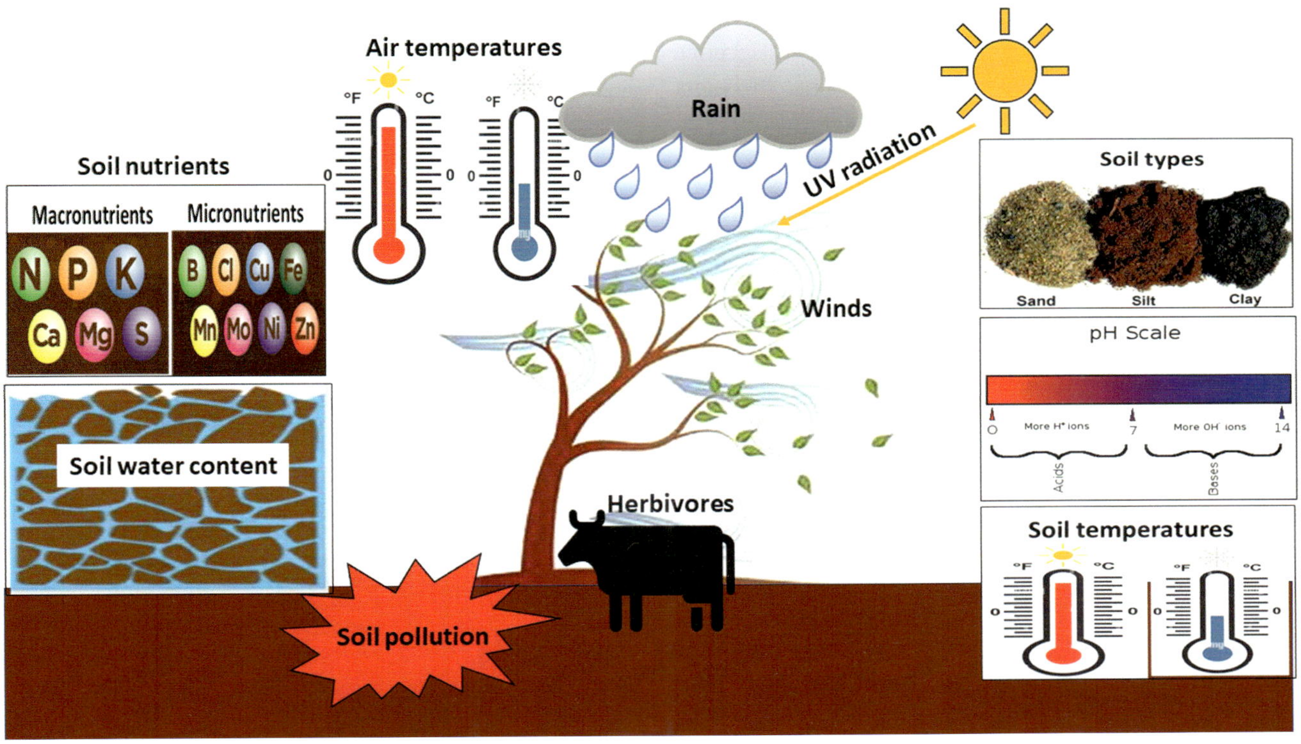

FIGURE 11.2 A representative diagram illustrates the abiotic and biotic factors influencing the dynamics of the rhizosphere microbiome.

plant-based factors affecting the rhizosphere microbiome dynamics due to their significant impacts on the taxonomical diversity and the constituents of the recruited microbes from the bulk soil toward the rhizosphere zones.

Terrestrial plants are sessile organisms in the soil, so they must establish interactions with existing surrounding organisms (microbes, animals, and even other plant species), whether these organisms are beneficial or harmful to these plants. Regarding the rhizosphere zone, terrestrial plants produce extracellular components that form rhizodeposition, which is responsible for shaping the plant's rhizosphere microbiome (Figure 11.3). In this context, the plant rhizodeposition's composition and functions are discussed along with the rhizodeposition's production mechanisms. Regarding the composition, the plant's rhizodeposition is composed of two major parts: (1) cell-based components and (2) chemical-based components (de Bruijn 2013). The cell-based components consist of the sloughed-off root cells and the root border cells. The sloughed-off root cells are detached from the root cap into the rhizosphere (de Bruijn 2013). Similarly, the border cells are detached from the root into the plant's rhizosphere, while these cells are alive just for weeks or even months (de Bruijn 2013). The chemical-based components consist of the mucilage and the root exudate, which is the most significant portion of the whole rhizodeposition (Koroney et al., 2016). As the root mucilage is an extracellular, viscus mixture secreted from the root cap-cells and stacked on the root tips (Koroney et al., 2016). The mucilage consists of polysaccharides, proteins, and polyphenols (Weiller et al., 2017). The root exudate is an extracellular liquid mixture released from the meristematic zone of the root tips into the rhizosphere (Bais et al., 2006; Badri and Vivanco, 2009). The root exudates are secreted into the rhizosphere via various mechanisms, including diffusion, ion channel pumping, and vesicle transport (P. Yu et al., 2021). The root exudate is a rich natural source of nitrogen and carbon compounds, such as amino acids, organic acids, aromatic acids, vitamins, phytohormones, volatile compounds, and sugars (Tian et al., 2020).

Collectively, all of these rhizodeposition components are responsible for the taxonomical diversity of the plant-rhizosphere microbiome through two significant strategies: one of them is known as positive interactions, and the other is the negative interactions (Bais et al., 2006; Philippot et al., 2013). Since some rhizodeposition components work as chemoattractants to the beneficial microbes, positive interactions with surrounding organisms regarding the symbiotic relationships with the host plants are established. These beneficial microorganisms, also known as mutualists, establish win-win situations with the host plants. For instance, the arbuscular mycorrhiza and the rhizobia assist the host plants in phosphate absorption and nitrogen fixation, respectively (Ezawa and Saito, 2018). In addition, different components of the plant rhizodeposition act as repellents to a broad spectrum of pathogens. As evidence, the sloughed-off root cells are considered the primary physical defense line against plant pathogens (Hirsch et al., 2013). Moreover, the border cells release an array of phytochemicals as natural antimicrobial agents repelling the plant pathogens (Sugimoto et al., 2010; Koroney et al., 2016). Among secreted phytochemicals from the plant root into its rhizosphere of *Arabidopsis thaliana* is rhizathalene A, responsible for plant protection against insect attack (Vaughan et al., 2013). Similarly, the antifungal activities of five phenylpropanoids secreted from the roots of barley against the soil-borne fungal pathogen *Fusarium graminearum* (Lanoue et al., 2010).

The constituent of the plant rhizodeposition is significantly impacted by extrinsic conditions (e.g., pollution and climatic conditions) affecting the plant's gene expressions and manipulating the pathways for rhizodeposition

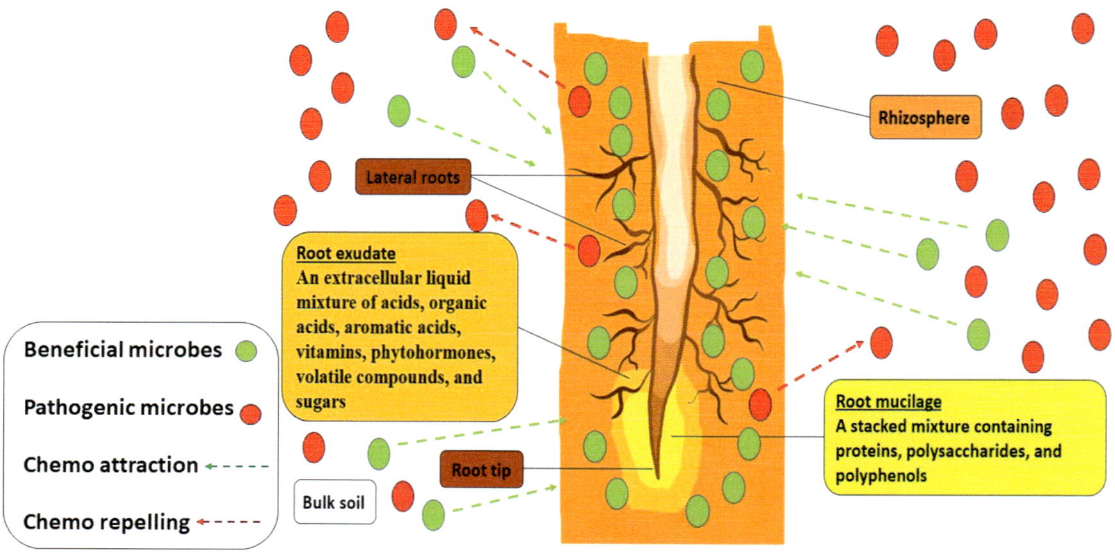

FIGURE 11.3 Diagram representing the effects of the plant rhizodeposition on re-shaping of the rhizosphere microbiome through secretion of the root mucilage and exudates to attract and/or repel the beneficial and/or pathogenic microbes, respectively.

production. The intrinsic conditions are related to the plant itself, regardless of the extrinsic conditions, which are (1) plant species, (2) plant genotypes, and (3) plant developmental stages. For evidence, there are C3 and C4 plant species according to the photosynthesis mechanisms, can recruit significant taxonomically different microbiomes in their rhizospheres even if these plant species are cultivated in the same field (Bledsoe et al., 2020). The constituent of the microbial rhizosphere of various genotypes of the same plant species also exhibits a significant taxonomical diversity, giving each plant species cultivar its fingerprint of the rhizospheric microbiome (Sivaram et al., 2020). That means the rhizospheric microbiome constituent can be considered as a heritable trait similar to other heritable characteristics, such as the morphological, physiological, and molecular traits (Horton et al., 2014). The plant developmental stages play critical roles in reshaping the plant rhizospheric microbiome composition as *Arabidopsis thaliana* recruits taxonomically different rhizospheric microbiomes at its various developmental stages (seedling, vegetative, bolting, and flowering) (Salas-González et al., 2021).

11.5 CONTROLLING RHIZOSPHERE MICROBIOME FOR NUTRIENT ACHIEVEMENTS

The microbial constituents of plant rhizosphere are a vital parameter in determining cultivated crop plant performance, development, fitness, and productivity; they assist in making the nutrients in the soil available and accessible, in addition to easing the process of plant absorption (Trivedi et al., 2020). Microbial species improve plant resilience against abiotic stress by regulating nutrient availability and growth, and help maintain plant health.

One of the primitive and conventional rhizosphere modeling and controlling techniques was the human practice of transplantation of fruitful soils to enhance cultivated crop productivity (Mougel et al., 2006). Crop plants shape the rhizosphere components by secreting root exudates that attract microorganisms. It has been displayed that the microbial structure of the plant rhizosphere is impacted by many physiological and chemical properties of the soil profile and organic matter accessibility, cultivated plant species, and even the phase of the plant's lifecycle. It has been proposed that functional features are the primary factor regulating the rhizosphere microbial structure (Zhang et al., 2019).

PGPR are essential components of the rhizosphere microbiome that not only facilitate organic matter accessibility and availability to the host plant but also help modify the plant root system structure and design. It was recognized that the better N_2 fixation and uptake among some maize species over others refers to the presence of nitrogen-fixing bacterial species in high amounts, like in the case of the *Oryza indica* rhizosphere compared to *Oryza japonica* species (Zhang et al., 2019). Thus, it was validated by studying the implantation outcome of 16 cultivated samples containing *O. indica* (associated with nitrogen-fixing bacterial strains). This resulted in the apparent efficient utilization of N_2 among *O. indica* species. It has been proven that some flavonoid compounds released from maize species' roots influence recruiting of bacterial members of the Oxalobacteraceae family in their rhizosphere, such as bacterial species involved in easy nitrogen fixation and absorption (P. Yu et al., 2021). Mainly, *Burkholderia*, *Rhizobium*, and *Sphingomonas* genera are the major nitrogen-assimilating bacterial groups that associate with the rhizosphere of sweetcorn (Walters et al., 2018).

Many microbial species impart to the nitrogen cycle in the rhizosphere; some bacteria are known for N_2 assimilation, like diazotrophs; others induce oxidation of NH_3 gas, including bacterial and archaeal species (Bargaz et al., 2018). It is worthy to mention that *Rhizobium leguminosarum* (Evans and Russell) paved the way for potential utilization and inoculation of N_2 assimilating bacterial strains in agricultural applications. Moreover, melted out a new concept of establishing root nodules' part, which is responsible for biological N_2 fixation, the evolved term of nitrogen assimilation biologically points to the involvement of beneficial microorganisms in transforming atmospheric nonreactive nitrogen to NH_3 utilizing nitrogenase enzyme (Ryan et al., 2009). This mutualistic beneficial relationship of host plant-rhizosphere microbiome facilitates the two-way pathway of movement of assimilated N_2 from microbes to the host plant, consecutively gaining plant organic carbons (Nelson and Mele 2007).

Azotobacter and *Clostridia* species are good examples of heterotrophic bacteria that assimilate the atmospheric N_2. These diazotrophs participate in fixing atmospheric nitrogen on behalf of plants in return for consuming carbon secreted from roots (Biswas and Gresshoff 2014). Yoneyama et al. (2017) recognized that the association between nitrogen-fixing microbiome and maize aerial roots in nitrogen-deficient soil results in increasing atmospheric N_2 assimilation by more than 80% and being acquired successfully by maize (Yoneyama et al., 2017).

In case of insufficient nutrient availability, the rhizosphere-inhabiting microorganisms contribute to regulating and gaining essential minerals like Fe^{+2} and P4. Under the circumstances, throughout the soil of phosphate deficiency, the A. thaliana plant species' phosphate starvation response (PSR) influences the microbial structure of the surrounding soil of the plant, in addition to inoculation with symbiotic microbes that altogether leads to PSR gene expression and improves plant absorption of the inorganic P_4 mineral (Castrillo et al., 2017). This microbial structure of the rhizosphere triggered by PSR gene expression contains microbes of symbiotic and rival opportunistic techniques that may support plant P4 acquisition or suppress the uptake process like what rival opportunistic does. The fungal composition in the rhizosphere microbiome is also shown to be significantly involved in nonmycorrhizal plant phosphorus absorption during phosphorus deficiency conditions (Finkel et al., 2019).

11.6 EFFECT OF ROOT EXUDATES RECRUITMENT (INDUCTION) OF RHIZOSPHERE MICROBIOME AND IMPROVEMENT OF NUTRIENT UPTAKE

The root exudates regulate the interactions of plants with other plants and rhizosphere microbes that benefit the plant, like nutrition and defense (Feng et al., 2021; Bhimani et al., 2024). The effect of root exudates is highly variable, complex, and dynamic in improving and impacting the composition and function of the rhizosphere microbiome (Jacoby et al., 2020; Yin et al., 2021). The composition and extent of root exudates depend on environmental conditions, such as pathogen invasion (Kong et al., 2021), and plant-specific characteristics, including genotype, developmental stage, health status, plant domestication, plant compartment, and mode of photosynthesis (Feng et al., 2021; Yue et al., 2023). The variety of primary and secondary compounds found in root exudates includes sugars, organic acids (long-chain fatty acids, short-chain organic acids, amino acids, and plant volatiles), metabolites, phytohormones, complex mucus-like polymers, phenolics, flavonoids, and auxins (Kong et al., 2021; Wu et al., 2023). They offer a highly bioavailable source of nutrients and energy for microbial growth and act as a signaling messenger to influence the rhizosphere microbiome (Koprivova and Kopriva 2022). Rhizosphere microbes can increase soil fertility and adjust soil pH by diverging root exudates, indirectly promoting plant growth (Wang et al., 2024).

Plants recruit soil-beneficial microbial communities after attack by pathogenic microorganisms without triggering a strong immune response to hold up their development and fitness (Yin et al., 2021). In order to support the growth and symbiosis of mycorrhizal fungi establishment as a mechanism for nutrient acquisition, the plant increases the production of strigolactones under nutrient-deficient conditions (Aliche et al., 2020). Different rhizosphere microbiomes can use chemical groups that the plant exudes as food sources and signals. Hida et al. (2020) showed that *Pseudomonas protegens* CHA0 has chemoreceptors for amino acids, enabling chemotaxis toward amino acids (Rizaludin et al., 2021). The root exudates from *Flaveria bidentis* can significantly increase the abundances of Bacillus frigoritolerans and *B. megaterium*, which can stimulate soil fertility by nitrogen-fixing and phosphate-solubilizing. The dominant chemoattractants (butyl-phenol, methyl stearate, and arginine) in the root exudates mediate the rhizosphere bacterial community assembly of *Casuarina equisetifolia* (Lin et al., 2022).

The allelochemicals such as benzoic acid and cinnamic acid derivatives secreted by rice roots can inhibit surrounding weeds (J. Li et al., 2022). Furthermore, the root exudates from *Rehmannia glutinosa* can enhance the growth of *Fusarium oxysporum*, which disrupts the growth and development of *R. glutinosa* by altering the expression of Leucine-rich repeat receptor-like protein kinases (RgLRRs), leading to replant disease (Yang et al., 2022). The four short-chain organic acids (citric acid, fumarate, succinic acid, and pyruvate acid) may be concerned for the enrichment of *Comamonadaceae*, a potentially beneficial microbial group (Wen et al., 2020). The sugars, phenolic acid, and *p*-coumaric acid in the root exudates can significantly affect the natural composition of soil microbiota (Zhou and Wu, 2012). Benzoxazinoid root exudates released by maize enhance plant defense and productivity by shaping rhizosphere microbiota (Hu et al., 2018). Flavonoids (a type of root exudate) can stimulate bacterial roots that form nodules for nitrogen fixation in legumes (Hassan and Mathesius 2012). Flavones can stimulate the growth of Oxalobacteraceae species, which can benefit maize growth by nitrogen acquisition and modulate developmental processes (H. Yu et al., 2021). Flavonoids can also increase phosphorus availability for plants by desorbing phosphates from soil minerals or dissolving mineral-phosphate complexes (Cesco et al., 2012).

The root exudates of *Arabidopsis*, including phenolics and triterpenes, can enhance the abundance of PGPR and influence the constituents of the microbial community around its roots (Huang et al., 2019). *Arabidopsis* rhizosphere exudates, such as malic acid, increased when plants were infected with *Pseudomonas syringae* pv. *tomato*, and attract beneficial *Bacillus subtilis* (Berendsen et al., 2012). *Arabidopsis* can specifically promote the growth of certain bacterial species, including *Microbacterium*, *Stenotrophomonas*, and *Xanthomonas* sp., in the rhizosphere after response to foliar defense activation by the pathogen *Hyaloperonospora arabidopsidis* (Berendsen et al., 2018). The root-secreted amino acids and long-chain fatty acids were enhanced by *P. syringae* pv. *Tomato* infection can attract soil-specific *Pseudomonas* populations, providing plant resistance against aboveground pathogen attack via the induction of plant ISR (Wen et al., 2021).

Some bacterial root microbiota are stimulated by coumarins, which are made by plants under nutrient-limiting soils and can stimulate iron nutrition for the host plant (Pantigoso et al., 2022). Phytohormones used by nitrogen-fixing rhizobacteria can improve biomass, tolerance to abiotic stressors, and resident bacteria colonization (Ryu et al., 2020). Plants can produce compounds that enhance or suppress QS-regulated responses in bacteria (Berendsen et al., 2012). Defense phenylpropanoids are synthesized in response to pathogen attacks, while the diterpene rhizathalene A is continuously secreted even without the pathogen. These compounds represent a constitutive direct defense mechanism and enhance plant resistance to specific pathogens (Wu et al., 2023). Plants may actively recruit disease-suppressive microbes in response to pathogen attacks, offering potential strategies for disease control and increased crop production (J. Li et al., 2021).

11.7 IMPACT OF STRESS CONDITIONS ON RHIZOSPHERE–PLANT ROOT RELATIONSHIPS

The rhizosphere microbiome structures and functions are significantly dynamic compared with the bulk soil microbiome, responding to different stressors. These structural and functional changes of the rhizosphere microbiomes are considered adaptive responses, affecting the host plants under stress. Therefore, it is mandatory to highlight stressor types herein, their impacts on the rhizosphere microbiome structures and functions, and their effect on different host plants' adaptation under these stresses. Different stressors are categorized mainly into abiotic and biotic stressors. Soil pollution, drought, submergence, and salinity are the primary abiotic stress sources affecting the rhizosphere microbiome structures and functions. Meanwhile, the biological threats of the host plants are significantly variable, ranging from micro-threats such as bacteria, viruses, and fungi to macro-threats such as competing plants (weeds) and herbivores (Huang et al., 2022).

Nowadays, there are significant pollution discharges into different environmental matrices due to the high rates. The soil/terrestrial matrix is a sink for many organic and inorganic pollutant discharges that disturb the soil microbiome, especially the rhizosphere microbiome. As evidence, the cultivation of ryegrass for 14 days reduces the phenanthrene pollution more than polluted non-cultivated soil by 8% (Jiang et al., 2022). Moreover, the pyrene removal efficiency was up to 20% higher in the polluted soil planted with *Festuca arundinacea* compared with the non-planted polluted soil (X. Chen et al., 2018). Such significant improvement in the remediation of organic pollutants in the planted soils is attributed to the enrichment of the resistant microbes to organic pollutants' toxicity. In contrast, those sensitive to the pollutants' toxicity are depleted (Dai et al., 2022). For instance, the rhizosphere microorganisms are responsible for the dissipation/degradation of organic pollutants discharged into soils, such as polycyclic aromatic hydrocarbons (PAHs), polychlorinated biphenyls (PCBs), and bis (2-ethylhexyl) phthalate (DEHP) more than the bulk soil microbiome (Huang et al., 2023). Knowing that the bacteria are better degraders of DEHP pollutants than fungi and the archaea (Zhu et al., 2023). Meanwhile, the bacterial communities are more sensitive to the toxicity of many other organic pollutants, such as PCBs and lindane (Y. Wang et al., 2022).

Like organic pollution, heavy metals are potential inorganic pollutants threatening life on the earth. These heavy metals are inorganic contaminants that can be categorized into essential and non-essential metals based on their biological significance (Theron et al., 2012). The essential heavy metals are used as cofactors for many enzymes, such as copper and iron (Theron et al., 2012). However, non-essential metals are potent toxins for living organisms such as arsenic and lead (Theron et al., 2012). So, soil pollution with these non-essential heavy metals gains excellent attention, especially their impact on the rhizosphere microbiome structure and function. As evidence, the plant-growth-promoting microorganisms are enriched in the rhizosphere in the heavy metal-polluted soils (Kumar and Verma 2018). These microbes have a significant role in the remediation of heavy metals and enhance the tolerance of the host plants as well *via* the formation of growth-promoting phytohormones and improving the availability of nutrients, including nitrogen and phosphorus (Kumar and Verma 2018).

Besides soil pollution, the moisture content of the soil is another essential factor shaping the soil microbiota. The soil drought crisis is defined as a significant decrease in the soil moisture, leading to a decrease in microbial respiration and, hence, their cellular activities (Denardin et al., 2020). However, drought-resistant microbial taxa have become the dominant taxa. Consequently, there are apparent variations in the microbial constituents as an adaptive response to soil drought, particularly in the plant rhizospheres. For example, the rhizosphere of *Sorghum bicolor* under drought stress is enriched with *Actinobacteria* species along with *Firmicutes* species (Wipf et al., 2021). Both genera can produce well-protected spores that can be resilient and thrive under drought (Beskrovnaya et al., 2021). Oppositely, the soil submergence floods the soil with water amounts exceeding the soil's need, leading to soil particles compacting (Hartman and Tringe, 2019). Such soil flooding leads to dramatic changes in gas exchange between the atmosphere, soil, microorganisms, soil pH, and nutrient status (Abdul Rahman et al., 2021). Such abiotic stress changes significantly the soil microflora composition. As evidence, the soil microflora community shifts toward the anaerobic microbes, especially anaerobic bacteria, while the obligate aerobic microbes, especially fungi, will gradually decrease (Abdul Rahman et al., 2021). For instance, there are bacterial species that represent the dominant organisms in flooded agricultural soils, such as *Anaeromyxobacter*, *Alphaproteobacteria*, *Betaproteobacteria*, *Deltaproteobacteria*, and *Malikia* (Sánchez-Rodríguez et al., 2018; Furtak et al., 2020).

The soil salinity is produced by the accumulation of Cl^- and Na^+ ions in the soil because of many reasons, such as soil drought and/or harmful agricultural practices, for example, soil irrigation with salty water and/or routine application of mineral/chemical fertilizers and plant protection agents like herbicides, insecticides, and fungicides (Holík et al., 2019; Abdul Rahman et al., 2021). According to the adaptive mechanisms under salinity stress, the microbes are categorized into halophiles and halotolerants. Regarding the halophiles, these microbes require salt to grow. Based on the required salt concentrations, there are four subtypes of the halophiles that are halotolerants, weak halophiles (1–3% of NaCI), intermediate halophiles (3–15% of NaCI), and intense halophiles (15–30% NaCI) (Abdul Rahman et al., 2021; Mainka et al., 2021). Firmicutes is the dominant species in the polar desert instead of Acidobacteria and Actinobacteria due to soil salinity increase

(Van Horn et al., 2014). In conclusion, several previous investigations have proved the significant effects of the abiotic stressors in reshaping the soil microbial community compositions by enriching the microbial taxa that are persistent to stressors.

11.8 PLANT GROWTH-PROMOTING MICROBES

11.8.1 Fungi and Oomycetes

11.8.1.1 Fungi in Plant Growth Promotion

Fungi, particularly mycorrhizal fungi, form symbiotic associations with plant roots, enhancing nutrient uptake, especially phosphorus, and increasing resistance to pathogens. AMF create extensive hyphal networks that extend the root absorption zone, improving water and nutrient availability (M. Chen et al., 2018). Fungi form symbiotic relationships with plants through structures like mycorrhizae, which enhance nutrient uptake, particularly phosphorus and nitrogen. In return, plants supply fungi with carbohydrates.

This mutualism increases plant growth and resilience (M. Chen et al., 2018). Certain fungi produce phytohormones such as auxins, gibberellins, and cytokinins that promote plant growth. For instance, *Trichoderma* spp. enhance root elongation through auxin-like compounds, improving stress tolerance and overall plant vigor (Goyal and Kalia, 2020).

Fungi, like *Trichoderma* spp. produce antifungal and antibacterial metabolites that suppress soilborne pathogens, indirectly promoting plant growth by stimulating its systemic resistance (SR) (khan et al 2020). *Trichoderma* species are well-known biocontrol agents, as they can directly inhibit the growth of harmful pathogens by forming antibiotics, lytic enzymes, and other antimicrobial compounds. This promotes plant growth by protecting the roots from pathogen attack (Benítez et al., 2004).

AMF contribute to soil fertility by facilitating the uptake of phosphorus and other vital nutrients, like zinc, copper, and magnesium, through their extensive hyphal networks. These fungi also influence soil pH and organic matter content, further benefiting plant growth (J. Li et al., 2021). Mycorrhizal fungi help plants manage biotic and abiotic stresses, such as drought, salinity, and soil pathogens. By stimulating nutrient absorption and producing stress-related hormones, these fungi aid in improving plant health (Rapparini and Peñuelas, 2014). Ectomycorrhizal fungi predominantly associate with trees, aiding nitrogen and phosphorus uptake while conferring resistance to heavy metals. Their ability to decompose organic material is critical for nutrient cycling in forest soils (Emamverdian et al., 2015).

11.8.1.2 Oomycetes and Their Agricultural Significance

Oomycetes are a diverse eukaryotic microorganism resembling fungi but differ in evolutionary lineage. Known for their association with plant diseases (e.g., *Phytophthora* species), many non-pathogenic oomycetes provide notable benefits to plant growth and resilience by affecting root-level interactions and nutrient uptake (Jiang and Tyler 2012). These oomycetes function like biocontrol agents, suppressing the growth of plant pathogens and boosting plant immune responses, thereby promoting healthy plant growth (Li et al., 2024).

Certain oomycetes, such as *Pythium oligandrum*, are recognized for stimulating plant growth and disease suppression. These organisms enhance root architecture, improve nutrient availability, and reduce the prevalence of harmful microbes through enzymatic activity and competitive exclusion (Bělonožníková et al., 2022). Certain beneficial oomycetes like *Pythium periplocum* promote positive microbial interactions in the rhizosphere. These interactions help suppress pathogenic organisms and boost the activity of beneficial microbes, improving soil health and plant growth (Mendes et al., 2013). This enhances the health of the soil and the plant (J. Li et al., 2021). Oomycetes, which form mutualistic relationships with plants, contribute to plant growth by promoting nutrient uptake and improving plant health. These interactions can also involve pathogen suppression, making these oomycetes beneficial for sustainable agriculture (J. Li et al., 2021).

Producing stable, effective oomycete-based products for large-scale agricultural use is challenging. Advances in biotechnology and nanotechnology provide opportunities to create innovative and reliable formulations (Boyetchko et al., 2020). Including oomycetes in IPM strategies has effectively suppressed pest populations while promoting crop yields. This aligns with sustainable agricultural practices and reduces the environmental impact (Galli et al., 2024). Some oomycetes activate systemic resistance in plants, enabling them to combat pathogens and environmental stress. This process implies the enhancement of defense-related genes and pathways mediated by salicylic acid and jasmonic acid (Raaymakers and Van den Ackerveken 2016).

In the rhizosphere, oomycetes interact with microorganisms to improve nutrient cycling. Their activity boosts phosphorus solubilization and nitrogen fixation, ensuring plants access essential nutrients that promote growth (Hakim et al., 2021). The interactions between plants and oomycetes are complex and require deeper exploration to maximize their agricultural potential. Unraveling the genetic and molecular basis of these relationships is essential for practical applications (Willsey et al., 2017).

Genomic and proteomic studies offer valuable insights into how oomycetes enhance plant growth. These advancements can support the breeding of resilient crops with enhanced resistance to stress and disease (Li et al., 2024). Exploring the diversity of oomycetes in natural ecosystems may reveal novel species with unique growth-promoting characteristics. Conservation of microbial biodiversity is critical for discovering and utilizing such resources (Nechwatal et al., 2008).

Beneficial oomycetes stimulate soil fertility by enhancing organic matter decomposition and stimulating microbial activity. These processes create healthier and more dynamic soil ecosystems (Frąc et al., 2018). The eco-friendly nature of oomycetes makes them well-suited for organic farming systems. Their application reduces dependency on synthetic fertilizers and pesticides, promoting sustainable practices (Blakney et al., 2023).

Oomycetes play a critical role in crop diversification efforts. They improve the productivity and resilience of non-traditional crops, promoting sustainable farming across diverse environmental conditions (Gahagan et al., 2023). Integrating oomycetes into crop breeding programs can lead to the development of resilient plants that are more able of withstanding biotic and abiotic stresses, addressing food security challenges (Jha et al., 2021).

The utilization of oomycetes in agriculture promotes sustainable resource management by decreasing chemical inputs, enhancing soil health, and supporting balanced ecosystems (Li et al., 2024). Oomycetes possess immense potential to revolutionize modern agriculture. By promoting plant growth, enhancing soil health, and improving stress tolerance, they offer sustainable solutions for global farming challenges. Continued research and technological advancements will play an important role in fully realizing their agricultural potential (L. Wang et al., 2023).

11.8.2 Nematodes and Their Role in Soil Ecosystems

Nematodes are microscopic, unsegmented worms found across various environments, from aquatic ecosystems to the terrestrial biosphere. They play pivotal roles in soil ecosystems in supporting soil health, enhancing nutrient cycling, and maintaining ecological balance. These organisms are vital components of the soil food web and contribute to multiple ecosystem services, such as decomposition and nutrient mineralization, while serving as reliable indicators of soil health. Their ubiquitous presence and diverse feeding habits underline their ecological significance (Neher 2001).

Soil nematodes are grouped into five primary trophic categories depending on their feeding behavior: bacterivores, fungivores, herbivores, omnivores, and predators. Bacterivorous nematodes primarily consume bacteria, accelerating nutrient turnover and improving nitrogen availability. Fungivorous nematodes feed on fungal hyphae, influencing the dynamics of fungal populations and facilitating fungal-driven nutrient pathways. Although often perceived as agricultural pests, herbivorous nematodes shape plant communities and soil-plant interactions. Predatory nematodes regulate populations of other nematodes and soil fauna, thereby maintaining ecological balance. Omnivorous nematodes exhibit feeding flexibility, adapting to variable food resources and contributing to soil resilience (Ferris et al., 2001).

The role of nematodes in nutrient cycling is well-established. By consuming soil microbes, such as bacteria and fungi, nematodes release nitrogen and other nutrients in bioavailable forms through their metabolic waste. This process is particularly critical in low-input agricultural systems, where external nutrient inputs are limited. Bacterivorous nematodes are especially effective in enhancing nitrogen mineralization, a key process for plant growth. Studies have demonstrated that soils with active nematode populations often exhibit improved crop yields and greater nitrogen efficiency, showcasing their contribution to sustainable agriculture (Ingham et al., 1985).

Nematodes actively decompose organic matter by fragmenting organic debris and enhancing microbial colonization. This process increases the surface area of organic material, enabling soil microbes to access and decompose it more effectively. Decomposition facilitated by nematodes contributes to the formation of humus, a stable organic compound essential for soil fertility. Furthermore, nematode activity influences the soil's physical properties, like aggregate stability, water retention, and porosity, all of which are essential for plant growth and soil health (Maina et al., 2021).

The community component of soil nematodes reflects soil conditions and disturbances, making them reliable bioindicators of soil health. Metrics like the maturity index (MI) and enrichment index (EI) provide valuable insights into soil ecosystems' functional and structural integrity. For instance, soils subjected to organic management often show higher nematode diversity and a balanced trophic structure, whereas degraded or polluted soils exhibit shifts toward opportunistic species. The ability of nematodes to respond rapidly to environmental divergences underscores their importance in soil monitoring and management (Neher 2001).

Plant-parasitic nematodes, like root-knot nematodes (*Meloidogyne* spp.) and cyst nematodes (*Heterodera* spp.), pose significant challenges to global agriculture by damaging crops and reducing yields. However, effective management practices can mitigate their impact, including crop rotation, cover cropping, and nematode-resistant plant varieties. Simultaneously, free-living nematodes contribute positively to agricultural productivity by promoting nutrient availability and suppressing soil-borne pathogens. Integrating nematode management with sustainable farming practices can optimize their benefits while minimizing their drawbacks (Jones et al., 2013).

Nematodes exert a significant influence on soil microbial communities through their grazing activities. By consuming bacteria and fungi, nematodes regulate microbial biomass and community composition, fostering a dynamic equilibrium within the soil ecosystem. This interaction enhances the functional diversity of soil microbes, supporting processes including organic matter decomposition, pathogen suppression, and nutrient mobilization. Nematodes also stimulate microbial activity by releasing nutrient-rich excretions, creating feedback loops that enhance soil fertility (Jiang et al., 2018).

The impact of nematodes on carbon dynamics in the soil is both direct and indirect. Through their feeding and metabolic activities, nematodes facilitate the decomposition of organic materials, stabilizing soil organic matter. Stable organic matter is a long-term carbon reservoir, contributing to carbon sequestration and mitigating climate change. Additionally, nematodes influence the microbial-mediated processes that govern the soil's turnover and storage of carbon compounds. Their contribution to the soil's carbon economy highlights their importance in global carbon cycling (Brussaard et al., 2007).

The study of soil nematodes presents numerous challenges, primarily due to their microscopic size, vast diversity, and the complexity of soil environments. Traditional methods of nematode identification, such as morphological analysis, are time-consuming and require specialized expertise. Recent advancements in molecular biology, including HTS and DNA metabarcoding, have revolutionized nematode research, enabling rapid and accurate recognition of species and their functional roles. However, the integration of molecular data with ecological context remains an ongoing challenge for researchers (Elhady et al., 2024).

Climate change has profound implications for soil nematodes and their ecological functions. Rising temperatures, altered precipitation patterns, and rising CO_2 levels can influence nematode populations, community structures, and interactions with other soil organisms. For instance, shifts in temperature-sensitive species may impact on nutrient cycling and organic matter dynamics. Understanding the responses of nematodes to climate change is essential for predicting and managing their roles in future soil ecosystems (Dutta and Phani 2023).

Nematodes are indispensable for soil ecosystems, providing nutrient cycling, organic matter decomposition, and soil health assessment. Their interactions with soil microbes and their sensitivity to environmental changes highlight their ecological and agricultural significance. Continued research into nematode biology and ecology, particularly in climate change and sustainable land management, will enhance our capacity to harness their benefits for ecosystem stability and productivity (Pinto et al., 2024).

11.8.3 Rhizobacteria as Plant Growth Promoters

Rhizobacteria, a group of beneficial microorganisms living in the rhizosphere, play a significant role in plant growth and development. These bacteria form complex interactions with plant roots, contributing to nutrient acquisition, disease suppression, and stress tolerance. PGPR offers an eco-friendly alternative to chemical fertilizers and pesticides in sustainable agriculture. Their multifunctional abilities underline their importance in modern crop management practices (Glick 1995).

PGPRs are classified depending on their association with plant roots and their mechanisms of promoting plant growth. They can be free-living, symbiotic, or endophytic. PGPR promotes plant growth via direct mechanisms, including nitrogen fixation, phosphate solubilization, and phytohormone production, as well as indirect mechanisms as biocontrol of pathogens and stimulation of systemic resistance. These multifaceted interactions enhance crop productivity and resilience (Vessey 2003; Lugtenberg and Kamilova 2009).

1. *Nitrogen Fixation*: Nitrogen-fixing rhizobacteria, like *Rhizobium* and *Azospirillum* species, convert atmospheric nitrogen into ammonia, making it available to plants. This symbiotic action is critical for legumes and other crops, reducing dependency on synthetic nitrogen fertilizers (Zahran 1999).
2. *Phosphate Solubilization*: Phosphorus, often immobilized in soil, becomes available to plants via the action of phosphate-solubilizing bacteria like *Pseudomonas* and *Bacillus*. These bacteria release organic acids and enzymes to mobilize phosphorus, enhancing nutrient uptake (Rodríguez and Fraga 1999).
3. *Phytohormone Production*: PGPR synthesizes phytohormones as auxins, gibberellins, and cytokinins, which regulate root growth, cell division, and stress responses. For example, *Azospirillum brasilense* produces indole-3-acetic acid (IAA), stimulating root elongation and nutrient absorption (Spaepen et al., 2007).
4. *Biocontrol of Pathogens*: PGPR suppresses plant pathogens by forming antibiotics, siderophores, and hydrolytic enzymes. For instance, *Pseudomonas fluorescens* produces 2,4-diacetyl phloroglucinol (DAPG), which inhibits fungal growth and protects plants from root diseases (Raaijmakers et al., 2002; Weller et al., 2002).
5. *Induced Systemic Resistance (ISR)*: PGPR activates plants' immune responses, enhancing their ability to withstand pathogen attacks. This phenomenon, known as ISR, involves signaling pathways mediated by jasmonic acid and ethylene (Pieterse et al., 2003).

PGPR contributes significantly to sustainable agriculture by stimulating soil fertility, enhancing crop yields, and reducing environmental pollution. Their application as biofertilizers and biopesticides minimizes the utilization of chemical inputs, promoting eco-friendly farming practices (Adesemoye et al., 2009).

Abiotic stresses, like drought, salinity, and heavy metal toxicity, adversely affect crop productivity. PGPR alleviates these stresses by producing stress-related enzymes, osmolytes, and exopolysaccharides. For example, *Bacillus subtilis* enhances drought tolerance in plants by modulating stress-responsive genes(Patel et al., 2023). The commercial utilization of PGPR, like biofertilizers and biopesticides, has gained momentum due to advancements in microbial biotechnology. Products such as *Azotobacter*, *Rhizobium*, and *Pseudomonas*-based formulations are widely used in agriculture. However, challenges related to field efficacy, formulation stability, and large-scale production remain (Bashan et al., 2014).

Despite their potential, PGPR research faces challenges, including variability in field performance, environmental interactions, and strain-specific effects. Integrating molecular tools, including genomics and proteomics, can enhance our understanding of PGPR functions and their optimization for agricultural applications (Vejan et al., 2016). The integration of PGPR into precision agriculture and its use with advanced technologies like nanotechnology and genetic engineering holds promise for improving crop productivity and resilience. Research focusing on multi-strain consortia and their synergistic effects will further enhance their application potential (Vessey 2003).

11.9 ENGINEERING OF MICROBIOME RHIZOSPHERE IN ASSOCIATION WITH PLANT ROOTS

The rhizosphere is a small soil area directly impacted by root secretions and related microbial activity. It is critical in plant health, nutrient acquisition, and stress tolerance. It is also an essential location for plant–microbe interactions and nutrient cycling (Vives-Peris et al., 2020).

Through many methods, the rhizosphere microbiome substantially affects plant health. Beneficial microbes that stimulate disease resistance, improve nutrient uptake and enhance plant development include diazotrophic bacteria, mycorrhizal fungi, and PGPR (Mendes et al., 2013).

The plant genotype strongly influences the rhizosphere microbiome, by which specific root exudates shape the microbial communities. Plant breeding and genetic engineering provide support to optimize the microbiome by selecting plant traits to promote associations with beneficial microbes and to support beneficial microbial populations rather than directly changing microbial genomes. Traits like root architecture, exudate composition, and rhizospheric signaling molecules are crucial targets (Bulgarelli et al., 2013).

11.9.1 Redirecting the Rhizosphere Microbiome by Plant Breeding and Genetic Engineering

Plant breeding, genomics and transcriptomics, genetic engineering, and gene-editing technologies are being utilized to modify the rhizosphere microbial community precisely, optimizing plant–microbe interactions for improved agricultural sustainability and productivity.

11.9.1.1 Plant Breeding

Traditional plant breeding has evolved, focusing on production, disease resistance, and abiotic stress tolerance by selecting genotypes that promote beneficial rhizosphere microbial communities. Some studies focus on selecting plant genotypes that promote the formation of advantageous rhizosphere microbial communities. Root exudation enhances plant resource use and promotes communication between soil bacteria and plants, reducing stress via the release of primary and secondary metabolites (Ali et al., 2022). To enhance nutrient uptake, some plants emit root exudates that draw phosphorus-solubilizing fungi or nitrogen-fixing bacteria (Pérez-Jaramillo et al., 2016).

Breeding programs have incorporated traits that improve root architecture, various exudates, and microbial colonization. Researchers bred wheat varieties with altered root architecture, which influenced the composition of the rhizosphere microbiome and improved nutrient uptake (York et al., 2016). (Mendes, Raaijmakers, et al., 2018) breeding *Phaseolus vulgares* for *Fusarium oxysporum* resistance led to significant changes in rhizosphere microbiome composition depending on the susceptible and resistant cultivars. Breeding of wheat for root exudates enriched with organic acids, flavonoids or sugars promoted beneficial microbes like nitrogen-fixing bacteria or phosphate-solubilizing fungi (Hakim et al., 2021).

11.9.1.2 Plant Genetic Engineering and Gene Editing

Genetic engineering allows for precise interventions in plant genomes, enabling the redirection of microbial populations in their rhizosphere. Icoz and Stotzky (2008) also discovered that the insect resistance-Bt crops genetically modified to express insecticidal proteins influence the variety and functionality of soil microbial communities. Makonde (2009) reported that *Bt* maize exudates Bt-endotoxin in the soil, reducing the rhizosphere diversity of Bradyrhizobium and *Rhizobium* species (Makonde, 2009). Di Giovanni et al. (1999) detected differences between the rhizosphere bacterial communities of the parental genotype and the two transgenic alfalfa plants, one with the alpha-amylase genes and the other with lignin peroxidase genes, suggesting changes in root exudation patterns, which in turn impact on microbial populations in the rhizosphere. Liu et al. (2024) reported that overexpression of the phosphate starvation response 2 (*OsPHR2*) gene in rice had demonstrated an improvement in phosphorus uptake by changing root exudation patterns, which in turn recruit beneficial P-solubilizing bacteria like *Pseudomonas* (Liu et al., 2024).

The recruitment of advantageous rhizosphere microorganisms is made easier by the precise alteration of important genes in plant genomes, which is made possible by sophisticated genome editing techniques like CRISPR/Cas systems and zinc finger nucleases. These technologies have the potential to improve the microbes and plants interactions, which would increase plant productivity and health (Bano et al., 2021).

The knowledge of nodulation genes in legumes was advanced using CRISPR/Cas9 genome editing. The leghemogloin gene was edited in *Lotus japonicus* to reveal their cooperative functions in symbiotic nitrogen fixation (Wang et al., 2019). Lebedeva et al. (2023) used CRISPR/Cas9 to knock out the *MtCLE35* gene, which is upregulated by rhizobia and nitrate treatments, revealing its role in nodulation suppression. Also, CRISPR/Cas9 was applied to form knockout mutants of Nodule-specific Cysteine-Rich (*NCR*) genes, revealing their specific roles in nodule development of *Medicago truncatula* (Güngör et al., 2023).

11.9.1.3 Genomics and Transcriptomics

Modern molecular techniques, as quantitative trait locus (QTL) and genome-wide association (GWA) investigations, have been used to explore the possible genetic basis for host plant selection of the rhizosphere microbiome (Park et al., 2023). Mendes et al. (2013) studied three quantitative trait loci (QTLs) linked to *Bacillus cereus*-induced disease suppression in tomato inbred lines(Mendes et al., 2013). This investigation revealed that the favorable relationships between plants and rhizosphere bacteria could be strengthened using genetic heterogeneity in host plants. GWA and QTL mapping identified genomic regions linked to root exudates, architecture, and microbial recruitment. Microbial assembly was influenced by genes controlling the synthesis and release of phenolics, terpenoids, and flavonoids (Huang et al., 2019). Also, Deng et al. (2021) used GWA to identify genetic loci in *Sorghum* genotypes, manage colonization, and predict rhizosphere microbiome structure, even under various environmental conditions (Deng et al., 2021).

11.9.2 Redirecting the Rhizosphere Microbiome by Introducing or Stimulating Microorganisms

Plant development and health are greatly influenced by the rhizosphere microbiome, which comprises bacteria, fungi, and archaea. It mediates nutrient uptake, enhances soil structure, and offers defense against pathogens (Mendes et al., 2013; Philippot et al., 2024). Intensive farming methods, such as the overuse of chemical pesticides and fertilizers, have upset these microbial communities' natural equilibrium, forming soil deterioration and lower crop productions (Kumar and Dubey 2020).

Research indicates that introducing beneficial microorganisms to the rhizosphere microbiome can restore soil health, improve plant resilience, decrease the need for chemical inputs in agriculture, and improve nutrient uptake. Beneficial microorganisms are crucial to the sustainability of agricultural systems because they improve soil structure, increase nutrient absorption, and protect plants from disease (Asghar et al., 2024). Beneficial microbes benefit plants under salt stress situations by enhancing soil fertility and structure, rising water absorption, and improving nutrient uptake and usage efficiency (Jalal et al., 2023). Strategies to introduce or stimulate beneficial microorganisms in the rhizosphere can be through microbial bioinoculants and microbiome engineering.

11.9.2.1 Microbial Bioinoculants

This method involves introducing beneficial microbes into the rhizosphere. A wide range of microbes are beneficial to the plant, including nitrogen-fixing bacteria, endo- and ectomycorrhizal fungi, and PGPR (Raaijmakers et al., 2009).

Certain microorganisms enhance plant nutrient uptake (Richardson et al., 2009). Di Salvo et al. (2018) showed that *Azospirillum brasilense* inoculation improves plant development and nutrient intake by increasing wheat growth and changing rhizosphere microbial populations. PGPR bacteria, like *Pseudomonas fluorescens*, enhance plant growth by forming phytohormones and facilitating nutrient uptake, leading to enhance biomass and yield in plants (Berendsen et al., 2012). Certain PGPR can fix nitrogen from the atmosphere and transform it into forms that plants can easily absorb. Joshi et al. (2019) revealed that using bioinoculants *Rhodococcus qingshengii* S10107 and *Pseudomonas jesenii* MP1 significantly improved soil nitrogen status and increased the yield in chickpeas. do Amaral Leite et al. (2024) inoculating phosphate-solubilizing bacteria improves maize phosphorus uptake from magnesium-enriched poultry manure biochar, enhancing soil availability, growth, and yield.

Bioinoculants can cause systemic resistance in plants that can be used as candidates in disease resistance and bioremediation applications (Chaudhary and Shukla, 2019). Manipulating the rhizosphere microbiome can significantly reduce disease incidence by introducing beneficial microorganisms, such as specific microbial consortia, which can outcompete or inhibit pathogenic microorganisms for plant health (Gao et al., 2019). Bakker et al. (2007) revealed that introducing fluorescent *Pseudomonas* species into crop roots can ISR against pathogens (Bakker et al., 2007). Also, Micallef et al. (2009) inoculated wheat varieties with the biocontrol *Pseudomonas fluorescens* strain WCS365 to affect the component of the rhizosphere bacterial community and the plant's defense mechanisms. They found that divergence in microbial diversity and abundance resulted after the introduction of WCS365, which changed the rhizosphere's native bacterial communities. Inoculation of *Arabidopsis* by *Pseudomonas syringae* increases malic acid (MA) in the rhizosphere. In response to MA, beneficial rhizobacteria *Bacillus subtilis* rose, enhancing induced systematic resistance in the plant and limiting infection (Lakshmanan et al., 2012).

Microbes thrive in extreme environments, developing resistance to environmental insults and improper soil conditions, making them a sustainable way to boost crop yield and soil health (S. Ali et al., 2023). Numerous investigations have used different microbes to mitigate abiotic stresses by promoting certain growth traits. Plants that have PGPR with ACC deaminase are better able to handle stress and flourish in unfavorable conditions (Saleem et al., 2007). SkZ et al. (2018) reported that inoculation of maize seedlings with *Pseudomonas putida* strain FBKV2 stimulated maize seedling growth, survival rates, and relative water content under drought conditions. They also found that this improvement was associated with enhanced stress protein expression, accumulation of osmoprotectants, antioxidant defense, and better regulation of stress-responsive genes. Inoculating wheat plants with *Burkholderia phytofirmans* (Naveed et al., 2014), *Bacillus safensis*, and *Ochrobactrum pseudogregnonense* (Chakraborty et al., 2013) improves drought tolerance by increasing photosynthetic activity, soil moisture, nutrient and water absorption, and antioxidant activity. Also, inoculating *Arabidopsis* with *Azospirillum brasilense* improved its drought tolerance by raising ABA levels (Cohen et al., 2015).

Salt stress was reported to alter the microbial populations in plant portions that are above and below ground (Hou et al., 2021). Salt stress significantly changes the structure of the microbiome and stimulates the abundance of *Enterobacter* spp. (Yaish et al., 2016). Salt stress reduced the amount of Actinobacteria and Chloroflexi in the groundnut rhizosphere while increasing the relative richness of Acidobacteria and Cyanobacteria (Xu et al., 2020).

Biofertilizers containing living microbes can stimulate the availability of primary nutrients, which can reroute the rhizosphere microbiota. Beneficial native microbes can flourish by changing farming techniques. Saharan et al. (2018) found that the presence of "biofertilization" with arbuscular mycorrhizal fungi (*Rhizophagus irregularis* strain BEG 75) and PGPR (*Pseudomonas fluorescens* R62 and R81) helps finger millet, a shallow-rooted cereal, to use deep-rooted neighboring pigeon pea and have better growth under drought conditions (Saharan et al., 2018). Qiao et al. (2019) utilized bio-organic fertilizer supplemented with *Trichoderma guizhouense* NJAU 4742 to boost the number of potentially helpful microbial groups in the rhizosphere, like the genera *Massilia* and *Cladorrhinum*, and maize and cabbage yields by studying their rotation systems.

11.9.2.2 Microbiome Engineering

This technique combines microbial genetic engineering to create chemicals that support advantageous plant-microbial interactions. Rodríguez-Salazar et al. (2009) genetically engineered *Azospirillum brasilense* to improve its stress tolerance by enhancing trehalose production. Inoculating it to maize showed improved drought resistance and growth performance compared to wild-type strains.

11.9.3 Transgenic Plants and Their Associated Rhizosphere

Genetic engineering can optimize rhizosphere interactions by precisely altering root traits. Transgenic plants with root-specific modifications can impact the properties and performance of above-ground plants through internal and rhizosphere-mediated effects. The microbiome of transgenic plants can be actively shaped, impacting microbial diversity and soil health. In order to maintain ecological balance and biosafety, it is crucial to comprehend how genetic alterations impact rhizosphere populations (Yang et al., 2024). Few researchers studied the impact of transgenic plants on altering the rhizosphere microbiota by changing or shifting their root exudation pattern, thereby reshaping the microbiome.

Overexpressing *ALMT1* (*aluminum-activated malate transporter 1*) genes in *Arabidopsis thaliana* that produce malic acid increases malic acid efflux and the microbial recruitment of *Bacillus subtilis*. This interaction also enhances plant resistance against pathogens like *Pseudomonas syringae* (Rudrappa et al., 2008). Transgenic potato plants expressing a lactonase gene demonstrated enhanced resistance to the bacterial pathogen *Pectobacterium carotovorum* by disrupting its communication mechanisms (Mendes et al., 2013). In addition, drought-resistant transgenic rice modified the profiles of root exudates, which in turn encourages advantageous microbial interactions (Lakshmanan et al., 2014). Geddes et al. (2019) genetically modified *Medicago truncatula* and *Hordeum vulgare* to produce rhizomes, which exude to promote beneficial microbes and so affect the microbial community composition.

Wei et al. (2021) found that intercropping the drought-tolerant Ea-DREB2B transgenic sugarcane with wild-type soybean established a beneficial rhizosphere environment that favored the growth of Actinobacteria, which positively affects plant health. Huang et al. (2024) studied the diversity and structure of rhizosphere microbial communities present with Bt-transgenic rice expressing the *cry1Ab/cry1Ac* genes (Huahui No. 1) and its non-transgenic parental line (Minghui63). They found shifts in the microbial composition in Bt rice, suggesting its impact on the diversity and structure of rhizosphere microbial communities. The CRISPR–Cas 9 editing of the *SlHyPRP1* gene in tomato enhanced multi-stress tolerance. These genetic changes in plants affect root exudation patterns, which can change rhizosphere microbial dynamics and increase plant survival in harsh environments (Tran et al., 2023).

Researchers have utilized CRISPR–Cas9 to edit genes implicated in secondary metabolites such as phenolics and carotenoids, which are crucial for plant defense and interactions with soil microbes. This modification can alter root exudation profiles, affecting the rhizosphere microbiome structure and function, thereby improving plant health and growth (Sarma, 2024).

Zhong et al. (2022) performed hairy root transformation in melon utilizing CRISPR-mediated gene editing and *Agrobacterium rhizogenes*-mediated to produce cucurbitacin B. They found that this root-secreted metabolite selectively enriches the rhizosphere with *Bacillus* and *Enterobacter*, which conferred resistance against the soilborne fungal pathogen *Fusarium oxysporum*. Their findings provided relevant evidence of the targeted manipulation of plant genomes for enhancing plant fitness by indirectly influencing the rhizosphere microbiome composition.

11.9.4 Interaction between Genetic Engineering Microbes and Transgenic Plants

The relationship between transgenic plants (TPs) and genetically modified microbes (GEMs) is a frontier in environmental science and agricultural biotechnology with enormous promises and substantial obstacles. Understanding and controlling their intricate interactions is the key to future investigations, even though present research frequently concentrates on changing only one partner at a time. GEM-TP interactions offer potential for sustainable agriculture but require ongoing research, interdisciplinary collaboration, and ethical considerations. By combining GE and multi-omics technologies with GEM-TP interactions, scientists can tackle environmental challenges such as (1) developing specialized GE microbes that can interact with specific transgenic plant traits, producing phytohormones, antimicrobial compounds, and stress-responsive metabolites to enhance growth, stress tolerance, and disease resistance. (2) Synthetic biology and microbial engineering can develop smart bioinoculants that respond dynamically to environmental cues, providing real-time adaptive support for transgenic crops. (3) GEM-TP interactions can optimize nutrient uptake and utilization, decreasing chemical fertilizer dependency and aligning with sustainable agricultural practices through improved nitrogen fixation, phosphorus solubilization, and iron chelation. (4) GEMs can boost stress tolerance in transgenic plants, ensuring drought, salinity, and heat stress mitigation and stable crop yields under challenging environmental conditions.

11.10 CONCLUSION

Rhizosphere microbiome is an active, complex system that is essential for plant well-being, nutrient uptake, and stress tolerance. Enhanced multi-omics technology has shed more light on microbial diversity, functional pathways, and microbe–plant interactions, allowing for the identification of beneficial taxa and their roles in nutrient cycling and plant defense. Root exudates are a critical regulation process, influencing microbial community structure and function and mediating plant adaptation to environmental stress.

Plant growth-promoting microbes (PGPMs) such as bacteria, fungi, and oomycetes, when harnessed, hold powerful means of augmenting nutrient uptake, stress tolerance, and disease management. Likewise, microbiome engineering through bioinoculants, host plant breeding, and gene manipulation presents a new horizon for sustainable agriculture. The novel applications of CRISPR–Cas technologies and synthetic biology present new horizons for enhancing both microbial and plant genomes, which can be used to build resilient agroecosystems.

Future research will need to pay attention to the incorporation of omics results with ecological and agronomic understanding to translate laboratory findings to field outcomes. Ecological safety, microbial persistence, and host–microbe co-adaptation will need careful attention for effective use. Shaping the rhizosphere microbiome, combined with transgenic plants and bioinoculants, provides a sustainable pathway to reduce chemical inputs, enhance crop productivity, and enhance food security against climate change.

REFERENCES

Abbaszadeh-Dahaji P, Baniasad-Asgari A, Hamidpour M. 2019. The effect of Cu-resistant plant growth-promoting rhizobacteria and EDTA on phytoremediation efficiency of plants in a Cu-contaminated soil. Environ Sci Pollut Res. 26(31):31822–31833.

Abdul Hamid NW, Nadarajah K. 2022. Microbe related chemical signalling and its application in agriculture. Int J Mol Sci. 23(16):8998.

Abdul Rahman NSN, Abdul Hamid NW, Nadarajah K. 2021. Effects of abiotic stress on soil microbiome. Int J Mol Sci. 22(16):9036.

Adams A, Landry D, Sykes V, Rickman T, Cham AK, Timling A, Kelly H, McBeath JH, Olukolu BA. 2024. Maize Kernel-associated metagenomes reveal potential microbe–microbe interactions that underlie fusarium ear rot disease. phytobiomes J.:PBIOMES–07.

Adesemoye AO, Torbert HA, Kloepper JW. 2009. Plant growth-promoting rhizobacteria allow reduced application rates of chemical fertilizers. Microb Ecol. 58:921–929.

Afridi MS, Ali S, Salam A, César Terra W, Hafeez A, Sumaira, Ali B, S. AlTami M, Ameen F, Ercisli S, et al. 2022. Plant microbiome engineering: Hopes or hypes. Biology (Basel). 11(12).

Ait-El-Mokhtar M, Meddich A, Baslam M. 2023. Plant-microbiome interactions under drought—insights from the molecular machinist's toolbox. Front Sustain Food Syst. 7:1253735.

Ali MH, Mandal S, Ghorai M, Lal MK, Tiwari RK, Kumar M, Ghosh A, Al-Tawaha AR, Gopalakrishnan AV, Shekhawat MS. 2023. Perspectives of omics and plant microbiome. In: Genomics, Transcr Proteomics Metabolomics Crop Plants. [place unknown]: Elsevier; p. 131–144.

Ali Q, Khan AR, Raza W, Bilal MS, Khalid S, Ayaz M, Khan A, Mundra S. 2025. Mechanisms of microbial VOC-mediated communication in plant ecosystems and agricultural applications. J Sustain Agric Environ. 4(1):e70044.

Ali S, Tyagi A, Mir RA, Rather IA, Anwar Y, Mahmoudi H. 2023. Plant beneficial microbiome a boon for improving multiple stress tolerance in plants. Front Plant Sci. 14:1266182.

Ali S, Tyagi A, Park S, Mir RA, Mushtaq M, Bhat B, Mahmoudi H, Bae H. 2022. Deciphering the plant microbiome to improve drought tolerance: Mechanisms and perspectives. Environ Exp Bot. 201:104933.

Aliche EB, Screpanti C, De Mesmaeker A, Munnik T, Bouwmeester HJ. 2020. Science and application of strigolactones. New Phytol. 227(4):1001–1011.

Allen R, Rivas-Ubach A, Borkum MI, Köberl M, Bilbao A, Colby SM, Hoyt DW, Bingol K, Kim Y-M, Wendler JP, et al. 2017. Rhizosphere the state of rhizospheric science in the era of multi-omics: A practical guide to omics technologies. [place unknown].

Alseekh S, Aharoni A, Brotman Y, Contrepois K, D'Auria J, Ewald J, C. Ewald J, Fraser PD, Giavalisco P, Hall RD. 2021. Mass spectrometry-based metabolomics: A guide for annotation, quantification and best reporting practices. Nat Methods. 18(7):747–756.

Alshareef SA. 2024. Metabolic analysis of the CAZy class glycosyltransferases in rhizospheric soil fungiome of the plant species Moringa oleifera. Saudi J Biol Sci. 31(4):103956.

Andargie YE, Lee G, Jeong M, Tagele SB, Shin J-H. 2023. Deciphering key factors in pathogen-suppressive microbiome assembly in the rhizosphere. Front Plant Sci. 14:1301698.

Arun KB, Madhavan A, Sindhu R, Emmanual S, Binod P, Pugazhendhi A, Sirohi R, Reshmy R, Awasthi MK, Gnansounou E. 2021. Probiotics and gut microbiome–Prospects and challenges in remediating heavy metal toxicity. J Hazard Mater. 420:126676.

Asghar W, Craven KD, Swenson JR, Kataoka R, Mahmood A, Farias JG. 2024. Enhancing the resilience of agroecosystems through improved rhizosphere processes: A strategic review. Int J Mol Sci. 26(1):109.

Azarias Guimarães A, Duque Jaramillo PM, Simão Abrahão Nóbrega R, Florentino LA, Barroso Silva K, de Souza Moreira FM. 2012. Genetic and symbiotic diversity of nitrogen-fixing bacteria isolated from agricultural soils in the western Amazon by using cowpea as the trap plant. Appl Environ Microbiol. 78(18):6726–6733.

Badri DV, Vivanco JM. 2009. Regulation and function of root exudates. Plant Cell Environ. 32(6):666–681.

Baetz U, Martinoia E. 2014. Root exudates: The hidden part of plant defense. Trends Plant Sci. 19(2):90–98.

Bais HP, Weir TL, Perry LG, Gilroy S, Vivanco JM. 2006. The role of root exudates in rhizosphere interactions with plants and other organisms. Annu Rev Plant Biol. 57(1):233–266.

Bakker PAHM, Pieterse CMJ, Van Loon LC. 2007. Induced systemic resistance by fluorescent Pseudomonas spp. Phytopathology. 97(2):239–243.

Bano S, Wu X, Zhang X. 2021. Towards sustainable agriculture: rhizosphere microbiome engineering. Appl Microbiol Biotechnol.1–20.

Bargaz A, Lyamlouli K, Chtouki M, Zeroual Y, Dhiba D. 2018. Soil microbial resources for improving fertilizers efficiency in an integrated plant nutrient management system. Front Microbiol. 9:1606.

Bashan Y, de-Bashan LE, Prabhu SR, Hernandez J-P. 2014. Advances in plant growth-promoting bacterial inoculant technology: formulations and practical perspectives (1998–2013). Plant Soil. 378:1–33.

Bělonožníková K, Hýsková V, Chmelík J, Kavan D, Čeřovská N, Ryšlavá H. 2022. Pythium oligandrum in plant protection and growth promotion: Secretion of hydrolytic enzymes, elicitors and tryptamine as auxin precursor. Microbiol Res. 258:126976.

Benítez T, Rincón AM, Limón MC, Codon AC. 2004. Biocontrol mechanisms of *Trichoderma* strains. Int Microbiol. 7(4):249–260.

Bentley DR, Balasubramanian S, Swerdlow HP, Smith GP, Milton J, Brown CG, Hall KP, Evers DJ, Barnes CL, Bignell HR. 2008. Accurate whole human genome sequencing using reversible terminator chemistry. Nature. 456(7218):53–59.

Berendsen RL, Pieterse CMJ, Bakker PAHM. 2012. The rhizosphere microbiome and plant health. Trends Plant Sci. 17(8):478–486.

Berendsen RL, Vismans G, Yu K, Song Y, de Jonge R, Burgman WP, Burmølle M, Herschend J, Bakker PAHM, Pieterse CMJ. 2018. Disease-induced assemblage of a plant-beneficial bacterial consortium. ISME J. 12(6):1496–1507.

Beskrovnaya P, Sexton DL, Golmohammadzadeh M, Hashimi A, Tocheva EI. 2021. Structural, metabolic and evolutionary comparison of bacterial endospore and exospore formation. Front Microbiol. 12:630573.

Bhaduri D, Pal S, Purakayastha TJ, Chakraborty K, Yadav RS, Akhtar MS. 2015. Soil quality and plant-microbe interactions in the rhizosphere. Sustain Agric Rev. 17:307–335.

Bhimani P, Mahavar P, Rajguru B, Bhatt VD, Nathani NM, Shri M. 2024. Unveiling the green dialogue: Advancements in omics technologies for deciphering plant–microbe interactions in soil. Discov Plants. 1(1).

Birhanu AG. 2023. Mass spectrometry-based proteomics as an emerging tool in clinical laboratories. Clin Proteomics. 20(1):32.

Biswas B, Gresshoff PM. 2014. The role of symbiotic nitrogen fixation in sustainable production of biofuels. Int J Mol Sci. 15(5):7380–7397.

Blakney AJC, Bainard LD, St-Arnaud M, Hijri M. 2023. Soil chemistry and soil history significantly structure oomycete communities in Brassicaceae crop rotations. Appl Environ Microbiol. 89(1):e01314–e01322.

Bledsoe RB, Goodwillie C, Peralta AL. 2020. Long-term nutrient enrichment of an oligotroph-dominated wetland increases bacterial diversity in bulk soils and plant rhizospheres. Msphere. 5(3):10–1128.

Boyetchko SM, Kiss J, Delos M, Grant W, Gwynn R, Quesada-Moraga E, Yousef-Naef M, Garrido-Jurado I, Rao T, Jurat-Fuentes JL. 2020. Biopesticides for Sustainable Agriculture. [place unknown]: Burleigh Dodds Science Publishing.

Brussaard L, De Ruiter PC, Brown GG. 2007. Soil biodiversity for agricultural sustainability. Agric Ecosyst Environ. 121(3):233–244.

Bulgarelli D, Rott M, Schlaeppi K, Ver Loren van Themaat E, Ahmadinejad N, Assenza F, Rauf P, Huettel B, Reinhardt R, Schmelzer E. 2012. Revealing structure and assembly cues for *Arabidopsis* root-inhabiting bacterial microbiota. Nature. 488(7409):91–95.

Bulgarelli D, Schlaeppi K, Spaepen S, Van Themaat EVL, Schulze-Lefert P. 2013. Structure and functions of the bacterial microbiota of plants. Annu Rev Plant Biol. 64(1):807–838.

Canarini A, Kaiser C, Merchant A, Richter A, Wanek W. 2019. Root exudation of primary metabolites: Mechanisms and their roles in plant responses to environmental stimuli. Front Plant Sci. 10:157.

Carvalhais LC, Rincon-Florez VA, Brewer PB, Beveridge CA, Dennis PG, Schenk PM. 2019. The ability of plants to produce strigolactones affects rhizosphere community composition of fungi but not bacteria. Rhizosphere. 9:18–26.

Castañeda-Gómez L, Walker JKM, Powell JR, Ellsworth DS, Pendall E, Carrillo Y. 2020. Impacts of elevated carbon dioxide on carbon gains and losses from soil and associated microbes in a Eucalyptus woodland. Soil Biol Biochem. 143:107734.

Castrillo G, Teixeira PJPL, Paredes SH, Law TF, De Lorenzo L, Feltcher ME, Finkel OM, Breakfield NW, Mieczkowski P, Jones CD. 2017. Root microbiota drive direct integration of phosphate stress and immunity. Nature. 543(7646):513–518.

Cesco S, Mimmo T, Tonon G, Tomasi N, Pinton R, Terzano R, Neumann G, Weisskopf L, Renella G, Landi L. 2012. Plant-borne flavonoids released into the rhizosphere: Impact on soil bio-activities related to plant nutrition: A review. Biol Fertil Soils. 48:123–149.

Chakraborty U, Chakraborty BN, Chakraborty AP, Dey PL. 2013. Water stress amelioration and plant growth promotion in wheat plants by osmotic stress tolerant bacteria. World J Microbiol Biotechnol. 29:789–803.

Chamkhi I, El Omari N, Balahbib A, El Menyiy N, Benali T, Ghoulam C. 2022. Is——The rhizosphere a source of applicable multi-beneficial microorganisms for plant enhancement? Saudi J Biol Sci. 29(2):1246–1259.

Chandok IK, Afreen H, Afreen R, Haider S, Moharana DP, Hussain T, Aamir M, Shanmugam V, Malik MZ. 2022. Functional genomics tools for studying microbe-mediated stress tolerance in plants. In: Mitig Plant Abiotic Stress by Microorg. [place unknown]: Elsevier; p. 175–204.

Chandra P, Rai AK, Sundha P, Basak N, Kaur H. 2022. Rhizospheric soil–plant-microbial interactions for abiotic stress mitigation and enhancing crop performance. In: Soil Heal Environ Sustain Appl geospatial Technol. [place unknown]: Springer; p. 593–614.

Chaudhary T, Shukla P. 2019. Bioinoculants for bioremediation applications and disease resistance: innovative perspectives. Indian J Microbiol. 59:129–136.

Chen M, Arato M, Borghi L, Nouri E, Reinhardt D. 2018. Beneficial services of arbuscular mycorrhizal fungi–from ecology to application. Front Plant Sci. 9:1270.

Chen X, Chen H, Yuan JS, Köllner TG, Chen Y, Guo Y, Zhuang X, Chen Xinlu, Zhang Y, Fu J. 2018. The rice terpene synthase gene Os TPS 19 functions as an (S)-limonene synthase in planta, and its overexpression leads to enhanced resistance to the blast fungus Magnaporthe oryzae. Plant Biotechnol J. 16(10):1778–1787.

Chen X, Sun M, Chong S, Si J, Wu L. 2022. Transcriptomic and metabolomic approaches deepen our knowledge of plant–endophyte interactions. Front Plant Sci. 12:700200.

Chenu C, Cosentino D. 2011. Microbial regulation of soil structural dynamics. In: Archit Biol soils life Inn Sp. [place unknown]: CABI Wallingford UK; p. 37–70.

Chernin L, Toklikishvili N, Ovadis M, Kim S, Ben-Ari J, Khmel I, Vainstein A. 2011. Quorum-sensing quenching by rhizobacterial volatiles. Environ Microbiol Rep. 3(6):698–704.

Cohen AC, Bottini R, Pontin M, Berli FJ, Moreno D, Boccanlandro H, Travaglia CN, Piccoli PN. 2015. *Azospirillum brasilense* ameliorates the response of *Arabidopsis thaliana* to drought mainly via enhancement of ABA levels. Physiol Plant. 153(1):79–90.

Contreras-Cornejo HA, Macías-Rodríguez L, Del-Val EK, Larsen J. 2016. Ecological functions of *Trichoderma* spp. and their secondary metabolites in the rhizosphere: Interactions with plants. FEMS Microbiol Ecol. 92(4):fiw036.

Contreras-Cornejo HA, Schmoll M, Esquivel-Ayala BA, González-Esquivel CE, Rocha-Ramírez V, Larsen J. 2024. Mechanisms for plant growth promotion activated by *Trichoderma* in natural and managed terrestrial ecosystem. Microbiol Res.:127621.

Crandall SG, Gold KM, Jiménez-Gasco M del M, Filgueiras CC, Willett DS. 2020. A multi-omics approach to solving problems in plant disease ecology. PLoS One. 15(9):e0237975.

Dai Y, Liu R, Chen J, Li N. 2022. Bioremediation of HMW-PAHs-contaminated soils by rhizosphere microbial community of fire phoenix plants. Chem Eng J. 432:134246. https://doi.org/10.1016/j.cej.2021.134246

Dalcin Martins P, Hoyt DW, Bansal S, Mills CT, Tfaily M, Tangen BA, Finocchiaro RG, Johnston MD, McAdams BC, Solensky MJ. 2017. Abundant carbon substrates drive extremely high sulfate reduction rates and methane fluxes in Prairie Pothole Wetlands. Glob Chang Biol. 23(8):3107–3120.

de Bruijn FJ. 2013. Molecular Microbial Ecology of the Rhizosphere. John Wiley & Sons, New Jersey, USA: Wiley Online Library.

De Faria MR, Simara L, Soares Costa A, Chiaramonte JB, Bettiol W, Mendes R. 2021. The rhizosphere microbiome: functions, dynamics, and role in plant protection. Trop Plant Pathol. 46:13–25.

Denardin LG de O, Alves LA, Ortigara C, Winck B, Coblinski JA, Schmidt MR, Carlos FS, Toni CAG de, Camargo FA de O, Anghinoni I. 2020. How different soil moisture levels affect the microbial activity. Ciência Rural. 50:e20190831.

Deng S, Caddell DF, Xu G, Dahlen L, Washington L, Yang J, Coleman-Derr D. 2021. Genome wide association study reveals plant loci controlling heritability of the rhizosphere microbiome. ISME J. 15(11):3181–3194.

De Vos RCH, Moco S, Lommen A, Keurentjes JJB, Bino RJ, Hall RD. 2007. Untargeted large-scale plant metabolomics using liquid chromatography coupled to mass spectrometry. Nat Protoc. 2(4):778–791.

Di Giovanni GD, Watrud LS, Seidler RJ, Widmer F. 1999. Comparison of parental and transgenic alfalfa rhizosphere bacterial communities using Biolog GN metabolic fingerprinting and enterobacterial repetitive intergenic consensus sequence-PCR (ERIC-PCR). Microb Ecol. 37:129–139.

Dijkstra FA, Carrillo Y, Pendall E, Morgan JA. 2013. Rhizosphere priming: A nutrient perspective. Front Microbiol. 4:216.

Di Salvo LP, Ferrando L, Fernández-Scavino A, García de Salamone IE. 2018. Microorganisms reveal what plants do not: wheat growth and rhizosphere microbial communities after *Azospirillum brasilense* inoculation and nitrogen fertilization under field conditions. Plant Soil. 424:405–417.

Diwan D, Rashid MM, Vaishnav A. 2022. Current understanding of plant-microbe interaction through the lenses of multi-omics approaches and their benefits in sustainable agriculture. Microbiol Res. 265:127180.

do Amaral Leite A, de Souza Cardoso AA, de Almeida Leite R, Barrera AMV, Queiroz DDL, Viana TC, de Oliveira-Longatti SM, Silva CA, de Souza Moreira FM, Lehmann J. 2024. Phosphate-solubilizing bacteria increase maize phosphorus uptake from magnesium-enriched poultry manure biochar. Biol Fertil Soils. 60(3):421–436.

Doddavarapu B, Lata C, Shah JM. 2024. Epigenetic regulation influenced by soil microbiota and nutrients: Paving road to epigenome editing in plants. Biochim Biophys Acta (BBA)-General Subj.:130580.

Dubey RK, Tripathi V, Prabha R, Chaurasia R, Singh DP, Rao CS, El-Keblawy A, Abhilash PC, Dubey RK, Tripathi V. 2020. Metatranscriptomics and metaproteomics for microbial communities profiling. Unravelling Soil Microbiome Perspect Environ Sustain.:51–60.

Dutta TK, Phani V. 2023. The pervasive impact of global climate change on plant-nematode interaction continuum. Front Plant Sci. 14:1143889.

Ehrmann J, Ritz K. 2014. Plant: Soil interactions in temperate multi-cropping production systems. Plant Soil. 376:1–29.

Eichmann R, Richards L, Schäfer P. 2021. Hormones as go-betweens in plant microbiome assembly. Plant J. 105(2):518–541.

El-ghfar MHAA, Ibrahim HM, Hassan IM, Abdel Fattah AA, Mahmoud MH. 2016. Peels of lemon and orange as value-added ingredients: Chemical and antioxidant properties.

Int J Curr Microbiol Appl Sci. 5(12):777–794. https://doi.org/10.20546/ijcmas.2016.512.089
Elhady A, Alghanmi L, Abd-Elgawad MMM, Heuer H, Saad MM, Hirt H. 2024. Plant-parasitic nematode research in the arid desert landscape: A systematic review of challenges and bridging interventions. Front Plant Sci. 15:1432311.
Emamverdian A, Ding Y, Mokhberdoran F, Xie Y. 2015. Heavy metal stress and some mechanisms of plant defense response. Sci World J. 2015:756120.
Etesami H, Beattie GA. 2018. Mining halophytes for plant growth-promoting halotolerant bacteria to enhance the salinity tolerance of non-halophytic crops. Front Microbiol. 9:148.
Ezawa T, Saito K. 2018. How do arbuscular mycorrhizal fungi handle phosphate? New insight into fine-tuning of phosphate metabolism. New Phytol. 220(4):1116–1121.
Feng H, Fu R, Hou X, Lv Y, Zhang N, Liu Y, Xu Z, Miao Y, Krell T, Shen Q, Zhang R. 2021. Chemotaxis of beneficial rhizobacteria to root exudates: The first step towards root–microbe rhizosphere interactions. Int J Mol Sci. 22(13). https://doi.org/10.3390/ijms22136655
Fernandes MLP, Bastida F, Jehmlich N, Martinović T, Větrovský T, Baldrian P, Delgado-Baquerizo M, Starke R. 2022. Functional soil mycobiome across ecosystems. J Proteomics. 252:104428.
Ferris H, Bongers T, de Goede RGM. 2001. A framework for soil food web diagnostics: Extension of the nematode faunal analysis concept. Appl soil Ecol. 18(1):13–29.
Fierer N. 2017. Embracing the unknown: Disentangling the complexities of the soil microbiome. Nat Rev Microbiol. 15(10):579–590.
Finkel OM, Salas-González I, Castrillo G, Spaepen S, Law TF, Teixeira PJPL, Jones CD, Dangl JL. 2019. The effects of soil phosphorus content on plant microbiota are driven by the plant phosphate starvation response. PLoS Biol. 17(11):e3000534.
Foster KR, Bell T. 2012. Competition, not cooperation, dominates interactions among culturable microbial species. Curr Biol. 22(19):1845–1850.
Frąc M, Hannula SE, Bełka M, Jędryczka M. 2018. Fungal biodiversity and their role in soil health. Front Microbiol. 9:707.
Furtak K, Grządziel J, Gałązka A, Niedźwiecki J. 2020. Prevalence of unclassified bacteria in the soil bacterial community from floodplain meadows (fluvisols) under simulated flood conditions revealed by a metataxonomic approachss. Catena. 188:104448.
Gahagan AC, Shi Y, Radford D, Morrison MJ, Gregorich E, Aris-Brosou S, Chen W. 2023. Long-term tillage and crop rotation regimes reshape soil-borne Oomycete communities in soybean, corn, and wheat production systems. Plants. 12(12):2338.
Galli M, Feldmann F, Vogler UK, Kogel K-H. 2024. Can biocontrol be the game-changer in integrated pest management? A review of definitions, methods and strategies. J Plant Dis Prot. 131(2):265–291.
Galloway-Peña J, Hanson B. 2020. Tools for analysis of the microbiome. Dig Dis Sci. 65(3):674–685.
Gamalero E, Bona E, Glick BR. 2022. Current techniques to study beneficial plant-microbe interactions. Microorganisms. 10(7):1380.
Gao C, Montoya L, Xu L, Madera M, Hollingsworth J, Purdom E, Hutmacher RB, Dahlberg JA, Coleman-Derr D, Lemaux PG. 2019. Strong succession in arbuscular mycorrhizal fungal communities. ISME J. 13(1):214–226.
Geddes BA, Paramasivan P, Joffrin A, Thompson AL, Christensen K, Jorrin B, Brett P, Conway SJ, Oldroyd GED, Poole PS. 2019. Engineering transkingdom signalling in plants to control gene expression in rhizosphere bacteria. Nat Commun. 10(1):3430.
Gkarmiri K, Mahmood S, Ekblad A, Alström S, Högberg N, Finlay R. 2017. Identifying the active microbiome associated with roots and rhizosphere soil of oilseed rape. Appl Environ Microbiol. 83(22).
Glick BR. 1995. The enhancement of plant growth by free-living bacteria. Can J Microbiol. 41(2):109–117.
Gonin M, Salas-González I, Gopaulchan D, Frene JP, Roden S, Van de Poel B, Salt DE, Castrillo G. 2023. Plant microbiota controls an alternative root branching regulatory mechanism in plants. Proc Natl Acad Sci. 120(15): e2301054120.
Goyal A, Kalia A. 2020. Fungal phytohormones: Plant growth-regulating substances and their applications in crop productivity. Agric Important Fungi Sustain Agric Vol 1 Perspect Divers Crop Product.:143–169.
Goyal RK, Habtewold JZ. 2023. Evaluation of legume–Rhizobial symbiotic interactions beyond nitrogen fixation that help the host survival and diversification in hostile environments. Microorganisms. 11(6):1454.
Guan N, Liu L. 2020. Microbial response to acid stress: mechanisms and applications. Appl Microbiol Biotechnol. 104(1):51–65.
Gullberg J, Jonsson P, Nordström A, Sjöström M, Moritz T. 2004. Design of experiments: An efficient strategy to identify factors influencing extraction and derivatization of *Arabidopsis thaliana* samples in metabolomic studies with gas chromatography/mass spectrometry. Anal Biochem. 331(2):283–295.
Güngör B, Biró JB, Domonkos Á, Horváth B, Kaló P. 2023. Targeted mutagenesis of Medicago truncatula Nodule-specific Cysteine-Rich (NCR) genes using the Agrobacterium rhizogenes-mediated CRISPR/Cas9 system. Sci Rep. 13(1):20676.
Hacquard S, Spaepen S, Garrido-Oter R, Schulze-Lefert P. 2017. Interplay between innate immunity and the plant microbiota. Annu Rev Phytopathol. 55(1):565–589.
Hakim S, Naqqash T, Nawaz MS, Laraib I, Siddique MJ, Zia R, Mirza MS, Imran A. 2021. Rhizosphere engineering with plant growth-promoting microorganisms for agriculture and ecological sustainability. Front Sustain Food Syst. 5:617157.
Han S, Na L, Rongchao Z, Xiuqin H, Wenyu Z, Bo Z, Xinpeng L, Zhen W, Jie X. 2023. Study on signal transmission mechanism of arbuscular mycorrhizal hyphal network against root rot of Salvia miltiorrhiza. Sci Rep. 13(1):16936.
Haraguchi S, Yoshiga T. 2020. Potential of the fungal feeding nematode Aphelenchus avenae to control fungi and the plant parasitic nematode Ditylenchus destructor associated with garlic. Biol Control. 143:104203.
Harman S, Ashbee HR, Evans EG V. 2004. Testing of antifungal combinations against yeasts and dermatophytes. J Dermatolog Treat. 15(2):104–107.
Hart MM, Gorzelak M, Ragone D, Murch SJ. 2014. Arbuscular mycorrhizal fungal succession in a long-lived perennial. Botany. 92(4):313–320.

Hartman K, Tringe SG. 2019. Interactions between plants and soil shaping the root microbiome under abiotic stress. Biochem J. 476(19):2705–2724.

Hartmann M, Six J. 2023. Soil structure and microbiome functions in agroecosystems. Nat Rev Earth Environ. 4(1):4–18.

Hassan S, Mathesius U. 2012. The role of flavonoids in root–rhizosphere signalling: Opportunities and challenges for improving plant–microbe interactions. J Exp Bot. 63(9):3429–3444.

Hida A, Oku S, Miura M, Matsuda H, Tajima T, Kato J. 2020. Characterization of Methyl-accepting Chemotaxis Proteins (MCPs) for amino acids in plant-growth-promoting rhizobacterium Pseudomonas protegens CHA0 and enhancement of amino acid chemotaxis by MCP genes overexpression. Biosci Biotechnol Biochem. 84(9):1948–1957.

Hirsch PR, Miller AJ, Dennis PG. 2013. Do root exudates exert more influence on rhizosphere bacterial community structure than other rhizodeposits? Mol Microb Ecol Rhizosph. 1:229–242.

Holík L, Hlisnikovský L, Honzík R, Trögl J, Burdová H, Popelka J. 2019. Soil microbial communities and enzyme activities after long-term application of inorganic and organic fertilizers at different depths of the soil profile. Sustainability. 11(12):3251.

Horton MW, Bodenhausen N, Beilsmith K, Meng D, Muegge BD, Subramanian S, Vetter MM, Vilhjálmsson BJ, Nordborg M, Gordon JI. 2014. Genome-wide association study of *Arabidopsis thaliana* leaf microbial community. Nat Commun. 5(1):5320.

Hou S, Wolinska KW, Hacquard S. 2021. Microbiota-root-shoot-environment axis and stress tolerance in plants. Curr Opin Plant Biol. 62:102028.

Hu L, Robert CAM, Cadot S, Zhang XI, Ye M, Li B, Manzo D, Chervet N, Steinger T, Van Der Heijden MGA. 2018. Root exudate metabolites drive plant-soil feedbacks on growth and defense by shaping the rhizosphere microbiota. Nat Commun. 9(1):2738.

Hua C, Wang Yonglin, Zheng X, Dou D, Zhang Z, Govers F, Wang Yuanchao. 2008. A Phytophthora sojae G-protein α subunit is involved in chemotaxis to soybean isoflavones. Eukaryot Cell. 7(12):2133–2140.

Huang AC, Jiang T, Liu Y-X, Bai Y-C, Reed J, Qu B, Goossens A, Nützmann H-W, Bai Y, Osbourn A. 2019. A specialized metabolic network selectively modulates *Arabidopsis* root microbiota. Science (80-). 364(6440):eaau6389.

Huang Q, Zhang Y, Tan Y, Kong H, Cao Y, Wang J, Yin G, Guo A. 2024. Bt-Modified transgenic rice may shift the composition and diversity of rhizosphere microbiota. Plants. 13(10):1300.

Huang X, Chaparro JM, Reardon KF, Zhang R, Shen Q, Vivanco JM. 2014. Rhizosphere interactions: Root exudates, microbes, and microbial communities 1. 275(February):267–275.

Huang Y-H, Liu Y, Geng J, Lü H, Zhao H-M, Xiang L, Li H, Mo C-H, Li Y-W, Cai Q-Y. 2022. Maize root-associated niches determine the response variation in bacterial community assembly and function to phthalate pollution. J Hazard Mater. 429:128280. https://doi.org/10.1016/j.jhazmat.2022.128280

Huang Y-H, Yang Y-J, Wu X, Zhu C-L, Lü H, Zhao H-M, Xiang L, Li H, Mo C-H, Li Y-W, et al. 2023. Adaptation of bacterial community in maize rhizosphere for enhancing dissipation of phthalic acid esters in agricultural soil. J Hazard Mater. 444:130292. https://doi.org/10.1016/j.jhazmat.2022.130292

Icoz I, Stotzky G. 2008. Fate and effects of insect-resistant Bt crops in soil ecosystems. Soil Biol Biochem. 40(3):559–586.

Ilangumaran G, Subramanian S, Smith DL. 2024. Complete genome sequences of Rhizobium sp. strain SL42 and Hydrogenophaga sp. strain SL48, microsymbionts of Amphicarpaea bracteata. Front Microbiomes. 3:1309947.

Ingham RE, Trofymow JA, Ingham ER, Coleman DC. 1985. Interactions of bacteria, fungi, and their nematode grazers: Effects on nutrient cycling and plant growth. Ecol Monogr. 55(1):119–140.

Jacoby RP, Chen L, Schwier M, Koprivova A, Kopriva S. 2020. Recent advances in the role of plant metabolites in shaping the root microbiome. F1000Research. 9.

Jain A, Chakraborty J, Das S. 2020. Underlying mechanism of plant–microbe crosstalk in shaping microbial ecology of the rhizosphere. Acta Physiol Plant. 42(1):8.

Jain A, Sarsaiya S, Singh R, Gong Q, Wu Q, Shi J. 2024. Omics approaches in understanding the benefits of plant-microbe interactions. Front Microbiol. 15.

Jain A, Singh HB, Das S. 2021. Deciphering plant-microbe crosstalk through proteomics studies. Microbiol Res. 242:126590.

Jalal A, Oliveira CE da S, Rosa PAL, Galindo FS, Teixeira Filho MCM. 2023. Beneficial microorganisms improve agricultural sustainability under climatic extremes. Life. 13(5):1102.

Jamil F, Mukhtar H, Fouillaud M, Dufossé L. 2022. Rhizosphere signaling: Insights into plant–rhizomicrobiome interactions for sustainable agronomy. Microorganisms. 10(5):899.

Jayaraman S, Naorem AK, Lal R, Dalal RC, Sinha NK, Patra AK, Chaudhari SK. 2021. Disease-suppressive soils—Beyond food production: a critical review. J Soil Sci Plant Nutr. 21:1437–1465.

Jha AB, Gali KK, Alam Z, Lachagari VBR, Warkentin TD. 2021. Potential application of genomic technologies in breeding for fungal and oomycete disease resistance in pea. Agronomy. 11(6):1260.

Jia B, Chang X, Fu Y, Heng W, Ye Z, Liu P, Liu L, Al Shoffe Y, Watkins CB, Zhu L. 2022. Metagenomic analysis of rhizosphere microbiome provides insights into occurrence of iron deficiency chlorosis in field of Asian pears. BMC Microbiol. 22:1–16.

Jiang L, Zhang D, Song M, Guan G, Sun Y, Li J, Cheng X, Luo C, Zhang G. 2022. The positive role of root decomposition on the bioremediation of organic pollutants contaminated soil: A case study using PCB-9 as a model compound. Soil Biol Biochem. 171:108726. https://doi.org/10.1016/j.soilbio.2022.108726

Jiang RHY, Tyler BM. 2012. Mechanisms and evolution of virulence in oomycetes. Annu Rev Phytopathol. 50(1):295–318.

Jiang Y, Zhou H, Chen L, Yuan Y, Fang H, Luan L, Chen Y, Wang X, Liu M, Li H. 2018. Nematodes and microorganisms interactively stimulate soil organic carbon turnover in the macroaggregates. Front Microbiol. 9:2803.

Jiao S, Lu Y. 2020. Abundant fungi adapt to broader environmental gradients than rare fungi in agricultural fields. Glob Chang Biol. 26(8):4506–4520.

Jones JT, Haegeman A, Danchin EGJ, Gaur HS, Helder J, Jones MGK, Kikuchi T, Manzanilla-López R, Palomares-Rius JE, Wesemael WML. 2013. Top 10 plant-parasitic nematodes in molecular plant pathology. Mol Plant Pathol. 14(9):946–961.

Jones OAH, Sdepanian S, Lofts S, Svendsen C, Spurgeon DJ, Maguire ML, Griffin JL. 2014. Metabolomic analysis of soil communities can be used for pollution assessment. Environ Toxicol Chem. 33(1):61–64.

Joshi D, Chandra R, Suyal DC, Kumar S, Reeta G. 2019. Impacts of bioinoculants Pseudomonas jesenii MP1 and Rhodococcus qingshengii S10107 on chickpea (Cicer arietinum L.) yield and soil nitrogen status. Pedosphere. 29(3):388–399.

Joubert PM, Krasileva K V. 2024. Distinct genomic contexts predict gene presence–absence variation in different pathotypes of Magnaporthe oryzae. Genetics. 226(4):iyae012.

Kajiura LN, Stewart SD, Dresios J, Uyehara CFT. 2015. Simultaneous extraction of viral and bacterial nucleic acids for molecular diagnostic applications. J Biomol Tech JBT. 26(4):118.

Kalita P, Mohapatra B, Maruthi M. 2024. Role of effectors in plant–pathogen interactions. In: Biotechnol Adv Dis Toler Plants. [place unknown]: Springer; p. 363–376.

Kehe J, Ortiz A, Kulesa A, Gore J, Blainey PC, Friedman J. 2021. Positive interactions are common among culturable bacteria. Sci Adv. 7(45):eabi7159.

Khan N, Ali S, Shahid MA, Mustafa A, Sayyed RZ, Curá JA. 2021. Insights into the interactions among roots, rhizosphere, and rhizobacteria for improving plant growth and tolerance to abiotic stresses: A review. Cells. 10(6):1551.

Khan RAA, Najeeb S, Hussain S, Xie B, Li Y. 2020. Bioactive secondary metabolites from *Trichoderma* spp. against phytopathogenic fungi. Microorganisms. 8(6):817.

Khatabi B, Gharechahi J, Ghaffari MR, Liu D, Haynes PA, McKay MJ, Mirzaei M, Salekdeh GH. 2019. Plant–microbe symbiosis: What has proteomics taught us? Proteomics. 19(16):1800105.

Kong HG, Song GC, Sim H-J, Ryu C-M. 2021. Achieving similar root microbiota composition in neighbouring plants through airborne signalling. ISME J. 15(2):397–408.

Koprivova A, Kopriva S. 2022. Plant secondary metabolites altering root microbiome composition and function. Curr Opin Plant Biol. 67:102227. https://doi.org/10.1016/j.pbi.2022.102227

Koroney AS, Plasson C, Pawlak B, Sidikou R, Driouich A, Menu-Bouaouiche L, Vicré-Gibouin M. 2016. Root exudate of Solanum tuberosum is enriched in galactose-containing molecules and impacts the growth of Pectobacterium atrosepticum. Ann Bot. 118(4):797–808.

Kruthika S, Ashu A, Anand A, Sammi Reddy K, Vara Prasad P V, Gurumurthy S. 2024. Unveiling the role of root exudates in plant adaptation to drought and heat stress. J Crop Heal. 76(5):941–955.

Kumar A, Dubey A. 2020. Rhizosphere microbiome: Engineering bacterial competitiveness for enhancing crop production. J Adv Res. 24:337–352.

Kumar A, Verma JP. 2018. Does plant—Microbe interaction confer stress tolerance in plants: A review? Microbiol Res. 207:41–52.

Kumari Ankita, Kumari Archana, Sharma H, Sharma P, Bhattacharya S, Mishra T, Al-Tawaha AR, Lal MK, Tiwari RK, Mandal S. 2023. Modern approaches in studying the role of plant-microbial interactions: A way towards the development of sustainable agriculture. In: New Front Plant-Environment Interact Innov Technol Dev. [place unknown]: Springer; p. 69–91.

Lakshmanan V, Kitto SL, Caplan JL, Hsueh Y-H, Kearns DB, Wu Y-S, Bais HP. 2012. Microbe-associated molecular patterns-triggered root responses mediate beneficial rhizobacterial recruitment in *Arabidopsis*. Plant Physiol. 160(3):1642–1661.

Lakshmanan V, Selvaraj G, Bais HP. 2014. Functional soil microbiome: Belowground solutions to an aboveground problem. Plant Physiol. 166(2):689–700.

Lan X, Ning Z, Jia Y, Lin W, Xiao E, Cheng Q, Cai Q, Xiao T. 2024. The rhizosphere microbiome reduces the uptake of arsenic and tungsten by Blechnum orientale by increasing nutrient cycling in historical tungsten mining area soils. Sci Total Environ. 924.

Lanoue A, Burlat V, Henkes GJ, Koch I, Schurr U, Röse USR. 2010. De novo biosynthesis of defense root exudates in response to Fusarium attack in barley. New Phytol. 185(2):577–588.

Larekeng SH, NGADIMAN YK, Khairina Y, Simarmata R, Christita M. 2024. Unraveling the potential of ACC Deaminase-producing microbes in various agricultural stresses: Current status, limitations, and recommendations. Pak J Bot. 56(2):791–805.

Lebedeva MA, Dobychkina DA, Lutova LA. 2023. CRISPR/Cas9-mediated knock-out of the MtCLE35 gene highlights its key role in the control of symbiotic nodule numbers under high-nitrate conditions. Int J Mol Sci. 24(23):16816.

Li H, La S, Zhang X, Gao L, Tian Y. 2021. Salt-induced recruitment of specific root-associated bacterial consortium capable of enhancing plant adaptability to salt stress. ISME J. 15(10):2865–2882.

Li J, Lin S, Ma H, Wang Y, He H, Fang C. 2022. Spatial-temporal distribution of allelopathic rice roots in paddy soil and its impact on weed-suppressive activity at the seedling stages. Front Plant Sci. 13:940218.

Li J, Wang C, Liang W, Liu S. 2021. Rhizosphere microbiome: The emerging barrier in plant-pathogen interactions. Front Microbiol. 12.

Li M, Song Z, Li Z, Qiao R, Zhang P, Ding C, Xie J, Chen Y, Guo H. 2022. Populus root exudates are associated with rhizosphere microbial communities and symbiotic patterns. Front Microbiol. 13:1042944.

Li Y, Ahmadi F, Wan S, Kariman K. 2024. Oomycetes as biocontrol agents: Unveiling their potential and mechanisms in plant disease control. Mycosphere. 15(1):5017–5037.

Lin Q, Li M, Wang Y, Xu Z, Li L. 2022. Root exudates and chemotactic strains mediate bacterial community assembly in the rhizosphere soil of Casuarina equisetifolia L. Front Plant Sci. 13:988442.

Liu J, Xu W, Zhang Q, Liao W, Li L, Chen S, Yang J, Wang Z, Xu F. 2024. OsPHR2-mediated recruitment of Pseudomonadaceae enhances rice phosphorus uptake. Plant Commun. 5(8).

Liu M, Adl S, Cui X, Tian Y, Xu X, Kuzyakov Y. 2020. In situ methods of plant-microbial interactions for nitrogen in rhizosphere. Rhizosphere. 13:100186.

Liu Y, Wang H, Qian X, Gu J, Chen W, Shen X, Tao S, Jiao S, Wei G. 2023. Metagenomics insights into responses of rhizobacteria and their alleviation role in licorice allelopathy. Microbiome. 11(1):109.

Liu Y, Xun W, Chen L, Xu Z, Zhang N, Feng H, Zhang Q, Zhang R. 2022. Rhizosphere microbes enhance plant salt tolerance: Toward crop production in saline soil. Comput Struct Biotechnol J. 20:6543–6551.

Liu Z, Ma A, Mathé E, Merling M, Ma Q, Liu B. 2021. Network analyses in microbiome based on high-throughput multiomics data. Brief Bioinform. 22(2):1639–1655.

Lugtenberg B, Kamilova F. 2009. Plant-growth-promoting rhizobacteria. Annu Rev Microbiol. 63(1):541–556.

Lundberg DS, Lebeis SL, Paredes SH, Yourstone S, Gehring J, Malfatti S, Tremblay J, Engelbrektson A, Kunin V, Rio TG del. 2012. Defining the core *Arabidopsis thaliana* root microbiome. Nature. 488(7409):86–90.

Maina S, Karuri H, Ng'endo RN. 2021. Free-living nematode assemblages associated with maize residues and their ecological significance. J Nematol. 53:e2021– e2038.

Mainka T, Weirathmüller D, Herwig C, Pflügl S. 2021. Potential applications of halophilic microorganisms for biological treatment of industrial process brines contaminated with aromatics. J Ind Microbiol Biotechnol. 48(1–2):kuab015.

Makonde HM. 2009. Assessment of Effects of Bacillus thuringiensis Cry1A (c) δ-endotoxin on nitrogen fixing bacteria and their host plants in clay soil.

Manickam S, Rajagopalan VR, Kambale R, Rajasekaran R, Kanagarajan S, Muthurajan R. 2023. Plant metabolomics: Current initiatives and future prospects. Curr Issues Mol Biol. 45(11):8894–8906.

Mansfield J, Genin S, Magori S, Citovsky V, Sriariyanum M, Ronald P, Dow MAX, Verdier V, Beer SV, Machado MA. 2012. Top 10 plant pathogenic bacteria in molecular plant pathology. Mol Plant Pathol. 13(6):614–629.

Masenya K, Manganyi MC, Dikobe TB. 2024. Exploring cereal metagenomics: Unravelling microbial communities for improved food security. Microorganisms. 12(3):510.

Meena M, Mehta T, Nagda A, Yadav G, Sonigra P. 2023. PGPR-mediated synthesis and alteration of different secondary metabolites during plant-microbe interactions. In: Plant-Microbe Interact Adv Mol Biochem Approaches. [place unknown]: Elsevier; p. 229–255.

Mendes LW, Kuramae EE, Navarrete AA, Van Veen JA, Tsai SM. 2014. Taxonomical and functional microbial community selection in soybean rhizosphere. ISME J. 8(8):1577–1587.

Mendes LW, Mendes R, Raaijmakers JM, Tsai SM. 2018. Breeding for soil-borne pathogen resistance impacts active rhizosphere microbiome of common bean. ISME J. 12(12): 3038–3042.

Mendes LW, Raaijmakers JM, de Hollander M, Mendes R, Tsai SM. 2018. Influence of resistance breeding in common bean on rhizosphere microbiome composition and function. ISME J. 12(1):212–224.

Mendes R, Garbeva P, Raaijmakers JM. 2013. The rhizosphere microbiome: Significance of plant beneficial, plant pathogenic, and human pathogenic microorganisms. FEMS Microbiol Rev. 37(5):634–663.

Mhlongo MI, Piater LA, Madala NE, Labuschagne N, Dubery IA. 2018. The chemistry of plant–microbe interactions in the rhizosphere and the potential for metabolomics to reveal signaling related to defense priming and induced systemic resistance. Front Plant Sci. 9:112.

Micallef SA, Shiaris MP, Colón-Carmona A. 2009. Influence of *Arabidopsis thaliana* accessions on rhizobacterial communities and natural variation in root exudates. J Exp Bot. 60(6):1729–1742.

Micheel CM, Nass SJ, Omenn GS. 2012. Omics-based clinical discovery: Science, technology, and applications. In: Evol Transl Omi Lessons Learn Path Forw. [place unknown]: National Academies Press (US).

Mishra AK, Sudalaimuthuasari N, Hazzouri KM, Saeed EE, Shah I, Amiri KMA. 2022. Tapping into plant–microbiome interactions through the lens of multi-omics techniques. Cells. 11(20):3254.

Mougel C, Offre P, Ranjard L, Corberand T, Gamalero E, Robin C, Lemanceau P. 2006. Dynamic of the genetic structure of bacterial and fungal communities at different developmental stages of Medicago truncatula Gaertn. cv. Jemalong line J5. New Phytol. 170(1):165–175.

Nakayasu ES, Nicora CD, Sims AC, Burnum-Johnson KE, Kim Y-M, Kyle JE, Matzke MM, Shukla AK, Chu RK, Schepmoes AA. 2016. MPLEx: A robust and universal protocol for single-sample integrative proteomic, metabolomic, and lipidomic analyses. MSystems. 1(3):10–1128.

Naveed M, Hussain MB, Zahir ZA, Mitter B, Sessitsch A. 2014. Drought stress amelioration in wheat through inoculation with Burkholderia phytofirmans strain PsJN. Plant Growth Regul. 73:121–131.

Nechwatal J, Wielgoss A, Mendgen K. 2008. Diversity, host, and habitat specificity of oomycete communities in declining reed stands (Phragmites australis) of a large freshwater lake. Mycol Res. 112(6):689–696.

Neher DA. 2001. Role of nematodes in soil health and their use as indicators. J Nematol. 33(4):161.

Nelson DR, Mele PM. 2007. Subtle changes in rhizosphere microbial community structure in response to increased boron and sodium chloride concentrations. Soil Biol Biochem. 39(1):340–351.

Nemergut DR, Schmidt SK, Fukami T, O'Neill SP, Bilinski TM, Stanish LF, Knelman JE, Darcy JL, Lynch RC, Wickey P. 2013. Patterns and processes of microbial community assembly. Microbiol Mol Biol Rev. 77(3):342–356.

Newman MM, Lorenz N, Hoilett N, Lee NR, Dick RP, Liles MR, Ramsier C, Kloepper JW. 2016. Changes in rhizosphere bacterial gene expression following glyphosate treatment. Sci Total Environ. 553:32–41.

Niazi P, Monib AW, Ozturk H, Mansoor M, Azizi A, Hassand MH. 2023. Review on surface elements and bacterial biofilms in plant-bacterial associations. J Res Appl Sci Biotechnol. 2(1):204–214.

Olanrewaju OS, Glick BR, Babalola OO. 2024. Metabolomics-guided utilization of beneficial microbes for climate-resilient crops. Curr Opin Chem Biol. 79.

Palmer JD, Foster KR. 2022. Bacterial species rarely work together. Science (80-). 376(6593):581–582.

Pantigoso HA, Newberger D, Vivanco JM. 2022. The rhizosphere microbiome: Plant–microbial interactions for resource acquisition. J Appl Microbiol. 133(5):2864–2876.

Park I, Seo Y-S, Mannaa M. 2023. Recruitment of the rhizo-microbiome army: Assembly determinants and engineering of the rhizosphere microbiome as a key to unlocking plant potential. Front Microbiol. 14:1163832.

Patel M, Islam S, Husain FM, Yadav VK, Park H-K, Yadav KK, Bagatharia S, Joshi M, Jeon B-H, Patel A. 2023. Bacillus subtilis ER-08, a multifunctional plant growth-promoting rhizobacterium, promotes the growth of fenugreek (Trigonella foenum-graecum L.) plants under salt and drought stress. Front Microbiol. 14:1208743.

Paul S, Parvez SS, Goswami A, Banik A. 2024. Exopolysaccharides from agriculturally important microorganisms: conferring soil nutrient status and plant health. Int J Biol Macromol.:129954.

Pérez-Jaramillo JE, Mendes R, Raaijmakers JM. 2016. Impact of plant domestication on rhizosphere microbiome assembly and functions. Plant Mol Biol. 90(6):635–644. https://doi.org/10.1007/s11103-015-0337-7

Philippot L, Chenu C, Kappler A, Rillig MC, Fierer N. 2024. The interplay between microbial communities and soil properties. Nat Rev Microbiol. 22(4):226–239.

Philippot L, Raaijmakers JM, Lemanceau P, Van Der Putten WH. 2013. Going back to the roots: The microbial ecology of the rhizosphere. Nat Rev Microbiol. 11(11):789–799.

Pieterse CMJ, Van Pelt JA, Verhagen BWM, Ton J, Van Wees SCM, Leon-Kloosterziel KM, Van Loon LC. 2003. Induced systemic resistance by plant growth-promoting rhizobacteria. Symbiosis. 35(1):39–54.

Pinto AG, Kos T, Puškarić J, Vrandečić K, Benković-Lačić T, Brmež M. 2024. Soil ecosystem functioning through interactions of nematodes and fungi *Trichoderma* sp. Sustainability. 16(7):2885.

Priya P, Aneesh B, Harikrishnan K. 2021. Genomics as a potential tool to unravel the rhizosphere microbiome interactions on plant health. J Microbiol Methods. 185:106215.

Qiao C, Penton CR, Xiong W, Liu C, Wang R, Liu Z, Xu X, Li R, Shen Q. 2019. Reshaping the rhizosphere microbiome by bio-organic amendment to enhance crop yield in a maize-cabbage rotation system. Appl Soil Ecol. 142:136–146.

Qin S, Zhang H, He Y, Chen Z, Yao L, Han H. 2023. Improving radish phosphorus utilization efficiency and inhibiting Cd and Pb uptake by using heavy metal-immobilizing and phosphate-solubilizing bacteria. Sci Total Environ. 868:161685.

Qu Q, Zhang Z, Peijnenburg W, Liu W, Lu T, Hu B, Chen Jianmeng, Chen Jun, Lin Z, Qian H. 2020. Rhizosphere microbiome assembly and its impact on plant growth. J Agric Food Chem. 68(18):5024–5038.

Raaijmakers JM, Paulitz TC, Steinberg C, Alabouvette C, Moënne-Loccoz Y. 2009. The rhizosphere: A playground and battlefield for soilborne pathogens and beneficial microorganisms.

Raaijmakers JM, Vlami M, De Souza JT. 2002. Antibiotic production by bacterial biocontrol agents. Antonie Van Leeuwenhoek. 81:537–547.

Raaymakers TM, Van den Ackerveken G. 2016. Extracellular recognition of oomycetes during biotrophic infection of plants. Front Plant Sci. 7:906.

Raimi AR, Ezeokoli OT, Adeleke RA. 2023. Soil nutrient management influences diversity, community association and functional structure of rhizosphere bacteriome under vegetable crop production. Front Microbiol. 14:1229873.

Rane NR, Tapase S, Kanojia A, Watharkar A, Salama E-S, Jang M, Yadav KK, Amin MA, Cabral-Pinto MMS, Jadhav JP. 2022. Molecular insights into plant–microbe interactions for sustainable remediation of contaminated environment. Bioresour Technol. 344:126246.

Rapparini F, Peñuelas J. 2014. Mycorrhizal fungi to alleviate drought stress on plant growth. Use Microbes Alleviation soil Stress Vol 1.:21–42.

Rath KM, Fierer N, Murphy D V, Rousk J. 2019. Linking bacterial community composition to soil salinity along environmental gradients. ISME J. 13(3):836–846.

Regalado J, Lundberg DS, Deusch O, Kersten S, Karasov T, Poersch K, Shirsekar G, Weigel D. 2020. Combining whole-genome shotgun sequencing and rRNA gene amplicon analyses to improve detection of microbe–microbe interaction networks in plant leaves. ISME J. 14(8):2116–2130.

Ren H, Gao G, Ma Y, Li Z, Wang S, Gu J. 2023. Shift of root nitrogen-acquisition strategy with tree age is mediated by root functional traits along the collaboration gradient of the root economics space. Tree Physiol. 43(8):1341–1353.

Richardson AE, Barea J-M, McNeill AM, Prigent-Combaret C. 2009. Acquisition of phosphorus and nitrogen in the rhizosphere and plant growth promotion by microorganisms.

Rivas-Ubach A, Pérez-Trujillo M, Sardans J, Gargallo-Garriga A, Parella T, Penuelas J. 2013. Ecometabolomics: Optimized NMR-based method. Methods Ecol Evol. 4(5):464–473.

Rivas-Ubach A, Sardans J, Pérez-Trujillo M, Estiarte M, Peñuelas J. 2012. Strong relationship between elemental stoichiometry and metabolome in plants. Proc Natl Acad Sci. 109(11):4181–4186.

Rizaludin MS, Stopnisek N, Raaijmakers JM, Garbeva P. 2021. The chemistry of stress: Understanding the 'cry for help'of plant roots. Metabolites. 11(6):357.

Roberts DP, Maul JE, McKenna LF, Emche SE, Meyer SLF, Collins RT, Bowers JH. 2010. Selection of genetically diverse *Trichoderma* spp. isolates for suppression of *Phytophthora capsici* on bell pepper. Can J Microbiol. 56(10):864–873.

Rodríguez H, Fraga R. 1999. Phosphate solubilizing bacteria and their role in plant growth promotion. Biotechnol Adv. 17(4–5):319–339.

Rodríguez-Salazar J, Suarez R, Caballero-Mellado J, Iturriaga G. 2009. Trehalose accumulation in *Azospirillum brasilense* improves drought tolerance and biomass in maize plants. FEMS Microbiol Lett. 296(1):52–59.

Rudrappa T, Czymmek KJ, Paré PW, Bais HP. 2008. Root-secreted malic acid recruits beneficial soil bacteria. Plant Physiol. 148(3):1547–1556.

Ryan J, Ibrikci H, Sommer R, McNeill A. 2009. Nitrogen in rainfed and irrigated cropping systems in the Mediterranean region. Adv Agron. 104:53–136.

Ryu M-H, Zhang J, Toth T, Khokhani D, Geddes BA, Mus F, Garcia-Costas A, Peters JW, Poole PS, Ané J-M. 2020. Control of nitrogen fixation in bacteria that associate with cereals. Nat Microbiol. 5(2):314–330.

Saharan K, Schütz L, Kahmen A, Wiemken A, Boller T, Mathimaran N. 2018. Finger millet growth and nutrient uptake is improved in intercropping with pigeon pea through "biofertilization" and "bioirrigation" mediated by arbuscular mycorrhizal fungi and plant growth promoting rhizobacteria. Front Environ Sci. 6:46.

Salas-González I, Reyt G, Flis P, Custódio V, Gopaulchan D, Bakhoum N, Dew TP, Suresh K, Franke RB, Dangl JL. 2021. Coordination between microbiota and root endodermis supports plant mineral nutrient homeostasis. Science (80-). 371(6525):eabd0695.

Saleem M, Arshad M, Hussain S, Bhatti AS. 2007. Perspective of Plant Growth Promoting Rhizobacteria (PGPR) containing ACC deaminase in stress agriculture. J Ind Microbiol Biotechnol. 34(10):635–648.

Samantara K, Shiv A, de Sousa LL, Sandhu KS, Priyadarshini P, Mohapatra SR. 2021. A comprehensive review on epigenetic mechanisms and application of epigenetic modifications for crop improvement. Environ Exp Bot. 188:104479.

Sánchez-Rodríguez AR, Chadwick DR, Tatton GS, Hill PW, Jones DL. 2018. Comparative effects of prolonged freshwater and saline flooding on nitrogen cycling in an agricultural soil. Appl Soil Ecol. 125:56–70.

Sanon A, Andrianjaka ZN, Prin Y, Bally R, Thioulouse J, Comte G, Duponnois R. 2009. Rhizosphere microbiota interfers with plant-plant interactions.

Sardans J, Penuelas J, Rivas-Ubach A. 2011. Ecological metabolomics: overview of current developments and future challenges. Chemoecology. 21:191–225.

Sarim KM, Srivastava R, Ramteke PW. 2020. Next-generation omics technologies for exploring complex metabolic regulation during plant-microbe interaction. In: Microb Serv Restor Ecol. [place unknown]: Elsevier; p. 123–138.

Sarma RK. 2024. Harnessing the rhizosphere soil microbiome for sustainable agriculture: Recent technological developments and future challenges. In: Prog Soil Microbiome Res. [place unknown]: Springer; p. 481–500.

Sarsaiya S, Jain A, Shu F, Yang M, Pu M, Jia Q, Gong Q, Wu Q, Qian X, Shi J. 2024. Enhancing dendrobine production in Dendrobium nobile through mono-culturing of endophytic fungi, *Trichoderma longibrachiatum* (MD33) in a temporary immersion bioreactor system. Front Plant Sci. 15:1302817.

Sasse J, Martinoia E, Northen T. 2018. Feed your friends: Do plant exudates shape the root microbiome? Trends Plant Sci. 23(1):25–41.

Schlechter RO, Miebach M, Remus-Emsermann MNP. 2019. Driving factors of epiphytic bacterial communities: A review. J Adv Res. 19:57–65.

Sharma M, Sudheer S, Usmani Z, Rani R, Gupta P. 2020. Deciphering the omics of plant-microbe interaction: Perspectives and new insights. Curr Genomics. 21(5):343–362.

Silva LI da, Pereira MC, Carvalho AMX de, Buttrós VH, Pasqual M, Dória J. 2023. Phosphorus-solubilizing microorganisms: A key to sustainable agriculture. Agriculture. 13(2):462.

Singh BJ, Chakraborty A, Sehgal R. 2023. A systematic review of industrial wastewater management: Evaluating challenges and enablers. J Environ Manage. 348:119230. https://doi.org/10.1016/j.jenvman.2023.119230

Singh K, Chandra R, Purchase D. 2022. Unraveling the secrets of rhizobacteria signaling in rhizosphere. Rhizosphere. 21:100484.

Sivaram AK, Subashchandrabose SR, Logeshwaran P, Lockington R, Naidu R, Megharaj M. 2020. Rhizodegradation of PAHs differentially altered by C3 and C4 plants. Sci Rep. 10(1):16109.

SkZ A, Vardharajula S, Vurukonda SSKP. 2018. Transcriptomic profiling of maize (Zea mays L.) seedlings in response to Pseudomonas putida stain FBKV2 inoculation under drought stress. Ann Microbiol. 68:331–349.

Song T, Sardar MF, Wang X, Li B, Zhang Z, Wu D, Zhu C, Li H. 2024. Distribution of antibiotic resistant bacteria in different soil types following manure application. Soil Ecol Lett. 6(2):230210.

Spaepen S, Vanderleyden J, Remans R. 2007. Indole-3-acetic acid in microbial and microorganism-plant signaling. FEMS Microbiol Rev. 31(4):425–448.

Spragge F, Bakkeren E, Jahn MT, BN Araujo E, Pearson CF, Wang X, Pankhurst L, Cunrath O, Foster KR. 2023. Microbiome diversity protects against pathogens by nutrient blocking. Science (80-). 382(6676):eadj3502.

Sugimoto K, Jiao Y, Meyerowitz EM. 2010. *Arabidopsis* regeneration from multiple tissues occurs via a root development pathway. Dev Cell. 18(3):463–471.

Sui X, Yan L, Jiang Y ying. 2017. The vaccines and antibodies associated with Als3p for treatment of Candida albicans infections. Vaccine. 35(43):5786–5793. https://doi.org/10.1016/J.VACCINE.2017.08.082

Sun L, Song K, Shi L, Duan D, Zhang H, Sun Y, Qin Q, Xue Y. 2021. Influence of elemental sulfur on cadmium bioavailability, microbial community in paddy soil and Cd accumulation in rice plants. Sci Rep. 11(1):11468.

Sun X, Huang D, Huang Y, Häggblom M, Soleimani M, Li J, Chen Zheng, Chen Zhenyu, Gao P, Li B. 2024. Microbial-mediated oxidative dissolution of orpiment and realgar in circumneutral aquatic environments. Water Res. 251:121163.

Świątczak J, Kalwasińska A, Brzezinska MS. 2024. Plant growth–promoting rhizobacteria: Peribacillus frigoritolerans 2RO30 and Pseudomonas Sivasensis 2RO45 for their effect on canola growth under controlled as well as natural conditions. Front Plant Sci. 14:1233237.

Szabo Z, Janaky T. 2015. Challenges and developments in protein identification using mass spectrometry. TrAC Trends Anal Chem. 69:76–87.

Tanna B, Mishra A. 2018. Metabolomics of seaweeds: Tools and techniques. In: Plant Metab Regul under Environ Stress. [place unknown]: Elsevier; p. 37–52.

Tedersoo L, Bahram M, Zobel M. 2020. How mycorrhizal associations drive plant population and community biology. Science (80-). 367(6480):eaba1223.

Tedersoo L, Magurno F, Alkahtani S, Mikryukov V. 2024. Phylogenetic classification of arbuscular mycorrhizal fungi: New species and higher-ranking taxa in Glomeromycota and Mucoromycota (class Endogonomycetes). MycoKeys. 107:249–271.

Thepbandit W, Athinuwat D. 2024. Rhizosphere microorganisms supply availability of soil nutrients and induce plant defense. Microorganisms. 12(3):558.

Theron AJ, Tintinger GR, Anderson R. 2012. Harmful interactions of non-essential heavy metals with cells of the innate immune system.

Thomas T, Gilbert J, Meyer F. 2012. Metagenomics-a guide from sampling to data analysis. Microb Inform Exp. 2:1–12.

Tian T, Reverdy A, She Q, Sun B, Chai Y. 2020. The role of rhizodeposits in shaping rhizomicrobiome. Environ Microbiol Rep. 12(2):160–172. https://doi.org/10.1111/1758-2229.12816

Tiwari P, Bose SK, Park K-I, Dufossé L, Fouillaud M. 2024. Plant-microbe interactions under the extreme habitats and their potential applications. Microorganisms. 12(3):448.

Tran MT, Son GH, Song YJ, Nguyen NT, Park S, Thach TV, Kim J, Sung YW, Das S, Pramanik D. 2023. CRISPR-Cas9-based precise engineering of SlHyPRP1 protein towards multi-stress tolerance in tomato. Front Plant Sci. 14:1186932.

Trivedi P, Leach JE, Tringe SG, Sa T, Singh BK. 2020. Plant–microbiome interactions: From community assembly to plant health. Nat Rev Microbiol. 18(11):607–621.

Turner TR, Ramakrishnan K, Walshaw J, Heavens D, Alston M, Swarbreck D, Osbourn A, Grant A, Poole PS. 2013. Comparative metatranscriptomics reveals kingdom level changes in the rhizosphere microbiome of plants. ISME J. 7(12):2248–2258.

Usman NI, Wali MM. 2024. Nitrogen fixation by rhizobacterial nif mechanism: an advanced genetic perspective. In: Updat Rhizobacteria. [place unknown]: IntechOpen.

Valente Lima J, Tinôco RS, Olivares FL, Chia GS, Melo Júnior JAG de, Silva GB da. 2021. Rhizobacteria modify root architecture and improve nutrient uptake in oil palm seedlings despite reduced fertilizer. Rhizosphere. 19.

van Dam NM, Bouwmeester HJ. 2016. Metabolomics in the rhizosphere: Tapping into belowground chemical communication. Trends Plant Sci. 21(3):256–265.

Van Horn DJ, Okie JG, Buelow HN, Gooseff MN, Barrett JE, Takacs-Vesbach CD. 2014. Soil microbial responses to increased moisture and organic resources along a salinity gradient in a polar desert. Appl Environ Microbiol. 80(10):3034–3043.

Vaughan MM, Wang Q, Webster FX, Kiemle D, Hong YJ, Tantillo DJ, Coates RM, Wray AT, Askew W, O'Donnell C. 2013. Formation of the unusual semivolatile diterpene rhizathalene by the *Arabidopsis* class I terpene synthase TPS08 in the root stele is involved in defense against belowground herbivory. Plant Cell. 25(3):1108–1125.

Vejan P, Abdullah R, Khadiran T, Ismail S, Nasrulhaq Boyce A. 2016. Role of plant growth promoting rhizobacteria in agricultural sustainability—A review. Molecules. 21(5):573.

Vessey JK. 2003. Plant growth promoting rhizobacteria as biofertilizers. Plant Soil. 255:571–586.

Větrovský T, Baldrian P. 2013. The variability of the 16S rRNA gene in bacterial genomes and its consequences for bacterial community analyses. PLoS One. 8(2):e57923.

Větrovský T, Kohout P, Kopecký M, Machac A, Man M, Bahnmann BD, Brabcová V, Choi J, Meszárošová L, Human ZR, et al. 2019. A meta-analysis of global fungal distribution reveals climate-driven patterns. Nat Commun. 10(1):5142. https://doi.org/10.1038/s41467-019-13164-8

Vinale F, Sivasithamparam K, Ghisalberti EL, Marra R, Woo SL, Lorito M. 2008. *Trichoderma*–plant–pathogen interactions. Soil Biol Biochem. 40(1):1–10.

Vives-Peris V, De Ollas C, Gómez-Cadenas A, Pérez-Clemente RM. 2020. Root exudates: From plant to rhizosphere and beyond. Plant Cell Rep. 39(1):3–17.

Walters WA, Jin Z, Youngblut N, Wallace JG, Sutter J, Zhang W, González-Peña A, Peiffer J, Koren O, Shi Q. 2018. Large-scale replicated field study of maize rhizosphere identifies heritable microbes. Proc Natl Acad Sci. 115(28):7368–7373.

Wang C, Wang X, Zhang Y, Morrissey E, Liu Y, Sun L, Qu L, Sang C, Zhang H, Li G. 2023. Integrating microbial community properties, biomass and necromass to predict cropland soil organic carbon. ISME Commun. 3(1):86.

Wang H, Shi C, Zhang Q, Guo D, Liu S, Yu Q. 2024. Biochar assists phytoremediation of cadmium by regulation of rhizosphere microbiome in paddy fields. Environ Technol Innov. 36.

Wang H-B, Zhang Z-X, Li H, He H-B, Fang C-X, Zhang A-J, Li Q-S, Chen R-S, Guo X-K, Lin H-F. 2011. Characterization of metaproteomics in crop rhizospheric soil. J Proteome Res. 10(3):932–940.

Wang L, Lin H, Dong Y, Li B, He Y. 2020. Effects of endophytes inoculation on rhizosphere and endosphere microecology of Indian mustard (Brassica juncea) grown in vanadium-contaminated soil and its enhancement on phytoremediation. Chemosphere. 240:124891.

Wang L, Rubio MC, Xin X, Zhang B, Fan Q, Wang Q, Ning G, Becana M, Duanmu D. 2019. CRISPR/Cas9 knockout of leghemoglobin genes in Lotus japonicus uncovers their synergistic roles in symbiotic nitrogen fixation. New Phytol. 224(2):818–832.

Wang L, Zhao F, Liu H, Chen H, Zhang F, Li S, Sun T, Nekrasov V, Huang S, Dong S. 2023. A modified Agrobacterium-mediated transformation for two oomycete pathogens. PLoS Pathog. 19(4):e1011346.

Wang R, Cui L, Li J, Li W. 2023. Factors driving the halophyte rhizosphere bacterial communities in coastal salt marshes. Front Microbiol. 14:1127958.

Wang S, Chen W, Gao Q, Zhou C. 2022. Dynamic changes of Rhizosphere soil microbiome and functional genes involved in carbon and nitrogen cycling in Chinese fir monoculture. Forests. 13(11):1906.

Wang Y, Li X-M, Yang S-Q, Zhang F-Z, Wang B-G, Li H-L, Meng L-H. 2022. Sesquiterpene and sorbicillinoid glycosides from the endophytic fungus *Trichoderma longibrachiatum* EN-586 derived from the Marine Red Alga *Laurencia obtusa*. Mar Drugs. 20(3):177.

Wei B, Zhang J, Wen R, Chen T, Xia N, Liu Y, Wang Z. 2021. Genetically modified sugarcane intercropping soybean impact on rhizosphere bacterial communities and co-occurrence patterns. Front Microbiol. 12:742341.

Weidemüller P, Kholmatov M, Petsalaki E, Zaugg JB. 2021. Transcription factors: Bridge between cell signaling and gene regulation. Proteomics. 21(23–24):2000034.

Weidenhamer JD, Cipollini D, Morris K, Gurusinghe S, Weston LA. 2023. Ecological realism and rigor in the study of plant-plant allelopathic interactions. Plant Soil. 489(1):1–39.

Weiller F, Moore JP, Young P, Driouich A, Vivier MA. 2017. The Brassicaceae species Heliophila coronopifolia produces root border-like cells that protect the root tip and secrete defensin peptides. Ann Bot. 119(5):803–813.

Weller DM, Raaijmakers JM, Gardener BBM, Thomashow LS. 2002. Microbial populations responsible for specific soil suppressiveness to plant pathogens. Annu Rev Phytopathol. 40(1):309–348.

Wen T, Yuan J, He X, Lin Y, Huang Q, Shen Q. 2020. Enrichment of beneficial cucumber rhizosphere microbes mediated by organic acid secretion. Hortic Res. 7.

Wen T, Zhao M, Yuan J, Kowalchuk GA, Shen Q. 2021. Root exudates mediate plant defense against foliar pathogens by recruiting beneficial microbes. Soil Ecol Lett. 3:42–51.

White III RA, Callister SJ, Moore RJ, Baker ES, Jansson JK. 2016. The past, present and future of microbiome analyses. Nat Protoc. 11(11):2049–2053.

Wiggins G, Thomas J, Rahmatallah Y, Deen C, Haynes A, Degon Z, Glazko G, Mukherjee A. 2022. Common gene expression patterns are observed in rice roots during associations with plant growth-promoting bacteria, Herbaspirillum seropedicae and *Azospirillum brasilense*. Sci Rep. 12(1):8827.

Willsey T, Chatterton S, Cárcamo H. 2017. Interactions of root-feeding insects with fungal and oomycete plant pathogens. Front Plant Sci. 8:1764.

Wilson SK, Pretorius T, Naidoo S. 2023. Mechanisms of systemic resistance to pathogen infection in plants and their potential application in forestry. BMC Plant Biol. 23(1):404.

Wipf HM-L, Bùi T-N, Coleman-Derr D. 2021. Distinguishing between the impacts of heat and drought stress on the root microbiome of Sorghum bicolor. Phytobiomes J. 5(2):166–176.

Wu B, Liu F, Fang W, Yang T, Chen G-H, He Z, Wang S. 2021. Microbial sulfur metabolism and environmental implications. Sci Total Environ. 778:146085.

Wu L, Weston LA, Zhu S, Zhou X. 2023. Editorial: Rhizosphere interactions: Root exudates and the rhizosphere microbiome. Front Plant Sci. 14.

Xu J, Zhang Y, Zhang P, Trivedi P, Riera N, Wang Y, Liu X, Fan G, Tang J, Coletta-Filho HD, et al. 2018. The structure and function of the global citrus rhizosphere microbiome. Nat Commun. 9(1).

Xu Y, Zhang G, Ding H, Ci D, Dai L, Zhang Z. 2020. Influence of salt stress on the rhizosphere soil bacterial community structure and growth performance of groundnut (Arachis hypogaea L.). Int Microbiol. 23(3):453–465.

Xue Z, Kable ME, Marco ML. 2018. Impact of DNA sequencing and analysis methods on 16S rRNA gene bacterial community analysis of dairy products. Msphere. 3(5):10–1128.

Yaish MW, Al-Lawati A, Jana GA, Vishwas Patankar H, Glick BR. 2016. Impact of soil salinity on the structure of the bacterial endophytic community identified from the roots of caliph medic (Medicago truncatula). PLoS One. 11(7):e0159007.

Yan Q, Zhang X, Li M, Zhang G, Huang L-F. 2024. Global climate change promotes aboveground growth and high-latitude species distribution in medicinal plants: Insights from the genus Panax. J Clean Prod. 478:143863.

Yang C, Xie Z, Qian S, Zhang J, Yu Z, Li M, Gu L, Qin S, Zhang Z. 2022. Functional analysis of Rehmannia Glutinosa key LRR-RLKs during interaction of root exudates with Fusarium oxysporum reveals the roles of immune proteins in formation of replant disease. Front Plant Sci. 13:1044070.

Yang L, Qian X, Zhao Z, Wang Y, Ding G, Xing X. 2024. Mechanisms of rhizosphere plant-microbe interactions: molecular insights into microbial colonization. Front Plant Sci. 15:1491495.

Yang L, Zhang Y, Li F. 2012. Soil enzyme activities and soil fertility dynamics. Adv citrus Nutr.:143–156.

Yang T, Adams JM, Shi Y, He J, Jing X, Chen L, Tedersoo L, Chu H. 2017. Soil fungal diversity in natural grasslands of the Tibetan Plateau: Associations with plant diversity and productivity. New Phytol. 215(2):756–765.

Yin C, Casa Vargas JM, Schlatter DC, Hagerty CH, Hulbert SH, Paulitz TC. 2021. Rhizosphere community selection reveals bacteria associated with reduced root disease. Microbiome. 9:1–18.

Yoneyama T, Terakado-Tonooka J, Minamisawa K. 2017. Exploration of bacterial N_2-fixation systems in association with soil-grown sugarcane, sweet potato, and paddy rice: A review and synthesis. Soil Sci Plant Nutr. 63(6):578–590.

York LM, Carminati A, Mooney SJ, Ritz K, Bennett MJ. 2016. The holistic rhizosphere: Integrating zones, processes, and semantics in the soil influenced by roots. J Exp Bot. 67(12):3629–3643.

Yu H, Zheng X, Weng W, Yan X, Chen P, Liu X, Peng T, Zhong Q, Xu K, Wang C. 2021. Synergistic effects of antimony and arsenic contaminations on bacterial, archaeal and fungal communities in the rhizosphere of Miscanthus sinensis: Insights for nitrification and carbon mineralization. J Hazard Mater. 411:125094.

Yu P, He X, Baer M, Beirinckx S, Tian T, Moya YAT, Zhang X, Deichmann M, Frey FP, Bresgen V. 2021. Plant flavones enrich rhizosphere Oxalobacteraceae to improve maize performance under nitrogen deprivation. Nat plants. 7(4):481–499.

Yu P, Hochholdinger F. 2018. The role of host genetic signatures on root–microbe interactions in the rhizosphere and endosphere. Front Plant Sci. 9:1896.

Yue H, Yue W, Jiao S, Kim H, Lee Y-H, Wei G, Song W, Shu D. 2023. Plant domestication shapes rhizosphere microbiome assembly and metabolic functions. Microbiome. 11(1):70.

Zahran HH. 1999. Rhizobium-legume symbiosis and nitrogen fixation under severe conditions and in an arid climate. Microbiol Mol Biol Rev. 63(4):968–989.

Zhang J, Liu Y-X, Zhang N, Hu B, Jin T, Xu H, Qin Y, Yan P, Zhang X, Guo X. 2019. NRT1. 1B is associated with root microbiota composition and nitrogen use in field-grown rice. Nat Biotechnol. 37(6):676–684.

Zhang S, Li C, Si J, Han Z, Chen D. 2022. Action mechanisms of effectors in plant-pathogen interaction. Int J Mol Sci. 23(12):6758.

Zhong Y, Xun W, Wang X, Tian S, Zhang Y, Li D, Zhou Y, Qin Y, Zhang B, Zhao G. 2022. Root-secreted bitter triterpene modulates the rhizosphere microbiota to improve plant fitness. Nat plants. 8(8):887–896.

Zhou X, Wu F. 2012. p-Coumaric acid influenced cucumber rhizosphere soil microbial communities and the growth of Fusarium oxysporum f. sp. cucumerinum Owen. PLoS One. 7(10):e48288.

Zhu C-L, Lü H-X, Huang Y-H, Cheng J, Li H, Li Y-W, Mo C-H, Zhao H-M, Xiang L, Cai Q-Y. 2023. Rice genotypes and root-associated niches shifted bacterial community in response to pollution of di-(2-ethylhexyl) phthalate (DEHP) for promoting DEHP removal. J Hazard Mater. 452:131227.

Zulfiqar M, Singh V, Steinbeck C, Sorokina M. 2024. Review on computer-assisted biosynthetic capacities elucidation to assess metabolic interactions and communication within microbial communities. Crit Rev Microbiol.:1–40.

12 Mechanism and Perspectives of Microbiome in Improvements of Salinity Tolerance

Bhramar Dutta[1], *and Rajib Bandopadhyay*[1]*

[1]Microbiology section, UGC-Centre for Advanced Studies, Department of Botany, The University of Burdwan, Golapbag-713104, Purba Bardhaman, West Bengal, India

12.1 INTRODUCTION

Soil salinization, a growing issue worldwide, has detrimental consequences on ecosystem. Excess of salt in soil adversely affected soil quality, induced land degradation, hindered plant development, resulting in low economic returns. Based on the drivers and mechanisms, there are two types of soil salinity: primary and secondary salinity. Primary salinization occurred due to natural causes such as climatic factors, shallow groundwater, and aeolian salt deposition (Hassani et al., 2021). While anthropogenic activities, irrigated cropping system, improper drainage, and runoffs from road salts accelerated secondary salinization. Similarly, aquifer draining, sea-water intrusion in coastal areas, and inadequate precipitation prevented the natural leaching of the salts, ultimately accumulated in the soil.

Taking India as an example, approximately 2.1% (~6.7 million ha) of its total geographic area is salt-affected. Irrigated agriculture covers ~17% of the land in the country. According to estimates, nearly 10% of the additional area is salinization every year, potentially affecting 50% (treble to 20 million ha) of all cultivable land by 2050 (Kumar and Sharma, 2020).

Excessive accumulation of ions (Na^+, Ca^{2+}, Mg^{2+}, Cl^-, HCO^{3-}, $SO^{2-}{}_4$) in the root zone restricted plant metabolism, impaired photosynthesis, limited natural growth, and reduced seed germination. High amounts of NaCl in irrigation water accelerated the developmental stages of terrestrial plants, such as leaf number, timing of leaf out and flowering, collar girth, and seedling biomass. Although rice, wheat, and maize are major field crops globally, they have experienced the most significant impact from salt stress (moderate to low). Some studies showed the impact of irrigation water salinity on crop yield. During spike developmental stages, rice plants exhibited increased apical shoot growth, as a result, the stem grew taller, but simultaneously, spikelet primordia number declined. These tend to shorten the duration of spikelet differentiation, resulting fewer spikes and fewer kernels (El Sabagh et al., 2021).

Plants harbor a wide variety of microscopic organisms from three primary domains of life, Bacteria, Archaea, and Eukarya (algae, fungi, and protista) residing inside or outside their tissues. This community of microorganisms (referred to as microbiota), along with their collective genetic material, inhabiting a particular environment, known as the plant microbiome or phytomicrobiome, has a shared governance of host and microbiota. Collectively, the host and symbiotic association of microbes, holobionts, was originally conceived by Adolf Meyer-Abich; later, Lynn Margulis described it with evolution theory (Baedke et al., 2020). In the past few years, microbiome research has potentially unveiled host–microbiota relationships and metabolites that aid in enhancing stress tolerance and growth features in plants (Berg et al., 2020). The utilization of robust and rapid technologies, advanced high-throughput sequencing (HTS) platforms, sophisticated next-generation sequencing (NGS) systems, Clustered Regularly Interspaced Short Palindromic Repeats (CRISPR)-mediated genetic manipulation, as well as improvement in bioinformatics software tools, has opened the possibilities of studying plant-associated microbes and their relationship with the host plant. Several studies have demonstrated the effect of microorganisms in safeguarding plant productivity (Qin et al., 2016). Recent evidence has revealed that root exudations are an integral part of attracting beneficial microbial strains and deterring pathogens (Chen and Liu, 2024). Although a massive explosion in plant microbiome research has been reported, deep insight into microbes that are involved in salinity tolerance, their synergistic activity, and the changes that they produce at the molecular level by such an approach is slow.

Plant-associated halotolerant microorganisms thrive under high saline conditions and cohabit with plants. Microbial diversity surrounding the plant roots is so rich that the rhizosphere is considered a hotspot. Rhizospheric microbial community consisted of plant growth-promoting (PGP) bacteria (PGPB), archaea, endophytic microbes, mycorrhizal fungi, algae, and other organisms (Thepbandit and Athinuwat, 2024). They are able to alleviate the salinity stress impact on plants. Among the reported halotolerant PGPB, dominant genera like *Bacillus*, *Pseudomonas*, *Pantonea*, *Rhizobium*, *Serratia*, and *Enterobacter* have been documented as effective bio-inoculants, triggering the growth of agricultural plants in extreme

***Corresponding author:** rajibindia@gmail.com (Dr Rajib Bandopadhyay)

DOI: 10.1201/9781003544425-12

salinity stress (Dutta et al., 2023; Zheng et al., 2024). Moreover, reports are available that members of arbuscular mycorrhizal fungi (AMF) orders Glomerales, Archaeosporales, Paraglomerales, and Diversisporales, in the subphylum Glomeromycotina, mediate salinity tolerance in plants (Orozco-Mosqueda et al., 2020). The extent to which various microbes influence salinity tolerance differed significantly due to certain factors like choice of host plant species, soil condition, severity of salt stress, environmental conditions, and their complex interaction (Romero-Munar et al., 2019). Therefore, a systematic understanding of the structure of phyto-microbiome (mainly bacteriobiome and to some extent mycobiome), function, and mechanism of the community under salinity stress is important for designing promising microbial inoculants or synthetic microbial communities (SynComs) in the future.

In this review, we emphasized three perspectives: (1) halotolerant microbiome community structure and their unique taxa, (2) the role of halotolerant bacteria (above ground and below ground) in plant resilience to salinity stress, and (3) mechanisms by which microbiomes enhance the growth of the plants in conditions of saline stress. Adding to these, we explored promising approaches coupling traditional culture-dependent approaches with high-throughput technologies in the development of beneficial microbial consortia for future agricultural systems. In recent times, climate change is invariably impacting natural cycles, and has lowered plant productivity. A triangular conjecture of soil, plant microbiome, and plant growth may hold tremendous potential for the development of stress-tolerant crops.

12.2 EXPLORATION OF PLANT-ASSOCIATED MICROBIOME

Plants harbor a variety of microbes that colonize across their entire structure, some of which, the rhizosphere inhabitants, are closest to the root system, while others stay on the above-ground organs like leaves, and are called phyllosphere inhabitants (Figure 12.1a). Microorganisms that are able to colonize on aerial surfaces of plant organs (epiphytes) or live within the plant tissue (endophyte) comprise diverse groups of microbes. Halo-tolerant PGPB support the growth and overall development of plants in harsh salinity stress or assist the plants in withstanding a saline environment. AMF symbionts help plants to combat abiotic stress and stimulate plant growth by enhancing the host's water absorption capacity, nutrient acquisition, and producing plant hormones.

The host-microbiome (including microbiota and their genomes) interaction differs rapidly due to environmental stressors like fluctuation of temperature, wind, ultraviolet (UV) radiation, water, and nutrient scarcity.

12.2.1 Rhizosphere

Microbial communities inhabiting three separate rhizo-compartments, namely, the root surface (rhizosphere), adjacent to the root (rhizoplane), and inner root tissue (root endosphere), are crucial for maintaining the soil ecosystem and plant development. Thus, tripartite interaction between plant, soil, and microbiome is essential for shaping the community structure and ecosystem functions. The association between rhizosphere-residing microorganisms and plant roots is vibrant and dynamic. Under salinity stress, plant roots released exudates that attract specific microbes. The composition of root exudates, namely, soluble and diffusible, mainly sugar, phenolic acid, amino acids, sterol, and organic acids, served as a chemoattractant for soil microorganisms. A recent study has shown that many strains of *Pseudomonas* spp. produced strain-specific metabolites, siderophores, purines, including osmolytes (trehalose) in salinized soil and developed plant species-specific microbial community (Zheng et al., 2024). Salt stress led to root-associated microbiome shift and enrichment of salinity-tolerant microbes. For instance, in an experiment, *Hibiscus hamabo* plants treated with salt (40 and 90mM NaCl) showed changes in the composition of microbial community and reduction in operational taxonomic unit (OTU) richness of the bacterial community (Yuan et al., 2019).

Plant roots are colonized by diverse bacteria. Particularly, plant-growth promoting rhizobacteria (PGPR) are capable of mineralization of inorganic nutrients into the soil (Figure 12.1b). Previous study has described the prevalence of certain halotolerant bacterial families—Firmicutes, Proteobacteria, Bacteroidetes, and Actinobacteria in the root microbiome under salt stress condition (Yadav et al., 2024).

Root-associated fungi such as endophytic mycorrhiza, *Piriformospora*, and free-living *Trichoderma* spp., and viruses have enhanced plant tolerance to salinity. *Trichoderma*, the rhizosphere fungus, possesses antagonistic activity toward soil-borne pathogens, and also promotes plant growth and resistance by reducing ion toxicity (Na^+ and/or Cl^-), the concentration in the soil (Liu et al., 2023). A study by Sabeem et al. (2022) demonstrated that through improvement of the antioxidant system, elevated levels of antioxidant enzymes, *Piriformospora indica* has reduced the Reactive Oxygen Species (ROS) in salt-stressed date palm (*Phoenix dactylifera*). Plants inoculated with AMF often showed increased accumulation of proline content in the chloroplast and cytoplast, enhanced strigolactones (SLs) level, synthesis of jasmonic acid, salicylic acid, sugar, and several inorganic nutrients, suggesting a symbiotic relation with the mycorrhiza and salt tolerance of plants (Begum et al., 2019). Rice (*Oryza sativa* L.) plant inoculated with certain *Glomus* spp., *Acaulospora*, has ameliorated salt stress by regulating root morphological characteristics, root system architecture, root hydraulic conductivities, and upregulating root plasma membrane H^+-ATPase activity (da Silva et al., 2022).

A previous study conferred that viral infection caused by plum pox virus (PPV) and potato virus X (PVX) in tobacco plant (*Nicotiana benthamiana*) supported plant defense and was adaptive to osmotic gradients (Hernández-Walias et al., 2022).

12.2.2 Phyllosphere

The phyllosphere consisted of both vegetative organs, such as leaves and stems, and reproductive organs, which includes flowers, fruits and seeds. Although microbes dwelling in leaves played crucial roles in a variety of biological processes, including

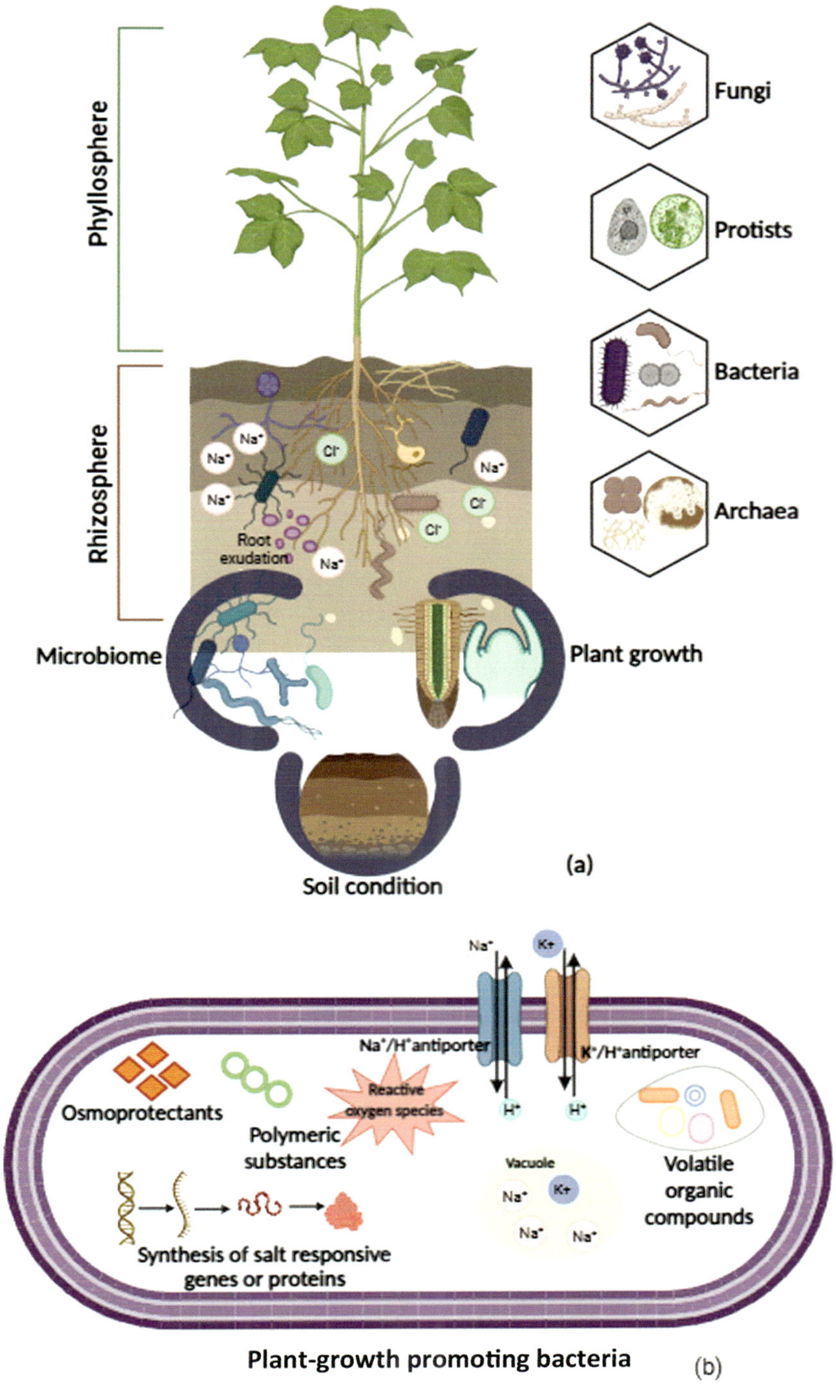

FIGURE 12.1 A pictorial representation of plant microbiome. (a) General overview of soil salinity, plant growth, and functioning of plant-associated microorganisms involved in enhancement of plant tolerance against salt stress. (b) Metabolic and genetic properties of halophilic plant growth-promoting bacteria in alleviating salinity stress in plants. Various mechanisms included the production of osmoprotectants, exopolysaccharides, volatile organic compounds; channels/transporters-mediated ion homeostasis; upregulation of reactive oxygen species; expression of stress-responsive genes, which produce ameliorated salinity stress in plants, e.g., ACC deaminase, phytohormones, siderophore, nitrogen uptake, phosphate solubilization.

health and vitality of plants, foliar functional traits, defense mechanism against pathogens, etc., bacterial community in leaf influenced ecosystem productivity through their role in defining host phenotypic plasticity (Wagner, 2021). In the last few years, phyllosphere microbiome has been studied extensively, but there is still knowledge gap of how and why microbial communities vary across spatio-temporal organization within the plant leaves. The epiphytic microbial communities, unlike rhizosphere are subjected to hostile environmental elements like swings of temperature, variation of humidity levels, exposure to sunlight solar radiation, UV radiation, CO_2 and often nutrient stress, etc., strongly affect phyllosphere microbiome. For instance, salinity affected microbial structure in *Tamarix chinensis*. Qu et al. (2024) demonstrated that salt stress reduced abundance of Actinobacteria and an increase in the abundance of Proteobacteria in *T. chinensis*. In addition, humidity served as major factor that shaped microbial diversity in *T. aphylla*. When humidity is low, bacterial population on the dry leaf surface exposed to extreme daytime desiccation, particularly in arid regions, but dew that condensed at night allows bacterial population to proliferate in a highly concentrated salt solution (Qvit-Raz et al., 2008). Ibekwe et al. (2021) observed significant effect of seasons (March, May, and June) in spinach (*Spinacia oleracea*) leaf surface-inhabiting bacterial community. Salinity significantly affected the abundance of bacterial family Halomonadaceae, followed by Weeksellaceae. In addition, harvesting months of May and June influenced the growth of individual bacterial species. In another study, richness of epiphytic phyllosphere bacteria and fungi in holm oak leaves (*Quercus ilex*) increased in summer, as compared to spring, due to increased colonization time in summer (Peñuelas et al., 2012).

The phyllosphere of mangrove vegetation is abundant in a diverse array of microorganisms that help in salt-stress adaptation, such as seawater splashing. Salt secretor mangroves (with salt glands) like *Acanthus ilicifolius*, *Avicennia marina*, and *Aegiceras corniculatum* had different community than nonsecretor mangroves (without salt glands) like, *Kandelia candel*, *Sonneratia apetala*, and *Excoecaria agallocha* (Yang et al., 2024). For instance, dominant bacterial species like *Salinisphaera*, *Halomonas*, *Kushneria*, *Fulvimarina*, etc. were obtained from salt-secreting plants. The enzymes that promoted nutrient cycling were associated with nitrate reduction (nitrate reductase, NADH), phosphorus utilization (3-phytase), and assisted plant stress resistance (catalase), and were abundant in this type of mangrove. In contrast, some PGPB, like *Bacillus*, *Pantonea*, and *Pseudomonas*, inhabiting nonsecretor plants have the ability to enhance growth and nutritional status of host by delivering essential nitrogen and phosphorus. Abundance of superoxide reductase, acid phosphatase, catalase, and nitrate reduction (nitrate reductase, NO-forming) was notably higher in nonsecretor mangroves. Therefore, microbial communities inhabiting mangroves that lack salt glands are better able to withstand salinity stress.

For fungal communities, *Cladosporium* is recognized as the most abundant taxon in the phyllosphere of *Avicennia marina*. Recent finding demonstrated that yeast (*Hortaea* sp.) population reduced with higher salt concentration and *Neodevriesia* population tends to increase at middle salinity (Yang et al., 2023). Similar to this study, *Cladosporium* was reported as the most abundant fungi in both secretor and nonsecretor mangroves (Yang et al., 2024).

12.3 MICROBES AND THEIR FUNCTIONS IN MANAGING SALINITY STRESS

Microorganisms isolated from salt-stress-prone areas are adapted to salt-tolerant traits and accelerate salinity-affected plant growth. The involvement of microorganisms related to plants, such as PGPB and AMF, can improve plant growth in a saline environment. PGPB help to mitigate salinity stress by modulating phytohormone levels, solubility of nutrients, producing protective bioactive components, and thereby keeping the host plant's ion-homoeostasis intact. Kaleh et al. (2023) have reported on the development of plants under stress conditions, highlighting the functions of rhizobacteria, *Pseudomonas* sp., and *Bacillus* sp. Recent research has shown that fungal microbiota in mulberry (*Morus* sp.) rhizosphere have improved the plant's response to a high-salinity environment. Dominant fungal genera belonged to the phylum Ascomycota and Mucoromycota (Lei et al., 2025). AMF symbionts, with their extensive mycelial network, increase plant resistance by promoting nutrient assimilation, water absorption, regulating plant hormones, ion balance, etc. Abundant occurrence of AMF like *Gigaspora* sp., *Diversispora* sp., *Rhizophagus irregularis, and Funneliformis mosseae* was found in salt-marsh plants (Wahab et al., 2023). Co-inoculation of *Glomus* sp. and *Streptomyces* sp. enhanced the uptake of nutrients and improved growth indices (Karimi and Noori, 2022). Some archaeal species live in the plant–soil interface-mediated nitrogen absorption and iron acquisition of plant roots. The halophilic archaea, *Halolamina* and *Haloferax*, that live in the saline Rann of Kutch desert in Gujarat, India, have lessened the effect of salinity stress in Banni grasses (*Dicanthium*, *Cenchrus*) (Yadav et al., 2019). The abundance of archaeal phyla, Crenarchaeota and Thermoplasmatota in the rhizosphere of mulberry (*Morus* sp.), helped the plant to grow in saline soil (Lei et al., 2025).

12.3.1 Bacterial Community Structure in Plants that Grew under Salinity

Rhizosphere bacteria deserve close attention because of their pivotal role in managing plant nutrients during various abiotic stresses. Water holding capacity, structural integrity, nutrient accessibility, pH levels, and microbial diversity are all diminished in saline soil. Soil salinity profoundly influenced the composition and activities of the rhizosphere microbiome. The makeup and functions of the microbial population are changed by the variation in salinity levels. Studies have shown that halotolerant bacterial genera belonging to Gamaproteobacteria, Alphaproteobacteria, and Bacteroids were associated with *Suaeda salsa* (Amaranthaceae) plant development. Kearl et al. (2019) observed a substantial effect of the rhizosphere bacterial community on the growth of forage crop, alfalfa (*Medicago sativa*), in contrast with plants that were not inoculated. Bacterial community succession in saline-alkaline land, such as Gamaproteobacteria (*Halomonas* and *Kushneria*), Firmicutes (*Bacillus* and *Virgibacillus*), and Actinobacteria (*Zhihengiluella*), enhanced the richness of raw alfalfa.

Teixeira et al. (2013) studied rhizomicrobiomes of Antarctic hair grass (*Deschampsia antarctica*) and Antarctic pearlwort (*Colobanthus quitensis*) and found Firmicutes (mainly Clostridia) as the most dominant phylum. However, Bacteroidetes, Proteobacteria (Gama and Alpha proteobacteria), and Actinobacteria were prevalent phyla, isolated from the plant rhizosphere in different parts of the Antarctic Peninsula (Prekrasna et al., 2022). Minor variations in the composition of bacterial communities may be attributed to changes in the root microbiome. The impact is primarily driven by two key factors—the type of soil and the limitation of the plant species diversity in the Antarctic soil. Similarly, rhizosphere bacterial communities found in Arctic vascular plants, namely, *Cassiope tetragona*, *Silene acaulis*, and *Dryas octopetala* exhibited dominance of Proteobacteria, followed by Acidobacteria and Bacteroidetes (Son and Lee, 2022). Relative abundance of Firmicutes and Bacteroidetes in both polar deserts positively correlated with increased salinity of the soil. These findings provide a helpful linkage among the rhizobacterial communities. Another study revealed the ACC deaminase enzyme-producing rhizobacteria, *Klebsiella* sp., associated with native shrub, *Parastrephia quadrangularis* (Asteraceae) from Atacama Desert, capacitated a notable increase (45–60%) in wheat (*Triticum aestivum*) seedling biomass (Acuña et al., 2019). Rhizosphere-associated genera of the perennial speargrass (*Stipagrostis ciliate*) and annual succulent, *Tetraena stapffii*, in the Namib Desert mostly belonged to Proteobacteria, Actinobacteria, and Firmicutes. Most of these members functioned as PGPB (Maphosa et al., 2025). The dominance of the Enterobacteriaceae family (mainly *Porphyromonas*) from the rhizospheric soil of desert salt grasses (*Distichlis spicata*) was shown in a recent investigation (Castro-Severyn et al., 2024). These data suggested that bacterial communities associated with desert plants (hot and cold) harbor a large bacterial biodiversity. Furthermore, it is evident that the host genotype of the pioneer plants possessed the ability to select the rhizobiome by secreting molecules in the rhizosphere. Rhizosphere bacterial communities of coastal halophyte, Mediterranean Sea lavender (*Limonium axillare*) were dominated by Firmicutes (*Bacillus* and *Staphylococcus*) population in the temperate region (Alhaddad et al., 2024). On the other hand, *Phragmites australis*, grown in the Subtropical salt marshes, high occurrence of Proteobacteria and Bacteroidetes, which were recorded by Wang et al. (2023). The findings suggested that the host plant's adaptation to salt stress is largely dependent on bacterial communities. Moreover, a corresponding increase in the number of bacterial phyla directed the importance of the specific edaphic factors (physical and chemical components) linked to each region as well.

12.4 MECHANISM UNDERLYING SALT STRESS MITIGATION BY THE PLANT-ASSOCIATED MICROBES

High salinity negatively affected plant growth by limiting nutrient uptake, causing ion imbalance, impairing photosynthesis, and reducing seed germination. Excessive salt build-up in the soil ultimately led to osmotic stress, oxidative stress, and mineral toxicity. Plant cells initiated metabolic processes in response to salt stress, which brought changes in the exudation of root and shoot. These exudates have an impact on the beneficial microbes associated with plants and their community composition. However, excessive salinity reduced the functioning of soil microbial communities. Plant microbiota, particularly rhizosphere resident, helped plants to cope with salinity through direct and indirect mechanisms.

12.4.1 Direct Action

Plant-associated microbes produced various compounds, extracellular polymeric substances (EPS), biofilm, siderophore, etc. Additionally, PGPB assisted in maintaining homeostasis in plant cells by increasing antioxidant activity during salinity stress (Siddika et al., 2024). EPS attached itself with soil cations like Na^+, Ca^{2+}; thus, the ions could not get to the stem. This way, EPS reduced the amount of bioavailable ions that plants can absorb. Additionally, EPS facilitated plant–microbe interaction, thereby imparting salt tolerance to plants (Alhoqail, 2025). Some microorganisms that are associated with plant growth and their various mechanisms are listed in Table 12.1.

Polyamines (PAs) are aliphatic amines, namely, spermidine (Spd), spermine (Spm), and putrescine (Put), which acted as physiological effectors and signal molecules in plant–microbe interactions. Polyamines functioned as ROS homeostasis, improved photosynthesis by decreasing the restrained impacts of salinity, overall, augmented growth and development of plants in stressful environments (Ahmed et al., 2023). Upregulation of spermidine synthase, spermine synthase, arginine decarboxylase, and putrescine synthase genes under salinity stress conditions increased polyamine content (Mahmud et al., 2024).

Plant-associated microbes generated volatile organic compounds (VOCs), which made plants more resistant to diseases and stimulated plant growth under salinity stress. Acetoin, a VOC produced by rhizobacteria, has been shown to protect plants against salt stress (Cappellari and Banchio, 2020). Microbial VOCs (mVOCs) helped plants in water acquisition by influencing root architecture, also increasing quantum yield by increasing chlorophyll content (Schenkel et al., 2018). Some mVOCs are able to exert induced resistance (IR) against NaCl stress (Gil et al., 2023).

Biofilm formation on the root surface conferred strong adherence and promoted microbial colonization. Thus, helping in microaggregation of soil, which improved soil structure (Anju et al., 2021). In addition, improved cation exchange capacity of soil, which indirectly altered soil salinity. Additionally, they have been shown to reduce sodium levels in saline water (Ahsan et al., 2023).

High Na^+ content in the saline hindered the plant's ability to take up macronutrients such as nitrogen (N).

TABLE 12.1
Potential Applications of the Halotolerant Microorganisms with Multiple PGP Traits on Improving Salinity Stress Tolerance in Plants

Direct Action	Host Plant	Halotolerant Bacteria	Fungi	Archaea	Reference
Iron solubilization/ Siderophore	*Avicennia marina*	*Oceanobacillus picturae*	-	-	El-Tarabily and Youssef, 2010
	-	-	-	*Halorubrum saccharovorum* *Halococcus* sp.	Naitam and Kaushik, 2021
	Tagetes patula nana	-	*Glomus etunicatum*	-	Winkelmann, 2017
Phosphate solubilization	*Avicennia marina*	*Oceanobacillus picturae*	-	-	El-Tarabily and Youssef, 2010
	Lotus japonicas	*Bacillus gibsonii*	-	-	Castaldi et al., 2023
	Acacia gerrardii	-	*Claroideoglomus etunicatum* *Funneliformis mosseae*	-	Hashem et al., 2016
	Triticum aestivum	-	-	*Halolamina* sp.	Naitam and Kaushik, 2021
Potassium solubilization	*Solanum lycopersicum*	*Bacillus licheniformis*	*Aspergillus violaceofuscus*	-	Muthuraja and Muthukumar, 2022
	Gossypium hirsutum	*Bacillus mucilaginosus*	*Aspergillus terreus*	-	Hamid and Bashir, 2019
	Brassica napus	*B. edaphicus*			
Nitrogen (N_2) fixation	*Medicago sativa*	*Kosakonia cowanii* *Sinorhizobium meliloti*	-	-	Noori et al., 2018
	Zea mays	*Citrobacter* sp. *Klebsiella* sp.	-	-	Ji et al., 2024
	Triticum durum	-	*Rhizophagus irregularis*, *Funneliformis mosseae*	-	Fileccia et al., 2017
ACC deaminase	*Vigna radiata*	*Enterococcus* sp., *Pantoea* sp.	-	-	Panwar et al., 2016
Phytohormone (IAA)	*Lotus japonicas*	*Bacillus amyloquefaciens*	-	-	Castaldi et al., 2023
	Malus domestica	-	*Funneliformis mosseae*	-	Gao et al., 2020
	Brassica oleracea	-	*Verticillium lecanii*	-	Taheri, 2025
	Rhizosphere of wild grass	-	-	*Halococcus hamelinensis*	Naitam and Kaushik, 2021
Exopolysaccharide	*Phragmites australis*	*Glutamicibacter* sp.	-	-	Hidri et al., 2023
Polyamine	*Salicornia bigelovii*	*Streptomyces euryhalinus*	-	-	El-Tarabily et al., 2020
	Hordeum brevisubulatum	-	*Epichloë bromicola*	-	Chen et al.,2019
	Zea mays	-	*Funneliformis costrictum*	-	Ahmed et al., 2023
Biofilm	*Lotus japonicas*	*Bacillus halotolerans*	-	-	Castaldi et al., 2023
	Oryza sativa	*Bacillus safensis*	-	-	Ahsan et al., 2023
Volatile organic compounds (VOCs)	*Arabidopsis thaliana*	*Pseudomonas simiae* *Paraburkholderia phytofirman*	-	-	Ledger et al., 2016
	Mentha piperita	*Pseudomonas putida*	-	-	Gil et al., 2023
	Eleusine coracana	-	*Aspergillus sojae*	-	Mahadik and Kumudini, 2024

Where "-" indicates not identified

Certain halotolerant bacteria secreted nitrogenase enzyme that transformed atmospheric nitrogen (N_2) into ammonia (NH_3) (Ji et al., 2024). In order to obtain nitrogen, the roots absorb ammonia. Although increased salinity significantly affected nitrogen fixation activity, cyanobacterial nitrogenase activity persisted. Cyanobacterial flavodoxin from *Anabaena variabilis* showed tolerance to salinity in transgenic *Medicago truncatula* lines (Coba de la Peña et al., 2010).

Insoluble iron (Fe), trivalent cation (Fe^{3+}) in soil is inaccessible to plants. Siderophore molecules are iron-chelating compounds, synthesized by bacteria and fungi, that help to sequester ferric ions present in the rhizosphere (Naitam and Kaushik, 2021). Siderophore scavenged ferric iron (Fe^{3+}) from extracellular environment and reduced it to ferrous iron (Fe^{2+}) and then transported into the cell through membrane receptor (Zamanzadeh-Nasrabadi et al., 2023). Hence, siderophore produced by PGPB or AMF enhances metal bioavailability.

Inorganic phosphorus (P_i) and potassium (K) solubilization ability of PGPB enhanced the ability to withstand salt stress in plants. Low bioavailability of phosphorus in soil hinders plant growth. Phosphate-solubilizing bacteria converted insoluble phosphorus to monobasic ($H_2PO_4^-$) and dibasic phosphate (HPO_4^{2-}) anions that are taken up by plants (Teles et al., 2024). Bacteria secreted organic acids (gluconic acid, butyric acid, 2,3-dimethyl fumaric acid, acetic acid) or by enzymes (phosphatase) and dissolved the mineral complexes through the release of H^+ ions.

Plants experienced sodium-induced potassium starvation in salt-affected soil due to imbalance of cytosolic K^+/Na^+ ratio. Since Na^+ interfered with potassium homeostasis, halotolerant microbes contributed to the solubilization of potassium (Muthuraja and Muthukumar, 2022). Bacteria can break down insoluble K-containing mineral sources like mica, biotite, potassium-feldspar, muscovite, and illite by proffering organic acids (succinic acid, oxalic acid, citric acid, tartaric acid) and liberated soluble K^+, making them bioavailable to plants (Hamid and Bashir, 2019). Furthermore, microbes sequestered other macronutrients like zinc, copper, and boron.

The phytohormones gibberellin, indole acetic acid, cytokinin, and abscisic acid are produced by rhizobacteria that can modulate endogenous hormone levels in plants. It has been well established that, as a consequence of environmental stresses, emission of stress hormone, ethylene, rises significantly, especially in stress-sensitive plants, which subsequently prevented the root elongation and inhibited shoot growth (Taheri, 2025). On the other hand, root-associated microbes, PGPB, like *Pseudomonas* spp., *Bacillus* spp., and AMF, like *Glomus* sp., *Trichoderma* sp., *Gigaspora* sp., contain 1-aminocyclopropane-1-carboxylate (ACC) deaminase, which regulates plant growth under stress by limiting ethylene concentrations within the plants (Orozco-Mosqueda et al., 2020).

Microbial ACC-deaminase cleaved the ACC (served as precursor of ethylene synthesis in plants) compound into ammonia and α-ketobutyrate, thus diminishing ethylene production.

Plant-associated bacteria possessed the capacity to synthesize indole-3-acetic acid (IAA) as a secondary metabolite (Kaleh et al., 2023). In addition, under stressful conditions, plant development was regulated by the interplay of ACC deaminase activity and IAA production.

IAA promoted ethylene biosynthesis via transcription of ACC synthase, resulting in an increase in ethylene level, which created inhibitory effects on plant growth (Taheri, 2025). This surge of ethylene led to initiate feedback mechanism that inhibited IAA production, which hinders the positive effects of IAA-catalyzed plant growth.

12.4.2 Indirect Action

Indirect mechanisms included hydrogen cyanide (HCN) production, induced systemic tolerance (IST) (Ryu et al., 2004), modulation of the genes that respond to salt stress, and suppression of lipid peroxidation or malonaldehyde (MDA) synthesis genes (Hidri et al., 2023).

12.5 UNRAVELING RHIZOSPHERE MICROBIOME ASSEMBLY IN PLANTS (*Oryza Sativa*)

Salt stress differentially affected microbiome assembly in plants and recruited a specific bacterial community under stress. Considerable attention on rice (*Oryza sativa*) microbiomes has increased in recent years because rice has served as a fundamental staple crop globally for human consumption. Salinity stress decreased the diversity in microbial communities associated with rice plants. The complexity in community structure depended on several factors. Hence, recognizing the network of rice microbiome across plant compartments (rhizosphere and phyllosphere) is essential for improving salt stress-resilient cultivars. A study, conducted in bulk soil and rhizosphere of Poaceae, showed a reduced trend in alpha diversity (e.g., Chao/Shannon index and Phylogenetic diversity) of bacteria with elevated salinity, but fungal community was less influenced (Li et al., 2024). The sequencing of 16S rRNA gene using MiSeq platform, and HTS methods revealed the abundance of Bacteroidetes with higher salinity, while a sharp decline in Choloflexi and Acidobacteria. Number of fungal species was lower compared to bacteria, yet they established a bipartite network. This finding indicated fungal richness had a strong influence on bacterial community assembly.

Before the emergence of HTS platform, culture-dependent approach recovered *Bacillus*, *Plseudomonas*, *Pantonea*, *Exiguobacterium*, and *Rhizobium* from rhizosphere; *Methylobacterium*, *Enterobacter*, and *Erwinia* from phyllosphere; *Micrococcus*, *Kocuria*, *Caulobacter* from seed microbiome (Ding et al., 2019). However, an amplicon-based study demonstrated that the bacterial community was largely dominated by Proteobacteria (mainly Alpha and Beta-proteobacteria). Actinobacteria, Acidobacteria, and Chloroflexi. Notably, the distribution of Deltaproteobacteria (*Geobacter* and *Desulfococcus*) pointed out oxic–anoxic condition in rice root environment. Anaerobic bacteria belonging to this class and their activity in ferric iron reduction (Fe(III) to ferrous iron, Fe(II)), and sulfate reduction ($[SO_4]_2$ to sulfide $[H_2S]$) provided more nutrients for rice (Ding et al., 2019). Metagenome and proteome-based analyses of above-and below-ground parts of rice made a clear distinction in community composition. The protein sequences in the phyllosphere sample matched (60%) mostly with Alphaproteobacteria (mainly *Methylobacterium* and *Agrobacterium*), whereas rhizosphere inhabitants were composed of 33% of Alphaproteobacteria (mainly *Bradyrhizobium* and *Azospirillum*). DNA-based study revealed dominance of Alphaproteobacteria (35%) in the phyllosphere. Remarkably, rhizosphere sample was composed of Alpha, Beta, and Deltaproteobacteria (Knief et al., 2012). Apart from bacteria, population of methanogenic archaea (*Methanococcus*, *Methanobacteria*, and *Methanomassiliicoccus*) and methanotrophic bacteria (*Methylococcus* and *Methylomonas*) from rice paddies were recorded as important players in methane oxidation (Du et al., 2023).

The fungal community in the rhizosphere sample was dominated by Ascomycota (*Aspergillus* and *Trichoderma*), Basidomycota (*Piriformospora*), and Mortierellomycota (*Mortierella*), was identified by 18S rRNA gene sequencing (Ding et al., 2019). Yin et al. (2023) identified *Papiliotrema*, *Khuskia*, and *Symmetrospora* resided in the rice phyllosphere. Further, the abundance of core fungal taxa varied in cultivated (*Oryza sativa* L.) and wild rice (*Oryza rufipogon*). Compared to the cultivated rice, wild rice has a more intricate and stable fungal network.

12.6 CONSTRUCTION OF SYNTHETIC MICROBIAL COMMUNITIES (SynCom)

Considering the underlying knowledge of how plant holobionts (the plant and its associated microorganisms that interact with each other) impact the plant performance, construction of synthetic microbial communities (SynCom) has become a necessity. SynComs are artificial combination of microbial taxa, culturing under controlled conditions. The goal is to completely imitate the functional and mechanistic insights of a natural microbiome. In order to preserve essential interactions between microorganisms and host, while producing a less complex, original microbial community, SynCom approach has been used to provide functional and mechanistic insight into native microbiota. Formulating inoculants through SynCom has emerged as a valuable tool for establishing synergistic interaction between the members (or their genes), and to perform in varying environmental circumstances (Jing et al., 2024). Numerous studies have found that SynCom inoculation ameliorated detrimental outcomes of salinity in crops. Schmitz et al. (2022) characterized root microbiome originating from Indigofera (*Indigofera argentea*) and constructed a SynCom consisting of *Ensifer* sp., *Bacillus* sp., *Massilia* sp., *Enterobacter* sp., and *Streptomyces* sp. that is able to combat salt stress in tomato (*Solanum lycopersicum*). A SynCom of bacterial endophyte, *Pseudomonas* sp. and *Ensifer* sp., isolated from nodules of *Medicago* spp. enhanced nodulation rate and growth under plethora of abiotic stresses in *Medicago sativa* (Flores-Duarte et al., 2023). The successful application of SynCom in *Vigna radiata* has been demonstrated to mitigate salinity stress (Dubey et al., 2025). Root microbiome analysis of *O. japonica* (Nipponbare variety) and *O. indica* (IR24 variety) revealed that rice genotype recruited specific microbiota. Furthermore, SynCom had improved organic nitrogen condition in *indica*-enriched SynCom, compared to *japonica*-enriched SynCom. Therefore, organic nitrogen is converted to nitrate and ammonium by SynCom bacteria more efficiently (Zhang et al., 2019).

12.7 CONCLUSION AND FUTURE PROSPECTS

The influence of beneficial microbes in the growth and development of plants, and their multifarious roles in combating salinity stress, are now well understood. Our finding is confined to only a certain group of microorganisms, specifically bacteria; the interplay between the microbiota contributing to the plant microbiome is still unknown. A growing number of research has explored microbes associated with plants, especially PGPR. However, diversity, assembly, and functionality of plant-associated microbial communities under variable environmental conditions, including salinity, climate change, and plant growth stages, are still unclear. As microbiome structure is influenced by general stoichiometry, understanding the ecological traits of a microbiome is crucial for sustaining ecosystem productivity or designing synthetic microbial consortia. Concentrated studies are required to characterize cross-kingdom interactions between AMF and PGPB, tailor SynComs, and carry out field trials to establish climate-change resilient plants. Moreover, multi-omics technologies, together with NGS, have ushered transformative possibilities for understanding salinity stress coping mechanisms in plants.

Despite the massive progress in phytomicrobiome studies, unresolved challenges persist due to lab-farm gap. More work is needed to transfer this knowledge into actionable farmer-centric information. A more comprehensive understanding of the microbial populations connected to the plants is required and how this is translated to plant health, and ecosystem as a whole. We anticipate a shift in practice in future, emphasizing precision microbiome engineering to increase yields under extreme conditions. One such fast-developing technology is smart agriculture. The integration of machine learning (ML) and deep learning (DL) can enhance overall productivity, real-time crop insights, and biodiversity monitoring. Artificial intelligence (AI) has been a promising avenue in uncovering the salinity-influenced soils. For instance, algorithms like Artificial Neural Network (ANN), random forest tree ensembles (RF), and soil exchangeable sodium percentage (ESP) can effectively detect plant lesions, classify plants, and evaluate salinity level. As a corollary, digital agriculture might potentially resolve the current challenges of sustainable plant production, reshaping the traditional farming practices.

ACKNOWLEDGMENTS

BD is appreciative of the financial support provided by the Department of Science and Technology (DST), New Delhi, India, for PURSE Phase II program. The authors thank the Department of Botany, University of Burdwan, for providing research infrastructure. Figures were created with BioRender [https://BioRender.com].

AUTHORSHIP CONTRIBUTION STATEMENT

All authors listed have made a substantial, direct, and intellectual contribution to the work. Authors have read and approved the submitted version.

DATA AVAILABILITY

No data was used for the review described in the article. The data that support the present study are available in the cited references.

FUNDING

This research received no external funding.

DECLARATION OF COMPETING INTEREST

The authors declare that they have no known competing financial interests or personal relationships that could be construed as a potential conflict of interest.

ETHICAL STATEMENT

This review does not include any ethical considerations.

ABBREVIATIONS

ACC	1-AMINOCYCLOPROPANE-1-carboxylate
AI	artificial intelligence
AMF	arbuscular mycorrhizal fungi
ANN	artificial neural network
CRISPR	Clustered Regularly Interspaced Short Palindromic Repeats
DL	deep learning
EPS	extracellular polymeric substances
ESP	soil exchangeable sodium percentage
HCN	hydrogen cyanide
HTS	high-throughput sequencing
IAA	indole-3-acetic acid
IST	induced systemic tolerance
MDA	malonaldehyde
ML	machine learning
mVOC	microbial VOCs
NGS	next-generation sequencing
OUT	operational taxonomic unit
PAs	polyamines
PGPB	plant growth promoting bacteria
PGPR	plant growth promoting rhizobacteria
PPV	plum pox virus
PVX	potato virus X
RF	random forest tree ensembles
ROS	reactive oxygen species
SL	strigolactone
SynComs	synthetic microbial communities
VOC	volatile organic compounds

REFERENCES

Acuña JJ, Campos M, de la Luz Mora M, Jaisi DP, Jorquera MA (2019) ACCD-producing rhizobacteria from an Andean Altiplano native plant (*Parastrephia quadrangularis*) and their potential to alleviate salt stress in wheat seedlings. Appl. Soil Ecol. 136:184–90. https://doi.org/10.1016/j.apsoil.2019.01.005

Ahmed M, Abdel-Fattah GG, Holford P, Korany SM, Alsherif EA, Abd Elgawad H, Ulhassan Z, Jośko I, Ali B, Sheteiwy MS (2023) *Funneliformis constrictum* modulates polyamine metabolism to enhance tolerance of *Zea mays* L. to salinity. Microbiol. Res. 266:127254. https://doi.org/10.1016/j.micres.2022.127254

Ahsan Y, Qurashi AW, Liaqat I, Latif A, Afzaal M, Khalid A, Sardar AA (2023) Efficacy of rice husk and halophilic bacterial biofilms in the treatment of saline water. J. Basic. Microbiol. 63:855–67. https://doi.org/10.1002/jobm.202300062

Alhaddad FA, Bitaar ZM, Abu-Dieyeh MH (2024) Diversity, characterization, and potential applications of bacterial endophytes isolated from the halophyte *Limonium axillare*. J. Plant Growth Regul. 1–8. https://doi.org/10.1007/s00344-024-11254-6

Alhoqail WA (2025) Exopolysaccharide-Pproducing PGPR: Mechanisms for alleviating salinity-induced plant stress. P. J Environ. Stud. XX: 1–18. https://doi.org/10.15244/pjoes/196747

Anju VT, Siddhardha B, Dyavaiah M (2021) Effect of soil biofilms on ecological function and impact on soil properties. In: Singh JS, Tiwari S, Singh AK (Eds), Microbes in land use change management. pp. 91–107. Elsevier. https://doi.org/10.1016/B978-0-12-824448-7.00006-1

Baedke J, Fábregas-Tejeda A, Nieves Delgado A (2020) The holobiont concept before Margulis. JEZ-B. 334:149–55. https://doi.org/10.1002/jez.b.22931

Begum N, Qin C, Ahanger MA, Raza S, Khan MI, Ashraf M, Ahmed N, Zhang L (2019) Role of arbuscular mycorrhizal fungi in plant growth regulation: Implications in abiotic stress tolerance. Front. Plant. Sc. 10:1068. https://doi.org/10.3389/fpls.2019.01068

Berg G, Rybakova D, Fischer D, Cernava T, Vergès MC, Charles T, Chen X, Cocolin L, Eversole K, Corral GH, Kazou M (2020) Microbiome definition re-visited: Old concepts and new challenges. Microbiome. 1–22. https://doi.org/10.1186/s40168-020-00875-0

Cappellari LD, Banchio E (2020) Microbial volatile organic compounds produced by *Bacillus amyloliquefaciens* GB03 ameliorate the effects of salt stress in *Mentha piperita* principally through acetoin emission. J Plant Growth Regul. 39:764–75. https://doi.org/10.1007/s00344-019-10020-3

Castaldi S, Valkov VT, Ricca E, Chiurazzi M, Isticato R (2023) Use of halotolerant *Bacillus amyloliquefaciens* RHF6 as a bio-based strategy for alleviating salinity stress in *Lotus japonicus* cv Gifu. Microbiol. Res. 268:127274. https://doi.org/10.1016/j.micres.2022.127274

Castro-Severyn J, Fortt J, Sierralta M, Alegria P, Donoso G, Choque A, Avellaneda AM, Pardo-Esté C, Saavedra CP, Stoll A, Remonsellez F (2024) Rhizospheric bacteria from the Atacama Desert hyper-arid core: Cultured community dynamics and plant growth promotion. Microbiol. Spectr. e00056–24. https://doi.org/10.1128/spectrum.00056-24

Chen L, Liu Y (2024) The function of root exudates in the root colonization by beneficial soil rhizobacteria. Biology. 13:95. https://doi.org/10.3390/biology13020095

Chen T, Li C, White JF, Nan Z (2019) Effect of the fungal endophyte *Epichloë bromicola* on polyamines in wild barley (*Hordeum brevisubulatum*) under salt stress. Plant. Soil. 436:29–48. https://doi.org/10.1007/s11104-018-03913-x

Coba de la Peña T, Redondo FJ, Manrique E, Lucas MM, Pueyo JJ (2010) Nitrogen fixation persists under conditions of salt stress in transgenic *Medicago truncatula* plants expressing a cyanobacterial flavodoxin. Plant. Biotechnol. J. 8:954–65. https://doi.org/10.1111/j.1467-7652.2010.00519.x

da Silva HF, Tavares OC, da Silva LD, Zonta E, da Silva EM, Júnior OJ, Nobre CP, Berbara RL, García AC (2022) Arbuscular mycorrhizal fungi and humic substances increased the salinity tolerance of rice plants. Biocat. Agri. Biotechnol. 44:102472. https://doi.org/10.1016/j.bcab.2022.102472

Ding LJ, Cui HL, Nie SA, Long XE, Duan GL, Zhu YG (2019) Microbiomes inhabiting rice roots and rhizosphere. FEMS Microbiol. Ecol. 95:fiz040. https://doi.org/10.1093/femsec/fiz040

Du X, Wang S, Xu Z, Liu J, Gao Z, Ren X, Ma K, Hu S (2023) Response of soils properties and archaea community to saline-sodic soils under long-term rice-based cropping system. Arch. Agron. Soil Sci. 69:533–49. https://doi.org/10.1080/03650340.2021.2013473

Dubey S, Bhattacharjee A, Oza Y, Saxena SS, Pradhan S, Sharma A, Sharma S (2025) Harnessing Syn Coms for rhizosphere engineering to alleviate salt stress in *Vigna radiata*: From lab experiments to the field. Plant. Physiol. Biochem. 110304. https://doi.org/10.1016/j.plaphy.2025.110304

Dutta B, Datta A, Dey A, Ghosh AK, Bandopadhyay R (2023) Establishment of seed biopriming in salt stress mitigation of rice plants by mangrove derived *Bacillus* sp. Biocat. Agri. Biotechnol. 48:102626. https://doi.org/10.1016/j.bcab.2023.102626

El Sabagh A, Islam MS, Skalicky M, Ali Raza M, Singh K, Anwar Hossain M, Hossain A, Mahboob W, Iqbal MA, Ratnasekera D, Singhal RK (2021) Salinity stress in wheat (*Triticum aestivum* L.) in the changing climate: Adaptation and management strategies. Front. Agron. 3:661932. https://doi.org/10.3389/fagro.2021.661932

El-Tarabily KA, El Baghdady KZ, Al Khajeh AS, Ayyash MM, Aljneibi RS, El-Keblawy A, Abu Qamar SF (2020) Polyamine-producing actinobacteria enhance biomass production and seed yield in *Salicornia bigelovii*. Biol. Fertil. Soils. 56:499–519. https://doi.org/10.1007/s00374-020-01450-3

El-Tarabily KA, Youssef T (2010) Enhancement of morphological, anatomical and physiological characteristics of seedlings of the mangrove *Avicennia marina* inoculated with a native phosphate-solubilizing isolate of *Oceanobacillus picturae* under greenhouse conditions. Plant. Soil. 332:147–62. https://doi.org/10.1007/s11104-010-0280-y

Fileccia V, Ruisi P, Ingraffia R, Giambalvo D, Frenda AS, Martinelli F (2017) Arbuscular mycorrhizal symbiosis mitigates the negative effects of salinity on durum wheat. PLoS One.12:e0184158. https://doi.org/10.1371/journal.pone.0184158

Flores-Duarte NJ, Navarro-Torre S, Mateos-Naranjo E, Redondo-Gómez S, Pajuelo E, Rodríguez-Llorente ID (2023) Nodule synthetic bacterial community as legume biofertilizer under abiotic stress in estuarine soils. Plants. 12:2083. https://doi.org/10.3390/plants12112083

Gao T, Liu X, Shan L, Wu Q, Liu Y, Zhang Z, Ma F, Li C (2020) Dopamine and arbuscular mycorrhizal fungi act synergistically to promote apple growth under salt stress. Environ. Exp. Bot. 178:104159. https://doi.org/10.1016/j.envexpbot.2020.104159

Gil SS, Cappellari LDR, Giordano W, Banchio E (2023) Antifungal activity and alleviation of salt stress by volatile organic compounds of native *Pseudomonas* obtained from *Mentha piperita*. Plants (Basel). 12:1488. https://doi.org/10.3390/plants12071488

Hamid B, Bashir Z (2019) Potassium solubilizing microorganisms: An alternative technology to chemical fertilizers. J. Res. Dev. 19:79–84.

Hashem A, Abd_ Allah EF, Alqarawi AA, Al-Huqail AA, Wirth S, Egamberdieva D (2016) The interaction between arbuscular mycorrhizal fungi and endophytic bacteria enhances plant growth of *Acacia gerrardii* under salt stress. Front. Microbiol. 7:1089. https://doi.org/10.3389/fmicb.2016.01089

Hassani A, Azapagic A, Shokri N (2021) Global predictions of primary soil salinization under changing climate in the 21st century. Nat. Commun. 12:6663. https://doi.org/10.1038/s41467-021-26907-3

Hernández-Walias FJ, García M, Moreno M, Giannoukos I, González N, Sanz-García E, Necira K, Canto T, Tenllado F (2022) Transgenerational tolerance to salt and osmotic stresses induced by plant virus infection. Int. J Mol. Sci. 23:12497. https://doi.org/10.3390/ijms232012497

Hidri R, Metoui-Ben Mahmoud O, Zorrig W, Azcon R, Abdelly C, Debez A (2023) The halotolerant rizhobacterium *Glutamicibacter* sp. alleviates salt impact on *Phragmites australis* by producing exopolysaccharides and limiting plant sodium uptake. Plant. Direct. 7:e535. https://doi.org/10.1002/pld3.535

Ibekwe AM, Ors S, Ferreira JF, Liu X, Suarez DL (2021) Influence of seasonal changes and salinity on spinach phyllosphere bacterial functional assemblage. PLoS One. 2021 16:e0252242. https://doi.org/10.1371/journal.pone.0252242

Ji C, Ge Y, Zhang H, Zhang Y, Xin Z, Li J, Zheng J, Liang Z, Cao H, Li K (2024) Interactions between halotolerant nitrogen-fixing bacteria and arbuscular mycorrhizal fungi under saline stress. Front. Microbio. 15:1288865. https://doi.org/10.3389/fmicb.2024.1288865

Jing J, Garbeva P, Raaijmakers JM, Medema MH (2024) Strategies for tailoring functional microbial synthetic communities. ISME J. 18:wrae049. https://doi.org/10.1093/ismejo/wrae049

Kaleh AM, Singh P, Mazumdar P, Wong GR, Chua KO, Harikrishna JA (2023) A halotolerant plant growth promoting consortium of *Bacillus* sp. RB3 and *Pseudomonas* sp. EB3 primes banana, *Musa acuminata* cv. Berangan, against salinity and Foc-TR4 stresses. Curr. Plant. Biol. 35:100294. https://doi.org/10.1016/j.cpb.2023.100294

Karimi R, Noori A (2022) *Streptomyces rimosus* rhizobacteria and *Glomus mosseae* mycorrhizal fungus inoculation alleviate salinity stress in grapevine through morphophysiological changes and nutritional balance. Sci. Hortic. 305:111433. https://doi.org/10.1016/j.scienta.2022.111433

Kearl J, McNary C, Lowman JS, Mei C, Aanderud ZT, Smith ST, West J, Colton E, Hamson M, Nielsen BL (2019) Salt-tolerant halophyte rhizosphere bacteria stimulate growth of alfalfa in salty soil. Front. Microbiol. 10:1849. https://doi.org/10.3389/fmicb.2019.01849

Knief C, Delmotte N, Chaffron S, Stark M, Innerebner G, Wassmann R, Von Mering C, Vorholt JA (2012) Metaproteogenomic analysis of microbial communities in the phyllosphere and rhizosphere of rice. ISME J. 6:1378–90. https://doi.org/10.1038/ismej.2011.192

Kumar P, Sharma PK (2020) Soil salinity and food security in India. Front. Sustain. Food Syst. 4:533781. https://doi.org/10.3389/fsufs.2020.533781

Ledger T, Rojas S, Timmermann T, Pinedo I, Poupin MJ, Garrido T, Richter P, Tamayo J, Donoso R (2016) Volatile-mediated effects predominate in *Paraburkholderia phytofirmans* growth promotion and salt stress tolerance of *Arabidopsis thaliana*. Front. Microbiol. 7:1838. https://doi.org/10.3389/fmicb.2016.01838

Lei J, Gu H, Liu Z, Hu X, Yu Z, Guan Q, Jin J, Liu X, Wang G, Liu J (2025) Recruitment of specific rhizosphere microorganisms in saline-alkali tolerant rice improves adaptation to saline-alkali stress. Sci. Total. Environ. 963:178413. https://doi.org/10.1016/j.scitotenv.2025.178413

Li R, Jiao H, Sun B, Song M, Yan G, Bai Z, Wang J, Zhuang X, Hu Q (2024) etee. Microorganisms. 12:683. https://doi.org/10.3390/microorganisms12040683

Liu Z, Xu N, Pang Q, Khan RA, Xu Q, Wu C, Liu T (2023) A salt-tolerant strain of *Trichoderma longibrachiatum* HL167 is effective in alleviating salt stress, promoting plant growth, and managing fusarium wilt disease in cowpea. J. Fungi. 9:304. https://doi.org/10.3390/jof9030304

Mahadik SP, Kumudini BS (2024) Halotolerant fungi secreting phytohormones and volatile organic compounds enhance growth and mineral content in finger millet under salinity stress. Plant Stress. 11:100426. https://doi.org/10.1016/j.stress.2024.100426

Mahmud MR, Islam MA, Hu Q, Zhang X, Wang W, Xu N, Yang C, Liu J (2024) Salt stress-induced polyamine biosynthesis contributes to blast resistance in rice. Phytopathol. Res. 6:64. https://doi.org/10.1186/s42483-024-00282-5

Maphosa S, Steyn M, Lebre PH, Gokul JK, Convey P, Marais E, Maggs-Kölling G, Cowan DA (2025) Rhizosphere bacterial communities of Namib Desert plant species: Evidence of specialised plant-microbe associations. Microbiol. Res. 293:128076. https://doi.org/10.1016/j.micres.2025.128076

Muthuraja R, Muthukumar T (2022) Co-inoculation of halotolerant potassium solubilizing *Bacillus licheniformis* and *Aspergillus violaceofuscus* improves tomato growth and potassium uptake in different soil types under salinity. Chemosphere. 294:133718. https://doi.org/10.1016/j.chemosphere.2022.133718

Naitam MG, Kaushik R (2021) Archaea: An agro-ecological perspective. Curr. Microbiol. 78:2510–21. https://doi.org/10.1007/s00284-021-02537-2

Noori F, Etesami H, Zarini HN, Khoshkholgh-Sima NA, Salekdeh GH, Alishahi F (2018) Mining alfalfa (*Medicago sativa* L.) nodules for salinity tolerant non-rhizobial bacteria to improve growth of alfalfa under salinity stress. Ecotoxicol. Environ. Saf. 162:129–38. https://doi.org/10.1016/j.ecoenv.2018.06.092

Orozco-Mosqueda M del Carmen, Glick BR, Santoyo G (2020) ACC deaminase in plant growth-promoting bacteria (PGPB): An efficient mechanism to counter salt stress in crops. Microbiol. Res. 235:126439. https://doi.org/10.1016/j.micres.2020.126439

Panwar M, Tewari R, Nayyar H (2016) Native halo-tolerant plant growth promoting rhizobacteria *Enterococcus* and *Pantoea* sp. improve seed yield of Mungbean (*Vigna radiata* L.) under soil salinity by reducing sodium uptake and stress injury. Physiol. Mol. Biol. Plants. 22:445–59. https://doi.org/10.1007/s12298-016-0376-9

Peñuelas J, Rico L, Ogaya R, Jump AS, Terradas JJ (2012) Summer season and long-term drought increase the richness of bacteria and fungi in the foliar phyllosphere of *Quercus ilex* in a mixed Mediterranean forest. Plant. Biol. 14:565–75. https://doi.org/10.1111/j.1438-8677.2011.00532.x

Prekrasna I, Pavlovska M, Miryuta N, Dzhulai A, Dykyi E, Convey P, Kozeretska I, Bedernichek T, Parnikoza I (2022). Antarctic hairgrass rhizosphere microbiomes: Microscale effects shape diversity, structure, and function. Microbes. Environ. 37:ME21069. https://doi.org/10.1264/jsme2.ME21069

Qin Y, Druzhinina IS, Pan X, Yuan Z (2016) Microbially mediated plant salt tolerance and microbiome-based solutions for saline agriculture. Biotechnol. Adv. 34:1245–59. https://doi.org/10.1016/j.biotechadv.2016.08.005

Qu X, Pan Y, Wang P, Ran L, Qin G, Li Q, Kang P (2024) Response of phyllosphere and rhizosphere microbial communities to salt stress of *Tamarix chinensis*. Plants. 13:1091. https://doi.org/10.3390/plants13081091

Qvit-Raz N, Jurkevitch E, Belkin S (2008) Drop-size soda lakes: Transient microbial habitats on a salt-secreting desert tree. Genetics. 178:1615–22. https://doi.org/10.1534/genetics.107.082164

Romero-Munar A, Baraza E, Gulías J, Cabot C (2019) Arbuscular mycorrhizal fungi confer salt tolerance in giant reed (*Arundo donax* L.) plants grown under low phosphorus by reducing leaf Na^+ concentration and improving phosphorus use efficiency. Front. Plant. Sci. 10:843. https://doi.org/10.3389/fpls.2019.00843

Ryu CM, Farag MA, Hu CH, Reddy MS, Kloepper JW, Paré PW (2004) Bacterial volatiles induce systemic resistance in *Arabidopsis*. Plant. Physiol. 134:1017–26. https://doi.org/10.1104/pp.103.026583

Sabeem M, Abdul Aziz M, Mullath SK, Brini F, Rouached H, Masmoudi K (2022) Enhancing growth and salinity stress tolerance of date palm using *Piriformospora indica*. Front. Plant. Sci. 13:1037273. https://doi.org/10.3389/fpls.2022.1037273

Schenkel D, Maciá-Vicente JG, Bissell A, Splivallo R (2018) Fungi indirectly affect plant root architecture by modulating soil volatile organic compounds. Front. Microbiol. 9:1847. https://doi.org/10.3389/fmicb.2018.01847

Schmitz L, Yan Z, Schneijderberg M, de Roij M, Pijnenburg R, Zheng Q, Franken C, Dechesne A, Trindade LM, Van Velzen R, Bisseling T (2022) Synthetic bacterial community derived from a desert rhizosphere confers salt stress resilience to tomato in the presence of a soil microbiome. ISME J. 16:1907–20. https://doi.org/10.1038/s41396-022-01238-3

Siddika A, Rashid AA, Khan SN, Khatun A, Karim MM, Prasad PV, Hasanuzzaman M (2024) Harnessing plant growth-promoting rhizobacteria, *Bacillus subtilis* and *B. aryabhattai* to combat salt stress in rice: A study on the regulation of antioxidant defense, ion homeostasis, and photosynthetic parameters. Front. Plant. Sci. 15:1419764. https://doi.org/10.3389/fpls.2024.1419764

Son D, Lee EJ (2022) Soil microbial communities associated with three arctic plants in different local environments in Ny–Ålesund, Svalbard. JMB. 32:1275. https://doi.org/10.4014/jmb.2208.08009

Taheri P (2025) Endophytic fungi as regulators of phytohormones production: Cytomolecular effects on plant growth, stress protection and importance in sustainable agriculture. Plant Stress. 17:100978. https://doi.org/10.1016/j.stress.2025.100978

Teixeira LC, Peixoto RS, Rosado AS (2013) Bacterial diversity in rhizosphere soil from antarctic vascular plants of admiralty bay in maritime Antarctica. In: Frans J.de Bruijn (Ed), Molecular microbial ecology of the rhizosphere. 1:1105–12. https://doi.org/10.1002/9781118297674.ch105

Teles EA, Xavier JF, Arcênio FS, Amaya RL, Gonçalves JV, Rouws LF, Zonta E, Coelho IS (2024) Characterization and evaluation of potential halotolerant phosphate solubilizing bacteria from *Salicornia fruticosa* rhizosphere. Front. Plant. Sci. 14:1324056. https://doi.org/10.3389/fpls.2023.1324056

Thepbandit W, Athinuwat D (2024) Rhizosphere microorganisms supply availability of soil nutrients and induce plant defense. Microorganisms. 12:558. https://doi.org/10.3390/microorganisms12030558

Wagner MR (2021) Prioritizing host phenotype to understand microbiome heritability in plants. New. Phytol. 232:502–9. https://doi.org/10.1111/nph.17622

Wahab A, Muhammad M, Munir A, Abdi G, Zaman W, Ayaz A, Khizar C, Reddy SP (2023) Role of arbuscular mycorrhizal fungi in regulating growth, enhancing productivity, and potentially influencing ecosystems under abiotic and biotic stresses. Plants. 12:3102. https://doi.org/10.3390/plants12173102

Wang R, Cui L, Li J, Li W (2023) Factors driving the halophyte rhizosphere bacterial communities in coastal salt marshes. Front. Microbiol. 14:1127958. https://doi.org/10.3389/fmicb.2023.1127958

Winkelmann G (2017) A search for glomuferrin: A potential siderophore of arbuscular mycorrhizal fungi of the genus *Glomus*. Biometals. 30:559–64. https://doi.org/10.1007/s10534-017-0026-x

Yadav AN, Gulati S, Sharma D, Singh RN, Rajawat MV, Kumar R, Dey R, Pal KK, Kaushik R, Saxena AK (2019) Seasonal variations in culturable archaea and their plant growth promoting attributes to predict their role in establishment of vegetation in Rann of Kutch. Biologia. 74:1031–43. https://doi.org/10.2478/s11756-019-00259-2

Yadav J, Srivastva AK, Singh R (2024) Diversity of halotolerant endophytes from wheat (*Triticum aestivum*) and their response to mitigate salt stress in plants. Biocatal. Agri. Biotechnol. 56:103000. https://doi.org/10.1016/j.bcab.2023.103000

Yang X, Dai Z, Yuan R, Guo Z, Xi H, He Z, Wei M (2023) Effects of salinity on assembly characteristics and function of microbial communities in the phyllosphere and rhizosphere of salt-tolerant *Avicennia marina* mangrove species. Microbiol. Spectr. 11:e03000–22. https://doi.org/10.1128/spectrum.03000-22

Yang X, Yuan R, Yang S, Dai Z, Di N, Yang H, He Z, Wei M (2024) A salt-tolerant growth-promoting phyllosphere microbial combination from mangrove plants and its mechanism for promoting salt tolerance in rice. Microbiome. 12:270. https://doi.org/10.1186/s40168-024-01969-9

Yin Y, Wang YF, Cui HL, Zhou R, Li L, Duan GL, Zhu YG (2023) Distinctive structure and assembly of phyllosphere microbial communities between wild and cultivated rice. Microbiol. Spectr. 11:e04371–22. https://doi.org/10.1128/spectrum.04371-22

Yuan Y, Brunel C, van Kleunen M, Li J, Jin Z (2019) Salinity-induced changes in the rhizosphere microbiome improve salt tolerance of *Hibiscus hamabo*. Plant and Soil. 443:525–37. https://doi.org/10.1007/s11104-019-04258-9

Zamanzadeh-Nasrabadi SM, Mohammadiapanah F, Hosseini-Mazinani M, Sarikhan S (2023) Salinity stress endurance of the plants with the aid of bacterial genes. Front. Genet. 14:1049608. https://doi.org/10.3389/fgene.2023.1049608

Zhang J, Liu YX, Zhang N, Hu B, Jin T, Xu H, Qin Y, Yan P, Zhang X, Guo X, Hui J (2019) NRT1. 1B is associated with root microbiota composition and nitrogen use in field-grown rice. Nat. Biotechnol. 37:676–84. https://doi.org/10.1038/s41587-019-0104-4

Zheng Y, Cao X, Zhou Y, Ma S, Wang Y, Li Z, Zhao D, Yang Y, Zhang H, Meng C, Xie Z (2024). Purines enrich root-associated *Pseudomonas* and improve wild soybean growth under salt stress. Nat. Commun. 15:3520. https://doi.org/10.1038/s41467-024-47773-9

13 Mechanism and Perspectives of the Microbiome in Improving Drought Tolerance

Everlon Cid Rigobelo
Graduate Program in Agricultural and Livestock Microbiology, São Paulo State University (UNESP), School of Agricultural and Veterinary Sciences, Jaboticabal, SP, Brazil

13.1 INTRODUCTION

Humans have dominated agriculture since ancient times. This development was significant, as it transformed the lifestyle of primitive civilizations. Previously, humans were nomadic and changed their habitat because of extractivism (Dong et al., 2022). The advent of agriculture allowed civilizations to establish permanent settlements, as it was no longer necessary to relocate (Dong et al., 2017). Agriculture allowed early human civilizations to transition from nomadic hunter-gatherer lifestyles to permanent settlements. This was a major development in the transformation of primitive societies. The ability to cultivate crops and produce food in a fixed location has been fundamental to the development of permanent settlements and cities. Through agriculture, societies are able to ensure food security and provide reliable sustenance for a growing population (Angelakis et al., 2021). As a result, supporting large populations depends on efficient and sustainable agricultural practices that meet the increasing demand for food.

The development of agriculture has led to food surpluses, which have allowed some members of society to pursue other specialized roles beyond food production. Agriculture forms the foundation of most economies, especially early civilizations. It provides raw materials, employment, and traded goods. Advances in agricultural techniques and technology have enabled massive increases in food production to support a growing global population. Agriculture shapes landscapes and ecosystems as land is cultivated for crops and livestock (Charania & Li, 2019). The rise of agriculture is linked to the development of new technologies, social structures, and cultural practices in human society.

Control over food production through agriculture has allowed for the rise of more complex social and political organizations. Agriculture remains vital for modern civilization, although its role has evolved with industrialization and globalization. It is essential for food security and economic stability in many parts of the world (Sheng et al., 2018). In summary, the transition to agriculture was a pivotal development that enabled the rise of permanent settlements, population growth, economic specialization, and increasingly complex human civilization. This fundamentally transformed human society and continued to play a crucial role in supporting the world's population.

Climate change poses significant challenges to agriculture because it affects production, costs, and the environment. Nanobiosensors and nanoformulations (NFs) offer promising solutions for enhancing farming practices and sustainability (Miguel-Rojas & Pérez-De-Luque, 2023). These technologies can detect plant stresses, optimize resource usage, and improve the efficiency of agrochemicals, potentially increasing crop protection and production while reducing their environmental impact.

Nonetheless, the impact of climate change on agriculture is multifaceted and varies significantly across regions. Rising global temperatures have resulted in uneven effects on cropland nitrogen usage and losses, with some countries benefiting from enhanced yields and improved nitrogen use efficiency, while others face declines in both areas (Ren et al., 2023). This regional disparity underscores the importance of developing tailored strategies to effectively address agricultural challenges in diverse environments.

Successfully adapting agriculture to climate change demands a comprehensive and integrated approach. Climate-smart agriculture (CSA) has demonstrated the ability to boost productivity in shifting climate scenarios while simultaneously reducing greenhouse gas emissions (De Pinto et al., 2020). Moreover, the adoption of advanced technologies—including precision farming and genetic improvement—can optimize cropping systems and refine management practices (Raihan et al., 2024). When combined with sustainable methods such as organic agriculture and environmentally responsible techniques, these innovations can help lower emissions and enhance soil organic carbon levels, fostering long-term agricultural resilience and sustainability (Holka et al., 2022).

Water scarcity presents considerable obstacles to farming and generates widespread issues that affect crop yields, food availability, and people's means of subsistence (Adhikari, 2018). Water scarcity is a key factor affecting plant growth and development, causing significant decreases in crop yields worldwide (Verma et al., 2023). This can result in reduced harvests, occasional total crop losses, and shorter

DOI: 10.1201/9781003544425-13

growing seasons owing to higher temperatures and erratic rainfall patterns (Lungu, 2014). For example, in Zambia, traditional farming methods have resulted in a 40% reduction in the national long-term average maize yield due to drought conditions (Lungu, 2014).

Water scarcity exerts a profound impact on plant physiology through multiple mechanisms that directly influence growth, development, and overall productivity (Fadiji et al., 2022). Drought conditions trigger morphoanatomical, physiological, and biochemical changes in plants, including reduced transpiration rates and enhanced water-use efficiency (Bijalwan et al., 2022). These adaptations, however, often result in leaf water deficits that can lead to oxidative stress and, in severe cases, cell death (Bijalwan et al., 2022). Under stress, plants undergo a range of physiological and molecular modifications, many of which serve as adaptive responses to challenging environments (Bijalwan et al., 2022).

Water shortages disrupt critical plant processes such as nutrient absorption, protein metabolism, photosynthesis, carbon allocation, and enzymatic and hormonal activities (Mahpara, 2019). In dry soils, limited nutrient uptake contributes to deficiencies that further hinder normal plant functions (Mahpara, 2019). Interestingly, elevated CO_2 levels can alleviate some negative effects of drought by enhancing photosynthetic rates and improving water-use efficiency (Wang et al., 2017).

The physiological response of plants to drought depends on the intensity of the stress. Moderately stressed plants may exhibit increased vigor and growth, while those experiencing severe stress often accumulate higher concentrations of amino acids (Sconiers et al., 2020). These differences in stress responses can also influence interactions between drought-stressed plants and herbivores (Sconiers et al., 2020). Water scarcity affects a wide range of plant functions, including water relations, nutrient uptake, photosynthetic activity, and metabolic pathways. A comprehensive understanding of these effects is crucial for developing effective strategies to enhance drought tolerance in crops and reduce the adverse impacts of water scarcity on global food production (Chandra et al., 2021).

The notion of "technological drought" has been proposed to explain crop failures and water shortages that occur due to insufficient irrigation technology or inadequate water management practices (Haque Mondol et al., 2022). This concept underscores how socioeconomic factors can amplify the effects of drought, particularly in developing economies, where agricultural production relies on small-scale farmers' ability to regulate the crop water supply (Haque Mondol et al., 2022). Drought poses significant threats to human welfare, including increased risks of illness, mortality, and large-scale population displacement (Verma et al., 2023). Its impact extends to various environmental and agricultural aspects, such as soil carbon depletion, reduced freshwater availability, diminished crop yields, decreased livestock productivity, and disrupted fish migration patterns (Bhattacharyya et al., 2020). These consequences go beyond immediate agricultural losses and affect food security, economic stability, and overall social well-being. As climate change exacerbates drought conditions, addressing these challenges is crucial to maintain sustainable agriculture and ensure global food security.

Drought poses significant challenges to food production and causes various problems in agricultural systems. The consequences of drought, including crop failure, reduced yields, and water scarcity, can have severe implications for food security (Haque Mondol et al., 2022). Drought negatively affects agricultural productivity by affecting crop physiology, morphology, and yield (Ray et al., 2020). In the United States, meteorological drought has been associated with 13% of the overall yield variability in maize and soybean production (Zipper et al., 2016). The impact of drought varies depending on factors such as crop type, soil texture, and timing. For example, lentils, groundnuts, and pigeon peas exhibit less drought-induced yield reduction than do cowpea and green grams (Daryanto et al., 2015). In legumes, drought during the reproductive stage typically results in greater yield reduction than that during the vegetative stage (Daryanto et al., 2015). Additionally, the concept of "technological drought" highlights how socioeconomic conditions can exacerbate the effects of water scarcity, particularly in developing economies (Haque Mondol et al., 2022).

Droughts significantly affect food production by reducing crop yields, altering cropping patterns, and threatening food security. It also affects soil health, leading to long-term sustainability issues in agricultural production (Cuhadar, 2024). The effects of drought on agriculture present major challenges in terms of both productivity and agricultural sustainability, necessitating adaptation measures to mitigate these effects (Cuhadar, 2024). Climate change is expected to intensify these challenges, potentially influencing global food prices and regional food security (Ejaz Qureshi et al., 2012). Addressing drought in modern agriculture requires a comprehensive approach that integrates various strategies and technologies. The approach to drought management has shifted from a reactive crisis response to proactive risk mitigation, necessitating a deeper understanding of agroecology to develop effective solutions (Rane & Minhas, 2017). Key strategies for preventing drought include watershed management, onsite water preservation, and integrated farming systems that incorporate resilient and contingency crop plans (Rane & Minhas, 2017). Advanced techniques for assessing agroecosystems, stress reduction methods, and genetic modifications of crops for drought resistance are essential (Rane & Minhas, 2017). Notably, while technological progress is important, conventional practices, such as contour bunding, continue to be widely utilized in rain-fed agricultural regions to reduce the impact of drought (Desalegn et al., 2006). Additionally, climate-smart farming techniques, crop diversification, and agroforestry have demonstrated efficacy in drought-prone areas (Adhikari, 2018). Small-scale water-harvesting structures, such as rainwater collection systems, conservation ponds, and drip irrigation, can also be advantageous (Adhikari, 2018).

Addressing droughts in contemporary agriculture requires a holistic strategy. This encompasses effective water management, enhanced irrigation methods, the utilization of drought-resistant crop varieties, soil conservation practices, and supportive agricultural policies (Cuhadar, 2024). Sustainable farming systems may also reduce vulnerability to climate risks by fostering adaptability among farmers (Knutson et al., 2011). The advancement of modern water-efficient agriculture is crucial for drought adaptation, with the ultimate goal of sustainable socioeconomic development (Yang et al., 2015). Table 13.1 illustrates the utilization of microorganisms to enhance plant growth, thereby aiding in the mitigation of abiotic stresses such as drought.

13.2 PROBLEMS CAUSED BY DROUGHT

Food security is profoundly affected by drought through various mechanisms at different levels. Common consequences of droughts include crop failure, pasture loss, and water source depletion, resulting in food shortages, water scarcity, and livestock losses (Mdemu, 2021). Agricultural regions experience severe reductions in crop yields and production during droughts. For example, extreme drought events in South Africa's Free State Province in 2015 and 2018 caused significant decreases in maize and sorghum production, jeopardizing food security (Orimoloye, 2022). The effects of drought on food security are not uniform and depend on socioeconomic factors. Households with superior social conditions, more resources, and diverse income sources generally show less vulnerability to drought (Mdemu, 2021). The relationship between drought and food security is multifaceted and involves several factors. For example, in the Middle East, food security is influenced by drought, livestock numbers, population growth, and agricultural production (Hameed et al., 2019).

Drought and climate change are intricately linked, with climate change worsening global drought conditions. Research indicates that climate change alters precipitation patterns, increases temperatures, and intensifies evaporation, contributing to more frequent and severe droughts (Patel & Patel, 2024). Semiarid regions are particularly susceptible to drought because of rising global temperatures and changing precipitation patterns (Patel & Patel, 2024). Climate change is anticipated to create more arid background conditions, intensifying land-atmosphere feedback and resulting in more frequent, severe, longer-lasting, and geographically extensive droughts (Han & Singh, 2023). Notably, some areas may experience increased average annual precipitation along with more erratic rainfall patterns. This shifts the nature of drought from historical patterns of reduced rainfall to more unpredictable precipitation that can threaten livelihoods (Memon et al., 2018). This underscores the complex relationship between climate change and drought, where changes in precipitation timing and intensity can be as impactful as overall rainfall.

Climate change has a profound impact on the frequency and severity of droughts. Human-driven global warming is responsible for more than 30% of increased drought intensity and geographic spread, primarily due to enhanced evaporation rates (Bevacqua et al., 2024). Significantly, these effects are not limited to immediate climate conditions; up to 41% of climate change's contribution to drought severity is linked to lingering soil dryness from previous years, underscoring the need to account for delayed and cumulative impacts when assessing drought risk and developing management strategies (Bevacqua et al., 2024).

Drought represents a major threat to food security, affecting its four fundamental pillars: availability, access,

TABLE 13.1
Illustrates the Utilization of Microorganisms to Enhance Plant Growth, Thereby Aiding in the Mitigation of Abiotic Stresses such as Drought

Microorganisms	Effect on Plants	Reference
Microbiome	Reduced diseases	Singh et al. (2020)
Ectomycorrhizal fungi	Increased nutrient uptake	Li et al. (2021)
Microbiome	Promoted plant growth	Pang et al. (2020)
Microorganisms	Drought resistance	Ebadi & Etesami (2025)
Firmicutes	Drought resistance	Balestrini et al. (2017)
Associated Microbes	Drought resistance	Xie et al. (2021)
SynComs	Drought resistance	Poudel et al. (2021)
Microbiome	Drought tolerance	Lopez et al. (2003)
Microbiome	Drought tolerance	Sato and Yokoya (2007)
Microbiome	Stress defense	Azri et al. (2021)
Microbiome	Stress defense	Vrantsi et al. (2021)
Microbiome	Environmental adaptation	Vigneron et al. (2024)
Microbiome	Environmental adaptation	Anderson et al. (1995)
Microbiome	Increased dry weight	Lalas et al. (2018)
Microbiome	Drought resistance	Orimoloye (2022)
Microbiome	Drought resistance	Afridi et al. (2022)
Microbiome	Drought resistance	Rossmann et al. (2017)

utilization, and stability (Mafongoya et al., 2022). To mitigate these challenges, a range of adaptive strategies have been proposed, including the construction of water reservoirs, deployment of temporary irrigation systems, provision of agricultural insurance (Pratiwi et al., 2020), and establishment of drought early warning systems (Akwango et al., 2017). Proactive adaptation to climate change and increased investment in agricultural productivity are essential for safeguarding food security, especially in regions that are highly susceptible to drought (Ejaz Qureshi et al., 2012).

13.3 PLANT MICROBIOME

The plant microbiome consists of a wide variety of microorganisms, such as bacteria, archaea, and fungi, which reside in both the internal (endosphere) and external (episphere) parts of the plant tissues (Singh et al., 2020). These microbial communities are considered essential extensions of the plant's genetic framework and have evolved alongside their host plants over time (Dastogeer et al., 2020). The plant microbiome is crucial for promoting plant growth, development, and health by aiding nutrient absorption, boosting resistance to biotic and abiotic stresses, and supporting various physiological functions (Hao et al., 2024). The composition and diversity of the plant microbiome are influenced by several factors, including the genetics of the host plant, soil properties, environmental conditions, and specific plant organs or tissues involved (Rossmann et al., 2017). Notably, recent progress in high-throughput community analysis and next-generation sequencing (NGS) technologies has significantly enhanced our ability to explore and comprehend the composition and dynamics of plant-associated microbial communities (Afridi et al., 2022). In conclusion, the plant microbiome represents a highly complex and dynamic ecosystem that has garnered significant scientific interest because of its potential to advance sustainable agriculture. By strategically managing these microbial communities, researchers and practitioners can tackle key challenges in crop production and protection, such as reducing the dependence on chemical pesticides, increasing agricultural yields, and promoting environmental sustainability (Singh et al., 2020; Hao et al., 2024). In particular, the rhizosphere microbiome is crucial for maintaining plant health, supporting plant development, and enhancing overall plant resilience. Through various mechanisms, these microorganisms bolster plant defenses against pests and diseases, facilitate the uptake and cycling of essential nutrients, and help plants endure abiotic stresses such as drought and soil salinity (Pérez-Jaramillo et al., 2015; Malviya et al., 2022). The specific roles of the rhizosphere microbiome include nutrient mineralization, hormone regulation, and suppression of plant pathogens (Hu et al., 2021). Importantly, the presence of beneficial microbial communities can help mitigate ecological stress in plants, potentially reducing the need for synthetic fertilizers and pesticides, while encouraging sustainable crop management (Malviya et al., 2022). Interestingly, the diversity and composition of the resident plant rhizosphere microbiome can have a more significant effect on promoting plant growth than direct introduction of beneficial plant traits through probiotic consortia (Hu et al., 2021). Furthermore, certain microbial volatile organic compounds (mVOCs) produced by bacteria isolated from agaves and cacti have shown promising effects on plant growth and development, even in arid and semiarid environments (Camarena-Pozos et al., 2018).

In summary, the plant microbiome substantially influences plant fitness through various mechanisms, including improved nutrient uptake, growth stimulation, stress tolerance, and pathogen resistance. Understanding and harnessing these plant–microbiome interactions can result in increased crop productivity and sustainability in agriculture (Xie et al., 2021).

13.4 RHIZOBIOME

The rhizobiome, also referred to as the root-associated or rhizosphere microbiome, comprises a diverse array of microorganisms residing in the narrow soil region surrounding plant roots known as the rhizosphere (Vukanti, 2020). This dynamic microbial community is fundamental to plant health, growth, and development, as it significantly influences nutrient uptake, stress resilience, and protection against pathogens (Haldar et al., 2022). Notably, the rhizobiome's composition differs substantially from that of the broader bulk soil microbiome. Plants actively shape their rhizobiome by releasing root exudates and interacting with their immune systems, selectively recruiting specific microbial species (Kumari et al., 2022).

Within the rhizobiome, beneficial microorganisms enhance plant growth and reinforce defense mechanisms, while some may act as potential pathogens (Moorthy & Balasundaram, 2021). Recent research has demonstrated that a distinct subset of the rhizobiome—strongly influenced by plant genetics—consistently assembles in the rhizosphere of particular plant genotypes, regardless of varying soil conditions (Xun et al., 2024).

Overall, rhizobiomes represent highly complex and ever-changing ecosystems that are vital for plant health and the advancement of sustainable agriculture. Gaining a deeper understanding of their structure, function, and interactions with host plants is crucial for developing innovative strategies to enhance crop productivity and bolster resistance against biological and environmental stresses (Moorthy & Balasundaram, 2021). Cutting-edge molecular techniques, including metagenomics and metatranscriptomics, have considerably improved our ability to investigate rhizobiome composition and functionality. Nevertheless, fully unraveling the intricacies of this essential microbial community remains an ongoing scientific challenge (Xun et al., 2024).

13.5 ENDOPHYTIC MICROBES

Endophytes are symbiotic microorganisms that inhabit plant tissues without causing harm to their hosts (Negi et al., 2023).

These organisms are mainly bacteria and fungi from various phyla, such as Euryarchaeota, Ascomycota, Basidiomycota, Mucoromycota, Firmicutes, Proteobacteria, and Actinobacteria (Azeem et al., 2020). Among these, Proteobacteria, which include α-, β-, γ-, and δ-Proteobacteria, are the most prevalent bacterial group, while Bacteroidetes, Deinococcus-Thermus, and Acidobacteria show the least diversity (Negi et al., 2023). Common bacterial genera found include *Achromobacter, Burkholderia, Bacillus, Enterobacter, Herbaspirillum, Pseudomonas, Pantoea, Rhizobium*, and *Streptomyces* (Kumar et al., 2022). Fungal endophytes frequently identified in plants are *Aspergillus, Alternaria, Bipolaris, Cladosporium, Diaporthe, Fusarium, Mucor, Nigrospora, Penicillium, Piriformospora, Trichoderma, Ulocladium*, and *Yarrowia* (Kumar et al., 2022).

Endophytic microbes form a diverse and widespread group that colonizes various plant tissues and organs across numerous host species (Stone et al., 2000). Their presence and diversity are shaped by a variety of biotic and abiotic factors that are crucial for plant growth, development, and defense against different stressors (Dutta et al., 2014). There are notable differences between endophytic and rhizospheric microbes. Endophytes reside within plant tissues, particularly in roots, stems, and leaves, whereas rhizospheric microbes are found in the soil around plant roots (Krishnaraj & Pasha, 2017). The internal colonization by endophytes offers them greater protection and adaptation to their host plants, potentially making them more beneficial than rhizospheric microbes for practical applications (Ali et al., 2017).

Host plants exert a stronger influence on the assembly of endophytic fungal communities compared to bacterial communities. Research on *Carex cepillacea* showed that all dominant rhizosphere bacterial taxa reached the root endosphere, while only 65.5% of the dominant rhizosphere fungal taxa did so (Xie et al., 2023). This indicates distinct community assembly patterns for fungi and bacteria in the root-soil continuum.

Notably, endophytic microbes are less affected by external environmental factors, such as temperature, rainfall, seasonal changes, and ultraviolet (UV) radiation, compared to rhizospheric and phylloplane microbes (Banik et al., 2018). This relative environmental stability may enhance their effectiveness in promoting plant growth and health.

In conclusion, while both endophytic and rhizospheric microbes are vital for plant growth and health, endophytes seem to have certain advantages due to their internal colonization and closer association with the host plant. These differences make endophytes particularly promising for applications in sustainable agriculture, bioremediation, and bioactive compound production (Muthusamy Pandian et al., 2024).

13.6 MECHANISMS OF MICROBIOME-MEDIATED DROUGHT TOLERANCE

Microbiomes play a crucial role in enhancing plant drought tolerance through a variety of integrated mechanisms that boost plant resilience under water-limited conditions. These beneficial microbes help plants adapt by regulating osmotic balance, elevating the activity of antioxidant enzymes, modulating phytohormone levels, forming protective biofilms, and optimizing the uptake of water and essential nutrients. Additionally, microbes produce volatile organic compounds (VOCs) and activate stress-responsive genes, both of which are essential for strengthening drought resistance. Collectively, these strategies support plants in coping with environmental stress by promoting physiological and biochemical changes that safeguard cellular function and sustain growth during periods of drought (Luu & Maurel, 2004).

Plants use numerous methods to regulate their internal water equilibrium and sustain their cellular functions when faced with environmental stressors. To maintain the water balance, plants employ osmotic adjustment by accumulating soluble compounds known as osmolytes. These substances, including proline, glycine–betaine, polyamines, and sugars, help to preserve cell turgor through osmotic regulation and shield against oxidative harm (Chakraborty & Kumari, 2023). Aquaporins, a group of membrane proteins that enable water diffusion across cell membranes, are crucial for regulating water balance. Plants modulate aquaporin gene expression, protein levels, subcellular redistribution, and channel activity in response to environmental cues (Luu & Maurel, 2004).

Plants also use hydrogel-like molecules, such as pectins, to control water uptake. In pollen grains, pectins function as "valves" for water entry, allowing for a regulated process of water absorption into dehydrated pollen (Vieira & Feijó, 2016). Moreover, plants can adapt to water stress through osmoregulation, as observed in cotton plants subjected to repeated water shortages. This adaptation enables plants to maintain photosynthesis and translocation at low leaf water potentials (Ackerson & Hebert, 1981). In conclusion, plants utilize a complex set of mechanisms to regulate their internal water balance, including the accumulation of osmolytes, control of aquaporins, use of hydrogel-like substances, and adaptation to stress. These processes modify the water and solute transport characteristics of plant membranes, ensuring proper cellular function under various environmental conditions (Chen et al., 2022).

Plants enhance their defenses against harmful molecules through multiple strategies. They synthesize a wide variety of small organic compounds known as phytochemicals that serve as chemical defenses against microbes, insects, and predators (Bhojwani & Dantu, 2013). When subjected to excessive ultraviolet B (UVB) radiation damage, plants increase the production of protective secondary metabolites and antioxidative enzymes (Kumar & Jose, 2020). These phytochemicals, particularly phenolic compounds, act as natural antioxidants and protect against reactive species and free radicals (Olszowy, 2019).

The plant defense system is highly complex, involving the coordinated regulation of numerous genes and transcription factors (Akbar et al., 2023). This intricate network manages several processes, including the activation of antioxidant defenses, the generation of reactive oxygen species, and the biosynthesis of secondary metabolites. Key antioxidants such as SOD and peroxidase (POD), alongside a broad range of secondary metabolites like phenolics, terpenes, and nitrogen-containing compounds, play crucial roles in protecting plants from various biotic stresses (Akbar et al., 2023).

To enhance their defense capabilities, plants employ multiple strategies to boost the production of protective substances. These strategies include the synthesis of diverse phytochemicals, the induction of antioxidative enzymes, and the fine-tuned regulation of complex defense pathways. The ability to generate these protective compounds is essential for plants to mitigate the detrimental effects of environmental stressors, thereby supporting their growth, development, and productivity (Singh et al., 2024).

Plant hormones are fundamental in enabling plants to adapt to fluctuating environmental conditions and to respond effectively to various stresses. By dynamically adjusting hormone levels and signaling pathways, plants optimize their stress responses and enhance their resilience.

Hormones play crucial roles in regulating plant growth and stress tolerance by modifying their production, distribution, and signal transduction (Colebrook et al., 2013). For example, salt stress induces a decrease in gibberellin (GA) levels and signaling in plants, leading to growth inhibition (Colebrook et al., 2013). In contrast, increased GA biosynthesis and signaling facilitate plant escape in response to shading and submergence (Colebrook et al., 2013). Abscisic acid (ABA) concentrations are continuously adjusted in response to fluctuating physiological, metabolic, and environmental conditions (Bansal et al., 2011).

Different hormones may have conflicting or complementary roles. Although ABA is the most extensively studied stress-responsive hormone, recent findings have suggested that cytokinins, brassinosteroids, and auxins also contribute significantly to plant responses to environmental stress (Peleg & Blumwald, 2011). The interaction between various plant hormones results in synergistic or antagonistic effects that are vital for plant responses to abiotic stresses (Peleg & Blumwald, 2011).

Plants employ sophisticated mechanisms, involving biosynthesis, signaling pathways, and hormone interactions, to control hormone levels. The transcriptional regulation of hormone metabolism plays a pivotal role in this process (Colebrook et al., 2013). Understanding these mechanisms is essential for improving stress tolerance in crops and could lead to the creation of genetically engineered plants with increased resistance to abiotic stress (Zhang et al., 2018).

To combat drought, plants develop protective surface layers, primarily through the formation of the cuticle, a specialized covering on the epidermis of terrestrial plants (Mcallister, 2022). This cuticle is composed of a resilient cross-linked polymer known as cutin and a protective wax layer that seals the plant surface (Samuels et al., 2008).

The cuticle serves as a crucial barrier between the plant and the atmosphere and consists of lipids, cutin, and intracuticular wax, with epicuticular wax deposited on the exterior (Aharoni et al., 2004). This outer waxy layer is vital for conserving water and shielding plants from environmental stressors, including dehydration and UV radiation damage (Aharoni et al., 2004).

Notably, certain grain crops, such as barley and wheat, develop distinct waxy blooms during their reproductive stages, which is linked to increased drought resistance (Mcallister, 2022). The makeup and organization of the cuticle can differ among plant species and in response to environmental cues and stressors (Borisjuk et al., 2014). For example, water scarcity can substantially increase cuticle thickness in some wheat varieties, such as Drysdale and RAC875 (Bi et al., 2017). Moreover, drought stress can result in increased alkane levels in the cuticular wax of wheat strains, potentially increasing drought tolerance (Bi et al., 2017).

In summary, plants establish protective barriers against drought through intricate cuticle formation, involving the production and accumulation of cutin, waxes, and other substances. The capacity to alter cuticle composition and structure in response to environmental stresses such as drought underscores the importance of this protective layer in plant adaptation to challenging environments. Plants use various strategies to increase their capacity for water and essential nutrient absorption under drought conditions.

Plants can enhance their ability to withstand drought by forming symbiotic relationships with beneficial microorganisms. Ectomycorrhizal fungi (EMFs) are known to boost nutrient and water absorption in host plants, particularly by enhancing calcium (Ca) uptake. Under drought conditions, EMFs significantly increase the availability of soil Ca and its uptake in seedlings, which is positively linked to water-use efficiency (Li et al., 2021). Similarly, certain soil bacteria, such as *Bacillus* species from the Atacama Desert, can enhance plant resilience to drought stress by improving water and nutrient absorption. These bacteria foster advantageous plant traits that aid in water and nutrient uptake, supporting their use as bioinoculants to boost plant production in the face of increasing drought conditions (Santander et al., 2024).

Importantly, the use of silicon (Si), a quasi-essential element, can alleviate the adverse effects of water scarcity. Si supports plant growth by improving seed germination, cell membrane stability, carbon assimilation, plant-water relations, and osmotic adjustment. It also enhances antioxidant activity, stimulates phytohormone biosynthesis, improves nutrient uptake, and regulates key enzyme activity in plants under drought stress (Irfan et al., 2023). Additionally, some plants have developed specific mechanisms to survive low-water conditions, which are categorized as tolerance, avoidance, and escape strategies (Farahani et al., 2009).

In summary, plants enhance their ability to absorb water and essential nutrients during drought through various strategies, including symbiotic relationships with microorganisms, the application of beneficial elements like silicon, and the development of specific drought tolerance mechanisms. These adaptations allow plants to sustain growth and productivity in water-limited environments, highlighting the importance of understanding and utilizing these mechanisms for sustainable agriculture in drought-prone areas. Recent research has shown that drought causes significant changes in the root microbiome, particularly in the enrichment of monodermal bacteria (Xu & Coleman-Derr, 2019). This enrichment is linked to changes in plant metabolism and may enhance drought tolerance in the host (Xu & Coleman-Derr, 2019). Moreover, the application of exogenous melatonin has been shown to improve plant performance under drought conditions by altering the microbiome composition and increasing the activities of nonstructural carbohydrate metabolism enzymes and antioxidant enzyme systems in plant roots (Ye et al., 2022).

In conclusion, microbiome-mediated drought tolerance involves intricate interactions between plants and their associated microorganisms. These mechanisms encompass physiological changes in plants, shifts in microbial community composition, and alterations in metabolic pathways. Understanding these mechanisms can guide strategies to enhance crop productivity and resilience to drought stress (Allison, 2023).

13.7 ROLE OF SPECIFIC MICROBIAL GROUPS IN DROUGHT TOLERANCE

Microbial communities significantly influence plant drought resilience, and certain groups play crucial roles. Studies have shown that Actinobacteria and Acidobacteria are more prevalent in peanut rhizospheres subjected to drought during the seedling and podding phases, whereas Cyanobacteria and Gemmatimonadetes are more prevalent during flowering (Dai et al., 2019). In Mediterranean oak species, fungi exhibit greater host specificity than prokaryotes do, with Quercus ilex supporting more diverse fungal communities than Quercus suber does (Rutten et al., 2020).

Notably, research has identified potential pathogens associated with drought tolerance. In Neotropical rainforests, fungal taxa, which are typically described as plant pathogens, proliferate with increased drought susceptibility, whereas the Strelitziana and Ochroconis fungal genera are recognized as potential candidates for enhancing drought tolerance (Cambon et al., 2022). Moreover, the diversity of arbuscular mycorrhizal fungi has been linked to microbially mediated drought tolerance. Nutrition enhancement relies heavily on diazotrophic activity and nitrogenase complexes in tree seedlings (Allsup et al., 2023).

To summarize, while various microbial groups contribute to drought tolerance, their functions can be intricate and dependent on context. Actinobacteria, Acidobacteria, and specific fungal taxa, including arbuscular mycorrhizal fungi, appeared to be particularly significant. However, the interplay between microbial communities and drought tolerance can fluctuate depending on factors such as plant species, ecosystem type, and substrate quality (Petrushin et al., 2023). Scientists have identified several key factors and strategies for selecting microbial groups that can increase plant drought tolerance. Isolation from drought-affected environments. Microorganisms isolated from water-stressed soils or plants often possess adaptive mechanisms for handling water scarcity. For example, bacteria and fungi extracted from upland rice roots in drought-prone regions have demonstrated improved plant growth-promoting capabilities under both well-watered and water-stressed conditions (Pang et al., 2020).

Assessing plant growth-enhancing characteristics: Microorganisms were tested for their capacity to generate growth hormones, siderophores, organic acids, and enzymes such as ACC deaminase, which are vital for the ability of plants to withstand stress. The synthesis of these substances, particularly under drought conditions, serves as a reliable indicator of their potential to increase plant drought resistance (Ebadi & Etesami, 2025).

Tolerance to osmotic stress: Microbes that can endure high osmotic stress in laboratory settings, such as those that grow in media with increasing concentrations of polyethylene glycol (PEG) or NaCl, are more likely to be effective under drought conditions. For example, *Pseudomonas putida* and *Bacillus megaterium* display high osmotic tolerance and elevated proline contents under stress (Marulanda et al., 2009).

Notably, the composition of root microbial communities can change during drought, with an increase in gram-positive bacteria, especially Firmicutes and Actinobacteria (Ebadi & Etesami, 2025). This natural selection process can inform the selection of drought-resistant microbial strains. In summary, choosing microbial groups to increase plant drought tolerance involves isolating strains from drought-affected environments, evaluating plant growth-promoting traits and osmotic stress tolerance, and identifying naturally drought-adapted microbial communities. Combining these approaches can result in the identification of effective microbial inoculants to improve plant drought tolerance (Balestrini et al., 2017).

13.8 MICROBIOME ENGINEERING AND SYNTHETIC COMMUNITIES FOR DROUGHT TOLERANCE

Recent advancements in microbiome manipulation and the creation of synthetic microbial communities have shown great potential in improving plant drought resistance. Host-mediated microbiome engineering (HMME) has proven effective in cultivating beneficial microbial assemblages that increase plant resilience to water scarcity (Styer et al., 2024). By repeatedly selecting microbiomes from drought-resistant plants and using them to seed subsequent generations, scientists have successfully developed simplified microbiomes that increase both growth and drought tolerance in crops, such as rice and wheat (Xie et al., 2021). Research has shown that drought stress can reduce the diversity of root-associated microbial communities, with shifts in composition driven by changes in the relative abundance of key phyla, such as Actinobacteria, Firmicutes, and

Proteobacteria (Xie et al., 2021). These findings suggest that certain microbial groups may play crucial roles in conferring drought tolerance. Specific bacterial and fungal strains isolated from drought-exposed rhizosphere soils have demonstrated positive effects on plant growth and drought resistance when applied in controlled inoculation experiments (Xie et al., 2021).

The application of synthetic microbial communities (SynComs) and HMME is a promising strategy for developing drought-resistant microbial communities in sustainable agriculture (Ali et al., 2022). These approaches, combined with high-throughput methods and multiomics technologies, can aid in identifying key microbial taxa and functions associated with improved drought tolerance (Ali et al., 2022). However, the transition from laboratory findings to field applications remains challenging, underscoring the need for additional research to fully leverage microbiome engineering to increase crop resilience to drought stress (Gu et al., 2024).

13.9 ENVIRONMENTAL AND ECOLOGICAL FACTORS IMPACTING MICROBIOME EFFICACY

The composition and function of microbiomes in various ecosystems are strongly influenced by environmental and ecological factors. For example, in the human gut, the initial microbiome is shaped by factors such as the delivery method, dietary habits, and infant surroundings (Donovan, 2020). Research has shown that ambient temperature is a crucial environmental factor that affects the structure and composition of the vaginal microbiome, which may have implications for preterm birth risk (Liu et al., 2024). Studies have revealed that glucocorticoids (GCs) affect gut microbiome diversity in wild animals, with higher GC levels correlating with lower diversity. However, this relationship is not straightforward, as ecological factors, such as food availability and population density, can independently affect both GCs and microbial diversity (Petrullo et al., 2022). In aquatic ecosystems, human-induced disturbances and geological events, such as earthquakes, can significantly alter microbiome composition and stability (Vigneron et al., 2024).

The formation of host-associated microbial communities is influenced by the interplay among host conditions, environmental factors, and microbial interactions. Large-scale disturbances affecting host populations, such as earthquake-induced changes in marine environments, can dramatically impact microbiome structure by restricting microbial dispersal opportunities (Pearman et al., 2024). In terrestrial ecosystems, habitat composition affects microbiome diversity, as evidenced by the breeding of blue tits in different forest types (Drobniak et al., 2021).

Understanding the intricate dynamics of microbiome diversity and its response to environmental and ecological factors is essential for maintaining ecosystem health and function. Future studies should concentrate on uncovering the mechanisms underlying these interactions and their implications for ecosystem services in the face of global pressure for change (Singh & Rai, 2024).

13.10 PRODUCTION OF STRESS PROTEINS TO PROMOTE TOLERANCE TO DROUGHT STRESS

When plants experience drought stress, they produce various proteins that are crucial for enhancing their drought tolerance. In wheat, researchers have observed the accumulation of dehydrins, a group of proteins thought to safeguard membranes and macromolecules from denaturation during drought (Lopez et al., 2003). Notably, the presence of a 24-kDa dehydrin has been linked to improved drought resistance in certain wheat varieties, as evidenced by their ability to maintain shoot dry matter production more effectively. Studies on rice have shown that increasing the expression of the small heat shock protein-encoding gene sHSP17.7 leads to increased drought tolerance in genetically modified seedlings (Sato & Yokoya, 2007). Rice plants with relatively high levels of the sHSP17.7 protein demonstrated a superior capacity to recover after water stress and exhibited increased survival rates under drought conditions simulated by polyethylene glycol treatment.

Remarkably, exposing rice seedlings to heat shock can also induce drought tolerance, indicating a potential interconnection between heat and drought stress responses (Sato & Yokoya, 2007). Furthermore, research on drought-stressed grapevines has revealed an increase in proteins associated with energy and carbohydrate metabolism as well as those linked to stress defense mechanisms (Azri et al., 2021). These proteins may contribute to active stress acclimation and ROS scavenging, thereby increasing drought tolerance.

In summary, the synthesis of stress proteins, including dehydrins, heat shock proteins, and proteins involved in energy metabolism and ROS scavenging, plays a vital role in increasing drought tolerance in various crop species. A deeper understanding of the mechanisms governing stress protein production and their functions could provide valuable insights for the development of drought-resistant crop varieties through genetic engineering or selective breeding programs.

13.11 HUMIC AND FULVIC ACIDS REDUCE DROUGHT DAMAGE

Humic (HA) and fulvic (FA) acids are organic compounds that naturally occur in soil solutions, surface water, and groundwater (Anderson et al., 1995). These acids are key components of humic substances, which are diverse mixtures of organic materials resulting from the decomposition of plant and animal remains in the environment (MacCarthy et al., 2015). The solubilities and structural characteristics of humic and FAs differ. HAs typically consist of larger molecules with higher molecular weights, whereas

FAs are smaller and more soluble (Vo-Minh Nguyen et al., 2022). Both acid types contain hydrophilic polar or ionic functional groups and hydrophobic structural units, which influence their behavior in aqueous environments (Anderson et al., 1995). Notably, HAs demonstrate greater sorption ability than FAs because of their higher molecular weight, specific UV absorbance (SUVA), and percentage of the total hydrophobic fraction (%THLF) (Vo-Minh Nguyen et al., 2022).

HA and FA are vital for various environmental processes, including soil fertility, plant growth, and water quality. They are widely distributed throughout the Earth's ecosystem and are found in surface soils, oceans, rivers, and other ecological environments (Yang & Antonietti, 2020). Understanding their properties and behavior is crucial for managing soil health, water quality, and agricultural productivity.

These acids positively affect plant growth and nutrient uptake, potentially alleviating drought damage. Growth chamber experiments revealed that low concentrations of HA or FA added to growth media increased the dry weight of roots and tops, plant height, and total content of essential nutrients (N, K, P, Ca, Mg, and Fe) in tobacco plants (Lalas et al., 2018). These acids aid in the adsorption and translocation of Ca, Mg, and Fe within plants, potentially increasing their drought resistance. Interestingly, while low concentrations of HA and FA are beneficial, high concentrations of HAs negatively affect plant growth and nutrient uptake (Mylonas & Mccants, 1980). These findings indicate that precise dosing is essential when these substances are used to increase plant resilience. Furthermore, the different molecular weights and structures of HA and FA influence their interactions with plant and soil components (Vrantsi et al., 2021). HAs generally have a more significant effect on enzyme activity than FAs do, possibly because of their higher molecular weights.

In conclusion, although they do not directly address drought damage, the ability of HA and FA to increase plant growth, nutrient uptake, and enzyme activity suggests that they could play a role in improving plant resilience to drought stress. However, further research specifically focusing on drought conditions is needed to confirm their effectiveness in reducing drought damage. The optimal concentration and application method need to be determined, as excessive amounts may have detrimental effects on plant growth and nutrient uptake. Figure 13.1 shows the effects of plant growth-promoting rhizobacteria (PGPR) on plants, enhancing drought resistance.

13.12 CONCLUSION

Microbiomes present substantial promise for boosting drought resistance in agriculture, but significant obstacles remain in both our understanding and practical application of this strategy. While research has explored how specific microbiome components can enhance plant growth during

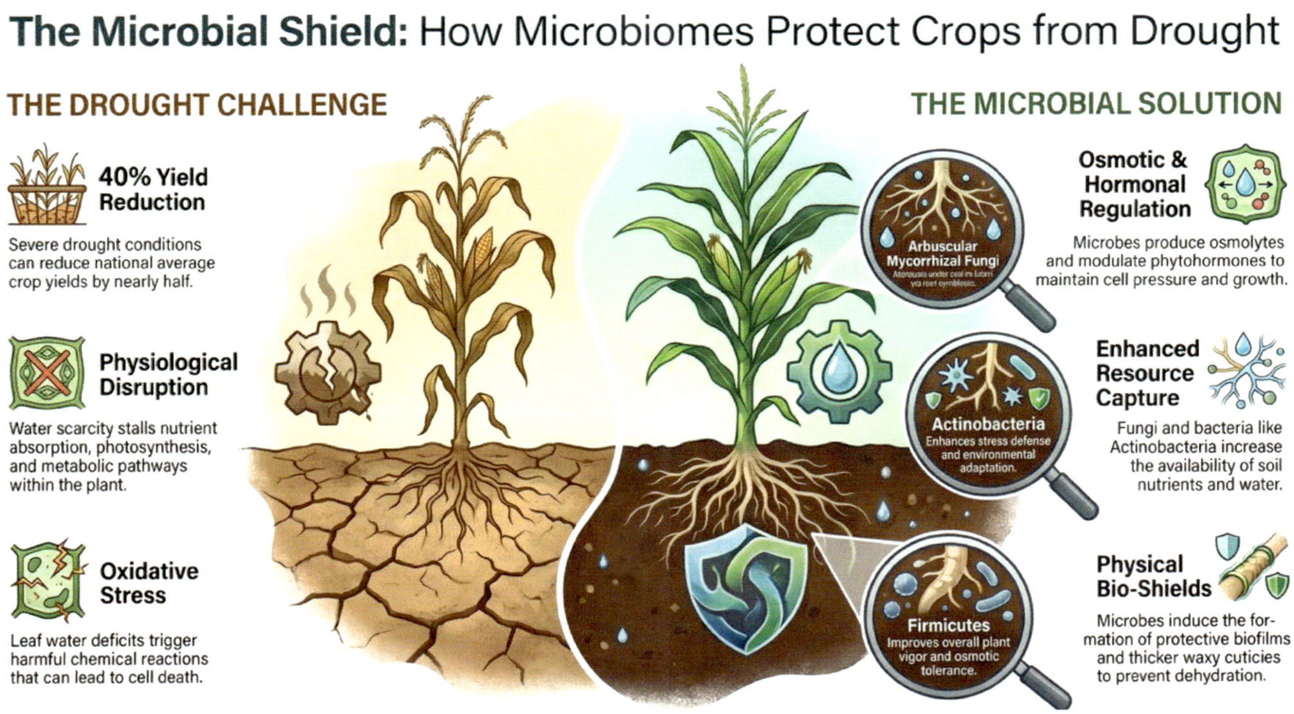

FIGURE 13.1 Effects of PGPR on plants, enhancing drought resistance. Observed benefits include improved shoot and root growth, induced systemic resistance, biocontrol, reduced fertilizer and pesticide use, and increased resistance to both biotic and abiotic stresses, including drought.

drought, our grasp of the broader root microbiome's impact on plant stress tolerance is still incomplete (Caddell et al., 2019). This highlights the necessity for more advanced and holistic approaches to studying plant–microbiome interactions under drought conditions.

To unlock the full potential of the rhizosphere microbiome, it is essential to further develop and scale DNA-SIP technologies into high-throughput methods. Such advancements would allow for deeper insights into the complex relationships between plants and their associated microbes, especially the noncultivable populations that may play critical roles in alleviating drought stress (Ebadi & Etesami, 2025). Currently, technological limitations hinder our ability to fully leverage these beneficial microbial communities.

Moreover, the molecular mechanisms governing interactions among drought, root-associated microorganisms, and host plants are not yet well characterized (Ait-El-Mokhtar et al., 2023). This lack of clarity limits the capacity to intentionally manipulate these relationships to improve drought tolerance. Interestingly, although organic and conventional farming foster different soil microbiomes, recent findings suggest that the resilience and resistance of these microbiomes to severe drought do not differ significantly (Kost et al., 2024). This indicates that current organic farming practices may not provide sufficient protection against severe drought, pointing to a need for innovative agricultural techniques that better utilize microbiomes for drought adaptation.

In conclusion, while the potential of microbiomes to enhance drought tolerance is clear, there are substantial gaps in our understanding of the underlying interactions and in translating this knowledge into effective, large-scale agricultural solutions. Future research should prioritize overcoming these challenges to fully take advantage of microbiome-based strategies for crop resilience in the face of increasing drought stress.

13.13 FUTURE AND PERSPECTIVES

This section examines the key challenges involved in leveraging microbiomes to enhance plant drought resistance. Currently, the precise mechanisms through which rhizosphere microbiota support plant stress tolerance remain only partially understood, highlighting the need for more thorough and integrated research. The complex interactions among drought conditions, soil microbes, and plant responses at the molecular and cellular levels are not yet fully elucidated, limiting our ability to effectively manipulate these relationships for agricultural benefit.

Advancing research in this area requires improved methodologies for studying plant–microbe interactions, particularly for microbes that are difficult to culture in laboratory settings. Translating laboratory discoveries to practical field applications remains a major obstacle, as environmental variability can significantly impact outcomes. Identifying optimal microbial consortia for boosting drought resilience is still an open question, and a deeper understanding of the environmental factors influencing microbial performance is essential.

For certain interventions, such as applying specific organic acids, it is critical to establish precise dosing guidelines, as excessive concentrations can harm plant growth and nutrient uptake. More comprehensive research is also needed to clarify the long-term impacts of microbial treatments on plant physiology and crop yield. Integrating microbiome-based approaches with other drought resistance strategies, including traditional breeding and genetic modification, presents additional complexity and requires coordinated, interdisciplinary efforts among microbiologists, plant scientists, ecologists, and agricultural experts.

Further investigation into the synergistic effects of microbial treatments combined with other drought resilience techniques could yield valuable insights for developing more robust and effective solutions. Long-term, multi-season studies conducted across diverse environments are necessary to better understand the sustained benefits of microbial interventions on plant health and productivity. Additionally, exploring the design and application of tailored microbial consortia to meet the specific needs of different crops and regional climates may lead to more targeted and efficient agricultural practices.

REFERENCES

Ackerson, R. C., & Hebert, R. R. (1981). Osmoregulation in cotton in response to water stress. *Plant Physiology*, *67*(3), 484–488. https://doi.org/10.1104/pp.67.3.484

Adhikari, S. (2018). Drought impact and adaptation strategies in the mid-hill farming system of Western Nepal. *Environments*, *5*(9), 101. https://doi.org/10.3390/environments5090101

Afridi, M. S., S Altami, M., Karunakaran, R., Hafeez, A., César Terra, W., Ercisli, S., Ali, S., Sumaira, S., Ali, B., Ameen, F., Medeiros, F. H. V., Marc, R. A., & Salam, A. (2022). Plant microbiome engineering: Hopes or hypes. *Biology*, *11*(12), 1782. https://doi.org/10.3390/biology11121782

Aharoni, A., Jetter, R., Dixit, S., Van Arkel, G., Thoenes, E., & Pereira, A. (2004). The SHINE clade of AP2 domain transcription factors activates wax biosynthesis, alters cuticle properties, and confers drought tolerance when overexpressed in *Arabidopsis*. *The Plant Cell*, *16*(9), 2463–2480. https://doi.org/10.1105/tpc.104.022897

Ait-El-Mokhtar, M., Baslam, M., & Meddich, A. (2023). Plant-microbiome interactions under drought—insights from the molecular machinist's toolbox. *Frontiers in Sustainable Food Systems*, *7*. https://doi.org/10.3389/fsufs.2023.1253735

Akbar, M. U., Aqeel, M., Shah, M. S., Jeelani, G., Iqbal, N., Latif, A., Elnour, R. O., Hashem, M., Alzoubi, O. M., Habeeb, T., Qasim, M., & Noman, A. (2023). Molecular regulation of antioxidants and secondary metabolites act in conjunction to defend plants against pathogenic infection. *South African Journal of Botany*, *161*, 247–257. https://doi.org/10.1016/j.sajb.2023.08.028

Akwango, D., Egeru, A., Obaa, B. B., Turyahabwe, N., & Baguma, Y. (2017). Effect of drought early warning system on household food security in Karamoja subregion, Uganda. *Agriculture & Food Security*, *6*(1). https://doi.org/10.1186/s40066-017-0120-x

Ali, S., Charles, T. C., & Glick, B. R. (2017). *Endophytic Phytohormones and Their Role in Plant Growth Promotion* (pp. 89–105). Springer. https://doi.org/10.1007/978-3-319-65897-1_6

Ali, S., Tyagi, A., Mir, R. A., Mushtaq, M., Park, S., Bhat, B., Bae, H., & Mahmoudi, H. (2022). Deciphering the plant microbiome to improve drought tolerance: Mechanisms and perspectives. *Environmental and Experimental Botany*, *201*, 104933. https://doi.org/10.1016/j.envexpbot.2022.104933

Allison, S. D. (2023). Microbial drought resistance may destabilize soil carbon. *Trends in Microbiology*, *31*(8), 780–787. https://doi.org/10.1016/j.tim.2023.03.002

Allsup, C. M., Lankau, R. A., & George, I. (2023). Shifting microbial communities can enhance tree tolerance to changing climates. *Science*, *380*(6647), 835–840. https://doi.org/10.1126/science.adf2027

Anderson, M. A., Hung, A. Y. C., Scott, M. S., & Mills, D. (1995). Factors affecting the surface tension of soil solutions and solutions of humic acids. *Soil Science*, *160*(2), 111–116. https://doi.org/10.1097/00010694-199516020-00004

Angelakis, A., Samonis, G., Nikolaidis, C., Vuorinen, H., Juuti, P., Katko, T., Zhang, J., & Juuti, R. (2021). Water quality and life expectancy: Parallel courses in time. *Water*, *13*(6), 752. https://doi.org/10.3390/w13060752

Azeem, F., Ali, M. A., Siddique, M. H., Moosa, A., Hussain, S., Abbas, A., Rashid, F., Shahzadi, M., Batool, R., & Nadeem, H. (2020). *Endophytes as Guardians of Plants Against Diseases* (pp. 221–242). Springer Nature Singapore. https://doi.org/10.1007/978-981-15-4439-2_10

Azri, W., Cosette, P., Jardak, R., Guillou, C., Mliki, A., & Riahi, J. (2021). Physiological and proteomic analyses of Tunisian local grapevine (Vitis vinifera) cultivar Razegui in response to drought stress. *Functional Plant Biology: FPB*, *49*(1), 25–39. https://doi.org/10.1071/fp21026

Balestrini, R., Fotopoulos, V., Ruocco, M., & Chitarra, W. (2017). *Potential Role of Beneficial Soil Microorganisms in Plant Tolerance to Abiotic Stress Factors* (pp. 191–207). Springer. https://doi.org/10.1007/978-3-319-63336-7_12

Banik, A., Dash, G. K., Swain, P., Kumar, U., Mukhopadhyay, S. K., & Dangar, T. K. (2018). Application of rice (*Oryza sativa* L.) root endophytic diazotrophic Azotobacter sp. strain Avi2 (MCC 3432) can increase rice yield under green house and field condition. *Microbiological Research*, *219*, 56–65. https://doi.org/10.1016/j.micres.2018.11.004

Bansal, K. C., Lenka, S. K., & Tuteja, N. (2011). *Abscisic Acid in Abiotic Stress Tolerance: An 'Omics' Approach* (pp. 143–150). Bentham Science. https://doi.org/10.2174/978160805058111101010143

Bevacqua, E., Rakovec, O., Schumacher, D. L., Kumar, R., Thober, S., Samaniego, L., Seneviratne, S. I., & Zscheischler, J. (2024). Direct and lagged climate change effects intensified the 2022 European drought. *Nature Geoscience*. https://doi.org/10.1038/s41561-024-01559-2

Bhattacharyya, P., Pal, S., & Pathak, H. (2020). *Impact of Climate Change on Agriculture: Evidence and Predictions* (pp. 17–32). Springer Singapore. https://doi.org/10.1007/978-981-15-9132-7_2

Bhojwani, S. S., & Dantu, P. K. (2013). *Production of Industrial Phytochemicals* (pp. 275–286). Springer India. https://doi.org/10.1007/978-81-322-1026-9_18

Bi, H., Lopato, S., Langridge, P., Kovalchuk, N., Borisjuk, N., & Tricker, P. J. (2017). The impact of drought on wheat leaf cuticle properties. *BMC Plant Biology*, *17*(1). https://doi.org/10.1186/s12870-017-1033-3

Bijalwan, P., Sharma, M., & Kaushik, P. (2022). *Review of the Effects of Drought Stress on Plants: A Systematic Approach*. mdpi ag. https://doi.org/10.20944/preprints202202.0014.v1

Borisjuk, N., Hrmova, M., & Lopato, S. (2014). Transcriptional regulation of cuticle biosynthesis. *Biotechnology Advances*, *32*(2), 526–540. https://doi.org/10.1016/j.biotechadv.2014.01.005

Caddell, D. F., Deng, S., & Coleman-Derr, D. (2019). *Role of the Plant Root Microbiome in Abiotic Stress Tolerance* (pp. 273–311). https://doi.org/10.1007/978-3-030-10504-4_14

Camarena-Pozos, D. A., López, M. G., López-Bucio, J., Flores-Núñez, V. M., & Partida-Martínez, L. P. (2018). Smells from the desert: Microbial volatiles that affect plant growth and development of native and nonnative plant species. *Plant, Cell & Environment*, *42*(4), 1368–1380. https://doi.org/10.1111/pce.13476

Cambon, M., Bonal, D., Stahl, C., Coste, S., Burban, B., Ziegler, C., Cartry, D., Cazal, J., Schimann, H., Goret, J., Buée, M., Ritter, Y., Heuret, P., Léger, P., Louisanna, E., Roy, M., Levionnois, S., Chancerel, É., Fort, T., . . . Vacher, C. (2022). Drought tolerance traits in neotropical trees correlate with the composition of phyllosphere fungal communities. *Phytobiomes Journal*, *7*(2), 244–258. https://doi.org/10.1094/pbiomes-04-22-0023-r

Chakraborty, P., & Kumari, A. (2023). Chapter 9—Role of compatible osmolytes in plant stress tolerance under the influence of phytohormones and mineral elements. In *Improving Stress Resilience in Plants* (pp. 165–201). Elsevier. https://doi.org/10.1016/b978-0-443-18927-2.00011-x

Chandra, P., Wunnava, A., Verma, P., Chandra, A., & Sharma, R. K. (2021). Strategies to mitigate the adverse effect of drought stress on crop plants—influences of soil bacteria: A review. *Pedosphere*, *31*(3), 496–509. https://doi.org/10.1016/s1002-0160(20)60092-3

Charania, I., & Li, X. (2019). Smart farming: Agriculture's shift from a labor intensive to technology native industry. *Internet of Things*, *9*, 100142. https://doi.org/10.1016/j.iot.2019.100142

Chen, S.-Y., Wang, L., Jia, P.-F., Yang, W.-C., Sze, H., & Li, H.-J. (2022). Osmoregulation determines sperm cell geometry and integrity for double fertilization in flowering plants. *Molecular Plant*, *15*(9), 1488–1496. https://doi.org/10.1016/j.molp.2022.07.013

Colebrook, E. H., Hedden, P., Thomas, S. G., & Phillips, A. L. (2013). The role of gibberellin signaling in plant responses to abiotic stress. *Journal of Experimental Biology*, *217*(1), 67–75. https://doi.org/10.1242/jeb.089938

Cuhadar, M. (2024). *The Impact of Drought on Agricultural Production and Agricultural Adaptation to Drought* (pp. 83–107). Nobel Tip Kitabevleri. https://doi.org/10.69860/nobel.9786053359432.5

Dai, L., Yu, Z., Xu, Y., Zhang, G., Ding, H., & Zhang, Z. (2019). Effect of drought stress and developmental stages on microbial community structure and diversity in peanut rhizosphere soil. *International Journal of Molecular Sciences*, *20*(9), 2265. https://doi.org/10.3390/ijms20092265

Daryanto, S., Jacinthe, P.-A., & Wang, L. (2015). Global synthesis of drought effects on food legume production. *PLoS One*, *10*(6), e0127401. https://doi.org/10.1371/journal.pone.0127401

Dastogeer, K. M. G., Tumpa, F. H., Sultana, A., Akter, M. A., & Chakraborty, A. (2020). Plant microbiome–an account of the factors that shape community composition and diversity. *Current Plant Biology*, *23*, 100161. https://doi.org/10.1016/j.cpb.2020.100161

De Pinto, A., Kwon, H.-Y., Koo, J., Cenacchi, N., & Dunston, S. (2020). Climate smart agriculture and global food-crop production. *PLoS One*, *15*(4), e0231764. https://doi.org/10.1371/journal.pone.0231764

Desalegn, C. E., Babel, M. S., Gupta, A. D., Seleshi, B. A., & Merrey, D. (2006). Farmers' perception of water management under drought conditions in the Upper Awash Basin, Ethiopia. *International Journal of Water Resources Development*, *22*(4), 589–602. https://doi.org/10.1080/07900620600779723

Dong, G., Huang, X., Lu, Y., Zhang, S., & Ma, M. (2022). Spatiotemporal variation in human settlements and their interaction with living environments in Neolithic and Bronze Age China. *Progress in Physical Geography: Earth and Environment*, *46*(6), 949–967. https://doi.org/10.1177/03091333221087992

Dong, G., Liu, F., & Chen, F. (2017). Environmental and technological effects on ancient social evolution at different spatial scales. *Science China Earth Sciences*, *60*(12), 2067–2077. https://doi.org/10.1007/s11430-017-9118-3

Donovan, S. M. (2020). Evolution of the gut microbiome in infancy within an ecological context. *Current Opinion in Clinical Nutrition & Metabolic Care*, *23*(3), 223–227. https://doi.org/10.1097/mco.0000000000000650

Drobniak, S. M., Gustafsson, L., Zagalska-Neubauer, M., Janas, K., Barczyk, J., & Cichoń, M. (2021). Habitat shapes diversity of gut microbiomes in a wild population of blue tits *Cyanistes caeruleus*. *Journal of Avian Biology*, *2022*(2). https://doi.org/10.1111/jav.02829

Dutta, D., Puzari, K. C., Gogoi, R., & Dutta, P. (2014). Endophytes: Exploitation as a tool in plant protection. *Brazilian Archives of Biology and Technology*, *57*(5), 621–629. https://doi.org/10.1590/s1516-8913201402043

Ebadi, A., & Etesami, H. (2025). Chapter 18—Monitoring changes in the rhizospheric microbiome under drought conditions: Approaches and challenges. In *Sustainable Agriculture under Drought Stress* (pp. 265–286). Elsevier. https://doi.org/10.1016/b978-0-443-23956-4.00018-1

Ejaz Qureshi, M., Hanjra, M. A., & Ward, J. (2012). Impact of water scarcity in Australia on global food security in an era of climate change. *Food Policy*, *38*, 136–145. https://doi.org/10.1016/j.foodpol.2012.11.003

Fadiji, A. E., Santoyo, G., Babalola, O. O., & Yadav, A. N. (2022). Efforts toward overcoming drought stress in crops: Revisiting the mechanisms employed by plant growth-promoting bacteria. *Frontiers in Microbiology*, *13*(520). https://doi.org/10.3389/fmicb.2022.962427

Farahani, H., Valadabadi, S., Shiranirad, A., Daneshian, J., & Khalvati, M. (2009). Medicinal and aromatic plants farming under drought conditions. *Journal of Horticulture and Forestry*, *1*(6), 86–92. https://doi.org/10.5897/jhf.9000075

Gu, Z., Tian, G., Gao, L., Hu, C., Gan, Y., & Zhou, J. (2024). Role of microbes in alleviating crop drought stress: A review. *Plants*, *13*(3), 384. https://doi.org/10.3390/plants13030384

Haldar, S., Mondal, S., Kumari, A., Ghosh, A., Chattopadhyay, D., & Ghosh, A. (2022). Chapter 16—Rhizosphere microbiome engineering. In *Current Developments in Biotechnology and Bioengineering* (pp. 377–396). Elsevier. https://doi.org/10.1016/b978-0-323-88504-1.00014-5

Hameed, M., Ahmadalipour, A., & Moradkhani, H. (2019). Drought and food security in the middle east: An analytical framework. *Agricultural and Forest Meteorology*, *281*, 107816. https://doi.org/10.1016/j.agrformet.2019.107816

Han, J., & Singh, V. P. (2023). A review of widely used drought indices and the challenges of drought assessment under climate change. *Environmental Monitoring and Assessment*, *195*(12). https://doi.org/10.1007/s10661-023-12062-3

Hao, J.-R., Li, Y., & Ge, Y. (2024). Harnessing the plant microbiome for environmental sustainability: From ecological foundations to novel applications. *Science of the Total Environment*, *951*, 175766. https://doi.org/10.1016/j.scitotenv.2024.175766

Haque Mondol, M. A., Zhu, X., Dunkerley, D., & Henley, B. J. (2022). Technological drought: A new category of water scarcity. *Journal of Environmental Management*, *321*, 115917. https://doi.org/10.1016/j.jenvman.2022.115917

Holka, M., Kowalska, J., & Jakubowska, M. (2022). Reducing carbon footprint of agriculture—can organic farming help to mitigate climate change? *Agriculture*, *12*(9), 1383. https://doi.org/10.3390/agriculture12091383

Hu, J., Kowalchuk, G. A., Jousset, A., Friman, V.-P., Li, M., Wei, Z., Yang, T., Hautier, Y., Xu, Y., & Shen, Q. (2021). Introduction of probiotic bacterial consortia promotes plant growth via impacts on the resident rhizosphere microbiome. *Proceedings of the Royal Society B: Biological Sciences*, *288*(1960). https://doi.org/10.1098/rspb.2021.1396

Irfan, M., Ercisli, S., Maqsood, M. A., Mahboob, W., Hussain, S., Hafeez, O. B. A., Rehman, H. U., Aziz, T., Sarwar, N., & Akhtar, M. (2023). Silicon nutrition in plants under water-deficit conditions: Overview and prospects. *Water*, *15*(4), 739. https://doi.org/10.3390/w15040739

Knutson, C. L., Hayes, M. J., Graf, L., Nothwehr, J., Haigh, T., Widhalm, M., & Kleinschmidt, M. (2011). Farmer perceptions of sustainable agriculture practices and drought risk reduction in Nebraska, USA. *Renewable Agriculture and Food Systems*, *26*(3), 255–266. https://doi.org/10.1017/s174217051100010x

Kost, E., Krause, H.-M., Mäder, P., Kundel, D., Six, J., Mayer, J., Conz, R. F., & Hartmann, M. (2024). *Microbial Resistance and Resilience to Drought under Organic and Conventional Farming*. Cold Spring Harbor Laboratory. https://doi.org/10.1101/2024.04.17.589021

Krishnaraj, P. U., & Pasha, M. A. (2017). *Metagenome of Rhizosphere and Endophytic Ecosystem* (pp. 125–156). Springer Singapore. https://doi.org/10.1007/978-981-10-5050-3_9

Kumar, A., Maurya, V. K., Susmita, C., Chuarasiya, U., Maurya, D. K., & Singh, S. K. (2022). Chapter 15—Environmental factors and plant–microbes (endophytes) interaction: An overview and future outlook. In *Microbial Endophytes and Plant Growth* (pp. 245–257). Elsevier. https://doi.org/10.1016/b978-0-323-90620-3.00009-x

Kumar, N., & Jose, J. (2020). Current developments in the nano-mediated delivery of photoprotective phytochemicals. *Environmental Science and Pollution Research*, *27*(31), 38446–38471. https://doi.org/10.1007/s11356-020-10100-y

Kumari, N., Uma, U., Jaiswal, S., Rai, A., Kumar, D., & Iquebal, M. A. (2022). Metatranscriptomic approach to understand the role of the rhizosphere microbiome: A review. *Bhartiya Krishi Anusandhan Patrika*, *Of*. https://doi.org/10.18805/bkap414

Lalas, S., Athanasiadis, V., & Dourtoglou, V. G. (2018). Humic and fulvic acids as potentially toxic metal reducing agents in water. *CLEAN—Soil, Air, Water*, *46*(2), 1700608. https://doi.org/10.1002/clen.201700608

Li, Y., Zhou, Y., Li, H., Yin, Y., Zou, X., Zhang, S., Zhang, T., & Liu, L. (2021). Ectomycorrhizal symbioses increase soil calcium availability and water use efficiency of Quercus acutissima seedlings under drought stress. *European Journal of Forest Research*, *140*(5), 1039–1048. https://doi.org/10.1007/s10342-021-01383-y

Liu, Z., Li, M., Qin, L., Hao, Y., Zhao, Q., Bian, Y., Zhao, S., Liu, X., Sun, T., Feng, Q., Zhao, H., & Li, X. (2024). *Ambient Temperature Affects the Composition of the Vaginal Microbiome and Temperature Sensitive Vaginal Microbes Influence Assisted Reproductive Technology Outcomes*. Springer Science Business Media Llc. https://doi.org/10.21203/rs.3.rs-4596870/v1

Lopez, C. G., Kronstad, W. E., Peterson, C. J., & Banowetz, G. M. (2003). Dehydrin expression and drought tolerance in seven wheat cultivars. *Crop Science*, *43*(2), 577–582. https://doi.org/10.2135/cropsci2003.5770

Lungu, O. I. (2014). *Use of Conservation Tillage and Cropping Systems to Sustain Crop Yields Under Drought Conditions in Zambia* (pp. 425–439). Springer. https://doi.org/10.1007/978-3-319-09360-4_23

Luu, D., & Maurel, C. (2004). Aquaporins in a challenging environment: Molecular gears for adjusting plant water status. *Plant, Cell & Environment*, *28*(1), 85–96. https://doi.org/10.1111/j.1365-3040.2004.01295.x

MacCarthy, P., Clapp, C. E., Malcolm, R. L., & Bloom, P. R. (2015). *An Introduction to Soil Humic Substances* (pp. 1–12). Soil Science Society of America. https://doi.org/10.2136/1990.humicsubstances.c1

Mafongoya, P., Mwadzingeni, L., Phophi, M. M., & Musokwa, M. (2022). *Climate Change Impacts on Food and Nutrition Security on Smallholder Farmers in Southern Africa* (pp. 233–249). Springer Nature Singapore. https://doi.org/10.1007/978-981-16-6771-8_14

Mahpara, S. (2019). Nitrogen fertilization induced drought tolerance in sunflower: A review. *Pure and Applied Biology*, *8*(2). https://doi.org/10.19045/bspab.2019.80110

Malviya, D., Chaurasia, R., Yadav, B., Sharma, S. K., Singh, U. B., Shafi, Z., Vishwakarma, S. K., Ilyas, T., Singh, H. V., & Shahid, M. (2022). *Engineering the Plant Microbiome for Biotic Stress Tolerance: Biotechnological Advances* (pp. 133–151). Springer Nature Singapore. https://doi.org/10.1007/978-981-19-5872-4_7

Marulanda, A., Azcón, R., & Barea, J.-M. (2009). Stimulation of plant growth and drought tolerance by native microorganisms (AM fungi and bacteria) from dry environments: Mechanisms related to bacterial effectiveness. *Journal of Plant Growth Regulation*, *28*(2), 115–124. https://doi.org/10.1007/s00344-009-9079-6

Mcallister, T. (2022, June 27). Identification of a key regulator controlling cuticular wax in barley. https://doi.org/10.48448/76pn-b060

Mdemu, M. V. (2021). Community's vulnerability to drought-driven water scarcity and food insecurity in central and northern semiarid areas of Tanzania. *Frontiers in Climate*, *3*. https://doi.org/10.3389/fclim.2021.737655

Memon, M. H., Aamir, N., & Ahmed, N. (2018). Climate change and drought: Impact of food insecurity on gender based vulnerability in district Tharparkar. *The Pakistan Development Review*, *57*(3), 307–321. https://doi.org/10.30541/v57i3pp.307-321

Miguel-Rojas, C., & Pérez-De-Luque, A. (2023). Nanobiosensors and nanoformulations in agriculture: New advances and challenges for sustainable agriculture. *Emerging Topics in Life Sciences*, *7*(2), 229–238. https://doi.org/10.1042/etls20230070

Moorthy, A., & Balasundaram, U. (2021). *Rhizosphere Metagenomics: Methods and Challenges* (pp. 1–20). Springer Singapore. https://doi.org/10.1007/978-981-16-0889-6_1

Muthusamy Pandian, T., Esakkimuthu, R., Rangasamy, A., Rengasamy, K., Alagesan, S., & Devasahayam, J. S. S. (2024). Exploring the potential of bacterial endophytes in plant disease management. *Current Microbiology*, *81*(12). https://doi.org/10.1007/s00284-024-03918-z

Mylonas, V. A., & Mccants, C. B. (1980). Effects of humic and fulvic acids on growth of tobacco. 2. Tobacco growth and ion uptake[1]. *Journal of Plant Nutrition*, *2*(3), 377–393. https://doi.org/10.1080/01904168009362785

Negi, R., Devi, R., Kour, D., Sharma, B., Yadav, A. N., Yadav, A., Kaur, T., Kumar, S., & Chaubey, K. K. (2023). Plant endophytes: Unveiling hidden applications toward agro-environment sustainability. *Folia Microbiologica*, *69*(1), 181–206. https://doi.org/10.1007/s12223-023-01092-6

Olszowy, M. (2019). What is responsible for antioxidant properties of polyphenolic compounds from plants? *Plant Physiology and Biochemistry*, *144*, 135–143. https://doi.org/10.1016/j.plaphy.2019.09.039

Orimoloye, I. R. (2022). Agricultural drought and its potential impacts: Enabling decision-support for food security in vulnerable regions. *Frontiers in Sustainable Food Systems*, *6*. https://doi.org/10.3389/fsufs.2022.838824

Pang, Z., Yu, D., Xu, P., & Zhao, Y. (2020). Microbial diversity of upland rice roots and their influence on rice growth and drought tolerance. *Microorganisms*, *8*(9), 1329. https://doi.org/10.3390/microorganisms8091329

Patel, R., & Patel, A. (2024). Evaluating the impact of climate change on drought risk in semiarid region using GIS technique. *Results in Engineering*, *21*, 101957. https://doi.org/10.1016/j.rineng.2024.101957

Pearman, W. S., Vaux, F., Morales, S. E., Gemmell, N. J., & Fraser, C. I. (2024). Host population crashes disrupt the diversity of associated marine microbiomes. *Environmental Microbiology*, *26*(3). https://doi.org/10.1111/1462-2920.16611

Peleg, Z., & Blumwald, E. (2011). Hormone balance and abiotic stress tolerance in crop plants. *Current Opinion in Plant Biology*, *14*(3), 290–295. https://doi.org/10.1016/j.pbi.2011.02.001

Pérez-Jaramillo, J. E., Raaijmakers, J. M., & Mendes, R. (2015). Impact of plant domestication on rhizosphere microbiome assembly and functions. *Plant Molecular Biology*, *90*(6), 635–644. https://doi.org/10.1007/s11103-015-0337-7

Petrullo, L., Ren, T., Wu, M., Boonstra, R., Palme, R., Boutin, S., Mcadam, A. G., & Dantzer, B. (2022). Glucocorticoids coordinate changes in gut microbiome composition in wild North American red squirrels. *Scientific Reports*, *12*(1). https://doi.org/10.1038/s41598-022-06359-5

Petrushin, I. S., Vasilev, I. A., & Markova, Y. A. (2023). Drought tolerance of legumes: Physiology and the role of the microbiome. *Current Issues in Molecular Biology*, *45*(8), 6311–6324. https://doi.org/10.3390/cimb45080398

Poudel, M., Garrett, K. A., Costa, L. A. S., Mendes, R., Bueno, C. G., Meng, Y., Martins, S. J., & Folimonova, S. Y. (2021). The role of plant-associated bacteria, fungi, and viruses in drought stress mitigation. *Frontiers in Microbiology*, *12*(108923). https://doi.org/10.3389/fmicb.2021.743512

Pratiwi, E. P. A., Ramadhani, E. L., Nurrochmad, F., & Legono, D. (2020). The impacts of flood and drought on food security in central java. *Journal of the Civil Engineering Forum*, *6*(1), 69. https://doi.org/10.22146/jcef.51872

Raihan, A., Ridwan, M., & Rahman, M. S. (2024). An exploration of the latest developments, obstacles, and potential future pathways for climate-smart agriculture. *Climate Smart Agriculture*, *1*(2), 100020. https://doi.org/10.1016/j.csag.2024.100020

Rane, J., & Minhas, P. S. (2017). *Agriculture Drought Management Options: Scope and Opportunities* (pp. 51–72). Springer Singapore. https://doi.org/10.1007/978-981-10-5744-1_3

Ray, R. L., Gao, M., & Ampim, P. A. Y. (2020). *Crop Protection Under Drought Stress* (pp. 145–170). Springer. https://doi.org/10.1007/978-3-030-46111-9_6

Ren, C., Jin, J., Zhang, X., Reis, S., Gu, B., Wang, S., & Xu, J. (2023). Climate change unequally affects nitrogen use and losses in global croplands. *Nature Food*, *4*(4), 294–304. https://doi.org/10.1038/s43016-023-00730-z

Rossmann, M., Mendes, R., Chiaramonte, J. B., Sarango-Flores, S. W., & Kmit, M. C. P. (2017). *Plant Microbiome: Composition and Functions in Plant Compartments* (pp. 7–20). Springer. https://doi.org/10.1007/978-3-319-59997-7_2

Rutten, G., Gómez-Aparicio, L., & Frey, B. (2020). *Mediterranean Oaks Harbor More Specific Soil Microbes at the Dry End of a Precipitation Gradient*. Cold Spring Harbor Laboratory. https://doi.org/10.1101/2020.05.14.095943

Samuels, L., Lam, P., Debono, A., Wen, M., Kunst, L., & Jetter, R. (2008). Use of *Arabidopsis eceriferum* mutants to explore plant cuticle biosynthesis. *Journal of Visualized Experiments*, *16*. https://doi.org/10.3791/709

Santander, C., González, F., Santos, C., Pérez, U., Vidal, G., Aroca, R., Cornejo, P., & Ruiz, A. (2024). Enhancing water status and nutrient uptake in drought-stressed lettuce plants (lactuca sativa L.) via inoculation with different bacillus spp. isolated from the atacama desert. *Plants (Basel, Switzerland)*, *13*(2), 158. https://doi.org/10.3390/plants13020158

Sato, Y., & Yokoya, S. (2007). Enhanced tolerance to drought stress in transgenic rice plants overexpressing a small heat-shock protein, sHSP17.7. *Plant Cell Reports*, *27*(2), 329–334. https://doi.org/10.1007/s00299-007-0470-0

Sconiers, W. B., Rowland, D. L., & Eubanks, M. D. (2020). Pulsed drought: The effects of varying water stress on plant physiology and predicting herbivore response. *Crop Science*, *60*(5), 2543–2561. https://doi.org/10.1002/csc2.20235

Sheng, P., Guo, X., Jones, M. K., Yang, L., Shang, X., & Sun, Z. (2018). North–south patterning of millet agriculture on the Loess Plateau: Late Neolithic adaptations to water stress, NW China. *The Holocene*, *28*(10), 1554–1563. https://doi.org/10.1177/0959683618782610

Singh, A., Verma, S., Choudhary, P., Kumar, M., & Chakdar, H. (2020). *Plant Microbiome: Trends and Prospects for Sustainable Agriculture* (pp. 129–151). Springer. https://doi.org/10.1007/978-3-030-36248-5_8

Singh, H., Mishra, A. K., Mohanto, S., Kumar, A., Mishra, A., Amin, R., Darwin, C. R., & Emran, T. B. (2024). A recent update on the connection between dietary phytochemicals and skin cancer: Emerging understanding of the molecular mechanism. *Annals of Medicine & Surgery*. https://doi.org/10.1097/ms9.0000000000002392

Singh, N. K., & Rai, A. K. (2024). Unraveling the complex dynamics of soil microbiome diversity and its implications for ecosystem functioning: A comprehensive review. *Microbiology Research Journal International*, *34*(3), 17–47. https://doi.org/10.9734/mrji/2024/v34i31434

Stone, J., Bacon, C., & White, J. (2000). *An Overview of Endophytic Microbes: Endophytism Defined* (pp. 17–44). https://doi.org/10.1201/9781482277302-1

Styer, A., Pettinga, D., Caddell, D., & Coleman-Derr, D. (2024). *Improving Rice Drought Tolerance Through Host-Mediated Microbiome Selection*. Elife Sciences. https://doi.org/10.7554/elife.97015

Verma, K., Augustine, A. A., Priya Soni, R., Chauhan, D., Kumar, P., & Pratibha, P. (2023). Drought; influence of drought in agriculture and management strategies for drought-a review. *Asian Journal of Microbiology, Biotechnology & Environmental Sciences*, *25*(4), 638–642. https://doi.org/10.53550/ajmbes.2023.v25i04.005

Vieira, A. M., & Feijó, J. A. (2016). Hydrogel control of water uptake by pectins during in vitro pollen hydration of Eucalyptus globulus. *American Journal of Botany*, *103*(3), 437–451. https://doi.org/10.3732/ajb.1500373

Vigneron, A., Agogué, H., Ader, M., Jézéquel, D., Leboulanger, C., Groleau, A., Carré, C., Hugoni, M., Got, P., Troussellier, M., Duperron, S., Cloarec, L. A., Oger, P. M., Bernard, C., & Roques, C. (2024). Seismic events as potential drivers of the microbial community structure and evolution in a paleo-ocean analog. *Communications Earth & Environment*, *5*(1). https://doi.org/10.1038/s43247-024-01615-0

Vo-Minh Nguyen, H., Hur, J., & Shin, H.-S. (2022). Humic acids and fulvic acids: Characteristics, sorption of hydrophobic organic contaminants, and formation of disinfection by-products during chlorination. https://doi.org/10.5772/intechopen.105518

Vrantsi, E., Athanasiadis, V., Papadaki, E. S., Lalas, S. I., Lakka, A., Bozinou, E., & Dourtoglou, V. G. (2021). Humic and fulvic acids as specific sorbents of herbicides in water. *CLEAN—Soil, Air, Water*, *49*(11), 2000467. https://doi.org/10.1002/clen.202000467

Vukanti, R. V. N. R. (2020). *Structure and Function of Rhizobiome* (pp. 241–261). Springer. https://doi.org/10.1007/978-3-030-36248-5_13

Wang, Y., Yan, D., Wang, J., Ding, Y., & Song, X. (2017). Effects of elevated CO_2 and drought on plant physiology, soil carbon and soil enzyme activities. *Pedosphere*, *27*(5), 846–855. https://doi.org/10.1016/s1002-0160(17)60458-2

Xie, H., Chai, Y., Liu, Z., Hao, W., & Gai, J. (2023). Community assembly of endophytic bacteria and fungi differs in soil-root continuum of Carex cepillacea. *Applied Soil Ecology*, *194*, 105206. https://doi.org/10.1016/j.apsoil.2023.105206

Xie, J., Sehim, A. E., Li, X., Zhang, D., Chen, S., Wu, J., & Dawwam, G. E. (2021). Drought stress triggers shifts in the root microbial community and alters functional categories in the microbial gene pool. *Frontiers in Microbiology*, *12*. https://doi.org/10.3389/fmicb.2021.744897

Xu, L., & Coleman-Derr, D. (2019). Causes and consequences of a conserved bacterial root microbiome response to drought stress. *Current Opinion in Microbiology*, *49*, 1–6. https://doi.org/10.1016/j.mib.2019.07.003

Xun, W., Ma, A., Liu, Y., Zhang, R., Zhang, N., Yan, H., Shao, J., Xu, Z., Shen, Q., Zhang, R., & Miao, Y. (2024). Dissection of rhizosphere microbiome and exploiting strategies for sustainable agriculture. *The New Phytologist*, *242*(6), 2401–2410. https://doi.org/10.1111/nph.19697

Yang, F., & Antonietti, M. (2020). Artificial humic acids: Sustainable materials against climate change. *Advanced Science*, *7*(5), 1902992. https://doi.org/10.1002/advs.201902992

Yang, J., Yang, Y., Chen, H., Tan, C., Wang, S., & Wang, S. (2015). Drought adaptation in the Ningxia Hui autonomous region, China: Actions, planning, pathways and barriers. *Sustainability*, *7*(11), 15029–15056. https://doi.org/10.3390/su71115029

Ye, F., Zhao, C., Zhang, P., Liu, S., Li, X., Liu, L., & Jiang, M. (2022). Exogenous melatonin reprograms the rhizosphere microbial community to modulate the responses of barley to drought stress. *International Journal of Molecular Sciences*, *23*(17), 9665. https://doi.org/10.3390/ijms23179665

Zhang, L., Yan, P., Li, X., Zhang, L., Ahammed, G. J., & Han, W.-Y. (2018). *Plant Hormones as Mediators of Stress Response in Tea Plants* (pp. 269–288). Springer Singapore. https://doi.org/10.1007/978-981-13-2140-5_12

Zipper, S. C., Kucharik, C. J., & Qiu, J. (2016). Drought effects on US maize and soybean production: Spatiotemporal patterns and historical changes. *Environmental Research Letters*, *11*(9), 094021. https://doi.org/10.1088/1748-9326/11/9/094021

14 Exploring the Use of Microbiome as Biocontrol Agents to Manage Pests and Diseases

Madhumitha L., Praveen Kumar M., Gokul V., Meenambiga Setti Sudharsan, Ivo Romauld, Vivek P., and Sowmya H.*
Department of Bioengineering, School of Engineering, Vels Institute of Science Technology and Advanced Studies, Chennai.

14.1 INTRODUCTION

14.1.1 Overview of Microbiome in Agriculture

The microbiome is an assemblage of all microbes that naturally exist on and within human bodies, such as viruses, fungi, bacteria, and their genes. Even though microorganisms are so small that they are only visible under a microscope, they have a big influence on people's health and well-being. They help our immune system grow, shield us from infections, and allow us to break down food so we can make energy. The impacts of these microbes are diverse on health and can even influence our reactions to specific environmental toxins because the microbiome is a crucial interface between the body and the environment. Certain microorganisms change environmental compounds in ways that increase their toxicity, while others serve as a buffer, reducing the harm that environmental substances cause. The complexity of microbial community organization has been better understood in recent decades because of microbiome research. Currently, research on the microbiome is just beginning to comprehend microbial complex community structure, intra- and intergeneric interactions, and the extension to biotic and abiotic variables (Sánchez-Cañizares et al., 2017). Our knowledge of the dynamics that take place inside microbiomes and how they interact with host species has increased as a result of the development of new instruments and the growing demand for various experimental techniques (Goodrich et al., 2017). Several researchers have concentrated on the complexity of soil and plant microbiomes, producing fresh findings for the impending Green Revolution in sustainable agricultural production (Cui et al., 2019).

It is essential to develop new analytical methods to fully grasp the extended potential of microbiomes in order to increase crop productivity by applying microbial communities more effectively (Bashiardes et al., 2018). The development of crops depends on the preservation of a diversified microbiome and a well-balanced rhizosphere. However, by maintaining soil fertility and availability of nutrients to the plant using different methods of solubilization, microbiomes have been used on farms to boost agricultural yields (Ab Rahman et al., 2018). The production of food, consumable resources, and biomaterials depends on the control of plant diseases. To handle the expanding concern for global food security, it is estimated that the most significant international issue of our time is the need to boost world food production by 70% by 2050 (Ingram, 2011). The existing food production system is to blame for 25% of global greenhouse gas emissions and 60% of the world's terrestrial biodiversity loss. Since crop protection is essential to preserving agricultural productivity from pathogen competition, the growing global population necessitates an effective management system and disease control in crop production (Oerke & Dehne, 2004). One of the few strategies that demonstrates the capacity to manage disease is beneficial biocontrol of microbiomes, which can work by either directly opposing plant pathogens through the release of antimicrobial chemical compounds or by competing with them (Chen et al., 2019). Although systemic acquired resistance (SAR) can result from localized phytopathogen infection. Additionally, the use of advantageous, nonpathogenic rhizobacteria is possible to bioprime plants, leading to induced systemic resistance (ISR) to infections. The total number of microorganisms and their biocontrol activity rose when compost was added to soil that contained cane molasses, citrus pulp, crab shell waste, and chitosan (Cheng et al., 2016).

14.1.2 Importance of Pest

Insects, an extremely diverse group of species, are essential to the ecosystem. Certain insects are classified as pests because they pose a threat to human health, stored goods, forestry, agriculture, and other industries. Insects, like all other creatures, coexist closely with microbes, which have a significant impact on their ecology and evolutionary history. Both permanent and temporary connections between microbes like bacteria, viruses, fungi, protozoa, and archaea and their insect host can be advantageous or detrimental to the fitness of the insects (Kaufman et al., 2000).

***Corresponding author:** meenambiga.se@vistas.ac.in (Dr. Meenambiga Setti Sudharsan)

DOI: 10.1201/9781003544425-14

For example, endosymbionts often rely on their insect hosts to receive nutrients, but they can also offer fitness benefits in terms of overcoming host defenses, providing nourishment, and protecting against viruses, parasites, and other environmental stressors. Microbes may, however, also be harmful, lowering viability and resulting in morbidity. Additionally, there may be physiological expenses associated with endosymbiont harboring (Krams et al., 2017).

Most studies on the interactions between insects and microbes up until 20 years ago concentrated on a single species of bacteria. The genus of aphids, among the most thoroughly studied models of pest insects, exhibits advantageous host–microbiome interactions mediated by the endosymbiotic bacterium *Buchnera*. genus of aphids, among the most thoroughly researched models of pest insects, advantageous host-microbiome interactions are *Buchnera* endosymbiotic bacteria (Douglas, 1998). They are inextricably linked to one another in terms of nutrition. A further well-researched insect–endosymbiont connection involves parasitic bacteria belonging to the genus *Wolbachia* and (many distinct) insect species. These parasites that reproduce can increase their own transmission by causing their hosts to become feminized, kill only males, become cytoplasmically incompatible, or undergo parthenogenesis (Kageyama et al., 2002). For many insects, *Wolbachia* is a mutualistic organism that can also help with nutritional provisioning and confer resistance to viruses, insecticides, or plant defenses. Even while these instances show the wide range of impacts that a single microbe species can have on its insect host, it is now widely acknowledged that microbiome is the collective term for the wide variety of microbes found in all living organisms (Nikoh et al., 2014).

Studies to describe the insect microbiome have increased dramatically during the last 10 years, primarily using methods for NGS. These investigations have been concentrating on characterizing the microbiome's makeup and its potential functional involvement in insect biology (Broderick & Lemaitre, 2012). Therefore, it is essential to expand our understanding to encompass the entire microbial community rather than focusing on a single microbial species. Through genetic and metabolic interactions, all elements of the microbiome have the capacity to collectively influence an insect's behavior and physiological characteristics. However, for historical or pragmatic reasons, all of this research concentrates on populations of bacteria with insufficient or the rest of the microbiome not being integrated. Nearly 1,000 articles since 2004 were found by searching for insects and microbiomes in the Web of Science. The great majority of these articles focused on the bacterial communities found in insects, and they frequently lacked information about the makeup or functions of other microorganisms, like viruses, fungi, and protozoa. These additional microbial members, however, can also affect the fitness and life history of their insect hosts. We contend that the insect host is also impacted by the coordinated actions and interactions of various microbial community members (Janson et al., 2008).

14.2 TYPES OF MICROBIOMES

A collection of microorganisms known as the microbiome resides in particular habitats. Microbiomes can be divided into different sorts according to the environment they live in. These are the primary kinds, as illustrated in Figure 14.1.

14.2.1 Marine Microbiome

Protists, bacteria, archaea, fungi, viruses, and other microbes are all associated with animals on Earth. Animal–microbial interactions in the ocean have traditionally been studied in systems with just one symbiont and host. However, new studies of the variety found in microbes that coexist with various hosts of marine animals are turning the field into one that examines relationships between a more varied ecology and the animal host. The possibility that microbes could impact the condition, physiology, ecology, and behavior of marine organisms could improve our understanding of the adaptation of marine creatures to change, particularly the increasing anthropogenic and climate-related changes that are currently affecting the ocean environment (Apprill, 2017). From simple species like sponges and ctenophores to more complicated ones like sharks and sea squirts, the microbiomes of a wide variety of marine animals are now being studied (Webster et al., 2016). One of the most well-studied symbiotic connections in the ocean is that between the bacterium that emits light, *Aliivibrio fischeri* and the *Hawaiian bobtail* squid. This interaction acts as a template system for studying symbiosis in general. This connection has shed light on basic mechanisms in microbial–animal symbioses, particularly the biochemical exchanges and communication between the bacterium and its host (McFall-Ngai, 2000).

The oligochaete worm *Olavius algarvensis*, which has no gut, is another extensively researched oceanic host of microbes. The Mediterranean Sea's shallow marine

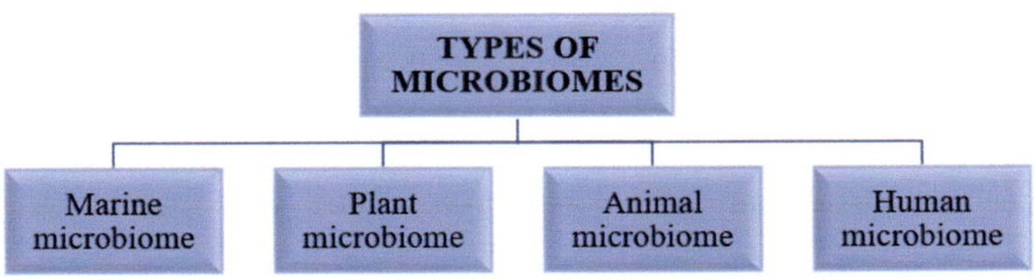

FIGURE 14.1 Types of microbiomes.

sediments are home to these three-centimeter-long worms. A range of extracellular bacterial endosymbionts reside there and collectively consume sulfur in the environment to feed the worms, which without having a digestive, excretory, or oral system (Dubilier et al., 2001). Some of the top tools for 'omics and visualization available have helped this system. Interactions between the host and microbiota, including energy transmission between the two and the worm's natural defenses, identifying the consortia, have been studied using transcriptomics and proteomics. Multi-labeled probing, for instance, has enhanced imaging of the microbiome. The main benefit of this system is that it offers a range of microbial and host genetic resources, as well as studying the relationship between the microbiome and the host using a microbial consortium with low diversity (Woyke et al., 2006).

The most typical instances of animals as hosts that can develop dysbiosis from a symbiotic relationship with microalgae, which manifests as bleaching, are corals. Research on coral microbiomes has demonstrated how changes in the ocean habitat, specifically with regard to inorganic nutrients, light, and temperature, all have an impact on the host's physiology and calcification as well as the quantity and functionality of the microalgal symbionts. One of the earliest indications of symbiosis between marine animals in multiple domains has been provided by studies that have revealed resident bacteria, archaea, and fungi also play a part in the coral's nutrient and organic matter cycling, with viruses potentially helping to structure the composition of these members. *Endozoicomonas*, a gamma proteobacterium with a versatile lifestyle, is becoming a key component of the coral's microbiome. Since reefs have recently experienced widespread bleaching, corals will probably remain a practical and well-liked system for studying symbiosis and dysbiosis (Neave et al., 2016). Sponges are widespread in the ocean's many benthic environments, and because of their ubiquity and capacity to filter vast amounts of saltwater, it is now known that they have a significant impact on pelagic as well as benthic processes. Among the earliest animal lineages, they have a comparatively simple body plan and are frequently associated with viruses, fungi, bacteria, archaea, and algal protists. The complexity of sponge microbiomes, which are made up of both specialists and generalists, seems to be influenced by host phylogeny (Webster & Thomas, 2016).

Assessing cetacean microbiomes might be challenging due to challenges in obtaining microbial samples. For instance, several of the rare whale species are deep divers. Different methods exist for collecting gut microbiota samples from cetaceans. The most often used technique involves obtaining samples of faeces from the surroundings and acquiring a clean probe from the facility (Suzuki et al., 2019). Marine animals are shielded from the outside world by their skin. Animal health is measured by the epidermal microbiome of the skin, which also acts as an ecological indication of the surrounding ecosystem's condition. Understanding how marine mammals' skin microbiota often looks helps one understand the differences between these populations and the sea's free microbial communities. A number of stresses put cetaceans at risk by increasing their susceptibility to specific diseases. Despite their considerable vulnerability, their respiratory microbiota is poorly understood in relation to airway infections. A sample of the "blow" can be used to assess their health. A range of chemical and microbiological substances, including proteins, lipids, and airway linings' cellular debris when released into the comparatively colder air outside condense into a discernible mound of collected vapor. There are several ways for obtaining exhaled breath samples. However, the deployment of aerial drones is among the most recent (Vendl et al., 2019).

14.2.2 Terrestrial

14.2.2.1 Plant Microbiome

Recent years have seen a surge in interest in the plant microbiome, which plays a role in food production and plant health. Plants cohabit with many microbial consortia known as the plant's microbiota. These bacteria reside in plant tissues' endosphere and episphere and are essential to plants' physiology and ecology (Dastogeer et al., 2020). In addition to environmental influences like management, land use, and climate, plant microbiomes are influenced by genotype, organ, species, and health state. According to certain studies, the microbiome of a plant can reveal information about or be linked to its general health (Berendsen et al., 2012). Plant tissue contains a variety of niches that are colonized by plant and plant-associated bacteria. Because of ultraviolet (UV) light and changing climate, the phyllosphere—the collective term for all above-ground plant components—is a constantly changing habitat. It is mostly made up of leaves. The characteristics of the soil typically affect the sections of plants that are below ground, primarily the roots. Through the pathogenic activities of certain microbiota members, harmful interactions affect the development of plants. Conversely, advantageous microbial interactions encourage the growth of plants (Shelake et al., 2019). Plant microbiomes are classified into four types, as illustrated in Figure 14.2.

14.2.2.1.1 Rhizosphere Microbiome

The soil's 1–10 mm layer that envelops the roots is known as the rhizosphere, and it is impacted by the plant through the deposits of mucilage, dead plant cells, and root exudates. Numerous specialised species can be found in the rhizosphere, including viruses, oomycetes, bacteria, fungi, nematodes, algae, protozoa, and archaea. It is estimated that mycorrhizal fungi, which are found in over 200,000 plant species and are common components of the rhizosphere ecosystem, are linked to over 80% of all plants. By controlling the carbon and nutrient cycles, relationships between mycorrhizae and roots have a big influence on land ecosystems (van Der Heijden et al., 2015). Plant health depends on mycorrhizae since they supply up to 80% of the phosphate and nitrogen needed by plants. In return, their host plants give the fungal lipids and carbohydrates. More variety within and between species is revealed by current studies

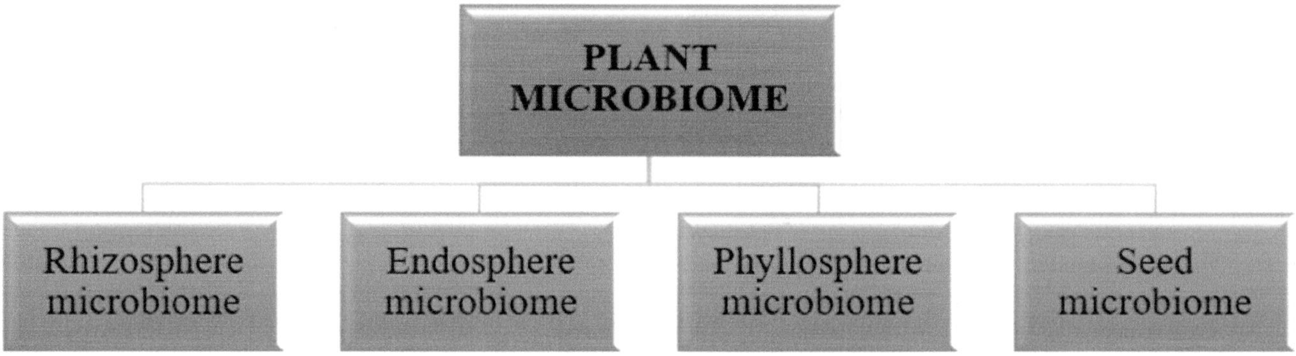

FIGURE 14.2 Plant microbiome.

employing sequencing technologies to study arbuscular mycorrhizal fungi than previously believed (Rich et al., 2017). The beneficial rhizosphere that is most commonly researched plant-growth-promoting rhizobacteria (PGPR), mycorrhizae, rhizobium bacteria, and biocontrol microorganisms are among the species. According to estimates, one gram of soil may contain more than one million unique bacterial genomes, and there are more than 50,000 Operational Taxonomic Units (OTUs) in the potato rhizosphere. The Bacteroidota, Actinomycetota, Planctomycetota, Pseudomonadota, Acidobacteriota, and Bacillota groups are the most prevalent bacteria among the prokaryotes found in the rhizosphere (İnceoğlu et al., 2011).

14.2.2.1.2 Endosphere Microbiome

Certain bacteria, known as endophytes, infiltrate and reside within the internal tissues of plants, forming the endospheric microbiome. The endosphere is mostly colonized by arbuscular mycorrhizal and other endophytic fungi. An essential component of endospore populations is bacteria and, to a lesser extent, archaea. Certain endophytic microorganisms engage with their host and clearly benefit plants. The microbial populations in the rhizosphere and rhizoplane are quite different from those in the endosphere. There may be significant differences between the root endophytic community and the nearby soil community. Outside of the plant, the microbial population is typically more diverse than the endophytic community. There may be differences in the endophytic microbiome's identity and variety between the plant's above and below-ground tissues (Bulgarelli et al., 2012).

14.2.2.1.3 Phyllosphere Microbiome

The term "phyllosphere" refers to a plant's above-ground terrain, which includes its stem, leaves, flowers, and fruits. It is believed to be slightly deficient in nutrients in contrast to the endosphere and rhizosphere. The ecosystem of the phyllosphere is further agile than that of the endosphere and rhizosphere. Seasonal and diurnal variations in radiation, moisture, and temperature affect microbial colonization. Furthermore, these environmental effects exert an unintended consequence on the composition of the microbiome and impact the physiology of plants (photosynthesis, respiration, water uptake, etc.). Additionally, wind and rain alter the phyllosphere microbiome over time (Lindow, 1996). In many of these microbiomes, interactions between the microbes and plants they are associated with can have an important effect on the health, function, and development of the host plant. The phyllosphere, or leaf surface, is home to a variety of bacterial, fungal, algal, archaeal, and viral microbiomes. Numerous facets of host plant physiology may be influenced by interactions between the phyllosphere bacteria and the host plant. But as of 2020, our understanding of these bacterial relationships in the phyllosphere is still somewhat limited, and we must improve our basic understanding of the phyllosphere microbiome's dynamics (Friesen et al., 2011).

14.2.2.1.4 Seed Microbiome

The beneficial endophytes of plants, such as those that give disease resistance, can be vertically transmitted through the seeds of plants. Instead of relying on the possibility that all of the right soil-dwelling microorganisms would be present at the germination location, it makes natural that seeds will vertically transmit their most important symbionts, according to a 2021 research study. Much of the study on agriculturally valuable plants has contributed to the recent paradigm shift of mutualistic fungi and transmission of bacteria through host plant seeds. High microbial diversity was discovered in rice seeds, with the embryo rather than the pericarp which is home to the most diverse microorganisms. It was discovered that the dominating microbiome members in maize stems were fungi of the genus *Fusarium* that were spread through seeds. The term "seed microbiome" refers to this aspect of the plant microbiome (Johnston-Monje et al., 2021).

14.2.2.2 Animal Microbiome

The coevolution of host and microbial lineages has been crucial in helping mammals adapt to their varied lifestyles, and the gut microbiota of mammals has been identified as a major regulator of host physiology. In mammals, diet—particularly herbivory—is a significant predictor of microbial diversity. Despite significant dietary changes, the majority of mammalian microbiomes also exhibit a strong correlation with host phylogeny. This implies that the structure of the gut microbiomes in mammals is significantly influenced by host characteristics, such as gut physiology, which vary throughout host phylogeny. It is even hypothesized that the vertebrate adaptive immune system originated as a selective mechanism for preserving symbiotic equilibrium (McFall-Ngai et al., 2013). The significance of phylogenetically connected variables for the overall vertebrate microbiome diversity is yet unknown. In several non-mammalian taxa, photosymbiotic, the finding that host species that are more closely related have more comparable microbiomes has been documented. Significant differences in phylosymbiotic signals between mammalian species have been discovered by other analyses, occasionally with contradictory findings. A strong phylosymbiotic association suggests that microbial assembly is regulated by host variables. Differences in the existence or strength of a detectable phylosymbiotic signal throughout host phylogeny may be helpful for discovering such mechanisms by means of comparative analysis, even if the precise mechanisms are unclear. However, as of 2020, the majority of research has concentrated on particular taxa and generalizations have been difficult because of different methods for evaluating host specificity and phylosymbiosis, as well as microbiome surveys (Song et al., 2020).

14.2.2.3 Human Microbiome

Microbiota are present in biofluids and human tissues such as the skin, ovarian follicles, uterus, lung, saliva, conjunctiva, biliary tract, gastrointestinal tract, oral mucosa, and seminal fluid. The human microbiome is the collective term for them. Protists, viruses, fungi, bacteria, and archaea are some of the various forms of the human microbiota. Usually not included in this description, the human body can also support microorganisms. In genomics, the aggregate genomes of resident microorganisms are frequently referred to as the human microbiome; the word human metagenome has the same meaning (Marchesi & Ravel, 2015). A large number of bacteria that have non-human cells, roughly on par with human cells, colonize humans. There is a relationship between mutualistic germs and their human hosts, whereas commensal bacteria live alongside people without damage or advantage. Nevertheless, the metabolites produced by certain non-pathogenic microorganisms, including trimethylamine, which is transformed by the body by Flavin-containing monooxygenase 3 (FMO3)-mediated oxidation into trimethylamine N-oxide can be harmful to human hosts. While the function of the majority of microorganisms is unclear, some of them carry out actions that are known to benefit the human host. Sometimes they are considered normal microbiota or normal flora because they are expected to exist and do not normally cause disease. The task of sequencing the human microbiota's genome was taken on by the Human Microbiome Project (HMP), which pays particular attention to the microbes that often reside on the skin, vagina, digestive tract, mouth, and nose. When it released its first findings in 2012, it accomplished a significant milestone (Willey et al., 2014). Human microbiomes are classified into four types, as illustrated in Figure 14.3.

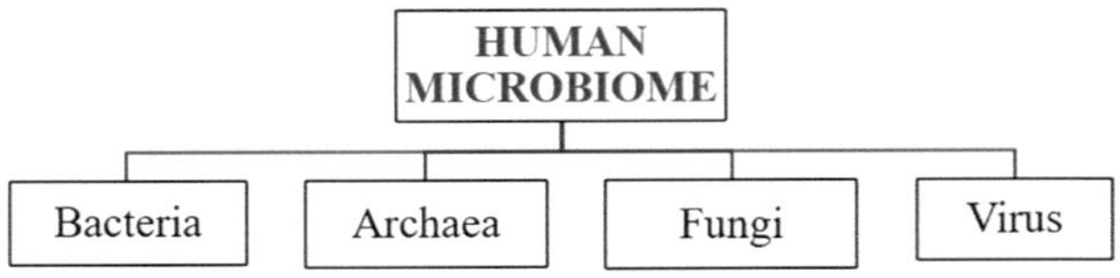

FIGURE 14.3 Human microbiome.

14.2.2.3.1 Bacteria

In different regions of the body, populations of microorganisms reside on the skin and mucosal surfaces, such as bacteria and yeasts. They play a part in normal, healthy human physiology, but disease can arise if the number of microbes increases beyond normal limits, if they take up residence in bodily regions that are typically sterile or not colonised. The 500–1,000 species of bacteria found in the human gut are believed to be made up of a few phyla, including Bacillota and Bacteroidota, as well as Pseudomonadota, Verrucomicrobiota, Actinobacteriota, Fusobacteriota, and "Cyanobacteria" (Kennedy et al., 2023). The sticky substance called plaque is made up of a variety of species of bacteria that are found in the oral cavity, including *Actinomyces viscosus* and *Actinomyces naeslundii*. This solidifies into calculus, commonly known as tartar, if it is not brushed away. Furthermore, dental disease is caused by the same bacteria's release of acids that erode tooth enamel. The vaginal microbiome is composed primarily of *Lactobacillus* species. It took a lot of time to consider that *Lactobacillus acidophilus* was one of these species, the most prevalent, but subsequent research has demonstrated that *Lactobacillus iners* is really the most prevalent. *Lactobacillus crispatus* came next, and *Lactobacillus gasseri, Lactobacillus jensenii, and Lactobacillus delbruekii* are other *lactobacilli* that are discovered in the vagina. Candidiasis and bacterial vaginosis are two illnesses that can result from disruption of the vaginal flora (Gajer et al., 2012).

14.2.2.3.2 Archaea

In the human gut, archaea are found, although their numbers are far smaller because of the diverse range of microbes found in this organ. The most prevalent groups are the methanogens, namely, *Methanosphaera stadtmanae* and *Methanobrevibacter smithii*. Methanogen colonization rates vary, nevertheless, and only roughly 50% of people have populations of these creatures that are easily observable. It has been proposed that certain methanogens are linked to human periodontal disease, but as of 2007, there were no well-defined examples of archaeal pathogens. The majority of methanogens, and specifically *Methanobrevibacter smithii*, are responsible for methane-dominant small intestine bacterial overgrowth (SIBO) (Eckburg et al., 2003).

14.2.2.3.3 Fungi

The human stomach contains fungi, especially yeasts. The *Candida* species that have been studied the most are those that can become harmful in both healthy and immunocompromised hosts. *Malassezia* species are among the yeasts that live on the skin and feed on the oils that the sebaceous glands release (Marcon & Powell, 1992).

14.2.2.3.4 Virus

Viruses, especially bacterial viruses (bacteriophages), colonize many regions of the body. The oral cavity, lungs, intestines, and skin are among these colonized areas. Virus communities are not just a reflection of bacterial communities; they have been linked to certain diseases. Researchers found a new class of viroid-like components called "obelisks" and a related class of proteins called "oblins" in the human microbiome (Zheludev et al., 2024).

14.2.3 Interaction of Microbiome with Plant Health

The variety of microorganisms in the rhizosphere and their structural community are necessary for the growth, development, and health of plants. The scientists called the plant-associated microbial community the "second genome of plants," which is significantly larger than the plant's genome, because of its crucial role in plant health. Fertile agricultural soils are home to tens of thousands of different species of microbes, which vary in number and diversity. Algae, bacteria, nematodes, fungi, protozoa, and microarthropods are all common members of soil microbial communities (Naseby & Lynch, 2002). As rhizodeposits, they consume most of the carbon produced by plant roots, these bacteria are regarded as significant participants in the food web even though the majority of them have no influence on plants. Less than 10% of all rhizosphere bacteria still exist, and they can either help or hurt plants. Algae, endophytic Arbuscular mycorrhiza fungus, and bacterial species that support plant growth and are free-living, associative, and symbiotic can all be considered beneficial microorganisms in soil, while pathogenic microorganisms include nematodes, oomycetes, and pathogenic fungi and bacteria. According to recent studies on plant–microbe interactions, different species of plants flourishing in the same conditions are linked to host-specific microbial species. The quantity and type of root exudates determine the number and diversity of both beneficial and harmful bacteria because these exudates, in conjunction with the physicochemical features of the soil, impact the structure of the rhizosphere's microbial community and, consequently, the general health of the plant (Somers et al., 2004).

The most significant of the beneficial microbes, plants, and soil interact symbiotically, which includes dynamic alterations in the interacting couples' genomes through the development of metabolic and signaling networks. During the past 20 years, two symbiotic associations in plant–microbe interactions—the arbuscular mycorrhizal fungal (AMF) association and the root nodule (RN) symbiosis—have been extensively studied (Kawaguchi & Minamisawa, 2010). A third category of microorganisms is referred to as endophytes. The "endosymbionts," or endophytes, reside inside plant tissues and do not cause illness. Among the novel bioactive metabolites, they offer quinones, tannins, saponins, terpenoids, alkaloids, phenolic acids, and steroids. Every one of these relationships is critical to better development of plants and growth (Gouda et al., 2016). Soil microorganisms are either directly or indirectly involved in enhancing the growth of plants and development via a range of growth-promoting methods. Certain germs have strain-specific processes, whereas others are highly common and typical among the microbes that can be cultured. The microbial population fluctuates greatly under natural agro ecosystems according to ambient and soil physicochemical conditions, temperature, soil moisture regimes, plant species, and vegetation cover. Microorganisms with growth-promoting traits and plant development are either stimulated or suppressed by changes in growth circumstances. The most common traits that aid in fixing biological nitrogen are part of plant growth and development, the creation of phytohormones, nutrient solubilization, the conduct of biocontrol, the hydrolytic enzymes' expulsion, and the initiation of tolerance for stress. Because beneficial soil bacteria are inexpensive, environmentally benign, and reduce reliance on non-renewable resources, their use is essential for agroecological systems to improve soil health and crop yield (Sathya et al., 2017). In the rhizosphere, edaphic characteristics influence plant–microbe synergism. One or both of the partners may benefit, gain, or suffer from such synergism. These interactions may lead to improved plant growth and productivity, healthy soil, and agricultural plants that are more resilient to stress. Using microbial inoculants and/or their metabolites, or rhizosphere engineering, to increase beneficial microbial populations can help alter the soil microbiome and boost crop yield and agroecosystem sustainability. Utilizing these advantageous microorganisms can help plants continue to grow and produce in a variety of settings while also enhancing soil health, which can support the sustainability of agroecosystems (Velmourougane et al., 2017).

14.3 MECHANISM OF MICROBIOME-BASED BIOCONTROL

Biological control of plant diseases involves the suppression of plant pathogen populations through the action of living organisms. Isolates of beneficial bacteria that are highly efficient against illnesses can be selected to be cultivated in artificial media. "Augmentative biological control" refers to the application of such mass-produced and selective antagonists at high concentrations once or multiple times over a season of growth. Growers use microbial biological control agents (MBCAs), which are approved plant protection products made by biocontrol businesses. These products include living bacteria. Certain products contain antimicrobial metabolites produced by specific microbial species, while other products merely contain antimicrobial compounds devoid of the antagonist's living cells. In the EU, these substances are regarded as chemical activities for legal purposes. Moreover, bacteriophages and mycoviruses may be useful MBCAs against plant diseases. In 2017, 101 MBCAs were registered in Europe, Japan, New Zealand, Australia, Brazil, and the United States for the purpose of disease control (Glare et al., 2012). Through a variety of mechanisms, agents of microbial biological control shield crops against disease damage. They can induce resistance or strengthen existing resistance against plant tissue diseases caused by pathogens without actively combating the pathogen. Competition with viruses for resources and space is another indirect connection. Furthermore, through antibiosis or hyperparasitism, MBCAs may have a direct interaction with the pathogen (Figure 14.4). Hyperparasites enter and kill the cells, spores, mycelium, and resting sites of bacterial and fungal pathogens. Secondary metabolite synthesis with antimicrobial qualities that can prevent infections is an additional direct method of action. Secondary metabolites that are secreted in situ in small amounts help antagonists have a competitive edge. During mass manufacture, biocontrol agents (BCA) that release secondary metabolites that are already effective have been selected in some cases. In the biological control product, these medicines can be delivered with or without living antagonist cells (Raaijmakers & Mazzola, 2012).

Even if there are additional known mechanisms of action, MBCA research frequently focuses on competition, hyperparasitism, antibiosis, and induced resistance and priming. For instance, the fungi that cause chestnut blight, *Cryphonectria parasitica*, are made hypervirulent by employing fungal viruses belonging to the Hypoviridae family. Some antagonists work by deactivating the enzymes that cause infections or by breaking down pathogen structures enzymatically. For instance, one strain of *Chryseobacterium* sp. can degrade a lectin that the pathogen known as rice blast, *Magnaporthe oryzae*, needs for spore adhesion on the surface of the leaf of the host (Ikeda et al., 2013). It is anticipated that the use of multi-omics will reveal numerous hitherto undiscovered mechanisms of microbial interaction. It is also recognized that the chemicals that MBCAs create, including secondary metabolites, can behave in a variety of ways. For instance, 2,4-diacetylphloroglucinol (DAPG) functions as both a microbial-associated molecular pattern (MAMP) and an antibiotic metabolite that can directly affect the pathogen. Therefore, it is nearly impossible to artificially separate the effects of DAPG's in-situ action on various ways of reaction and the simultaneous action of both antibiosis and induced resistance. Another

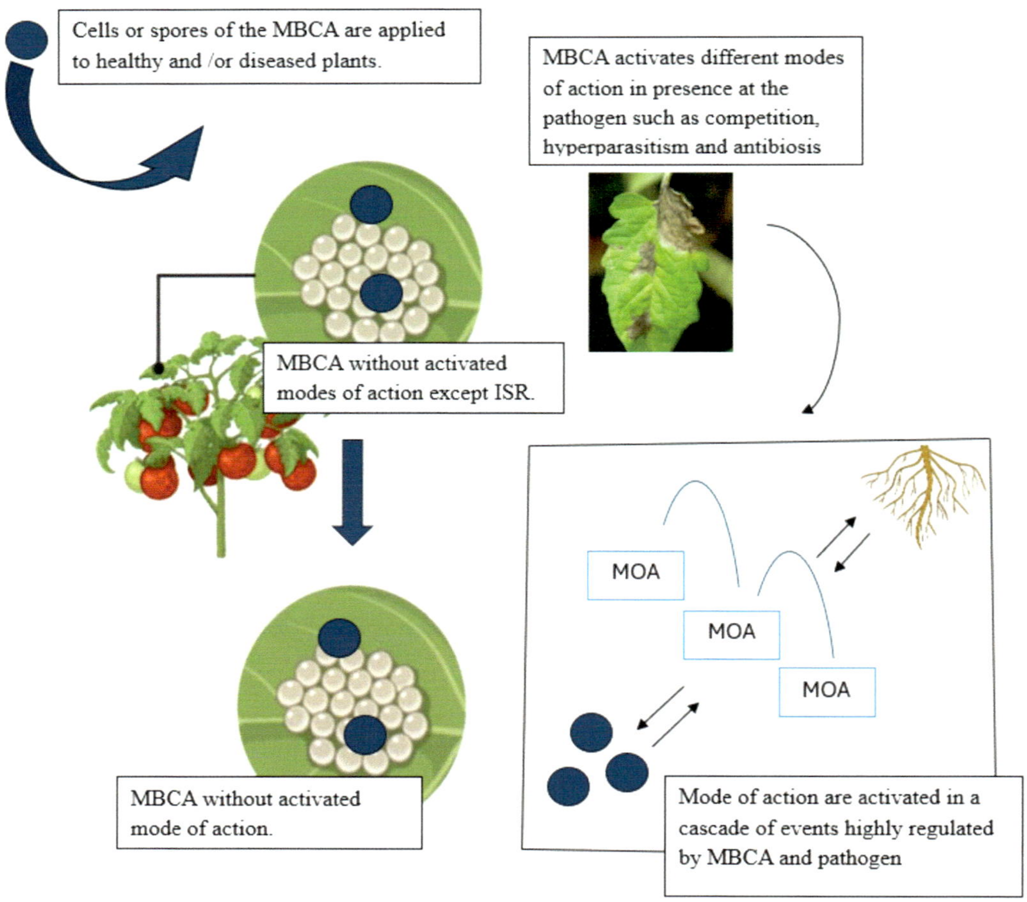

FIGURE 14.4 The targeted pathogen and the MBCA interact momentarily in situ, triggering various modes of action in a series of events.

illustration is the generation of siderophores that bind iron to compete with the pathogen for nutrients, which the plants also identify as MAMPs that cause resistance (Höfte & Bakker, 2007).

To understand how MBCAs work, scientists are methodically differentiating their modes of action. In addition to being required for enrolment, where the method of action must be specified, this information is crucial for maximizing the usage of MBCAs, as illustrated in Table 14.1. Nevertheless, microbial interactions have a more complex nature and defy easy classification into risk managers, scientists, and regulators. Results show that antagonistic interactions are fueled by many modes of action in numerous instances when the method has been carefully examined for a single biocontrol strain. Additionally, differentiation into distinct modes of activity is not always evident and appears to be fabricated. One of the essential components of the fungal hosts' successful attack is the production of hydrolytic enzymes by *Trichoderma* spp., which permeabilize and break down the fungal cell wall. Secondary antimicrobial metabolites are entering the cell more easily as a result of the cell wall's increased permeability (Karlsson et al., 2017). *Pseudoperonospora cubensis*, *Sclerotinia sclerotiorum*, and *Sphaerotheca fusca* are foliar pathogens that are controlled by isolate T39 of *Trichoderma harzianum*, which was first chosen to combat *Botrytis cinerea*. It is well known that antagonistic *Trichoderma* spp. isolates generate antimicrobial metabolites and work through hyperparasitism. Comprehensive research on *T. harzianum* T39 showed that the targeted pathogens are not being hampered by any antibiotic metabolites. There is no connection between the isolated capacity to manufacture chitinases and the biocontrol activity of other, non-antagonistic strains of *T. harzianum*. On bean leaves, *T. harzianum* T39 produces a number of proteases that inhibit *B. cinerea* enzymes. Pathogen enzymes that are vital to the pathogen during host infection, In the presence of proteases, enzymes such as β-1,3-glucanase, cutinase, pectin methyl esterase, chitinase, pectate lyase, and exo- and endo polygalacturonase were less active. Experiments using protease inhibitors either completely or in part eliminated the effect of biocontrol. The *T. harzianum* T39 biocontrol impact can, therefore, be partially attributed to the enzymes it produces that inhibit pathogen enzymes. The additional established mechanisms of action of *T. harzianum* T39 were locally induced resistance, and nutritional competition. And concluded that *T. harzianum* T39 controls biotrophic and necrotrophic foliar pathogens through a variety of modes of action. He also assumed that other biocontrol systems also involve multiple mechanisms, but in most cases, only a portion of the potential mechanisms have been clarified (Elad, 2000).

14.4 KEY MICROBIAL AGENTS FOR BIOCONTROL

Industry and academic research activities have accelerated due to the growing market demand for chemical pesticide substitutes. Increasing the competitiveness of these innovations against synthetic pesticides and making them more predictable is the objective. A synthetic microbiological material is made up of components to increase its efficiency and survivability in addition to a biocontrol agent (Schisler et al., 2004). The BCA development will assist with product handling and application, stabilize organisms during production, distribution, and storage, agent protection from damaging environmental elements, and increase organism activity (Jones & Burges, 1998). Gram-positive or Gram-negative bacteria, fungi, actinomycetes, protozoa, or viruses can all be considered BCAs. Planning the strategy to develop stable formulations might be aided by having a thorough knowledge of the organism. Various genera of bacteria have varying degrees of difficulties in their formulation. Spore-producing bacteria often have a higher survival rate than non-spore-producing bacteria (McQuilken et al., 1998). *Bacillus* and *Streptomyces* are examples of sporulating Gram-positive microbes that produce spores that are resistant to heat and desiccation. Since it is easy to synthesize these spores into stable compounds (Emmert & Handelsman, 1999). Novel approaches to enhancing the biocontrol product's effectiveness can result from the knowledge about the biocontrol agent. Different microbial BCAs are represented in Figure 14.5. A formulation that breaks the exogenous dormancy of the biocontrol agent's spores after application might be able to increase the agent's rate of action (Jones & Burges, 1998).

TABLE 14.1
Modes of Action in Connection with the Creation and Application of Biological Control Agents that Are Microbial

Achievement Mode	Screening Method	Pathogen Influence	Resistance Risk	Dependency of Environmental Conditions
Resistance induced	Complex plant bioassay	Specific to board	Low	Low
Hyper parasitism	Simplified bioassay	Pathogen-specific communication	Low	High
In situ-produced metabolites of antimicrobials	Simplified bioassay	Specific to board	Low	Moderate
Antimicrobial byproduct metabolites	In-vitro assay	Board	Moderate	Low

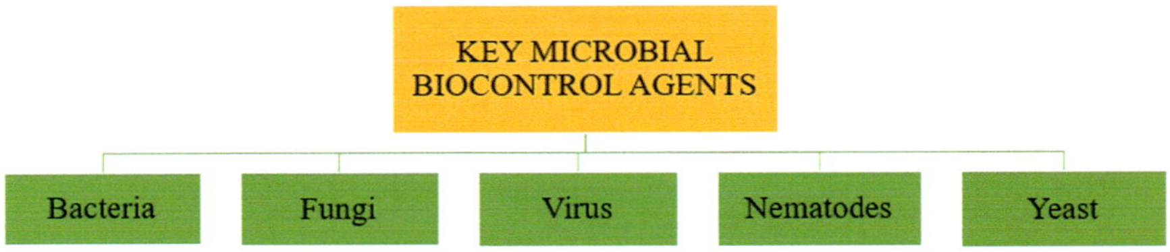

FIGURE 14.5 Key microbial BCA.

14.4.1 Bacterial Biocontrol Agent

14.4.1.1 *Bacillus thuringiensis* as Biocontrol Agent

Bacillus thuringiensis (Bt) is one of the most abundant entomopathogenic microbes for the biological management of pest insects (Samsonov et al., 1997). Fascinatingly, past accounts indicate that Bt's insecticidal qualities were known for many years prior to the bacterium's identification; some even speculate that ancient Egypt might have been using Bt spores. The Japanese biologist Shigetane Ishiwatari discovered the contemporary Bt bacterium in 1901 while studying silkworm wilt disease. He called it *Bacillus sotto*. Ten years later, Ernst Berliner in the Thuringia region of Germany identified the same bacteria from a Mediterranean flour moth that was ill (*Ephestia kuehniella*) and called it *Bacillus thuringiensis* (Siegel, 2000). *Bacillus cereus* (Bc), *Bacillus mycoides*, and *Bacillus anthracis* are all closely related to Bt, a spore-forming, Gram-positive soil bacterium. The only way to differentiate Bt from *Bacillus cereus* is that one or more bodies of inclusion are produced during the sporulation phase, which are poisonous, mostly insects belonging to the Dipteran, Lepidopteran, and Coleopteran orders (de Barjac, 1981b; Andrews et al., 1987). The capacity of Bt to form crystalline proteinaceous inclusions is its distinguishing characteristic. Several insecticidal crystal proteins (ICPs) make up the distinctive crystalline proteinaceous parasporal inclusions that Bt produces. In 1915, Bt parasporal inclusions were initially noticed (Berliner, 1915).

ICPs, also referred to as δ-endotoxins, can be spherical, cuboidal, bipyramidal, or flat rhomboid, with two crystal shapes based on the makeup of their ICP. A partial association exists between ICP content, crystal structure, and bioactivity against the intended insects (Bulla et al., 1977). The *Cry* and *Cyt* genes encode δ-endotoxins. Because they are regulated by a special RNA polymerase that is also produced during the spore-forming process, these genes become active during sporulation. These *Cry /Cyt* poisons can account for up to 20% of the protein content of the spores (Aronson, 2002). Endotoxins are ecologically appropriate approaches to controlling insects that affect major crops because of their unique poisonous effect on specific species. ICPs unique to Lepidoptera (cryI), Diptera and Lepidoptera (cryII), Coleoptera (cryIII), Diptera (cryIV), and Lepidoptera (cryV) are encoded by the two crystal (cry) gene types now known (Höfte & Whiteley, 1989). Bt strains, which can be introduced to an ecosystem to decrease insects, naturally supply most ecological niches (Andrews et al., 1987). They have isolated a number of Bt subspecies from dead or dying insect larvae; these isolates usually exhibit lethal effects on the insect from which they were extracted (Goldberg & Margalit, 1977). Both conventional and organic farming employ bioinsecticides generated from *B. thuringiensis* (Gustafson et al., 1997). Commercial preparations of Microbial Pest Control Agents (MPCAs) include Bt that contain formulation agents, particularly ICPs, and usually live spores. They are fermented products that have been treated. In addition to being biodegradable, which creates a sizable and competitive market, they are superior to chemical sprays in terms of safety, specificity, and efficacy. The advantages are contingent, though, as Bt alone functions when it is present on the plant components that insects consume. In the presence of early instar larvae, Bt is usually administered since larvae that are older are more resilient. Bt sprays only stay on the leaf surface for a few days because Cry proteins are degraded by UV light, meteorological conditions, and the leaf surface's chemical environment (McGaughey & Whalon, 1992).

14.4.1.2 *Pseudomonas fluorescens* as Biocontrol Agent

P. fluorescens species can endure a range of climatic conditions due to their versatile utilization of organic and inorganic substances. Large populations of this genus can be found in all three of the main natural habitats—terrestrial, freshwater, and marine—and they also have close relationships with both plants and animals. This extensive dissemination indicates a high degree of genetic and physiological adaptability (Nowak-Thompson et al., 1997). Because of their numerous uses as bioremediators and agents that promote plant growth, the bacteria in the genus *Pseudomonas* are ecologically significant and functionally diverse. Gram-negative, chemoheterotrophic rods that move and have polar flagella are known as *pseudomonads* (Palleroni, 1984). *Fluorescent pseudomonads* are gaining special attention as the preferred biocontrol agent among the many *Pseudomonas* species. By actively combating phytopathogens and fostering disease resistance in the host plant, Pseudomonas carries out its biocontrol function (Cartieaux et al., 2003). Numerous characteristics of pseudomonads make them ideal as growth-promoting and BCAs (Weller, 1988). The use of *P. fluorescens* strains, as well as associated pseudomonads for biological control, has important applications in two contexts. The initial example concerns the use of BCAs as soil or plant inoculants. It has been effectively used in field research related to agronomy (Karthikeyan & Gnanamanickam, 2008). The second context, when it comes to disease-suppressive soils, is that biological control by *P. fluorescens* strains and allied pseudomonads is crucial. Plants that are prone to disease can flourish without experiencing severe damage from virulent root infections (Janvier et al., 2007). Among closely related species, *P. fluorescens*' BCAs have evolved into key models for research on plant defense systems as well as secondary metabolism. As a result, their direct impacts on the pathogens and their activation of the ISR pathways in plants are now well understood. By using antimicrobial compounds or by altering rhizodeposition patterns, these BCAs most likely have the capacity to interfere with the activities of the remaining community of microbes in the rhizosphere (Phillips et al., 2004).

14.4.1.3 *A. radiobacter* as a Biocontrol Agent

A. radiobacter is an inexpensive yet effective means to stop the growth of crown gall disease because it produces the antibiotic agrocin 84 (K84 strain). It was successful to suppress the bacteriocin-sensitive pathogenic strains (Kerr & Htay, 1974). Also, it was found that the crown gall is capable of being physiologically regulated by a bacteriocin that *A. radiobacter* (84 strain), a non-pathogen found in the rhizosphere of vulnerable plants. In vitro experiments revealed that strain 84's live cells reduced the growth of galls in cherry leaf scars and had an inhibitory impact on *A. tumefaciens*. In a field trial, the crown gall was prevented by dipping strain 84's new rootstock. The heat-killed bacterial cells, however, had no impact. Young Chrysanthemum plants may also be protected from *Agrobacterium* pathogens by using *A. radiobacter* (K84 strain). According to the trials, cuttings treated with a suspension of strain K84 were effective in reducing the number of pathogen strains (Amiot et al., 1982). Strains K84, 0341, and the mutant that does not produce agrocin (K84 Agr) of *A. radiobacter* were successful in controlling crown galls in the rootstocks of stone fruits. K84 and 0341 strains of *A. tumefaciens*, which are agrocin-resistant 84, were able to manage plum tree crown galls when they were inserted into the soil. Strain K84 was able to suppress the illness that affects peach plants when *A. tumefaciens* strains that are either resistant to or susceptible to agrocin 84 were added to the soil. The strain K84 was more efficient than the delicate tension. K84 and K84 Agr strains may be used to manage plum crown galls cultivated in soil treated with *A. tumefaciens* strains that are resistant to or susceptible to agrocin 84 (López et al., 1989).

A. radiobacter strains K84 and K1026 were used to control the peach seedlings' delicate strains. Galls were absent from plants treated with the K1026 strain, while they were present in some plants that were exposed to the K84 strain. *Agrobacterium* strains isolated from galls showed that all of the K1026 strain and control bacterial isolates, or the majority of plant isolates treated with the K84 strain, displayed the same traits as inoculation strains in tests utilizing both resistant and sensitive strains (Vicedo et al., 1993). In comparison to native Tunisian strains (C58 and B6), the effectiveness of the genetically modified *A. radiobacter* (GEM) strain, the K84 strain, and the K1026 strain was evaluated. The K84 strain worked well against all tomato and target crown gall isolates, with the exception of the B6 strain. K1026 and GEM were successful in warding off every isolate. In the field, the antagonists outperformed Peach almond GF677 and quince BA29 rootstocks in preventing illness of the crown gall on the rootstocks of bitter almond trees (Rhouma et al., 2004). When inoculated with *A. tumefaciens* biovar 3, the grapevine or K84 strain of bacteria (*A. radiobacter* biovar 2) inhibited the galls in tomato seedling stems. Compared to the pathogen alone, all strains of *A. radiobacter* decreased the incidence and size of galls. Strain VAR03–1 decreased the size and production of galls on grapevine roots (Nalini et al., 2016). Gall formation on tomato stems was completely inhibited by the production of agrocin by *A. radiobacter* isolates UHFBA-8, UHFBA-11, and UHFBA-12. Crown gall is less common in peach rootstocks treated with UHFBA-11 isolated by root dip than when they are not treated. Isolate UHFBA-8, a root dipping therapy, reduced the incidence of rootstock cherry comparing colt crown galls to untreated plants. During the growth season, *A. radiobacter* isolates UHFBA-8 and -11 successfully colonize the root systems of peaches and colts as rifampicin-resistant mutants. *A. radiobacter* developed a variety of biological control strategies, including efficient root colonization, physical obstructions, binding of infection sites, and synthesis of agrocin (Gupta et al., 2010).

14.4.2 Fungal Biocontrol Agent

14.4.2.1 *Trichoderma* spp. as Biocontrol Agent

Numerous facultative, anaerobic, and cosmopolitan fungi, *Trichoderma* spp., are present in several substrates, including decomposing wood and agricultural soils (Irina & Christian, 2004; Samuels, 1996). *Trichoderma* species serve a variety of purposes beyond restricting the growth of harmful bacteria. These include (1) stimulating the colonization of rhizospheres, (2) facilitating the growth of plants and roots, and (3) strengthening defense systems in plants (Harman, 2004a; Vinale et al., 2008). *Trichoderma* species are effective biocontrol chemicals that fight phytopathogens caused by fungi. They may work directly through processes like mycoparasitism or indirectly through altering environmental circumstances, boosting plant growth, defense mechanisms, and antibiosis, or by competing for nutrients and space (Howell, 2003; Papavizas, 1985; Vinale et al., 2008). *Trichoderma* species have innate resistance to a wide range of harmful substances, like phenolic compounds, fungicides, and herbicides. Thus, they have the ability to proliferate quickly and affect infections by generating metabolic chemicals that either kill the cells (antibiosis), prevent spore germination (fungistasis), or may impact the rhizosphere (for example, by making the soil more acidic to prevent the growth of diseases). Since microorganisms most frequently die from starvation, competition for scarce resources is crucial to the biocontrol of phytopathogens. Siderophores are chelators specific to ferric iron that have a low molecular weight that filamentous fungi release when iron is in short supply. Strong siderophores produced by *Trichoderma* species chelate iron and inhibit the growth of other fungi (Benitez et al., 2004).

The intricate process of mycoparasitism often entails the synthesis of an enzyme that breaks down cell walls. The four successive steps that comprise the process of mycoparasitism are chemotropism and recognition, adhesion and coiling, penetration of the cell wall, and digestion of the host cell. *Trichoderma* strains identify other fungi, move directly in their direction, and then create enzymes that hydrolyze and break down their cell walls one after the other. *Trichoderma* coils its hyphae around the host after adhering to them, develops appressoria on their surface, and collapses the hyphae after entering the host cell (Steyaert et al., 2003). Apart from controlling infections, *Trichoderma* spp. promote plant defense mechanisms and improve root formation and growth (biofertilizer). It has been demonstrated that certain strains of *Trichoderma* can firmly and persistently colonize root surfaces and pierce the epidermis. Additionally, pre-treating plants with *Trichoderma* spp. has been demonstrated to boost their resistance to pathogen attack (Harman, 2004a). In addition, *Trichoderma* species generate plants' induced systemic resistance (ISR), hypersensitive response, and systemic acquired resistance (SAR) (Vinale et al., 2008; Benitez et al., 2004). Control of disease by root colonization, the pathogen, in *Trichoderma* spp., the biocontrol agent, the host plant, and other environmental factors interact in a complex way (Hoitink et al., 2006; Harman, 2004a; Alfano et al., 2007). *Trichoderma* spp. generate secondary chemicals and antibiotics, which are essential for antagonistic biocontrol activities (Ajitha & Lakshmidevi, 2010; Vinale et al., 2008). Various *Trichoderma* species are still used commercially to protect and encourage the growth of different crops (Lumsden et al., 1993; Samuels, 1996). The establishment of a biocontrol program is contingent upon the biocontrol agent's ability to effectively regulate the host plant–pathogen interaction. *Trichoderma* has a proven track record of effectively handling this interaction. Additionally, it has been shown that fungi improve plants' defense mechanisms. *Trichoderma* is an efficient biocontrol agent; therefore, using it will undoubtedly guarantee long-term disease control.

14.4.2.2 *Gliocladium* spp. as Biocontrol Agent

The soil fungus *Gliocladium* is found in almost every soil and other natural environments, particularly those that contain organic matter. It is a saprophytic species that may grow promptly and is easily identified due to its abundant sporulation. It can convert a wide range of natural and xenobiotic substances and produce a large number of secondary metabolites. It also produces a lot of degradative enzymes, including chitinase. As *Gliocladium* grows around the pathogen, it releases an enzyme that can break down the pathogen's chitin. Additionally, *Gliocladium virens* produces antibiotics called glicotoxin. *G. virens* uses a variety of approaches, including parasitism, antibiosis, competition, and cell death, to combat plant pathogens (Kyeong et al., 2000). These qualities make it an easy colonizer, and it has been extensively investigated as a biocontrol agent and a stimulant for plant growth. Because of its widespread function as a soil decomposer, it benefits the entire ecosystem and most likely helps to increase soil fertility (Kubicek & Harman, 1998). One of the most widely utilized biocontrol strains, formerly designated as *G. virens*, was discovered to differ from the species belonging to this genus, including the type, both morphologically and physiologically.

Under typical exposure conditions, *Gliocladium* can be regarded as non-pathogenic for healthy persons. In rare cases, it can infect immunocompromised persons, just like other saprophytic fungi (Nevalainen et al., 1998). *Gliocladium* species are well known for working as bio-fungicides against a variety of pathogenic mycoflora. *Trichoderma virens (*syn. *Gliocladium virens), Clonostachys rosea f. rosea (*syn. *Gliocladium roseum),* and *Clonostachys rosea f. catenulata (*syn. *Gliocladium catenulatum)* are the most often used species as BCAs. Many fungal mycopathogens, including *P. ultimum*, *B. cinerea*, *F. graminearum*, *F. udum*, *Phytophthora cinnamomi*, *P. citricola*, *Alternaria alternata*, *Verticillium* spp., and *Chaetomium* spp., are strongly hostile to *Gliocladium* spp. (Agarwal et al., 2011; Jabnoun-Khiareddine et al., 2009; Hassine et al., 2017). *Gliocladium* species are known to generate a variety of antagonistic metabolites, such as gliotoxin and viridin (Ayent et al., 1992). Based on the type of antibiotic they generate, the strains of *T. virens* can be separated into P and Q groupings. Gliovirin is produced by Group P members, which have a narrow range of antifungal activity and mostly target oomycetes (Howell, 2006). Nevertheless, gliotoxin, which possesses a number of antifungal and antibacterial qualities, is produced by group Q members (Anitha & Murugesan, 2005). Reports state that a different species of Gliocladium generates certain volatile antifungal compounds that prevent *P. ultimum* and *V. dahliae* from growing. Among these, annulene, an antifungal antibiotic, was found (Stinson et al., 2003). In addition, it has been found that *Gliocladium* spp. may have a biocontrol mechanism of action through mycoparasitism against a variety of fungal infections (Sun et al., 2015; Sun et al., 2015a). According to the latest research, in southwest Greece, 199 potential mycoparasites were discovered in agricultural soils. Of these, the strongest and the most potent antagonist against *S. sclerotia* of *S. sclerotiorum* was *Gliocladium* sp. G21–3 (Tsapikounis, 2015).

14.4.3 Viral Biocontrol Agent

14.4.3.1 Baculoviruses as Biocontrol Agents

Managing pests is significant for the production systems of both forestry and agriculture. In the past, chemicals were utilized as a common control measure. However, other techniques, such as biological control measures, are currently emerging as an attractive substitute for the use of chemicals since synthetic pesticides affect both the environment and organisms that are not their intended targets, as well as increasing public knowledge of these impacts (Herzfeld & Kay, 2011). There have been reports of virus infections in over 800 insect and mite species. The most prevalent and well-researched family of insect pathogenic viruses is the *baculoviridae*. Baculoviruses are diseases peculiar to invertebrates that were occasionally created as biopesticides to control insects, especially Lepidopteran species (Black et al., 1997). Baculoviruses and cypo viruses are two groups of cytoplasmic polyhedrosis viruses that are advantageous in integrated pest management (IPM) schemes as microbial control agents because of their biological characteristics. For instance, these viruses are protected and made more sustainable by their protective occlusion bodies, or polyhedra. Throughout their replicative life, baculoviruses exhibit two distinct phenotypes: the budded virus (BV), which is necessary for the infection to propagate within the host, and the occlusion-derived virus (ODV), which is necessary for the infection to transfer between larvae (Toprak et al., 2005). They only affect a single variety of insect and are very clear about their host range systems. Members of arthropod species have been identified to have different kinds of viruses. In Chinese wild insect populations, it has been shown that over 200 insect virus isolates from many viral families, including Baculoviridae, Reoviridae, Densovirinae, and Entomopoxvirinae, can produce epizootics (Sun, 2015). Baculoviruses (*nucleopolyhedrovirus* and *granuloviruses*) are the most commonly used and preferred microbial control agents (Hunter-Fujita et al., 1998). Because of its efficacy and the huge variety of viruses that have been isolated from economically significant pests, Baculoviridae is the focus of most viral bioinsecticide research and development activities. More than 1690 viruses identified in more than 1100 insect and mite species are classified as baculoviruses (Grzywacz & Moore, 2017; Onstad, 1998). Baculoviruses are the most often used viral type for biocontrol since they are more distinct from viruses that affect vertebrate species and are taken into account for appropriate use as biopesticides.

14.4.4 Nematodes as Biocontrol Agents

14.4.4.1 *Steinernema* spp. and Heterorhabditidae spp.

Depending on the family, entomopathogenic viruses have different modes of pathogenesis and reproduction, but the infection almost always results from the host ingesting the virus. Infections caused by baculoviruses often begin in the hemocoel and progress to other important organs and tissues, particularly fat bodies. Within 5–14 days, acute infections cause the host to die. Baculoviruses are classified into two genera: granuloviruses (GVs) and nucleopolyhedroviruses (NPV). Baculoviruses can only infect hosts that belong to the same order and, typically, family. Commercial baculovirus production as biopesticides is thought to pose the least amount of harm to humans and wildlife (Grzywacz & Moore, 2017). Because of their specificity and ability to produce secondary inoculum, baculoviruses and other insect viruses are desirable substitutes for chemical insecticides and perfect parts of insect pest management systems because they do not harm insects that are not the intended target, such as other biological control organisms, or have any detrimental effects on the environment or ecosystems (Cunningham, 1995; Groner, 1990; Huber, 1986). Due to their non-toxicity to consumers, relative specificity to their target pests, and ease of application using standard pesticide equipment, entomopathogenic nematodes are a useful biopesticide that can be included in IPM programs. The most successful biological control organisms have been found to be entomopathogenic nematodes. Because they have little effect on non-target insects and are thought to be extremely harmless to the environment, entomopathogenic nematodes have been widely distributed in agricultural fields. In this work, we reviewed the pathogenicity and mechanism of nematodes, as well as the phylogeny of Heterorhabditidae and Steinernematidae. The genera *Neosteinernema* and *Steinernema* represent the family Steinernematidae, while the genus *Heterorhabditis* represents the family Heterorhabditidae. Beneficial nematodes can be employed as biological control organisms because they combat soil-borne insect pests without endangering people, animals, plants, or earthworms. They can be utilized to efficiently keep soil-borne pests under control because they are inhabitants that live in the soil. Understanding their life cycles and roles, connecting the right species of nematodes with the species of pest, distributing them in the right conditions (sunlight, soil moisture, and soil temperature), and only utilizing insecticides that are compatible with them are all crucial to success. Entomopathogenic nematodes must be handled carefully during storage and transit, in addition to having an appropriate environment to survive in the soil following administration (Berry, 2007).

We suggest that cross-generational partner interaction is crucial for determining whether the bacterium–nematode complex or just one of the partners is adverse, using Koch's postulates (Lacey, 1997). There are numerous examples of symbiotic interactions between worms and bacteria in different areas of the body, even though the bulk of Heterorhabditis and Steinernema symbionts are located in the nematode stomach (Polz et al., 2000). Interestingly, *Paenibacillus nematophilus* is associated with the cuticle of Heterorhabditis species, whereas *O. carolinensis* is associated with the insect pathogen *Serratia marcescens* on its outer cuticle. Furthermore, according to orthodoxy, these connections are required since the symbionts of Heterorhabditis and Steinernema are usually not living freely, and *S. carpocapsae's* symbiont is an auxotroph for nicotinic acid, a compound that is absent from the surrounding environment (Orchard and Goodrich-Blair, 2004). Additionally, the biological traits of the two nematode taxa differ from one another. Steinernema have specialized host structures that they use to convey their symbionts, whereas Heterorhabditis use a complex multistep mechanism to transmit symbionts maternally

(Enright, 2004). Together, these findings show that, despite having very distinct biological bases for their symbiotic relationship with bacteria, Steinernema and Heterorhabditis are greatly adjusted to entomopathogenic and exhibit modifications likely to result from an extended commitment to the entomopathogenic lifestyle (Chaston & Goodrich-Blair, 2010). Applying data from a small number of representative members to entire clades can be problematic, as demonstrated by the exceptions and variances that have been noted for these complete signature features. Specialisations should not be used to dismiss recently identified correlations. We warn against presuming that these specializations apply to all or even the majority of Steinernematids or Heterorhabditids because, in fact, just a few species in either genus have been fully investigated.

14.4.5 Yeast as Biocontrol Agent

14.4.5.1 *Candida oleophila* spp. as BCAs

Yeast strain, *O. Candida oleophila*, has been isolated from the surfaces of Golden Delicious apples and chosen because of its strong and consistent biocontrol ability toward *Penicillium expandum* and *Botrytis cinerea*, two significant global infections of preserved apples and pears. All stages required for developing a biofungicide, including formulation, mechanism of action, observing, and massive usage, were examined following the isolation of *Candida oleophila* strain O from apple fruit surfaces (Jijakli et al., 1993). Nevertheless, no prior research has been done on the application of UV protectants to preharvest antagonistic yeast. The effects of UV light on BCA survival have not been extensively studied. However, it has been suggested that UV rays may decrease the efficiency of BCA when used prior to harvest. Therefore, in order to shield BCA compositions from the damaging effects of sunshine, several sunscreens have been chosen and incorporated. The following are among them: p-amino benzoic acid (PABA), tyrosine, tryptophan, ascorbic acid, tinpol LPW, Congo red, silicate, talc, inositol, folic acid, zinc oxide, titanium oxide, silicate, talc, and riboflavin. There is currently not much knowledge on how direct sunlight affects antagonistic yeasts. Finding out how ultraviolet-B (UV-B) radiation affects *C. oleophila* longevity both in vitro and in vivo, as well as screening and comparing the efficacy of the most successful chemicals employed in earlier research to increase strain O's survival, were the objectives of the current study.

When administered before harvest, the BCAs used to treat postharvest fruit disease are known to become susceptible to environmental factors, despite being extremely effective in postharvest settings. Therefore, temperature and sunlight factors are the abiotic elements that have the greatest effects on BCA viability in field settings (mainly ultraviolet A [UV-A] [320–400 nm] and UV-B [280–320 nm]), pH, relative humidity, and water availability (Ippolito & Nigro, 2000; Teixido et al., 1999; Magan, 2001). Numerous microbial fungicides have been shown to photodegrade and become inactive when exposed to different wavelengths of sunlight (Hadapad et al., 2009). Both direct and indirect DNA damage are caused by UV-B rays. Direct damage causes cyclobutane pyrimidine dimers to form in DNA, which can lead to transcription failure or mutation (Griffiths et al., 1998). *Saccharomyces cerevisiae* and *Pichia guilliermondii* are two examples of yeast cells which rapidly produce reactive oxygen species (ROS) when exposed to high temperatures. Oxidative stress to the mitochondria causes a major drop in intracellular ATP levels, cell viability, and eventually yeast survival (Sui & Liu, 2014; Kim et al., 2013).

The yeast cells were harmed by the high temperature. Both lactose and maltose were successful in increasing *C. oleophila's* resistance under stress at high temperatures (45°C, 10 min). Temperature is a key environmental factor that significantly influences the effectiveness of antagonistic yeasts used as biological control agents against postharvest diseases in fruit crops. It is well known that sugar can protect against the negative effects of both high and low temperatures. In the current work, sugar shielded the cells from heat shock, an undesirable stressor. The heat shock also triggered a resistance response, which made the yeast cells more resilient overall to the osmotic and oxidative stressors prevalent in fruit wound sites (Liu et al., 2011b; Zhang et al., 2017; Cañamás et al., 2007; Sui et al., 2012). Additionally, sugars have the potential to serve as a protective element to formulate dry and liquid biocontrol yeast formulations (Stephan et al., 2016). Since yeast antagonists compete for space and nutrients, which is one of the fundamental methods of action of postharvest biological control systems, their usage is advantageous for postharvest biocontrol (Droby et al., 2016; Wilson & Wisniewski, 1989). The biocontrol efficacy of heat-stressed cells of *C. oleophila* with maltose and lactose was generally lower than that of fresh cells against Alternaria rot, blue mold, and gray mold, but it was higher than that of yeast cells exposed to heat stress without sugar. Yeast suspensions were made to achieve the same cell counts; however, the proportion of live cells varied. For every treatment, varying amounts of live yeasts were used. Given that infections require living cells to be controlled, this may be the reason why heat-treated *C. oleophila* exhibits reduced biocontrol activity. Three bio-factors of the postharvest biocontrol system are used to identify the infection symptom on the fruit target—the pathogen, the biocontrol agent, and the host fruit—as well as an environmental factor were employed (Liu et al., 2013a).

14.5 APPLICATIONS OF MICROBIOME IN PEST AND DISEASE MANAGEMENT

A wide range of resources that can be used to control pests is represented by the contribution of microbiomes to several features associated with insect invasiveness. Using biochemicals to eradicate or interfere with insect symbiosis is a straightforward tactic (Baumann, 2005). Although they are more frequently used to combat insect-borne plant or human infections, antimicrobial peptides (AMPs) have also been investigated for manipulating insect symbionts (Carter et al., 2013). Rhizosphere microbes can form relationships with their host plants that are advantageous, neutral, or harmful, depending on the specifics of the relationship. In particular relationships between microorganisms and model plants, like symbioses between Rhizobium and legume (Nair, 1993), roots release 5–20% of photosynthate, the byproducts of photosynthesis, mainly into the rhizosphere, or soil–root interface (Ray et al., 2020). Among the many benefits of the plant microbiome, symbiotic relationships between phosphate-solubilizing bacteria, arbuscular mycorrhizal fungi, and nitrogen-fixing bacteria—primarily rhizobia—are common instances of how plants get phosphorus and nitrogen, respectively (Kohl et al., 2014). When legume crops and nitrogen-fixing bacteria interact with phosphate-solubilizing bacteria and arbuscular mycorrhizal fungi, plants have greater access to both nitrogen and phosphorus. This is because bacteria fix phosphorus and nitrogen ions, whereas fungi deliver them to plants (Kumar et al., 2015). Another technique is phytostimulation, also known as biostimulation, which is derived from the microbiome's synthesis of phytohormones and is inherently connected to plant development (Hashem et al., 2017). In addition, plants' resistance to adverse conditions like drought, salt, and exposure to heavy metals depends on their microbiome. Plant growth and productivity were interrupted due to soil salinity. However, the production of phytohormones by the microbiome may lessen the negative consequences of high soil salinity, enhancing plant resistance to these adverse circumstances (Gaiero et al., 2013). In contaminated environments, four strategies were found to manage the microbial community: feeding of supply lines, disruption, root exudate interference, and microbiome-based plant selection. For survival, growth optimization, and biomass output, plants in the soil contaminated by oil rely on their microbiome. The following are a few applications of microbes in pest and disease management given in Figure 14.6 (Stringlis et al., 2018).

14.6 CASE STUDY OF SUCCESSFUL MICROBIOME-BASED BIOCONTROL

A brief summary of microbiome-based agrifood success stories that satisfy a few of the specified requirements is provided in this compilation of case studies: (1) Commercialization-related microbiome-based advancements within the agrifood network; (2) Industry-based microbiome-based mitigation strategies to improve product quality and yield or reduce production-related monetary damage; (3) Agrifood-related microbiome-based innovations that improved the health of plants, animals, and people; (4) Knowledge based on microbiomes that resulted in improved management of food safety systems; or (5) Microbiome-based amendments to laws and regulations. The overview of the main findings of some of the case studies mentioned here is represented in Table 14.2.

14.6.1 Case Study 1

The first success story presented is the identification of the very first seed-endophytic bacteria that are capable of entirely preventing disease. This has been made possible by comprehensive methods, such as both large-scale disease occurrence monitoring and traditional microbiome analyses. According to preliminary evaluations, a variety of *Sphingomonas* bacteria were accumulated and passed down through the generations in rice seeds that were resistant to disease. These bacteria also conferred resistance to

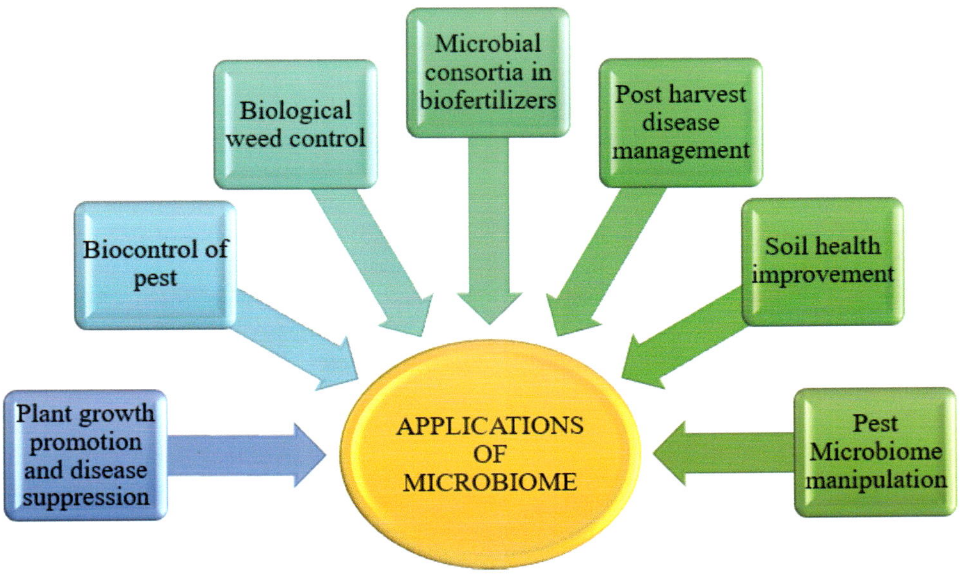

FIGURE 14.6 Applications of microbiome.

TABLE 14.2
Key Insights from Case Studies

Case Study	Key Insights	References
Resistance to disease provided by the microbiota in plant seeds	It has been demonstrated that *Sphingomonas melonis* ZJ26, which naturally exists in rice seeds as an endophyte, promotes resistance to disease to the newly emerging plant pathogen, *Burkholderia*, hence influencing the host's morphology.	Matsumoto et al., 2021
Using nitrogen-fixing microorganisms to increase sustainable agricultural productivity: the Brazilian example	By means of biological nitrogen fixation (BNF), diazotrophic bacteria provide nitrogen to the plants. In agricultural contexts, chemical N fertilizers have been essentially or entirely substituted by these microorganisms. Brazilian soybean plantations show how BNF can be used to produce food in a more sustainable manner.	Santos et al., 2021; Hungria et al., 2015; Rondina et al., 2020,
BioAg products derived from fungi for enhanced plant growth	An inoculum carried in seeds of the filamentous fungus *Penicillium bilaiae* is employed to increase the accessibility and availability of soil-derived phosphorus to plant roots, hence improving growth performance through colonization of the root system and the formation of organic acid exudates.	Gómez-Muñoz et al., 2018
From the laboratory to the field, *Bacillus* sp. S4	A commercial product for enhancing drought-stress tolerance in maize was developed from a strain of the endophytic bacteria *Bacillus simplex* that was obtained during basic research.	Naveed et al., 2014; Mitter et al., 2016

genotypes that were vulnerable to disease against the plant pathogen *Burkholderia plantarii*. Using culture-based techniques, it has been demonstrated that in resistance to disease genotypes, endophytic strains constituted the majority of the seed-endophytic bacterial species of *Sphingomonas melonis* (ZJ26). More importantly, S. melonis ZJ26 restored the phenotypic plasticity in disease-prone rice seeds following their exposure to *B. plantarii*. After *S. melonis* ZJ26 was identified as the factor that confers resistance by integrating molecular interaction tests, gene mutagenesis, and High Throughput Sequencing (HTS) data, it became easier to identify its primary mechanism of action, which underlies the suppression of pathogens by interfering with the pathogen's virulence signaling cascade. The methodology used here shows that investigations of microbiomes that combine several methods and data sources can yield significantly more mechanistic information than those that only use HTS. Even though this discovery is the first of its kind, it is reasonable to believe that a similar strategy will make it easier to find bacteria in other plant species that perform comparable tasks and open the door for the use of prevention of disease in microbes in sustainable agriculture (Matsumoto et al., 2021).

14.6.2 Case Study 2

Case Study 2 shows how N-fixing microorganisms can be applied effectively to increase enduring crop production in Brazil (Rondina et al., 2020; Hungria et al., 2015; Santos et al., 2021). Brazil's 37 million hectares of soybeans yielded a mediocre output of 3.4 tonnes/ha during the 2019–2020 crop season. Given that soybeans need about 80 kg of nitrogen per tonne of grain, and that leaching, volatilization, and denitrification processes rarely result in N fertilizer use efficiencies more than 50% (Lassaletta et al., 2014; Hungria et al., 2007; Zhang et al., 2015). Without Biological Nitrogen Fixation (BNF), Brazilian soybean crops would require about 20 million tonnes of nitrogen, or 43 million tonnes of urea. Greenhouse gas (GHG) emissions can also be significantly decreased by using N_2-fixing bacteria in place of chemical fertilizers. Considering that approximately 10 kg of CO_2-equivalent GHG emissions are associated with every kilogram of N fertilizer (Brock et al., 2012). The absence of N_2-fixing microorganisms in Brazilian soybean farms would result in the annual release of almost 430 million tonnes of CO_2-equivalent gases. However, environmental damage and production costs were significantly reduced since Brazil created the most effective program for employing inoculants with Bradyrhizobium. In recent years, 79% of soybeans have adopted inoculants containing N2-fixing microorganisms (Santos et al., 2021). which amounts to about 10.2 billion US$ in N fertilizer savings. The creation of inoculants with biostimulants, including N_2-fixing bacteria for grasses, has also been aided by the Bradyrhizobium BNF program's strong scientific and research efforts. Brazil registered the first *Azospirillum brasilense*-based inoculant for grasses in 2009 and was initially meant to be used with maize, wheat, and rice. The benefits of co-inoculating legumes with strains of *Bradyrhizobium* spp. and *A. brasilense* have been shown by recent studies and field data. Between 2017 and 2019, farmers adopted co-inoculation of soybeans with both species, making up 25% of Brazil's total planted soybean area. The 2019–2020 crop season saw more than 73 million inoculant doses overall (most farmers apply multiple doses per hectare), with a notable rise in *A. brasilense*-based inoculants over the previous 4 years, indicating an increasing trend of co-inoculation in soybeans. The development and implementation of regulations and methods that ultimately led to the acceptance of inoculants in Brazil took nearly 40 years of sporadic usage of inoculants of soybean (Chibeba et al., 2015; Hungria et al., 2013; Santos et al., 2021).

14.6.3 Case Study 3

One popular and common restricting factor for agricultural cultivation and growth is the roots' ability to get enough phosphorus (P) in the soil (Lambers et al., 2006). It has been observed that phosphate-solubilizing bacteria can dissolve inorganic P in soil that would otherwise be unavailable (Richardson and Simpson, 2011). In regard to this, a novel biological technique has been developed to supply more P that is available to plants, supporting the seedling stage and lowering the need for P fertilizers. It deepens our understanding about specific filamentous fungi (*Penicillium* genus) that release enough organic acids into the soil to break down inorganic P (Dutton and Evans, 1996). Thus, it is possible to alter and enhance the rhizosphere microbiome in terms of organismal content and function by using a seed-borne infection that includes these fungi. This serves as the foundation for case study 3, which demonstrates how incorporating a specific species of Penicillium into the seeds of plants can increase the vigor of plant seedlings, enhancing their developmental efficiency and resistance to insects and disease (Raymond et al., 2021; Gómez-Muñoz et al., 2018). A PGP product that is sold commercially is the result of this innovation. Making this a reality requires an understanding of the biology of fungal species of the genus *Penicillium*. Because of their quick development, strong colonization, and abundant secreted metabolome—which often contains antimicrobial compounds—on the surface of the roots and in the rhizosphere, they embrace an advantage in competition. In addition, a sizable proportion of conidia that are typical of *Penicillium* spp. germinate quickly, allowing them to avoid the susceptible period (between the inoculum and the rhizosphere's active development). Because *Penicillium bilaiae* does not create any metabolites that are toxic to humans or other animals, the biotech industry discovered that it performed well for this kind of usage (Currrie and Thom, 1915).

14.6.4 Case Study 4

Other microbial endophytes that were isolated during scholarly study have produced commercially viable products to increase crop plants' resistance to drought stress. The following success story case study 4 provides an example. When it comes to crop productivity and global warming, drought stress is the biggest obstacle. More than 50% of arable land is predicted to experience severe plant growth issues by 2050 as a result of droughts becoming more frequent and severe in the future (Vinocur & Altman, 2005). There is increasing attention to using PGP microbes to ensure crop yields under drought stress circumstances, in addition to conventional breeding and gene manipulation to provide crop plants greater drought resistance (Ullah et al., 2019; Naylor & Coleman-Derr, 2018). In 2014, the US startup INDIGO Ag (US) teamed up with the Austrian Institute of Technology GmbH (AIT, Austria) to search for microbial endophytes that might improve plant growth. Plant-associated microorganisms were routinely isolated by AIT researchers for use in various plant microbiome studies. These microbes were then tested for a variety of plant-beneficial traits, including biocontrol and PGP. They produced a group that included multiple hundred isolates of bacteria from various plant species and divisions by completing this task. From the roots or seeds of maize, 10 strains of bacteria were identified and put through a rigorous biochemical testing process based on their capacity to make indole-3-acetic acid (IAA) and their activity as 1-aminocyclopropane-carboxylate (ACC) deaminase. Furthermore, greenhouse studies were conducted to assess the effects of the selected strains on maize seedlings (Mitter et al., 2016; Naveed et al., 2014). Large-scale testing of the strains' interactions with various crops using hyperspectral imaging revealed that one strain of both repeatable PGP in maize under N stress and seedling vigor in winter wheat under freshwater and N deprivation were consistently improved by *Bacillus simplex* (*Bacillus* sp. S4). Concurrently, the genome sequencing of the B. simplex strain revealed several genes that are essential for plant association and potential PGP, such as the pathway for the production of methyl thiolated cytokinins or the siderophore aerobactin. The B. simplex strain underwent rigorous field testing for 3 years after the in vitro investigation, during which time it demonstrated efficacy in improving root architecture and crop resilience to drought stress. In order to treat corn seed in the US, INDIGO Ag introduced a product called Indigo™ 30 in 2018 that contained the strain as a flowable powder. The strain is being utilized for maize, wheat, and sorghum and is sold in the US, Brazil, Argentina, and numerous EU nations under the trade name biotrinsicTM W10.

14.7 ADVANCEMENT OF MICROBIOME IN PEST AND DISEASE CONTROL

The demand for sustainable agricultural methods has fueled the development of microbiomes as BCAs for disease and pest management in recent years. By increasing plant resilience and directly reducing pest populations, microbiomes—especially endophytic microbes—offer creative solutions. This comprehensive strategy incorporates a number of tactics that take advantage of the organic relationships that exist between bacteria, plants, and pests.

14.7.1 Role of Endophytic Microbes

Endophytic entomopathogenic microbes, like bacteria and fungi, inhabit plant tissues, promoting plant vitality and providing an internal shield against pests. These microbes can enhance growth, improve nutrient absorption, and activate plant defenses, all contributing to the overall robustness of the plant and its ability to resist pests (Sanjay Kumar et al., 2024).

14.7.2 New Developments in Biocontrol Techniques

The modification of biocontrol organisms through genetic engineering aims to enhance their performance and precision in combating target pests, representing a novel advancement. To boost the application and success of BCAs, innovations such as RNA interference (RNAi) and nanoparticle-based delivery mechanisms are under examination (Tiwari and Awanindra Kumar, 2024).

14.7.3 Engaging in Pest Management

The incorporation of microbiomes into IPM systems enables the development of a holistic approach that merges biological, cultural, and chemical methods. This synthesis promotes soil vitality and diversity while reducing reliance on synthetic pesticides (Kumar et al., 2024).

14.7.4 Technological Innovations

Delivery technologies utilizing nanoparticles enhance the application of biocontrol medications, while progress in genomics has deepened our comprehension of microbial interactions. By employing genetic modification to boost the synthesis of antimicrobial compounds, present-day BCAs are gaining efficacy. (Montoya-Martínez et al., 2024). As sustainable farming methods become imperative, the application of BCAs to combat plant ailments has progressed remarkably. Leading this initiative are bacteria like *Bacillus subtilis* and *P. fluorescens*, along with fungi like *Trichoderma*. Serving as an adaptable element of integrated pest control, *Trichoderma* species improve plant development and resilience in addition to combating infections. The antibiotics produced by *Bacillus subtilis* and *P. fluorescens* contribute to crop vitality by curtailing various plant diseases. Fungal endophytes and bacteriophage treatments represent innovative strategies for plant disease prevention that reduce adverse effects while targeting specific pathogens. These advancements highlight the shift toward biological solutions that foster ecological balance and decrease reliance on chemical pesticides, ultimately championing an environmentally friendly agriculture (Warrior et al., 2002; Majumder et al., 2014).

14.8 CHALLENGES OF USING MICROBIOME AS BIOCONTROL AGENT

Even with their promise, microorganisms, or Microbial Biological Management Agents (MBMAs), present certain limitations that could impede their application in studies of plant microbiomes. To guarantee the complete dependability of NGS, it is essential to address these obstacles.

1. NGS methods are fundamentally dependent on data interpretation to uncover insights, no matter how extensive the data may be. The bioinformatics evaluation of high-throughput sequencing output is currently evolving. NGS is increasingly accessible and easier to use, aided by the emergence of commercial platforms such as CLC Bio, Geneious, and Galaxy. In the past, it was predominantly confined to bioinformaticians (Meyers et al., 2008). This evolution is facilitated by enhanced processing power and affordable cloud computing solutions. The prevailing movement focuses on streamlining and parameterizing these tools, rendering them usable for individuals lacking advanced bioinformatics expertise. Such initiatives simplify the management and examination of generated sequences, opening avenues for the utilization of NGS in biocontrol research, including the investigation of interactions between BCAs and local microbial communities. NGS demands specialized knowledge and skills to derive meaningful insights and mitigate biases in interpretation. Biases can be introduced at various stages, encompassing experimental design, sample preparation, and bioinformatics processing (Buttigieg and Ramette, 2014). Slants can emerge in every step of bioinformatics procedures, and an infallible protocol does not exist. Easing up on protocols may result in an increase of false positives, while more rigid protocols could omit valuable information. Analyzing data necessitates meticulous attention to parameters, as outcomes are not always straightforward. Biocontrol researchers, particularly those pursuing a PhD, require instruction in creating NGS data, oversight, and statistical evaluation prior to investigating microbiomes. It is also wise to continue working closely with bioinformatics teams (Shakya et al., 2013).
2. Environmental parameters like temperature, humidity, soil acidity, and water content are essential to determining the effectiveness of BCAs. For example, *P. fluorescens* shows diminished efficacy at higher temperatures due to a drop in the production of antifungal compounds, while *Trichoderma* species thrive best within a temperature span of

20–30°C. Similarly, high humidity levels can encourage specific fungal pathogens, which weakens the effectiveness of BCAs like *Trichoderma* spp. and *Bacillus subtilis*, particularly when soil moisture levels are excessively elevated. In addition, the survival and operational capability of these agents are influenced by soil acidity and the makeup of the microbial community, making it challenging to forecast their performance in various agricultural environments (Babbal et al., 2017).

3. The ability of metatranscriptomics to discern between microorganisms that are alive and those that are dead, and to examine the microbial population at its actual functioning level, makes it very intriguing. To guarantee a genuinely accurate illustration of the genes that are expressed beneath a particular condition, however, a number of parts of the scientific procedures preceding it are necessary to re-examine RNA extraction and purification. One way to cut down on washing time is to alter the procedure used to remove the plant's rhizosphere and phyllosphere bacteria. The majority of regimens currently use a 15-minute to over an hour-long washing stage, while this is not always required. As seen in deep-sea water metatranscriptomics research, this sampling technique can severely alter microbial gene expression and skew the study (Feike et al., 2012).
4. The efficiency of microorganisms' biocontrol potential is significantly influenced by the genotypes and species of plants. Numerous investigations have been carried out that demonstrate how plants impact the biocontrol capabilities of BCAs. Various plant species display differing degrees of rhizosphere colonization, antibiotic production by BCAs, and the triggering of induced systemic resistance. The relationship between BCAs and pathogens in the rhizosphere zone is largely influenced by the secretion of root exudates, the absorption of nutrients and water, and the presence of other microorganisms in the vicinity. Reports indicate that the rhizosphere enrichment process, which enhances plant vitality by attracting bacteria from the surrounding environment, is affected by the expression of plant genes. To effectively tackle a variety of plant diseases, it is essential to undertake comprehensive research to unravel the complexities behind the specificity of BCAs for certain hosts (Choudhary et al., 2008; Umer et al., 2021).
5. The main factor influencing the low field efficiency of BCAs lies in their ability in order to adjust to local environmental conditions, both biotic and abiotic. Thus, it becomes essential to examine the rhizosphere's BCA distribution patterns to obtain a deeper understanding of this issue. To achieve pertinent biocontrol outcomes, it is important to collect and analyze more suitable native BCA strains. The success of BCAs significantly hinges on their mechanisms of action, selectivity toward plant diseases, and tolerance to extreme climatic conditions. When an optimal relationship exists, the efficacy of the BCAs is enhanced. For biocontrol strategies to succeed, BCAs must be deployed at the right moment. Biocontrol will work if the antagonist is administered prior to the infection being established. Future research should look into untapped microbial diversity in order to find novel strains with strong biocontrol properties. To identify microbes that have broad-spectrum antagonistic action against a variety of phytopathogens, this may require a thorough sampling and screening of varied habitats (Sehrawat et al., 2019).
6. Despite the extensive characterization of gene families believed to play vital roles in biocontrol traits or interactions between plants and pathogens, genomic and genetic sequences of plant-associated microbial communities remain largely underrepresented in databases. A staggering 50% of sequences in a metagenomic dataset may still be uncharacterized, as the function of many sequences is typically inferred through homology with databases like Swissprot. Nonetheless, the availability of genomes for pathogens, Biological Control Agents (BCAs), and epiphytic strains is rapidly on the rise. This influx will enhance future data interpretation by generating catalogs of genes and genomes from rhizosphere and phyllosphere bacteria, complemented by metagenomic insights. Nevertheless, defining a core microbiome for every crop species, either at the taxonomic level, remains a significant unresolved challenge despite the growing number of genes and genomes that are available. Ultimately, these scientific advancements will empower researchers to discover innovative strategies for improving pathogen management through optimal timing of applications, refined BCA formulations, recognition of supportive microbial strains for BCAs, or the engineering of compounds that modify microbial community structure in favor of "BCA-friendly" or "pathogen-antagonistic" microbiomes (Massart et al., 2015).
7. Despite host specificity, which can prevent non-target effects, it restricts the use of biocontrol drugs. Bacteriophages are very specialized, requiring the creation of phage cocktails to target various disease strains. This adds complexity and cost to biocontrol efforts. *Trichoderma* species are efficient against certain infections, such as Fusarium, but may not act against others, necessitating the use of numerous treatments in combination or rotation. Additionally, there is a risk of harming beneficial

bacteria and disrupting the soil's microbial ecology (Shaviv & Mikkelsen, 1993).

8. Ensuring vitality and activity during mass manufacturing and storage is one of the major issues in the formulation and manufacture of BCAs. Product quality can fluctuate when production is scaled up, and BCAs must be carefully formulated to maintain stability and efficacy because they are frequently sensitive to environmental factors like temperature and moisture. Whether for foliar sprays, soil applications, or seed treatments, the formulation must also be customized for the delivery mechanism and take into account possible interactions with other formulation ingredients because they might be expensive to produce and formulate. BCAs are less competitive when compared to chemical pesticides (Shaviv & Mikkelsen, 1993).
9. Commercialization and regulatory issues are two of the biggest obstacles preventing biocontrol products from being widely used. In nations such as the US and the EU, where stringent rules prevail, the process of registering BCAs is often challenging, expensive, and lengthy. This especially poses a barrier to innovation and delays the introduction of new biocontrol products for small- and medium-sized enterprises (SMEs). Additionally, the limited market for biocontrol solutions and farmers' doubts about the effectiveness of these agents compared to chemical pesticides further curtail investment in research and development. To tackle these challenges, it is essential to establish public-private collaborations, create standardized regulatory systems, and implement educational programs aimed at fostering the use and acceptance of BCAs (Sharifzadeh et al., 2017).

14.9 FUTURE PROSPECTS AND RESEARCH DIRECTIONS

As agriculture pivots toward more eco-friendly practices, BCAs are becoming increasingly vital in tackling pests and diseases. To unlock their complete potential, ongoing innovation and research are essential. Future explorations will include the discovery of novel BCAs, harnessing biotechnology to boost effectiveness, adopting sustainable practices, and establishing frameworks for favorable policies. Upcoming research will focus on the complex interactions between plant pathogens and BCAs. This encompasses examining how diverse modes of action can simultaneously suppress diseases, leading to more effective integrated disease management strategies.

1. *Genetic Engineering and Synthetic Science*: Subsequent investigations ought to focus on employing genetic modification and synthetic biology to boost the efficacy of existing BCAs. This entails augmenting the production of antimicrobial compounds and improving stress resilience to develop more robust biocontrol strategies (Kumar et al., 2024).
2. *Formation of Synthetic Microbial Consortia*: The creation of engineered microbial communities is suggested as a promising area of study. These communities can optimize the stability and efficacy of BCAs in agricultural settings, allowing for tailored approaches that specifically address certain pathogens (El-Husseini & Monir Mohamed, 2006).
3. *Integration with IPM*: Future investigations should focus on integrating BCAs into existing IPM frameworks. This holistic approach has the potential to reduce reliance on chemical pesticides while improving pest management strategies.
4. *Advanced Delivery Systems*: Nanotechnology and encapsulation represent innovative delivery methods that warrant additional investigation. These advances can enhance the stability and efficacy of BCAs, increasing their utility for agricultural producers (Tomar et al., 2024).
5. *Discovery of New Agents*: Creating innovative BCAs is crucial for enhancing the arsenal available to tackle pathogens and pests. An abundance of potential agents is found within the plant microbiome, especially in the rhizosphere, phyllosphere, and endosphere. Metagenomics and high-throughput sequencing have uncovered novel strains of *Bacillus* and *Pseudomonas* with unique antibiotic characteristics. Marine environments are also showing promise, where *Streptomyces* derived from these conditions demonstrate antifungal properties against *Rhizoctonia* and *Fusarium*. The capability of bacteriophages and entomopathogenic fungi such as *Beauveria* and *Metarhizium* in precise biocontrol is currently under investigation (Galli et al., 2024).
6. *Advancing Biotechnology:* Biotechnology presents techniques to enhance the efficiency of BCAs. Genetic modification has been employed to amplify the production of lipopeptides in *Bacillus* and antifungal compounds in *Pseudomonas* to bolster pathogen suppression. Engineered microbial communities deliver a more integrated approach to disease management by fusing various agents with supportive roles. Additionally, attributes such as pathogen specificity and stress resilience in BCAs are being advanced through molecular breeding and CRISPR-driven genome editing (Fira et al., 2018).

Upcoming research should delve into the societal impacts and economic feasibility of broader biocontrol technology application. Evaluating the financial viability and possible advantages for farmers and the agriculture industry as a whole would be beneficial. It is crucial to inform farmers, extension services, and policymakers about biocontrol strategies to ensure their widespread implementation. Additional efforts could be directed toward developing outreach programs and educational initiatives to enhance awareness and practicality of biocontrol methods. Moreover, further inquiries might support the creation of regulatory structures and policies that promote the safe and efficient utilization of BCAs in farming, considering that biocontrol could diminish reliance on chemical pesticides. Additionally, exploring how biocontrol strategies can bolster agricultural resilience against climate change effects might present a fruitful direction. Investigating the relationships between diseases, BCAs, and shifting environmental conditions could pave the way for future studies. By focusing on these identified research needs, scientists can propel the biocontrol field forward and create more effective and sustainable strategies for managing crop diseases (Berger et al., 2018; Mohanram & Kumar, 2019).

14.10 CONCLUSION

BCAs that leverage microbial antagonists offer a more secure alternative to or enhancement alongside chemical products. Consequently, in recent decades, the study of the identification, characterization, and market development of these BCAs has seen a steady expansion. Filamentous fungi, yeasts, and bacteria are some of the most researched microbial antagonists. The biocontrol microorganisms' most studied genera and species are listed in this review, along with an explanation of their key traits for application as BCAs in the field and post-harvest to prevent fungal infections of vegetables. Additionally covered are a few potential crucial tactics to maximize the choice and application of novel BCAs.

Although a lot of research has been done on microbial antagonists, there are currently not enough BCA formulations to meet the increasing demand for their application in sustainable agriculture systems. By offering a thorough understanding of their methods of action, which is essential to maximizing BCA activity, and by streamlining the registration processes using omic and molecular technologies, it can increasingly help to choose microbial antagonists more quickly and efficiently. Despite the technological and administrative challenges related to creating microbial BCAs, there is a notable shift toward managing plant diseases that present reduced risks to human well-being and the ecosystem. Furthermore, there is increasing governmental backing to seek resolutions and financial support for investigations into innovative technologies that tackle broader challenges linked to climate change, biodiversity, and environmental preservation. Moreover, by fostering cooperative microbial interactions, enhancing the modularity of bacterial networks, creating a more complex ecological niche for Bacillus, and boosting the photosynthetic efficiency of plants, biocontrol bacteria uphold the equilibrium of the phyllosphere microbial community and encourage plant development. In light of these findings, it is recommended that pesticides and biological control agents be thoughtfully integrated into vineyard practices in this grape-growing region. This strategy proves more effective in preserving grape quality and maintaining the stability of the phyllosphere microbiota. Consequently, there will be a growing emphasis on research aimed at crafting and developing more potent bioproducts, including microbial formulations to combat fungal diseases (Bashan et al., 2014; He et al., 2024).

REFERENCES

Ab Rahman SF, Singh E, Pieterse CM, Schenk PM. Emerging microbial biocontrol strategies for plant pathogens. Plant Science. 2018 Feb 1;267:102–111.

Agarwal T, Malhotra A, Trivedi PC, et al. Biocontrol potential of Gliocladium virens against fungal pathogens isolated from chickpea, lentil and black gram seeds. Journal of Agricultural Technologies. 2011;7(6):1833–1839.

Ajitha PS, Lakshmedevi N. Effect of volatile and von-volatile compounds from *Trichoderma* spp. against *Colletotrichum capsici* incitant of anthracnose on Bell peppers. Nature and Science. 2010;8:265–296.

Alfano G, Lewis Ivey LM, Cakir C, Bos JIB, Miller SA, Madden Kamoun VL, Hoitink JAH. Systemic modulation of gene S. expression in tomato by *Trichoderma hamatum* 382. Biological Control. 2007;97:429–437.

Amiot AF, Róux J, Faivre M. Biological control of Agrobacterium tumefaciens (Schmit et Townsend) Conn on Chrysanthemum with K84 Agrobacterium radiobacter (Beijerinck et Var Delder) Conn strain. In: ISHS Acta Horticulturae, 125: Symposium on Chrysanthemum, 1982.

Andrews RE, Jr., Faust RM, Wabiko H, Raymond KC, Bulla LA, Jr. The biotechnology of Bacillus thuringiensis. CRC Critical Reviews in Biotechnology. 1987;6:163–232.

Anitha R, Murugesan K. Production of gliotoxin on natural substrates by *Trichoderma virens*. Journal of Basic Microbiology. 2005;45(1):12–19.

Apprill A. Marine animal microbiomes: Toward understanding host–microbiome interactions in a changing ocean. Frontiers in Marine Science. 2017 Jul 18;4:222.

Aronson A. Sporulation and deltaendotoxin synthesis by Bacillus thuringiensis. Cellular and Molecular Life Sciences. 2002; 59:417–425.

Ayent AG, Hanson JR, Truneh A. Metabolites of Gliocladium flavofuscum. Phytochemistry. 1992;32(1):197–198.

Babbal A, Khasa YP. Microbes as biocontrol agents. Probiotics and Plant Health. 2017;507–552.

Bashan Y, de-Bashan LE, Prabhu SR, Hernandez J-P. Advances in plant growth-promoting bacterial inoculant technology: Formulations and practical perspectives (1998–2013). Plant Soil. 2014;378:1–33.

Bashiardes S, Godneva A, Elinav E, Segal E. Towards utilization of the human genome and microbiome for personalized nutrition. Current Opinion in Biotechnology. 2018 Jun 1;51:57–63.

Baumann P. Biology bacteriocyte-associated endosymbionts of plant sap-sucking insects. Annual Review of Microbiology. 2005;59:155–189. https://doi.org/10.1146/annurev.micro.59.030804.121041

Benitez T, Rincon AM, Limon MC, Codon AC. Biocontrol mechanism of *Trichoderma* strains. International Microbiology. 2004;7:249–260.

Berendsen RL, Pieterse CM, Bakker PA. The rhizosphere microbiome and plant health. Trends in Plant Science. 2012 Aug 1;17(8):478–486.

Berger B, Patz S, Ruppel S, Dietel K, Faetke S, Junge H, Becker M. Successful formulation and application of plant growth-promoting Kosakonia radicincitans in Maize cultivation. BioMed Research International. 2018;2018:6439481.

Berliner E. [About the sleep sickness of the Ephestia kühniella Zell. and its vector Bacillus thuringiensis.] Zeitschrift für Angewandte Entomologie. 1915;2:29–56 (in German).

Berry R. Application and evaluation of entomopathogens for control of pest insects in mint. In: Field Manual of Techniques in Invertebrate Pathology: Application and Evaluation of Pathogens for Control of Insects and Other Invertebrate Pests, Edited by LA Lacey and HK Kaya (2nd ed., pp. 599–608). 2007.

Black BC, Brennan LA, Dierks PM, Gard IE. Commercialization of baculoviral insecticides. In: The Baculoviruses (pp. 341–387). Berlin: Springer, 1997.

Brock P, Madden P, Schwenke G, Herridge D, Brock P, Madden P, et al. Greenhouse gas emissions profile for 1 tonne of wheat produced in central zone (east) New South Wales: A life cycle assessment approach. Crop Pasture Science. 2012;63:319–329. https://doi.org/10.1071/CP11191

Broderick NA, Lemaitre B. Gut-associated microbes of Drosophila melanogaster. Gut microbes. 2012 Jul 14;3(4):307–321.

Bulgarelli D, Rott M, Schlaeppi K, Ver Loren van Themaat E, Ahmadinejad N, Assenza F, Rauf P, Huettel B, Reinhardt R, Schmelzer E, Peplies J. Revealing structure and assembly cues for *Arabidopsis* root-inhabiting bacterial microbiota. Nature. 2012 Aug 2;488(7409):91–95.

Bulla LA, Jr., Kramer KJ, Davidson LI. Characterization of the entomocidal parasporal crystal of Bacillus thuringiensis. Journal of Bacteriology. 1977;130:375–383.

Buttigieg PL, Ramette A. A guide to statistical analysis in microbial ecology: A community-focused, living review of multivariate data analyses. FEMS Microbiology Ecology. 2014;90(3):543–550.

Cañamás TP, Viñas I, Usall J, Magan N, Morelló JR, Teixidó N. Mechanisms of glycine betaine enhancing oxidative stress tolerance and biocontrol efficacy of Pichia caribbica against blue mold on apples. Letters in Applied Microbiology. 2007;45:6–12. https://doi.org/10.1111/j.1472-765X.2007.02156.x

Carter V, Underhill A, Baber I, Sylla L, Baby M, Larget-Thiery I, et al. Killer bee molecules: Antimicrobial peptides as effector molecules to target sporogonic stages of plasmodium. PLoS Pathogens. 2013;9:e1003790. https://doi.org/10.1371/journal.ppat.1003790

Cartieaux F, Thibaud MC, Zimmerli L, Lessard P, Sarrobert C, David P, Gerbaud A, Robaglia C, Somerville S, Nussaume L. Transcriptome analysis of *Arabidopsis* colonized by a plant-growth promoting rhizobacterium reveals a general effect on disease resistance. Plant Journal. 2003;36:177–188.

Chaston J, Goodrich-Blair H. Common trends in mutualism revealed by model associations between invertebrates and bacteria. FEMS Microbiology Review. 2010;34:41–58.

Chen QL, Cui HL, Su JQ, Penuelas J, Zhu YG. Antibiotic resistomes in plant microbiomes. Trends in Plant Science. 2019 Jun 1;24(6):530–541.

Cheng X, van de Mortel JE, Etalo DW, Dekkers E, Raaijmakers JM. Bacterial determinants of plant growth promotion and induced systemic resistance by Pseudomonas fluorescens. Plant Growth Promotion By Pseudomonas Fluorescens. 2016:127.

Chibeba AM, de Fátima Guimarães M, Brito OR, Nogueira MA, Araujo RS, Hungria M, et al. Co-inoculation of soybean with *Bradyrhizobium* and *Azospirillum* promotes early nodulation. American Journal of Plant Sciences. 2015;6:1641–1649. https://doi.org/10.4236/AJPS.2015.610164

Choudhary DK, Johri BN. Interactions of Bacillus spp. and plants–with special reference to Induced Systemic Resistance (ISR). Microbiological Research. 2008;164:493–513.

Cui E, Fan X, Li Z, Liu Y, Neal AL, Hu C, Gao F. Variations in soil and plant-microbiome composition with different quality irrigation waters and biochar supplementation. Applied Soil Ecology. 2019 Oct 1;142:99–109.

Cunningham JC. Baculoviruses as microbial insecticides [cited 2019 Apr 12], 1995. https://www.cfs.nrcan.gc.ca/publications?id=21623

Currie JN, Thom C. An oxalic acid producing Penicillium. Journal of Biological Chemistry. 1915;22:287–293. https://doi.org/10.1016/S0021-9258(18)87646-3

Dastogeer KM, Tumpa FH, Sultana A, Akter MA, Chakraborty A. Plant microbiome–an account of the factors that shape community composition and diversity. Current Plant Biology. 2020 Sep 1;23:100161.

de Barjac H. Insect pathogens in the genus Bacillus. In: The Aerobic Endosporeforming Bacteria: Classification and Identification, Edited by RCW Berkley and M Goodfellow (pp. 241–250). New York, London: Academic Press Inc., 1981b.

Douglas AE. Nutritional interactions in insect-microbial symbioses: Aphids and their symbiotic bacteria Buchnera. Annual Review of Entomology. 1998 Jan;43(1):17–37.

Droby S, Wisniewski M, Teixidó N, Spadaro D, Jijakli MH. The science, development, and commercialization of postharvest biocontrol products. Postharvest Biology and Technology. 2016;122:22–29. https://doi.org/10.1016/j.postharvbio.2016.04.006

Dubilier N, Mülders C, Ferdelman T, de Beer D, Pernthaler A, Klein M, Wagner M, Erséus C, Thiermann F, Krieger J, Giere O. Endosymbiotic sulphate-reducing and sulphide-oxidizing bacteria in an oligochaete worm. Nature. 2001 May 17;411(6835):298–302.

Dutton MV, Evans CS. Oxalate production by fungi: Its role in pathogenicity and ecology in the soil environment. Canadian Journal of Microbiology. 1996;42:881–895. https://doi.org/10.1139/m96-114

Eckburg PB, Lepp PW, Relman DA. Archaea and their potential role in human disease. Infection and Immunity. 2003 Feb;71(2):591–596.

Elad Y. Biological control of foliar pathogens by means of *Trichoderma harzianum* and potential modes of action. Crop Protection. 2000 Sep 12;19(8–10):709–714.

El-Husseini MM. Microbial control of insect pests: Is it an effective and environmentally safe alternative. Arab Journal of Plant Protection. 2006;24(2):162–169.

Emmert E, Handelsman J. Biocontrol of plant disease: A (Gram-) positive perspective. FEMS Microbiology Letters. 1999;171:1–9.

Enright MR, Griffin CT. Specificity of association between Paenibacillus spp. and the entomopathogenic nematodes, Heterorhabditis spp. Microbial Ecology. 2004;48:414–423.

Feike J, Jürgens K, Hollibaugh JT, Krüger S, Jost G, Labrenz M. Measuring unbiased metatranscriptomics in suboxic waters of the central Baltic Sea using a new in situ fixation system. The ISME Journal. 2012;6(2):461–470.

Fira D, Dimkić I, Berić T, Lozo J, Stanković S. Biological control of plant pathogens by Bacillus species. Journal of Biotechnology. 2018;285:44–55.

Friesen ML, Porter SS, Stark SC, von Wettberg EJ, Sachs JL, Martinez-Romero E. Microbially mediated plant functional traits. Annual Review of Ecology, Evolution, and Systematics. 2011 Dec 1;42(1):23–46.

Gaiero JR, McCall CA, Thompson KA, Day NJ, Best AS, Dunfield KE. Inside the root microbiome: Bacterial root endophytes and plant growth promotion. Americal Journal of Botany. 2013;100:1738–1750. https://doi.org/10.3732/ajb.1200572

Gajer P, Brotman RM, Bai G, Sakamoto J, Schütte UM, Zhong X, Koenig SS, Fu L, Ma Z, Zhou X, Abdo Z. Temporal dynamics of the human vaginal microbiota. Science Translational Medicine. 2012 May 2;4(132):132ra52.

Galli M, Feldmann F, Vogler UK, Kogel KH. Can biocontrol be the game changer in integrated pest management? A review of definitions, methods and strategies. Journal of Plant Diseases and Protection. 2024;131(2):265–291.

Glare T, Caradus J, Gelernter W, Jackson T, Keyhani N, Köhl J, Marrone P, Morin L, Stewart A. Have biopesticides come of age? Trends in Biotechnology. 2012 May 1;30(5):250–258.

Goldberg LJ, Margalit J. A bacterial spore demonstrating rapid larvicidal activity against Anopheles sergentii, Uranotaenia unguiculata, Culex univitattus, Aedes aegypti and Culex pipiens. Mosq News. 1977;37:355–358.

Gómez-Muñoz B, Jensen LS, de Neergaard A, Richardson AE, Magid J. Effects of Penicillium bilaii on maize growth are mediated by available phosphorus. Plant Soil. 2018;431:159–173. https://doi.org/10.1007/s11104-018-3756-

Goodrich JK, Davenport ER, Clark AG, Ley RE. The relationship between the human genome and microbiome comes into view. Annual Review of Genetics. 2017 Nov 27;51(1): 413–433.

Gouda S, Das G, Sen SK, Shin HS, Patra JK. Endophytes: A treasure house of bioactive compounds of medicinal importance. Frontiers in Microbiology. 2016 Sep 29;7:1538.

Griffiths HR, Mistry P, Herbert KE, Lunec J. Molecular and cellular effects of ultraviolet light-induced genotoxicity. Critical Reviews in Clinical Laboratory Sciences. 1998;35:189–237.

Groner A. Safety to nontarget invertebrates of baculoviruses. In: Safety of Microbial Insecticides (pp. 135–147). 1990.

Grzywacz D, Moore S. Production, formulation, and bioassay of baculoviruses for pest control. In: Microbial Control of Insect and Mite Pests (Chapter 7), Edited by LA Lacey (pp. 109–124). Academic Press, 2017 [cited 2019 Apr 16]. http://www.sciencedirect.com/science/article/pii/B978012803527600007X

Gupta AK, Khosla K, Bhardwaj SS, Thakur A, Devi S, Jarial RS, Sharma C, Singh KP, Srivastava DK, Lal R. Biological control of crown gall on peach and cherry rootstock colt by native Agrobacterium radiobacter isolates. Open Horticulture Journal. 2010;3:1–10.

Gustafson ME, Clayton RA, Lavrik PB, Jonhson GV, Leimgruber RM, Sims SR, Bartnicki DE. Largescale production and characterization of Bacillus thuringiensis subsp. tenebrionis insecticidal protein from Escherichia coli. Applied Microbiology and Biotechnology. 1997;47:255–261.

Hadapad AB, Hire RS, Vijayalakshmi N, Dongre TK. UV protectants for the biopesticide based on Bacillus sphaericus Neide and their role in protecting the binary toxins from UV radiation. Journal of Invertebrate Pathology. 2009;100:147–152.

Harman GE, Howell CR, Viterbo A, Chet I, Lorito M. *Trichoderma* species-opportunistic, avirulent plant symbionts. Nature Reviews Microbiology. 2004a;2:43–56.

Hashem A, Abd Allah EF, Alqarawi AA, Radhakrishnan R, Kumar A. Plant defense approach of bacillus subtilis (Bera 71) against *Macrophomina phaseolina* (tassi) Goid in mung bean. Journal of Plant Interactions. 2017;12:390–401. https://doi.org/10.1080/17429145.2017.1373871

Hassine M, Jabnoun-Khiareddine H, Aydi Ben Abdallah R, et al. In vitro and in vivo antifungal activity of culture filtrates and organic extracts of Penicillium sp. and Gliocladium spp. against Botrytis cinerea. Journal of Plant Pathology & Microbiology. 2017;8(12):427.

He T, Yang M, Du H, Du R, He Y, Wang S, Deng W, et al. Biocontrol agents transform the stability and functional characteristics of the grape phyllosphere microenvironment. Frontiers in Plant Science. 2024;15:1439776.

Herzfeld D, Kay S. Integrated pest management. In: Private Pesticide Applicator Study Manual (19th ed.). Minnesota: University of Minnesota Extension, 2011. http://apps.extension.umn.edu/agriculture/pesticide-safety/ppat_manual/Intro.pdf

Höfte H, Whiteley HR. Insecticidal crystal proteins of Bacillus thuringiensis. Microbiology Review. 1989;53:242–255.

Höfte M, Bakker PA. Competition for iron and induced systemic resistance by siderophores of plant growth promoting rhizobacteria. In: Microbial Siderophores (pp. 121–133). Berlin, Heidelberg: Springer Berlin Heidelberg, 2007.

Hoitink HAJ, Madden LV, Dorrance AE. Systemic resistance induced by *Trichoderma* spp.: Interactions between the host, the pathogen, the biocontrol agent, and soil organic matter quality. Phytopathology. 2006;96:186–189.

Howell CR. Mechanisms employed by *Trichoderma* species in the biological control of plant diseases: The history and evolution of current concepts. Plant Disease. 2003;87:4–10.

Howell CR. Understanding the mechanisms employed by *Trichoderma virens* to effect biological control of cotton diseases. Phytopathology. 2006;96(2):178–180.

Huber J. Use of baculoviruses in pest management programs [cited 2019 Apr 12], 1986. http://agris.fao.org/agris-search/search.do?recordID=US19880053890

Hungria M, Campo RJ, Mendes IC. A importância do processo de fixação biológica do nitrogênio para a cultura da soja: Componente essencial para a competitividade do produto brasileiro. Documentos 283–Londrina, Brazil: Embrapa Soja, 2007.

Hungria M, Nogueira MA, Araujo RS. Co-inoculation of soybeans and common beans with rhizobia and azospirilla: Strategies to improve sustainability. Biology and Fertility Soils. 2013;49:791–801. https://doi.org/10.1007/s00374-012-0771-5

Hungria M, Nogueira MA, Araujo RS. Soybean seed coinoculation with *Bradyrhizobium* spp. and *Azospirillum brasilense*: A new biotechnological tool to improve yield and sustainability. American Journal of Plant Sciences. 2015;6:811–817. https://doi.org/10.4236/AJPS.2015.66087

Hunter-Fujita FR, Entwistle PF, Evans HF, Crook NE. Insect viruses and pest management. In: Insect Viruses and Pest Management [cited 2019 Apr 12], 1998.

Ikeda K, Kitagawa H, Shimoi S, Inoue K, Osaki Y, Nakayashiki H. Attachment of airborne pathogens to their host: A potential target for disease control. Acta Phytopathologica Sinica. 2013;43:18.

İnceoğlu Ö, Al-Soud WA, Salles JF, Semenov AV, van Elsas JD. Comparative analysis of bacterial communities in a potato field as determined by pyrosequencing. PLoS One. 2011 Aug 19;6(8):e23321.

Ingram J. A food systems approach to researching food security and its interactions with global environmental change. Food Security. 2011 Dec; 3:417–431.

Ippolito A, Nigro F. Impact of preharvest application of biological control agents on postharvest diseases of fresh fruits and vegetables. Crop Protection. 2000;19:610–619.

Irina D, Christian PK. Species and biodiversity in *Trichoderma* and *Hypocera*: From aggregate species to species clusters. Journal of Zhejiang University Science. 2004;6:100–112.

Jabnoun-Khiareddine H, Daami-Remadi M, Ayed F, et al. Biocontrol of tomato verticillium wilt by using indigenous Gliocladium spp. and Penicillium sp. isolates. Dyn Soil Dyn Plant. 2009;3(1):70–79.

Janson EM, III Stireman JO, Singer MS, Abbot P. Phytophagous insect–microbe mutualisms and adaptive evolutionary diversification. Evolution. 2008 May 1;62(5):997–1012.

Janvier C, Villeneuve F, Alabouvette C, Edel-Hermann V, Mateille T, Steinberg C. Soil health through soil disease suppression: Which strategy from descriptors to indicators? Soil Biology and Biochemistry. 2007;39:1–23.

Jijakli MH, Lepoivre P, Tossut P, Thonard P. Biological control of Botrytis cinerea and Penicillium sp. on post-harvest apples by two antagonistic yeasts. Mededelingen van de Faculteit Landbowwetenschappen University of Gent. 1993;58/3b:1349–1358.

Johnston-Monje D, Gutiérrez JP, Lopez-Lavalle LA. Seed-transmitted bacteria and fungi dominate juvenile plant microbiomes. Frontiers in Microbiology. 2021 Oct 22;12:737616.

Jones KA, Burges HD. Technology of formulation and application. In Formulation of Microbial Pesticides—Beneficial Microorganisms, Nematodes and Seed Treatments, Edited by HD Burges (pp. 7–30). Dordrecht: Kluwer Academic.

Kageyama D, Nishimura G, Hoshizaki S, Ishikawa Y. Feminizing wolbachia in an insect, Ostrinia furnacalis (Lepidoptera: Crambidae). Heredity. 2002 Jun;88(6):444–449.

Karlsson M, Atanasova L, Jensen DF, Zeilinger S. Necrotrophic mycoparasites and their genomes. Microbiology Spectrum. 2017 Apr 30;5(2):10–128.

Karthikeyan V, Gnanamanickam SS. Biological control of Setaria blast (Magnaporthe grisea) with bacterial strains. Crop Protectection. 27:263–267.

Kaufman MG, Walker ED, Odelson DA, Klug MJ. Microbial community ecology & insect nutrition. American Entomologist. 2000 Jul 1;46(3):173–185.

Kawaguchi M, Minamisawa K. Plant–microbe communications for symbiosis. Plant and Cell Physiology. 2010 Sep 1;51(9):1377–1380.

Kennedy KM, de Goffau MC, Perez-Muñoz ME, Arrieta MC, Bäckhed F, Bork P, Braun T, Bushman FD, Dore J, de Vos WM, Earl AM. Questioning the fetal microbiome illustrates pitfalls of low-biomass microbial studies. Nature. 2023 Jan 26;613(7945):639–649.

Kerr A, Htay K. Biological control of crown gall through bacteriocin production. Physiological Plant Pathology. 1974;41:37–40.

Kim IS, Kim YS, Kim H, Jin I, Yoon HS. Saccharomyces cerevisiae KNU5377 stress response during high-temperature ethanol fermentation. Molecular Cells. 2013;35:210–218. https://doi.org/10.1007/s10059-013-2258-0

Kohl L, Oehl F, Van Der Heijden MGA. Agricultural practices indirectly influence plant productivity and ecosystem services through effects on soil biota. Ecological Applications. 2014;24:1842–1853. https://doi.org/10.1890/13-1821.1

Krams IA, Kecko S, Jõers P, Trakimas G, Elferts D, Krams R, Luoto S, Rantala MJ, Inashkina I, Gudrā D, Fridmanis D. Microbiome symbionts and diet diversity incur costs on the immune system of insect larvae. Journal of Experimental Biology. 2017 Nov 15;220(22):4204–4212.

Kubicek CP, Harman GE. (Eds.). *Trichoderma* and *Gliocladium*, Vol. 1 and 2. Taylor & Francis Ltd., 1998.

Kumar A, Sharma S, Mishra S, Dames JF. Arbuscular mycorrhizal inoculation improves growth and antioxidative response of *Jatropha curcas* (L.) under Na_2SO_4 salt stress. Plant Biosystems. 2015;149:260–269. https://doi.org/10.1080/11263504.2013.845268

Kumar S, Chavda N, Rathore A, Nigam R, Arvind M, Nilesh R, Yuvarani R. Role of biocontrol agents in suppressing plant pathogen: A compressive analysis. Journal of Scientific Research and Reports. 2024;30(10):1004–1015.

Kumar S, Chavda N, Rathore A, Nigam R, Shivam S, Arvind M, Nilesh R, Yuvarani R. Role of biocontrol agents in suppressing plant pathogen: A compressive analysis. 2024 Oct 21 (Sciencedomain International);30(10):1004–1015.

Kyeong SJ, Hong MK, Bong KC. Purification and antifungal activities of an antibiotic produced by Gliocladium virens G1 against plant patogen. Plant Pathology Journal. 2000;17(1):53–56.

Lacey LA. Manual of Techniques in Insect Pathology. San Diego: Academic Press, 1997.

Lambers H, Shane MW, Cramer MD, Pearse SJ, Veneklaas EJ. Root structure and functioning for efficient acquisition of phosphorus: Matching morphological and physiological traits. Annals of Botany. 2006;98:693–713. https://doi.org/10.1093/aob/mcl114

Lassaletta L, Billen G, Grizzetti B, Anglade J, Garnier J. 50 year trends in nitrogen use efficiency of world cropping systems: The relationship between yield and nitrogen input to cropland. Environmental Research Letters. 2014;9:105011. https://doi.org/10.1088/1748-9326/9/10/105011

Lindow SE. Role of immigration and other processes in determining epiphytic bacterial populations: Implications for disease management. In: Aerial Plant Surface Microbiology (pp. 155–168). Boston, MA: Springer US, 1996.

Liu J, Sui Y, Wisniewski M, Droby S, Liu Y. Review: Utilization of antagonistic yeasts to manage postharvest fungal diseases of fruit. International Journal of Food Microbiology. 2013a;167:153–160. https://doi.org/10.1016/j.ijfoodmicro.2013.09.004

Liu J, Wisniewski M, Droby S, Vero S, Tian S, Hershkovitz V. Glycine betaine improves oxidative stress tolerance and biocontrol efficacy of the antagonistic yeast Cystofilobasidium infirmominiatum. International Journal of Food Microbiology. 2011b;146:76–83. https://doi.org/10.1016/j.ijfoodmicro.2011.02.007

López MM, Gorris MT, Salcedo CI, Montojo AM, Miró M. Evidence of biological control of Agrobacterium tumefaciens strains sensitive and resistant to agrocin 84 by different Agrobacterium radiobacter strains on stone fruit trees. Applied and Environmental Microbiology. 1989;55:741–746.

Lumsden RD, Lewis JA, Lock JC. Managing soilborne plant pathogens with fungal antagonists. In: Pest Management: Biology Based Technologies, Edited by RD Lumsden and JL Vaughn (pp. 196–203). Washington: American Chemical Society Publishers, 1993.

Magan N. Physiological approaches to improving the ecological fitness of fungal biocontrol agents. In: Fungi as Biocontrol Agents: Progress, Problems and Potential, Edited by TM Butt, C Jackson and N Magan (p. 390). Wallingford, Bristol, UK: CABI Publishing.

Majumder D, Kongbrailatpam JD, Suting EG, Kangjam B, Lyngdoh D. Pseudomonas fluorescens: A potential biocontrol agent for management of fungal diseases of crop plants. Future Challenges in Crop Protection against Fungal Pathogens. 2014:317–342.

Marchesi JR, Ravel J. The vocabulary of microbiome research: A proposal. Microbiome. 2015 Dec;3:1–3.

Marcon MJ, Powell DA. Human infections due to Malassezia spp. Clinical Microbiology Reviews. 1992 Apr;5(2):101–119.

Massart S, Martinez-Medina M, Haissam Jijakli M. Biological control in the microbiome era: Challenges and opportunities. Biological Control. 2015;89:98–108.

Matsumoto H, Fan X, Wang Y, Kusstatscher P, Duan J, Wu S, et al. Bacterial seed endophyte shapes disease resistance in rice. Nature Plants. 2021;7:60–72. https://doi.org/10.1038/s41477-020-00826-5

McFall-Ngai MJ. Negotiations between animals and bacteria: The 'diplomacy of the squid-vibrio symbiosis. Comparative Biochemistry and Physiology Part A: Molecular & Integrative Physiology. 2000 Aug 1;126(4):471–480.

McFall-Ngai MJ, Hadfield MG, Bosch TC, Carey HV, Domazet-Lošo T, Douglas AE, Dubilier N, Eberl G, Fukami T, Gilbert SF, Hentschel U. Animals in a bacterial world, a new imperative for the life sciences. Proceedings of the National Academy of Sciences. 2013 Feb 26;110(9):3229–3236.

McGaughey WH, Whalon ME. Managing insect resistance to Bacillus thuringiensis toxins. Science, Wash. 1992; 258:1451–1455.

Mcquilken MP, Halmer P, Rhodes DJ. Application of microorganisms to seeds. In: Formulation of Microbial Pesticides—Beneficial Microorganisms, Nematodes and Seed Treatments, Edited by HD Burges (pp. 255–286). Dordrecht: Kluwer Academic, 1998.

Meyers BC, Axtell MJ, Bartel B, Bartel DP, Baulcombe D, Bowman JL, Cao X, et al. Criteria for annotation of plant MicroRNAs. The Plant Cell. 2008;20(12): 3186–3190.

Mitter B, Pastar M, Sessitsch A. Plant-Endophyte Combinations and Uses Therefore. U.S. Patent No US 9364005 B2. Alexandria, VA: U.S. Patent and Trademark Office, 2016.

Mohanram S, Kumar P. Rhizosphere microbiome: Revisiting the synergy of plant-microbe interactions. Annals of Microbiology. 2019;69:307–320.

Montoya-Martínez AC, Valenzuela Ruiz V, Chávez-Luzanía RA, Villa-Rodríguez ED, de los Santos Villalobos S. Biological control agents for mitigating plant diseases. In: New Insights, Trends, and Challenges in the Development and Applications of Microbial Inoculants in Agriculture (pp. 27–35). Academic Press, 2024.

Nair P, K R. An Introduction to Agroforestry. Dordrecht: Kluwer Academic Publishers Netherlands, 1993, p. 499.

Nalini S, Parthasarathi R, Prabudoss V. Production and characterization of lipo-peptide from Bacillus SNAU01 under solid state fermentation and its potential application as an antibiofilm agent. Biocatalysis and Agricultural Biotechnology. 2016;5:123–132.

Naseby DC, Lynch JM. Enzymes and microorganisms in the rhizosphere. Enzymes in the environment. Activity, ecology and applications. Marcel Dekker, New York. 2002 Jan;29:109–123.

Naveed M, Mitter B, Yousaf S, Pastar M, Afzal M, Sessitsch A. The endophyte Enterobacter sp. FD17: A maize growth enhancer selected based on rigorous testing of plant beneficial traits and colonization characteristics. Biology and Fertility of Soils. 2014;50:249–262. https://doi.org/10.1007/s00374-013-0854-y

Naylor D, Coleman-Derr D. Drought stress and root-associated bacterial communities. Frontiers in Plant Science. 2018; 8:2223. https://doi.org/10.3389/fpls.2017.02223

Neave MJ, Apprill A, Ferrier-Pagès C, Voolstra CR. Diversity and function of prevalent symbiotic marine bacteria in the genus Endozoicomonas. Applied Microbiology and Biotechnology. 2016 Oct;100:8315–8324.

Nevalainen H, Neethling D. In *Trichoderma* and *Gliocladium*. In: Basic Biology, Taxonomy and Genetics, Edited by CP Kubicek and GE Harman (Vol. 1, pp. 193–205). London: Taylor & Francis Ltd., 1998.

Nikoh N, Hosokawa T, Moriyama M, Oshima K, Hattori M, Fukatsu T. Evolutionary origin of insect–Wolbachia nutritional mutualism. Proceedings of the National Academy of Sciences. 2014 Jul 15;111(28):10257–10262.

Nowak-Thompson B, Gould SJ, Loper JE. Identification and sequence analysis of the genes encoding a polyketide synthase required for pyoluteorin biosynthesis in Pseudomonas fluorescens Pf-5. Gene. 1997;204:17–24.

Oerke EC, Dehne HW. Safeguarding production—losses in major crops and the role of crop protection. Crop Protection. 2004 Apr 1;23(4):275–285.

Onstad D. Two databases provide information about insect pathogens. ASM News. 1998;64:9.

Orchard SS, Goodrich-Blair H. Identification and functional characterization of aXenorhabdus nematophila oligopeptide permease. Applied and Environmental Microbiology. 2004;70:5621–5627.

Palleroni NJ. Genus I Pseudomonas nigula. In: Bergey's Manual of Systematic Bacteriology, Edited by NR Krig and JG Holt (pp. 141–199). Baltimore: Williams and Wilkins, 1984.

Papavizas GC. Trichodema and Gliocladium: Biology, ecology and potential for biocontrol. Annual Review of Phytopathology. 1985;22:23–54.

Phillips DA, Fox TC, King MD, Bhuvaneswari TV, Teuber LR. Microbial products trigger amino acid exudation from plant roots. Plant Physiology. 2004;136:2887–2894.

Polz MF, Felbeck H, Novak R, Nebelsick M, Ott JA. Chemoautotrophic, sulfur-oxidizing symbiotic bacteria on marine nematodes: Morphological and biochemical characterization. Microbial Ecology. 2000;24:313–329.

Raaijmakers JM, Mazzola M. Diversity and natural functions of antibiotics produced by beneficial and plant pathogenic bacteria. Annual Review of Phytopathology. 2012 Sep 8;50(1):403–424.

Ray P, Lakshmanan V, Labbe JL Craven KD. Microbe to microbiome: A paradigm shift in the application of microorganisms for sustainable agriculture. Frontiers in Microbiology. 2020;11:622926. https://doi.org/10.3389/fmicb.2020.622926

Raymond NS, Gómez-Muñoz B, van der Bom FJT, Nybroe O, Jensen LS, Müller-Stöver DS, et al. Phosphate-solubilising microorganisms for improved crop productivity: A critical assessment. New Phytologist. 2021;229:1268–1277. https://doi.org/10.1111/nph.169249

Rhouma A, Ferchichi A, Hafsa M, Boubaker A. Efficacy of the non-pathogenic Agrobacterium strains K84 and K1026 against crown gall in Tunisia. Phytopathologia Mediterranea. 2004;43:167–176.

Rich MK, Nouri E, Courty PE, Reinhardt D. Diet of arbuscular mycorrhizal fungi: Bread and butter? Trends in Plant Science. 2017 Aug 1;22(8):652–660.

Richardson AE, Simpson RJ. Soil microorganisms mediating phosphorus availability update on microbial phosphorus. Plant Physiology. 2011;156:989–996. https://doi.org/10.1104/pp.111.175448

Rondina ABL, dos Santos Sanzovo AW, Guimarães GS, Wendling JR, Nogueira MA, Hungria M. Changes in root morphological traits in soybean co-inoculated with *Bradyrhizobium* spp. and *Azospirillum brasilense* or treated with *A. brasilense* exudates. Biology and Fertility of Soils. 2020;56:537–549. https://doi.org/10.1007/s00374-020-01453-0

Samsonov P, Padrón RI, Pardo C, Cabrera J, De al Riva GA. Bacillus thuringiensis from biodiversity to biotechnology. Journal of Industrial Microbiology and Biotechnology. 1997;19:202–219.

Samuels GJ. *Trichoderma*: A review of biology and systematics of the genus. Mycological Research. 1996;100:923–935.

Sánchez-Cañizares C, Jorrín B, Poole PS, Tkacz A. Understanding the holobiont: The interdependence of plants and their microbiome. Current Opinion in Microbiology. 2017 Aug 1; 38:188–96.

Santos MS, Nogueira MA, Hungria M. Outstanding impact of *Azospirillum brasilense* strains ab-V5 and ab-V6 on the Brazilian agriculture: Lessons that farmers are receptive to adopt new microbial inoculants. Revista Brasileira de Ciencia do Solo. 2021;45:e0200128. https://doi.org/10.36783/18069657rbcs20200128

Sathya A, Vijayabharathi R, Gopalakrishnan S. Plant growth-promoting actinobacteria: A new strategy for enhancing sustainable production and protection of grain legumes. 3 Biotech. 2017 Jun;7(2):102.

Schisler DA, Slininger PJ, Behle RW, Jackson MA. Formulation of Bacillus spp. for biological control of plant diseases. Phytopathology. 2004;94:1267–1271.

Sehrawat A, Sindhu SS. Potential of biocontrol agents in plant disease control for improving food safety. Defence Life Science Journal. 2019;4:220–225.

Shakya M, Quince C, Campbell JH, Yang ZK, Schadt CW, Podar M. Comparative metagenomic and rRNA microbial diversity characterization using archaeal and bacterial synthetic communities. Environmental Microbiology. 2013;15(6):1882–1899.

Sharifzadeh MS, Damalas CA, Abdollahzadeh G, Ahmadi-Gorgi H. Predicting adoption of biological control among Iranian rice farmers: An application of the extended technology acceptance model (TAM2). Crop Protection. 2017;96:88–96.

Shaviv A, Mikkelsen RL. Controlled- release fertilizers to increase efficiency of nutrient use and minimize environmental degradation-a review. Fertilizer Research. 1993;35:1–12.

Shelake RM, Pramanik D, Kim JY. Exploration of plant-microbe interactions for sustainable agriculture in CRISPR era. Microorganisms. 2019 Aug 17;7(8):269.

Siegel JP. Bacteria. In: Field Manual of Techniques in Invertebrate Pathology, Edited by LL Lacey and HK Kaya (pp. 209–230). Dordrecht, Netherlands: Kluwer Scientific Publishers. 2000.

Somers E, Vanderleyden J, Srinivasan M. Rhizosphere bacterial signalling: A love parade beneath our feet. Critical Reviews in Microbiology. 2004 Jan 1;30(4):205–240.

Song SJ, Sanders JG, Delsuc F, Metcalf J, Amato K, Taylor MW, Mazel F, Lutz HL, Winker K, Graves GR, Humphrey G. Comparative analyses of vertebrate gut microbiomes reveal convergence between birds and bats. MBio. 2020 Feb 25;11(1):10–128.

Stephan D, Silva APMD, Bisutti IL. Optimization of a freeze-drying process for the biocontrol agent Pseudomonas spp. and its influence on viability, storability and efficacy. Biological Control. 2016;94:74–81. https://doi.org/10.1016/j.biocontrol.2015.12.004

Steyaert JM, Ridgway HJ, Elad Y, Stewart A. Genetic basis of mycoparasitism: A mechanism of biological control by species of *Trichoderma*. Journal of Crop and Horticultural Science. 2003;31:281–291.

Stinson M, Ezra D, Hess WM, Sears J, Strobel G. An endophytic Gliocladium sp. of Eucryphiacordifolia producing selective volatile antimicrobial compounds. Plant Science. 2003;165(4):913–922.

Stringlis IA, Zhang H, Pieterse CM, Bolton MD, de Jonge R. Microbial small molecules-weapons of plant subversion. Natural Product Reports. 2018;35:410–433. https://doi.org/10.1039/C7NP00062F

Sui Y, Liu J. Effect of glucose on thermotolerance and biocontrol efficacy of the antagonistic yeast Pichia guilliermondii. Biological Control. 2014;74:59–64. https://doi.org/10.1016/j.biocontrol.2014.04.003

Sui Y, Liu J, Wisniewski M, Droby S, Norelli J, Hershkovitz V. Pretreatment of the yeast antagonist, Candida oleophila, with glycine betaine increases oxidative stress tolerance in the microenvironment of apple wounds. International Journal of Food Microbiology. 2012;157:45–51. https://doi.org/10.1016/j.ijfoodmicro.2012.04.010

Sun X. History and current status of development and use of viral insecticides in China. Viruses. 2015;7:306–319.

Sun ZB, Li SD, Zhong ZM, et al. A perilipin gene from Clonostachys rosea f. catenulata HL-1-1 is related to sclerotial parasitism. International Journal of Molecular Sciences. 2015;16:5347–5362.

Sun ZB, Sun MH, Li SD. Identification of mycoparasitism-related genes in Clonostachys rosea 67–1 active against Sclerotinia sclerotiorum. Scientific Reports. 2015a;5:18169.

Suzuki A, Ueda K, Segawa T, Suzuki M. Fecal microbiota of captive Antillean manatee Trichechus manatus manatus. FEMS Microbiology Letters. 2019 Jun;366(11):fnz134.

Teixido N, Usall J, Vinas I. Efficacy of preharvest and postharvest Candida sake biocontrol treatments to prevent blue mould on apples during cold storage. International Journal of Food Microbiology. 1999;50:203–210.

Tiwari AK. Advances in biological control strategies for sustainable pest management. Uttar Pradesh Journal of Zoology. 2024;45(3):214–232.

Tomar P, Thakur N, Jhamta S, Chowdhury S, Kapoor M, Singh S, Shreaz S, et al. Bacterial biopesticides: Diversity, role in pest management and beneficial impact for agricultural and environmental sustainability. *Heliyon*. 2024.

Toprak U, Bayram S, Gürkan MO. Gross pathology of SpliNPVs and alterations in Spodoptera littoralis Boisd. (Lepidoptera: Noctuidae) morphology due to baculoviral infection. Tarim Bilimleri Dergisi. 2005;11:65–71.

Tsapikounis FA. An integrated evaluation of mycoparasites from organic culture soils as biological control agents of sclerotia of Sclerotinia sclerotiorum in the laboratory. BAOJ Microbiology. 2015;1(1):001.

Ullah A, Nisar M, Ali H, Hazrat A, Hayat K, Keerio AA, et al. Drought tolerance improvement in plants: An endophytic bacterial approach. Applied Microbiology and Biotechnology. 2019;103:7385–7397. https://doi.org/10.1007/s00253-019-10045-4

Umer M, Mubeen M, Iftikhar Y, Shad MA, Usman HM, Sohail MA, Atiq MN, Abbas A, Ateeq M. Role of rhizobacteria on plants growth and biological control of plant diseases: A review. Plant Protein. 2021;5:59–73.

van Der Heijden MG, Martin FM, Selosse MA, Sanders IR. Mycorrhizal ecology and evolution: The past, the present, and the future. New Phytologist. 2015 Mar;205(4): 1406–1423.

Velmourougane K, Saxena G, Prasanna R. Plant-microbe interactions in the rhizosphere: Mechanisms and their ecological benefits. Plant-Microbe Interactions in Agro-Ecological Perspectives: Volume 2: Microbial Interactions and Agro-Ecological Impacts. 2017:193–219.

Vendl C, Ferrari BC, Thomas T, Slavich E, Zhang E, Nelson T, Rogers T. Interannual comparison of core taxa and community composition of the blow microbiota from East Australian humpback whales. FEMS Microbiology Ecology. 2019 Aug;95(8):fiz102.

Vicedo B, Penalver R, Asins MJ, Lopez MM. Biological control of Agrobacterium tumefaciens, colonization, and pAgK84 transfer with Agrobacterium radiobacter K84 and the Tra-mutant strain K1026. Applied and Environmental Microbiology. 1993;59(1):309–315.

Vinale F, Sivasithamparam K, Ghisalberti LE, Marra R, Woo LS, Lorito M. *Trichoderma*-plant–pathogen interactions. Soil Biology and Biochemistry. 2008;40:1–10.

Vinocur B, Altman A. Recent advances in engineering plant tolerance to abiotic stress: Achievements and limitations. Current Opinion in Biotechnology. 2005;16:123–132. https://doi.org/10.1016/j.copbio.2005.02.001

Warrior P, Konduru K, Vasudevan P. Formulation of biological control agents for pest and disease management. In: Biological Control of Crop Diseases (pp. 435–456). CRC Press, 2002.

Webster NS, Negri AP, Botté ES, Laffy PW, Flores F, Noonan S, Schmidt C, Uthicke S. Host-associated coral reef microbes respond to the cumulative pressures of ocean warming and ocean acidification. Scientific Reports. 2016 Jan 13;6(1): 19324.

Webster NS, Thomas T. The sponge hologenome. MBio. 2016 May 4;7(2):10–128.

Weller DM. Biological control of soil borne plant pathogens in the rhizosphere with bacteria. Annual Review of Phytopathology. 1988;26:379–407.

Willey JM, Sherwood LM, Woolverton CJ. Prescott's Microbiology. McGraw-Hill, 2014.

Wilson C, Wisniewski M. Biological control of postharvest diseases of fruits and vegetables: An emerging technology. Annual Review of Phytopathology. 1989;27:425–441. https://doi.org/10.1146/annurev.py.27.090189.002233

Woyke T, Teeling H, Ivanova NN, Huntemann M, Richter M, Gloeckner FO, Boffelli D, Anderson IJ, Barry KW, Shapiro HJ, Szeto E. Symbiosis insights through metagenomic analysis of a microbial consortium. Nature. 2006 Oct 26;443(7114):950–955.

Zhang X, Davidson EA, Mauzerall DL, Searchinger TD, Dumas P, Shen Y. Managing nitrogen for sustainable development. Nature. 2015;528:51–59. https://doi.org/10.1038/nature15743

Zhang X, Zhang G, Li P, Yang Q, Chen K, Zhao L, et al. Mechanisms of glycine betaine enhancing oxidative stress tolerance and biocontrol efficacy of Pichia caribbica against blue mold on apples. Biological Control. 2017;108:55–63. https://doi.org/10.1016/j.biocontrol.2017.02.011

Zheludev IN, Edgar RC, Lopez-Galiano MJ, De la Pena M, Babaian A, Bhatt AS, Fire AZ. Viroid-like colonists of human microbiomes. Cell. 2024 Nov 14;187(23): 6521–6536.

15 CRISPR–Cas-Based Microbiome Editing

Tarek A. A. Moussa[1]*, *Nevien M. Sabry*[2], *Ayat M. Zedan*[3], *and Rehab M. Hafez*[1]

[1]Botany and Microbiology Department, Faculty of Science, Cairo University, Giza 12613, Egypt

[2]Cell Biology Department, National Research Center, Giza, Egypt

[3]College of Medicine, Al-Iraqia University, Baghdad, Iraq

15.1 INTRODUCTION

CRISPR–Cas systems are a groundbreaking genome editing technology that is used to modify the genomes with high precision. CRISPR–Cas systems are considered immunity present in prokaryotes that utilize CRISPR and CRISPR-associated (Cas) endonuclease proteins to battle invading viruses and plasmids, and have been adopted for several applications in molecular biology and genetic engineering (Doudna and Charpentier 2014). Researchers can achieve double-strand breaks at defined genomic areas by designing guide RNAs that lead Cas enzymes, such as Cas9, to target sequences (Jinek et al., 2012).

CRISPR has transformed into agriculture, enabling the enhancement of agricultural characteristics, and medicine, where it offers significant promise for addressing genetic abnormalities (Knott and Doudna 2018). Nonetheless, the technology has engendered significant ethical dilemmas about matters such as germline editing and the production of "designer babies" (Lander et al., 2019). As CRISPR advances and discovers novel applications, continuous research and public dialogue will be essential to harness its transformational potential effectively. Notwithstanding these obstacles, CRISPR continues to be a powerful technique that revolutionizes our understanding of genetics and enhances our ability to manipulate biological systems for the advancement of humanity.

The straightforwardness and adaptability of CRISPR have enabled its extensive application across various domains, including fundamental research, biotechnology, and medicine. It has been employed to establish animal models of human diseases, devise innovative medicines for genetic abnormalities, enhance crop characteristics, and potentially eliminate invasive species or disease vectors (Hsu et al., 2014; Gao 2021).

15.2 DISCOVERY OF CRISPR

Since its discovery in 1987, the CRISPR–Cas system has undergone significant advancements. From its initial discovery as a repeating DNA sequence, it has undergone substantial changes to its current position as a potent genome editing method. If research into its processes and use continues, the CRISPR–Cas system can revolutionize many academic disciplines and enterprises.

15.2.1 Evolution of CRISPR–Cas System

The discovery of the CRISPR–Cas system happened in 1987, when Yoshizumi Ishino and associates discovered a peculiar repetitive DNA sequence in *Escherichia coli*, which they named "CRISPR" (Ishino et al., 1987). Originally, the purpose of CRISPR was unclear, and it was perceived as merely a repeated DNA sequence. Subsequent studies have indicated its existence in many bacteria and archaea, suggesting a more complex function.

In the early 2000s, researchers initiated the exploration of CRISPR's function. In 2005, Francis Mojica and associates elucidated CRISPR's function in bacterial defense against viral infections (Mojica et al., 2005). This advancement stimulated additional investigation into the mechanics of CRISPR. The CRISPR–Cas system provides adaptive immunity by incorporating viral DNA into the host genome, enabling the host to identify and combat subsequent infections.

The CRISPR array comprises spacer sequences originating from viral DNA, whereas the Cas enzymes function to cleave the viral DNA (Barrangou et al., 2007). In 2010, researchers identified that the CRISPR–Cas system was developed to target specific DNA sequences, serving as a potent instrument for genome editing (Marraffini and Sontheimer, 2010).

The CRISPR–Cas9 system was initially delineated (Jinek et al., 2012). The programmability of CRISPR–Cas9 was demonstrated, facilitating precise genome editing. This advancement converted CRISPR into a formidable instrument for biotechnology and medicine. Since that time, CRISPR–Cas9 has been extensively utilized for various applications, including gene editing, gene regulation, and gene therapy.

CRISPR-Cpf1 is an alternative to CRISPR–Cas9, which has been discovered (Zetsche et al., 2015). These systems have expanded the range of genome editing methods available, offering innovative opportunities for fundamental research, biotechnology, and medicine. CRISPR–Cas technology has transformed genetics and possesses the potential to disrupt multiple sectors.

***Corresponding author:** tarekmoussa@cu.edu.eg (Dr. Tarek A. A. Moussa)

DOI: 10.1201/9781003544425-15

15.2.2 Function of CRISPR–Cas System

The CRISPR–Cas system is exceptionally flexible, enabling bacteria to develop resistance to novel viral infections swiftly (Sorek et al., 2013). The system has been demonstrated to be instrumental in preserving bacterial genome stability and inhibiting the uptake of exogenous DNA (Fonfara et al., 2016).

The CRISPR–Cas9 system is a type II variant of the CRISPR–Cas systems that is being extensively used for genome editing (Jinek et al., 2012). The CRISPR–Cas9 system comprises the Cas9 enzyme and a guide RNA molecule that guides the enzyme to cleave particular DNA regions (Cong et al., 2013). The CRISPR–Cas9 system is exceptionally efficient and specialized, facilitating accurate genome editing across several organisms (Mali et al., 2013).

Besides its function in bacterial defense, the CRISPR–Cas system is also pivotal in controlling bacterial gene expression (Bikard et al., 2013). The system has been demonstrated to modulate the gene expression involved in several physiological functions, including metabolism, DNA repair, and stress response (Qi et al., 2013). The CRISPR–Cas system is a sophisticated and versatile bacterial defense mechanism that provides adaptive immunity against viral infections. The mechanism is pivotal in controlling bacterial gene expression and preserving bacterial genome stability. The CRISPR–Cas9 technology has been extensively employed for genome editing for various purposes and is remarkably efficient and precise.

15.2.3 Structure of CRISPR Locus and Classification of Cas Proteins

The principal function of the adaptive immunity system CRISPR–Cas is to safeguard archaea and bacteria from viruses and other mobile genetic elements (MGE), which constitute a distinctive genome segment of the bacteria (Barrangou et al., 2007; Hille and Charpentier, 2016; Mohanraju et al., 2016; Barrangou and Horvath, 2017; Hille et al., 2018). A typical CRISPR–Cas locus comprises a CRISPR array and adjacent cas genes that constitute one or more operons. It contains a succession of repeated DNA sequences, termed CRISPR repeats, interspersed by brief viral DNA segments referred to as spacers (Mojica et al., 2005). Spacers separate several direct repeats, typically ranging from 25 to 36 base pairs, which are highly conserved within a specific CRISPR locus that constitutes CRISPR arrays (Jansen et al., 2002); specific repetitions are identical to segments of viral or plasmid genomes (Mojica et al., 2005). Conversely, the spacers generally range from 20 to 60 base pairs in length and originate from viral DNA (Pourcel et al., 2005). The CRISPR repeats and spacers are organized distinctly, with the repeats interspersed by the spacers (Bolotin et al., 2005).

Numerous CRISPR–Cas systems encompass other accessory genes that modify essential immune responses (Shmakov et al., 2018; Makarova et al., 2020). The architectural genomic loci structure and the composition of Cas genes within CRISPR–Cas systems display significant variability. The CRISPR locus is translated into an RNA molecule precursor, pre-crRNA, which is then processed into mature crRNA molecules (Deltcheva et al., 2011). The crRNA molecules direct the Cas proteins to the target viral DNA (Jinek et al., 2012).

15.2.4 Classification of the CRISPR–Cas System

The Cas (CRISPR-associated) proteins constitute a protein family integral to the CRISPR–Cas system (Haft et al., 2005). The Cas proteins are classified into various categories: type I, II, and III (Makarova et al., 2011), along with the less prevalent but distinct type IV (Koonin and Makarova, 2013; Makarova et al., 2013). With rare exceptions, all three CRISPR–Cas variants fully fulfill the components required for the key operations of the defensive system. Recent comprehensive sequence and structural analyses of effector complexes from diverse CRISPR–Cas system variations have revealed fundamental principles governing the organization and function of these complexes (Staals et al., 2013; Zhang et al., 2013).

Type I Cas proteins are the most extensively researched and are responsible for cleaving target viral DNA in a process termed interference (Hille et al., 2018). Type II Cas proteins are tasked with cleaving target viral DNA by a mechanism referred to as primed adaptation (Makarova et al., 2013; Makarova et al., 2018). Type III Cas proteins modulate gene expression and are not directly involved in the cleavage of viral DNA (Makarova et al., 2018).

The Cas9 protein belongs to the Type II Cas protein family and is among the most extensively studied Cas proteins. Prokaryotes employ CRISPR and its associated protein (Cas9) as an adaptive immune mechanism to defend against bacteriophages and viruses (Hille and Charpentier, 2016; Hille et al., 2018) while investigating an alkaline phosphatase gene in *E. coli*.

15.2.5 Structure and Function of Effector Complexes

Effector complexes are essential for the functionality of the CRISPR–Cas system, facilitating the identification and cleavage of target DNA (Marraffini and Sontheimer 2010). The effector complex consists of a Cas protein and a guide RNA molecule that collectively identify and attach to the target DNA sequence (Jinek et al., 2012). The most extensively researched effector complex is the Cas9-gRNA complex, known for its excellent specificity and efficiency in targeting and DNA cleavage regions (Cong et al., 2013).

The Cas9-gRNA complex structure has been analyzed using X-ray crystallography and cryo-electron microscopy (Nishimasu et al., 2014). The complex comprises a Cas9 protein and a gRNA molecule, which are interconnected by several interactions (Fonfara et al., 2016). The gRNA molecule identifies the sequences targeted in the DNA via base pairing, wherein the nucleotides of the gRNA establish hydrogen bonds with the nucleotides of the target DNA sequence (Jiang et al., 2016).

Upon recognition and binding of the Cas9-gRNA to the target DNA sequence, it cleaves the DNA through hydrolysis (Zetsche et al., 2015). The Cas9 protein comprises two distinct nuclease domains that facilitate the cleavage of the target DNA sequence (Yamano et al., 2017). The initial nuclease domain, known as the RuvC-like domain, cleaves the target DNA sequence on the strand corresponding to the gRNA molecule (Joung and Kleinstiver, 2017). The second nuclease domain, known as the HNH-like domain, cleaves the target DNA sequence on the strand that is similar to the gRNA molecule (Kim et al., 2016).

Alongside the Cas9-gRNA complex, various additional effector complexes have been identified inside the CRISPR–Cas system (Shmakov et al., 2017). These comprise the Cas12a-gRNA complex, the Cas13-gRNA complex, and the Cpf1-gRNA complex, among others (Bharathkumar et al., 2022). Each effector complex possesses a distinct structure and function, all contributing to the targeting and cleavage of invading viral DNA (Kato et al., 2022). Additional investigation into the effector complexes of the CRISPR–Cas system is expected to provide an innovative understanding of bacterial defense mechanisms and the advancement of new biotechnological techniques.

15.2.5.1 Effector Complexes of Class 1

Class 1 CRISPR–Cas systems are characterized by a multi-subunit effector complex that identifies and cleaves the target DNA sequence (Brouns et al., 2008). The Cascade complex, the most extensively researched Class 1 effector complex, consists of five distinct Cas proteins: CasA, CasB, CasC, CasD, and CasE (Jore et al., 2011).

The Cascade complex can identify the target DNA sequence via base pairing, wherein the nucleotides of the crRNA molecule establish hydrogen bonds with the nucleotides of the target DNA sequence (Wiedenheft et al., 2011). The Cascade complex necessitates the existence of a protospacer adjacent motif (PAM) sequence, a brief nucleotide sequence located adjacent to the target DNA sequence (Deltcheva et al., 2011). The Cascade complex identifies the PAM sequence and is essential for cleaving the target DNA sequence.

Upon recognition and binding to the target DNA sequence, the Cascade complex recruits the Cas3 protein, which is responsible for cleaving the target DNA sequence (Mulepati and Bailey 2011). The Cas3 protein functions as a helicase-nuclease, unwinding the target DNA sequence and fragmenting it into smaller pieces (Sinkunas et al., 2013). The cleavage of the target DNA sequence is a crucial step in the CRISPR–Cas defense system, as it inhibits the replication of invading viral DNA, thereby protecting the host cell from potential damage.

Multiple Class 1 effector complexes, such as the CMR complex, CSM complex, and Cascade complex (Zhang et al., 2012), have been found. These complexes resemble the Cascade complex in their ability to detect and cleave the target DNA sequence via base pairing (Staals et al., 2014). Nonetheless, they diverge from the Cascade complex in terms of their composition and mode of action.

15.2.5.2 Effect Complexes of Class 2

Conversely, Class 2 systems, shown by Cas9, function with a more straightforward architecture. Cas9, directed by a single-guide RNA (sgRNA), attaches to target DNA and induces double-strand breaks with exceptional accuracy, making it a preferred option for gene-editing applications (Xue and Greene 2021). Class 2 effector complexes play a crucial role in numerous biological processes, particularly in cellular response mechanisms and gene regulation. These complexes, primarily of proteins and nucleic acids, are essential in facilitating the effects of signaling pathways vital for cell differentiation, development, and immunological responses (Aljabali et al., 2024). A significant characteristic of Class 2 effector complexes is their modularity, facilitating dynamic interactions with diverse cellular components (Hille et al., 2018). Research indicates that the transcription factors in these complexes can markedly affect gene expression patterns, reacting to external stimuli and internal signals (Nidhi et al., 2021).

A notable instance of Class 2 effector complexes is the assembly of nuclear factor kappa B (NF-κB) with its inhibitors. NF-κB is crucial in modulating immunological responses and inflammation (Guo et al., 2024). Upon activation by diverse cytokines, NF-κB translocates to the nucleus, where it binds to DNA sequences to initiate the transcription of target genes implicated in immunological and inflammatory responses. Research indicates that altering the regulatory mechanisms of NF-κB can lead to uncontrolled inflammatory responses, highlighting the significance of comprehending Class 2 effector complexes in disease pathophysiology (Schreiter et al., 2014; Guo et al., 2024).

Furthermore, Class 2 effector complexes are essential for chromatin remodeling and modification. These complexes often contain histone-modifying enzymes that are crucial for regulating gene expression at the epigenetic level. The engagement of Class 2 effectors with chromatin may lead to transcriptional activation or repression. This signifies a meticulously regulated interaction that guarantees appropriate gene expression is essential for cellular function and organismal development (De Majo and Calore 2018). Understanding this regulatory mechanism clarifies the role of epigenetic modifications in various diseases.

Grasping Class 2 effector complexes is crucial for understanding the cellular mechanisms that regulate both normal physiological states and pathological circumstances. Ongoing research into these complexes reveals that their manipulation presents intriguing therapeutic possibilities, particularly in disorders characterized by dysregulated signaling pathways or gene expression. As our understanding of these complexes expands, novel techniques for modulating their actions for disease intervention will arise (Braicu et al., 2019; Bhol et al., 2024).

15.3 MECHANISM OF CRISPR

Three phases can be identified in the operational mechanism of CRISPR–Cas systems: interference, maturation, expression, and adaptation. Cas proteins recognize the target DNA during the adaptation phase and acquire a new spacer sequence. The immunological memory is established by incorporating this spacer region into the CRISPR array. The CRISPR array is transcribed into a precursor RNA transcript during the expression and maturation phase, which is subsequently processed into smaller RNA fragments known as crRNAs, each containing a single spacer adjacent to a segment of the repeat sequence. The active Cas–crRNA complex is formed when these crRNAs engage with one or more Cas proteins. The Cas–crRNA complex identifies foreign nucleic acid targets within the cell during the interference phase. These targets are recognized through base pairing with complementary crRNA sequences. The target nucleic acid is cleaved and destroyed following successful recognition (Rath et al., 2015; Loureiro and da Silva, 2019; Asmamaw and Zawdie, 2021).

15.4 ROLE OF CRISPR–CAS SYSTEMS IN GENOME EDITING OF MICROBIOMES

CRISPR–Cas systems have revolutionized the field of genome editing, offering unprecedented precision and efficiency in modifying microbial genomes (Jinek et al., 2012). One of the primary applications of CRISPR–Cas in microbiome editing is the targeted modification of specific microbial genes, enabling researchers to dissect the complex interactions within microbiomes and identify key microbial players (Hsu et al., 2014). For instance, CRISPR–Cas has been used to edit the genomes of gut microbiome bacteria, such as Bacteroides thetaiotaomicron, to study their roles in nutrient metabolism and immune system regulation (Rahmati et al., 2025).

CRISPR–Cas systems have also been employed to engineer microbiomes for improved human health, such as the development of novel probiotics and microbiome-based therapies (Xiang et al., 2025). For example, CRISPR–Cas-mediated editing of the gut microbiome has been shown to alleviate symptoms of diseases such as inflammatory bowel disease and obesity (L. Wang et al., 2024). Additionally, CRISPR–Cas has been used to engineer microbiomes for environmental sustainability, such as the development of novel bioremediation strategies for pollutant degradation (Cardi et al., 2023).

The CRISPR–Cas system's ability to target specific genes and edit microbiome genomes has also enabled the development of novel diagnostic tools and biomarkers for microbiome-related diseases (Amen et al., 2025). For instance, CRISPR–Cas-based diagnostics have been developed to detect specific microbial pathogens, such as *Mycobacterium tuberculosis*, in clinical samples (Huang et al., 2023). Furthermore, CRISPR–Cas has been used to engineer microbiomes for biotechnological applications, including the production of novel enzymes, biofuels, and bioproducts (Ansori et al., 2023).

CRISPR–Cas systems are also used to study the microbiome's role in various diseases, including diabetes, neurological disorders, and cancer (Amen et al., 2025). For example, CRISPR–Cas-mediated editing of the gut microbiome has been shown to influence the development of colorectal cancer (Hu et al., 2023). Additionally, CRISPR–Cas has been used to modify microbiomes to produce novel compounds, such as antibiotics and anticancer agents (Kadkhoda et al., 2024).

The CRISPR–Cas systems' development has also enabled the creation of novel gene editing tools, such as base editors and prime editors, which have improved the efficiency and precision of microbiome editing (Komor et al., 2016; Anzalone et al., 2019). These advances have expanded the CRISPR–Cas's applications in microbiome research, enabling researchers to study the complex interactions within microbiomes and develop novel microbiome-based therapies.

CRISPR–Cas systems have transformed the field of microbiome editing, offering unprecedented precision and efficiency in modifying microbiome genomes. The CRISPR–Cas's applications in microbiome editing are vast, ranging from basic research to translational applications, and are expected to continue to grow in the coming years (Sadhu et al., 2025).

15.5 APPLICATIONS OF CRISPR–CAS SYSTEMS

15.5.1 In Plant Biotechnology

The CRISPR–Cas system functions by a straightforward yet effective mechanism. The crRNA guide identifies specific DNA sequences, whereas the Cas enzyme cleaves the target DNA, facilitating accurate editing (Wang et al., 2013). This technology has significant implications for the treatment of genetic disorders, the improvement of agricultural productivity, and the progression of synthetic biology (Long et al., 2014; Dever et al., 2016). The adaptability of CRISPR has prompted substantial research, resulting in various versions, including CRISPR–Cas9 and CRISPR-Cpf1 (Cai et al., 2015; Zetsche et al., 2015).

Researchers have been utilizing plant biotechnology transformed by CRISPR–Cas systems to explore promising routes for enhancing crop productivity and global food security (Figure 15.1). This technique created crops with improved climate resiliency, stress tolerance, nutritional profiling, and better traits (Ndudzo et al., 2024).

15.5.1.1 Prospects in Plant Gene Editing

15.5.1.1.1 Plant Nutritional Values

Some crops have been biofortified with CRISPR/Cas9 to alter their nutritional value, either by enhancing or reducing it. CRISPR can knock out specific genes related to allergenic proteins in particular food crops, making them safer for individuals with food allergies to consume. This

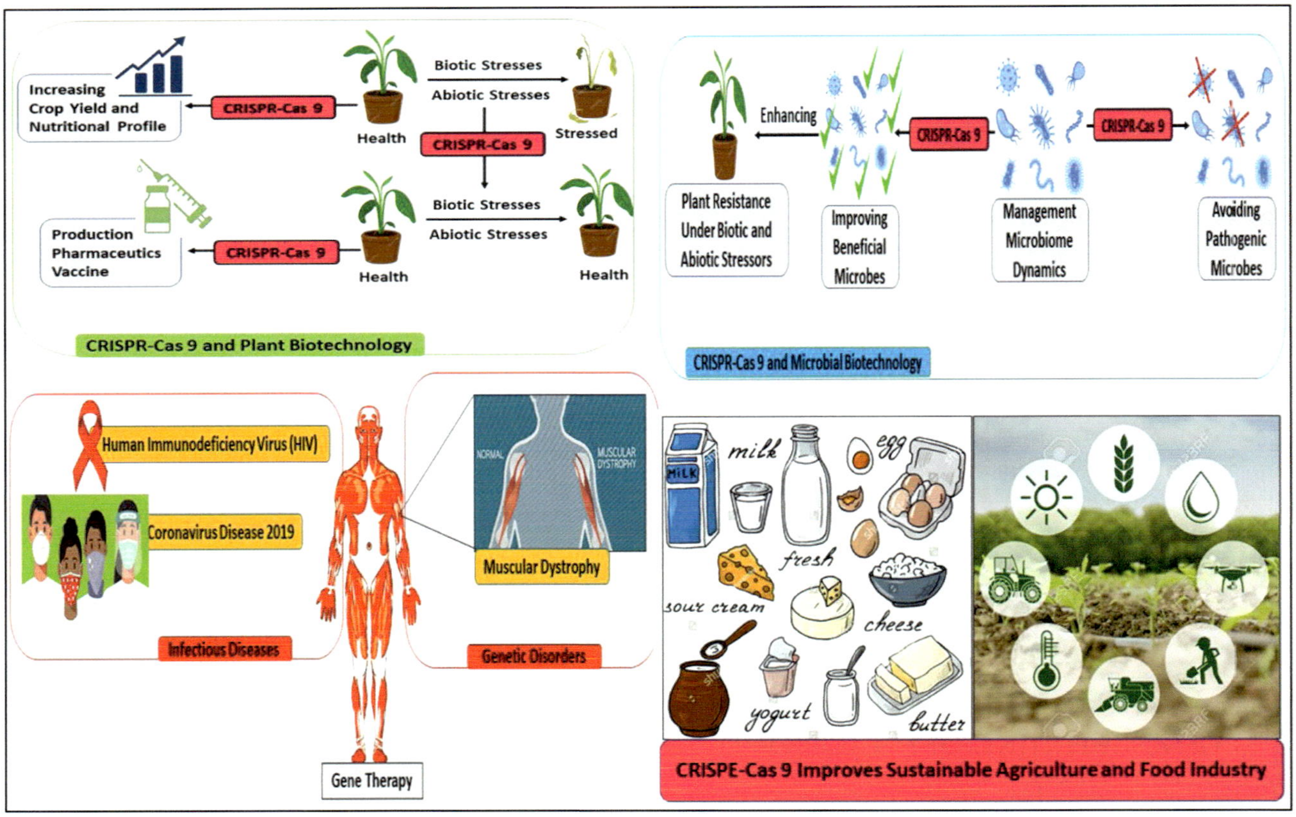

FIGURE 15.1 Applications for the CRISPR–Cas9 system in plant biotechnology, microbial biotechnology, medical biotechnology (gene therapy), and food biotechnology.

technique was used to lower gluten in wheat by targeting specific sequences in gluten-encoding genes (*α-gliadin* genes), which reduces allergenic epitopes in gluten proteins for celiac patients (Sánchez-León et al., 2018). Similarly, Herman et al. (2003) knocked out Gly m proteins (Gly m Bd 30K, Gly m 4) in soybean-producing hypoallergenic soybeans suitable for broader consumption. Also, CRISPR has been applied by Camerlengo et al. (2020). To edit ATI subunits WTAI-CM3 and WTAI-CM16 in the Italian durum wheat cultivar Svevo, aiming to reduce allergens. Moreover, genes encoding vicilin (7S globulin) and legumin (11S globulin) proteins have been edited to minimize allergenicity in cowpeas and lentils (Bhowmik et al., 2021). Researchers have used CRISPR to knock out Ara h allergenic proteins in peanut varieties, leading to a reduction in its allergenicity without significantly affecting its taste or nutritional content (Rajyaguru et al., 2023; Singh et al., 2024; Wang et al., 2024). Also, Nakayasu et al. (2018) have created potato cultivars with the complete elimination of steroidal glycoalkaloid (SGA) accumulation in their hairy roots by knocking out the St16DOX genes, which encode steroid 16α-hydroxylase in SGA biosynthesis, thereby guaranteeing safer consumption and increasing marketability.

The CRISPR/Cas9 technique has been utilized to enhance the nutritional profile of specific crops. It was employed to mitigate nutritional deficits in rice populations. Trijatmiko et al. modified both the *OsNAS2* and *OsYSL2* genes to enhance the transport and accumulation of iron and zinc in their grains (Trijatmiko et al., 2016). The OsNAS2 genes encode nicotine amine synthase, and their modification increases the bioavailability of iron and zinc in rice grains (Johnson et al., 2011; Lee et al., 2023).

CRISPR/Cas9 has been utilized to enhance provitamin A levels by modifying specific genes involved in the carotenoid biosynthesis pathway, such as the *IbOr* gene (Orange gene) in sweet potatoes (*Ipomoea batatas*) (Park et al., 2015), and in maize, where it stabilizes the phytoene synthase (*PSY*), lycopene epsilon cyclase (*LcyE*), and beta-carotene hydroxylase 1 (*crtRB1*) genes (TRAN et al., 2017; Thanh, 2019), as well as the Zmpsy1 (Phytoene Synthase 1) gene in maize kernels (Niaz et al., 2023; Lorenzo et al., 2024). Furthermore, the amino acid composition of rice (*Oryza sativa*) was enhanced through the modification of the *OsAAP6* (Amino Acid Permease 6) gene, essential for the plant's amino acid transport and distribution (Mondal et al., 2020; Huang et al., 2021), and resulted in beta-carotene accumulation in cassava (*Manihot esculenta*) via the editing of lycopene epsilon-cyclase genes involved in carotenoid biosynthesis (Kaur et al., 2020).

CRISPR–Cas9 was utilized to knock out genes associated with the negative regulation of tomato carotenoid biosynthesis pathways, thereby enhancing lycopene content by targeting the *CrtR-b2* gene, which inhibits lycopene production (Huang et al., 2021).

15.5.1.1.2 Phenotypic Traits

CRISPR–Cas9 has been utilized to stimulate the phenotypic traits of some crops. Lin et al. (2007) identified the *qSH1* gene that encodes a *BEL1-type homeobox transcription factor* as a major QTL for grain detachment in wild rice, which regulates the development of the abscission layer at the base of the rice spikelet. Also, the wild rice (*Oryza rufipogon*) productivity was increased by editing a group of specific key genes, such as DEP1, Gn1a, and GS3, that collectively improve panicle structure and grain characteristics (number, density, and size). Editing the *DEP1* gene modifies panicle structure by increasing the density of the clusters, resulting in more efficient flower clusters with improved light capture and photosynthetic efficiency. Knocking out the *Gn1a* gene increases cytokinin levels and boosts panicle grain production. The *GS3* gene controls grain size, so its editing leads to larger grains by altering their development (Li et al., 2016; Wang et al., 2018). In addition, Li et al. (2018) knocked out the *IPA1* gene and regulating genes (*SH4* and *qSH1*) for seed shattering in *Oryza rufipogon*, which involved the formation of healthy plant architecture with better grain yield by preventing premature grain loss. In addition, a new domestication strategy was developed for *Oryza alta*, an allotetraploid rice with high biomass (Yu et al., 2021). They edited genes related to seed shattering (*qSH1* gene) and awn length (*An-1* gene orthologues), resulting in lower seed shattering rates and shorter awn lengths. In addition, CRISPR–Cas9 was explored in wild rice to edit multiple domestication traits, including *SH4* and *qSH1* genes, to reduce grain shattering and other yield-related genes (Zhang et al., 2022). They emphasized the potential of multi-gene targeting to improve plant architecture and grain productivity, making the edited wild rice suitable for cultivation. The quantitative trait locus *KRN2* gene studied in maize and its rice ortholog, *OsKRN2*, which regulates the grain number of those crops (Chen et al., 2022). They knocked out both genes that encode WD40 proteins. *KRN2* in maize increased grain yield by 10%, while knocking out the OsKRN2 gene in rice resulted in an 8% increase.

Similarly, CRISPR–Cas9 has been responsible for developing varieties of tomato plants with compact canopies and more synchronized fruit ripening, resulting in better fruit size and enhanced yield. These varieties are ideal for high-density planting and mechanical harvesting, which is necessary for pruning in commercial cultivation. Researchers edited a group of genes, including SP (self-pruning), SlCLV3, and SP5G, to control the growth habits and flowering cycle of wild tomatoes. Knocking out the *SP* gene leads to determinate growth. It produces shorter plantlets by controlling the transition from vegetative to reproductive growth, then editing *SP5G*, a paralog of *SP*, which further accelerates flowering and synchronizes fruit ripening. Additionally, editing the *SlCLV3* gene regulates tomato fruit size and controls genes in the fw2.2 and fw3.2 loci that enhance fruit weight, size, and production (Zsögön et al., 2018).

Similarly, it was worked with the CRISPR technique to edit genes related to *Physalis pruinosa* (Groundcherry), plant growth habits (*SP* gene), and fruit size (*CLV1* gene). They produced more compact ground cherry plants with bigger fruits, which gave them a more appropriate architecture for farming (Lemmon et al., 2018).

Also, the CRISPR-engineered wild maize relatives were produced by editing combined genes, *CLE* and *FON* genes, to grow maize with more prominent ears and more kernels per ear, as well as the *IPA1* gene to improve the angle of leaves and plant height, by which light capture can be optimized, and planting density can be increased (Jiang et al., 2022). In addition, introduced CRISPR-edited *BABY BOOM1* (*BBM1*) genes, key regulators of embryogenesis, in maize, allowing the formation of clonal seeds to ensure uniformity and stable hybrid formations (Khanday et al., 2019).

Moreover, CRISPR/Cas9 can also facilitate hybrid breeding by inducing seedless fruit development in *Solanum peruvianum* by targeting the *SELF-INCOMPATIBILITY1* gene (S-RNase) (Lin et al., 2022).

15.5.1.1.3 Metabolic Characters

CRISPR methods have been utilized to regulate specific metabolic pathways, thereby enhancing certain plant capabilities. It improved biological nitrogen fixation and reduced reliance on chemical fertilizers that cause ecological divergence. NIN-like protein 7 (*NLP7*) was knocked out to boost the nitrogen-fixing efficiency, which is critical for nodule formation in soybeans (Jiang et al., 2021).

The CRISPR–Cas9 genome editing technique facilitates accurate and economical sequence modification of specific genomic loci, rendering phytoremediation an environmentally friendly, conservation-focused, and sustainable approach for the bioremediation of toxic heavy metals (THMs) without modifying soil physicochemical properties. This novel methodology facilitates the creation of new phytoremediators through the modification of *THM*-responsive genes that encode the zinc-iron permease (ZIP) protein family, natural resistance-associated macrophage proteins (NRAMP), multidrug and toxin efflux proteins (MATE), P-type P1B-ATPases, cation diffusion facilitator (CDF) proteins, HM-responsive miRNAs, among others (Pandita, 2021). Gomez et al. (2019) employed CRISPR–Cas9 gene editing to target the cytochrome P450 genes (*CYP79D1* and *CYP79D2*) in Manihot esculenta, which facilitates the manufacture of hazardous cyanogenic glucosides that yield lethal cyanide upon hydrolysis.

15.5.1.2 Resistivity to Stress

15.5.1.2.1 Climate-Resiliency and Stress Tolerance

Biotic challenges, resulting from infections by bacteria, fungi, and viruses, combined with abiotic stresses such as drought, salinity, heat, and cold, pose substantial threats to global agricultural output, leading to yield reductions (Figure 15.1).

CRISPR techniques have been utilized to inhibit or eradicate susceptibility genes associated with various diseases and to develop crops that exhibit climate resilience and stress tolerance. Researchers have employed multiplex CRISPR systems to concurrently edit multiple genes, either knocking out or altering specific susceptibility genes in crops, thereby improving their resistance to diseases without detriment to other plant characteristics.

The *MLO* (Mildew Locus O) gene family is linked to susceptibility to powdery mildew, and its disruption may confer resistance. Nekrasov et al. (2017) utilized CRISPR–Cas9 to knock out the *SlMLO1* gene (*Solanum lycopersicum MLO1*, Mildew Locus O1) in tomatoes, resulting in plants devoid of functional *SlMLO1*. The *TaMLO* (*Triticum aestivum*) *MLO* (Mildew Locus O) gene was modified in wheat to confer resistance to powdery mildew (Wang et al., 2014). The multiplex CRISPR–Cas9 technology was utilized to modify two critical susceptibility genes, *Xa13* (*Os8N3*) and *OsSWEET14*, in rice, enhancing resistance to bacterial blight induced by *Xanthomonas oryzae* pv. *oryzae*. The *OsSWEET14* gene is associated with the sugar transporter that the pathogen activates to promote sugar efflux from plant cells for nutrition acquisition (Bastet et al., 2018; Li et al., 2018). The Xa13 gene is essential for bacterial proliferation and the manifestation of illness symptoms. A receptor-like cytosolic kinase ortholog (*SlRIPK*) was found in tomatoes, and knockdown mutants were produced using CRISPR/Cas9 technology (Wang et al., 2022). Transgenic *Solanum lycopersicum* plants that overexpress *SlRIPK* exhibit enhanced resistance to harmful bacteria and fungi. The *ZmFER1* gene was edited to confer resistance to *Fusarium verticillioides* in maize (Liu et al., 2022).

CRISPR–Cas systems were utilized to identify and degrade viral genomes, providing robust protection against viral infections. CRISPR/Cas9 has been employed to knockout the eIF4E (Eukaryotic Translation Initiation Factor 4E) gene in potatoes to enhance resistance against potato virus Y (Noureen et al., 2022), in particular *Brassica* species to confer resistance to the Potyvirus TuMV (Illing 2019), in cherry tomatoes to develop resistance to the pepper veinal mottle virus (PVMV) (Kuroiwa et al., 2022) and in cucumber plants to provide resistance against the Cucumber Vein Yellowing Virus (CVYV) and other potyviruses without compromising plant growth (Wang et al., 2014).

Global agriculture confronts issues stemming from abiotic stresses, notably climate change. Gene editing of *OsNRT1*.1A in rice enhances nitrogen use efficiency by facilitating nitrate absorption and the translocation of nitrate ions from the soil to plant roots, enabling growth in drought-prone and nitrogen-deficient soils (Li et al., 2018). CRISPR–Cas technology facilitates accurate modification of stress-responsive genes in crop varieties that excel under adverse conditions (Khan et al., 2019).

The modification of the OsSPL14 gene in indica rice "MH86" enhanced its root architecture, hence augmenting its drought tolerance and resulting in more resilient plants under water scarcity conditions (Lian et al., 2020). The consequences of overexpressing a constitutively active version of DREB2A in *Arabidopsis* demonstrate increased tolerance to drought and heat stress, but notable growth inhibition, dwarfism, and diminished reproductive potential, highlighting the necessity for stringent regulation (Kim et al., 2012). The dehydration-responsive element-binding protein 2 (DREB2) transcription factors in tomato (SlDREB2) were modified to modulate abiotic stress responses in both tomato and *Arabidopsis* (Hichri et al., 2016). Moreover, CRISPR is employed to alter stress-responsive genes within the CBF/DREB1 pathways, thereby enhancing adaptation responses to cold and drought in barley and wheat crops (Caccialupi et al., 2023).

The *SlCBF1* gene was deleted to examine its function in the chilling tolerance of tomato plants. They created *slcbf1* mutants that exhibited pronounced chilling injury symptoms, reduced proline and protein levels, and elevated hydrogen peroxide concentrations (Li et al., 2018). CRISPR-mediated modification of the *SlMAPK3* gene was conducted in *Solanum lycopersicum* to encode a mitogen-activated protein kinase (*MAPK*) linked to stress signaling for heat tolerance (Yu et al., 2019). The *SlHyPRP1* gene was modified in *Solanum lycopersicum*, producing multi-stress-tolerant alleles that exhibit improved heat resistance but heightened vulnerability to *Fusarium oxysporum* f. sp. *lycopersici* (Tran et al., 2023).

15.5.2 In Microbes Biotechnology

The presence of CRISPR–Cas systems in diverse microbiomes underscores their crucial role in microbial ecology and evolution. These systems regulate microbial populations by providing immunity against bacteriophages and plasmids. Furthermore, they influence gene flow and horizontal gene transfer among microbial communities, demonstrating the importance of CRISPR in shaping microbiome dynamics (Pinilla-Redondo et al., 2022) (Figure 15.1).

Researchers used CRISPR–Cas systems to improve crop stress tolerance by modifying microbiomes to boost their beneficial functions for plants to withstand environmental stresses like extreme temperatures, drought, or salinity (Hsu et al., 2014), and confer resistance to diseases, reduce chemical pesticide reliance, and stimulate sustainable agriculture (Borrelli et al., 2018; Shelake et al., 2019; Han et al., 2025).

Targeting pathogenic microbes directly or modifying beneficial microbes to produce antimicrobial compounds can protect crops from various diseases. Microbes were designed to target antibiotic resistance and virulence factors, such as *bla NDM-1* and *gyrAD87G* in *E. coli* and *Staphylococcus aureus*, using phage-based vectors like M13 and phiNM1 phagemids, aiming to modulate microbiomes and enhance microbial properties, such as sensitizing antibiotics, degrading harmful compounds, or improving plant health (Bikard and Barrangou 2017).

CRISPR–Cas9 optimizes metabolic pathways in *E. coli* and *Saccharomyces cerevisiae* for biofuel and bioplastic production: *E. coli* has several genes that can be modified to redirect carbon flux toward ethanol production. *PykF* and *Ppc* genes are used for glycolysis and anaplerotic reactions. *AceE* and *AceF* genes control acetyl-CoA production as a precursor for bioplastic synthesis, while adhE overexpression enhances ethanol production. Deletion of ldhA prevents pyruvate from being diverted into lactate. *Saccharomyces cerevisiae* genes *PDC1* and *ADH1* are essential for ethanol fermentation, while the *ACS2* gene enhances acetyl-CoA synthesis for bioplastic precursor production. *LAT1* and *DLD1* genes in the mitochondrial lactate pathway reduce byproduct formation, increasing metabolic output efficiency (Shelake et al., 2019).

15.5.3 In Therapeutics for Management of Infectious Diseases

CRISPR-Cas systems are explored as innovative tools that can improve the ability to fight off viral and microbial infectious diseases (Serajian et al., 2021). It can help in diagnostics, treatment, and antimicrobial therapies targeting resistant strains (Jolany Vangah et al., 2020).

15.5.4 In the Food Industry

15.5.4.1 Prospects in the Food Industry

Applications for CRISPR–Cas systems in the food industry are also being explored to enhance their sustainability, safety, and nutritional value (Figure 15.1).

CRISPR in the food industry targets consumer demand for natural and minimally processed foods by producing safer, higher-quality products and more sustainable food options. CRISPR–Cas9 systems were utilized to modify phage genomes by adding or removing sequences that enhance their effectiveness and reduce pathogen loads in food. CRISPR-engineered phages targeting virulence genes in *Salmonella* to reduce the foodborne illness worldwide, and those in *Listeria* strains to eliminate bacteria present in food-processing environments without affecting non-pathogenic bacteria (Khambhati et al., 2023).

CRISPR-edited genomes are used to remove harmful genes in probiotic strains to obtain safe strains to be applied in fermented foods, dairy products, and dietary supplements (Bikard et al., 2014). By enabling precise modifications to the genomes of yeast, bacteria, and other fermentative organisms in product manufacturing like wine, beer, and cheese, CRISPR–Cas9 enhances metabolic pathways to improve the fermentation efficiency, flavor, and sensory qualities of these products (Vilela 2021).

Additionally, CRISPR is being utilized to modify the DNA of microbes that produce food, enhancing product quality, extending shelf life, and reducing food waste. It has been used to improve the resistance of *Lactobacillus* strains against bacteriophage attacks in dairy production, resulting in more desirable flavor profiles, longer-lasting products, and efficient fermentation in dairy manufacture, such as cheese and yogurt (Xie et al., 2024). In addition, CRISPR has been utilized to create mushrooms that do not brown by blocking the *polyphenol oxidase* (*PPO*) gene, thereby reducing food waste and increasing shelf life for consumers and sailors (Waltz 2016). *Saccharomyces cerevisiae* strains have been genetically modified to enhance ethanol formation in beer and wine manufactories by targeting glycolysis and flavor compound biosynthesis pathways (Liu et al., 2019).

15.6 FUTURE PERSPECTIVES

As CRISPR technology continues to advance, researchers are investigating new frontiers, including gene regulation, synthetic biology, and genome engineering (Zhu 2022; Ansori et al., 2023; Aljabali et al., 2024). Developing more precise and efficient CRISPR variants, such as prime editing, will further expand the applications of this technology. With its vast potential, CRISPR technology is poised to transform various fields, improving human health, agriculture, and biotechnology (Liu et al., 2021).

15.7 CONCLUSION

The advances in CRISPR–Cas technology have enabled researchers to dissect complex interactions within microbiomes, identify key microbial players, and engineer microbiomes for improved human health, environmental sustainability, and biotechnological applications. The CRISPR–Cas system's ability to target specific genes and edit microbiome genomes has opened new avenues for understanding microbiome function, identifying novel therapeutic targets, and developing microbiome-based therapies. Despite the challenges and limitations associated with CRISPR–Cas, ongoing research is focused on improving the specificity, efficiency, and delivery of CRISPR–Cas systems, paving the way for the widespread adoption of this technology in microbiome research and applications. As CRISPR–Cas continues to evolve, it is poised to transform our understanding of microbiomes and unlock the full potential of microbiome editing, enabling the development of novel strategies for improving human health, environmental sustainability, and biotechnological innovation.

REFERENCES

Aljabali AAA, El-Tanani M, Tambuwala MM. 2024. Principles of CRISPR-Cas9 technology: Advancements in genome editing and emerging trends in drug delivery. J Drug Deliv Sci Technol [Internet]. 92:105338. https://doi.org/https://doi.org/10.1016/j.jddst.2024.105338

Amen RA, Hassan YM, Essmat RA, Ahmed RH, Azab MM, Shehata NR, Elgazzar MM, El-Sayed WM. 2025. Harnessing the microbiome: CRISPR-based gene editing and antimicrobial peptides in combating antibiotic resistance and cancer. Probiotics Antimicrob Proteins. 17(4):1938–1968. https://doi.org/10.1007/s12602-025-10573-8

Ansori AN, Antonius Y, Susilo RJ, Hayaza S, Kharisma VD, Parikesit AA, Zainul R, Jakhmola V, Saklani T, Rebezov M, et al. 2023. Application of CRISPR-Cas9 genome editing technology in various fields: A review. Narra J. 3(2):e184. https://doi.org/10.52225/narra.v3i2.184

Anzalone A V, Randolph PB, Davis JR, Sousa AA, Koblan LW, Levy JM, Chen PJ, Wilson C, Newby GA, Raguram A, Liu DR. 2019. Search-and-replace genome editing without double-strand breaks or donor DNA. Nature. 576(7785):149–157. https://doi.org/10.1038/s41586-019-1711-4

Asmamaw M, Zawdie B. 2021. Mechanism and applications of CRISPR/Cas-9-mediated genome editing. Biologics. 15:353–361. https://doi.org/10.2147/BTT.S326422

Barrangou R, Fremaux C, Deveau H, Richards M, Boyaval P, Moineau S, Romero DA, Horvath P. 2007. CRISPR provides acquired resistance against viruses in prokaryotes. Science (80-). 315(5819):1709–1712.

Barrangou R, Horvath P. 2017. A decade of discovery: CRISPR functions and applications. Nat Microbiol. 2(7):1–9.

Bastet A, Lederer B, Giovinazzo N, Arnoux X, German-Retana S, Reinbold C, Brault V, Garcia D, Djennane S, Gersch S. 2018. Trans-species synthetic gene design allows resistance pyramiding and broad-spectrum engineering of virus resistance in plants. Plant Biotechnol J. 16(9):1569–1581.

Bharathkumar N, Sunil A, Meera P, Aksah S, Kannan M, Saravanan KM, Anand T. 2022. CRISPR/Cas-based modifications for therapeutic applications: A review. Mol Biotechnol. 64(4):355–372. https://doi.org/10.1007/s12033-021-00422-8

Bhol NK, Bhanjadeo MM, Singh AK, Dash UC, Ojha RR, Majhi S, Duttaroy AK, Jena AB. 2024. The interplay between cytokines, inflammation, and antioxidants: Mechanistic insights and therapeutic potentials of various antioxidants and anti-cytokine compounds. Biomed Pharmacother [Internet]. 178:117177.https://doi.org/https://doi.org/10.1016/j.biopha.2024.117177

Bhowmik P, Konkin D, Polowick P, Hodgins CL, Subedi M, Xiang D, Yu B, Patterson N, Rajagopalan N, Babic V. 2021. CRISPR/Cas9 gene editing in legume crops: Opportunities and challenges. Legum Sci. 3(3):e96.

Bikard D, Barrangou R. 2017. Using CRISPR-Cas systems as antimicrobials. Curr Opin Microbiol. 37:155–160.

Bikard D, Euler CW, Jiang W, Nussenzweig PM, Goldberg GW, Duportet X, Fischetti VA, Marraffini LA. 2014. Exploiting CRISPR-Cas nucleases to produce sequence-specific antimicrobials. Nat Biotechnol. 32(11):1146–1150.

Bikard D, Jiang W, Samai P, Hochschild A, Zhang F, Marraffini LA. 2013. Programmable repression and activation of bacterial gene expression using an engineered CRISPR-Cas system. Nucleic Acids Res. 41(15):7429–7437.

Bolotin A, Quinquis B, Sorokin A, Ehrlich SD. 2005. Clustered regularly interspaced short palindrome repeats (CRISPRs) have spacers of extrachromosomal origin. Microbiology. 151(8):2551–2561.

Borrelli VMG, Brambilla V, Rogowsky P, Marocco A, Lanubile A. 2018. The enhancement of plant disease resistance using CRISPR/Cas9 technology. Front Plant Sci. 9:1245.

Braicu C, Buse M, Busuioc C, Drula R, Gulei D, Raduly L, Rusu A, Irimie A, Atanasov AG, Slaby O, et al. 2019. A comprehensive review on MAPK: A promising therapeutic target in cancer. Cancers (Basel). 11(10). https://doi.org/10.3390/cancers11101618

Brouns SJJ, Jore MM, Lundgren M, Westra ER, Slijkhuis RJH, Snijders APL, Dickman MJ, Makarova KS, Koonin E V, Van Der Oost J. 2008. Small CRISPR RNAs guide antiviral defense in prokaryotes. Science (80-). 321(5891):960–964.

Caccialupi G, Milc J, Caradonia F, Nasar MF, Francia E. 2023. The triticeae CBF gene cluster—to frost resistance and beyond. Cells. 12(22):2606.

Cai Y, Chen L, Liu X, Sun S, Wu C, Jiang B, Han T, Hou W. 2015. CRISPR/Cas9-mediated genome editing in soybean hairy roots. PLoS One. 10(8):e0136064. https://doi.org/10.1371/journal.pone.0136064

Camerlengo F, Frittelli A, Sparks C, Doherty A, Martignago D, Larré C, Lupi R, Sestili F, Masci S. 2020. CRISPR-Cas9 multiplex editing of the α-amylase/trypsin inhibitor genes to reduce allergen proteins in durum wheat. Front Sustain Food Syst. 4:104.

Cardi T, Murovec J, Bakhsh A, Boniecka J, Bruegmann T, Bull SE, Eeckhaut T, Fladung M, Galovic V, Linkiewicz A, et al. 2023. CRISPR/Cas-mediated plant genome editing: Outstanding challenges a decade after implementation. Trends Plant Sci. 28(10):1144–1165. https://doi.org/10.1016/j.tplants.2023.05.012

Chen W, Chen L, Zhang X, Yang N, Guo J, Wang M, Ji S, Zhao X, Yin P, Cai L. 2022. Convergent selection of a WD40 protein that enhances grain yield in maize and rice. Science (80-). 375(6587):eabg7985.

Cong L, Ran FA, Cox D, Lin S, Barretto R, Habib N, Hsu PD, Wu X, Jiang W, Marraffini LA. 2013. Multiplex genome engineering using CRISPR/Cas systems. Science (80-). 339(6121):819–823.

Deltcheva E, Chylinski K, Sharma CM, Gonzales K, Chao Y, Pirzada ZA, Eckert MR, Vogel J, Charpentier E. 2011. CRISPR RNA maturation by trans-encoded small RNA and host factor RNase III. Nature. 471(7340):602–607.

De Majo F, Calore M. 2018. Chromatin remodelling and epigenetic state regulation by non-coding RNAs in the diseased heart. Non-Coding RNA Res [Internet]. 3(1):20–28. https://doi.org/10.1016/j.ncrna.2018.02.003

Dever DP, Bak RO, Reinisch A, Camarena J, Washington G, Nicolas CE, Pavel-Dinu M, Saxena N, Wilkens AB, Mantri S, et al. 2016. CRISPR/Cas9 β-globin gene targeting in human haematopoietic stem cells. Nature. 539(7629):384–389. https://doi.org/10.1038/nature20134

Doudna JA, Charpentier E. 2014. The new frontier of genome engineering with CRISPR-Cas9. Science (80–). 346(6213):1258096.

Fonfara I, Richter H, Bratovič M, Le Rhun A, Charpentier E. 2016. The CRISPR-associated DNA-cleaving enzyme Cpf1 also processes precursor CRISPR RNA. Nature. 532(7600):517–521.

Gao C. 2021. Genome engineering for crop improvement and future agriculture. Cell. 184(6):1621–1635.

Gomez MA, Lin ZD, Moll T, Chauhan RD, Hayden L, Renninger K, Beyene G, Taylor NJ, Carrington JC, Staskawicz BJ. 2019. Simultaneous CRISPR/Cas9-mediated editing of cassava eIF 4E isoforms nCBP-1 and nCBP-2 reduces cassava brown streak disease symptom severity and incidence. Plant Biotechnol J. 17(2):421–434.

Guo Q, Jin Y, Chen X, Ye X, Shen X, Lin M, Zeng C, Zhou T, Zhang J. 2024. NF-κB in biology and targeted therapy: New insights and translational implications. Signal Transduct Target Ther [Internet]. 9(1):53. https://doi.org/10.1038/s41392-024-01757-9

Haft DH, Selengut J, Mongodin EF, Nelson KE. 2005. A guild of 45 CRISPR-associated (Cas) protein families and multiple CRISPR/Cas subtypes exist in prokaryotic genomes. PLoS Comput Biol. 1(6):e60.

Han X, Li S, Zeng Q, Sun P, Wu D, Wu J, Yu X, Lai Z, Milne RJ, Kang Z. 2025. Genetic engineering including genome editing for broad-spectrum disease resistance in crops. Plant Commun. 6(2):101195.

Herman EM, Helm RM, Jung R, Kinney AJ. 2003. Genetic modification removes an immunodominant allergen from soybean. Plant Physiol. 132(1):36–43.

Hichri I, Muhovski Y, Clippe A, Žižková E, Dobrev PI, Motyka V, Lutts S. 2016. SlDREB2, a tomato dehydration-responsive element-binding 2 transcription factor, mediates salt stress tolerance in tomato and A rabidopsis. Plant Cell Environ. 39(1):62–79.

Hille F, Charpentier E. 2016. CRISPR-Cas: Biology, mechanisms and relevance. Philos Trans R Soc B Biol Sci. 371(1707):20150496.

Hille F, Richter H, Wong SP, Bratovič M, Ressel S, Charpentier E. 2018. The biology of CRISPR-Cas: Backward and forward. Cell. 172(6):1239–1259.

Hsu PD, Lander ES, Zhang F. 2014. Development and applications of CRISPR-Cas9 for genome engineering. Cell. 157(6):1262–1278. https://doi.org/10.1016/j.cell.2014.05.010

Hu Y, Liu L, Jiang Q, Fang W, Chen Y, Hong Y, Zhai X. 2023. CRISPR/Cas9: A powerful tool in colorectal cancer research. J Exp Clin Cancer Res. 42(1):308. https://doi.org/10.1186/s13046-023-02901-z

Huang T, Zhang R, Li J. 2023. CRISPR-Cas-based techniques for pathogen detection: Retrospect, recent advances, and future perspectives. J Adv Res [Internet]. 50:69–82. https://doi.org/https://doi.org/10.1016/j.jare.2022.10.011

Huang Y, Dong H, Shang M, Wang K. 2021. CRISPR/Cas systems: The link between functional genes and genetic improvement. Crop J. 9(3):678–687.

Illing LJC. 2019. The optimisation of CRISPR-Cas9 transformation in brassica species using the eIEF4E resistance gene as a model target. PhD thesis, University of Warwick.

Ishino Y, Shinagawa H, Makino K, Amemura M, Nakata A. 1987. Nucleotide sequence of the iap gene, responsible for alkaline phosphatase isozyme conversion in Escherichia coli, and identification of the gene product. J Bacteriol. 169(12):5429–5433.

Jansen R, Embden JDA van, Gaastra W, Schouls LM. 2002. Identification of genes that are associated with DNA repeats in prokaryotes. Mol Microbiol. 43(6):1565–1575.

Jiang F, Taylor DW, Chen JS, Kornfeld JE, Zhou K, Thompson AJ, Nogales E, Doudna JA. 2016. Structures of a CRISPR-Cas9 R-loop complex primed for DNA cleavage. Science (80-). 351(6275):867–871.

Jiang S, Jardinaud M-F, Gao J, Pecrix Y, Wen J, Mysore K, Xu P, Sanchez-Canizares C, Ruan Y, Li Q. 2021. NIN-like protein transcription factors regulate leghemoglobin genes in legume nodules. Science (80-). 374(6567):625–628.

Jiang Y, Sun K, An X. 2022. CRISPR/Cas system: Applications and prospects for maize improvement. ACS Agric Sci Technol. 2(2):174–183.

Jinek M, Chylinski K, Fonfara I, Hauer M, Doudna JA, Charpentier E. 2012. A programmable dual-RNA–guided DNA endonuclease in adaptive bacterial immunity. Science (80-). 337(6096):816–821.

Johnson AAT, Kyriacou B, Callahan DL, Carruthers L, Stangoulis J, Lombi E, Tester M. 2011. Constitutive overexpression of the OsNAS gene family reveals single-gene strategies for effective iron-and zinc-biofortification of rice endosperm. PLoS One. 6(9):e24476.

Jolany Vangah S, Katalani C, Boone HA, Hajizade A, Sijercic A, Ahmadian G. 2020. CRISPR-based diagnosis of infectious and noninfectious diseases. Biol Proced Online. 22:1–14.

Jore MM, Lundgren M, Van Duijn E, Bultema JB, Westra ER, Waghmare SP, Wiedenheft B, Pul Ü, Wurm R, Wagner R. 2011. Structural basis for CRISPR RNA-guided DNA recognition by Cascade. Nat Struct Mol Biol. 18(5):529–536.

Joung JK, Kleinstiver B. 2017. Engineered CRISPR-CAS9 nucleases with altered PAM specificity. Nature. 523(7561): 481–485.

Kadkhoda H, Gholizadeh P, Samadi Kafil H, Ghotaslou R, Pirzadeh T, Ahangarzadeh Rezaee M, Nabizadeh E, Feizi H, Aghazadeh M. 2024. Role of CRISPR-Cas systems and anti-CRISPR proteins in bacterial antibiotic resistance. Heliyon [Internet]. 10(14):e34692. https://doi.org/https://doi.org/10.1016/j.heliyon.2024.e34692

Kato K, Zhou W, Okazaki S, Isayama Y, Nishizawa T, Gootenberg JS, Abudayyeh OO, Nishimasu H. 2022. Structure and engineering of the type III-E CRISPR-Cas7-11 effector complex. Cell [Internet]. 185(13):2324–2337.e16. https://doi.org/https://doi.org/10.1016/j.cell.2022.05.003

Kaur, N, Alok A, Kumar P, Kaur Navjot, Awasthi P, Chaturvedi S, Pandey P, Pandey A, Pandey AK, Tiwari S. 2020. CRISPR/Cas9 directed editing of lycopene epsilon-cyclase modulates metabolic flux for β-carotene biosynthesis in banana fruit. Metab Eng. 59:76–86.

Khambhati K, Bhattacharjee G, Gohil N, Dhanoa GK, Sagona AP, Mani I, Bui N Le, Chu D, Karapurkar JK, Jang SH. 2023. Phage engineering and phage-assisted CRISPR-Cas delivery to combat multidrug-resistant pathogens. Bioeng Transl Med. 8(2):e10381.

Khan Y, Bin Q, Hassan T. 2019. The impact of climate changes on agriculture export trade in Pakistan: Evidence from time-series analysis. Growth Change. 50(4):1568–1589.

Khanday I, Skinner D, Yang B, Mercier R, Sundaresan V. 2019. A male-expressed rice embryogenic trigger redirected for asexual propagation through seeds. Nature. 565(7737):91–95.

Kim D, Kim Sojung, Kim Sunghyun, Park J, Kim J-S. 2016. Genome-wide target specificities of CRISPR-Cas9 nucleases revealed by multiplex Digenome-seq. Genome Res [Internet]. 26(3):406–415. https://doi.org/10.1101/gr.199588.115

Kim J-S, Mizoi J, Kidokoro S, Maruyama K, Nakajima J, Nakashima K, Mitsuda N, Takiguchi Y, Ohme-Takagi M, Kondou Y. 2012. *Arabidopsis* GROWTH-REGULATING FACTOR7 functions as a transcriptional repressor of abscisic acid–and osmotic stress–responsive genes, including DREB2A. Plant Cell. 24(8):3393–3405.

Knott GJ, Doudna JA. 2018. CRISPR-Cas guides the future of genetic engineering. Science (80-). 361(6405):866–869.

Komor AC, Kim YB, Packer MS, Zuris JA, Liu DR. 2016. Programmable editing of a target base in genomic DNA without double-stranded DNA cleavage. Nature [Internet]. 533(7603):420–424. https://doi.org/10.1038/nature17946

Koonin E V, Makarova KS. 2013. CRISPR-Cas: Evolution of an RNA-based adaptive immunity system in prokaryotes. RNA Biol. 10(5):679–686.

Kuroiwa K, Thenault C, Nogué F, Perrot L, Mazier M, Gallois J-L. 2022. CRISPR-based knock-out of eIF4E2 in a cherry tomato background successfully recapitulates resistance to pepper veinal mottle virus. Plant Sci. 316:111160.

Lander ES, Baylis F, Zhang F, Charpentier E, Berg P, Bourgain C, Friedrich B, Joung JK, Li J, Liu D. 2019. Adopt a moratorium on heritable genome editing. Nature. 567(7747):165–168.

Lee S, Rahman MM, Nakanishi H, Nishizawa NK, An G, Nam HG, Jeon J-S. 2023. Concomitant activation of OsNAS2 and OsNAS3 contributes to the enhanced accumulation of iron and zinc in rice. Int J Mol Sci. 24(7):6568.

Lemmon ZH, Reem NT, Dalrymple J, Soyk S, Swartwood KE, Rodriguez-Leal D, Van Eck J, Lippman ZB. 2018. Rapid improvement of domestication traits in an orphan crop by genome editing. Nat Plants. 4(10):766–770.

Li M, Li X, Zhou Z, Wu P, Fang M, Pan X, Lin Q, Luo W, Wu G, Li H. 2016. Reassessment of the four yield-related genes Gn1a, DEP1, GS3, and IPA1 in rice using a CRISPR/Cas9 system. Front Plant Sci. 7:377.

Li R, Zhang L, Wang L, Chen L, Zhao R, Sheng J, Shen L. 2018. Reduction of tomato-plant chilling tolerance by CRISPR–Cas9-mediated SlCBF1 mutagenesis. J Agric Food Chem. 66(34):9042–9051.

Lian L, Xu H, Zhang H, He W, Cai Q, Lin Y, Wei L, Pan L, Xie X, Zheng Y. 2020. Overexpression of OsSPL14 results in transcriptome and physiology changes in indica rice 'MH86'. Plant Growth Regul. 90:265–278.

Lin C-S, Hsu C-T, Yuan Y-H, Zheng P-X, Wu F-H, Cheng Q-W, Wu Y-L, Wu T-L, Lin S, Yue J-J. 2022. DNA-free CRISPR-Cas9 gene editing of wild tetraploid tomato Solanum peruvianum using protoplast regeneration. Plant Physiol. 188(4):1917–1930.

Lin Z, Griffith ME, Li X, Zhu Z, Tan L, Fu Y, Zhang W, Wang X, Xie D, Sun C. 2007. Origin of seed shattering in rice (Oryza sativa L.). Planta. 226:11–20.

Liu C, Kong M, Zhu J, Qi X, Duan C, Xie C. 2022. Engineering null mutants in ZmFER1 confers resistance to ear rot caused by Fusarium verticillioides in maize. Plant Biotechnol J. 20(11):2045.

Liu K, Yuan X, Liang L, Fang J, Chen Y, He W, Xue T. 2019. Using CRISPR/Cas9 for multiplex genome engineering to optimize the ethanol metabolic pathway in Saccharomyces cerevisiae. Biochem Eng J. 145:120–126.

Liu W, Li L, Jiang J, Wu M, Lin P. 2021. Applications and challenges of CRISPR-Cas gene-editing to disease treatment in clinics. Precis Clin Med. 4(3):179–191. https://doi.org/10.1093/pcmedi/pbab014

Long C, McAnally JR, Shelton JM, Mireault AA, Bassel-Duby R, Olson EN. 2014. Prevention of muscular dystrophy in mice by CRISPR/Cas9-mediated editing of germline DNA. Science. 345(6201):1184–1188. https://doi.org/10.1126/science.1254445

Lorenzo CD, Blasco-Escámez D, Beauchet A, Wytynck P, Sanches M, Garcia del Campo JR, Inzé D, Nelissen H. 2024. Maize mutant screens: From classical methods to new CRISPR-based approaches. New Phytol. 244(2):384–393.

Loureiro A, da Silva GJ. 2019. CRISPR-Cas: Converting a bacterial defence mechanism into a state-of-the-art genetic manipulation tool. Antibiot (Basel, Switzerland). 8(1). https://doi.org/10.3390/antibiotics8010018

Makarova KS, Haft DH, Barrangou R, Brouns SJJ, Charpentier E, Horvath P, Moineau S, Mojica FJM, Wolf YI, Yakunin AF. 2011. Evolution and classification of the CRISPR–Cas systems. Nat Rev Microbiol. 9(6):467–477.

Makarova KS, Wolf YI, Iranzo J, Shmakov SA, Alkhnbashi OS, Brouns SJJ, Charpentier E, Cheng D, Haft DH, Horvath P. 2020. Evolutionary classification of CRISPR–Cas systems: A burst of class 2 and derived variants. Nat Rev Microbiol. 18(2):67–83.

Makarova KS, Wolf YI, Koonin E V. 2013. The basic building blocks and evolution of CRISPR–Cas systems. Biochem Soc Trans. 41(6):1392–1400.

Makarova KS, Wolf YI, Koonin E V. 2018. Classification and nomenclature of CRISPR-Cas systems: Where from here? Cris J [Internet]. 1(5):325–336. https://doi.org/10.1089/crispr.2018.0033

Mali P, Yang L, Esvelt KM, Aach J, Guell M, DiCarlo JE, Norville JE, Church GM. 2013. RNA-guided human genome engineering via Cas9. Science (80-). 339(6121):823–826.

Marraffini LA, Sontheimer EJ. 2010. CRISPR interference: RNA-directed adaptive immunity in bacteria and archaea. Nat Rev Genet. 11(3):181–190.

Mohanraju P, Makarova KS, Zetsche B, Zhang F, Koonin E V, Van der Oost J. 2016. Diverse evolutionary roots and mechanistic variations of the CRISPR-Cas systems. Science (80-). 353(6299):aad5147.

Mojica FJM, Díez-Villaseñor C sar, García-Martínez J, Soria E. 2005. Intervening sequences of regularly spaced prokaryotic repeats derive from foreign genetic elements. J Mol Evol. 60:174–182.

Mondal S, Gayen D, Karmakar S. 2020. Improvement of nutritional quality of rice seed through classical breeding and advance genetic engineering. Rice Res Qual Improv Genomics Genet Eng Vol 2 Nutr Biofortification Herbic Biot Stress Resist Rice.:541–562.

Mulepati S, Bailey S. 2011. Structural and biochemical analysis of nuclease domain of clustered regularly interspaced short palindromic repeat (CRISPR)-associated protein 3 (Cas3). J Biol Chem. 286(36):31896–31903. https://doi.org/10.1074/jbc.M111.270017

Nakayasu M, Akiyama R, Lee HJ, Osakabe K, Osakabe Y, Watanabe B, Sugimoto Y, Umemoto N, Saito K, Muranaka T. 2018. Generation of α-solanine-free hairy roots of potato by CRISPR/Cas9 mediated genome editing of the St16DOX gene. Plant Physiol Biochem. 131:70–77.

Ndudzo A, Sibanda Makuvise A, Moyo S, Bobo ED. 2024. CRISPR-Cas9 genome editing in crop breeding for climate change resilience: Implications for smallholder farmers in Africa. J Agric Food Res [Internet]. 16:101132. https://doi.org/10.1016/j.jafr.2024.101132

Nekrasov V, Wang C, Win J, Lanz C, Weigel D, Kamoun S. 2017. Rapid generation of a transgene-free powdery mildew resistant tomato by genome deletion. Sci Rep. 7(1):1–6.

Niaz M, Zhang B, Zhang Y, Yan X, Yuan M, Cheng Y, Lv G, Fadlalla T, Zhao L, Sun C. 2023. Genetic and molecular basis of carotenoid metabolism in cereals. Theor Appl Genet. 136(3):63.

Nidhi S, Anand U, Oleksak P, Tripathi P, Lal JA, Thomas G, Kuca K, Tripathi V. 2021. Novel CRISPR–Cas systems: An updated review of the current achievements, applications, and future research perspectives. Int J Mol Sci. 22(7). https://doi.org/10.3390/ijms22073327

Nishimasu H, Ran FA, Hsu PD, Konermann S, Shehata SI, Dohmae N, Ishitani R, Zhang F, Nureki O. 2014. Crystal structure of Cas9 in complex with guide RNA and target DNA. Cell. 156(5):935–949.

Noureen A, Khan MZ, Amin I, Zainab T, Mansoor S. 2022. CRISPR/Cas9-mediated targeting of susceptibility factor eIF4E-enhanced resistance against potato virus Y. Front Genet. 13:922019.

Pandita D. 2021. Cas9 technology: An innovative approach to enhance phytoremediation. In: Pirzadah TB, Malik B, Hakeem KR (eds.), *Plant-Microbe Dynamics Recent Advances for Sustainable Agriculture*. CRC Press; pp. 221–248.

Park S-C, Kim SH, Park S, Lee H-U, Lee JS, Park WS, Ahn M-J, Kim Y-H, Jeong JC, Lee H-S. 2015. Enhanced accumulation of carotenoids in sweetpotato plants overexpressing IbOr-Ins gene in purple-fleshed sweetpotato cultivar. Plant Physiol Biochem. 86:82–90.

Pinilla-Redondo R, Russel J, Mayo-Muñoz D, Shah SA, Garrett RA, Nesme J, Madsen JS, Fineran PC, Sørensen SJ. 2022. CRISPR-Cas systems are widespread accessory elements across bacterial and archaeal plasmids. Nucleic Acids Res. 50(8):4315–4328. https://doi.org/10.1093/nar/gkab859

Pourcel C, Salvignol G, Vergnaud G. 2005. CRISPR elements in Yersinia pestis acquire new repeats by preferential uptake of bacteriophage DNA, and provide additional tools for evolutionary studies. Microbiology. 151(3):653–663.

Qi LS, Larson MH, Gilbert LA, Doudna JA, Weissman JS, Arkin AP, Lim WA. 2013. Repurposing CRISPR as an RNA-guided platform for sequence-specific control of gene expression. Cell. 152(5):1173–1183.

Rahmati R, Zarimeidani F, Ghanbari Boroujeni MR, Sadighbathi S, Kashaniasl Z, Saleh M, Alipourfard I. 2025. CRISPR-assisted probiotic and in situ engineering of gut microbiota: A prospect to modification of metabolic disorders. Probiotics Antimicrob Proteins.:1–17.

Rajyaguru R, Maheshala N, K G. 2023. Genetic improvement in peanut: Role of Genetic Engineering. In: Tiwari S, Koul B (eds.), *Genetic Engineering of Crop Plants for Food and Health Security. Volume 1*. Springer; pp. 271–288.

Rath D, Amlinger L, Rath A, Lundgren M. 2015. The CRISPR-Cas immune system: Biology, mechanisms and applications. Biochimie. 117:119–128.

Sadhu S, Paul T, Yadav N. 2025. Therapeutic engineering of the gut microbiome using synthetic biology and metabolic tools: A comprehensive review with E. coli Nissle 1917 as a model case study. Arch Microbiol. 207(9):213. https://doi.org/10.1007/s00203-025-04417-w

Sánchez-León S, Gil-Humanes J, Ozuna C V, Giménez MJ, Sousa C, Voytas DF, Barro F. 2018. Low-gluten, nontransgenic wheat engineered with CRISPR/Cas9. Plant Biotechnol J. 16(4):902–910.

Schreiter S, Ding G-C, Heuer H, Neumann G, Sandmann M, Grosch R, Kropf S, Smalla K. 2014. Effect of the soil type on the microbiome in the rhizosphere of field-grown lettuce. Front Microbiol. 5:144.

Serajian S, Ahmadpour E, Oliveira SMR, Pereira M de L, Heidarzadeh S. 2021. CRISPR-Cas technology: Emerging applications in clinical microbiology and infectious diseases. Pharmaceuticals. 14(11):1171.

Shelake RM, Pramanik D, Kim J-Y. 2019. Exploration of plant-microbe interactions for sustainable agriculture in CRISPR era. Microorganisms. 7(8):269.

Shmakov SA, Makarova KS, Wolf YI, Severinov K V, Koonin E V. 2018. Systematic prediction of genes functionally linked to CRISPR-Cas systems by gene neighborhood analysis. Proc Natl Acad Sci. 115(23):E5307–E5316.

Shmakov SA, Smargon A, Scott D, Cox D, Pyzocha N, Yan W, Abudayyeh OO, Gootenberg JS, Makarova KS, Wolf YI, et al. 2017. Diversity and evolution of class 2 CRISPR-Cas systems. Nat Rev Microbiol. 15(3):169–182. https://doi.org/10.1038/nrmicro.2016.184

Singh S, Sangh C, Kona P, Bera SK. 2024. Genome editing in peanuts: Advancements, challenges and applications. Nucl. 67(1):127–139.

Sinkunas T, Gasiunas G, Waghmare SP, Dickman MJ, Barrangou R, Horvath P, Siksnys V. 2013. In vitro reconstitution of cascade-mediated CRISPR immunity in Streptococcus thermophilus. EMBO J. 32(3):385–394. https://doi.org/10.1038/emboj.2012.352

Sorek R, Lawrence CM, Wiedenheft B. 2013. CRISPR-mediated adaptive immune systems in bacteria and archaea. Annu Rev Biochem. 82:237–266. https://doi.org/10.1146/annurev-biochem-072911-172315

Staals RHJ, Agari Y, Maki-Yonekura S, Zhu Y, Taylor DW, Van Duijn E, Barendregt A, Vlot M, Koehorst JJ, Sakamoto K. 2013. Structure and activity of the RNA-targeting Type III-B CRISPR-Cas complex of Thermus thermophilus. Mol Cell. 52(1):135–145.

Staals RHJ, Zhu Y, Taylor DW, Kornfeld JE, Sharma K, Barendregt A, Koehorst JJ, Vlot M, Neupane N, Varossieau K, et al. 2014. RNA targeting by the type III-A CRISPR-Cas Csm complex of Thermus thermophilus. Mol Cell. 56(4):518–530. https://doi.org/10.1016/j.molcel.2014.10.005

Thanh ND. 2019. Provitamin a biofortification in maize through genetic engineering and marker-assisted selection. Tap Chi Sinh Hoc. 41(4).

Tran MT, Son GH, Song YJ, Nguyen NT, Park S, Thach TV, Kim J, Sung YW, Das S, Pramanik D, Lee J, Son K-H, Kim SH, Vu TV, Kim J-Y. 2023. CRISPR-Cas9-based precise engineering of SlHyPRP1 protein towards multi-stress tolerance in tomato. Front Plant Sci. 14:1186932.

Tran T, Ho T, Nguyen D. 2017. Overexpression of the IbOr gene from sweet potato (Ipomea batatasHoang Long') in maize increases total carotenoid and ß-carotene contents. Turkish J Biol. 41(6):1003–1010.

Trijatmiko KR, Dueñas C, Tsakirpaloglou N, Torrizo L, Arines FM, Adeva C, Balindong J, Oliva N, Sapasap M V, Borrero J. 2016. Biofortified indica rice attains iron and zinc nutrition dietary targets in the field. Sci Rep. 6(1):19792.

Vilela A. 2021. An overview of CRISPR-based technologies in wine yeasts to improve wine flavor and safety. Fermentation. 7(1):5.

Waltz E. 2016. Gene-edited CRISPR mushroom escapes US regulation. Nature. 532(7599):293.

Wang H, Yang H, Shivalila CS, Dawlaty MM, Cheng AW, Zhang F, Jaenisch R. 2013. One-step generation of mice carrying mutations in multiple genes by CRISPR/Cas-mediated genome engineering. Cell. 153(4):910–918. https://doi.org/10.1016/j.cell.2013.04.025

Wang L, Hu J, Li K, Zhao Y, Zhu M. 2024. Advancements in gene editing technologies for probiotic-enabled disease therapy. Iscience. 27(9).

Wang M, Schedel M, Gelfand EW. 2024. Gene editing in allergic diseases: Identification of novel pathways and impact of deleting allergen genes. J Allergy Clin Immunol.154(1):51–58.

Wang R, Li C, Li Q, Ai Y, Huang Z, Sun X, Zhou J, Zhou Y, Liang Y. 2022. Tomato receptor-like cytosolic kinase RIPK confers broad-spectrum disease resistance without yield penalties. Hortic Res. 9:uhac207.

Wang S, Ma B, Gao Q, Jiang G, Zhou L, Tu B, Qin P, Tan X, Liu P, Kang Y. 2018. Dissecting the genetic basis of heavy panicle hybrid rice uncovered Gn1a and GS3 as key genes. Theor Appl Genet. 131:1391–1403.

Wang Y, Cheng X, Shan Q, Zhang Y, Liu J, Gao C, Qiu J-L. 2014. Simultaneous editing of three homoeoalleles in hexaploid bread wheat confers heritable resistance to powdery mildew. Nat Biotechnol. 32(9):947–951.

Wiedenheft B, Lander GC, Zhou K, Jore MM, Brouns SJJ, Van Der Oost J, Doudna JA, Nogales E. 2011. Structures of the RNA-guided surveillance complex from a bacterial immune system. Nature. 477(7365):486–489.

Xiang X, Zhou Z, Rao X, Jiang X-R. 2025. Recent advances of engineered bacteria for therapeutic applications. Mol Ther. 34(2):714–733.

Xie Z, McAuliffe O, Jin Y-S, Miller MJ. 2024. Genomic modifications of lactic acid bacteria and their applications in dairy fermentation. J Dairy Sci. 107(11):8749–8764.

Xue C, Greene EC. 2021. DNA repair pathway choices in CRISPR-Cas9-mediated genome editing. Trends Genet. 37(7):639–656. https://doi.org/10.1016/j.tig.2021.02.008

Yamano T, Zetsche B, Ishitani R, Zhang F, Nishimasu H, Nureki O. 2017. Structural basis for the canonical and non-canonical PAM recognition by CRISPR-Cpf1. Mol Cell. 67(4):633–645.e3. https://doi.org/10.1016/j.molcel.2017.06.035

Yu H, Lin T, Meng X, Du H, Zhang J, Liu G, Chen M, Jing Y, Kou L, Li X. 2021. A route to de novo domestication of wild allotetraploid rice. Cell. 184(5):1156–1170.

Yu W, Wang L, Zhao R, Sheng J, Zhang S, Li R, Shen L. 2019. Knockout of SlMAPK3 enhances tolerance to heat stress involving ROS homeostasis in tomato plants. BMC Plant Biol. 19:1–13.

Zetsche B, Gootenberg JS, Abudayyeh OO, Slaymaker IM, Makarova KS, Essletzbichler P, Volz SE, Joung J, Van Der Oost J, Regev A. 2015. Cpf1 is a single RNA-guided endonuclease of a class 2 CRISPR-Cas system. Cell. 163(3):759–771.

Zhang J, Rouillon C, Kerou M, Reeks J, Brugger K, Graham S, Reimann J, Cannone G, Liu H, Albers S-V, et al. 2012. Structure and mechanism of the CMR complex for CRISPR-mediated antiviral immunity. Mol Cell. 45(3):303–313. https://doi.org/10.1016/j.molcel.2011.12.013

Zhang Y, Heidrich N, Ampattu BJ, Gunderson CW, Seifert HS, Schoen C, Vogel J, Sontheimer EJ. 2013. Processing-independent CRISPR RNAs limit natural transformation in Neisseria meningitidis. Mol Cell. 50(4):488–503.

Zhang Y-L, Xia Q-Y, Jiang X-Q, Hu W, Ye X-X, Huang Q-X, Yu S-B, Guo A-P, Lu B-R. 2022. Reducing seed shattering in weedy rice by editing SH4 and qSH1 genes: Implications in environmental biosafety and weed control through transgene mitigation. Biology (Basel). 11(12):1823.

Zhu Y. 2022. Advances in CRISPR/Cas9. Biomed Res Int. 2022:9978571.

Zsögön A, Čermák T, Naves ER, Notini MM, Edel KH, Weinl S, Freschi L, Voytas DF, Kudla J, Peres LEP. 2018. De novo domestication of wild tomato using genome editing. Nat Biotechnol. 36(12):1211–1216.

16 Microbiome in Crop Improvements at High Altitude

Ayon Pal and Madhumita G. Chaki*
Microbiology and Computational Biology Laboratory, Department of Botany, Raiganj University, Raiganj – 733134, West Bengal, India

16.1 INTRODUCTION

High-altitude agronomy experiences several constraints. Crop production in hilly and mountainous terrains is distinctive in many aspects from that of the plains concerning the farming pattern and system. The plains adopt various farming systems such as check basin, border and intercropping, while terrace farming is the most adopted cultivation practice in the elevational zones (Mishra and Rai, 2013). The harsh terrain, hostile soil, and climatic conditions of the arable farm at high altitudes pose economic pressures on the farmers (Hossain et al., 2020). Agronomic lands are vulnerable to soil loss, erosion, undulating topography, and rainfall in the hills. Access to adequate groundwater storage is scarce due to land relinquishment, heavy mechanization, and hastening land deprivation in steep landscapes (Yosef and Asmamaw, 2015). Despite many complications and challenges, hill agriculture embraces the cultivation of paddy, rice cereals, natural products, vegetables, blooms, restorative plants, apples, oranges, plums, banana, pineapple, papaya, mango, litchi, jackfruit, and guava (Sati 2024). Northeastern hill regions of India commonly follow a practice known as shifting cultivation or slash and burn cultivation, or Jhum cultivation, wherein land is cleared, burnt, and then recultivated (Nath et al., 2020). However, some standard farming practices globally include vertical farming (frequently seen as an emerging practice in developing countries, growing crops using aquaponics and hydroponics) (Gumisiriza et al., 2022), silvopasture (practiced in Europe and North and South America wherein trees, livestock and forage crops are cultivated on the same land) (Orefice et al., 2017), and arable farming (here traditional techniques such as sowing, ploughing, and harvesting are used to cultivate staple food crops most commonly done in Asia, North America, and Europe) (Fischer and Connor, 2018).

The terms "microbiota" and "microbiome" are distinctly related terms wherein the microbiota relates to a microbial ecosystem's living or biotic component, and the microbiome constitutes the abiotic factors in their physical habitat (Trivedi et al., 2022). Microbial consortia is an assemblage of microorganisms working synergistically with a symbiotic dependency (Lupatini et al., 2017). Microbes play an indispensable role in promoting plant growth since many soil-borne organisms possess growth-promoting ability through nitrogen fixation, phytohormone production, siderophore production, and resistance to biotic and abiotic stresses (Souza et al., 2015). Thus, deciphering the native microbes could serve as an avenue for the upliftment of high-altitude farming. Microbial diversity along the elevational gradients changes with a predominance of psychrophilic and psychotropic bacteria (Suyal et al., 2021). These microbes possess significant potential to strive and thrive in cold climate conditions and augment plant growth by serving as potent bioinoculants. This chapter explores the challenges and prospects accompanying high-altitude agriculture. It discusses the issues of mountain and hill agroecosystems, providing a comprehensive overview of the various factors that control the microbiome at high altitudes and emphasizing their role in crop improvements at high altitudes.

16.2 CHALLENGES OF HIGH-ALTITUDE AGRICULTURE

Agriculture at high altitudes is complex in itself. Apart from environmental, climatic, and topographical undulations, some significant constraints of high-altitude terrain regions include short growing seasons, temperature variations, water scarcity, high ultraviolet (UV) radiation, strong winds, limited crop varieties, frost risk, low oxygen levels, and poor soil quality. A diagrammatic representation of the challenges associated with agriculture at high altitudes is illustrated in Figure 16.1.

16.2.1 Frost Risk and Chilling Stress

Hillsides experience cold air and terrain curvature, imposing potential chilling stress upon the plants and restraining plant growth (Geitner et al., 2021). Distinctive survival strategies emerge within the plants and the associated microbiota to counter these stressful conditions, which otherwise would be detrimental or lethal to the plant. Chilling and freezing conditions pose significant threats to plant growth, causing physiological and mechanical damage and disrupting metabolic pathways (Zahra et al., 2021). A detailed elaboration of the role of microbiota in addressing frost damage is discussed later in this chapter.

*Corresponding author: ayonpal.ruc@gmail.com (Ayon Pal)

DOI: 10.1201/9781003544425-16

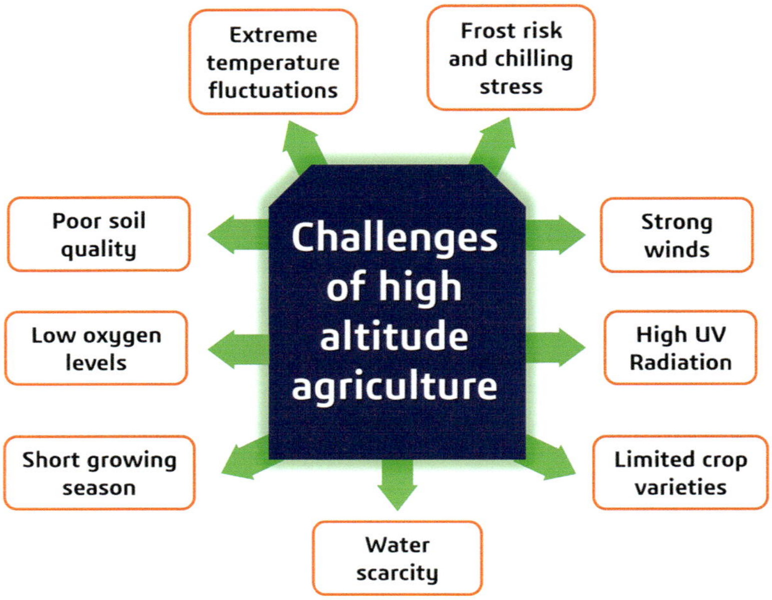

FIGURE 16.1 The challenges associated with agriculture at high altitudes.

16.2.2 Short Growing Season

High altitudes often experience shorter growing seasons due to prolonged winters and delayed soil thawing (Jaganathan et al., 2020). High altitudes cause physiological desiccation and freezing injury; thus, thriving at such high altitudes requires many backup adaptations to cope with such challenges.

16.2.3 Extreme Temperature Fluctuations

Significant day-night temperature differences can stress plants and reduce yields. Hill agronomy is prone to challenges such as climatic fluctuations leading to rising temperatures, decreasing rainfall, landslides, earthquakes, osmotic stress due to low water potential, and other natural disasters affecting crop yield and productivity. Weather and soil conditions curtail the growth of plants and are very labor-intensive; thus, hill farming is demanding and challenging (Maity et al., 2016).

16.2.4 Water Scarcity

Deforestation is a significant problem in hilly areas, resulting in landslides, soil erosion, earthquakes, and, thereby, increasing the risk of ecological impairment. Although some steep mountainous regions experience high rainfall, the water runs off the steep slopes instead of seeping under the ground, resulting in water scarcity. At high altitudes, atmospheric pressure declines along with low rainfall, lowering the water table in the foothills. High altitudes where cultivation was done previously but is limited currently due to climatic disruptions include the high-altitude Himalayan regions, which face the utmost challenging weather conditions, shortage of water supply, and frequent risks of natural calamities (Chakraborty and Saikia 2023; Dimri et al., 2021). The inconsistent snowfall, disappearing glaciers, and a declining water supply in the Lahaul-Spiti district of Himachal Pradesh also face agricultural distress (Bajpai, 2002; Manglani, 2023). In the Chumchaur region of Jumla, Nepal, at an elevation of 3,050 meters, rice cultivation is not possible due to climatic fluctuations (Bhattarai et al., 2020).

16.2.5 High UV Radiation

Extreme UV exposure at higher elevations significantly appears due to thinning atmospheric layers, high reflective surfaces such as snow, and less ozone. The mountains are more prone to UV-induced damage, triggering oxidative base damage, protein-DNA crosslinks, DNA single-strand breaks, and pyrimidine dimers, impeding plant growth and development (Prado et al., 2012; Sedej et al., 2020).

16.2.6 Strong Winds

Temperature and pressure gradients at high altitudes reduce friction, leading to jet streams and strong winds above sea level. Such strong, catastrophic winds loosen the topsoil, posing significant mechanical damage to the plant canopy (Moore et al., 2018). Wilting, drooping down of shoots, uprooting of trees and plants, and breaking of crown branches of older trees occur upon exposure to such high-frequency winds. All these cause major plant damage and injuries, the volume of which depends upon the direction, persistence, frequency, and turbulence of the winds (Quine et al., 2021).

16.2.7 Limited Crop Varieties

Steep-sloped terrains limit agricultural productivity, mostly preferring perennial plantation crops such as tea and coffee. Although canopy gaps, deforestation, and land burning lead to havoc, the deprivation of considerable areas of plain land for extensive cultivation has forced indigenous hill tribes to assume these traditional agricultural practices (Kurien et al., 2019).

16.2.8 Low Oxygen Levels

A shift occurs from normoxic to hypoxic conditions predominantly at high altitudes due to a decreased oxygen saturation with the elevational increase (Abbas et al., 2022). Deprivation of plant growth occurs as the oxygen levels drop. Respiration rate diminishes, and photosynthesis is inhibited, with a significant decline in nutrient uptake and augmented vulnerability to disease susceptibility due to less dissolved oxygen in the soil. This also causes leaf wilting and slows root growth (Saiz-Fernández et al., 2017).

16.2.9 Poor Soil Quality

Dry air and low temperature decrease the supply of nutrients in the soil, thereby restricting plant growth at colder extremities (Bhattacharya 2022). Reduced amount of dissolved oxygen in the soil affects the microbial community composition, which in turn affects plant growth and development. Since the soil has reduced water content and irrigation is also an issue at high altitudes, selective crops must be cultivated to grow sustainably with minimal water and nutrient necessities (Normatov et al., 2022).

16.3 FARMING PRACTICES COMMON TO HIGH ALTITUDES

The farming systems practiced in mountainous regions are intended to avert soil erosion and minimize water scarcity problems. Meanwhile, high-altitude agriculture also experiences space limitations, and to achieve a trade-off between limiting land and crop productivity, appropriate farming practices have been adopted to address these issues. Terrace farming, wherein the mountain slopes are cut out along the sides in a staircase pattern, may be of three types: bench-terracing, contour-terracing, and parallel-terracing (Arora et al., 2023). To retain the nutrient content of the soil, mixed cropping (where two or more crops with different nutritional requirements and maturation time are grown at once in the same field), intercropping, mulching (covering the topsoil with a layer of materials such as grass clippings, straw, stones, brick chips, or plastic to protect the soil from erosion or weathering), crop rotation, and agroforestry, tillage (loosening the soil surface thereby helping seed propagation) are standard agronomical soil routines in undulating areas (Khokhar et al., 2021; Upadhaya et al., 2020).

16.4 MICROBIOME AND POPULATION DIVERSITY OF HIGH ALTITUDES

The microbial assemblage exhibits a dynamic state change to acclimatize to the climatic variations. Microbial consortia differ according to environmental conditions. As the ecological settings fluctuate significantly, so do the plants, animals, and microbial distribution. The microbial community exhibits variability, which needs to be unveiled. High altitudes generally display distinctly variable conditions, with a unique population of psychrophilic (cold-loving) organisms being predominant (Dasauni and Nailwal 2020; Pearce 2017). Nevertheless, detailed information on the fluctuations of the bacterial diversity with elevation is inadequate. The climatic conditions, plant succession, and the vegetation types within the different regions of a mountain, starting from the valley to the foot of a hillside or skirting up to the summit, vary, as does the microbiome's phylogenetic diversity (Ciccazzo et al., 2016). A significant negative correlation is prevalent between elevation and microbial diversity (Aqeel et al., 2024; Lyngwi et al., 2013). The mean annual temperature (MAT), mean annual precipitation (MAP), and soil pH are the primary determinants of this variability (Chen et al., 2013). Disproportion in soil microbial community distribution along elevational gradients relies on abiotic and biotic factors, such as vegetation (Zhang et al., 2019). Thus, richness and phylogenetic diversity start to contract near higher elevational gradients (Zhang et al., 2019). However, phylogenetic clustering shows uniformity and is positively correlated with elevation, unlike the dispersive pattern displayed by macroorganisms. Temperature fluctuations are crucial for altering phylogenetic distribution patterns. Microbial consortia of diverse ecosystems exhibit dynamicity due to spatial divergence (Fu et al., 2023). However, no constant pattern of soil bacterial richness or diversity can be determined amongst the world's mountain systems as a general trend since it encompasses a dynamic population even across different transects on the same mountain (Dhakal and Kattel, 2019). Different elevational zones display inconsistent vegetation cover and microbial diversity (Peng et al., 2022). However, the elevation diversity trend is ambiguous and inconsistent. Exploring the community composition of high altitudinal elevations is not possible without metagenomics and a multivariate statistical determination (Wu, 2018).

16.5 ROLE OF MICROBIOMES IN ADDRESSING HIGH-ALTITUDE AGRICULTURE CHALLENGES

To alleviate cold stress, bioprospecting high-altitude environments is essential for discovering microorganisms with excellent cold-tolerant adaptability. Microorganisms with the potential to provide an enormous boost to agronomic development are referred to as bio-inoculants (Kumawat et al., 2021). Hilly agriculture is particularly uplifted by cold-adapted microbes serving as potent bio-inoculants for augmenting the hill agronomy. Native microbes possessing multileveled traits could be used as bioinoculants to promote hill agronomy. Psychrophiles are cold-loving organisms with growth optima at 15°C, while psychrotolerant or psychrotrophs (cold-tolerant) prefer temperatures between 20°C and 40°C with tolerance up to 0°C (Pathania et al.,

2022). Reports of seed bacterization have been documented through repeated field trials, significantly demonstrating the role of some potent isolates having a noteworthy role in improving plant growth, upregulating mineral deficiencies and nutrient uptake, and providing an impetus for enhancement to sustainable agriculture at high altitudes (Rizvi et al., 2021). The different facets of the microbiome in addressing the challenges related to high-altitude crop production are discussed in the succeeding sections.

16.5.1 Soil Microbiome as the Key Role Player in Combating Frost Damage

The soil harbors a plethora of microorganisms with dynamic properties. Microbes are indispensable in ameliorating a wide range of stress responses. The role of microorganisms and the microbiome in preventing ice nucleation and synthesis of anti-freeze compounds, accumulation of osmolyte (osmoprotectants), synthesis of stress proteins, and assistance in genetic adaptation is discussed in this section.

16.5.1.1 Ice-Nucleation Prevention and Synthesis of Anti-Freeze Compounds

In high-altitude regions, low temperature, drought, salinity, low soil water potential, and chilling stress are chief factors limiting the growth and productivity of major crops (Zahra et al., 2021). In response to supercooling conditions, the composition of membrane lipid changes, the content of compatible osmolytes such as proline is enhanced, and an increase in the number of specific proteins that enable the lowering of freezing stress. Microbes resilient to freezing conditions develop tolerance to chilling injuries through cold acclimation, wherein microbes cope by regulating specific physiological and metabolic responses. Considering the concept of coevolution, the psychrophilic bacteria of the high-altitude regions also develop equally compatible mechanisms to aid in combating chilling injuries. The high-altitude microbiota increases the freezing tolerance in plants by preventing the formation of membrane-damaging ice crystals due to their ice-nucleating inhibition activity (Esitken 2011). Stress responses cause molecular, biochemical, physiological, and morphological instabilities. Cold acclimatization is significantly achieved by different mechanisms to thrive through the abiotic chilling stressors. Adaptation to the cold environment in bacteria is carried out by certain mechanistically distinct responses, such as the upregulation of several classes of proteins, stress proteins, osmoprotectants, and genetic upregulation that counter the frost damage and chilling stress (Abdullah et al., 2022).

Exposure to extreme cold conditions, i.e., freezing or sub-zero temperatures, initiates the process of phase change (or polymerization) from liquid to solid (ice) form, facilitated by proteins called ice nucleation proteins and the phenomenon is termed ice nucleation (Lin et al., 2023). The cellular localization of these ice-nucleating proteins (INPs) is the outer membrane of some Gram-negative bacteria. Thus, it can be said that the key players of frost damage are Gram-negative bacterial outer membrane INP (Kassmannhuber et al., 2017). In the liquid state, water molecules are held by relatively weak hydrogen bonds with less intermolecular space between them. However, in the ice form, the nucleation proteins enable expansion into a hexagonal crystal lattice held together by relatively stable hydrogen bonds upon freezing (Cascajo Castresana 2017). Frost damage hinders plant growth by causing crystallization of cellular water, leading to dehydration and xylem vessel embolism (clogging of xylem vessels leading to blockage of water transport), which causes vascular system damage and thus lethality to plants (Hammond and Morris 2022). Some of the bacterial genera that cause frost damage through INP production are *Pseudomonas syringae, Erwinia herbicola, Pantoea agglomerans,* and *Xanthomonas campestris* (Cid et al., 2016; Rustama et al.). Ice nucleation inhibitor proteins are a class of proteins that avert frost damage by adsorption onto the ice surface. Inhibition of the action of INP is the mode of action exhibited by INP inhibitor proteins, such as anti-freeze proteins (AFPs, recombinant ice-binding proteins reported in *Pseudomonas putida* GR12–2 (expressing AfpA), *Pseudomonas syringae*, and *Xanthomonas campestris* inhibiting crystallization of ice by glycosylated AFP production (Uko et al., 2024). Cold-adapted (psychrophilic) microbes mechanistically display certain traits to thrive through harsh cold stress conditions. To sustain the vital metabolic functions, they constitutively synthesize several proteins such as cold-shock proteins, cryoprotectants, cold acclimation proteins (Caps), ice nucleation factors, and AFPs (Mishra et al., 2020) along with some other phenomena such as cell membrane-associated changes, which enable extreme cold adaptation and sustenance (Sinha 2019). In general, chilling-resistant organisms contrastingly demonstrate a distinct structure of the membrane in comparison to the chilling-sensitive ones (Bustamante et al., 2016). The former displays a high content of unsaturated fatty acids and a lower ratio of saturated to unsaturated fatty acids (Liu et al., 2018).

16.5.2 Osmolyte (Osmoprotectants) Accumulation

Low temperatures mainly distress the cellular protein stability, so cryoprotectants such as glycine betaine, glycerol, sucrose/bovine serum albumin, and trehalose/dextran, sorbitol, mannitol, glucose, and fructose are accumulated in the cells, which ensure adjustment under osmotically stressed conditions (Rawat et al., 2020).

16.5.3 Stress Proteins

In response to cold shock at ~10–15°C temperature range, synthesis of a distinct class of proteins CspA, CspC, and CspE with distinct stable secondary RNA structures (Yair et al., 2022) has been reported. Impairment of cellular freezing and ice formation or nucleation is prevented by the synthesis of a distinct class of ice nucleator inhibitors and AFPs that prevent the freezing of the cellular contents (Bagwell and Ricker, 2019). Countering such hostile and detrimental weather conditions is exhibited by significant strategic responses, including Csps and Caps proteins, altering membrane fluidity by increasing the unsaturated fatty-acid content, thus compensating for the loss and enabling endurance (Zhang and Gross 2021).

16.5.4 Genetic Adaptation Assistance

The entire gene family actively participating in restraining cold shock constitutes the *csp* gene cluster comprising *cspA*, *cspB*, *cspC*, *cspD*, and *cspE* genes synthesizing CspA, CspB, CspC, CspD, and CspE proteins, respectively. Besides cold shock tolerance, the *csp* gene family maintains other functions like regulating pathogenicity, motility, and biofilm formation (Liu et al., 2020; Yair et al., 2022). A comprehensive insight into the profile of chilling-resistant versus chilling-sensitive plants, non-inoculated controls, and bacterized plantlets demonstrates noteworthy distinctions. Stien Beirinckx et al. deciphered the effect of chilling temperatures, its consequent ability to impede the growth of *Zea mays* L. in the Northern Hemisphere (Beirinckx et al., 2020), and its abolition through the analysis of the associated root microbiome. Relieving chilling stress for thermophilic crops like maize is of utmost importance to impede the physiological disruptions affecting seed germination, root architecture, stunted growth, seedling emergence, and agronomic abnormalities. Therefore, the selection of potential PGPR strains to cope with chilling injuries to promote maize growth under chilling conditions has revealed the dominance of major plant-growth promoting rhizobacteria (PGPR) phyla in root endosphere with plant growth promoting competencies such as *Pseudoduganella* sp. RHG12 (Oxalobacteraceae) and *Bosea* sp. RHG5 (Bradyrhizobiaceae) (Beirinckx et al., 2020). The microbial consortia of the root microbiome of the maize plant and the probable functional role of the *csp* gene cluster in combating chilling stress are diagrammatically depicted in Figure 16.2. *Bradyrhizobium japonicum* in nodulated soybean plants (*Glycine max*) has been identified as a growth-promoting strain under chilling temperatures (Schmidt et al., 2015).

16.5.5 Supporting Local Crop Varieties

Microbiomes adapted to local high-altitude conditions can create bioinoculants, supporting indigenous crop varieties. *Pseudomonas fragi* CS11RH1, a high-altitude psychrotolerant, is used as a bioinoculant in the wheat seedlings of the Indian Himalayas. It augments wheat seedlings' plant biomass and nutrient uptake by increasing seed germination, germination rate, phosphate solubilization, and indole acetic acid (IAA) production ability (Selvakumar et al., 2009). High-altitude wheat rhizospheric soil from the Uttarakhand Himalayas has demonstrated the occurrence of a potent, phosphate-solubilizing bio-inoculant and plant growth-promoting isolate, *Pseudomonas lurida* M2RH3 (Selvakumar et al., 2011).

16.5.6 Enhancing Soil Fertility

Nitrogen-fixing bacteria (e.g., *Rhizobium* and *Azotobacter*) can improve nitrogen availability in nutrient-poor soils. Phosphate-solubilizing microbes can make phosphorus more accessible to plants, thereby enhancing growth. Different plant growth-promoting microbes display distinct features and help in the upregulation of crop growth.

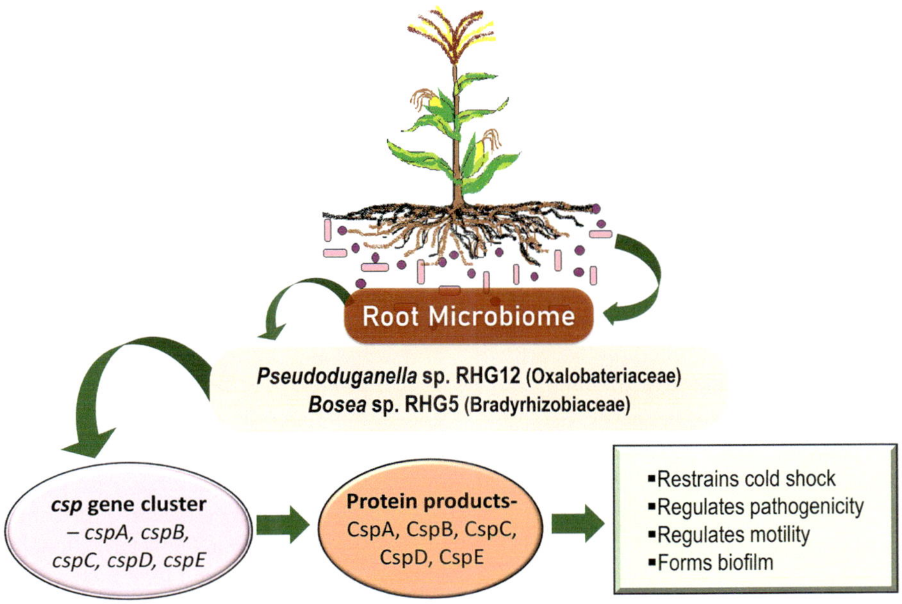

FIGURE 16.2 The microbial consortia of the root microbiome of the maize plant and the functional role of the *csp* gene cluster in combating chilling stress.

Bacillus subtilis, *Bacillus licheniformis*, *Bacillus pumilus*, *Bacillus methylotrophicus*, *Bacillus firmus*, and *Bacillus altitudinis* from Himachal Pradesh, India, have demonstrated phosphate-solubilization and other PGP traits significantly promoting the growth of tomato (*Solanum lycopersicum)* (Rawat et al., 2020). *Pseudomonas jesenii* MP1 displays phosphate-solubilization and N_2-fixation ability, isolated from the Indian Himalayan region. It significantly promotes growth in *Cicer arietinum* (Chickpea), *Vigna mungo*, and *Vigna radiata (L.)* (Rawat et al., 2020). *Lysinibacillus macrolides* ST-30 isolated from Uttarakhand (Munsyari, Kandakhal, and Nainital) demonstrated phosphate-solubilization ability and growth enhancement ability in *Cicer arietinum* (Chickpea) (Tomer et al., 2024).

16.5.7 Improving Plant Resilience to Stress

PGPB produce hormones like auxins and gibberellins, boosting plant tolerance to extreme temperatures and low oxygen levels. Endophytes (microbes living inside plants) can enhance stress tolerance by producing protective compounds. The growth of *Triticum aestivum* has shown a pronounced increase in the presence of the bacterium *Stenotrophomonas rhizophila*, with significant abilities like nitrogen fixation, IAA, siderophore, and ammonia production (Sharma et al., 2020; Soni et al., 2017). PGPB such as *Enterobacter asburae* RCE 2, *Enterobacter ludwogii* RCE 5, and *Klebsiella pneumonaiae* RCE 7 isolated from hills of Northeast India have demonstrated nitrogen fixation, IAA, phosphate solubilization, and biocontrol effects on Mandarin orange (Thokchom et al., 2014).

16.5.8 Stress Protein Production

Low temperature, variable rainfall, reduced atmospheric pressure, and soil nutritional stress are principal ecological conditions of a high altitudinal region. Adaptive responses at the cellular and molecular levels enable better adaptability through several cold tolerance mechanisms exhibited by the microbes from colder regions. The upsurge of proteins critically involved in countering chilling stress conditions includes the cold shock protein CapB, cold shock-like protein CspD, chaperone modulatory protein CbpM, chaperone protein ClpB, chaperone protein DnaJ, glycine betaine/choline transport system, stress response kinase A, stress response protein SCP 2, chaperone protein HtpG, chaperone SurA, co-chaperone protein HscB, chaperone modulatory protein CbpM, Tig, SmpB, and CspA. Cold habitats are hot spots of a huge microbial diversity of psychrophilic, psychrotolerant, and psychrotrophic microbiomes encompassing many cold adaptation genes (Sehgal and Chaturvedi 2022).

16.6 CARBON SEQUESTRATION

Some soil microbes can capture and store carbon, improving soil structure and fertility and mitigating climate change. Biotic soil components comprise micro, meso and macro biota (Samal and Mishra). An increased root-to-shoot ratio and biomass production are witnessed by the occurrence of opportunistic enrichment nematodes (*Rhabditis* and *Cervidellus)* (Laasli et al., 2022). A metagenomic and metaproteomic insight into the soil profile along an elevational gradient has deciphered an upsurge in the prevalence of genes and their corresponding enzymes involved in the carbon sequestration and degradation harbored by both bacterial and fungal microbiota such as carbon fixation genes (*pcc/acc, cbbL*, and *rbcL*) and carbon decomposition genes (*amyA, lcc, mnp, pulA, vdh*) of carbon sequestration microorganisms (Rai et al., 2021; Trivedi et al., 2016). The carbon fixation genes in microbes imply a positive correlation with their carbon sequestration potential (Zhang et al., 2023a). The *rbcL* gene encodes the large subunit of Ribulose bisphosphate carboxylase oxygenase (RuBisCo) enzyme, and the *cbbL* encodes for the large subunit of Ribulose bisphosphate carboxylase. *amyA* is an important gene implicated in carbon degradation. Chemical exudates, including primary (glucose, fructose, galactose, glycine, proline, and phenylalanine) and secondary metabolites (flavonoids, terpenoids, alkaloids, phenolics and glucosinolates) produced by roots, reframe the PGPR or other microbes that indirectly stimulate plant growth (Karnwal and Dohroo 2018).

16.6.1 Improving Crop Yield in Poor Soils and Accelerating Organic Matter Decomposition

Microbial consortia tailored to high-altitude conditions can create a favorable rhizosphere, thereby improving yield in rocky and shallow soils, alongside the decomposer microbial population, facilitating organic matter breakdown and enriching the soil with nutrients. Generally, high altitudes are habitats with extreme environmental settings, so the microbiota inhabiting therein are equally potent, resilient, and consistent with external instabilities. A phylogenetic diversity alteration amongst the low, medium and high-altitude zones is exhibited and is considered the essential criterion for selecting an organism as an inoculant for crop improvement at high altitudes. Metagenomic analysis of intestinal microbiota of low-altitude zones revealed an increase in members of Bacteroidetes and a decrease in Firmicutes with an increase in elevation (Liu et al., 2024). However, an increase in Bacteroidetes was consistent with a reduction in the population of Proteobacteria. A metagenomic insight into the gut microflora of extreme hypoxic zones revealed an increase in the Firmicutes/Bacteroidetes ratio to counteract the low oxygen levels by maintaining energy homeostasis (Lan et al., 2017). Microbial diversity and community composition are directly proportional to increasing altitude. Bacteria from high altitudes are more tolerant of stressful conditions, thus maintaining their stability, adaptation, and acclimatization. They are resilient to heat and pH changes and thereby boost the growth of plants upon inoculation by potent high-altitude bacteria following microbiome engineering, thus aiding in sustainable development (Joshi et al., 2024).

Microbiome engineering is a promising biotechnological process that assists in the upliftment of agricultural productivity and the restoration of ecological balance. It has expanded in almost every sphere, from agriculture to medicine to disease biology, the leading concerns of recent times. It emphasizes engineering the genetic makeup of microbiota, incorporating novel genes with potential applications to address challenging issues. High-altitude groups significantly display upregulated metabolic pathways and more intense quorum sensing (Gao et al., 2022). A higher proportion of Gram-negative to Gram-positive bacteria is prevalent at high-elevation localities (Kumar et al., 2019). A decline in aerobic bacteria and an upsurge in anaerobic bacteria is evident, corresponding to a drift from normoxic to hypoxic conditions. At high altitudes, the Gangotri soil of the Western Indian Himalayan region encompasses potent psychrotrophic diazotrophs. Predominant genera include members of *Cytophaga*, *Arthrobacter humicola, Brevibacillus invocatus, Pseudomonas mandelii*, and *Pseudomonas helmanticensis* (Kumar et al., 2019). Low-altitude regions, in turn, have a greater abundance of Betaproteobacteria, Bacteroidetes, and Firmicutes (Zou et al., 2022).

16.6.2 Physiological Adaptations for Cold Endurance in Bacteria

Global warming can lead to key temperature fluctuations such as extremely long winters or shortening winters in some areas, drastically affecting crop productivity (Baum 2024). Global climatic disruptions paradoxically cause significant impacts on food production (Sekar et al., 2024). Plants are susceptible to cold injury during the winters and are, therefore, endowed with specific response mechanisms such as reducing stomatal opening and mesophyll conductance, phytohormone biosynthesis, reactive oxygen species (ROS), nitric oxide (NO), as well as protein kinases, that ensure cold tolerance of plants (Abhijith Shankar et al., 2024). Additionally, the plant microbiome comprises specific PGPB with many physiological adaptations for cold endurance (Borker et al., 2024). These comprise phosphate solubilization, siderophore production, IAA production, biofertilizers, biocontrol agents, and bioremediation. *Acinetobacter*, *Bacillus*, *Enterobacter*, *Klebsiella*, *Pseudomonas*, and *Staphylococcus* produce plant growth enhancers such as ammonia, HCN, zinc solubilizing and hydrolytic enzymes (Ramírez-Cariño et al., 2024). *Pseudomonas koreensis* P2 is known to augment plant growth with growth enhancers such as IAA, siderophore, HCN, and siderophores (Ayangbenro et al., 2024). *Pseudomonas mosselii* increases proline and decreases glutathione production, thereby improving the cold tolerance of Asian rice (*Oryza sativa* L.) (Balasjin et al., 2023)

16.6.3 Bioremediation of Polluted Soils

Certain microbes decontaminate affected areas and tend to detoxify pollutants or heavy metals, thereby making soil safe for farming. Tolerance to toxic concentrations of Cr(III) and Pb(II) has been elucidated by some extremophilic fungal genera like *Stemphylium*, *Cladosporium*, and *Penicillium* from three high-altitude glaciers, Citlaltépetl and Iztaccíhuatl volcanoes in Mexico (Calvillo-Medina et al., 2020). The Arsenic-rich area of Salar de Huasco, an area located on the Chilean Altiplano, harbors genes related to arsenic tolerance, expulsion, and reduction (*arsR*, *arsC*, *arsB*, *arsP*, *arsK*, *arsJ*, and *acr3*). The *arsC* gene encodes an arsenate reductase protein that converts arsenate to arsenite, while *arsB* is responsible for removing the arsenite, thereby reducing arsenic toxicity. The *ars* operon genes are regulated in the presence of a repressor protein encoded by the *arsR* gene (Mohsin et al., 2021). Major bacterial genera carrying these arsenic-resistant gene clusters that possess mechanisms responsible for resisting arsenic are *Exiguobacterium*, *Acidithiobacillus*, *Desulfovibrio*, *Rhodohalobacter*, *Marinobacter*, *Psychroflexus*, and *Brumimicrobium* (Castro-Severyn et al., 2021).

16.6.4 Suppressing Pests and Diseases

Beneficial microbes can outcompete harmful pathogens or produce antimicrobial compounds, reducing crop losses. Examples include bacterial and fungal microbiota-mediated eradication of the tea green leafhopper (*Matsumurasca onukii*) population in high-altitude regions. Predominant bacterial genera include *Wolbachia*, *Rickettsia*, *Serratia*, *Arsenophonus*, and *Cardinium*, and the corresponding fungal genera include *Cladosporium*, *Fusarium*, *Alternaria*, and *Gibberella* (Zhang et al., 2023b).

16.6.5 Mitigating Soil Erosion

Microbial biofilms can bind soil particles together, reducing erosion caused by strong winds and runoff. Phosphate-solubilizing psychrophilic strains from the Himalaya, belonging to *Kluyvera intermedia*, *Pseudomonas corrugata*, *Pseudomonas putida*, *Pseudomonas fragi*, *Pantoea dispersa*, *Exiguobacterium acetylicum*, and *Serratia marcescens* convert phosphate from an insoluble to soluble form, which is in turn directly correlated to the amount of biofilm formation, an imperative eradicator of soil erosion (Ciccazzo et al., 2016).

16.6.6 Protection against UV Damage

Microbes producing pigments and antioxidants can help protect plant tissues from intense UV radiation at high altitudes. Shigenobu Yoshida et al. demonstrated the occurrence of UVA-absorbing *Methylobacterium* bacteria from the phyllosphere (entire above-ground plant surface) (Yoshida et al., 2017). Another extremophilic bacterium thriving in the hostile and most damaging UV radiation of Salar de Huasco, a high-altitude saline wetland in the Chilean Altiplano, is the *Rhodobacter* sp. Osmolytes such as glycine betaine and inositol function as efficient reactive oxygen scavengers to cope with UV-induced DNA damage (Pérez et al., 2017).

TABLE 16.1
The Major PGPR Acting on Specific Host Plants and Their Mechanism of Action in Upregulating Plant Growth

Sl. No.	Major PGPR	Host Plants	Mechanism of Action
1.	*Pseudomonas fragi* CS11RH1, *Pseudomonas lurida* M2RH3	*Triticum aestivum* (Wheat)	Phosphate solubilization and indole acetic acid (IAA) production ability.
2.	*Pseudomonas jesenii* MP1, *Lysinibacillus macrolides* ST-30	*Cicer arietinum* (Chickpea)	Phosphate-solubilization and N_2-fixation ability.
3.	*Stenotrophomonas rhizophila*	*Triticum aestivum* (Wheat)	IAA, siderophore, and ammonia production.
4.	*Enterobacter asburae* RCE 2, *Enterobacter ludwogii* RCE 5, and *Klebsiella pneumonaiae* RCE 7	Mandarian orange	Nitrogen fixation, IAA, phosphate solubilization.
5.	*Pseudomonas mosselii*	Asian rice (*Oryza sativa* L.)	Increases proline and decreases glutathione production.

The major PGPR with significant plant growth promoting abilities acting upon specific host plants is depicted in Table 16.1

16.7 GENETIC ENGINEERING AND GENOME EDITING IN MICROBIOME ENGINEERING FOR IMPROVING MICROBIAL STRAINS

With the advent of population expansion, the issue of food insecurity arises. Pests, diseases, and other biotic and abiotic agents negatively impact the environment. Maintaining sustainable development (to use the resources and preserve them for future generations) is essential to ensure future genetic improvements (Pathirana and Carimi 2022). Therefore, it is vital to address these hurdles and augment the development of an economy. In this context, the role of novel genome editing tools, genetic engineering, microbiome engineering, and synthetic biology would be a possible approach for bioprospecting potent microbial strains (Villena and Samolski 2022). Unveiling microbial diversity, exploiting potent microbial genera with significant combating properties to address biotic and abiotic stresses, mainly (chilling stress discussed here), is imperative to tackle the challenges (Akhtar et al., 2024). The implications of restructuring the microflora at high altitudes entail the creation of a tailored microbial consortium.

Conventional plant breeding techniques and genetic engineering are inadequate in addressing the swiftly evolving challenges. Recent advancements in novel genome editing platforms such as clustered regularly interspaced short palindromic repeats associated with CRISPR-associated protein 9 (CRISPR/Cas9), zinc finger nucleases (ZFNs) and transcription activator-like effector nucleases (TALENs) play an imperative role in mitigating significant threats posed to agriculture and thereby provide tremendous opportunities and open up to newer solutions (Pal et al., 2024). Although CRISPR/Cas9 evolved as the adaptive immune system of bacteria and archaea, it is not limited to prokaryotes. A guide RNA (gRNA) directs the Cas (CRISPR-associated) nuclease to create double-strand breaks in the DNA. In the clustered regularly interspaced short palindromic repeats (CRISPR) system, gRNA is activated by the crRNA (CRISPR RNA) and tracrRNA (trans-activating crRNA) (Liao and Beisel 2021). CRISPR/Cas technology aids in precise genetic modifications to improve crop stress tolerance and resilience. Different types of Cas nucleases initiate different targets. CRISPR/Cas9 and CRISPR/Casl2a target gRNA-initiating DSBs, while CRISPR/Casl3 targets RNA instead of DNA (Chen et al., 2024; Liao and Beisel, 2021). Chilling-sensitive plants contain saturated membrane lipids, while chilling-resistant plants acclimatize to the chilling temperatures by the change in membrane lipid composition from saturated to unsaturated fatty acids. A comprehensive approach in conferring cold tolerance includes CRISPR/Cas9-mediated upregulation of three genes of the rice plant, namely, OsPIN5b, GS3, and OsMYB30, which are altered to increase spike length and grain size and provide flexibility to cold stress (Yadav et al., 2023). C-repeat binding factor 1 (CBF1) is a novel transcription factor in tomato, and OsKASI-2 in rice confers tolerance to cold by increasing the content of unsaturated fatty acids by activating desaturase genes (Heidari, 2019; Zhang et al., 2024). AtCBF1, AtCBF2, UGT79B2, and UGT79B3 proteins were actively translated in *Arabidopsis* to provide resilience against frost damage (Caccialupi et al., 2023). Major staple crops such as rice, wheat, and maize are impaired by slight abruptions in climate change at high altitudes. CRISPR/Cas9-mediated gene editing has emerged as one of the most versatile tools in creating gene knock-out and targeted protein insertion, which provides several agronomical benefits, such as drought tolerance, salinity tolerance, nutritional enrichment, yield enhancement, and disease resistance (Hui et al., 2022). *OsBADH2* gene inhibits the synthesis of 2-acetyl-1-pyrroline (2-AP). CRISPR/Cas9-mediated knock out strategies have led to the development of aromatic rice, which is an important agronomic characteristic of rice. (Hui et al., 2022). Knock-out effects of the *TaSAL1* gene in wheat lead to improved tolerance to drought stress, since the *TaSAL1* gene is a negative regulator of drought stress (Mohr et al., 2022).

16.8 CONCLUSION

Issues pertaining to climatic disruptions are currently emerging at an extreme pace due to anthropogenic reasons and natural disasters, which necessitate immediate attention. The role of certain psychrophilic microbes in addressing the high-altitude farming problems is remarkable. Specific proteins, such as ice nucleation inhibitors, can reduce frost damage by altering ice formation on plant surfaces, thereby augmenting survival strategies in such cold niches. Cold-adapted (psychrophilic) microbes mechanistically display certain traits to thrive through harsh, cold-stressed conditions. Microbes resilient to freezing conditions develop tolerance to chilling injuries through cold acclimation, wherein microbes cope by upregulating specific physiological and metabolic responses. Thus, exploiting these indigenous microbes and generating customized tailored gene products through novel gene editing methods and modifying the microbes following a synthetic biology approach would lead toward a viable and feasible approach for sustainable agricultural improvement in the future to mitigate the issues at high altitudes.

REFERENCES

Abbas M, Sharma G, Dambire C, Marquez J, Alonso-Blanco C, Proaño K, Holdsworth MJ (2022) An oxygen-sensing mechanism for angiosperm adaptation to altitude. Nature 606(7914):565–569.

Abdullah SNA, Azzeme AM, Yousefi K (2022) Fine-tuning cold stress response through regulated cellular abundance and mechanistic actions of transcription factors. Frontiers in Plant Science 13:850216.

Abhijith Shankar P, Parida P, Bhardwaj R, Yadav A, Swapnil P, Seth CS, Meena M (2024) Deciphering molecular regulation of Reactive Oxygen Species (ROS) and Reactive Nitrogen Species (RNS) signalling networks in *Oryza* genus amid environmental stress. Plant Cell Reports 43(7):185.

Akhtar N, Shahzad A, Ilyas N, Bostan N, Jameel MA, Mukhtar S, Qin M, Li HY, Alhewairini SS (2024) Nanobiotechnology and microbial influence on cold adaptation in plants. Nanotechnology Reviews 13(1):20240059.

Aqeel M, Khalid N, Noman A, Ran J, Manan A, Hou Q, Dong L, Sun Y, Deng Y, Lee SS (2024) Interplay between edaphic and climatic factors unravels plant and microbial diversity along an altitudinal gradient. Environmental Research 242:117711

Arora S, Bhatt R, Sharma V, Hadda MS (2023) Indigenous practices of soil and water conservation for sustainable hill agriculture and improving livelihood security. Environmental Management 72(2):321–332.

Ayangbenro AS, Olanrewaju OS, Amoo AE, Babalola OO (2024) *Pseudomonas koreensis* AB36 from a gold mine: Genomic insights into adaptation to extreme conditions and potential plant growth promotion. Biologia:1–18.

Bagwell SN, Ricker JV (2019) Antifreeze proteins: Effective adaptations of organisms for low temperature survival. Bios 90(3):158–170.

Bajpai SC (2002) Lahaul-Spiti: A Forbidden Land in the Himalayas. Indus Publishing.

Balasjin NM, Maki JS, Schläppi MR (2023) *Pseudomonas mosselii* improves cold tolerance of Asian rice (*Oryza sativa* L.) in a genotype-dependent manner by increasing proline in japonica and reduced glutathione in indica varieties. Canadian Journal of Microbiology 70(1):15–31.

Baum SD (2024) Climate change, uncertainty, and global catastrophic risk. Futures 162:103432.

Beirinckx S, Viaene T, Haegeman A, Debode J, Amery F, Vandenabeele S, Nelissen H, Inzé D, Tito R, Raes J (2020) Tapping into the maize root microbiome to identify bacteria that promote growth under chilling conditions. Microbiome 8:1–13.

Bhattacharya A (2022) Low temperature stress and plant-water relationship: A review. Physiological Processes in Plants Under Low Temperature Stress:107–197.

Bhattarai B, Yadav BK, Roshan B, Chapagain S (2020) In vitro evaluation of botanical extracts against brown leaf spot (*Bipolaris oryzae*) of rice at Paklihawa, Rupandehi, Nepal. Frontiers in Plant Science 5:309–312.

Borker SS, Sharma P, Thakur A, Kumar A, Kumar A, Kumar R (2024) Physiological and genomic insights into a psychrotrophic drought-tolerant bacterial consortium for crop improvement in cold, semiarid regions. Microbiological Research 286:127818.

Bustamante CA, Monti LL, Gabilondo J, Scossa F, Valentini G, Budde CO, Lara MV, Fernie AR, Drincovich MF (2016) Differential metabolic rearrangements after cold storage are correlated with chilling injury resistance of peach fruits. Frontiers in Plant Science 7:1478.

Caccialupi G, Milc J, Caradonia F, Nasar MF, Francia E (2023) The triticeae CBF gene cluster—to frost resistance and beyond. Cells 12(22):2606.

Calvillo-Medina RP, Gunde-Cimerman N, Escudero-Leyva E, Barba-Escoto L, Fernández-Tellez EI, Medina-Tellez AA, Bautista-de Lucio V, Ramos-López MÁ, Campos-Guillén J (2020) Richness and metallo-tolerance of cultivable fungi recovered from three high altitude glaciers from Citlaltépetl and Iztaccíhuatl volcanoes (Mexico). Extremophiles 24:625–636.

Cascajo Castresana M (2017) Morphology and dynamics of ice crystals and the effect of proteins. Doctoral thesis.

Castro-Severyn J, Pardo-Esté C, Mendez KN, Fortt J, Marquez S, Molina F, Castro-Nallar E, Remonsellez F, Saavedra CP (2021) Living to the high extreme: Unraveling the composition, structure, and functional insights of bacterial communities thriving in the arsenic-rich Salar de Huasco altiplanic ecosystem. Microbiology Spectrum 9(1). https://doi.org/10.1128/spectrum.00444-00421

Chakraborty A, Saikia P (2023) Climate change impacts on the higher altitude forests of Indian Himalayan regions: Nature-based solutions for climate resilience and disaster risk reduction. The Palgrave Handbook of Socio-Ecological Resilience in the Face of Climate Change: Contexts from a Developing Country, Springer, pp. 313–328.

Chen F, Chen L, Yan Z, Xu J, Feng L, He N, Guo M, Zhao J, Chen Z, Chen H (2024) Recent advances of CRISPR-based genome editing for enhancing staple crops. Frontiers in Plant Science 15:1478398.

Chen Z, Yu G, Ge J, Sun X, Hirano T, Saigusa N, Wang Q, Zhu X, Zhang Y, Zhang J (2013) Temperature and precipitation control of the spatial variation of terrestrial ecosystem carbon exchange in the Asian region. Agricultural and Forest Meteorology 182:266–276.

Ciccazzo S, Esposito A, Borruso L, Brusetti L (2016) Microbial communities and primary succession in high altitude mountain environments. Annals of Microbiology 66:43–60.

Cid FP, Rilling JI, Graether SP, Bravo LA, Mora MdLL, Jorquera MA (2016) Properties and biotechnological applications of ice-binding proteins in bacteria. FEMS Microbiology Letters 363(11):fnw099.

Dasauni K, Nailwal TK (2020) Biodiversity of microbial life: Indian Himalayan region. Recent Advancements in Microbial Diversity, Elsevier, pp. 1–17.

Dhakal B, Kattel RR (2019) Effects of global changes on ecosystems services of multiple natural resources in mountain agricultural landscapes. Science of the Total Environment 676:665–682.

Dimri A, Allen S, Huggel C, Mal S, Ballesteros-Canovas J, Rohrer M, Shukla A, Tiwari P, Maharana P, Bolch T (2021) Climate change, cryosphere and impacts in the Indian Himalayan Region. Current Science 120:774–790.

Esitken A (2011) Use of plant growth promoting rhizobacteria in horticultural crops. Bacteria in Agrobiology: Crop Ecosystems, Springer, pp. 189–235.

Fischer R, Connor D (2018) Issues for cropping and agricultural science in the next 20 years. Field Crops Research 222:121–142.

Fu F, Li J, Li S, Chen W, Ding H, Xiao S, Li Y (2023) Elevational distribution patterns and drivers of soil microbial diversity in the Sygera Mountains, southeastern Tibet, China. Catena 221:106738.

Gao W, Liu P, Ye Z, Zhou J, Wang X, Huang X, Deng X, Ma L (2022) Divergent prokaryotic microbial assembly, coexistence patterns and functions in surrounding river sediments of a Cu-polymetallic deposit in Tibet. Science of The Total Environment 851:158192.

Geitner C, Mayr A, Rutzinger M, Löbmann MT, Tonin R, Zerbe S, Wellstein C, Markart G, Kohl B (2021) Shallow erosion on grassland slopes in the European Alps–Geomorphological classification, spatio-temporal analysis, and understanding snow and vegetation impacts. Geomorphology 373:107446.

Gumisiriza MS, Ndakidemi P, Nalunga A, Mbega ER (2022) Building sustainable societies through vertical soilless farming: A cost-effectiveness analysis on a small-scale non-greenhouse hydroponic system. Sustainable Cities and Society 83:103923.

Hammond W, Morris H (2022) The global vulnerability of plant xylem.

Heidari P (2019) Comparative analysis of C-repeat binding factors (CBFs) in tomato and *Arabidopsis*. Brazilian Archives of Biology and Technology 62:e19180715.

Hossain A, Krupnik TJ, Timsina J, Mahboob MG, Chaki AK, Farooq M, Bhatt R, Fahad S, Hasanuzzaman M (2020) Agricultural land degradation: Processes and problems undermining future food security. Environment, Climate, Plant and Vegetation Growth, Springer, pp. 17–61.

Hui S, Li H, Mawia AM, Zhou L, Cai J, Ahmad S, Lai C, Wang J, Jiao G, Xie L (2022) Production of aromatic three-line hybrid rice using novel alleles of BADH2. Plant Biotechnology Journal 20(1):59–74.

Jaganathan GK, Dalrymple SE, Pritchard HW (2020) Seed survival at low temperatures: A potential selecting factor influencing community level changes in high altitudes under climate change. Critical Reviews in Plant Sciences 39(6):479–492.

Joshi D, Suyal DC, Singh J, Maithani D, Rajwar J, Preeti, Singh M, Singh D (2024) Plant growth-promoting psychrotrophic microbes: Current research and future challenges. Microbial Inoculants: Applications for Sustainable Agriculture, Springer, pp. 247–280.

Karnwal A, Dohroo A (2018) Effect of maize root exudates on indole-3-acetic acid production by rice endophytic bacteria under influence of L-tryptophan. F1000Research 7.

Kassmannhuber J, Rauscher M, Schöner L, Witte A, Lubitz W (2017) Functional display of ice nucleation protein InaZ on the surface of bacterial ghosts. Bioengineered 8(5):488–500.

Khokhar A, Yousuf A, Singh M, Sharma V, Sandhu PS, Chary GR (2021) Impact of land configuration and strip-intercropping on runoff, soil loss and crop yields under rainfed conditions in the Shivalik foothills of north-west, India. Sustainability 13(11):6282.

Kumar S, Suyal DC, Yadav A, Shouche Y, Goel R (2019) Microbial diversity and soil physiochemical characteristic of higher altitude. PLoS One 14(3):e0213844.

Kumawat KC, Sharma P, Nagpal S, Gupta R, Sirari A, Nair RM, Bindumadhava H, Singh S (2021) Dual microbial inoculation, a game changer?–bacterial biostimulants with multifunctional growth promoting traits to mitigate salinity stress in spring mungbean. Frontiers in Microbiology 11:600576.

Kurien AJ, Lele S, Nagendra H (2019) Farms or forests? Understanding and mapping shifting cultivation using the case study of West Garo hills, India. Land 8(9):133.

Laasli S-E, Mokrini F, Lahlali R, Wuletaw T, Paulitz T, Dababat AA (2022) Biodiversity of nematode communities associated with wheat (*Triticum aestivum* L.) in Southern Morocco and their contribution as soil health bioindicators. Diversity 14(3):194.

Lan D, Ji W, Lin B, Chen Y, Huang C, Xiong X, Fu M, Mipam TD, Ai Y, Zeng B (2017) Correlations between gut microbiota community structures of Tibetans and geography. Scientific Reports 7(1):16982.

Liao C, Beisel CL (2021) The tracrRNA in CRISPR biology and technologies. Annual Review of Genetics 55(1):161–181.

Lin M, Cao H, Li J (2023) Control strategies of ice nucleation, growth, and recrystallization for cryopreservation. Acta Biomaterialia 155:35–56.

Liu D, Chen D, Xiao J, Wang W, Zhang L-J, Peng H, Han C, Yao H (2024) High-altitude-induced alterations in intestinal microbiota. Frontiers in Microbiology 15:1369627.

Liu X, Zhou Y, Xiao J, Bao F (2018) Effects of chilling on the structure, function and development of chloroplasts. Frontiers in Plant Science 9:1715.

Liu Y, Tan X, Cheng H, Gong J, Zhang Y, Wang D, Ding W (2020) The cold shock family gene cspD3 is involved in the pathogenicity of *Ralstonia solanacearum* CQPS-1 to tobacco. Microbial Pathogenesis 142:104091.

Lupatini M, Korthals GW, De Hollander M, Janssens TK, Kuramae EE (2017) Soil microbiome is more heterogeneous in organic than in conventional farming system. Frontiers in Microbiology 7:2064.

Lyngwi NA, Koijam K, Sharma D, Joshi S (2013) Cultivable bacterial diversity along the altitudinal zonation and vegetation range of tropical Eastern Himalaya. Revista de Biología Tropical 61(1):467–490.

Maity A, Vijay D, Mukherjee A, Lamichaney A (2016) Potential impacts of climate change on quality seed production: A perspective of hill agriculture. Conservation Agriculture: An

Approach to Combat Climate Change in Indian Himalaya, Springer, pp. 459–485.

Manglani O (2023) Cultivating hope: Situating place-based education, health sovereignty, and climate resiliency in Spiti Valley, India.

Mishra P, Rai S (2013) Use of indigenous soil and water conservation practices among farmers in Sikkim Himalaya.

Mishra PK, Bisht SC, Pandey B, Meena V, Parihar M, Mahanta D, Bisht J, Pattanayak A (2020) Plant growth promoting rhizobacteria: Mechanisms and alleviation of cold stress in plants. Plant Stress Biology: Strategies and Trends:185–226.

Mohr T, Horstman J, Gu YQ, Elarabi NI, Abdallah NA, Thilmony R (2022) CRISPR-Cas9 gene editing of the Sal1 gene family in wheat. Plants 11(17):2259.

Mohsin H, Shafique M, Rehman Y (2021) Genes and biochemical pathways involved in microbial transformation of arsenic. Arsenic Toxicity: Challenges and Solutions:391–413.

Moore J, Gardiner B, Sellier D (2018) Tree mechanics and wind loading. Plant Biomechanics: From Structure to Function at Multiple Scales:79–106.

Nath AJ, Sahoo U, Giri K, Sileshi G, Das A (2020) Incentivizing hill farmers for promoting agroforestry as an alternative to shifting cultivation in Northeast India. Agroforestry for Degraded Landscapes: Recent Advances and Emerging Challenges 1:425–444.

Normatov PI, Normatov I, Karimzoda A, Eshankulova R, Normatov A, Shermatov N (2022) Hydrochemistry of high-mountain Pamir rivers: Formation, changes and influence on the chemical composition of irrigation water. Journal of Environmental Science and Sustainable Development 5(1):19–47.

Orefice J, Carroll J, Conroy D, Ketner L (2017) Silvopasture practices and perspectives in the Northeastern United States. Agroforestry Systems 91:149–160.

Pal N, Nayak JK, Yadav RR, Tyagi S, Joshi P, Kumar S (2024) Potential applications of CRISPR/Cas9 mediated genome-editing approaches in improvements of cereals. Omics and System Biology Approaches for Delivering Better Cereals, CRC Press, pp. 111–132.

Pathania S, Solanki P, Putatunda C, Bhatia RK, Walia A (2022) Adaptation to cold environment: The survival strategy of Psychrophiles. Survival Strategies in Cold-Adapted Microorganisms:87–111.

Pathirana R, Carimi F (2022) Management and utilization of plant genetic resources for a sustainable agriculture. Plants 11(15):2038.

Pearce DA (2017) Extremophiles in antarctica: Life at low temperatures. Adaption of Microbial Life to Environmental Extremes: Novel Research Results and Application, Springer, pp. 99–131.

Peng S-l, Ge Z-w, Liu G-c, Mao L-f (2022) Environmental drivers of soil microbial activity and diversity along an elevational gradient. Journal of Mountain Science 19(5):1336–1347.

Pérez V, Hengst M, Kurte L, Dorador C, Jeffrey WH, Wattiez R, Molina V, Matallana-Surget S (2017) Bacterial survival under extreme UV radiation: A comparative proteomics study of *Rhodobacter* sp., isolated from high altitude wetlands in Chile. Frontiers in Microbiology 8:1173.

Prado FE, Rosa M, Prado C, Podazza G, Interdonato R, González JA, Hilal M (2012) UV-B radiation, its effects and defense mechanisms in terrestrial plants. Environmental Adaptations and Stress Tolerance of Plants in the Era of Climate Change, pp. 57–83.

Quine CP, Gardiner BA, Moore J (2021) Wind disturbance in forests: The process of wind created gaps, tree overturning, and stem breakage. Plant Disturbance Ecology, Elsevier, pp. 117–184.

Rai S, Solanki AC, Anal AKD (2021) Modern biotechnological tools: An opportunity to discover complex phytobiomes of horticulture crops. Microbiomes and Plant Health:85–124.

Ramírez-Cariño HF, Morales I, Guadarrama-Mendoza PC, Tlilayatzi-Bello AA, Valadez-Blanco R (2024) Beneficial traits of crop-associated rhizobacteria toward plant growth promotion, biofertilization, and nutrient enhancement of harvested agro-products. Bacterial Secondary Metabolites, Elsevier, pp. 299–334.

Rawat J, Yadav N, Pande V (2020) Role of rhizospheric microbial diversity in plant growth promotion in maintaining the sustainable agrosystem at high altitude regions. Recent Advancements in Microbial Diversity, Elsevier, pp. 147–196.

Rizvi A, Ahmed B, Khan MS, Umar S, Lee J (2021) Psychrophilic bacterial phosphate-biofertilizers: A novel extremophile for sustainable crop production under cold environment. Microorganisms 9(12):2451.

Rustama R, Hussaina S, Rustamb T, Ishtiaqc R Importance of pseudomonas syringae as biological aerosol in clouds and its environmental effect.

Saiz-Fernández I, De Diego N, Brzobohatý B, Muñoz-Rueda A, Lacuesta M (2017) The imbalance between C and N metabolism during high nitrate supply inhibits photosynthesis and overall growth in maize (Zea mays L.). Plant Physiology and Biochemistry 120:213–222.

Samal S, Mishra C Soil micro and Meso Fauna as effective tools for diagnosis of biotic and abiotic stress. Emerging Issues in Environmental Science:37.

Sati VP (2024) Farming Systems and Sustainable Agriculture in the Himalaya. Springer.

Schmidt J, Messmer M, Wilbois K-P (2015) Beneficial microorganisms for soybean (*Glycine max* (L.) Merr), with a focus on low root-zone temperatures. Plant and Soil 397:411–445.

Sedej TT, Erznožnik T, Rovtar J (2020) Effect of UV radiation and altitude characteristics on the functional traits and leaf optical properties in *Saxifraga hostii* at the alpine and montane sites in the Slovenian Alps. Photochemical & Photobiological Sciences 19:180–192.

Sehgal P, Chaturvedi P (2022) Omic technologies and cold adaptations. Survival Strategies in Cold-Adapted Microorganisms, Springer, pp. 253–284.

Sekar A, Sharma R, Mahajan A (2024) Synergistic impact of air pollution and heat on health and economy in India. The Climate-Health-Sustainability Nexus: Understanding the Interconnected Impact on Populations and the Environment, Springer, pp. 19–50.

Selvakumar G, Joshi P, Nazim S, Mishra PK, Bisht JK, Gupta HS (2009) Phosphate solubilization and growth promotion by *Pseudomonas fragi* CS11RH1 (MTCC 8984), a psychrotolerant bacterium isolated from a high altitude Himalayan rhizosphere. Biologia 64:239–245.

Selvakumar G, Joshi P, Suyal P, Mishra PK, Joshi GK, Bisht JK, Bhatt JC, Gupta HS (2011) *Pseudomonas lurida* M2RH3 (MTCC 9245), a psychrotolerant bacterium from the Uttarakhand Himalayas, solubilizes phosphate and promotes wheat seedling growth. World Journal of Microbiology and Biotechnology 27:1129–1135.

Sharma V, Salwan R, Al-Ani LKT (2020) Molecular aspects of plant beneficial microbes in agriculture. Academic Press.

Sinha N (2019) Psychrotrophic bacteria: A Boonto higher altitude agriculture. Indian Journal of Agricultural Biochemistry 32(1):10–22.

Soni R, Kumar A, Kanwar SS, Pabbi S (2017) Efficacy of liquid formulation of versatile rhizobacteria isolated from soils of the North-Western Himalayas on Solanum lycopersicum. Indian Journal of Experimental Biology (IJEB) 55:690–698.

Souza RD, Ambrosini A, Passaglia LM (2015) Plant growth-promoting bacteria as inoculants in agricultural soils. Genetics and Molecular Biology 38:401–419.

Suyal DC, Joshi D, Kumar S, Bhatt P, Narayan A, Giri K, Singh M, Soni R, Kumar R, Yadav A (2021) Himalayan microbiomes for agro-environmental sustainability: Current perspectives and future challenges. Microbial Ecology:1–33.

Thokchom E, Kalita MC, Talukdar NC (2014) Isolation, screening, characterization, and selection of superior rhizobacterial strains as bioinoculants for seedling emergence and growth promotion of Mandarin orange (*Citrus reticulata* Blanco). Canadian Journal of Microbiology 60(2):85–92.

Tomer S, Khati P, Suyal DC, Perveen K, Khan F, Barasarathi J (2024) Relative multi-beneficial effect of MOs on plant health of chickpea (*Cicer arietinum* L. var. PG-186). Frontiers in Microbiology 15:1452553.

Trivedi P, Batista BD, Bazany KE, Singh BK (2022) Plant–microbiome interactions under a changing world: Responses, consequences and perspectives. New Phytologist 234(6): 1951–1959.

Trivedi P, Trivedi C, Grinyer J, Anderson IC, Singh BK (2016) Harnessing host-vector microbiome for sustainable plant disease management of phloem-limited bacteria. Frontiers in Plant Science 7:1423.

Uko MP, Umana SI, Iwatt IJ, Udoekong NS, Mgbechidinma CL, Adie FU, Akan OD (2024) Microbial ice-binding structures: A review of their applications. International Journal of Biological Macromolecules:133670.

Upadhaya K, Barik S, Kharbhih VM, Nongbri G, Debnath G, Gupta A, Ojha A (2020) Traditional bun shifting cultivation practice in Meghalaya, Northeast India. Energy, Ecology and Environment 5:34–46.

Villena GK, Samolski I (2022) Bioprospecting and genomic-based biotechnology for economic use of Agro-Industrial Wastes. Sustainable Microbial Technologies for Valorization of Agro-Industrial Wastes, CRC Press, pp. 31–49.

Wu J (2018) Environmental gradients drive biogeographic patterns in soil microbial communities. ResearchSpace@ Auckland.

Yadav RK, Tripathi MK, Tiwari S, Tripathi N, Asati R, Chauhan S, Tiwari PN, Payasi DK (2023) Genome editing and improvement of abiotic stress tolerance in crop plants. Life 13(7):1456.

Yair Y, Michaux C, Biran D, Bernhard J, Vogel J, Barquist L, Ron EZ (2022) Cellular RNA targets of cold shock proteins CspC and CspE and their importance for serum resistance in septicemic *Escherichia coli*. Msystems 7(4):e00086–e00022.

Yosef BA, Asmamaw DK (2015) Rainwater harvesting: An option for dry land agriculture in arid and semi-arid Ethiopia. International Journal of Water Resources and Environmental Engineering 7(2):17–28.

Yoshida S, Hiradate S, Koitabashi M, Kamo T, Tsushima S (2017) Phyllosphere *Methylobacterium* bacteria contain UVA-absorbing compounds. Journal of Photochemistry and Photobiology B: Biology 167:168–175.

Zahra N, Shaukat K, Hafeez MB, Raza A, Hussain S, Chaudhary MT, Akram MZ, Kakavand SN, Saddiq MS, Wahid A (2021) Physiological and molecular responses to high, chilling, and freezing temperature in plant growth and production: Consequences and mitigation possibilities. Harsh Environment and Plant Resilience: Molecular and Functional Aspects, Springer, pp. 235–290.

Zhang L, Wang S, Bai B, Chen Y, Xiang Z, Chen C, Kuang X, Yang Y, Fu J, Chen L (2024) OsKASI-2 is required for the regulation of unsaturation levels of membrane lipids and chilling tolerance in rice. Plant Biotechnology Journal 22(8):2157–2172.

Zhang N, Chen K, Wang S, Qi D, Zhou Z, Xie C, Liu X (2023a) Dynamic response of the cbbL carbon sequestration microbial community to Wetland type in Qinghai Lake. Biology 12(12):1503.

Zhang Y, Gross CA (2021) Cold shock response in bacteria. Annual Review of Genetics 55(1):377–400.

Zhang Y, Liu S, Huang X-y, Zi H-b, Gao T, Ji R-j, Sheng J, Zhi D, Zhang Y-l, Gong C-m (2023b) Altitude as a key environmental factor shaping microbial communities of tea green leafhoppers (*Matsumurasca onukii*). Microbiology Spectrum 11(6):e01009–01023.

Zhang Y-Y, Wu W, Liu H (2019) Factors affecting variations of soil pH in different horizons in hilly regions. PLoS One 14(6):e0218563.

Zou Z, Yuan K, Ming L, Li Z, Yang Y, Yang R, Cheng W, Liu H, Jiang J, Luan T (2022) Changes in alpine soil bacterial communities with altitude and slopes at mount Shergyla, Tibetan Plateau: Diversity, structure, and influencing factors. Frontiers in Microbiology 13:839499.

17 Microbiome in Crop Improvements at High Temperature

Rehab M. Hafez[1]*, *Aya A. Mostafa*[2], *Rana Y. Maher*[1], *and Mohamed S. Zaki*[1]
[1]Botany and Microbiology Department, Faculty of Science, Cairo University, Giza 12613, Egypt
[2]Biotechnology Department, Faculty of Science, Cairo University, Giza 12613, Egypt

17.1 INTRODUCTION

Climate change and rising greenhouse gas (GHG) emissions are unprecedentedly threatening agricultural productivity, food security, and environmental sustainability (Gourdji et al., 2013; Calvin et al., 2023). Among all climatic stresses, high temperature is most damaging because it disrupts the physiological processes of plants, causes pollens to become non-viable, promotes senescence, and impairs symbiotic processes between plants and their respective microbiomes (Higashi et al. 2015; Lamaoui et al. 2018). The plant microbiome consists of intricate communities of fungi, bacteria, archaea, protists, and viruses that are tasked with playing a pivotal role in plant survival under such stress by stimulating nutrient uptake, phytohormone balance, and the triggering of antioxidant processes (Berg et al., 2016).

The holobiont worldview emphasizes that plant health and productivity cannot be explained by host genes in isolation but need to be evaluated in conjunction with the microbiota and their interactions (Vandenkoornhuyse et al., 2015; Sánchez-Cañizares et al., 2017). Heat stress alters endophytic and rhizospheric microbial composition and function, typically suppressing beneficial taxa and promoting opportunists and hence compromising resilience and nutrient cycling (Corné M.J. Pieterse et al., 2014a; Hassani et al., 2018). Nevertheless, certain microbial populations produce heat shock proteins (HSPs), exopolysaccharides (EPS), volatile organic compounds (VOCs), and secondary metabolites that facilitate plant resistance against heat and oxidative stress (Canarini et al., 2019; L. Wu et al., 2023) Microbiomes can be utilized recently for thermotolerance through: (1) bioinoculant application, thermotolerant strains of *Bacillus*, *Pseudomonas*, and *Azospirillum* with improved antioxidant activity, nutrient provision, and gene expression of stress-related activities (Abd El-Daim et al., 2019; Ali et al., 2021); and (2) plant microbiome engineering, the use of omics tools, synthetic biology, and CRISPR–Cas technology to design host–microbe interfaces for successful adaptation against stress (Trivedi et al., 2020; Asghar et al., 2024).

17.2 PLANT MICROBIOME COMMUNITY

Plant health and fitness are the capabilities of the plant as a living organism to survive, endure, and procreate in a particular habitat, and this fitness is closely linked with the plant-tissue physiological process, which is strongly correlated to the plant microbiota (da Silva Cruz et al., 2024). Plant microbiota refers to the reservoir of microorganisms that inhabit the plant environment. Plants do not exist as separate entities; they are present in association with a wide range of microorganisms, whether inside (endosphere) or surrounding (episphere) of plant tissues (Berendsen et al., 2012; Bulgarelli et al., 2012; Zhang et al., 2023). These microorganisms, as constituents of microbiota, are in continuous interaction with one another and the host plant itself (Palmer and Foster 2022). This reservoir comprises many living organisms of different kingdoms, like fungi, bacteria, protozoa, algae, and even viruses; among these microorganisms, bacteria and fungi are the most predominant. Thus, the host plant's tendency to acquire microorganisms into its microbiome is taxon-specific, as it is a multifactorial process and often is governed by specific factors like sharing a typical evolutionary trajectory between the two organisms (Mishra et al., 2020). There are many ways to detect, observe, identify, and investigate the existence of such microorganisms in plants, either by isolation and culturing or by detecting signals like many compounds (e.g., hormones such as salicylic acid, jasmonate, and ethylene) as an indicator of plant and microbes' integration (Robert-Seilaniantz et al., 2011; Corné M J Pieterse et al., 2014b).

Out of the strong interdependence, coevolutionary process, and correlation between plants and their microbiota, which even extends to their genomic level, both plants and their microbiome are treated as meta-organisms (Berg et al., 2016). The term "holobiont" was coined to describe this complex symbiotic relation; since plant fitness could not be determined only by the plant itself, there are still other factors, like its microbiota and their interactions, which both have a solid influence on determining the plant's fitness and health (Vandenkoornhuyse et al., 2015). So, the host plant with its microbiome (whole gene composition of the plant

***Corresponding author:** rehabhafez@sci.cu.edu.eg (Rehab M. Hafez)

DOI: 10.1201/9781003544425-17

microbiota) constitutes the plant holobiont as a new entity (Sánchez-Cañizares et al., 2017). However, a comprehensive understanding of plant physiological processes, performance, and proliferation needs to be considered concerning the new holobiont entity (Hacquard, 2016). Numerous studies and research have recently confirmed the evolution and adaptation of the symbiotic relationships between holobiont constituents, highlighting the strong association between the host plant and the microbiome (Uroz et al., 2019).

Generally, both components of the holobiont, which are the host plant and its microbiome, are responsible for shaping the plant community's performance and abundance (Hassani et al., 2018; Simon et al., 2019). Though recruiting microorganisms into the plant ecosystem is beneficial to plant functioning and fitness, as the microbiome is recognized to improve plant response toward biotic and abiotic conditions, in addition to enhancing plant development and proliferation (Compant et al., 2019).

There are many factors affecting the plant microbiome community structure formulation. Soil serves as the primary reservoir and source of the microbial community in the plant microbiome ecosystem, so it is responsible for shaping the plant microbiome community; despite admitting that plant microbiota is mainly derived from soil microbial reservoir, it takes the host plant a long time to recruit the soil microbiota to the desired abundance and prevalence (Yurgel et al., 2018; Rodriguez et al., 2019). The microorganisms themselves are a decisive parameter in influencing the microbiome structure that inhabits the plant based on microbial aspects and attributes. Certain organic matters that are secreted through root exudates surrounding the plant have a crucial role in attracting and recruiting specific species of the population. This process is highly reliant on several factors, including the plant's genetic makeup, the diversity and performance of the microbial population, and biotic and abiotic stressors (Trivedi et al., 2020; Park et al., 2023). Intensifying the studies and research conducted to investigate microbiome diversity with a few plants revealed some precise and improved figures of microbiome associated with some selected plant species, such as *Arabidopsis thaliana*, *Zea mays*, *Oryza sativa*, *Picea abies*, and *Populus deltoides* (Cregger et al., 2018; Terhonen et al., 2019).

Although many parameters are involved in microbiome structuring, many researchers assumed the presence of plant core microbiomes, which refers to the standard microbial community of plants. For instance, groups of Proteobacteria, Actinobacteria, and Acidobacteria are the most abundant phyla that constitute the core microbiome of the rhizosphere of grapevine plants (Xu et al., 2018), similar to the core microbiome of the poplar Plant and the core microbiome of *Citrus* (Zhang et al., 2021). Like the core microbiome concept, the microbiome functional core concept evolved as a term that refers to microbes performing the same specific function regardless of taxonomic similarity. The cumulative impact of microbial features and characteristics led to any microorganism of the community carrying out the same functions for the plant, irrespective of the microbial species type (Lemanceau et al., 2017).

Our understanding of plant microbiome structure, composition, and function has significantly risen owing to the massive advancements in revealing knowledge of gene levels, capitalizing on molecular techniques, sequencing technologies, bioinformatics, and metagenomics. Plant microbiomes' phylogenetic abundance and biodiversity could be comprehensively understood by applying metagenomics by sequencing specific marker genes from the whole microbiota structure. Additionally, it advances our understanding and clarifies the variables and parameters that influence and drive microbiome structure, compositions, and functions, whether they are biotic factors like organisms' prevalence and interactions or they are abiotic variables like temperatures, pH, salinity, drought, etc. (Bulgarelli et al., 2013). The metagenomic analysis reveals more insight into the microbiome of the holobiont (host plant and microbes), which is shown to be more adaptive and reactive to ecosystem challenges than the sole plant genome (Rosenberg and Zilber-Rosenberg 2018). While metagenomics reveals a comprehensive understanding of the community phenotypes compared to the conventional culturable way, metatranscriptomics, metaproteogenomics, and metabolomics give a strong picture of microbial community functioning and performance (Lebeis 2015).

17.3 CLIMATE CHANGE AND THE GLOBAL WARMING PHENOMENA

The climate of a given area is defined as the long-term normal weather conditions of this place. For instance, Hawaii, as a specific region, has hot and sunny weather, while Antarctica experiences bitterly frigid temperatures. Based on that, any alteration in the defined weather condition related to a specific region is defined as a change in climate, though the Intergovernmental Panel on Climate Change (IPCC) reported climate change as "altering in the typical condition of the weather that can be noticed and recognized by changes in the average and/or the variance of its attributes and this change last for a long time" (Allen et al., 2018; Santos and Bakhshoodeh, 2021).

As a result of human accumulative industrial works over centuries, mainly since the Industrial Revolution until now, with excessive dependence on fossil fuels, in addition to other many factors like deforestation and decreasing the green covered land, altogether led to a steady rise in the amount of GHGs (mainly carbon dioxide, and nitrous oxide, methane, chlorofluorocarbons) released into the atmosphere while primarily increasing the carbon dioxide amount that has been immitted to the earth atmosphere act as insulator of heat and radiation that have been reflected from earth after sunlight absorbance. This process resulted in a gradual increase in land and ocean temperatures on Earth, which is being dealt with as a frightening pattern of global warming. As a result of global warming, the earth's weather is dramatically altered, a phenomenon known as "climate change." This shift in the worldwide climate turned out to increase the potentiality, intensity, incidences, and magnitudes of heat waves and many abiotic stressors like droughts, salinity, and freezing stresses (Bigot et al., 2018; Bailey-Serres et al., 2019; Anderson and Song, 2020; Grossiord et al., 2020).

The scientific modules set to figure out climate change warned of the potential of temperature rise in the range of 2–4°C over the next several decades (Solomon, 2007).

It is worth mentioning that the global temperature range from the 30s to the 60s of the last century was reported to have declined, while it significantly climbed from the 70s of the previous century to the 14th of the current century. According to the IPCC, the Earth's mean surface temperature has increased by 1.1°C only in the second decade of this century compared to the last 50 years of the 19th century (Kogan, 2023). The standard worldwide temperature is expected to increase by 1.5°C between the 20s and the 40s of this century, according to the most recent assessment from the Intergovernmental Panel on Climate Change (Kogan 2023). The global temperature increase is driven by GHS emissions, mainly carbon dioxide and nitrous oxide, which are crucial pillars. In the first decade of this century and until the very beginning of the second decade, forest restoration activities in the tropical region controlled the amount of carbon dioxide emitted. Friedlingstein et al. (2020) recognized the annual carbon exchange flux not to exceed one billion metric tons annually, considering 3 PgC/yr^1 emission volume out of deforestation countered by the amount of 2 PgC/yr^1 produced from reforestation according to Pachauri et al. (2014).

Although approximately 6% of global warming is attributed to nitrous oxide, it contributes a comparatively smaller amount than carbon dioxide to the atmosphere's composition. Compared to carbon dioxide, only one nitrous oxide molecule has 265 times more heating effect than the carbon dioxide molecule in the atmosphere (Change, 2013). Moreover, nitrous oxide plays a significant role in decomposing O_3 molecules in the atmosphere's stratospheric layer (Change, 2013; Abalos et al., 2018). Although the majority of studies focus on plant development and proliferation periods to study the GHG emission rate and percentages, the time of the freeze–thaw cycle (FTC) plays a substantial role in drawing a comprehensive picture of GHG emission patterns and influencing crop plant growth in the pre-culturing season (Urakawa et al., 2014), as releasing both gases during FTC time represents a reasonable portion of the total annual emission of both gases from the Earth. The spring season of the FTC reported an increase in the percentage of released nitrous oxide gas in temperate forest and grassland ecosystems (Wu et al., 2020; Ejack and Whalen, 2021; Mander et al., 2021). Twenty to ninety percent of total nitrous oxide gas annually released during the FTC period comes from the soil of cultivated land (Jiang et al., 2021). On the other side, only 1–5% of the total annual carbon released comes from the same land during the same period (Yan et al., 2021).

17.4 IMPACTS OF GLOBAL WARMING ON THE QUALITY AND YIELDS OF CROPS

Global warming has a harsh effect on plants as the core constituent of the food cycle. Global warming leads to high deposition of N_2, rising Earth temperature, and CO_2 trapping in the atmosphere. In addition to changing the rain pattern, it alters the cultivated plants' growth and fitness by disturbing their physiological process and proliferation (Higashi et al., 2018). It is proven that the addition of N_2 supports the vegetative biomass of cultivated crop plants (Feng et al., 2023), while heating promotes woody plant development and inhibits the non-vascular small flowerless green plants (Bao et al., 2022). Global warming significantly impacts the cultivated crop plant physiological process, crop quality and yields, and the quality of human life and health (W. Wang et al., 2021; J. Wu et al., 2023).

17.4.1 Food Crops

According to IPCC forecasting, worldwide environmental warming is going to continue, leading to a rise in the probability of extreme climate incidents, notably the scarcity of water supply, and extreme global temperature increases, which will have significant negative consequences on crop yields in most places globally (Lamaoui et al., 2018; Lee et al., 2023). The alteration in the global climate, mainly the drought and heat waves, would reduce the yield of cultivated food crops, threatening the global food supply and availability (Figure 17.1).

The term "food security" refers to the human accessibility and affordability toward adequate, needed food that is available all of the time with enough amount to leverage a good quality of life quality (Nguyen, 2002). The adverse events of global environmental alteration resulted in negative impacts on the food crop yield, which risked food security globally (Zhu et al., 2022). Commonly, the dried and semi-arid locations of water supply have insufficient supply, resulting in reductions in cultivated crop quality and quantities that lead to negative impacts on people settling in such areas (McGuire 2017). So, securing and supporting an incremental rate of crop yield in agricultural land is becoming critical to ensure food security. It is worth mentioning that climate change poses significant negative impacts on arid areas, while it is anticipated to enhance agricultural production in the middle and northern latitude areas of the Northern Earth (Yang et al., 2015; Zhu et al., 2022).

The global warming and environmental condition alteration that negatively impacted the crop plants influence the plant fitness and growth through altering the physiological functions; for instance, drought is considered among the most notable abiotic stressors that affect crop plants, as it poses devastating impacts on yield quality and quantity (Zhang et al., 2022). The past 10 years have witnessed a 15% decline in grain crop yield due to drought conditions. While in Europe, drought conditions are projected to cause a 30% decline in wheat crop outcomes (Khan et al., 2019; Zahra et al., 2023). Shortage of water supply causes severe damage across the seedling phase, disrupting the photosynthesis functions, shrinking leaves, closing stomata, reducing the chlorophyll contents, decomposing cell components, and decreasing the plant's total weight content (Khan et al., 2019; Khan et al., 2021).

The most water-sensitive cellular function is photosynthesis because of the electron transport process, impacting, in addition to the physiological response of closing the stomata,

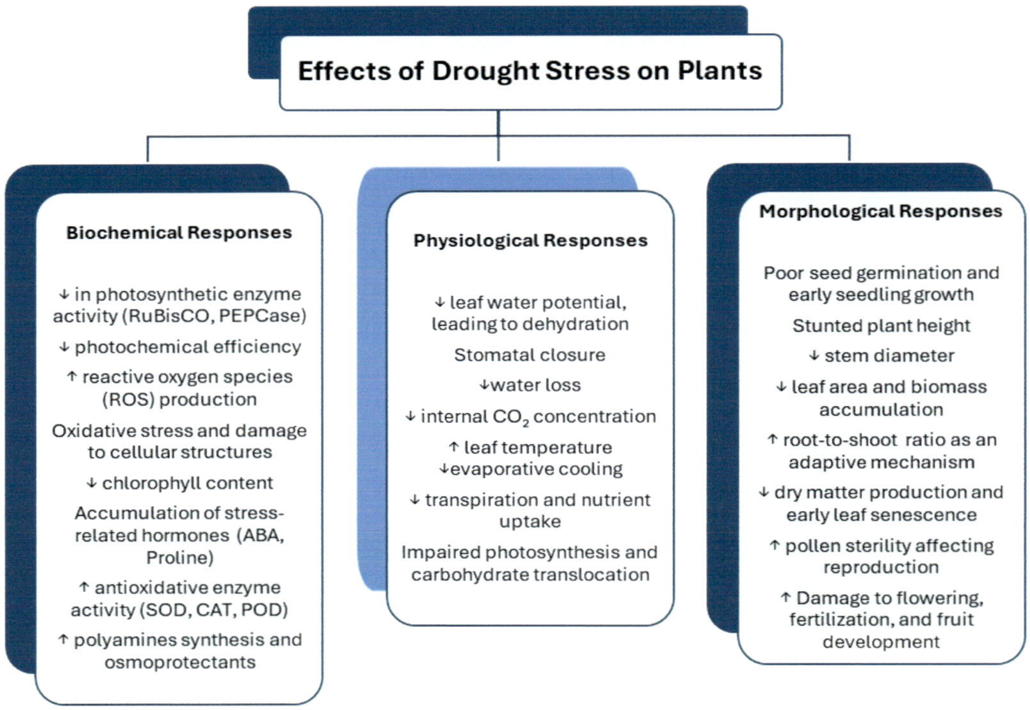

FIGURE 17.1 The biochemical, physiological and morphological changes in plants are caused by drought stress.

Source: Adapted from de Faria Melo et al. (2025).

that lowers the capability of photosynthesis (Ahmad et al., 2021; Wang et al., 2021). Water shortages stimulate the formation of reactive oxygen species (ROS), which will disturb the cellular balance state between ROS and the antioxidant molecules, thus leading to the degradation of the cell membrane and total cellular organelles (Afridi et al., 2022).

The cereal food crop of *Triticum aestivum* L. comes in the second rank of the most fundamental crop globally. The wheat crop occupies vast land on earth, exceeding 200 million hectares and yielding more than 780 million metric tons. It is almost being cultivated everywhere on the planet, including temperate to semi-arid climatic conditions (Shahbandeh and Elhag 2024). It contributes 20% of the worldwide consumption of essential human nourishment constituents, carbohydrates, and protein. Around 70% of the wheat grains are polysaccharides, and 7–20% consist of amino acids, giving wheat crops great economic importance. The concentration of the grains' total protein and carbohydrate content is correlated to inherent traits, growing patterns, and climatic factors (Lin et al., 2015).

Both stressors, extreme heating and water shortage, severely impact the wheat crops (Qaseem et al., 2019). The effect of extreme temperature and water shortage combined is more harmful and severe than the sole stressor alone (Gobin, 2018). Many investigations to evaluate cultivated crops' exposure to multi-extreme stressors resulted in unusual reactions and responses (Dixit et al., 2024). Wheat crops potentially face dual stressors of extreme heat and water shortage throughout the breeding phase, negatively influencing crop quality and quantity (Wheeler and Von Braun, 2013; Zhou et al., 2017).

Exposure to extreme heating waves degrades the thylakoid membrane's composition and disturbs the photosynthetic electron transportation process, suppressing the total photosynthesis function. Photosynthesis is the primary affected function in wheat under extreme heat and water shortage stress, leading to the declining total wheat crop yield. Investigations revealed that closing leaves' stomata, besides inhibiting the secretion of enzymes responsible for carbon dioxide assimilation, is the root cause of photosynthesis decline (Sharifi and Mohammadkhani 2016; Zahra et al., 2023). Both stressors cause disturbance in cell metabolism, development rate, leaf shrinkage, and alteration in plant morphology (Bu et al., 2022). Moreover, both stressors affect wheat's capturing and transporting of nitrogen. Only scarcity of water suppresses the translocation of nitrogen inside the plant, mainly in aboveground parts, resulting in shortening the spikes, in addition to inhibiting carbohydrate production and the function of adenosine diphosphate (ADP)-glucose pyrophosphorylase (AGPase), resulting in shortening the time of filling the grains. In summary, both factors impede the rate of photosynthesis, N_2 uptake and translocation, and CO_2 assimilation and suppress the development of winter wheat, resulting in a considerable loss in output (Ullah et al., 2023; Ulfat et al., 2024).

Likewise, drought has a devastating impact on *O. sativa* crop yield, which was expected to harm 2 million hectares of cultivated rice crops in the Asian continent during the water shortage over the next few years, and is expected to result in a 19% total loss of rice crop yield on the continent within this century. This incidence that affects rice crops is noticed in regions where the cultivation needs more irrigation patterns than other cereal crops (Leng and Hall 2019; Lal et al., 2022). Rice is a thermal-sensitive crop (Shi et al., 2016); when the field temperature reaches 38°C during the flowering and grain filling stage, it results in significant crop quantity loss (Tian et al., 2010). Furthermore, it is forecasted that heat rise and frequent water supply shortage in the upcoming years, in addition to low-tech irrigation equipment altogether, will induce scientists to come up with new techniques that help in the sufficient usage of water resources, such as direct-seeded arid rice cultivation in monsoon seasons and aerobically in dry seasons of low rainfall environments. Many studies reported a significant drop in the crop yield when the direct-seeded rice experienced frequent drought period conditions (Quinones et al., 2017). Thus, when the field experiences a combination of both stressors of warming and water supply shortage in different areas globally, which is reasonably sure to occur, recorded internal plant tissue temperature rises and drought's harshness increases (Khan et al., 2021).

Moreover, predictive modeling applied to investigate and forecast the potential impacts of rice crops during breeding phase exposure to extreme heating for a minimum 5 days, turned out that more than a quarter of earth-cultivated land with rice is potentially going to be negatively impacted by such abiotic stress by the end of the year 2050, comparing to only 8% in the early beginning of this century (Gourdji et al., 2013). Field studies proved that only a few hours of extreme warming exposure in the flowering phase would potentially lead to infertility of rice spikelets, while exceeding 24 hours resulted in several negative impacts on yield quantity and quality (Rang et al., 2011).

17.4.2 Feed Crops

Z. mays, or corn as a feed crop, is widely used for many objectives, like feeding cattle, pets, and animals. Also, it is utilized in producing food grains and contributes to biofuel formation under different environmental conditions (Dustgeer et al., 2021). Corn is considered one of the three major worldwide cultivated crops that are being consumed by livestock, and even in the energy production process as a biological fuel (Godfray et al., 2010). The corn crops cover around 200 million hectares globally, resulting in more than 1 billion tons of corn grains annually (Canton, 2021). Both alterations in climate, adverse incidence on cultivated crops, and incremental global need for consumed feed and food materials (due to the dramatic increase in the world population) pose a significant challenge facing agriculture globally. By the middle of this century, the need for consumed agricultural-driven food is expected to grow by 60% because of the world population expansion (Guo et al., 2023). Over 183 million hectares of cultivated land were covered by corn globally in 2019, yielding 350 million tons of corn grains. The ratio may shift in the coming years as consumption rates rise and cropland areas record a decline for many reasons (Mekouar 2020).

In recent systems of farming bovine and cattle, whether for the sake of milk or/and meat production, corn silage is a substantial constituent of the feeding system for livestock, mainly silage reported to be preferable to the livestock in comparison to other coarser grains owing to its high carbohydrate concentrations that turn into high-calorie energy sufficient to livestock's dietary requirements. Consequently, the silage cropland has been negatively affected by extreme heating and drought stress due to climate change, which has been reported frequently in recent years. These stressors ended up declining the quality and quantity of the yielded grains (Forzieri et al. 2014). Because of this, the quality of yield output from drought-affected crops is compromised not only by a smaller percentage of carbohydrates in yield but also by a general decrease in ease of swallowing and digestion, in addition to a drop in milk production capacity and the number of dairy cows (Crasta and Cox, 1996).

Another example of a feed crop is "Queen of Forages" or *Medicago sativa* L., commonly known as the alfalfa plant, which is among the top feeding crops livestock consume globally. These feeding crops are recognized for their substantial production potential and ability to thrive in various climatic adversities. Besides containing reasonable calories, around 20% of the straw of alfalfa encompasses essential elements and amino acids like lysine, methionine, and threonine, in addition to 10 types of vitamins. In addition, alfalfa contains a small amount of fiber, which makes it preferable to livestock. For soil, alfalfa is being developed in summer, with below-ground parts going deeply into the soil to minimize the destruction of the soil and require more irrigation requirements (Tanweer Kumar et al. 2018; Caiserman and Faour 2021). The rotational cultivation process was significantly impacted by the irrigation shortage, reporting that the crop yield declined annually by more than 600 kilograms per hectare and water loss by 33 millimeters in half a meter of the soil profile (Li and Huang 2008). Water shortage, extreme temperature, and salt concentration are among the abiotic stressors resulting from climate change that threaten alfalfa crops (Song et al. 2019). Shortage of water supply stress disrupts typical metabolic, biological, and physiological functions in alfalfa, like photosynthesis, assimilation, and proliferation; it also poses significant effects on the alfalfa crop's yield; the damages may reach half of the cultivated crop quantity (Jaleel et al. 2008).

17.4.3 Medicinal Crops

Being an essential pillar of conventional human medicine for hundreds of years, medicinal plants act as the main source for various disease therapies in different civilizations and nations. The notion of the cure in nature is the catalyst that drove the communities to extract the medication from medicinal plants. Throughout the decades, these plants and their crops have played a significant role in medical procedures, despite their minimal adverse events, domestically and globally; therefore, the need for plant-based medications has increased. Thus, it was fostered by being the sole source of active extracted molecules, which were extracted directly or passively from medicinal plants until contemporary medicine. It has been suggested that the existence of different bioactive compounds in drugs derived from herbal plants has therapeutic potential. So, this global acceptance of therapeutic potentiality came from the presence of many promising bioactive molecules in medicinal plants (Rani et al. 2023).

Substantial impacts of global climate disturbance are threatening the availability, quality, and yields of medicinal plant crops; the alteration in world weather imposes a negative influence on both human and plant species. Nevertheless, a comprehensive understanding of these impacts on cultivated medicinal plants remains uncertain (Yan et al. 2024). Being a core player in the development of the pharmaceutical industry, medicinal plant crops' yield and quality have been grabbing scientific attention for a long time, taking into consideration the major variables that affect the growth and fitness of the crop yield, like heating conditions, light intensity, water content, and soil profile (Park et al. 2023; Zheng et al. 2023). Many researchers surveyed the influence of global warming on medicinal plant yield. However, the applicability to generalize the results was limited according to the diversity of development conditions of different plant species and their growth patterns, in addition to differentiation in secondary metabolites and their correlation to cultivation conditions (Sun et al. 2023). Furthermore, many studies have already been conducted for the sake of deep investigation of single species to define parameters of climate disturbances; for instance, low water supply circumstances in which liquorice species were exposed, resulting in increased phenol derivatives in leaves and a decline in roots (Chang et al. 2016). Moreover, the presence of alkali compounds and the content of alkali compounds in periwinkle species strongly correlated with nitrogen availability (Singh and Agrawal 2015). All these examples foster the difficulty of generalizing the standard impact toward specific conditions. Therefore, for a reasonable understanding of each environmental variable on medicinal plants, it is proposed to specify the medicinal plant species and define which environmental parameter will be examined.

Panax is considered one of the promising medicinal plant families of worldwide usage. Known for its reputational residues, the promising pharmaceutical compound that the whole genus is distinguished by. Saponins, as steroidal molecules, have many medical applications and are being utilized now in the drug industry. Not only the steroids but also many bioactive compounds of antimicrobial and anti-inflammatory effects were extracted from the same genus of ginsenoside efficacy, in addition to a group of compounds with human immuno-supportive effects (Farh et al. 2020; Hou et al. 2021). The *Panax* genus encompasses around 14 species, mainly *Panax ginseng* (*Pg*), *Panax quinquefolius* (*Pq*), and *Panax notoginseng* (*Pn*), which are commonly used because of their pharmaceutical potential. Numerous research studies have demonstrated how the alteration in global climate caused by abiotic stressors could affect the crop yield of the *Panax* genus. Drought stressors, as an example, lead to an increase in the content of osmotic regulator agents, abscisic acid (ABA), and malondialdehyde (MDA) compounds accumulated in the root of *Panax ginseng* (Lei et al. 2023). Disturbance of N_2 supplement will impact subsequent growth of the belowground part in the same species (*Pg*) (Geem et al. 2023). Extensive research is required, especially considering the significance of the lower part of the undersoil of these plant genus components for medical purposes (Feng et al. 2023).

Yan et al. (2024)[98] combined meta-analyses with MaxEnt to investigate over 700 observation pairs and 588 sites of genus *Panax* to assess the global climate change effects on the genus's yield and allocation as a representative of common medicinal crops. Meta-analysis has been recognized as an evolved and scientifically credited technique that provides a detailed and in-depth evaluation of plant reactions to worldwide climate change (Adekanmbi et al. 2023). It has been used to assess the consequences of significant incidents like herbivorous plants' effects on cultivated crops' development, 108, and the impact of global lockdowns in 2019 because of the coronavirus pandemic on worldwide living organisms (Tucker et al. 2023). Conversely, the MaxEnt model provides a precise tool for forecasting plant prevalence and diversity alteration owing to global climate change (Bao and Yang 2022; Chen et al. 2022). Applying MaxEnt modeling, Yan et al. (2024) could demonstrate thoroughly the *Panax* genus as a typical medicinal plant representative—reaction toward any alteration or disturbance in the global climate.

Meta-analysis revealed that physiological processes have been promoted by climate alteration; mainly, the vegetative parts were developed in contrast to the roots. Regarding bioactive compound contents, saponin is primarily negatively impacted only on the vegetative mass. Meta-analysis confirmed that global climate disturbance possesses modest and multifactorial impacts on medicinal plants. MaxEnt indicates that the *Panax* genus's response to climate change throughout conservative dispersal, where *Panax ginseng* (*Pg*) and *Panax quinquefolius* (Pq) prevailed in the northern upper regions, yet *Panax notoginseng* (Pn) leveraged the warmed regions (Yan et al. 2024).

One of the most common genera and economic interests in the family Lamiaceae is "Tulsi" or *Ocimum*. The financial interest of the *Ocimum* genus is attributed to producing bioactive compounds of much pharmaceutical importance. This genus encompasses more than 64 species globally. The global economic importance of this commonly used genus comes from many aromatic bioactive compounds (including volatile oils or non-volatile ones) that are being produced, which are involved in many medicinal implications. The monoterpenes and sesquiterpenes are the major constituents of *Ocimum* volatile oils; the total amount of extracted acids from vegetative parts of the plant varies up to 1% of weight (Gurav et al. 2022; Zahran et al. 2022; Du et al. 2023). The vegetative part of the *Ocimum* genus is very rich in economically essential bioactive acids like labiatenic acid, hydroxycinnamic acid, hydroxy-methoxy benzoic acid or anisic acid derivatives, carnosic acid of benzenediol abietane diterpene, caffeoylquinic acid, pentacyclic triterpenoid acid, and other phenolic benzoic acid derivatives, besides containing other essential oils like clove oil (Hussain et al. 2017). The secondary metabolites of *Ocimum* plants are recognized for their pharmaceutical importance because of having inhibitory effects against many microbes (Mounira et al. 2022; Qasem et al. 2023), free radical formation (Ben Abdallah et al. 2020), allergic agents (Zengin et al. 2019), parasites and worms (Osei Akoto et al. 2020), viruses (Goel and Bhatia 2022), cytotoxicity, and tumor (Qamar et al. 2020), metastatic agents (Manaharan et al. 2014), bronchial-asthma (Costa et al. 2012), arthropod larvae (Ntonga et al. 2014), diabetes mellitus, and hyperglycemia (Raja et al. 2016), human convulsions and seizures (Medeiros Venancio et al. 2016), in addition to showing pain-killing efficacy and tissue protection effects on human skin and kidney (Almalki 2019; Bae et al. 2020).

The scarcity of water supply as a drought stressor has been reported to have significant impacts on the growth and development of *Ocimum* plant species and other plants of pharmaceutical importance, as well as on the secondary metabolite yield. Due to its medicinal importance, several studies worldwide have examined and reported the influence of drought stress on *Ocimum* species. When *Ocimum basilicum* species (sweet basil) was exposed to drought conditions in the field, it yielded around a 50% decrease in both fresh and dry mass of the species above ground parts. Furthermore, decreases in cultivar height and stem girth were also reported in field studies, while the whole plant height increased. The crop yield quantity of *Ocimum americanum* lowered when exposed to drought stress, and total essential oil concentration reduction was recognized when both *O. basilicum* "Genovese" and *O. americanum* were examined under severe drought conditions (Damalas 2019; Radácsi et al. 2020; Mulugeta and Radácsi 2022). Two years of investigation of the effect of drought stress on *O. basilicum* "Ohře" and *O. americanum* species, Mulugeta et al. (2023), ended up with clear negative impacts on plant fitness and growth, leading to changes in the secondary metabolites' compositions, including essential oil concentrations. On the contrary, it has been proven that a water shortage would induce more synthesis of bioactive molecules like alkaline compounds, phenolics, and isoprenoids, even essential oils in the case of marjoram (Farouk and Al-Huqail 2020; Abouzeid et al. 2022). Cultivar types are a crucial factor in determining the impact on essential oil concentration in the plant under drought stress, as it is reported to increase in the presence of "Kasia" and "Keskenylevelű" (Radácsi et al. 2020).

17.5 MICROBES UNDER HS AND RHIZOSPHERE ENRICHMENT

Microbial communities are integral to ecosystem functionality and essential in nutrient cycling, organic matter decomposition, and carbon fixation. As climate change accelerates, rising global temperatures have become a significant stressor for microbial ecosystems. High temperatures can shift microbial community composition, favoring thermophilic organisms while disadvantaging mesophilic species, leading to an overall decrease in biodiversity (Kumar et al. 2021). These changes can negatively impact key ecosystem services, including nutrient cycling and soil fertility. Moreover, the decline in microbial diversity caused by temperature stress can result in the loss of critical functional groups, which are essential for maintaining ecosystem stability (García et al. 2018).

In extreme environments, such as geothermal springs or deep-sea hydrothermal vents, microbial communities have evolved remarkable adaptations to survive high temperatures. These extremophiles exhibit unique biochemical properties that allow them to thrive under conditions lethal to most other organisms. For instance, thermophilic bacteria and archaea produce heat-stable enzymes that are highly sought after for industrial applications such as biofuel production and waste management. Studying these organisms provides insights into microbial adaptation mechanisms and expands our understanding of life's potential in extreme conditions, which has implications for astrobiology and the search for extraterrestrial life (Shu and Huang 2022).

17.5.1 The Impact of High Temperature on Microbial Community Assembly

Temperature profoundly influences microbial diversity, dictating which species can thrive in an environment. Elevated temperatures often favor heat-resistant thermophilic organisms, such as certain bacteria, archaea, and fungi, while mesophilic microbes may be outcompeted or die off. Direct physiological effects and temperature-induced changes in other environmental factors, such as nutrient availability and oxygen levels, influence these shifts in community composition. As microbial diversity decreases, the loss of functional groups that drive essential processes such as decomposition and nutrient cycling can have cascading effects on ecosystem health (Mironov et al., 2021).

High temperatures also have a significant impact on microbial functionality. Enzyme activity, crucial for microbial metabolism and various ecosystem processes, is often temperature-dependent (Dutta and Dutta 2016). Elevated temperatures can increase enzymatic rates to an optimal point; however, enzyme function can deteriorate beyond this threshold, slowing down essential processes like nutrient cycling and organic matter decomposition. As a result, ecosystems may experience disruptions in nutrient fluxes, with possible consequences for soil fertility, greenhouse gas emissions, and broader ecosystem stability. For instance, higher temperatures can impair soil microbial functions related to nitrogen fixation and carbon sequestration, exacerbating environmental issues such as soil degradation and climate change (Jansson and Hofmockel 2020).

The ecological consequences of high temperatures on microbial communities extend beyond shifts in diversity. As temperature increases, microbial-driven processes like nutrient cycling and organic matter decomposition can be significantly altered, potentially impacting ecosystem functioning. Loss of microbial diversity can reduce an ecosystem's resilience to climate change, impairing its ability to adapt to further environmental stresses. Reduced microbial diversity may also diminish ecosystem services such as soil fertility, carbon sequestration, and water purification. These changes emphasize the importance of understanding microbial community responses to temperature to predict and mitigate the potential impacts of climate change on ecosystem stability (Fierer and Jackson 2006).

Soil microbial communities are vital for nutrient cycling, decomposition of organic matter, and plant growth support. High temperatures can significantly alter the structure and function of these communities. As temperature increases, thermophilic organisms often take over, while mesophilic species, which prefer moderate conditions, are reduced or eliminated. These changes can disrupt microbial processes in the soil, such as nitrogen cycling and decomposition, leading to reduced soil fertility and a buildup of organic matter. Elevated temperatures can cause soil destruction, further stressing microbial populations and impairing nutrient availability. Such changes in soil microbial dynamics can have profound effects on agricultural productivity and ecosystem resilience (Condron et al., 2010).

Microorganisms have evolved various adaptive mechanisms to survive high temperatures. HSPs, which aid in protein refolding and prevent aggregation of denatured proteins, are one of the primary responses to temperature stress. Additionally, many thermophilic organisms possess heat-stable enzymes that continue functioning at high temperatures, facilitating essential metabolic processes. To protect cellular structures, some microbes also alter the composition of their cell membranes, increasing the concentration of saturated fatty acids to maintain membrane stability at elevated temperatures. These strategies allow microbes to thrive in extreme thermal environments, such as hot springs and geothermal vents and offer insights into potential biotechnological applications for heat-tolerant enzymes (Jeyachandran et al., 2023).

17.5.2 The Impact of High Temperature on Plant–Microbe Interactions

The relationship between plants and microbes is crucial for ecosystem function, and their interactions play significant roles in nutrient cycling, disease suppression, and overall plant health. However, rising global temperatures, driven by climate change, substantially threaten plant–microbe symbiosis. High temperatures can alter microbial communities in soil, shift microbial activity, and disrupt mutualistic interactions, particularly those beneficial to plant growth, such as nitrogen fixation and phosphorus uptake. Furthermore, temperature-induced stress can trigger plant defense mechanisms, which may affect microbial populations and influence the balance of microbial community dynamics. Understanding how elevated temperatures influence plant–microbe interactions is critical to predicting the long-term health of ecosystems and agricultural productivity (HAWKES et al., 2007).

Soil microbial communities form the foundation of plant–microbe interactions by influencing nutrient availability, decomposition, and organic matter turnover. High temperatures significantly impact soil microbial diversity and functional composition. For example, elevated temperatures can reduce the abundance of soil bacteria in organic matter decomposition and nutrient cycling. As soil warms, thermophilic microbes tend to become more dominant, while mesophilic organisms that participate in critical processes such as nitrogen cycling may be outcompeted. The resulting shift in microbial community structure can directly affect plant health by altering nutrient availability and potentially increasing plant susceptibility to disease. In addition, higher temperatures can increase microbial competition for resources, further destabilizing plant–microbe interactions (Rousk et al., 2012).

Mycorrhizal fungi form a symbiotic relationship with plant roots, aiding in the uptake of nutrients and water. During HS, these fungi play a crucial role in maintaining plant resilience (Wahab et al., 2023).

Mycorrhizal associations enhance plant nutrient acquisition, particularly phosphorus and nitrogen, which are essential for stress tolerance and metabolic function (Smith et al., 2011). Additionally, these fungi increase root surface area, improving water absorption from deeper soil layers, which is vital during periods of drought or high temperatures (Bhantana et al., 2021).

HS may impair fungal metabolism, reduce symbiotic functioning, and limit plant access to essential nutrients. Furthermore, HS can shift the balance of fungal species within the mycorrhizal community, favoring those that are more heat-tolerant but less efficient in nutrient uptake. Such changes can have severe implications for plant growth and survival, especially in nutrient-poor soils (Aroca et al., 2012).

Mycorrhizal fungi also contribute to plant thermotolerance by improving root health and enhancing stress signaling pathways. Studies have shown that plants inoculated with mycorrhizal fungi exhibit improved growth, better stress responses, and higher yields under HS than non-inoculated plants. The symbiotic relationship also helps the plant maintain a better balance of essential nutrients, ensuring that metabolic processes continue functioning efficiently during periods of thermal stress (Watts-Williams et al., 2015).

Rhizobia form a symbiotic relationship with legumes, providing them with nitrogen in exchange for carbon compounds. However, elevated temperatures can significantly affect this interaction by reducing the efficiency of nitrogen fixation. High temperatures can impair the enzymatic activity of nitrogenase, the enzyme responsible for converting atmospheric nitrogen into a form usable by plants. As a result, plants may experience nitrogen deficiency, leading to reduced growth, productivity, and overall health. Additionally, increased temperatures may alter the composition of rhizobia communities, selecting strains that are more heat-resistant but potentially less efficient at nitrogen fixation. This could lead to an imbalance in the symbiosis, further limiting plant access to this crucial nutrient (Zahran, 1999).

Plants possess intricate defense mechanisms to cope with abiotic stress, including elevated temperatures. One such response is the activation of HSPs, which help in protecting plant cells from thermal damage. However, the activation of these defense pathways can also influence plant–microbe interactions. Heat-induced stress responses can alter the chemical signaling between plants and their associated microbes, potentially suppressing beneficial microbial activities or enhancing pathogenic interactions. For example, specific plant defense responses may produce antimicrobial compounds that inhibit beneficial microbes like mycorrhizal fungi or rhizobia. This disruption in beneficial symbioses could compromise plant growth and resistance to pathogens (Bhantana et al., 2021).

Plant–microbe interactions are often mediated by complex chemical signaling mechanisms, which can be sensitive to environmental factors like temperature. High temperatures can disrupt the normal functioning of signaling molecules such as plant hormones (e.g., auxins, gibberellins [GAs]) and microbial signals (e.g., quorum-sensing molecules) (Jha and Saraf, 2012).

For example, the production of plant hormones involved in root development and microbial colonization may be altered under HS, affecting the establishment of beneficial microbial communities. Similarly, microbial signaling pathways that regulate the formation of symbiotic structures, such as nodules in legumes, may be impaired by elevated temperatures, reducing the effectiveness of plant–microbe interactions. These disruptions can limit nutrient uptake and negatively impact plant health (Shaffique et al., 2022).

The broader ecological implications of temperature-induced changes in plant–microbe interactions are profound. As temperatures rise, the altered dynamics of plant–microbe symbiosis can lead to reduced ecosystem productivity and increased vulnerability to pests and diseases (Wahid et al., 2020). For instance, the loss of beneficial microbes such as nitrogen-fixing bacteria and mycorrhizal fungi can lead to nutrient imbalances, affecting plant health and overall ecosystem functioning. Moreover, these disruptions can affect soil carbon dynamics, potentially leading to increased GHS emissions from the soil as microbial processes involved in carbon sequestration are compromised. In agricultural systems, the impact of high temperatures on plant–microbe interactions can lead to reduced crop yields, decreased food security, and greater reliance on chemical fertilizers and pesticides (Kumari et al., 2023).

17.5.3 Microbial-Mediated Thermotolerance in Plants

Thermotolerance is the ability of plants to withstand and adapt to high temperatures that would typically harm cellular processes and metabolic functions. As global temperatures rise due to climate change, thermotolerance has become an essential adaptive trait for maintaining plant productivity and survival, particularly for crops and native vegetation. HS can severely affect plant growth, leading to wilting, reduced photosynthesis, protein denaturation, and oxidative damage. This phenomenon disrupts normal metabolic processes and diminishes plant growth, resulting in significant agricultural losses (Kaushal et al., 2016).

Given these challenges, researchers have increasingly focused on the mechanisms that enhance plant thermotolerance, especially the role of plant-associated microbes. Microbial communities, including rhizobacteria, fungi, and endophytes, have been found to play a crucial role in enhancing plant resilience to high temperatures. These microbes can help plants mitigate the adverse effects of HS by influencing various physiological processes such as water retention, nutrient uptake, antioxidant defense, and hormonal regulation. Understanding how these microbial partners contribute to thermotolerance is a promising strategy for improving agricultural practices and food security in the face of climate change (Iqbal et al., 2023).

Plant-associated microbes significantly impact various physiological processes, enhancing the plant's ability to cope with thermal stress. Rhizobacteria, which live in the plant's root zone, can help plants endure HS by activating stress-responsive genes. These genes encode proteins that help plants manage oxidative stress and stabilize cellular membranes. Additionally, these microbes can produce phytohormones such as auxins, GAs, and cytokinins (CKs), which regulate essential plant functions under HS, including photosynthesis, water uptake, and stomatal conductance (Prerostova and Vankova, 2023).

By modulating these physiological processes, microbes ensure that plants can maintain better hydration, nutrient uptake, and photosynthetic efficiency, even during periods of high heat. These changes are critical for plant survival, as they help mitigate water loss and ensure that plants continue to produce energy through photosynthesis. Studies have shown that plant–microbe interactions also enhance chlorophyll content and leaf area, which are vital for photosynthetic activity under HS (Chouhan et al., 2023). Therefore, understanding the intricate relationship between microbes and plants at the physiological level offers a promising approach to enhancing thermotolerance in crops.

EPS are complex carbohydrate compounds produced by microorganisms that reside in the plant rhizosphere. These EPS play a significant role in enhancing plants' thermotolerance by improving soil moisture retention and stabilizing soil structure. The soil can dry out quickly during HS, which exacerbates plant water loss. However, EPS act as a barrier that retains moisture in the soil, reducing the impact of high temperatures on plant roots (Morcillo and Manzanera 2021). Additionally, EPS improve soil aggregation, improving nutrient uptake and root growth.

The increased water and nutrient availability provided by EPS helps maintain plant physiological functions during HS. Furthermore, EPS can help strengthen the plant's cell wall integrity, making it more resistant to the mechanical stress induced by heat. This mechanism allows plants to manage environmental stress better and improve their resilience to high temperatures (Maitra et al., 2021).

17.5.4 Antioxidant Enzyme Activity, Total Proteins, and Amino Acids under High-Temperature Stress

One of the most harmful effects of HS is the generation of ROS, which can cause oxidative damage to cellular structures like proteins, lipids, and DNA. To counter this, plants have evolved antioxidant systems that neutralize ROS. Beneficial microbes, particularly those in the root zone, enhance the plant's antioxidant defense system. Microbes can stimulate the synthesis of antioxidant enzymes, such as superoxide dismutase (SOD), catalase (CAT), and peroxidases (PODs), which actively scavenge ROS during HS (Belhadj Slimen et al., 2014).

SOD converts superoxide radicals into hydrogen peroxide, which CAT then breaks down into harmless water and oxygen molecules. Similarly, PODs play a crucial role in degrading hydrogen peroxide. The activity of these enzymes is vital for preventing oxidative damage, maintaining cellular integrity, and improving survival rates under HS. Additionally, increased antioxidant activity can also promote plant growth and physiological stability during extreme heat (Ikwegbue et al., 2017).

Amino acids are essential in mitigating the effects of high-temperature stress by stabilizing proteins, protecting against oxidative damage, maintaining energy homeostasis, and regulating stress-related gene expression. Their role in heat tolerance is crucial for both prokaryotic and eukaryotic organisms, making them key targets in research on stress adaptation and resilience in various biological systems (Peer et al., 2020).

Amino acids play a crucial role in cellular adaptation to high-temperature stress. Elevated temperatures can lead to protein denaturation, increased metabolic rates, oxidative stress, and disruptions in cellular homeostasis. To counteract these adverse effects, organisms utilize amino acids in various protective and adaptive mechanisms (Raza et al., 2023).

Certain amino acids, like proline, function as osmoprotectants, maintaining cellular hydration and preventing protein aggregation under HS. Glutamate and aspartate are involved in stress signaling pathways, activating responses that help cells survive elevated temperatures (Li et al., 2016).

17.5.4.1 HS Proteins

HSPs are a group of proteins produced in response to environmental stress, including high temperatures. These proteins act as molecular chaperones, assisting in the correct folding of denatured proteins and preventing their aggregation. During HS, proteins within plant cells can become denatured, disrupting cellular functions and leading to cell death. HSPs are critical in mitigating this damage by ensuring that proteins remain functional or are refolded correctly. Microbial interactions can induce the expression of HSPs in plants, enhancing their capacity to survive thermal stress. The production of these proteins is an adaptive mechanism that helps protect plant cells from heat-induced protein damage. Research has shown that plants with higher levels of HSPs exhibit better growth and stress tolerance under high-temperature conditions. By promoting the synthesis of these proteins, microbes help enhance plant survival and recovery after HS (ul Haq et al., 2019).

17.5.4.2 Secondary Metabolites and VOCs

Secondary metabolites are organic compounds produced by plants that do not directly participate in essential metabolic functions but play a crucial role in stress responses. These metabolites, such as flavonoids, phenolics, and alkaloids, have strong antioxidant properties that help plants cope with oxidative stress during heat exposure. Beneficial microbes contribute to the synthesis of these metabolites by stimulating plant biochemical pathways (Bartwal et al., 2013).

Additionally, microbial VOCs are produced by microbes in the rhizosphere and can trigger stress-resistance pathways in plants. These VOCs act as signaling molecules, inducing plant defense mechanisms that enhance heat tolerance. Research has shown that VOCs released by beneficial microbes can alter plant gene expression and enhance the synthesis of stress-related proteins, improving the plant's overall resilience to HS (Chatterjee and Niinemets 2022).

17.6 CROPS IMPROVEMENT UNDER THERMO-ADAPTATION/TOLERANCE MICROBIOMES

17.6.1 Nutrient Uptake

The microbial constituents of the plant rhizosphere are vital in determining cultivated crop plant performance, development, fitness, and productivity; they contribute to making the nutrients in the soil available and accessible and easing the process of plant absorption (Trivedi et al., 2020). Microbial species improve plant resilience against abiotic stress by regulating nutrient availability and growth and helping maintain plant health.

One of the primitive and conventional rhizosphere modeling and controlling techniques was the human practice of transplantation of fruitful soils to enhance cultivated crop productivity (Mougel et al., 2006). Crop plants shape the rhizosphere components by secreting root exudates that attract microorganisms. It has been demonstrated that the microbial structure of the plant rhizosphere is influenced by many physiological and chemical properties of the soil profile and organic matter accessibility, cultivated plant species, and even the phase of the plant's lifecycle. It has been proposed that functional features are the primary factor regulating the rhizosphere microbial structure (Castrillo et al., 2017; Zhang et al., 2019). Many symbiotic microbes inhibit the rhizosphere, promoting plant capabilities and accessibility to uptake nutrients by transferring phosphorus from complex organic form to digestible form and nitrogen assimilation.

Plant growth-promoting rhizobacteria are essential components of the rhizosphere microbiome that not only facilitate organic matter accessibility and availability to the host plant but also contribute to modifying the plant root system structure and design. It was recognized that the better N_2 fixation and uptake among some maize species over others refers to the presence of nitrogen-fixing bacterial species in high amounts, like in the case of the *Oryza indica* rhizosphere compared to *Oryza japonica* species (Zhang et al., 2019). Thus, it was validated by studying the implantation outcome of sixteen cultivated samples containing *O. indica* (associated with nitrogen-fixing bacterial strains). This resulted in the apparent efficient utilization of N_2 among *O. indica* species. It has been proven that some flavonoid compounds released from maize species' roots influence recruiting of bacterial members of the Oxalobacteraceae family in their rhizosphere, such as bacterial species involved in ease nitrogen fixation and absorption (Yu et al., 2021). Mainly, *Burkholderia*, *Rhizobium*, and *Sphingomonas* genera are the major nitrogen-assimilating bacterial groups that associate the rhizosphere of Sweetcorn (Walters et al., 2018).

Many microbial species contribute to the nitrogen cycle in the rhizosphere; some bacteria are known for N_2 assimilation, like diazotrophs; others induce oxidation of NH_3 gas, including bacterial and archaeal species (Bargaz et al., 2018). It is worth mentioning that *Rhizobium leguminosarum* (Evans & Russell) paved the way for potential utilization and inoculation of N_2 assimilating bacterial strains in field agricultural applications. Moreover, the new concept of establishing root nodules, which is responsible for biological N_2 fixation, the evolved term of nitrogen assimilation biologically points to the involvement of beneficial microorganisms in transforming atmospheric nonreactive nitrogen to NH_3 utilizing nitrogenase enzyme (Ryan et al., 2009). This mutualistic beneficial relationship of the host plant-rhizosphere microbiome facilitates the two-way movement of assimilated N_2 from microbes to the host plant in return for gaining plant organic carbons (Nelson and Mele 2007).

In cereal crops, a specific pathway evolved because of a partnership with arbuscular mycorrhizal fungi (AMF) in beneficial mutualistic relationships; this pathway is called the symbiotic signaling pathway, where the AMF induces the root cells to trigger this pathway. The cereal plant root cell also has differentiated a significant organelle where the transcription of the *nif* gene occurs; these developed cells are called nitroplasts, which are responsible for producing nitrogen-fixing enzymes or nitrogenase enzymes (Mahmud et al., 2021).

Azotobacter and *Clostridia* species are good examples of heterotrophic bacteria that assimilate the atmospheric N_2. These diazotrophs participate in fixing atmospheric nitrogen on behalf of plants in return for consuming carbon secreted from roots (Biswas and Gresshoff, 2014). Yoneyama et al. (2017) recognized that the association between nitrogen-fixing microbiome and maize aerial roots in nitrogen-deficient soil results in increasing atmospheric N_2 assimilation by more than 80% and being acquired successfully by maize. When Alfalfa yield was examined with nitrogen assimilating microbes and plant growth promoting ones, the yield measured when double inoculated and single microbe inoculated resulted in double inoculation improving not only the root system but the upper ground vegetative part as well (Nacamulli et al., 1997). Metagenomic analysis of rhizosphere soil sample of the Taoyuan site in China, where the *Oryza sativa* crops got more than 70% quantity of crop yield compared to cultivated *Oryza* crops in different sites of similar condition, revealed a high abundance of bacterial species over fungi in Taoyuan sample, specifically the prevalence of nitrogen-fixing bacterial species, which relates the higher yield production to the associated nitrogen-fixing bacteria in the rhizosphere (Guiñazú et al., 2010).

In insufficient nutrient availability, the rhizosphere-inhabiting microorganisms contribute to regulating and gaining essential minerals like Fe^{2+} and P4. Throughout the soil of phosphate deficiency, *A. thaliana* plant species' phosphate starvation response (PSR) influences the microbial structure of the surrounding soil of the plant, in addition to inoculation with symbiotic microbes that altogether lead to PSR gene expression and improve plant absorption of the inorganic P4 mineral (Castrillo et al., 2017). This microbial structure of the rhizosphere triggered by PSR gene expression contains microbes of symbiotic and rival opportunistic techniques that may support plant P4 acquisition or suppress the uptake process, like what rival opportunistic does. The fungal composition in the rhizosphere microbiome is also shown to be significantly involved in nonmycorrhizal plant phosphorus absorption during phosphorus deficiency conditions (Finkel et al., 2019).

17.6.2 Production and Regulation of Phytohormone

One of the significant mitigation techniques to hinder thermal stress is the release of phytohormones. Phytohormones are small chemical substances that, at low concentrations (less than 1 mM), stimulate and manage different physiological and biochemical functions that modulate and enhance crop plant fitness and proliferation even under unfavorable environmental conditions. Many plant growth-promoting bacteria (PGPBs) are known for their ability to release phytohormones in addition to inducing the associated plant to secrete them. Certain species of rhizobacteria induce *T. aestivum* L. plant development's features and yield quantity; this favorable impact of rhizobacteria is related to their function in enhancing phytohormone synthesis (García et al., 2017). PGPB employs multiple methods, such as phytohormone secretion, nutrient absorption, and N_2 assimilation, to enhance plant development and mitigate the fatal effects of abiotic stress (Sen and Chandrasekhar, 2014).

It has been confirmed that inoculation of PGPB in the rhizosphere of the crop plants led to improvement in cultivated crop plant development and fitness in addition to utilizing the available water content efficiently under unfavorable environmental stress, thus due to PGPB's distinguished capabilities to release phytohormones like indole-3-acetic acid (IAA), which is known to raise the height and biomass of the root system (Naz et al., 2009). Also, PGPB's favorable influence on plants may be ascribed to the generation of these organic substances of phytohormones that help in regulating and modulating crop plant development; some extracellular macromolecules of polysaccharides (EPSs) were noticed to be released by PGPR and the rise of ABA percent, which all led to reducing the detrimental effects on physiological functions like photosynthesis (Shilev 2020). There are six main categories of phytohormones: GA, CKs, ABA, auxins, brassinosteroids (BRs), and ethylene (C_2H_4), which are considered crucial signals in the resilience of environmental adversities (Akbar et al., 2022).

Phytohormones induce and manage cellular division and multiplication and coordinate complicated growth processes to attain plant health and breeding success (Li et al., 2021; EL Sabagh et al., 2022). Recent research has also demonstrated the significance of the influence of phytohormones in crop plant fitness under extreme heat conditions, including auxin, CK, BRs, gibberellic acid, and strigolactones. Auxins are known for their role in cell lengthening of the apex, signal transmission, and plant response to light and gravity (Du et al., 2020; Gomes and Scortecci, 2021). CK generally enhances plant development under normal and abnormal conditions by inducing cellular multiplication in addition to plant organ distinctions (Fathy et al., 2022; Sabagh et al., 2022). Like other phytohormones, GAs stimulate plant stem development and seed growth and regulate blooming processes, besides being recognized for safeguarding crop plants from thermal stress. Furthermore, transferring the genes responsible for GA production to plants of non-GA leads to proper accommodation and survival under thermal stress (Ikeda et al., 2012). Ethylene has a significant role in determining the plant response to aging factors, maturation of fruits, and physical standing (Iqbal et al., 2017). ABA contributes to plant adaptation to water shortage and salt stress (Chen et al. 2020). Jasmonate (JA), BRs, and SA are also among the phytohormones that contribute to crop plant development and resilience to abiotic stress (Wang et al. 2021; Manghwar and Li 2022).

Melatonin was recently recognized as a newly stated phytohormone due to its controlling involvement in unfavorable environmental stress adaptation. Melatonin operates with other phytohormones like auxin, GA, CK, ABA, ET, BRs, JA, SA, and other plant growth promoters like polyamines, NO, γ-aminobutyric acid, and serotonin to alleviate stress from unfavorable environmental incidences (Arnao and Hernández-Ruiz, 2018). Melatonin modulates several physiological functions: plant cell response to light, underground system development, seed germination, and leaf aging, in addition to acting as an ROS scavenging agent, so it induces many crop plants' tolerance against HS (Haskirli et al., 2021; Samanta et al., 2021; Wang et al., 2021; Yan et al., 2021; Samanta et al., 2022). According to Byeon and Back (2014), the melatonin concentration increased in the *O. sativa* crop when exposed to thermal stress, which indicates the melatonin function in thermal defense mechanisms. The damages that occurred in *Phacelia tanacetifolia* because of exposure to extreme heat are healed when treated with melatonin (Tiryaki and Keles 2012). External sources of melatonin, such as phytohormones, hinder the aggregation of hydrogen peroxide and MDA and induce the secretion of proline and scavenging agents. These all lead to lessening the devastating impact of thermal stress on crop plants. Also, adding melatonin to the seeds of *Arabidopsis* that are exposed to extreme temperatures stimulates the scavenging agents to reduce the ROS (Buttar et al., 2020).

Auxins are a big set of phytohormones that control a wide range of physiological functions in plants. IAA is among the widely investigated and studied auxins that initially existed in plants (Casanova-Sáez et al., 2021). It has been proven that melatonin and auxin are generated from the exact origin; also, tryptophan shares a similar structure with both substances. IAA is one of the most essential auxins produced from L-tryptophan digestion by many microorganisms, including plant growth-promoting rhizobacteria (PGPR). This auxin type facilitates the development of underground plant systems with numerous root hairs and laterals (Li et al., 2022). IAA induces a rise in the cellular osmotic concentration, enhances water diffusion inside the cell, and then increases the deposition and thickening of the cell wall, in addition to producing the protein. Also, IAA suppresses plant leaf cutting and wilting and promotes blossoming roses and fruit ripening. *Azatobacter* sp., *Azospirillum* sp., *Bacillus* sp., *Pseudomonas* sp., *Enterobacter* sp., etc. are groups of PGPR that are proven to release plant growth-promoting molecules like IAA (Park et al., 2021; Bag et al., 2022).

GAs are a set of phytohormones composed of tetracyclic diterpenoid acids. The GA group contains a variety of more than 125 compounds that have been extracted from plants. There are two enzymes in charge of the generation of GAs in the cell, which are gibberellin 3-oxidase (GA3ox) and gibberellin 20-oxidase (GA20ox) (Ayano et al., 2014). GAs exist in higher plants and contribute to managing and stimulating many developmental and metabolic functions during the crop plants' lifespan. Many plants' physiological processes, such as crop plant proliferation, sprouting of seeds, development of vegetative aboveground parts, blossoming of flowers, and ripening of fruits, in addition to neutralizing the suppression impact of ABA on sprouting the plant seed, are all among the physiological functions in which GAs are involved (Urbanova and Leubner-Metzger, 2016). Only a tiny number of GAs are found to be functionally active, like GAs GA1 and GA2 in *O. sativa*, while the most common form is gibberellic acid GA3 (Davies, 2010). Plant growth-promoting Rhizobacteria *Rhizobium* and *Azospirillum* were proven to generate GA phytohormones when inoculated in the rhizosphere of *T. aestivum* L., which improve plant growth under unfavorable conditions, especially extreme HS (Ilyas et al., 2020).

CKs are plant development modulators or phytohormones that participate in the development of plant physiological and biochemical processes, besides having crucial functions in crop plant survival and tolerating unfavorable environmental circumstances. CKs are natively generated in plants that perform essential functions like regulating the development, proliferation, and fitness of crop plants and ameliorating environmental stress (Kaur et al., 2022). CK has been recognized for enhancing the sprouting of seeds, growth of leaves, budding proliferation, and increasing chloroplast contents. It also facilitates cell multiplication, the development of crops, and quality and quantity, in addition to stimulating the eradication of ROS as scavenging agents (Hönig et al. 2018). Furthermore, these phytohormones are key factors in crop plant resilience to many extremely unfavorable environmental stressors like HS throughout, strengthening plants' scavenging systems (Pavlů et al., 2018; Cortleven et al., 2019).

CK supports the vitality and expansion of *Agrostis stolonifera* leaves; as a resilience to high temperatures, it triggers the scavenging functions and inhibits lipid oxidation and spoilage (Liu and Huang, 2002). In grains of *O. sativa*, CK regulates the cellular multiplication of embryos in the grain-filling phase and controls the grain capsule size (Yang et al., 2003). CK preserves the growth of the leaves under unfavorable conditions throughout, inducing the enzymes that are responsible for the eradication of ROS, like POD, SOD, and CAT (Kumari et al., 2018). CK is also recognized to enhance the yield of *T. aestivum* cultivated in winter and exposed to thermal stress by sustaining photosynthesis functions through the grain-filling phase and aiding the transfer of fixed carbon and nitrogen substances to grain (Yang et al., 2016). Additionally, external CK dosage was recognized to enhance the yield under abiotic stress, mainly CK ameliorating the negative impact of extreme temperature on genital organs in crops like *O. sativa* and *T. aestivum*, which led to improving the quality and quantity of their yield (Wu et al., 2017).

17.6.3 HS Proteins

The crop plant that experiences abiotic stress applies many defensive techniques to counter the threats of harsh conditions and survive. Extreme heat conditions force plants to take precautionary measures. One of the primary defensive mechanisms that is the most fundamental and consistent thermal adaptation reaction of plant cells to prevent aggregation of non-native functional proteins is the transcription of heat-responsive genes (HRGs), which are responsible for the generation of specific proteins and contribute to plant cells counteracting the thermal stress. Head shock proteins (HSPs) are among these specific proteins (Driedonks et al., 2015). HS proteins, or HSPs, are a big group of proteins that exist in the plant cell and are triggered by many abiotic stresses, mainly the thermal ones, as a practice to ameliorate stress and even autotoxicity. HSPs have been observed to be omnipresent molecules of chaperone proteins of molecular weights that vary from 10 to 200 kilo-daltons. According to the molecular weights, the HS proteins are categorized into groups like HSP101, HSP90, HSP70, HSP60, HSP23, and sHSP, which have all been reported in many crop plants. These HSPs were found to increase in the plant under thermal stress, as HSPs' primary function is to protect the native protein from damage and denaturation. Also, HSPs are responsible for transmitting signals and stabilizing cellular protein homeostasis by facilitating proper folding and assembling of partially folded and/or misfolded proteins in the cell and hindering the irreversible accumulation of proteins. Therefore, HSPs preserve and sustain cellular vitality, stability, and balance under thermal stress by performing an essential function in safeguarding cultivated crop plant cells from the destruction caused by thermal stress (Wu et al., 2022; Mumtaz et al., 2023).

Besides the molecular mechanisms HSPs play in plant molecular level to tolerate thermal stress, they have been proven to enhance the plant cell physiological functions like the photosynthesis process, carbon fixation, membrane permeability, and stability, in addition to stimulating utilizing the available water and nutrient contents efficiently, which all turned out proper crop plant amelioration to extreme heat waves (Ding et al., 2021; Zhou et al., 2022). Bi et al. (2021) recognized a decrease in the ROS concentration in the grass after exposure to thermal stress owing to an increase in the frequency of transcription of *FaHSP17.8-CII*. Not only that, but this HSP also leads to enhanced photosynthesis performance and chlorophyll content, in addition to minimizing chloroplast destruction.

PGPR contributes to tolerating thermal stress in crop plants like *T. aestivum* L. when inoculated with *Bacillus amyloliquefaciens* for whole-day exposure to 45°C. The microorganisms induce HS proteins HSP17.8, HSP26, and HsfA3, as well as ascorbate-glutathione's enzyme activation (Abd El-Daim et al., 2014). Other factors influencing plant growth, such as promoting rhizobacteria and impacting plant cells, such as the generation of osmolytes and improved metabolic activities through thermal stress, are still not thoroughly investigated. Furthermore, in this regard, the involvement of important HSP101, HSP90, and HSP70 HS proteins, which have been widely characterized for bestowing thermal resistance in diverse heat-tolerant *T. aestivum* L. cultivars, was still unknown in PGPR-inducing crop plants during extreme temperatures.

Bacillus safensis as PGPR was recognized by its alleviation of thermal stress effect on cellular membrane and ROS aggregation when *T. aestivum* L. was exposed to 40°C over 12 hours (Ghosh et al., 2018). Yet, HS factors that support *B. safensis* in ameliorating extreme heat conditions, like the antioxidant activities, thermal tolerability genes, and involvement of HSPs, are still not fully comprehended.

The incremental rate of HSP101C, HSP26, and HSP23.5 protein expression in cultivars of *T. aestivum* L. was recognized when the plants were exposed to thermal stress conditions for more than 24 hours. While HSPs of the HSP90 and HSP20 families were reported in *Glycine max* under thermal stress of 38°C for 12 hours, the HSPs of these families are essential in adapting to the HS for a long time (Song et al., 2017). Besides their fundamental function in folding proteins strongly dependent on adenosine triphosphate molecules, HS proteins of HSP70 and HSP90 are extremely important in grain proliferation and flower efflorescence during thermal stress (Płażek et al., 2020). Various heat shock factors (HSF) of Class A actively contribute to ameliorating the environmental incidences in *T. aestivum* L. and other *Arabidopsis* throughout modulating the encoding of response genes (Zhang et al., 2018). Transcription of HSP70 has been proven to increase *Panicum virgatum* when inoculated with PGPB under abiotic stress. HSP70 was noticed to facilitate protein assembly and protect the cell from grouping of emerging newly transcribed proteins, in addition to ensuring correct folding and delivery of proteins (Begum et al., 2019).

17.7 THE CROP-MICROBIOME CROSS-TALKS UPON RAISING TEMPERATURES

Global warming can be like a coin, having negative and positive effects on a broad spectrum of living organisms, including plants. This part of the chapter reviews three sub-sections addressing the dual impacts of global warming on different plant species: (1) the adverse impacts and (2) the influence of microbes on these different impacts at morphological, physiological, and molecular levels. The plants' morphological traits are visible structural traits of the vegetative and reproductive parts of the plant. The vegetative traits focus on the qualitative and quantitative parameters of leaves, stems, and roots, whereas the reproductive traits are related to the development of the reproductive organs and seeds. Physiological traits regulate these traits. Physiological traits are the functional characteristics of plants associated with the metabolic processes controlling plant growth, development, and adaptation to variable environmental conditions. The molecular levels depend on the genetic material and the differential expressions of genes regulating physiological and morphological traits. Thus, the molecular level can be considered the master key of plants' responses and adaptations toward varying circumstances (Hasanuzzaman et al., 2013; Lachowiec et al., 2016).

17.7.1 Ecological Structure and Function of Phyto-Microbiome

Plants have robust adaptive responses to significantly fluctuating environmental conditions, including extreme soil and ambient air temperatures. In general, extreme temperatures, whether excessively cold or hot, trigger many adverse impacts at the morphological and physiological levels (Ahmad et al., 2022). In contrast, at the molecular levels, stressed plants try to restore intracellular chaos due to temperature stresses, exhibiting adaptive mechanisms to tolerate such harsh conditions (Bäurle, 2016). Naturally, all plants are colonized with microbial communities that play crucial roles in plant adaptations to fluctuating biotic and abiotic environmental factors. As evidence, many studies have proven the vital roles of different species of microbes (bacteria and/or fungi) in increasing the host plant's fitness to HS (Nievola et al., 2017). Despite being sessile organisms, plants have robust adaptive responses to significant fluctuating environmental conditions, including extreme soil and ambient air temperatures. In general, extreme temperatures, whether excessively cold or hot, trigger many adverse impacts at morphological and physiological levels. In contrast, at the molecular levels, stressed plants try to restore intracellular chaos due to temperature stresses, exhibiting adaptive mechanisms to tolerate such harsh conditions.

17.7.2 Physiological and Molecular Response of Phyto-Microbiome against High Temperatures

Some significantly affected morphological traits in the context of vegetative and reproductive characteristics. According to the vegetative traits, there are signature phytoresponses toward elevating soil and/or air temperatures, such as reduction of stomata numbers on leaves, decrease of the leaf surface area, along with high susceptibility of leaf rolling and leaf color bleaching (Goufo et al., 2017). Moreover, the development of the plant reproductive organs is impaired significantly under higher temperatures, especially the male reproductive tissues (Giorno et al., 2013). Among the most significant phyto-physiological deteriorations under high temperatures are reductions in stomatal conductance, transpiration rate, and photosynthetic efficiency. These responses are exacerbated by the elevation of the cellular ROS content, leading to structural–functional deteriorations of cellular membranes and many proteins, accelerating leaf senescence and decreasing plant viability.

The molecular scale comprehensively explains such adaptive responses under higher temperatures. According to much of the previous literature, the heat shock transcription factors (HSFs) and HSPs are master molecular players regulating the plant morpho-physiological responses to HS (Haider et al., 2022; Kang et al., 2022; Kan et al., 2023). For example, the reduction of a plant's stomatal index during HS is attributed to HSP90-activation, which subsequently inhibits the stomatal development by phosphorylating and inactivating SPEECHLESS (SPCH), a key transcription factor for initiation of the stomatal formation (Lampard 2009). Another class of HSPs, known as small heat shock proteins (sHSPs), is involved in the leaf morphological traits such as leaf surface area, especially HSP26.8 in *A. stolonifera* L. (Sun et al., 2021). Since the overexpression of HSP26.8 in the transgenic *A. stolonifera* L. has twisted leaf blades with reduced leaf surface areas compared to the wild-type leaves (Sun et al., 2021). Furthermore, HSFs play master roles in plant responses to HS, especially HSFA1 and HSFA2. Class A1-HSFs (such as HSFA1a, HSFA1b, HSFA1d, *and* HSFA1e) are master regulators of heat shock-responsive gene expression, including class A2-HSFs and HSPs (Wunderlich et al., 2007). HSFA2 acts as a regulatory enhancer for specific HRGs related to metabolism and cellular redox balance, particularly during extended periods of HS (Charng et al., 2007). Nevertheless, severe heat waves or prolonged exposure to HS accumulate significantly high ROS contents, impairing photosynthesis efficiency due to structural malformation of chloroplasts, the disintegration of the chloroplast thylakoids, and chlorophyll degradation (Medina et al., 2021). These dramatic effects of HS finally accelerate the leaf senescence and reduce the whole plant's viability. One of the plant defense mechanisms against environmental stressors, including HS, is the phyto-microbiome, including the rhizosphere and endophytes.

17.7.3 Phyto-Microbiome in Combating High-Temperature Stress

Naturally, all plants are colonized with a wide range of different microbial communities that play crucial roles in plant adaptations to the fluctuating biotic and abiotic environmental factors (Ismail et al., 2020; X. Li et al., 2021). As evidence, there are many studies that have proved the vital roles of different species of microbes (bacteria and/or fungi) in increasing the host plant's fitness to HS, such as *T. aestivum* (Abd El-Daim et al., 2018; Abd El-Daim et al., 2019), *Sorghum bicolor* (Bruno et al., 2020), *Z. mays* (Mathur and Jajoo, 2020), *Solanum lycopersicum* (Mukhtar et al., 2020; Sarim et al., 2020), *and O. sativa* (Sangamesh et al., 2018). A comprehensive review re-analyzed metadata from previous studies to highlight the most consistent plant responses to HS when inoculated with microbes, regardless of the plant–microbe species. This analysis aimed to identify generalized plant responses as key fitness pathways influenced by microbial inoculation under HS conditions (Dastogeer et al., 2022). These consistent responses can be categorized into five major classes based on the measured plant traits, which are (1) plant growth, (2) photosynthesis efficiency, (3) antioxidant system, (4) phytohormones, and (5) oxidative stress signs. According to the generalized responses, the studied parameters of plant growth, photosynthesis efficiency, and antioxidant system are generally increased by microbial mediation. In contrast, the oxidative stress signs exhibit significant microbial-mediated reductions compared to the non-inoculated plants. Plant-microbe interactions can modulate plant growth regulators, resulting in microbially mediated up-regulate certain phytohormones while down-regulate others (Kumar and Verma 2018).

The plant growth parameters are positively regulated upon microbial inoculations regarding the shoot and root dry masses compared to the un-inoculated plants (Dastogeer et al., 2022). These observations may be linked to the significant increases in the photosynthetic pigments' levels (chlorophylls and carotenoids), which subsequently enhance the photosynthetic efficiency in the inoculated plants compared to those not inoculated (Dastogeer et al., 2022). The antioxidant system's parameters are summarized into two sub-groups of the expression levels of antioxidant compounds (phenolics and flavonoids) and antioxidant enzymes (glutathione reductase [GR], POD dismutase, and CAT). Like the plant growth parameters results, the antioxidant system parameters are notably enhanced due to microbial inoculations, increasing the plant's fitness toward HS (Dastogeer et al., 2022). Consequently, the oxidative stress signs decreased significantly in the inoculated plants under heat compared with the non-inoculated ones. As evidence, there are apparent decreases in the cellular contents of hydrogen peroxide (H_2O_2) and associated lipid peroxidation fingerprint compound known as MDA (Dastogeer et al., 2022).

The phytohormones have regulatory actions in many pathways and, hence, regulate the plant's fitness to stressful conditions, including HS. There are three differentially expressed phytohormones concluded from the meta-analysis of the previously conducted works of plant–microbial systems compared to non-inoculated ones: ABA, SA, and JA (Dastogeer et al., 2022). The low level of ABA improves plant fitness under HS through many pathways, including (1) plant growth promotion (Vysotskaya et al., 2023), (2) enhancing the photosynthesis efficiency through improving the gas exchange via stomata (Munemasa et al., 2015), and (3) reduction of oxidative stress via improving the antioxidant system (Ibrahim and Jaafar, 2013). In contrast, the SA and JA at high concentrations improve plant health under high temperatures through (1) induction of HSPs responsible for cellular proteins' integrity and functionality under stress exceptionally high temperatures (Sangwan et al., 2022), (2) reduction of ROS levels through stimulation of phenolic production (Rehman et al., 2023), and (3) efficient photosynthesis by maintaining the chlorophyll contents (Su et al., 2021; Das et al., 2024).

In conclusion, plant responses to environmental stresses, including global warming, are highly complex phenomena with interconnected pathways. However, key players—such as specific proteins and transcription factors—are primarily responsible for enhancing plant fitness under HS. Thus, it is mandatory to understand the roles of key regulatory elements and their relationships with microbial inoculation to comprehend how microbes enhance plant thermotolerance. This understanding is crucial for improving bio-based strategies to ensure food security under global warming threats (Carus 2017; Dammer et al., 2023; Demirel et al., 2024).

17.8 IMPACT OF MICROBIOMES-MEDIATED THERMO-ADAPTATION/TOLERANCE ON CROP IMPROVEMENT

Global climate change, characterized by rising temperatures and heat waves, seriously threatens agricultural productivity. Traditional breeding and genetic modification methods for HS resilience are time-consuming and may not fully address plant responses. Emerging strategy uses plant-associated microbiomes for thermo-adaptation and tolerance (Khan, 2023; Soares et al., 2024). The plant microbiome is a varied collection of microbes that live in the endosphere and rhizosphere of the plant. Plant-microbiome interaction is crucial in ecosystem stability, promoting plant health, nutrient cycling, resilience, and adaptation to climate-induced stresses (Petersen et al., 2023; Anas et al., 2025).

These microbes, representing bacteria, fungi, and archaea, can modulate plant stress responses through different mechanisms: (1) some can modulate the plant defense pathways and stress-responsive genes in plants, which increases their tolerance to HS (Branco et al., 2022; Parasar et al., 2024); (2) others produce phytohormones like ABA, SA, and JA which regulate plant responses to abiotic stresses, such as heat, by reducing water loss and promoting stomatal closure (Dastogeer et al., 2022); (3) some microbes associated with plants can boost the antioxidant defenses of the plant, to mitigate oxidative damage produced by HS and so protect the plants (Parasar et al., 2024); (4) they improve water uptake and nutrient acquisition in plants, improving their energy transfer and metabolic processes (Hajiboland, 2013); and (5) microbial interactions increase HSP expression in plants, which enhances their ability to withstand heat (Dastogeer et al., 2022).

This part emphasized the potential of plant microbiome engineering to mitigate HS, the role of bioinoculants in promoting tolerance to high temperatures, and the effects of microbiome-mediated thermo-adaptation on crop improvement.

17.8.1 Bioinoculants for Inducing High-Temperature Stress Tolerance in Plants

Bioinoculants of beneficial microorganisms were used to enhance the heat tolerance of plants by controlling their physiological and molecular responses. Bioinoculants enhance the thermotolerance of plants through various mechanisms: Ali et al. (2009) found that *Pseudomonas* sp. strain AKM-P6 induced thermotolerance of *S. bicolor* seedlings by enhancing the antioxidant enzyme activities. Inoculated seedlings exhibited increased survival and growth with higher roots and shoot biomass under elevated temperatures. Then, Ali et al. (2011) studied the impact of bioinoculating the thermotolerant *Pseudomonas putida* strain AKM-P7 on wheat under HS. They found a reduction in the expression of ROS response genes and increased growth parameters such as spikelet formation, grain yield, dry biomass, and shoot and root length. In addition, Abd El-Daim et al. (2014) revealed that rhizosphere bacterial treatment with *B. amyloliquefaciens* UCMB5113 or *Azospirillum brasilense* NO40 improved HS management of wheat seeds of the cultivars Olivin and Sids1. The bacterial inoculation enhanced growth parameters and reduced cellular damage under elevated temperatures. Later, Abd El-Daim et al. (2018) reported that *B. amyloliquefaciens* 5113 acquired HS tolerance to wheat. The study found that bacterial treatment induced stress-responsive gene expression and upregulated the activities of antioxidant enzymes, contributing to improved plant thermotolerance. Moreover, Bruno et al. (2020) found amelioration of HS response in Sorghum bicolor when co-inoculated with Cr^{6+} reducing-thermotolerant PGPB. The inoculation of *Bacillus cereus* TCR17, *Providencia rettgeri* TCR21, and *Myroides odoratimimus* TCR22 enhanced plant growth, antioxidant status, and decreased proline and MDA contents, indicating reduced oxidative damage under increased atmospheric temperature conditions. Bioinoculation of plant growth-promoting *B. cereus* SA1 to soybean plants enhanced their thermotolerance by promoting biologically active metabolites. This inoculation improved soybean plant biomass, chlorophyll content, and defense enzymes (Khan et al., 2020).

17.8.2 Plant Microbiome Engineering to Combat High-Temperature Stress

Using different molecular mechanisms, plant microbiome engineering provides promising approaches to enhance plant tolerance against high-temperature stress.

Some research focused on overexpressing *HSP* genes to confer heat tolerance to transgenic plants (Dixit and Dhankher 2011). AtSAP10, a stress-associated protein from *A. thaliana* ecotype Columbia, enhances plant resilience to high-temperature stress. They also reported that overexpression of *AtSAP10* gene in the transgenic plants permitted them to grow green and healthy under these conditions. Transgenic tobacco overexpressing constitutive expression of a tomato sHSP gene, *LeHSP21*, exhibited increased thermotolerance by enhancing its antioxidation abilities (Zhang et al., 2016). Huo et al. (2020) overexpressed the autophagy-related protein 18 (MdATG18a) gene in transgenic apples (*Malus domestica*) to ameliorate thermotolerance by reducing chloroplast stress damage. Also, Kumar et al. (2020) found that transgenic *Arabidopsis* with overexpressing *AtHsp101* genes had increased tolerance to elevated temperatures. In addition, Zhang et al. (2022) concluded that overexpressing the HS tolerance 2 (*TaHST2*) gene provided heat tolerance, faster grain filling, and higher yield potential in allohexaploid *T. aestivum* L.

Other investigations highlighted the biostimulant ability of melatonin to mediate temperature stress tolerance in genetically engineered plants. A new heat shock factor (HSF3) from *A. thaliana* plays a crucial role in the heat shock response and confers thermotolerance when overexpressed in transgenic plants (Prändl et al., 1998). Also, Byeon et al. (2014) overexpressed the rice gene encoding tryptophan decarboxylase (*TDC*). They reported that transgenic rice seeds exhibited an elevation in melatonin contents associated with enhanced tolerance to extreme temperatures (Xu et al., 2016). The acetyl serotonin O-methyltransferase (*ASMT*) gene was inserted into the tomato. They found that the transgenic tomatoes exhibited an increase in melatonin levels, upregulated HSPs, and so enhanced their thermotolerance. In addition, Wang et al. (2020) revealed that overexpression of the serotonin N-acetyltransferase (*SlSNAT*) gene improves thermotolerance in tomato plants. They also discovered that the molecular chaperone HSP40 enhanced the stability of *SlSNAT*, leading to increased melatonin synthesis that boosts the plant defense against ROS accumulation under high-temperature stress.

Some researchers tried to acquire the thermotolerance ability of genetically modified plants by accumulating osmolytes. Expression of choline oxidase gene (*codA*) from the bacteria *A. globiformis*, responsible for glycinebetaine synthesis, in transgenic *Araidopsis thaliana plants* conferred their thermotolerance (Alia Hayashi et al., 1998). In addition, Yang et al. (2007) genetically engineered tobacco plants to overproduce glycinebetaine, a compatible solute that protects plants from HS. Overexpressing *codA*- and *BADH* genes in transgenic tomato plants improved thermotolerance by producing glycinebetaine for reducing oxidative stress and enhancing photosynthesis (Zhang et al., 2020).

Other investigators worked to enhance the antioxidant defense systems of transgenic plants. Wang et al. (2006) overexpressed the cytosolic ascorbate peroxidase (*cAPX*) gene in transgenic tomatoes (*Lycopersicon esculentum*) to improve tolerance to HS by scavenging ROS and detoxifying them. Also, Shiraya et al. (2015) overexpressed Golgi/plastid-type manganese superoxide dismutase 1 (MSD1) in rice (*O. sativa*), stimulated heat tolerance, and improved grain quality under HS conditions. In addition, Chiang et al. (2015) studied the overexpression of the genes coding for ascorbate POD from Chinese cabbage (*BcAPX*) to enhance heat tolerance in transgenic *A. thaliana*. They also found that the APX activity in transgenic *Arabidopsis lines was* higher with elevated chlorophyll content and reduced oxidative damage under HS. Wang et al. (2019) identified a novel DnaJ protein that plays a vital role in tomatoes' thermotolerance. They overexpressed the gene of this protein in transgenic tomato plants, enhancing thermotolerance by reducing oxidative stress and improving photosynthesis.

One available report used fatty acid desaturase manipulation for membrane fluidity adjustments. Murakami et al. (2000) worked on silencing the chloroplast omega-3 fatty acid desaturase gene in tobacco (*Nicotiana tabacum*), which synthesizes trienoic fatty acids to increase membrane lipid saturation and enhance heat tolerance. Also, one available report is used to modulate the calcium signaling pathways. Finka et al. (2012) studied the overexpression of the plasma membrane cyclic nucleotide-gated calcium channel (CNGCs) gene in *A. thaliana* to modulate the calcium signaling pathways, enhancing thermal sensing and thermotolerance.

17.9 CONCLUSION

Global warming, especially, poses a cardinal danger to plant development and productivity and to food security. HS destabilizes physiologic processes such as photosynthesis, nutrient acquisition, and reproductive growth, and destabilizes symbiotic plant–microbe interactions. The rhizospheric microbiota is capable of developing an efficient biological buffer through the utilization of a wide array of mechanisms of plant thermotolerance, including mobilization of nutrients, EPS secretion, phytohormone regulation, HSP induction, antioxidant defense system, and VOC and secondary metabolite emission.

Plant growth-promoting rhizobacteria, mycorrhizal fungi, and diazotrophic bacteria are all beneficial microorganisms that do the fundamental things to mitigate high-temperature stress and encourage climate-resilient crop production. From here, the concept of the plant holobiont just emphasizes the fact that crop fitness does not depend on host genetics alone, but also on the colonizing microbiome. The recent developments in the areas of omics, synthetic biology, and CRISPR–Cas systems provide unparalleled opportunities for rhizosphere microbiome engineering and bioinoculant design under climate resilience.

Additional research must prioritize the extrapolation from laboratory and greenhouse findings to field-scale models in terms of microbial persistence, ecological safety, and host plant compatibility. By converging microbiome-based solutions with plant breeding and biotechnology, chemical input use can be reduced, crop resilience can be optimized, and food production can be sustained under predicted global warming temperatures. Exploiting the full potential of microbiome-mediated thermo-adaptation is, therefore, an environmentally friendly and promising path for future agriculture.

REFERENCES

Abalos D, van Groenigen JW, De Deyn GB. 2018. What plant functional traits can reduce nitrous oxide emissions from intensively managed grasslands? Glob Chang Biol. 24(1):e248–e258.

Abd El-Daim IA, Bejai S, Fridborg I, Meijer J. 2018. Identifying potential molecular factors involved in Bacillus amyloliquefaciens 5113 mediated abiotic stress tolerance in wheat. Plant Biol. 20(2):271–279.

Abd El-Daim IA, Bejai S, Meijer J. 2014. Improved heat stress tolerance of wheat seedlings by bacterial seed treatment. Plant Soil. 379:337–350.

Abd El-Daim IA, Bejai S, Meijer J. 2019. Bacillus velezensis 5113 induced metabolic and molecular reprogramming during abiotic stress tolerance in wheat. Sci Rep. 9(1):16282.

Abouzeid S, Lewerenz L, Yahyazadeh M, Radwan A, Hijazin T, Kleinwächter M, Selmar D. 2022. Favorable impacts of drought stress on the quality of medicinal Plants: Improvement of composition and content of their natural products. In: Environ Challenges Med Plants Sustain Prod Solut under Advers Cond. [place unknown]: Springer; pp. 105–131.

Adekanmbi O, Alao A, Ibisola B, Odekunle I. 2023. A comparative analysis of laboratory parameters of hospitalized COVID-19 patients by disease severity and mortality at a facility in Ibadan, Nigeria. Niger Med J. 64(2):243–250.

Afridi MS, Javed MA, Ali S, De Medeiros FHV, Ali B, Salam A, Sumaira, Marc RA, Alkhalifah DHM, Selim S. 2022. New opportunities in plant microbiome engineering for increasing agricultural sustainability under stressful conditions. Front Plant Sci. 13:899464.

Ahmad I, Kamran M, Meng X, Ali S, Ahmad S, Gao Z, Liu T, Han Q. 2021. Hormonal changes with uniconazole trigger canopy apparent photosynthesis and grain filling in wheat crop in a semi-arid climate. Protoplasma. 258:139–150.

Ahmad M, Imtiaz M, Shoib Nawaz M, Mubeen F, Imran A. 2022. What did we learn from current progress in heat stress tolerance in plants? Can microbes be a solution? Front Plant Sci. 13:794782.

Akbar A, Han B, Khan AH, Feng C, Ullah A, Khan AS, He L, Yang X. 2022. A transcriptomic study reveals salt stress alleviation in cotton plants upon salt tolerant PGPR inoculation. Environ Exp Bot. 200:104928.

Ali S, Tyagi A, Kumar R, Zhou G, Hassan MI. 2021. Plant growth-promoting rhizobacteria and thermotolerance: Mechanisms and applications. Front Microbiol. 12:619196. https://doi.org/10.3389/fmicb.2021.619196

Ali SZ, Sandhya V, Grover M, Kishore N, Rao LV, Venkateswarlu B. 2009. Pseudomonas sp. strain AKM-P6 enhances tolerance of sorghum seedlings to elevated temperatures. Biol Fertil Soils. 46:45–55.

Ali SZ, Sandhya V, Grover M, Linga VR, Bandi V. 2011. Effect of inoculation with a thermotolerant plant growth promoting Pseudomonas putida strain AKMP7 on growth of wheat (Triticum spp.) under heat stress. J Plant Interact. 6(4):239–246.

Alia Hayashi H, Sakamoto A, Murata N. 1998. Enhancement of the tolerance of *Arabidopsis* to high temperatures by genetic engineering of the synthesis of glycinebetaine. Plant J. 16(2):155–161.

Allen M, Dube OP, Solecki W, Aragón-Durand F, Cramer W, Humphreys S, Kainuma M. 2018. Special report: Global warming of 1.5 C. Intergov Panel Clim Chang. 27:677.

Almalki DA. 2019. Renoprotective effect of Ocimum Basilicum (basil) against diabetes-induced renal affection in albino rats. Mater Sociomed. 31(4):236.

Anas M, Khalid A, Saleem MH, Ali Khan K, Ahmed Khattak W, Fahad S. 2025. Symbiotic synergy: Unveiling plant-microbe interactions in stress adaptation. J Crop Heal. 77(1):1–21.

Anderson JT, Song B. 2020. Plant adaptation to climate change—where are we? J Syst Evol. 58(5):533–545.

Arnao MB, Hernández-Ruiz J. 2018. Melatonin and its relationship to plant hormones. Ann Bot. 121(2):195–207.

Aroca R, Porcel R, Ruiz-Lozano JM. 2012. Regulation of root water uptake under abiotic stress conditions. J Exp Bot. 63(1):43–57. https://doi.org/10.1093/jxb/err266

Asghar W, Craven KD, Swenson JR, Kataoka R, Mahmood A, Farias JG. 2024. Enhancing the resilience of agroecosystems through improved rhizosphere processes: A strategic review. Int J Mol Sci. 26(1):109.

Ayano M, Kani T, Kojima M, Sakakibara H, Kitaoka T, Kuroha T, Angeles-Shim RB, Kitano H, Nagai K, Ashikari M. 2014. Gibberellin biosynthesis and signal transduction is essential for internode elongation in deepwater rice. Plant Cell Environ. 37(10):2313–2324.

Bae AH, Kim G, Seol GH, Lee SB, Lee JM, Chang W, Min SS. 2020. Delta-and mu-opioid pathways are involved in the analgesic effect of Ocimum basilicum L in mice. J Ethnopharmacol. 250:112471.

Bag S, Mondal A, Majumder A, Mondal SK, Banik A. 2022. Flavonoid mediated selective cross-talk between plants and beneficial soil microbiome. Phytochem Rev. 21(5):1739–1760.

Bailey-Serres J, Parker JE, Ainsworth EA, Oldroyd GED, Schroeder JI. 2019. Genetic strategies for improving crop yields. Nature. 575(7781):109–118.

Bao S, Yang F. 2022. Influences of climate change and land use change on the habitat suitability of Bharal in the Sanjiangyuan district, China. Int J Environ Res Public Health. 19(24):17082.

Bao T, Jia G, Xu X. 2022. Warming enhances dominance of vascular plants over cryptogams across northern wetlands. Glob Chang Biol. 28(13):4097–4109.

Bargaz A, Lyamlouli K, Chtouki M, Zeroual Y, Dhiba D. 2018. Soil microbial resources for improving fertilizers efficiency in an integrated plant nutrient management system. Front Microbiol. 9:1606.

Bartwal A, Mall R, Lohani P, Guru SK, Arora S. 2013. Role of secondary metabolites and brassinosteroids in plant defense against environmental stresses. J Plant Growth Regul. 32(1):216–232. https://doi.org/10.1007/s00344-012-9272-x

Bäurle I. 2016. Plant heat adaptation: Priming in response to heat stress. F1000Research. 5.

Begum N, Hu Z, Cai Q, Lou L. 2019. Influence of PGPB inoculation on HSP70 and HMA3 gene expression in switchgrass under cadmium stress. Plants. 8(11):504.

Belhadj Slimen I, Najar T, Ghram A, Dabbebi H, Ben Mrad M, Abdrabbah M. 2014. Reactive oxygen species, heat stress and oxidative-induced mitochondrial damage. A review. Int J Hyperth. 30(7):513–523. https://doi.org/10.3109/02656736.2014.971446

Ben Abdallah F, Lagha R, Gaber A. 2020. Biofilm inhibition and eradication properties of medicinal plant essential oils against methicillin-resistant Staphylococcus aureus clinical isolates. Pharmaceuticals. 13(11):369.

Berendsen RL, Pieterse CMJ, Bakker PAHM. 2012. The rhizosphere microbiome and plant health. Trends Plant Sci. 17(8):478–486.

Berg G, Rybakova D, Grube M, Köberl M. 2016. The plant microbiome explored: Implications for experimental botany. J Exp Bot. 67(4):995–1002.

Bhantana P, Rana MS, Sun X, Moussa MG, Saleem MH, Syaifudin M, Shah A, Poudel A, Pun AB, Bhat MA, et al. 2021. Arbuscular mycorrhizal fungi and its major role in plant growth, zinc nutrition, phosphorous regulation and phytoremediation. Symbiosis. 84(1):19–37. https://doi.org/10.1007/s13199-021-00756-6

Bi S-S, Jin H-T, Talukder M, Ge J, Zhang C, Lv M-W, Ismail MAY, Li J-L. 2021. The protective effect of nano-selenium against cadmium-induced cerebellar injury via the heat shock protein pathway in chicken. Food Chem Toxicol. 154:112332.

Bigot S, Buges J, Gilly L, Jacques C, Le Boulch P, Berger M, Delcros P, Domergue J, Koehl A, Ley-Ngardigal B. 2018. Pivotal roles of environmental sensing and signaling mechanisms in plant responses to climate change. Glob Chang Biol. 24(12):5573–5589.

Biswas B, Gresshoff PM. 2014. The role of symbiotic nitrogen fixation in sustainable production of biofuels. Int J Mol Sci. 15(5):7380–7397.

Branco S, Schauster A, Liao H, Ruytinx J. 2022. Mechanisms of stress tolerance and their effects on the ecology and evolution of mycorrhizal fungi. New Phytol. 235(6):2158–2175.

Bruno LB, Karthik C, Ma Y, Kadirvelu K, Freitas H, Rajkumar M. 2020. Amelioration of chromium and heat stresses in Sorghum bicolor by Cr6+ reducing-thermotolerant plant growth promoting bacteria. Chemosphere. 244:125521.

Bu QR, Bao MY, Yang Y, Wang TM, Wang CZ. 2022. Targeting virulence factors of candida albicans with natural products. Foods. 11(19). https://doi.org/10.3390/foods11192951

Bulgarelli D, Rott M, Schlaeppi K, Ver Loren van Themaat E, Ahmadinejad N, Assenza F, Rauf P, Huettel B, Reinhardt R, Schmelzer E. 2012. Revealing structure and assembly cues for *Arabidopsis* root-inhabiting bacterial microbiota. Nature. 488(7409):91–95.

Bulgarelli D, Schlaeppi K, Spaepen S, Van Themaat EVL, Schulze-Lefert P. 2013. Structure and functions of the bacterial microbiota of plants. Annu Rev Plant Biol. 64(1):807–838.

Buttar ZA, Wu SN, Arnao MB, Wang Chaojie, Ullah I, Chengshe W. 2020. Melatonin suppressed the heat stress-induced damage in wheat seedlings by modulating the antioxidant machinery. Plants. 9(7):809.

Byeon Y, Back K. 2014. An increase in melatonin in transgenic rice causes pleiotropic phenotypes, including enhanced seedling growth, delayed flowering, and low grain yield. J Pineal Res. 56(4):408–414.

Byeon Y, Park S, Lee HY, Kim Y, Back K. 2014. Elevated production of melatonin in transgenic rice seeds expressing rice tryptophan decarboxylase. J Pineal Res. 56(3):275–282.

Caiserman A, Faour G. 2021. Spatial variability of evapotranspiration and pressure on groundwater resources: Remote sensing monitoring by crop type in the Bekaa plain, Lebanon. J Appl Remote Sens. 15(1):14517.

Calvin K, Dasgupta D, Krinner G, Mukherji A, Thorne PW, Trisos C, Romero J, Aldunce P, Barrett K, Blanco G, et al. 2023. IPCC, 2023: Climate Change 2023: Synthesis report. Contribution of working groups I, II and III to the sixth assessment report of the intergovernmental panel on climate change. In: Core Writing Team, H. Lee and J. Romero (eds.). IPCC, Geneva, Switzerland. Arias P, Bustamante M, Elgizouli I, Flato G, Howden M, Méndez-Vallejo C, Pereira JJ, Pichs-Madruga R, Rose SK, Saheb Y, et al., editors. [place unknown]. https://doi.org/10.59327/IPCC/AR6-9789291691647

Canarini A, Kaiser C, Merchant A, Richter A, Wanek W. 2019. Root exudation of primary metabolites: Mechanisms and their roles in plant responses to environmental stimuli. Front Plant Sci. 10:157.

Canton H. 2021. Food and agriculture organization of the United Nations—FAO. In: Eur Dir Int Organ 2021. [place unknown]: Routledge; pp. 297–305.

Carus M. 2017. Biobased economy and climate change—important links, pitfalls, and opportunities. Ind Biotechnol. 13(2): 41–51.

Casanova-Sáez R, Mateo-Bonmatí E, Ljung K. 2021. Auxin metabolism in plants. Cold Spring Harb Perspect Biol. 13(3):a039867.

Castrillo G, Teixeira PJPL, Paredes SH, Law TF, De Lorenzo L, Feltcher ME, Finkel OM, Breakfield NW, Mieczkowski P, Jones CD. 2017. Root microbiota drive direct integration of phosphate stress and immunity. Nature. 543(7646): 513–518.

Chang CL-T, Yang C-Y, Muthamilselvan T, Yang W-C. 2016. Field trial of medicinal plant, Bidens pilosa, against eimeriosis in broilers. Sci Rep. 6(1):24692.

Change IC. 2013. The physical science basis. (No Title).

Charng Y, Liu H, Liu N, Chi W, Wang C, Chang S, Wang T. 2007. A heat-inducible transcription factor, HsfA2, is required for extension of acquired thermotolerance in *Arabidopsis*. Plant Physiol. 143(1):251–262.

Chatterjee P, Niinemets Ü. 2022. Improving plant stress resistance by growth-promoting bacteria and evaluating the improvements by volatile emissions. Plant Soil. 476(1–2):403–419. https://doi.org/10.1007/s11104-022-05576-1

Chen K, Li G, Bressan RA, Song C, Zhu J, Zhao Y. 2020. Abscisic acid dynamics, signaling, and functions in plants. J Integr Plant Biol. 62(1):25–54.

Chen K, Wang B, Chen C, Zhou G. 2022. MaxEnt modeling to predict the current and future distribution of Pomatosace filicula under climate change scenarios on the Qinghai–Tibet plateau. Plants. 11(5):670.

Chiang CM, Chien HL, Chen LFO, Hsiung TC, Chiang MC, Chen SP, Lin KH. 2015. Overexpression of the genes coding ascorbate peroxidase from Brassica campestris enhances heat tolerance in transgenic *Arabidopsis thaliana*. Biol Plant. 59:305–315.

Chouhan D, Mathur P, Choudhuri C. 2023. Plant-microbe interaction and their role in mitigation of heat stress. In: [place unknown]; pp. 127–147. https://doi.org/10.1007/978-981-99-0030-5_6

Compant S, Samad A, Faist H, Sessitsch A. 2019. A review on the plant microbiome: Ecology, functions, and emerging trends in microbial application. J Adv Res. 19:29–37.

Condron L, Stark C, O'Callaghan M, Clinton P, Huang Z. 2010. The role of microbial communities in the formation and decomposition of soil organic matter. In: Soil Microbiol Sustain Crop Prod. Dordrecht: Springer Netherlands; pp. 81–118. https://doi.org/10.1007/978-90-481-9479-7_4

Cortleven A, Leuendorf JE, Frank M, Pezzetta D, Bolt S, Schmülling T. 2019. Cytokinin action in response to abiotic and biotic stresses in plants. Plant Cell Environ. 42(3):998–1018.

Costa RS, Carneiro TCB, Cerqueira-Lima AT, Queiroz NV, Alcântara-Neves NM, Pontes-de-Carvalho LC, da Silva Velozo E, Oliveira EJ, Figueiredo CA. 2012. Ocimum gratissimum Linn. and rosmarinic acid, attenuate eosinophilic airway inflammation in an experimental model of respiratory allergy to Blomia tropicalis. Int Immunopharmacol. 13(1):126–134.

Crasta OR, Cox WJ. 1996. Temperature and soil water effects on maize growth, development yield, and forage quality. Crop Sci. 36(2):341–348.

Cregger MA, Veach AM, Yang ZK, Crouch MJ, Vilgalys R, Tuskan GA, Schadt CW. 2018. The Populus holobiont: Dissecting the effects of plant niches and genotype on the microbiome. Microbiome. 6:1–14.

Damalas CA. 2019. Improving drought tolerance in sweet basil (Ocimum basilicum) with salicylic acid. Sci Hortic (Amsterdam). 246:360–365.

Dammer L, Carus M, Porc O. 2023. The use of food and feed crops for bio-based materials and the related effects on food security: Promoting evidence-based debates and recognizing potential benefits. Ind Biotechnol. 19(5):257–271.

Das AK, Ghosh PK, Nihad SAI, Sultana S, Keya SS, Rahman MA, Ghosh TK, Akter M, Hasan M, Salma U. 2024. Salicylic acid priming improves cotton seedling heat tolerance through photosynthetic pigment preservation, enhanced antioxidant activity, and osmoprotectant levels. Plants. 13(12):1639.

da Silva Cruz FV, da Costa NB, Juneau P. 2024. Non-pathogenic microbiome associated to aquatic plants and anthropogenic impacts on this interaction. Sci Total Environ.:174663.

Dastogeer KMG, Zahan MI, Rhaman MS, Sarker MSA, Chakraborty A. 2022. Microbe-mediated thermotolerance in plants and pertinent mechanisms-a meta-analysis and review. Front Microbiol. 13:833566.

Davies PJ. 2010. Regulatory factors in hormone action: Level, location and signal transduction. In: Plant Horm Biosynthesis, Signal Transduction, Action! [place unknown]: Springer; pp. 16–35.

de Faria Melo CC, Amaral DS, de Mello Prado R. 2025. The role of mineral nutrients in plant growth under drought stress. In: Sustain Agric under Drought Stress. [place unknown]: Elsevier; pp. 195–207.

Demirel Ö, Güneş H, Can C. 2024. Sustainable and modern biobased technologies: New approachs to food safety and security. Environ Dev Sustain.:1–28.

Ding J, Zheng L, Chu J, Liang X, Wang J, Gao X, Wu F, Sheng S. 2021. Characterization, and functional analysis of Hsp70 and Hsp90 gene families in Glyphodes pyloalis Walker (Lepidoptera: Pyralidae). Front Physiol. 12:753914.

Dixit AR, Dhankher OP. 2011. A novel stress-associated protein 'AtSAP10'from *Arabidopsis thaliana* confers tolerance to nickel, manganese, zinc, and high temperature stress. PLoS One. 6(6):e20921.

Dixit S, Sivalingam PN, Baskaran RKM, Senthil-Kumar M, Ghosh PK. 2024. Plant responses to concurrent abiotic and biotic stress: Unravelling physiological and morphological mechanisms. Plant Physiol Reports. 29(1):6–17.

Driedonks N, Xu J, Peters JL, Park S, Rieu I. 2015. Multi-level interactions between heat shock factors, heat shock proteins, and the redox system regulate acclimation to heat. Front Plant Sci. 6:999.

Du M, Spalding EP, Gray WM. 2020. Rapid auxin-mediated cell expansion. Annu Rev Plant Biol. 71(1):379–402.

Du P, Yuan H, Chen Y, Zhou H, Zhang Y, Huang M, Jiangfang Y, Su R, Chen Q, Lai J. 2023. Identification of key aromatic compounds in basil (Ocimum L.) using sensory evaluation, metabolomics and volatilomics analysis. Metabolites. 13(1):85.

Dustgeer Z, Seleiman MF, Imran K, Chattha MU, Alhammad BA, Jalal RS, Refay Y, Hassan MU. 2021. Glycine-betaine induced salinity tolerance in maize by regulating the physiological attributes, antioxidant defense system and ionic homeostasis. Not Bot Horti Agrobot Cluj-Napoca. 49(1):12248.

Dutta H, Dutta A. 2016. The microbial aspect of climate change. Energy, Ecol Environ. 1(4):209–232. https://doi.org/10.1007/s40974-016-0034-7

Ejack L, Whalen JK. 2021. Freeze-thaw cycles release nitrous oxide produced in frozen agricultural soils. Biol Fertil Soils. 57(3):389–398.

EL Sabagh A, Islam MS, Hossain A, Iqbal MA, Mubeen M, Waleed M, Reginato M, Battaglia M, Ahmed S, Rehman A. 2022. Phytohormones as growth regulators during abiotic stress tolerance in plants. Front Agron. 4:765068.

Farh ME-A, Kim Yu-Jin, Abbai R, Singh P, Jung K-H, Kim Yeon-Ju, Yang D-C. 2020. Pathogenesis strategies and regulation of ginsenosides by two species of Ilyonectria in Panax ginseng: Power of speciation. J Ginseng Res. 44(2):332–340.

Farouk S, Al-Huqail AA. 2020. Sodium nitroprusside application regulates antioxidant capacity, improves phytopharmaceutical production and essential oil yield of marjoram herb under drought. Ind Crops Prod. 158:113034.

Fathy M, Saad Eldin SM, Naseem M, Dandekar T, Othman EM. 2022. Cytokinins: Wide-spread signaling hormones from plants to humans with high medical potential. Nutrients. 14(7):1495.

Feng G, Xiong Y-J, Wei H-Y, Li Y, Mao L-F. 2023. Endemic medicinal plant distribution correlated with stable climate, precipitation, and cultural diversity. Plant Divers. 45(4):479–484.

Fierer N, Jackson RB. 2006. The diversity and biogeography of soil bacterial communities. Proc Natl Acad Sci. 103(3):626–631. https://doi.org/10.1073/pnas.0507535103

Finka A, Cuendet AFH, Maathuis FJM, Saidi Y, Goloubinoff P. 2012. Plasma membrane cyclic nucleotide gated calcium channels control land plant thermal sensing and acquired thermotolerance. Plant Cell. 24(8):3333–3348.

Finkel OM, Salas-González I, Castrillo G, Spaepen S, Law TF, Teixeira PJPL, Jones CD, Dangl JL. 2019. The effects of soil phosphorus content on plant microbiota are driven by the plant phosphate starvation response. PLoS Biol. 17(11):e3000534.

Forzieri G, Feyen L, Rojas R, Flörke M, Wimmer F, Bianchi A. 2014. Ensemble projections of future streamflow droughts in Europe. Hydrol Earth Syst Sci. 18(1):85–108.

Friedlingstein P, O'sullivan M, Jones MW, Andrew RM, Hauck J, Olsen A, Peters GP, Peters W, Pongratz J, Sitch S. 2020. Global carbon budget 2020. Earth Syst Sci Data Discuss. 2020:1–3.

García FC, Bestion E, Warfield R, Yvon-Durocher G. 2018. Changes in temperature alter the relationship between biodiversity and ecosystem functioning. Proc Natl Acad Sci. 115(43):10989–10994. https://doi.org/10.1073/pnas.1805518115

García JE, Maroniche G, Creus C, Suárez-Rodríguez R, Ramirez-Trujillo JA, Groppa MD. 2017. In vitro PGPR properties and osmotic tolerance of different *Azospirillum* native strains and their effects on growth of maize under drought stress. Microbiol Res. 202:21–29.

Geem KR, Kim J, Bae W, Jee M-G, Yu J, Jang I, Lee D-Y, Hong CP, Shim D, Ryu H. 2023. Nitrate enhances the secondary growth of storage roots in Panax ginseng. J Ginseng Res. 47(3):469–478.

Ghosh S, Sarkar P, Basak P, Mahalanobish S, Sil PC. 2018. Role of heat shock proteins in oxidative stress and stress tolerance. Heat Shock Proteins Stress.:109–126.

Giorno F, Wolters-Arts M, Mariani C, Rieu I. 2013. Ensuring reproduction at high temperatures: The heat stress response during anther and pollen development. Plants. 2(3):489–506.

Gobin A. 2018. Weather related risks in Belgian arable agriculture. Agric Syst. 159:225–236.

Godfray HCJ, Beddington JR, Crute IR, Haddad L, Lawrence D, Muir JF, Pretty J, Robinson S, Thomas SM, Toulmin

C. 2010. Food security: The challenge of feeding 9 billion people. Science (80-). 327(5967):812–818.

Goel A, Bhatia AK. 2022. Ocimum sanctum: In vitro antiviral potential against animal viruses.

Gomes GLB, Scortecci KC. 2021. Auxin and its role in plant development: Structure, signalling, regulation and response mechanisms. Plant Biol. 23(6):894–904.

Goufo P, Moutinho-Pereira JM, Jorge TF, Correia CM, Oliveira MR, Rosa EAS, António C, Trindade H. 2017. Cowpea (*Vigna unguiculata* L. Walp.) metabolomics: Osmoprotection as a physiological strategy for drought stress resistance and improved yield. Front Plant Sci. 8:586.

Gourdji SM, Sibley AM, Lobell DB. 2013. Global crop exposure to critical high temperatures in the reproductive period: Historical trends and future projections. Environ Res Lett. 8(2):24041.

Grossiord C, Buckley TN, Cernusak LA, Novick KA, Poulter B, Siegwolf RTW, Sperry JS, McDowell NG. 2020. Plant responses to rising vapor pressure deficit. New Phytol. 226(6):1550–1566.

Guiñazú LB, Andrés JA, Del Papa MF, Pistorio M, Rosas SB. 2010. Response of alfalfa (Medicago sativa L.) to single and mixed inoculation with phosphate-solubilizing bacteria and Sinorhizobium meliloti. Biol Fertil Soils. 46:185–190.

Guo A, Yue W, Yang J, Xue B, Xiao W, Li M, He T, Zhang M, Jin X, Zhou Q. 2023. Cropland abandonment in China: Patterns, drivers, and implications for food security. J Clean Prod. 418:138154.

Gurav TP, Dholakia BB, Giri AP. 2022. A glance at the chemodiversity of Ocimum species: Trends, implications, and strategies for the quality and yield improvement of essential oil. Phytochem Rev. 21(3):879–913.

Hacquard S. 2016. Disentangling the factors shaping microbiota composition across the plant holobiont. New Phytol. 209(2):454–457.

Haider S, Raza A, Iqbal J, Shaukat M, Mahmood T. 2022. Analyzing the regulatory role of heat shock transcription factors in plant heat stress tolerance: A brief appraisal. Mol Biol Rep. 49(6):5771–5785.

Hajiboland R. 2013. Role of arbuscular mycorrhiza in amelioration of salinity. Salt Stress Plants Signalling, Omi Adapt.:301–354.

Hasanuzzaman M, Nahar K, Alam MM, Roychowdhury R, Fujita M. 2013. Physiological, biochemical, and molecular mechanisms of heat stress tolerance in plants. Int J Mol Sci. 14(5):9643–9684.

Haskirli H, Yilmaz O, Ozgur R, Uzilday B, Turkan I. 2021. Melatonin mitigates UV-B stress via regulating oxidative stress response, cellular redox and alternative electron sinks in *Arabidopsis thaliana*. Phytochemistry. 182:112592.

Hassani MA, Durán P, Hacquard S. 2018. Microbial interactions within the plant holobiont. Microbiome. 6:1–17.

Hawkes LA, Broderick AC, Godfrey MH, Godley BJ. 2007. Investigating the potential impacts of climate change on a marine turtle population. Glob Chang Biol. 13(5):923–932. https://doi.org/10.1111/j.1365-2486.2007.01320.x

Higashi Y, Okazaki Y, Myouga F, Shinozaki K, Saito K. 2015. Landscape of the lipidome and transcriptome under heat stress in *Arabidopsis thaliana*. Sci Rep. 5(1):10533. https://doi.org/10.1038/srep10533

Higashi Y, Okazaki Y, Takano K, Myouga F, Shinozaki K, Knoch E, Fukushima A, Saito K. 2018. HEAT INDUCIBLE LIPASE1 remodels chloroplastic monogalactosyldiacylglycerol by liberating α-linolenic acid in *Arabidopsis* leaves under heat stress. Plant Cell. 30(8):1887–1905.

Hönig M, Plíhalová L, Husičková A, Nisler J, Doležal K. 2018. Role of cytokinins in senescence, antioxidant defence and photosynthesis. Int J Mol Sci. 19(12):4045.

Hou M, Wang R, Zhao S, Wang Z. 2021. Ginsenosides in Panax genus and their biosynthesis. Acta Pharm Sin B. 11(7):1813–1834.

Huo L, Sun X, Guo Z, Jia X, Che R, Sun Y, Zhu Y, Wang P, Gong X, Ma F. 2020. MdATG18a overexpression improves basal thermotolerance in transgenic apple by decreasing damage to chloroplasts. Hortic Res. 7.

Hussain AI, Chatha SAS, Kamal GM, Ali MA, Hanif MA, Lazhari MI. 2017. Chemical composition and biological activities of essential oil and extracts from Ocimum sanctum. Int J food Prop. 20(7):1569–1581.

Ibrahim MH, Jaafar HZE. 2013. Abscisic acid induced changes in production of primary and secondary metabolites, photosynthetic capacity, antioxidant capability, antioxidant enzymes and lipoxygenase inhibitory activity of Orthosiphon stamineus Benth. Molecules. 18(7):7957–7976.

Ikeda M, Fujiwara S, Mitsuda N, Ohme-Takagi M. 2012. A triantagonistic basic helix-loop-helix system regulates cell elongation in *Arabidopsis*. Plant Cell. 24(11):4483–4497.

Ikwegbue P, Masamba P, Oyinloye B, Kappo A. 2017. Roles of heat shock proteins in apoptosis, oxidative stress, human inflammatory diseases, and cancer. Pharmaceuticals. 11(1):2. https://doi.org/10.3390/ph11010002

Ilyas N, Mazhar R, Yasmin H, Khan W, Iqbal S, Enshasy HE, Dailin DJ. 2020. Rhizobacteria isolated from saline soil induce systemic tolerance in wheat (Triticum aestivum L.) against salinity stress. Agronomy. 10(7):989.

Iqbal N, Khan NA, Ferrante A, Trivellini A, Francini A, Khan MIR. 2017. Ethylene role in plant growth, development and senescence: Interaction with other phytohormones. Front Plant Sci. 8:475.

Iqbal S, Iqbal MA, Li C, Iqbal A, Abbas RN. 2023. Overviewing drought and heat stress amelioration—from plant responses to microbe-mediated mitigation. Sustainability. 15(2):1671. https://doi.org/10.3390/su15021671

Ismail, Hamayun M, Hussain A, Iqbal A, Khan SA, Lee I-J. 2020. Aspergillus niger boosted heat stress tolerance in sunflower and soybean via regulating their metabolic and antioxidant system. J Plant Interact. 15(1):223–232.

Jaleel CA, Manivannan P, Lakshmanan GMA, Gomathinayagam M, Panneerselvam R. 2008. Alterations in morphological parameters and photosynthetic pigment responses of Catharanthus roseus under soil water deficits. Colloids Surfaces B Biointerfaces. 61(2):298–303.

Jansson JK, Hofmockel KS. 2020. Soil microbiomes and climate change. Nat Rev Microbiol. 18(1):35–46. https://doi.org/10.1038/s41579-019-0265-7

Jeyachandran S, Chellapandian H, Park K, Kwak I-S. 2023. A review on the involvement of heat shock proteins (extrinsic chaperones) in response to stress conditions in aquatic organisms. Antioxidants. 12(7):1444. https://doi.org/10.3390/antiox12071444

Jha CK, Saraf M. 2012. Hormonal signaling by PGPR improves plant health under stress conditions. In: Bact Agrobiol Stress Manag. Berlin, Heidelberg: Springer Berlin Heidelberg; pp. 119–140. https://doi.org/10.1007/978-3-642-23465-1_7

Jiang S, Jardinaud M-F, Gao J, Pecrix Y, Wen J, Mysore K, Xu P, Sanchez-Canizares C, Ruan Y, Li Q. 2021. NIN-like protein transcription factors regulate leghemoglobin genes in legume nodules. Science (80-). 374(6567):625–628.

Kan Y, Mu X-R, Gao J, Lin H-X, Lin Y. 2023. The molecular basis of heat stress responses in plants. Mol Plant. 16(10):1612–1634.

Kang Y, Lee K, Hoshikawa K, Kang M, Jang S. 2022. Molecular bases of heat stress responses in vegetable crops with focusing on heat shock factors and heat shock proteins. Front Plant Sci. 13:837152.

Kaur S, Samota MK, Choudhary M, Choudhary M, Pandey AK, Sharma A, Thakur J. 2022. How do plants defend themselves against pathogens-biochemical mechanisms and genetic interventions. Physiol Mol Biol plants an Int J Funct plant Biol. 28(2):485–504. https://doi.org/10.1007/s12298-022-01146-y

Kaushal N, Bhandari K, Siddique KHM, Nayyar H. 2016. Food crops face rising temperatures: An overview of responses, adaptive mechanisms, and approaches to improve heat tolerance. Cogent Food Agric. 2(1):1134380.

Khan A, Ahmad M, Ahmed M, Gill KS, Akram Z. 2021. Association analysis for agronomic traits in wheat under terminal heat stress. Saudi J Biol Sci. 28(12):7404–7415.

Khan AL. 2023. The phytomicrobiome: Solving plant stress tolerance under climate change. Front Plant Sci. 14:1219366.

Khan I, Lei H, Shah AA, Khan Inayat, Muhammad I. 2021. Climate change impact assessment, flood management, and mitigation strategies in Pakistan for sustainable future. Environ Sci Pollut Res. 28:29720–29731.

Khan MA, Asaf S, Khan AL, Jan R, Kang S-M, Kim K-M, Lee I-J. 2020. Thermotolerance effect of plant growth-promoting Bacillus cereus SA1 on soybean during heat stress. BMC Microbiol. 20:1–14.

Khan Y, Bin Q, Hassan T. 2019. The impact of climate changes on agriculture export trade in Pakistan: Evidence from time-series analysis. Growth Change. 50(4):1568–1589.

Kogan F. 2023. The IPCC reports on global warming and land changes. In: Remote Sens L Surf Chang 1981–2020 Intensive Glob Warm. [place unknown]: Springer; pp. 67–79.

Kumar A, Verma JP. 2018. Does plant—microbe interaction confer stress tolerance in plants: A review? Microbiol Res. 207:41–52.

Kumar R, Khungar L, Shimphrui R, Tiwari LD, Tripathi G, Sarkar NK, Agarwal S-K, Agarwal M, Grover A. 2020. AtHsp101 research sets course of action for the genetic improvement of crops against heat stress. J Plant Biochem Biotechnol. 29:715–732.

Kumar R, Neerudu UK, Gothalwal R, Mohapatra S, Deshpande PK, Vani MM, Merugu R. 2021. Bioprocess parameters for thermophilic and mesophilic biogas production: Recent trends and challenges. In: [place unknown]; pp. 225–256. https://doi.org/10.1007/978-981-33-4611-6_8

Kumari A, Dash M, Singh SK, Jagadesh M, Mathpal B, Mishra PK, Pandey SK, Verma KK. 2023. Soil microbes: A natural solution for mitigating the impact of climate change. Environ Monit Assess. 195(12):1436. https://doi.org/10.1007/s10661-023-11988-y

Kumari S, Kumar S, Prakash P. 2018. Exogenous application of cytokinin (6-BAP) ameliorates the adverse effect of combined drought and high temperature stress in wheat seedling. J Pharmacogn Phytochem. 7(1):1176–1180.

Lachowiec J, Queitsch C, Kliebenstein DJ. 2016. Molecular mechanisms governing differential robustness of development and environmental responses in plants. Ann Bot. 117(5):795–809.

Lal MK, Tiwari RK, Kumar A, Dey A, Kumar R, Kumar D, Jaiswal A, Changan SS, Raigond P, Dutt S. 2022. Mechanistic concept of physiological, biochemical, and molecular responses of the potato crop to heat and drought stress. Plants. 11(21):2857.

Lamaoui M, Jemo M, Datla R, Bekkaoui F. 2018. Heat and drought stresses in crops and approaches for their mitigation. Front Chem. 6:26.

Lampard GR. 2009. The missing link? *Arabidopsis* SPCH is a MAPK specificity factor that controls entry into the stomatal lineage: *Arabidopsis* SPCH is a MAPK specificity factor that controls entry into the stomatal lineage. Plant Signal Behav. 4(5):425–427.

Lebeis SL. 2015. Greater than the sum of their parts: Characterizing plant microbiomes at the community-level. Curr Opin Plant Biol. 24:82–86.

Lee H, Calvin K, Dasgupta D, Krinner G, Mukherji A, Thorne P, Trisos C, Romero J, Aldunce P, Barret K. 2023. IPCC, 2023: Climate change 2023: Synthesis report, summary for policymakers. Contribution of Working Groups I, II and III to the Sixth Assessment Report of the Intergovernmental Panel on Climate Change [Core Writing Team, H. Lee and J. Romero (eds.)]. IPC.

Lei H, Zhang H, Zhang Z, Sun H, Li M, Shao C, Liang H, Wu H, Zhang Y. 2023. Physiological and transcriptomic analyses of roots from Panax ginseng CA Meyer under drought stress. Ind Crops Prod. 191:115858.

Lemanceau P, Blouin M, Muller D, Moënne-Loccoz Y. 2017. Let the core microbiota be functional. Trends Plant Sci. 22(7):583–595.

Leng G, Hall J. 2019. Crop yield sensitivity of global major agricultural countries to droughts and the projected changes in the future. Sci Total Environ. 654:811–821.

Li N, Euring D, Cha JY, Lin Z, Lu M, Huang L-J, Kim WY. 2021. Plant hormone-mediated regulation of heat tolerance in response to global climate change. Front Plant Sci. 11:627969.

Li Q, Li H, Yang Z, Cheng X, Zhao Y, Qin L, Bisseling T, Cao Q, Willemsen V. 2022. Plant growth-promoting rhizobacterium Pseudomonas sp. CM11 specifically induces lateral roots. New Phytol. 235(4):1575–1588.

Li X, Zhao C, Zhang T, Wang G, Amombo E, Xie Y, Fu J. 2021. Exogenous Aspergillus aculeatus enhances drought and heat tolerance of perennial ryegrass. Front Microbiol. 12:593722.

Li Y, Huang M. 2008. Pasture yield and soil water depletion of continuous growing alfalfa in the Loess Plateau of China. Agric Ecosyst Environ. 124(1–2):24–32.

Li Z, Yu J, Peng Y, Huang B. 2016. Metabolic pathways regulated by γ-aminobutyric acid (GABA) contributing to heat tolerance in creeping bentgrass (Agrostis stolonifera). Sci Rep. 6(1):30338. https://doi.org/10.1038/srep30338

Lin Y-S, Medlyn BE, Duursma RA, Prentice IC, Wang H, Baig S, Eamus D, De Dios VR, Mitchell P, Ellsworth DS. 2015. Optimal stomatal behaviour around the world. Nat Clim Chang. 5(5):459–464.

Liu X, Huang B. 2002. Cytokinin effects on creeping bentgrass response to heat stress: II. Leaf senescence and antioxidant metabolism. Crop Sci. 42(2):466–472.

Mahmud AA, Upadhyay SK, Srivastava AK, Bhojiya AA. 2021. Biofertilizers: A Nexus between soil fertility and crop productivity under abiotic stress. Curr Res Environ Sustain. 3:100063.

Maitra S, Pramanick B, Dey P, Bhadra P, Shankar T, Anand K. 2021. Thermotolerant Soil Microbes and Their Role in Mitigation of Heat Stress in Plants. In: [place unknown]; pp. 203–242. https://doi.org/10.1007/978-3-030-73507-4_8

Manaharan T, Thirugnanasampandan R, Jayakumar R, Ramya G, Ramnath G, Kanthimathi MS. 2014. Antimetastatic and anti-inflammatory potentials of essential oil from edible Ocimum sanctum leaves. Sci World J. 2014(1):239508.

Mander Ü, Krasnova A, Escuer-Gatius J, Espenberg M, Schindler T, Machacova K, Pärn J, Maddison M, Megonigal JP, Pihlatie M. 2021. Forest canopy mitigates soil N_2O emission during hot moments. NPJ Clim Atmos Sci. 4(1):39.

Manghwar H, Li J. 2022. Endoplasmic reticulum stress and unfolded protein response signaling in plants. Int J Mol Sci. 23(2):828.

Mathur S, Jajoo A. 2020. Arbuscular mycorrhizal fungi protects maize plants from high temperature stress by regulating photosystem II heterogeneity. Ind Crops Prod. 143:111934.

McGuire MR. 2017. Introduction: The interface of climate change and food security. In: Routledge Handb Technol Crime Justice. [place unknown]: Routledge; pp. 1–14.

Medeiros Venancio A, Ferreira-da-Silva FW, da Silva-Alves KS, de Carvalho Pimentel H, Macêdo Lima M, Fraga de Santana M, Barreto Alves P, Batista da Silva G, Leal-Cardoso JH, Marchioro M. 2016. Essential oil of Ocimum basilicum L. and (–)-Linalool blocks the excitability of Rat Sciatic Nerve. Evidence-Based Complement Altern Med. 2016(1):9012605.

Medina E, Kim S-H, Yun M, Choi W-G. 2021. Recapitulation of the function and role of ROS generated in response to heat stress in plants. Plants. 10(2):371.

Mekouar MA. 2020. 15. Food and agriculture organization of the United Nations (FAO). Yearb Int Environ Law. 31(1):326–340.

Mironov V, Vanteeva A, Merkel A. 2021. Microbiological activity during co-composting of food and agricultural waste for soil amendment. Agronomy. 11(5):928. https://doi.org/10.3390/agronomy11050928

Mishra S, Hättenschwiler S, Yang X. 2020. The plant microbiome: A missing link for the understanding of community dynamics and multifunctionality in forest ecosystems. Appl Soil Ecol. 145:103345.

Morcillo R, Manzanera M. 2021. The effects of plant-associated bacterial exopolysaccharides on plant abiotic stress tolerance. Metabolites. 11(6):337. https://doi.org/10.3390/metabo11060337

Mougel C, Offre P, Ranjard L, Corberand T, Gamalero E, Robin C, Lemanceau P. 2006. Dynamic of the genetic structure of bacterial and fungal communities at different developmental stages of Medicago truncatula Gaertn. cv. Jemalong line J5. New Phytol. 170(1):165–175.

Mounira D, Abdelouahab D, Widad F, Riadh BM, Farid B, Ahmed B. 2022. Cytotoxic, antioxidant and antimicrobial activities of Peganum harmala L. extracts. Biotechnol Acta. 15(1):61–71.

Mukhtar T, Rehman SU, Smith D, Sultan T, Seleiman MF, Alsadon AA, Amna, Ali S, Chaudhary HJ, Solieman THI. 2020. Mitigation of heat stress in Solanum lycopersicum L. by ACC-deaminase and exopolysaccharide producing Bacillus cereus: Effects on biochemical profiling. Sustainability. 12(6):2159.

Mulugeta SM, Radácsi P. 2022. Influence of drought stress on growth and essential oil yield of Ocimum species. Horticulturae. 8(2):175.

Mulugeta SM, Sárosi S, Radácsi P. 2023. Physio-morphological trait and bioactive constituents of Ocimum species under drought stress. Ind Crops Prod. 205:117545.

Mumtaz MA, Zhou Y, Gao C, Kamran HM, Altaf MA, Hao Y, Shu H, Zhang Y, Lu X, Abbas HMK. 2023. Interaction between transcriptional activator BRI1-EMS-SUPPRESSOR 1 and HSPs regulates heat stress tolerance in pepper. Environ Exp Bot. 211:105341.

Munemasa S, Hauser F, Park J, Waadt R, Brandt B, Schroeder JI. 2015. Mechanisms of abscisic acid-mediated control of stomatal aperture. Curr Opin Plant Biol. 28:154–162.

Murakami Y, Tsuyama M, Kobayashi Y, Kodama H, Iba K. 2000. Trienoic fatty acids and plant tolerance of high temperature. Science (80-). 287(5452):476–479.

Nacamulli C, Bevivino A, Dalmastri C, Tabacchioni S, Chiarini L. 1997. Perturbation of maize rhizosphere microflora following seed bacterization with Burkholderia cepacia MCI 7. FEMS Microbiol Ecol. 23(3):183–193.

Naz N, Hameed M, Ashraf M, Ahmad R, Arshad M. 2009. Eco-morphic variation for salt tolerance in some grasses from Cholistan Desert, Pakistan. Pak J Bot. 41(4):1707–1714.

Nelson DR, Mele PM. 2007. Subtle changes in rhizosphere microbial community structure in response to increased boron and sodium chloride concentrations. Soil Biol Biochem. 39(1):340–351.

Nguyen NV. 2002. Global climate changes and rice food security. Rome FAO. 625.

Nievola CC, Carvalho CP, Carvalho V, Rodrigues E. 2017. Rapid responses of plants to temperature changes. Temperature. 4(4):371–405.

Ntonga PA, Baldovini N, Mouray E, Mambu L, Belong P, Grellier P. 2014. Activity of Ocimum basilicum, Ocimum canum, and Cymbopogon citratus essential oils against Plasmodium falciparum and mature-stage larvae of Anopheles funestus ss. Parasite. 21.

Osei Akoto C, Acheampong A, Boakye YD, Naazo AA, Adomah DH. 2020. Anti-inflammatory, antioxidant, and anthelmintic activities of Ocimum basilicum (Sweet Basil) fruits. J Chem. 2020(1):2153534.

Pachauri RK, Allen MR, Barros VR, Broome J, Cramer W, Christ R, Church JA, Clarke L, Dahe Q, Dasgupta P. 2014. Climate change 2014: Synthesis report. Contribution of Working Groups I, II and III to the Fifth Assessment Report of the Intergovernmental Panel on Climate Change. [place unknown]: Ipcc.

Palmer JD, Foster KR. 2022. Bacterial species rarely work together. Science (80-). 376(6593):581–582.

Parasar BJ, Kashyap S, Sharma I, Marme SD, Das P, Agarwala N. 2024. Microbe mediated alleviation of drought and heat stress in plants-current understanding and future prospects. Discov Plants. 1(1):26.

Park I, Seo Y-S, Mannaa M. 2023. Recruitment of the rhizo-microbiome army: Assembly determinants and engineering of the rhizosphere microbiome as a key to unlocking plant potential. Front Microbiol. 14:1163832.

Park S, Kim A-L, Hong Y-K, Shin J-H, Joo S-H. 2021. A highly efficient auxin-producing bacterial strain and its effect on plant growth. J Genet Eng Biotechnol. 19:1–9.

Pavlů J, Novák J, Koukalová V, Luklová M, Brzobohatý B, Černý M. 2018. Cytokinin at the crossroads of abiotic stress signalling pathways. Int J Mol Sci. 19(8):2450.

Peer LA, Dar ZA, Lone AA, Bhat MY, Ahamad N. 2020. High temperature triggered plant responses from whole plant to cellular level. Plant Physiol Reports. 25(4):611–626. https://doi.org/10.1007/s40502-020-00551-3

Petersen C, Hamerich IK, Adair KL, Griem-Krey H, Torres Oliva M, Hoeppner MP, Bohannan BJM, Schulenburg H. 2023. Host and microbiome jointly contribute to environmental adaptation. ISME J. 17(11):1953–1965.

Pieterse Corné MJ, Zamioudis C, Van Does der D, Van Wees SCM. 2014a. Signalling networks involved in induced resistance. Induc Resist plant Def.:58–80.

Pieterse Corné MJ, Zamioudis C, Berendsen RL, Weller DM, Van Wees SCM, Bakker PAHM. 2014b. Induced Systemic Resistance by Beneficial Microbes. Annu Rev Phytopathol. 52(1):347–375. https://doi.org/10.1146/annurev-phyto-082712-102340

Płażek A, Hura K, Hura T, Słomka A, Hornyák M, Sychta K. 2020. Synthesis of heat-shock proteins HSP-70 and HSP-90 in flowers of common buckwheat (Fagopyrum esculentum) under thermal stress. Crop Pasture Sci. 71(8):760–767.

Prändl R, Hinderhofer K, Eggers-Schumacher G, Schöffl F. 1998. HSF3, a new heat shock factor from *Arabidopsis thaliana*, derepresses the heat shock response and confers thermotolerance when overexpressed in transgenic plants. Mol Gen Genet MGG. 258:269–278.

Prerostova S, Vankova R. 2023. Phytohormone-mediated regulation of heat stress response in plants. In: Plant Horm Clim Chang. Singapore: Springer Nature Singapore; p. 167–206. https://doi.org/10.1007/978-981-19-4941-8_8

Qamar KA, Farooq AD, Siddiqui BS, Kabir N, Begum S. 2020. Antiproliferative effects of Ocimum basilicum methanolic extract and fractions, Oleanolic acid and 3-epi-Ursolic acid. Curr Tradit Med. 6(2):134–146.

Qaseem MF, Qureshi R, Shaheen H. 2019. Effects of pre-anthesis drought, heat and their combination on the growth, yield and physiology of diverse wheat (Triticum aestivum L.) genotypes varying in sensitivity to heat and drought stress. Sci Rep. 9(1):6955.

Qasem A, Assaggaf H, Mrabti HN, Minshawi F, Rajab BS, Attar AA, Alyamani RA, Hamed M, Mrabti NN, Baaboua AE. 2023. Determination of chemical composition and investigation of biological activities of Ocimum basilicum L. Molecules. 28(2):614.

Quinones C, Mattes N, Faronilo J, Jagadish KSV. 2017. Drought stress reduces grain yield by altering floral meristem development and sink size under dry-seeded rice cultivation. Crop Sci. 57(4):2098–2108.

Radácsi P, Inotai K, Sárosi S, Seidler-Łożykowska K, Musie S, Zámboriné É. 2020. Effect of irrigation on the production and volatile compounds of sweet basil cultivars (Ocimum basilicum L.). Herba Pol. 66(4):14–24.

Raja TAR, Reddy RVRN, Buchineni M. 2016. An evaluation of anti-hyperglycemic activity of Ocimum sanctum Linn (leaves) in Wister rats. Pharma Innov. 5(1, Part A):1.

Rang ZW, Jagadish SVK, Zhou QM, Craufurd PQ, Heuer S. 2011. Effect of high temperature and water stress on pollen germination and spikelet fertility in rice. Environ Exp Bot. 70(1):58–65.

Rani A, Guleria M, Sharma Y, Sharma S, Chaudhary A, Sharma R, Kumar P. 2023. Insights into elicitor's role in augmenting secondary metabolites production and climate resilience in genus Ocimum–A globally important medicinal and aromatic crop. Ind Crops Prod. 202:117078.

Raza A, Charagh S, Abbas S, Hassan MU, Saeed F, Haider S, Sharif R, Anand A, Corpas FJ, Jin W, Varshney RK. 2023. Assessment of proline function in higher plants under extreme temperatures. Plant Biol. 25(3):379–395. https://doi.org/10.1111/plb.13510

Rehman M, Saeed MS, Fan X, Salam A, Munir R, Yasin MU, Khan AR, Muhammad S, Ali B, Ali I. 2023. The multifaceted role of jasmonic acid in plant stress mitigation: An overview. Plants. 12(23):3982.

Robert-Seilaniantz A, Grant M, Jones JDG. 2011. Hormone crosstalk in plant disease and defense: More than just jasmonate-salicylate antagonism. Annu Rev Phytopathol. 49:317–343.

Rodriguez PA, Rothballer M, Chowdhury SP, Nussbaumer T, Gutjahr C, Falter-Braun P. 2019. Systems biology of plant-microbiome interactions. Mol Plant. 12(6):804–821.

Rosenberg E, Zilber-Rosenberg I. 2018. The hologenome concept of evolution after 10 years. Microbiome. 6:1–14.

Rousk J, Frey SD, Bååth E. 2012. Temperature adaptation of bacterial communities in experimentally warmed forest soils. Glob Chang Biol. 18(10):3252–3258. https://doi.org/10.1111/j.1365-2486.2012.02764.x

Ryan J, Ibrikci H, Sommer R, McNeill A. 2009. Nitrogen in rainfed and irrigated cropping systems in the Mediterranean region. Adv Agron. 104:53–136.

Samanta S, Banerjee A, Roychoudhury A. 2021. Exogenous melatonin regulates endogenous phytohormone homeostasis and thiol-mediated detoxification in two indica rice cultivars under arsenic stress. Plant Cell Rep. 40(8):1585–1602.

Samanta S, Banerjee A, Roychoudhury A. 2022. Arsenic toxicity is counteracted by exogenous application of melatonin to different extents in arsenic-susceptible and arsenic-tolerant rice cultivars. J Plant Growth Regul. 41(6):2210–2231.

Sánchez-Cañizares C, Jorrín B, Poole PS, Tkacz A. 2017. Understanding the holobiont: The interdependence of plants and their microbiome. Curr Opin Microbiol. 38:188–196.

Sangamesh MB, Jambagi S, Vasanthakumari MM, Shetty NJ, Kolte H, Ravikanth G, Nataraja KN, Uma Shaanker R. 2018. Thermotolerance of fungal endophytes isolated from plants adapted to the Thar Desert, India. Symbiosis. 75:135–147.

Sangwan S, Shameem N, Yashveer S, Tanwar H, Parray JA, Jatav HS, Sharma S, Punia H, Sayyed RZ, Almalki WH. 2022. Role of salicylic acid in combating heat stress in plants: Insights into modulation of vital processes. Front Biosci. 27(11):310.

Santos RM, Bakhshoodeh R. 2021. Climate change/global warming/climate emergency versus general climate research: Comparative bibliometric trends of publications. Heliyon. 7(11).

Sarim KM, Srivastava R, Ramteke PW. 2020. Next-generation omics technologies for exploring complex metabolic regulation during plant-microbe interaction. In: Microb Serv Restor Ecol. [place unknown]: Elsevier; pp. 123–138.

Sen S, Chandrasekhar CN. 2014. Effect of PGPR on growth promotion of rice (Oryza sativa L.) under salt stress. Asian J Plant Sci Res. 4(5):62–67.

Shaffique S, Khan MA, Wani SH, Pande A, Imran M, Kang S-M, Rahim W, Khan SA, Bhatta D, Kwon E-H. 2022. A review on the role of endophytes and plant growth promoting rhizobacteria in mitigating heat stress in plants. Microorganisms. 10(7):1286.

Shahbandeh M, Elhag M. 2024. Microclimate changes and trend analysis of remotely sensed environmental parameters in West Asia Semi-arid region. Environ Dev Sustain. 26(6):14433–14447.

Sharifi P, Mohammadkhani N. 2016. Effects of drought stress on photosynthesis factors in wheat genotypes during anthesis. Cereal Res Commun. 44(2):229–239.

Shi P, Zhu Y, Tang L, Chen J, Sun T, Cao W, Tian Y. 2016. Differential effects of temperature and duration of heat stress during anthesis and grain filling stages in rice. Environ Exp Bot. 132:28–41.

Shilev S. 2020. Plant-growth-promoting bacteria mitigating soil salinity stress in plants. Appl Sci. 10(20):7326.

Shiraya T, Mori T, Maruyama T, Sasaki M, Takamatsu T, Oikawa K, Itoh K, Kaneko K, Ichikawa H, Mitsui T. 2015. Golgi/plastid-type manganese superoxide dismutase involved in heat-stress tolerance during grain filling of rice. Plant Biotechnol J. 13(9):1251–1263.

Shu W-S, Huang L-N. 2022. Microbial diversity in extreme environments. Nat Rev Microbiol. 20(4):219–235. https://doi.org/10.1038/s41579-021-00648-y

Simon J-C, Marchesi JR, Mougel C, Selosse M-A. 2019. Host-microbiota interactions: From holobiont theory to analysis. Microbiome. 7:1–5.

Singh A, Agrawal M. 2015. Responses of periwinkle (Catharanthus roseus L.) to CO2 enrichment: Chlorophyll florescence, pigments, metabolites, growth and biomass. Clim Chang Environ Sustain. 3(1):51–57.

Smith SE, Jakobsen I, Grønlund M, Smith FA. 2011. Roles of Arbuscular Mycorrhizas in plant phosphorus nutrition: Interactions between pathways of phosphorus uptake in Arbuscular Mycorrhizal roots have important implications for understanding and manipulating plant phosphorus acquisition. Plant Physiol. 156(3):1050–1057. https://doi.org/10.1104/pp.111.174581

Soares EV, Egamberdieva D, Smith DL, Khan AL. 2024. The phytomicrobiome: Solving plant stress tolerance. Bio-Based Solut Sustain Dev Agric. 2.

Solomon S. 2007. IPCC (2007): Climate change the physical science basis. In: Agu Fall Meet Abstr. Vol. 2007. [place unknown]; pp. U43D–01.

Song K, Yim WC, Lee B-M. 2017. Expression of heat shock proteins by heat stress in soybean. Plant Breed Biotechnol. 5(4):344–353.

Song Y, Lv J, Ma Z, Dong W. 2019. The mechanism of alfalfa (Medicago sativa L.) response to abiotic stress. Plant Growth Regul. 89:239–249.

Su Y, Huang Y, Dong X, Wang R, Tang M, Cai J, Chen J, Zhang X, Nie G. 2021. Exogenous methyl jasmonate improves heat tolerance of perennial ryegrass through alteration of osmotic adjustment, antioxidant defense, and expression of jasmonic acid-responsive genes. Front Plant Sci. 12:664519.

Sun X, Huang N, Li X, Zhu J, Bian X, Li H, Wang L, Hu Q, Luo H. 2021. A chloroplast heat shock protein modulates growth and abiotic stress response in creeping bentgrass. Plant Cell Environ. 44(6):1769–1787.

Sun Y, Alseekh S, Fernie AR. 2023. Plant secondary metabolic responses to global climate change: A meta-analysis in medicinal and aromatic plants. Glob Chang Biol. 29(2):477–504.

Tanweer Kumar TK, Bao AiKe BA, Bao ZhuLaTai BZ, Wang Fei WF, Gao Li GL, Wang SuoMin WS. 2018. The progress of genetic improvement in alfalfa (Medicago sativa L.).

Terhonen E, Blumenstein K, Kovalchuk A, Asiegbu FO. 2019. Forest tree microbiomes and associated fungal endophytes: Functional roles and impact on forest health. Forests. 10(1):42.

Tian X, Matsui T, Li S, Yoshimoto M, Kobayasi K, Hasegawa T. 2010. Heat-induced floret sterility of hybrid rice (Oryza sativa L.) cultivars under humid and low wind conditionsin the field of Jianghan Basin, China. Plant Prod Sci. 13(3):243–251.

Tiryaki I, Keles H. 2012. Reversal of the inhibitory effect of light and high temperature on germination of Phacelia tanacetifolia seeds by melatonin. J Pineal Res. 52(3):332–339.

Trivedi P, Leach JE, Tringe SG, Sa T, Singh BK. 2020. Plant–microbiome interactions: From community assembly to plant health. Nat Rev Microbiol. 18(11):607–621.

Tucker MA, Schipper AM, Adams TSF, Attias N, Avgar T, Babic NL, Barker KJ, Bastille-Rousseau G, Behr DM, Belant JL. 2023. Behavioral responses of terrestrial mammals to COVID-19 lockdowns. Science (80-). 380(6649):1059–1064.

Ulfat A, Aslam A, Mehmood A, Wazarat A. 2024. Variation in nonstructural carbohydrates and antioxidant metabolism in wheat leaf and spike under changing CO2 and nitrogen supply. Acta Physiol Plant. 46(7):75.

ul Haq S, Khan A, Ali M, Khattak AM, Gai W-X, Zhang H-X, Wei A-M, Gong Z-H. 2019. Heat shock proteins: Dynamic biomolecules to counter plant biotic and abiotic stresses. Int J Mol Sci. 20(21):5321. https://doi.org/10.3390/ijms20215321

Ullah A, Ali I, Noor J, Zeng F, Bawazeer S, Eldin SM, Asghar MA, Javed HH, Saleem K, Ullah S. 2023. Exogenous γ-aminobutyric acid (GABA) mitigated salinity-induced impairments in mungbean plants by regulating their nitrogen metabolism and antioxidant potential. Front Plant Sci. 13:1081188.

Urakawa R, Shibata H, Kuroiwa M, Inagaki Y, Tateno R, Hishi T, Fukuzawa K, Hirai K, Toda H, Oyanagi N. 2014. Effects of freeze–thaw cycles resulting from winter climate change on soil nitrogen cycling in ten temperate forest ecosystems throughout the Japanese archipelago. Soil Biol Biochem. 74:82–94.

Urbanova T, Leubner-Metzger G. 2016. Gibberellins and seed germination. Annu Plant Rev. 49 Gibberellins,.:253–284.

Uroz S, Courty PE, Oger P. 2019. Plant symbionts are engineers of the plant-associated microbiome. Trends Plant Sci. 24(10):905–916.

Vandenkoornhuyse P, Quaiser A, Duhamel M, Le Van A, Dufresne A. 2015. The importance of the microbiome of the plant holobiont. New Phytol. 206(4):1196–1206.

Vysotskaya L, Martynenko E, Ryabova A, Kuzmina L, Starikov S, Chetverikov S, Gaffarova E, Kudoyarova G. 2023. The growth-inhibitory effect of increased planting density can be reduced by abscisic acid-degrading bacteria. Biomolecules. 13(11):1668.

Wahab A, Muhammad M, Munir A, Abdi G, Zaman W, Ayaz A, Khizar C, Reddy SPP. 2023. Role of Arbuscular Mycorrhizal fungi in regulating growth, enhancing productivity, and potentially influencing ecosystems under abiotic and biotic stresses. Plants. 12(17):3102. https://doi.org/10.3390/plants12173102

Wahid F, Sharif M, Ali A, Fahad S, Adnan M, Noor M, Mian IA, Khan IA, Alam M, Saeed M, et al. 2020. Plant-microbes interactions and functions in changing climate. In: Environ Clim Plant Veg Growth. Cham: Springer International Publishing; pp. 397–419. https://doi.org/10.1007/978-3-030-49732-3_16

Walters WA, Jin Z, Youngblut N, Wallace JG, Sutter J, Zhang W, González-Peña A, Peiffer J, Koren O, Shi Q. 2018. Large-scale replicated field study of maize rhizosphere identifies heritable microbes. Proc Natl Acad Sci. 115(28):7368–7373.

Wang G, Cai G, Xu N, Zhang L, Sun X, Guan J, Meng Q. 2019. Novel DnaJ protein facilitates thermotolerance of transgenic tomatoes. Int J Mol Sci. 20(2):367.

Wang H, Ren C, Cao L, Jin X, Wang M, Zhang M, Zhao Q, Li H, Zhang Y, Yu G. 2021. The mechanisms underlying melatonin improved soybean seedling growth at different nitrogen levels. Funct Plant Biol. 48(12):1225–1240.

Wang S, Zheng J, Wang Y, Yang Q, Chen T, Chen Y, Chi D, Xia G, Siddique KHM, Wang T. 2021. Photosynthesis, chlorophyll fluorescence, and yield of peanut in response to biochar application. Front Plant Sci. 12:650432.

Wang W, Yuan S, Wu C, Yang S, Zhang W, Xu Y, Gu J, Zhang H, Wang Z, Yang J. 2021. Field experiments and model simulation based evaluation of rice yield response to projected climate change in Southeastern China. Sci Total Environ. 761:143206.

Wang X, Zhang H, Xie Q, Liu Y, Lv H, Bai R, Ma R, Li X, Zhang X, Guo Y-D. 2020. SlSNAT interacts with HSP40, a molecular chaperone, to regulate melatonin biosynthesis and promote thermotolerance in tomato. Plant Cell Physiol. 61(5):909–921.

Wang Y, Wisniewski M, Meilan R, Cui M, Fuchigami L. 2006. Transgenic tomato (Lycopersicon esculentum) overexpressing cAPX exhibits enhanced tolerance to UV-B and heat stress. J Appl Hortic. 8:87–90.

Watts-Williams SJ, Smith FA, McLaughlin MJ, Patti AF, Cavagnaro TR. 2015. How important is the mycorrhizal pathway for plant Zn uptake? Plant Soil. 390(1–2):157–166. https://doi.org/10.1007/s11104-014-2374-4

Wheeler T, Von Braun J. 2013. Climate change impacts on global food security. Science (80-). 341(6145):508–513.

Wu C, Cui K, Wang W, Li Q, Fahad S, Hu Q, Huang J, Nie L, Mohapatra PK, Peng S. 2017. Heat-induced cytokinin transportation and degradation are associated with reduced panicle cytokinin expression and fewer spikelets per panicle in rice. Front Plant Sci. 8:371.

Wu J, Gao T, Hu J, Zhao L, Yu C, Ma F. 2022. Research advances in function and regulation mechanisms of plant small heat shock proteins (sHSPs) under environmental stresses. Sci Total Environ. 825:154054.

Wu J, Gu Y, Sun K, Wang N, Shen H, Wang Y, Ma X. 2023. Correlation of climate change and human activities with agricultural drought and its impact on the net primary production of winter wheat. J Hydrol. 620:129504.

Wu L, Weston LA, Zhu S, Zhou X. 2023. Editorial: Rhizosphere interactions: Root exudates and the rhizosphere microbiome. Front Plant Sci. 14.

Wu X, Li T, Wang D, Wang F, Fu B, Liu G, Lv Y. 2020. Soil properties mediate the freeze-thaw-related soil N_2O and CO_2 emissions from temperate grasslands. Catena. 195:104797.

Wunderlich M, Doll J, Busch W, Kleindt CK, Lohmann C, Schöffl F. 2007. Heat shock factors: Regulators of early and late functions in plant stress response. Plant Stress. 1(1):16–22.

Xu J, Zhang Y, Zhang P, Trivedi P, Riera N, Wang Y, Liu X, Fan G, Tang J, Coletta-Filho HD. 2018. The structure and function of the global citrus rhizosphere microbiome. Nat Commun. 9(1):4894.

Xu W, Cai S, Zhang Y, Wang Y, Ahammed GJ, Xia X, Shi K, Zhou Y, Yu J, Reiter RJ. 2016. Melatonin enhances thermotolerance by promoting cellular protein protection in tomato plants. J Pineal Res. 61(4):457–469.

Yan F, Wei H, Ding Y, Li W, Liu Z, Chen L, Tang S, Ding C, Jiang Y, Li G. 2021. Melatonin regulates antioxidant strategy in response to continuous salt stress in rice seedlings. Plant Physiol Biochem. 165:239–250.

Yan Q, Zhang X, Li M, Zhang G, Huang L-F. 2024. Global climate change promotes aboveground growth and high-latitude species distribution in medicinal plants: Insights from the genus Panax. J Clean Prod. 478:143863.

Yang D, Li Yong, Shi Y, Cui Z, Luo Y, Zheng M, Chen J, Li Yanxia, Yin Y, Wang Z. 2016. Exogenous cytokinins increase grain yield of winter wheat cultivars by improving stay-green characteristics under heat stress. PLoS One. 11(5):e0155437.

Yang JC, Zhang JH, Wang ZQ, Zhu QS, Liu LJ. 2003. Involvement of abscisic acid and cytokinins in the senescence and remobilization of carbon reserves in wheat subjected to water stress during grain filling. Plant Cell Environ. 26(10):1621–1631.

Yang X, Chen F, Lin X, Liu Z, Zhang H, Zhao J, Li K, Ye Q, Li Y, Lv S. 2015. Potential benefits of climate change for crop productivity in China. Agric For Meteorol. 208:76–84.

Yang X, Wen X, Gong H, Lu Q, Yang Z, Tang Y, Liang Z, Lu C. 2007. Genetic engineering of the biosynthesis of glycinebetaine enhances thermotolerance of photosystem II in tobacco plants. Planta. 225:719–733.

Yoneyama T, Terakado-Tonooka J, Minamisawa K. 2017. Exploration of bacterial N_2-fixation systems in association with soil-grown sugarcane, sweet potato, and paddy rice: A review and synthesis. Soil Sci Plant Nutr. 63(6):578–590.

Yu P, He X, Baer M, Beirinckx S, Tian T, Moya YAT, Zhang X, Deichmann M, Frey FP, Bresgen V. 2021. Plant flavones enrich rhizosphere Oxalobacteraceae to improve maize performance under nitrogen deprivation. Nat plants. 7(4):481–499.

Yurgel SN, Douglas GM, Dusault A, Percival D, Langille MGI. 2018. Dissecting community structure in wild blueberry root and soil microbiome. Front Microbiol. 9:1187.

Zahra SM, Shahid MA, Misaal MA, Zaman M, Imran M, Azam S, Hussain F. 2023. Sustainable water management in indus basin and vulnerability due to climate change. Environ Sci Proc. 25(1):36.

Zahran EM, Abdel-Maqsoud NMR, Tammam OY, Abdel-Rahman IM, Elrehany MA, Bakhsh HT, Altemani FH, Algehainy NA, Alzubaidi MA, Abdelmohsen UR. 2022. Scabicidal potential of coconut seed extract in rabbits via downregulating inflammatory/immune cross talk: A comprehensive phytochemical/GC-MS and in silico proof. Antibiotics. 12(1):43.

Zahran HH. 1999. Rhizobium-legume symbiosis and nitrogen fixation under severe conditions and in an arid climate. Microbiol Mol Biol Rev. 63(4):968–989.

Zengin G, Ferrante C, Gnapi DE, Sinan KI, Orlando G, Recinella L, Diuzheva A, Jekő J, Cziáky Z, Chiavaroli A. 2019. Comprehensive approaches on the chemical constituents and pharmacological properties of flowers and leaves of American basil (Ocimum americanum L). Food Res Int. 125:108610.

Zhang J, Chen H, Wang H, Li B, Yi Y, Kong F, Liu J, Zhang H. 2016. Constitutive expression of a tomato small heat shock protein gene LeHSP21 improves tolerance to high-temperature stress by enhancing antioxidation capacity in tobacco. Plant Mol Biol Report. 34:399–409.

Zhang J, Liu W, Bu J, Lin Y, Bai Y. 2023. Host genetics regulate the plant microbiome. Curr Opin Microbiol. 72:102268.

Zhang J, Liu Y-X, Zhang N, Hu B, Jin T, Xu H, Qin Y, Yan P, Zhang X, Guo X. 2019. NRT1. 1B is associated with root microbiota composition and nitrogen use in field-grown rice. Nat Biotechnol. 37(6):676–684.

Zhang L, Zhang Z, Tao F, Luo Y, Zhang J, Cao J. 2022. Adapting to climate change precisely through cultivars renewal for rice production across China: When, where, and what cultivars will be required? Agric For Meteorol. 316:108856.

Zhang R, Liu G, Xu H, Lou H, Zhai S, Chen A, Hao S, Xing J, Liu J, You M. 2022. Heat Stress Tolerance 2 confers basal heat stress tolerance in allohexaploid wheat (Triticum aestivum L.). J Exp Bot. 73(19):6600–6614.

Zhang T, Li Z, Li D, Li C, Wei D, Li S, Liu Y, Chen THH, Yang X. 2020. Comparative effects of glycinebetaine on the thermotolerance in codA-and BADH-transgenic tomato plants under high temperature stress. Plant Cell Rep. 39:1525–1538.

Zhang Y, Chen F-S, Wu X-Q, Luan F-G, Zhang L-P, Fang X-M, Wan S-Z, Hu X-F, Ye J-R. 2018. Isolation and characterization of two phosphate-solubilizing fungi from rhizosphere soil of moso bamboo and their functional capacities when exposed to different phosphorus sources and pH environments. PLoS One. 13(7):e0199625.

Zhang Y, Trivedi P, Xu J, Roper MC, Wang N. 2021. The citrus microbiome: From structure and function to microbiome engineering and beyond. Phytobiomes J. 5(3):249–262.

Zheng H, Fu X, Shao J, Tang Y, Yu M, Li L, Huang L, Tang K. 2023. Transcriptional regulatory network of high-value active ingredients in medicinal plants. Trends Plant Sci. 28(4):429–446.

Zhou R, Yu X, Ottosen C-O, Rosenqvist E, Zhao L, Wang Y, Yu W, Zhao T, Wu Z. 2017. Drought stress had a predominant effect over heat stress on three tomato cultivars subjected to combined stress. BMC Plant Biol. 17:1–13.

Zhou Y, Wang Y, Xu F, Song C, Yang X, Zhang Z, Yi M, Ma N, Zhou X, He J. 2022. Small HSPs play an important role in crosstalk between HSF-HSP and ROS pathways in heat stress response through transcriptomic analysis in lilies (Lilium longiflorum). BMC Plant Biol. 22(1):202.

Zhu X, Yuan X, Zhang Y, Liu H, Wang J, Sun B. 2022. The global concern of food security during the COVID-19 pandemic: Impacts and perspectives on food security. Food Chem. 370:130830.

18 Microbiome
Current Challenges and Opportunities

Carolina E. Demaman Oro[1], *João H. C. Wancura*[2], *Dinalva Schein*[2], *Rogério M. Dallago*[1], *and Marcus V. Tres*[3]*

[1]Department of Food Engineering, Integrated Regional University of Alto Uruguay and Missions, 1621 Sete de Setembro Av., 99709-910, Erechim, RS, Brazil

[2]Department of Chemical Engineering, Federal University of Santa Maria, 1000 Roraima Av., Building 9B, 97105-900, Santa Maria, RS, Brazil

[3]Laboratory of Agroindustrial Processes Engineering (LAPE), Federal University of Santa Maria (UFSM), 3013 Taufik Germano Rd., University II DC, Cachoeira do Sul 96503-205, RS, Brazil

18.1 INTRODUCTION

Over the course of millions of years, microorganisms have adapted to form complex ecological networks on Earth, collectively known as microbiomes. These microbial assemblages can inhabit both living hosts—including humans, animals, and plants—and non-living environments such as soil, oceans, and the atmosphere. Within microbiomes, microorganisms engage in dynamic interactions with their hosts and with each other, fostering synergistic relationships that shape microbial community structure and influence the characteristics of their surrounding ecosystems. These interactions have significant implications, affecting both host physiology and environmental conditions. For example, the gut microbiome in humans has been associated with various aspects of physical and mental health, while microbial communities in soil play a fundamental role in agricultural productivity. Given the extensive influence of microbiomes on ecosystems, research into their composition, dynamics, and the factors driving their variations has expanded considerably in recent years (George et al., 2022; Grice & Segre, 2012; Shayanthan et al., 2022; Su et al., 2023).

The fundamental characteristics of ecosystems are closely linked to the core microbiomes they support. These distinctive features arise from intricate interaction networks within microbial communities, as well as between microorganisms and their hosts, influenced by metabolites produced by microbiota. Disruptions to these microbiomes can result in profound changes, including declines in host health or reductions in soil fertility. This highlights the crucial role of microbiome stability in sustaining the resilience of both organisms and their surrounding environments. Among various microbiomes, the gut microbiome stands out as the most diverse and densely populated, exerting a vital influence on host well-being. Alterations in its composition have been associated with a range of gastrointestinal disorders, such as ulcerative colitis, *Clostridium difficile* infections, and inflammatory bowel disease (Angima et al., 2024; Bolsega et al., 2021; Woo & Kim, 2024). These associations emphasize the essential role of gut microbial balance in maintaining overall health and preventing disease. Similarly, animals host diverse microbiomes, whose composition is directly linked to their growth and physiological health (Foo et al., 2017).

The plant microbiome, primarily represented by root-associated microbial communities, is equally complex. This microbiome includes microbes within the plant roots (endophytic compartment) and those in the rhizosphere. Root-associated microbial communities, composed of thousands of bacterial and fungal taxa, are influenced by environmental conditions, plant genetics, and microbe-microbe interactions. Functionally, these communities enhance plant physiology, provide protection against pathogens, and contribute to crop health and productivity (Brereton et al., 2020; dos Santos et al., 2023; Frąc et al., 2022; Kotoky et al., 2018; Nie et al., 2023; Su et al., 2023).

In contrast, the soil microbiome extends beyond the rhizosphere to encompass microbial communities in bulk soil. Agricultural management practices heavily influence these communities, and engineering soil microbiomes has shown promise in establishing more diverse and balanced microbial populations. Such efforts improve soil health, enhance nutrient cycling, and boost plant fertility (George et al., 2022; Gupta et al., 2024; Q. Liu et al., 2024; Singh et al., 2022; Yuan et al., 2021).

The significance of microbiomes in shaping host physiology and ecosystem characteristics has gained widespread recognition due to advances in microbiome research. Harnessing this knowledge, microbiome studies have emerged as a promising field with applications ranging from disease treatment to enhancing agricultural productivity. Despite its success in these areas, the study of the microbiome remains

***Corresponding author:** marcus.tres@ufsm.br (Prof. Marcus V. Tres)

DOI: 10.1201/9781003544425-18

in its early stages, offering immense potential for future exploration and innovation.

This chapter explores the multifaceted roles of microbial communities across human health, environmental sustainability, and agricultural productivity. It aims to provide an overview of the challenges associated with studying and utilizing microbiota and microbiomes, highlighting technological and methodological barriers. By addressing current challenges in microbiome research, such as technological limitations, data integration, and the lack of standardization, it seeks to provide a comprehensive understanding of the potential field. Additionally, it examines opportunities in personalized medicine, bioremediation, and sustainable agriculture, emphasizing the transformative potential of microbiome research to advance global health and environmental resilience.

18.2 CHALLENGES

18.2.1 Complexity and Diversity

The human body hosts a vast array of microorganisms—collectively referred to as the microbiota—that colonize and adapt to different regions within the organism. The term "microbiota" describes the living microorganisms residing in specific environments, such as the gut, skin, lungs, oral cavity, and urogenital tract. In contrast, "microbiome" encompasses the collective genomes of these microorganisms, along with structural components, metabolites, and environmental factors influencing their community (Hou et al., 2022). Bacteria constitute over 90% of these microbial populations, coexisting with viruses, yeasts, and archaea. Together, these microorganisms create a genetic repository that exceeds the human genome by more than 150-fold (Grice & Segre, 2012; Hou et al., 2022). The human microbiota plays an essential role in sustaining physiological functions, contributing to various biological processes, including nutrient metabolism, interactions with the immune, endocrine, and nervous systems, biosynthesis of vitamins and amino acids, and defense against pathogenic colonization (Filardo et al., 2024). Figure 18.1 illustrates the dominant phyla of the human microbiota.

Among microbial communities, the gut microbiota is the most complex (Filardo et al., 2024), accounting for approximately 10^{13} bacteria. The dominant phyla are *Firmicutes* (60–80%) and *Bacteroidetes* (20–40%), with smaller populations of *Proteobacteria*, *Actinobacteria*, *Fusobacteria*, and

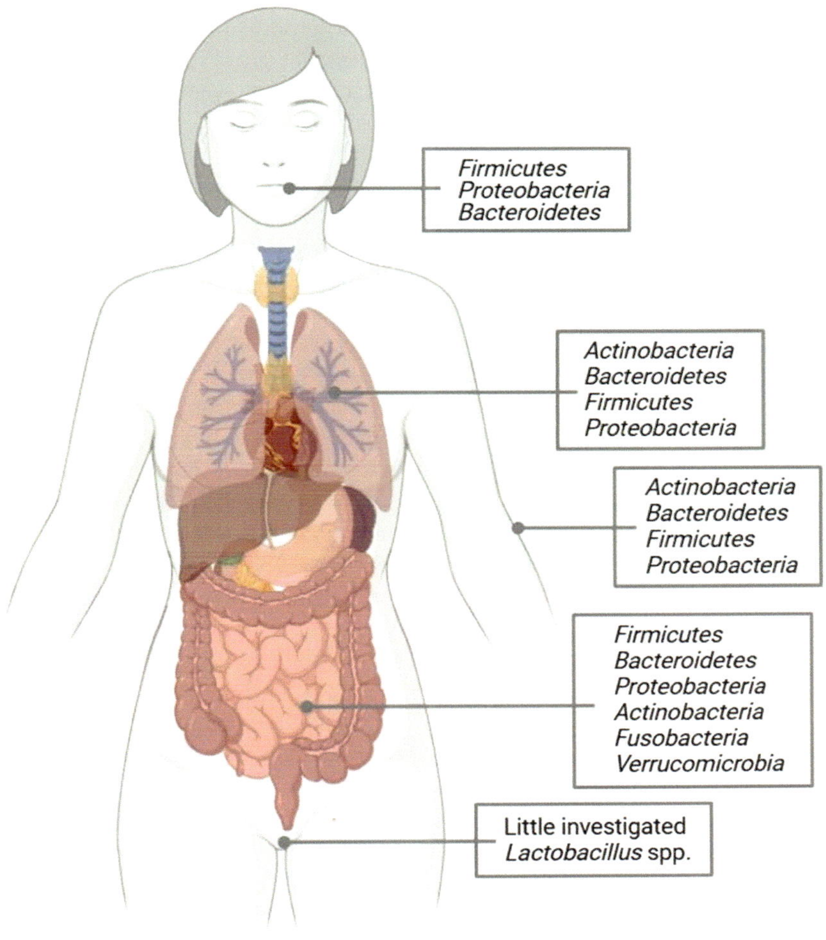

FIGURE 18.1 Dominant phyla of the human microbiota.

Verrucomicrobia (Filardo et al., 2022). Most microorganisms are obligate anaerobes, such as *Bacteroides*, *Clostridium*, and *Fusobacterium*, contrasting with facultative anaerobes like *Lactobacillus*, *Escherichia*, and *Enterococcus* (Filardo et al., 2022). The most extensively studied fungal species within the gut microbiota include *Candida*, *Saccharomyces*, *Malassezia*, and *Cladosporium* (Hou et al., 2022). The microorganisms in the gut microbiota perform three fundamental functions: host protection, metabolic activities, and contribution to immune system regulation (Salem et al., 2018). The composition of the gut microbiota may vary depending on the characteristics of different intestinal regions, such as pH, nutrient gradients, antimicrobial factors, and the presence of other microorganisms (Shealy et al., 2024). Dysbiosis within the gut microbiota has been associated with impaired nutrient absorption, immune system dysfunction, and increased intestinal permeability (Takiishi et al., 2017).

The oral microbiota constitutes the second largest microbial community in humans (D. Verma et al., 2018). The predominant bacterial phyla in this environment are *Firmicutes*, *Proteobacteria*, and *Bacteroidetes* (Segata et al., 2012). In cases of microbial imbalance, *Porphyromonas gingivalis* predominates (Demkovych et al., 2023). The oral microbiota inhabits various environments, such as saliva, tongue, tooth surfaces, and gums, and is susceptible to rapid alterations in its composition, influenced by pH variations and interactions with other bacteria (Segata et al., 2012). In contrast, the cervicovaginal microbiota remains one of the least explored microbial communities. In healthy women, it consists of a relatively simple microbial composition, predominantly featuring *Lactobacillus* spp., which account for over 80% of the total population (France et al., 2022). Additionally, the lung microbiota in healthy individuals primarily includes bacteria from the phyla *Actinobacteria*, *Bacteroidetes*, *Firmicutes*, and *Proteobacteria* (Yu et al., 2016), while the skin microbiota is generally composed of *Actinobacteria*, *Bacteroidetes*, *Firmicutes*, and *Proteobacteria* (Woo & Kim, 2024).

Microbiological niches within the human body exhibit substantial complexity and diversity, with microbiotas playing crucial roles in host health. Each region of the body hosts a unique microbiota, adapted to local conditions such as pH, nutrient availability, and microbial interactions. These diverse microbial ecosystems perform essential functions, including pathogen defense and immune system modulation. However, the balance of these communities is dynamic and can be easily disrupted by external factors, especially considering the use of medication, directly impacting the host's health. Diseases caused by microbial imbalance include inflammatory bowel disease, respiratory diseases, diabetes, and liver diseases, among others. Despite significant advancements, understanding the intricate interactions between host and resident microorganisms remains a challenge due to the complexity of these ecosystems. However, ongoing research on the human microbiota is crucial to comprehend this complexity and develop strategies to restore and maintain microbial balance (Filardo et al., 2024).

18.2.2 Technological Limitations, Data Integration, and Standardization

The identification and quantification of microorganisms in environmental, clinical, or food samples still face significant challenges, even with the use of modern technologies (Filardo et al., 2024). In distinct microbial communities, certain microorganisms are present in extremely low concentrations, making their detection and quantification difficult, as sequencing may end up prioritizing more abundant organisms (Bonk et al., 2018). Contextually, these low-abundance microorganisms remain underestimated or invisible in analyses, although they have the potential to play crucial ecological roles, such as regulating microbial interactions or maintaining specific ecosystem functions (Bartsch et al., 2023). Since the genomes of microorganisms vary widely in size, organization, and composition, such variability can make it difficult to uniformly recover DNA from different microorganisms in the same sample (Carabeo-Pérez et al., 2019). Furthermore, very small genomes, such as those of viruses and microorganisms with a high degree of simplification, may be disregarded in certain analyses.

The vast majority of microorganisms present in natural environments, in the human body, or in other ecosystems remain unknown or poorly studied due to the impossibility of cultivating them in the laboratory using traditional methods. Microorganisms considered "uncultivable" are those that cannot grow in standardized culture media available in laboratories. The claim that over 99% of microorganisms present in different natural environments have not yet been cultivated is widely accepted in microbiology, limiting the ability to study them in detail (Vartoukian et al., 2010). The difficulty in the cultivation process lies in the complex physical–chemical conditions required by them, where each microorganism has unique environmental requirements of temperature, pH, salinity, osmotic pressure, oxygen availability, etc. to develop (Kapinusova et al., 2023; Loftus et al., 2020; Ortúzar et al., 2024). Accordingly, in extreme ecosystems (such as hot springs, deep oceans, or deserts), microorganisms adapt to highly specific conditions that are difficult to reproduce in the laboratory. Moreover, some microorganisms require specific nutrients or highly specialized chemical compounds, which may not be present in conventional culture media (Kapinusova et al., 2023). Another issue concerns symbiotic relationships, where several microorganisms depend on interaction with other organisms (bacteria, fungi, plants, or animals) to develop. A well-known example is certain gut microbes, which require metabolites produced by other microorganisms to grow, making it impossible to cultivate them in isolation (Wu et al., 2023).

Another important issue is the limited availability of references in databases. The identification of microorganisms depends on the comparison of their DNA sequences with available cataloged genome data. However, several microorganisms have not yet been sequenced or catalogued, especially those from unexplored ecosystems such as deep soils, oceans, or the digestive tracts of some organisms. As a result, sequences obtained in microbiome studies are often classified as "unidentified" or assigned to generic categories such as *operational taxonomic units (OTUs)* or *unclassified bacteria* (Furtak et al., 2020; Wang et al., 2023).

Different sequencing platforms (such as Illumina, PacBio, and Oxford Nanopore) have varying levels of accuracy, sensitivity, and ability to detect full-length sequences. Short sequences (such as those obtained with Illumina) can make it difficult to distinguish between microorganisms that are very closely related phylogenetically. On the other hand, methods based on amplification of target genes (such as 16s rRNA) can introduce biases, since the primers used may preferentially amplify certain microbial groups and ignore others (Bailén et al., 2020; Filardo et al., 2024).

The integration of microbiomic data with other areas of biology is essential to understand the interactions between microorganisms and their environment, however, it is undeniable that it also presents considerable challenges. One of the main problems is the heterogeneity of the data, which have different scales, formats, and levels of complexity (Filardo et al., 2024). Microbiomic data is often sparse and dimensional, while metabolomic information varies according to environmental conditions. Furthermore, identifying causal relationships among these data requires sophisticated statistical approaches, such as mediation analyses and causal models (Samarrai et al., 2022).

Practical examples of this integration show important advances. In the DINE-CD study, metagenomic and metabolomic data were combined to investigate changes in the gut microbiome of Crohn's disease patients treated with a specific diet (Deek et al., 2024). The authors reported associations between microbial alterations as well as fecal and plasma metabolites, suggesting potential biomarkers of the disease. Another example is multiomics analysis in colorectal cancer, which integrates gut microbiomes with genomics and metabolomics to identify how *Fusobacterium nucleatum* species affect fatty acid metabolism and tumor progression (Kong et al., 2021; Pani, 2023). This microorganism has been linked to significant alterations in lipid metabolism and cancer-associated inflammation. *F. nucleatum*, for example, activates the TLR4/Keap1/NRF2 pathway, resulting in increased production of 12,13-epoxyoctadecenoic acid (12,13-EpOME), a compound that promotes cell migration and epithelial-mesenchymal transition, processes associated with metastasis in colorectal cancer (Pani, 2023). Data integration is also applied in the study of the gut–brain axis, where the metabolomic and transcriptomic approach explores the impact of the gut microbiome on mental health, in conditions such as anxiety and depression (Tan, 2023; Xiong et al., 2023). To facilitate these advances, tools such as Multi-Omics Factor Analysis (MOFA) and coassociative networks help map interactions and shared patterns across data (Zhang et al., 2024). Machine learning models are also critical to identifying complex relationships and predicting responses to therapies. In the work of Yu et al. (2024), the authors collected fecal samples from more than 160 cows, performed a 16S amplicon sequencing, and subsequently developed an interpretable machine learning framework to classify individuals based on their milk urea nitrogen concentrations. In the study, the authors were able to identify nine microorganisms that were strongly correlated with a key parameter. Despite technical challenges, such as the lack of standardization in data collection and the need for high computational capacity, multiomics integration offers a powerful approach to unravel the role of the microbiome in health and disease.

The lack of standardization in the collection, processing, and analysis of collected samples is another key challenge in microbiome analysis that limits the comparison of results between different studies. Each step, from sampling to data analysis, can be influenced by methodological decisions that vary widely among researchers (Filardo et al., 2024; Samarrai et al., 2022). Furthermore, the form as the samples are stored prior to analysis (temperature, storage time, or use of preservatives) can alter the integrity of the microbiota, generating inconsistencies in the data. Research highlights that samples refrigerated at 4°C exhibit relative stability for short periods, while immediate freezing at −80°C is often recommended to minimize degradation. Additionally, choosing preservatives such as RNAlater can better preserve RNA integrity for specific analyses, although it also incurs costs (Moossavi et al., 2019; Nagata et al., 2019). In complex environments, such as soil samples or human guts, there is a mixture of microbial DNA, host, and other organisms. Such complexity can generate noise in the data and interfere with accurate species identification. Accordingly, small contaminations, either during sample collection or sequencing, can lead to the detection of microorganisms that were not originally present, confusing the analyses (Moossavi et al., 2019).

Another crucial point is sample processing, which involves the use of different protocols for DNA and RNA extraction, directly influencing the representation of the microbiome. For example, DNA extraction protocols can be optimized for different tissue types or to better preserve certain bacterial species; however, this can lead to a lack of consistency between studies (Silverman et al., 2021). Comparative research demonstrates that variables such as the number of polymerase chain reaction (PCR) cycles or the choice of primers for genetic amplification can introduce bias into the data, affecting the diversity and abundance of the detected species (Silverman et al., 2021). Typically, increasing the number of PCR cycles can amplify sequences from low-abundance organisms, but at the cost of increased formation of chimeric sequences, which compromise data quality (Nearing et al., 2021). Furthermore, the choice of primers may generate a selective bias in the amplification of certain species, resulting in under- or overrepresentation of specific microbial communities (Nearing et al., 2021; Sipos et al., 2007).

In data analysis, the lack of standardization in the choice of bioinformatics methods also contributes to variability between studies. DNA sequencing analysis tools such as QIIME, Mothur, or DADA2 may produce slightly different results depending on how parameters are set for sequencing analysis, such as read quality or taxonomic key filtering (Govender & Ghai, 2025; Mancin et al., 2024; Navas-Molina et al.,

2013). Furthermore, data normalization approaches, such as abundance ratio filtering or sequencing depth standardization, vary among researchers, which can lead to divergent results on microbiome composition (Govender & Ghai, 2025).

Studies on the standardization of methodologies are essential to advance in this field (Filardo et al., 2024; Govender & Ghai, 2025; Samarrai et al., 2022). The *Consortium for Microbiome Studies* has promoted the standardization of sample collection and processing methodologies to minimize variability in results. Initiatives such as the *Microbiome Quality Control Project* are focused on establishing best practices for the analysis and interpretation of microbiome data with the aim of improving reproducibility and comparability between studies. Additionally, efforts such as those of the *Human Microbiome Project* aim to create more uniform collection and analysis protocols to ensure more consistent and reproducible results. Such initiatives have proven essential to facilitate the comparison of microbiomic data from different studies, providing a more solid basis for the discovery of universal patterns and the identification of biomarkers associated with diseases (Sinha et al., 2015).

18.3 OPPORTUNITIES

18.3.1 Personalized Medicine, Disease Prevention, and Treatment

The expansion of microbiome research in pre-clinical and clinical contexts represents a significant step forward in fully harnessing its potential (Aggarwal et al., 2023). Among disease prevention strategies, lifestyle modifications are prominent, and microbiome-related data can enhance precision in predicting responses to these alterations. Personalized medicine, in turn, leverages microbiome sequencing to identify microorganisms and their interactions with patient health. There is an established correlation between microbial dysbiosis and conditions that predispose individuals to diseases such as obesity, insulin sensitivity, and hypertension (Ratiner et al., 2024).

The gut microbiota has been recognized as the most significant microbial community in preserving overall health. Experimental strategies targeting the microbiome include microorganism-based approaches, such as dietary supplementation with probiotics (Ratiner et al., 2024). Additionally, the transfer of complex microbial communities seeks to establish a stable and optimized microbiota capable of colonizing dysbiotic environments, thereby restoring microbial balance (Junca et al., 2022). In this context, dietary prebiotics are defined as selectively fermentable ingredients that induce specific modifications in the composition and activity of the gastrointestinal microbiota (Bevilacqua et al., 2024). These compounds offer various health benefits, including local immune modulation, maintenance of intestinal wall integrity, and enhancement of immune function, elevating levels of immunoglobulins IgG, IgA, and IgM (Hao et al., 2011).

In this scenario, postbiotics, which are metabolites derived from microorganisms, are bioactive molecules produced or modulated by the microbiome, directly influencing the host's cells and tissues. The supplementation of these metabolites has being considered as a potential strategy in treating microbiome-associated diseases (Sorbara & Pamer, 2022; Vallianou et al., 2020). Fermented infant formulas, often supplemented with lactic acid-producing bacteria, typically lack live microorganisms (Lin et al., 2012). Moreover, postbiotics have been investigated for their potential role in obesity treatment, as they may induce thermogenesis and promote the browning of adipose tissue (Moreno-Navarrete et al., 2018).

Evidence collected by Bevilacqua et al. (2024) shows the promising use of probiotics in alleviating symptoms in various pathological conditions and improving quality of life, with positive effects on cognitive functions and the reduction of inflammatory and pro-inflammatory processes. Another clinical outcome observed with the use of specific probiotics is a reduction in the incidence of upper respiratory tract diseases (Bevilacqua et al., 2024; Hao et al., 2011).

Another procedure of great scientific interest is fecal microbiota transplantation, which has proven effective in modifying the microbiome by transferring a complex microbial community from a healthy donor (Sorbara & Pamer, 2022). The fungal microbiome has gained attention as a potential biomarker for various diseases, including cancer (Ratiner et al., 2024). The fecal microbiome has shown promise in detecting colorectal cancer through algorithms that generate classification models with high precision and specificity (Liu et al., 2022). The integration of metagenomic and metabolomic data may further enhance the diagnosis of microbiome-related diseases (Yachida et al., 2019). This effort aims to develop new therapies or preventive strategies based on microbiome manipulation (Filardo et al., 2024).

18.3.2 Environmental and Agricultural Applications

Over the past three decades, advancements in biological processes have significantly broadened the scope of bioremediation, facilitating the breakdown of hazardous pollutants into less toxic or non-toxic forms, often reducing them to concentrations below regulatory limits (Chauhan et al., 2024; Curiel-Alegre et al., 2024; Gupta et al., 2024; Kumar & Shukla, 2024; Lu et al., 2024; Tang et al., 2021; Vale et al., 2023). Bioremediation can be implemented either ex situ, involving costly dredging and transportation of contaminated materials, or in situ, which is more economical and relies on native or engineered microbial communities to degrade pollutants while fostering ecosystem resilience. The two principal strategies in bioremediation are bioaugmentation, which involves introducing hydrocarbon-degrading microorganisms, and biostimulation, which enhances the metabolic activity of native microbial populations through nutrient supplementation (Oro et al., 2024).

Soil health, a cornerstone of agricultural productivity, depends on the interaction of climate, management practices, physicochemical soil properties, and biological interactions. These factors collectively regulate crop yields and contribute to the sustainability of agricultural systems. As the central component of agroecosystems, soil plays a crucial role in maintaining fertility (Karimi et al.,

2022; Mohapatra et al., 2022; Su et al., 2023). Healthy soils function as dynamic systems composed of organic matter, mineral particles, and diverse microbial communities, including bacteria, fungi, archaea, nematodes, and protozoa. This microbial diversity supports essential processes such as nutrient cycling, organic matter decomposition, and pathogen suppression, thereby enhancing agricultural resilience against environmental challenges such as drought and disease (Al-Shammary et al., 2024; dos Santos et al., 2023; Tripathi & Gaur, 2021).

Beyond sustaining agricultural productivity, soil quality is also pivotal in regulating the biogeochemical cycles of carbon, nitrogen, sulfur, and phosphorus, playing a vital role in climate change mitigation. These cycles contribute to soil fertility and influence atmospheric composition through nitrogen fixation and carbon sequestration. Soil microorganisms, whether free-living or in association with plant roots, establish the rhizosphere—a specialized microenvironment where plants and microbes engage in mutualistic interactions. Plants release metabolites that signal and nourish microorganisms, while the rhizobiome reciprocates by recycling nutrients, fixing nitrogen, and protecting plants from environmental stressors. These interactions bolster plant growth and competitiveness, critical for sustainable agricultural systems (Chamkhi et al., 2022; Custódio et al., 2022; Hemkemeyer et al., 2021; Orozco-Mosqueda et al., 2022; Tripathi & Gaur, 2021).

Key microbial players in this system include arbuscular mycorrhizal fungi (AMF), ectomycorrhizal fungi (ECM), and plant growth-promoting rhizobacteria (PGPR). AMFs establish symbiotic relationships with most terrestrial plants, enhancing nutrient exchange, recycling resources, retaining water, and shielding plants from pathogens. ECMs are particularly important in temperate and boreal forests, contributing to nutrient and phosphorus uptake while providing physical barriers against pathogens. PGPRs, widely used in agroecosystems, stimulate plant immunity, recycle nutrients, and promote growth, with notable genera such as *Azospirillum*, *Bacillus*, and *Rhizobium* achieving commercial success. Some PGPRs also facilitate soil bioremediation by degrading hydrocarbons and pesticides, supporting ecosystem restoration (Alrajhi et al., 2024; Andrew et al., 2022; Nabi, 2023; Su et al., 2023; Vasco-Palacios et al., 2018; Xiao et al., 2022).

However, intensive agricultural practices and unsustainable land management have disrupted soil microbiota, often necessitating chemical fertilizers and agrochemicals to compensate for nutrient deficiencies. While effective in the short term, these inputs can lead to significant environmental issues, including soil and water contamination, eutrophication, greenhouse gas emissions, and toxicity to non-target organisms. In this context, bioremediation offers a sustainable alternative. By balancing microbial communities and employing techniques like bioaugmentation and biostimulation, bioremediation not only removes pollutants but also restores soil fertility. Advances in genetic engineering have further enhanced microbial capabilities for pollutant degradation, paving the way for more sustainable agriculture (Mącik et al., 2020; Omotayo & Omotayo, 2024; Saravanan et al., 2022).

For instance, recent studies have shown that microbial consortia can significantly enhance pollutant degradation. In one case, a consortium comprising *Brucella intermedia*, *Alcaligenes faecalis*, and *Aquamicrobium terrae* effectively degraded lambda-cyhalothrin in contaminated soils, improving both pollutant removal and plant agronomic outcomes. Such findings highlight the potential of bioremediation to address contamination challenges in agricultural systems (Anwar et al., 2023).

Urbanization, industrialization, and heavy metal contamination remain critical threats to soil quality and agricultural sustainability. Urban expansion compromises soil health, while industrial activities introduce toxic substances that degrade soil quality and pose risks to plants and humans. Despite ongoing monitoring and mitigation efforts, effective and scalable solutions are needed. Bioremediation has emerged as a cost-effective technology for addressing these challenges. By leveraging microbial populations to degrade pollutants in situ or ex situ, bioremediation restores soil health and enhances nutrient availability. Continued advancements in genetic engineering and process optimization further bolster its efficiency, making it a vital tool for environmental restoration and sustainable agriculture (Omotayo & Omotayo, 2024; Stoichev et al., 2023; Verma & Kuila, 2019).

Bioremediation has also been successfully applied to oil-contaminated shorelines, with biostimulation and bioaugmentation strategies proving effective. While biostimulation enhances native microbial activity through nutrient supplementation, bioaugmentation introduces specialized microbial strains when native populations are insufficient. Autochthonous microorganisms are often preferred to maintain local microbial diversity and ecological balance. As this field evolves, bioremediation continues to demonstrate its versatility and potential as a sustainable approach to combating environmental degradation (Chen et al., 2024; Ma et al., 2022; Oro et al., 2024; Zhang et al., 2024).

In this sense, the integration of bioremediation into agricultural practices offers a promising solution to the environmental challenges posed by intensive land use, urbanization, and industrialization. Techniques such as bioaugmentation and biostimulation, combined with microbial engineering, optimize pollutant degradation while maintaining soil balance and microbial diversity. Moreover, the synergy between plants and soil microorganisms, particularly in the rhizosphere, underscores the importance of leveraging natural processes to enhance soil health and crop resilience.

As global pressures on agricultural production intensify, the role of bioremediation in sustainable land management becomes increasingly vital. By addressing soil contamination, improving nutrient availability, and reducing the reliance on chemical inputs, bioremediation contributes to more sustainable and productive agricultural systems. Future research and innovation in this field will further refine these processes, paving the way for broader applications and reinforcing bioremediation's role as a cornerstone of environmental restoration and sustainable development.

18.3.3 Challenges and Opportunities

Microbiomes play a crucial role in agricultural systems, influencing plant health, soil fertility, and crop productivity. The plant-associated microbiome, particularly the rhizosphere—the zone surrounding plant roots—is a hotspot for microbial interactions that impact nutrient availability, disease resistance, and stress tolerance. Beneficial rhizosphere microorganisms, including plant growth-promoting bacteria and mycorrhizal fungi, contribute to improved plant resilience by enhancing nutrient uptake, producing phytohormones, and inducing systemic resistance against pathogens. However, harnessing these microbial communities for sustainable agriculture presents both challenges and opportunities.

One of the major challenges lies in the complexity and variability of soil microbial ecosystems. Soil microbiomes are highly dynamic and influenced by environmental factors such as soil type, climate conditions, and agricultural practices. Disturbances caused by excessive chemical fertilizers, pesticides, and monoculture farming can disrupt microbial diversity, reducing the effectiveness of beneficial microorganisms. Additionally, the interactions between microbes and plant hosts are often species-specific, making it difficult to generalize findings across different crops and soil conditions.

Despite these challenges, advances in microbiome research offer promising opportunities for sustainable agriculture. The ability of soil microorganisms to suppress plant pathogens through biocontrol mechanisms presents an alternative to chemical pesticides, reducing environmental impact and promoting soil health. Certain bacteria and fungi produce antimicrobial compounds or outcompete harmful pathogens for resources, thereby enhancing plant disease resistance. Moreover, microbial inoculants and biofertilizers, derived from naturally occurring beneficial microorganisms, are being developed to enhance soil fertility and crop yield.

Another emerging opportunity lies in precision agriculture, where high-throughput sequencing and metagenomic analyses enable the identification of key microbial species that contribute to plant health. By manipulating microbiomes through targeted interventions, such as tailored microbial consortia or soil amendments, it may be possible to optimize plant–microbe interactions for improved agricultural productivity. Furthermore, research into plant-microbiome engineering and synthetic biology may lead to the development of crops with enhanced microbial symbioses, further strengthening resilience against biotic and abiotic stresses.

In conclusion, while challenges remain in understanding and managing agricultural microbiomes, ongoing research continues to uncover innovative strategies to leverage microbial diversity for sustainable crop production. By integrating microbiome-based solutions into modern agricultural practices, it is possible to enhance food security, reduce environmental impact, and promote long-term soil health.

18.4 CONCLUSIONS

Microbiome research has ushered in a transformative era, demonstrating the pivotal role of microbial communities in health, environmental, and agricultural contexts. While substantial progress has been made, significant challenges, including the detection of low-abundance microorganisms and the cultivation of uncultivable species, continue to impede comprehensive understanding and utilization. However, advancements in sequencing technologies and initiatives to standardize methodologies have opened new avenues for discovery. The applications of microbiome research, ranging from personalized medicine and disease prevention to bioremediation and sustainable agriculture, highlight its potential to address critical global issues. Future efforts must focus on fostering interdisciplinary collaborations, enhancing data integration, and developing innovative strategies to overcome current barriers. By doing so, we can fully harness the power of microbial ecosystems to improve health outcomes, restore environmental balance, and ensure food security in the face of growing global challenges.

REFERENCES

Aggarwal, N., Kitano, S., Puah, G. R. Y., Kittelmann, S., Hwang, I. Y., & Chang, M. W. (2023). Microbiome and human health: Current understanding, engineering, and enabling technologies. *Chemical Reviews*, *123*(1), 31–72. https://doi.org/10.1021/acs.chemrev.2c00431

Alrajhi, K., Bibi, S., & Abu-Dieyeh, M. (2024). Diversity, distribution, and applications of arbuscular mycorrhizal fungi in the Arabian Peninsula. *Saudi Journal of Biological Sciences*, *31*(2). https://doi.org/10.1016/j.sjbs.2023.103911

Al-Shammary, A. A. G., Al-Shihmani, L. S. S., Fernández-Gálvez, J., & Caballero-Calvo, A. (2024). Optimizing sustainable agriculture: A comprehensive review of agronomic practices and their impacts on soil attributes. *Journal of Environmental Management*, *364*, 121487. https://doi.org/10.1016/J.JENVMAN.2024.121487

Andrew, C., Mueller, G. M., & Avis, P. G. (2022). Do local-scale context dependencies shape how ectomycorrhizal fungal diversity structures with reduced or sustained experimental N addition? *Pedobiologia*, *90*, 150791. https://doi.org/10.1016/J.PEDOBI.2022.150791

Angima, G., Qu, Y., Park, S. H., & Dallas, D. C. (2024). Prebiotic strategies to manage lactose intolerance symptoms. *Nutrients*, *16*(7), 1–14. https://doi.org/10.3390/nu16071002

Anwar, S., Liaqat, A., Munir, A., Ashraf, M. F., & Iqbal, S. (2023). Bioaugmentation of a novel bacterial consortium in cotton-planted soil for degradation of lambda-cyhalothrin. *Pedosphere*. https://doi.org/10.1016/J.PEDSPH.2023.07.019

Bailén, M., Bressa, C., Larrosa, M., & González-Soltero, R. (2020). Bioinformatic strategies to address limitations of 16rRNA short-read amplicons from different sequencing platforms. *Journal of Microbiological Methods*, *169*, 105811. https://doi.org/10.1016/j.mimet.2019.105811

Bartsch, M., Hahn, A., & Berkemeyer, S. (2023). Bridging the gap from enterotypes to personalized dietary recommendations: A metabolomics perspective on microbiome research. *Metabolites*, *13*(12), 1182. https://doi.org/10.3390/metabo13121182

Bevilacqua, A., Campaniello, D., Speranza, B., Racioppo, A., Sinigaglia, M., & Corbo, M. R. (2024). An update on prebiotics and on their health effects. *Foods*, *13*(3), 446. https://doi.org/10.3390/foods13030446

Bolsega, S., Bleich, A., & Basic, M. (2021). Synthetic microbiomes on the rise—application in deciphering the role of microbes in host health and disease. *Nutrients*, *13*(11). https://doi.org/10.3390/nu13114173

Bonk, F., Popp, D., Harms, H., & Centler, F. (2018). PCR-based quantification of taxa-specific abundances in microbial communities: Quantifying and avoiding common pitfalls. *Journal of Microbiological Methods*, *153*, 139–147. https://doi.org/10.1016/j.mimet.2018.09.015

Brereton, N. J. B., Gonzalez, E., Desjardins, D., Labrecque, M., & Pitre, F. E. (2020). Co-cropping with three phytoremediation crops influences rhizosphere microbiome community in contaminated soil. *Science of the Total Environment*, *711*, 135067. https://doi.org/10.1016/j.scitotenv.2019.135067

Carabeo-Pérez, A., Guerra-Rivera, G., Ramos-Leal, M., & Jiménez-Hernández, J. (2019). Metagenomic approaches: Effective tools for monitoring the structure and functionality of microbiomes in anaerobic digestion systems. *Applied Microbiology and Biotechnology*, *103*(23–24), 9379–9390. https://doi.org/10.1007/s00253-019-10052-5

Chamkhi, I., El Omari, N., Balahbib, A., El Menyiy, N., Benali, T., & Ghoulam, C. (2022). Is—the rhizosphere a source of applicable multi-beneficial microorganisms for plant enhancement? *Saudi Journal of Biological Sciences*, *29*(2), 1246–1259. https://doi.org/10.1016/J.SJBS.2021.09.032

Chauhan, R., Awasthi, S., Tiwari, P., Upadhyay, M. K., Srivastava, S., Dwivedi, S., Dhankher, O. P., & Tripathi, R. D. (2024). Biotechnological strategies for remediation of arsenic-contaminated soils to improve soil health and sustainable agriculture. *Soil and Environmental Health*, *2*(1), 100061. https://doi.org/10.1016/j.seh.2024.100061

Chen, T., Dong, Y., Huang, W., & Ma, Y. (2024). Dynamics of microbial community and functional genes during bioremediation of PAHs-contaminated soil by two biostimulants. *Biochemical Engineering Journal*, *208*(April), 109356. https://doi.org/10.1016/j.bej.2024.109356

Curiel-Alegre, S., de la Fuente-Vivas, D., Khan, A. H. A., García-Tojal, J., Velasco-Arroyo, B., Rumbo, C., Soja, G., Rad, C., & Barros, R. (2024). Unveiling the capacity of bioaugmentation application, in comparison with biochar and rhamnolipid for TPHs degradation in aged hydrocarbons polluted soil. *Environmental Research*, *252*(P2), 118880. https://doi.org/10.1016/j.envres.2024.118880

Custódio, V., Gonin, M., Stabl, G., Bakhoum, N., Oliveira, M. M., Gutjahr, C., & Castrillo, G. (2022). Sculpting the soil microbiota. *Plant Journal*, *109*(3), 508–522. https://doi.org/10.1111/tpj.15568

Deek, R. A., Ma, S., Lewis, J., & Li, H. (2024). Statistical and computational methods for integrating microbiome, host genomics, and metabolomics data. *ELife*, *13*. https://doi.org/10.7554/eLife.88956

Demkovych, A., Kalashnikov, D., Hasiuk, P., Zubchenko, S., & Vorobets, A. (2023). The influence of microbiota on the development and course of inflammatory diseases of periodontal tissues. *Frontiers in Oral Health*, *4*(August), 6–11. https://doi.org/10.3389/froh.2023.1237448

dos Santos, M. S. N., Ody, L. P., Kerber, B. D., Araujo, B. A., Oro, C. E. D., Wancura, J. H. C., Mazutti, M. A., Zabot, G. L., & Tres, M. V. (2023). New frontiers of soil fungal microbiome and its application for biotechnology in agriculture. *World Journal of Microbiology and Biotechnology*, *39*(11). https://doi.org/10.1007/s11274-023-03728-8

Filardo, S., Di Pietro, M., Protano, C., Antonucci, A., Vitali, M., & Sessa, R. (2022). Impact of air pollution on the composition and diversity of human gut microbiota in general and vulnerable populations: A systematic review. *Toxics*, *10*(10). https://doi.org/10.3390/toxics10100579

Filardo, S., Di Pietro, M., & Sessa, R. (2024). Current progresses and challenges for microbiome research in human health: A perspective. *Frontiers in Cellular and Infection Microbiology*, *14*. https://doi.org/10.3389/fcimb.2024.1377012

Foo, J. L., Ling, H., Lee, Y. S., & Chang, M. W. (2017). Microbiome engineering: Current applications and its future. *Biotechnology Journal*, *12*(3), 1–11. https://doi.org/10.1002/biot.201600099

Frąc, M., Hannula, E. S., Bełka, M., Salles, J. F., & Jedryczka, M. (2022). Soil mycobiome in sustainable agriculture. *Frontiers in Microbiology*, *13*(November), 1–9. https://doi.org/10.3389/fmicb.2022.1033824

France, M., Alizadeh, M., Brown, S., Ma, B., & Ravel, J. (2022). Towards a deeper understanding of the vaginal microbiota. *Nature Microbiology*, *7*(3), 367–378. https://doi.org/10.1038/s41564-022-01083-2

Furtak, K., Grządziel, J., Gałązka, A., & Niedźwiecki, J. (2020). Prevalence of unclassified bacteria in the soil bacterial community from floodplain meadows (fluvisols) under simulated flood conditions revealed by a metataxonomic approachss. *CATENA*, *188*, 104448. https://doi.org/10.1016/j.catena.2019.104448

George, S. E., James, J., Devereux, R., Wan, Y., Diamond, G. L., Bradham, K. D., Scheckel, K. G., & Thomas, D. J. (2022). Ingestion of remediated lead-contaminated soils affects the fecal microbiome of mice. *Science of the Total Environment*, *837*(April), 155797. https://doi.org/10.1016/j.scitotenv.2022.155797

Govender, P., & Ghai, M. (2025). Population-specific differences in the human microbiome: Factors defining the diversity. *Gene*, *933*, 148923. https://doi.org/10.1016/j.gene.2024.148923

Grice, E. A., & Segre, J. A. (2012). The human microbiome: Our second genome. *Annual Review of Genomics and Human Genetics*, *13*, 151–170. https://doi.org/10.1146/annurev-genom-090711-163814

Gupta, P. K., Nair, V. K., Dalvi, V., Dhali, S., Malik, A., & Pant, K. K. (2024). Field-scale assessment of soil, water, plant, and soil microbiome in and around Rania-Khan Chandpur Chromium contaminated site, India. *Journal of Hazardous Materials*, *467*(December 2023), 133747. https://doi.org/10.1016/j.jhazmat.2024.133747

Hao, Q., Lu, Z., Dong, B. R., Huang, C. Q., & Wu, T. (2011). Probiotics for preventing acute upper respiratory tract infections. In B. R. Dong (Ed.), *Cochrane Database of Systematic Reviews*. Issue 9. Art. No.: CD006895. John Wiley & Sons, Ltd. https://doi.org/10.1002/14651858.CD006895.pub2

Hemkemeyer, M., Schwalb, S. A., Heinze, S., Joergensen, R. G., & Wichern, F. (2021). Functions of elements in soil microorganisms. *Microbiological Research*, *252*, 126832. https://doi.org/10.1016/J.MICRES.2021.126832

Hou, K., Wu, Z. X., Chen, X. Y., Wang, J. Q., Zhang, D., Xiao, C., Zhu, D., Koya, J. B., Wei, L., Li, J., & Chen, Z. S. (2022). Microbiota in health and diseases. *Signal Transduction and Targeted Therapy*, *7*(1). https://doi.org/10.1038/s41392-022-00974-4

Junca, H., Pieper, D. H., & Medina, E. (2022). The emerging potential of microbiome transplantation on human health interventions. *Computational and Structural Biotechnology Journal*, *20*, 615–627. https://doi.org/10.1016/j.csbj.2022.01.009

Kapinusova, G., Lopez Marin, M. A., & Uhlik, O. (2023). Reaching unreachables: Obstacles and successes of microbial cultivation and their reasons. *Frontiers in Microbiology*, *14*. https://doi.org/10.3389/fmicb.2023.1089630

Karimi, H., Mahdavi, S., Asgari, B., Rajput, V. D., Minkina, T., Astatkie, T., & Moghiseh, E. (2022). Insights on the bioremediation technologies for pesticide- contaminated soils. *Environmental Geochemistry and Health*, *44*(4), 1329–1354. https://doi.org/10.1007/s10653-021-01081-z

Kong, C., Yan, X., Zhu, Y., Zhu, H., Luo, Y., Liu, P., Ferrandon, S., Kalady, M. F., Gao, R., He, J., Yin, F., Qu, X., Zheng, J., Gao, Y., Wei, Q., Ma, Y., Liu, J.-Y., & Qin, H. (2021). Fusobacterium nucleatum promotes the development of colorectal cancer by activating a cytochrome P450/Epoxyoctadecenoic acid axis via TLR4/Keap1/NRF2 signaling. *Cancer Research*, *81*(17), 4485–4498. https://doi.org/10.1158/0008-5472.CAN-21-0453

Kotoky, R., Rajkumari, J., & Pandey, P. (2018). The rhizosphere microbiome: Significance in rhizoremediation of polyaromatic hydrocarbon contaminated soil. *Journal of Environmental Management*, *217*, 858–870. https://doi.org/10.1016/j.jenvman.2018.04.022

Kumar, N., & Shukla, P. (2024). Microalgal multiomics-based approaches in bioremediation of hazardous contaminants. *Environmental Research*, *247*(October 2023), 118135. https://doi.org/10.1016/j.envres.2024.118135

Lin, H. V., Frassetto, A., Kowalik Jr, E. J., Nawrocki, A. R., Lu, M. M., Kosinski, J. R., Hubert, J. A., Szeto, D., Yao, X., Forrest, G., & Marsh, D. J. (2012). Butyrate and propionate Protect against diet-induced obesity and regulate gut hormones via free fatty acid receptor 3-independent mechanisms. *PLoS One*, *7*(4), e35240. https://doi.org/10.1371/journal.pone.0035240

Liu, N.-N., Jiao, N., Tan, J.-C., Wang, Z., Wu, D., Wang, A.-J., Chen, J., Tao, L., Zhou, C., Fang, W., Cheong, I. H., Pan, W., Liao, W., Kozlakidis, Z., Heeschen, C., Moore, G. G., Zhu, L., Chen, X., Zhang, G.,. . . Wang, H. (2022). Multi-kingdom microbiota analyses identify bacterial–fungal interactions and biomarkers of colorectal cancer across cohorts. *Nature Microbiology*, *7*(2), 238–250. https://doi.org/10.1038/s41564-021-01030-7

Liu, Q., Chen, Z., He, D., Pan, A., yuan, J., Liu, Y., Huang, L., & Feng, Y. (2024). Assembly of root-associated bacterial community and soil health in cadmium-contaminated soil affected by nano/bulk-biochar compost associations. *Environmental Pollution*, *360*(April), 124619. https://doi.org/10.1016/j.envpol.2024.124619

Loftus, S. E., Hunt, D. E., & Johnson, Z. I. (2020). Reused cultivation water from a self-inhibiting alga does not inhibit other algae but alters their microbiomes. *Algal Research*, *51*, 102067. https://doi.org/10.1016/j.algal.2020.102067

Lu, C. W., Lo, K. H., Wang, S. C., Kao, C. M., & Chen, S. C. (2024). An innovative permeable reactive bio-barrier to remediate trichloroethene-contaminated groundwater: A field study. *Science of the Total Environment*, *920*(October 2023), 170885. https://doi.org/10.1016/j.scitotenv.2024.170885

Ma, H., Zhao, Y., Yang, K., Wang, Y., Zhang, C., & Ji, M. (2022). Application oriented bioaugmentation processes: Mechanism, performance improvement and scale-up. *Bioresource Technology*, *344*(PB), 126192. https://doi.org/10.1016/j.biortech.2021.126192

Mącik, M., Gryta, A., & Frąc, M. (2020). Biofertilizers in agriculture: An overview on concepts, strategies and effects on soil microorganisms. *Advances in Agronomy*, *162*, 31–87. https://doi.org/10.1016/BS.AGRON.2020.02.001

Mancin, L., Paoli, A., Berry, S., Gonzalez, J. T., Collins, A. J., Lizarraga, M. A., Mota, J. F., Nicola, S., & Rollo, I. (2024). Standardization of gut microbiome analysis in sports. *Cell Reports Medicine*, *5*(10), 101759. https://doi.org/10.1016/j.xcrm.2024.101759

Mohapatra, D., Rath, S. K., & Mohapatra, P. K. (2022). Soil fungi for bioremediation of pesticide toxicants: A perspective. *Geomicrobiology Journal*, *39*(3–5), 352–372. https://doi.org/10.1080/01490451.2021.2019855

Moossavi, S., Engen, P. A., Ghanbari, R., Green, S. J., Naqib, A., Bishehsari, F., Merat, S., Poustchi, H., Keshavarzian, A., & Malekzadeh, R. (2019). Assessment of the impact of different fecal storage protocols on the microbiota diversity and composition: A pilot study. *BMC Microbiology*, *19*(1), 145. https://doi.org/10.1186/s12866-019-1519-2

Moreno-Navarrete, J. M., Serino, M., Blasco-Baque, V., Azalbert, V., Barton, R. H., Cardellini, M., Latorre, J., Ortega, F., Sabater-Masdeu, M., Burcelin, R., Dumas, M., Ricart, W., Federici, M., & Fernández-Real, J. M. (2018). Gut microbiota interacts with markers of adipose tissue browning, insulin action and plasma acetate in morbid obesity. *Molecular Nutrition & Food Research*, *62*(3). https://doi.org/10.1002/mnfr.201700721

Nabi, M. (2023). Role of microorganisms in plant nutrition and soil health. *Sustainable Plant Nutrition: Molecular Interventions and Advancements for Crop Improvement*, 263–282. https://doi.org/10.1016/B978-0-443-18675-2.00016-X

Nagata, N., Tohya, M., Takeuchi, F., Suda, W., Nishijima, S., Ohsugi, M., Ueki, K., Tsujimoto, T., Nakamura, T., Kawai, T., Miyoshi-Akiyama, T., Uemura, N., & Hattori, M. (2019). Effects of storage temperature, storage time, and Cary-Blair transport medium on the stability of the gut microbiota. *Drug Discoveries & Therapeutics*, *13*(5), 256–260. https://doi.org/10.5582/ddt.2019.01071

Navas-Molina, J. A., Peralta-Sánchez, J. M., González, A., McMurdie, P. J., Vázquez-Baeza, Y., Xu, Z., Ursell, L. K., Lauber, C., Zhou, H., Song, S. J., Huntley, J., Ackermann, G. L., Berg-Lyons, D., Holmes, S., Caporaso, J. G., & Knight, R. (2013). Advancing our understanding of the human microbiome using QIIME (pp. 371–444). https://doi.org/10.1016/B978-0-12-407863-5.00019-8

Nearing, J. T., Comeau, A. M., & Langille, M. G. I. (2021). Identifying biases and their potential solutions in human microbiome studies. *Microbiome*, *9*(1), 113. https://doi.org/10.1186/s40168-021-01059-0

Nie, M., Wu, C., Tang, Y., Shi, G., Wang, X., Hu, C., Cao, J., & Zhao, X. (2023). Selenium and Bacillus proteolyticus SES synergistically enhanced ryegrass to remediate Cu–Cd–Cr contaminated soil. *Environmental Pollution*, *323*(November 2022). https://doi.org/10.1016/j.envpol.2023.121272

Omotayo, A. O., & Omotayo, O. P. (2024). Potentials of microbe-plant assisted bioremediation in reclaiming heavy metal polluted soil environments for sustainable agriculture. *Environmental and Sustainability Indicators*, *22*(April), 100396. https://doi.org/10.1016/j.indic.2024.100396

Oro, C. E. D., Puton, B. M. S., Venquiaruto, L. D., Dallago, M., & Tres, M. V. (2024). Effective microbial strategies to remediate contaminated agricultural soils and conserve functions. *Agronomy*, *14*(11), 2637. https://doi.org/https://doi.org/10.3390/agronomy14112637

Orozco-Mosqueda, M. del C., Fadiji, A. E., Babalola, O. O., Glick, B. R., & Santoyo, G. (2022). Rhizobiome engineering: Unveiling complex rhizosphere interactions to enhance plant growth and health. *Microbiological Research*, *263*, 127137.

Ortúzar, M., Riesco, R., Criado, M., Alonso, M. del P., & Trujillo, M. E. (2024). Unraveling the dynamic interplay of microbial communities associated to Lupinus angustifolius in response to environmental and cultivation conditions. *Science of The Total Environment*, *946*, 174277. https://doi.org/10.1016/j.scitotenv.2024.174277

Pani, G. (2023). Fusobacterium & Co. at the stem of cancer: Microbe–cancer stem cell interactions in colorectal carcinogenesis. *Cancers*, *15*(9), 2583. https://doi.org/10.3390/cancers15092583

Ratiner, K., Ciocan, D., Abdeen, S. K., & Elinav, E. (2024). Utilization of the microbiome in personalized medicine. *Nature Reviews Microbiology*, *22*(5), 291–308. https://doi.org/10.1038/s41579-023-00998-9

Salem, I., Ramser, A., Isham, N., & Ghannoum, M. A. (2018). The gut microbiome as a major regulator of the gut-skin axis. *Frontiers in Microbiology*, *9*. https://doi.org/10.3389/fmicb.2018.01459

Samarrai, R., Frank, S., Lum, A., Woodis, K., Weinstock, G., & Roberts, D. (2022). Defining the microbiome of the head and neck: A contemporary review. *American Journal of Otolaryngology*, *43*(1), 103224. https://doi.org/10.1016/j.amjoto.2021.103224

Saravanan, A., Kumar, P. S., Jeevanantham, S., Anubha, M., & Jayashree, S. (2022). Degradation of toxic agrochemicals and pharmaceutical pollutants: Effective and alternative approaches toward photocatalysis. *Environmental Pollution*, *298*, 118844. https://doi.org/10.1016/J.ENVPOL.2022.118844

Segata, N., Haake, S. K., Mannon, P., Lemon, K. P., Waldron, L., Gevers, D., Huttenhower, C., & Izard, J. (2012). Composition of the adult digestive tract bacterial microbiome based on seven mouth surfaces, tonsils, throat and stool samples. *Genome Biology*, *13*(6). https://doi.org/10.1186/gb-2012-13-6-r42

Shayanthan, A., Ordoñez, P. A. C., & Oresnik, I. J. (2022). The role of Synthetic Microbial Communities (SynCom) in sustainable agriculture. *Frontiers in Agronomy*, *4*(June), 1–13. https://doi.org/10.3389/fagro.2022.896307

Shealy, N. G., Baltagulov, M., & Byndloss, M. X. (2024). A long journey to the colon: The role of the small intestine microbiota in intestinal disease. *Molecular Microbiology*. https://doi.org/10.1111/mmi.15270

Silverman, J. D., Bloom, R. J., Jiang, S., Durand, H. K., Dallow, E., Mukherjee, S., & David, L. A. (2021). Measuring and mitigating PCR bias in microbiota datasets. *PLOS Computational Biology*, *17*(7), e1009113. https://doi.org/10.1371/journal.pcbi.1009113

Singh, R., Pal, D. B., Alkhanani, M. F., Almalki, A. H., Areeshi, M. Y., Haque, S., & Srivastava, N. (2022). Prospects of soil microbiome application for lignocellulosic biomass degradation: An overview. *Science of the Total Environment*, *838*(February). https://doi.org/10.1016/j.scitotenv.2022.155966

Sinha, R., Abnet, C. C., White, O., Knight, R., & Huttenhower, C. (2015). The microbiome quality control project: Baseline study design and future directions. *Genome Biology*, *16*(1), 276. https://doi.org/10.1186/s13059-015-0841-8

Sipos, R., Székely, A. J., Palatinszky, M., Révész, S., Márialigeti, K., & Nikolausz, M. (2007). Effect of primer mismatch, annealing temperature and PCR cycle number on 16S rRNA gene-targetting bacterial community analysis. *FEMS Microbiology Ecology*, *60*(2), 341–350. https://doi.org/10.1111/j.1574-6941.2007.00283.x

Sorbara, M. T., & Pamer, E. G. (2022). Microbiome-based therapeutics. *Nature Reviews Microbiology*, *20*(6), 365–380. https://doi.org/10.1038/s41579-021-00667-9

Stoichev, T., Mucha, A. P., Bernabeu, A., & Almeida, C. M. R. (2023). Buried and surface oil degradation—evaluating bioremediation to increase PAHs removal through linear mathematical models. *Marine Pollution Bulletin*, *189*(December 2022), 114727. https://doi.org/10.1016/j.marpolbul.2023.114727

Su, Y., Wang, J., Gao, W., Wang, R., Yang, W., Zhang, H., Huang, L., & Guo, L. (2023). Dynamic metabolites: A bridge between plants and microbes. *Science of The Total Environment*, *899*, 165612. https://doi.org/10.1016/J.SCITOTENV.2023.165612

Takiishi, T., Fenero, C. I. M., & Câmara, N. O. S. (2017). Intestinal barrier and gut microbiota: Shaping our immune responses throughout life. *Tissue Barriers*, *5*(4), e1373208. https://doi.org/10.1080/21688370.2017.1373208

Tan, H.-E. (2023). The microbiota-gut-brain axis in stress and depression. *Frontiers in Neuroscience*, *17*. https://doi.org/10.3389/fnins.2023.1151478

Tang, F., Gao, M., Zeng, F., Xu, Z., & Tian, G. (2021). An old story with new insights into an ignored issue of metabolites in biochar-amended soil: Effect of biochar on dissipation of carbosulfan as an example. *Science of the Total Environment*, *790*, 148100. https://doi.org/10.1016/j.scitotenv.2021.148100

Tripathi, M., & Gaur, R. (2021). Bioactivity of soil microorganisms for agriculture development. *Microbes in Land Use Change Management*, 197–220. https://doi.org/10.1016/B978-0-12-824448-7.00012-7

Vale, K. do V., Soares, D. W. F., da Silva, A. L. B., Abreu, F. O. M. da S., de Oliveira Almeida, J. L. I., Oliveira, M. R. F., Reck, L., Melo, V. M. M., Alves, D. R., & Alves, C. R. (2023). Biossurfactant production by isolated bacteria from petroleum industry's effluent targeting bioremediation. *Results in Chemistry*, *6*(March), 0–9. https://doi.org/10.1016/j.rechem.2023.101038

Vallianou, N., Stratigou, T., Christodoulatos, G. S., Tsigalou, C., & Dalamaga, M. (2020). Probiotics, prebiotics, synbiotics, postbiotics, and obesity: Current evidence, controversies, and perspectives. *Current Obesity Reports*, *9*(3), 179–192. https://doi.org/10.1007/s13679-020-00379-w

Vartoukian, S. R., Palmer, R. M., & Wade, W. G. (2010). Strategies for culture of 'unculturable' bacteria. *FEMS Microbiology Letters*, no-no. https://doi.org/10.1111/j.1574-6968.2010.02000.x

Vasco-Palacios, A. M., Hernandez, J., Peñuela-Mora, M. C., Franco-Molano, A. E., & Boekhout, T. (2018). Ectomycorrhizal fungi diversity in a white sand forest in western Amazonia. *Fungal Ecology*, *31*, 9–18. https://doi.org/10.1016/J.FUNECO.2017.10.003

Verma, D., Garg, P. K., & Dubey, A. K. (2018). Insights into the human oral microbiome. *Archives of Microbiology*, *200*(4), 525–540. https://doi.org/10.1007/s00203-018-1505-3

Verma, S., & Kuila, A. (2019). Bioremediation of heavy metals by microbial process. *Environmental Technology and Innovation*, *14*, 100369. https://doi.org/10.1016/j.eti.2019.100369

Wang, M., Lei, M., & He, H. (2023). Catalog of operational taxonomic units and unified amplicon sequencing data for the microbiomes of medicinal plant roots. *Engineering Microbiology*, *3*(3), 100087. https://doi.org/10.1016/j.engmic.2023.100087

Woo, Y. R., & Kim, H. S. (2024). Interaction between the microbiota and the skin barrier in aging skin: A comprehensive review. *Frontiers in Physiology*, *15*(January), 1–14. https://doi.org/10.3389/fphys.2024.1322205

Wu, Y., Peng, L., Feng, P., Han, R., Khan, A., Kulshreshtha, S., Ling, Z., Liu, P., & Li, X. (2023). Gut microbes consume host energy and reciprocally provide beneficial factors to sustain a symbiotic relationship with the host. *Science of The Total Environment*, *904*, 166773. https://doi.org/10.1016/j.scitotenv.2023.166773

Xiao, L., Lai, S., Chen, M., Long, X., Fu, X., & Yang, H. (2022). Effects of grass cultivation on soil arbuscular mycorrhizal fungi community in a tangerine orchard. *Rhizosphere*, *24*, 100583. https://doi.org/10.1016/J.RHISPH.2022.100583

Xiong, R.-G., Li, J., Cheng, J., Zhou, D.-D., Wu, S.-X., Huang, S.-Y., Saimaiti, A., Yang, Z.-J., Gan, R.-Y., & Li, H.-B. (2023). The role of gut microbiota in anxiety, depression, and other mental disorders as well as the protective effects of dietary components. *Nutrients*, *15*(14), 3258. https://doi.org/10.3390/nu15143258

Yachida, S., Mizutani, S., Shiroma, H., Shiba, S., Nakajima, T., Sakamoto, T., Watanabe, H., Masuda, K., Nishimoto, Y., Kubo, M., Hosoda, F., Rokutan, H., Matsumoto, M., Takamaru, H., Yamada, M., Matsuda, T., Iwasaki, M., Yamaji, T., Yachida, T., . . . Yamada, T. (2019). Metagenomic and metabolomic analyses reveal distinct stage-specific phenotypes of the gut microbiota in colorectal cancer. *Nature Medicine*, *25*(6), 968–976. https://doi.org/10.1038/s41591-019-0458-7

Yu, G., Gail, M. H., Consonni, D., Carugno, M., Humphrys, M., Pesatori, A. C., Caporaso, N. E., Goedert, J. J., Ravel, J., & Landi, M. T. (2016). Characterizing human lung tissue microbiota and its relationship to epidemiological and clinical features. *Genome Biology*, *17*(1), 163. https://doi.org/10.1186/s13059-016-1021-1

Yu, Q., Wang, H., Qin, L., Wang, T., Zhang, Y., & Sun, Y. (2024). Interpretable machine learning reveals microbiome signatures strongly associated with dairy cow milk urea nitrogen. *IScience*, *27*(6), 109955. https://doi.org/10.1016/j.isci.2024.109955

Yuan, X., Hong, S., Xiong, W., Raza, W., Shen, Z., Wang, B., Li, R., Ruan, Y., Shen, Q., & Dini-Andreote, F. (2021). Development of fungal-mediated soil suppressiveness against Fusarium wilt disease via plant residue manipulation. *Microbiome*, *9*(1), 1–15. https://doi.org/10.1186/s40168-021-01133-7

Zhang, M., Liu, X., Zhu, W., Hu, S., Yan, X., & Hong, Q. (2024). Remediation of isoproturon-contaminated soil by Sphingobium sp. strain YBL2: Bioaugmentation, detoxification and community structure. *Journal of Hazardous Materials*, *476*(April), 134968. https://doi.org/10.1016/j.jhazmat.2024.134968

Zhang, Y., Thomas, J. P., Korcsmaros, T., & Gul, L. (2024). Integrating multi-omics to unravel host-microbiome interactions in inflammatory bowel disease. *Cell Reports Medicine*, *5*(9), 101738. https://doi.org/10.1016/j.xcrm.2024.101738

Index

Note: Page numbers in *italics* indicate a figure and page numbers in **bold** indicate a table on the corresponding page.

D

E

F

G

H

S

T

U

V

W

X

Y

Z

For Product Safety Concerns and Information please contact our
EU representative GPSR@taylorandfrancis.com Taylor & Francis
Verlag GmbH, Kaufingerstraße 24, 80331 München, Germany